NATURAL SCIENCES IN AMERICA

NATURAL SCIENCES IN AMERICA

Advisory Editor
KEIR B. STERLING

Editorial Board
EDWIN H. COLBERT
EDWARD GRUSON
ERNST MAYR
RICHARD G. VAN GELDER

CONTRIBUTIONS TO AMERICAN SYSTEMATICS

Introduction by
Keir B. Sterling

ARNO PRESS
A New York Times Company
New York, N. Y. • 1974

Reprint Edition 1974 by Arno Press Inc.

"Where Are We?" by Ernst Mayr was reprinted by permission of Cold Spring Harbor Laboratories.

Reprinted from copies in The American Museum of Natural History Library

NATURAL SCIENCES IN AMERICA
ISBN for complete set: 0-405-05700-8
See last pages of this volume for titles.

Manufactured in the United States of America

Publisher's Note: The articles in this compilation have been reprinted from the best available copies.

Library of Congress Cataloging in Publication Data

Main entry under title:

Contributions to American systematics.

    (Natural sciences in America)
    CONTENTS: Gill, T. Systematic zoology: its progress and purpose.--Mayr, E. Where are we?--Gulick, J. T. On diversity of evolution under one set of external conditions. [etc.]
    1. Zoology--Classification--Addresses, essays, lectures. I. Series.
QL351.C66   591'.01'2   73-17807
ISBN 0-405-05724-5

CONTENTS

Gill, Theodore
Systematic Zoology: Its Progress and Purpose (Reprinted from *Annual Report of the Smithsonian Institution for 1907*), Washington, D. C., 1907

Mayr, Ernst
Where Are We? (Reprinted from *Cold Spring Harbor Symposium on Quantitative Biology*, Vol. XXIV), Cold Spring Harbor, N. Y., 1959

Gulick, John T.
On Diversity of Evolution Under One Set of External Conditions (Reprinted from *Linnaean Society Journal, Zoology*, Vol. XI), London, 1871

Baldwin, J. Mark
A New Factor in Evolution (Reprinted from *The American Naturalist*, Vol. XXX), New York, 1896

Jordan, K[arl]
On Mechanical Selection and Other Problems (Reprinted from *Novitates Zoologicae British Museum*, Vol. III), London, 1896

Hyatt, Alpheus
Cycle in the Life of the Individual (Ontogeny) and in the Evolution of Its Own Group (Phylogeny) (Reprinted from *Science*, Vol. V), New York, 1897

Wright, Sewall
Evolution in Mendelian Populations (Reprinted from *Genetics,* Vol. XVI), Princeton, N. J., 1931

Jordan, David Starr
The Law of Geminate Species (Reprinted from *The American Naturalist,* Vol. XLII), New York, 1908

Taverner, P. A.
The Test of the Subspecies (Reprinted from *Journal of Mammalogy,* Vol. I), Baltimore, 1920

Osborn, Henry Fairfield
The Origin of Species, V: Speciation and Mutation (Reprinted from *The American Naturalist,* Vol. LXI), New York, 1927

Kinsey, Alfred C.
Supra-Specific Variation in Nature and in Classification from the View-Point of Zoology (Reprinted from *The American Naturalist,* Vol. LXXI), New York, 1937

Jordan, Karl
Where Subspecies Meet (Reprinted from *Novitates Zoologicae British Museum,* Vol. XLI), London, 1938

Simpson, George Gaylord
Criteria For Genera, Species, and Subspecies in Zoology and Paleozoology (Reprinted from *Annals of the New York Academy of Sciences,* Vol. XLIV), New York, 1943

Mayr, Ernst
Karl Jordan's Contribution to Current Concepts in Systematics and Evolution (Reprinted from *Transactions of the Royal Entomological Society,* Vol. 107), London, 1955

Seton, Ernest E. T.
Letter to the Editors: The Popular Names of Birds (Reprinted from *The Auk,* Vol. II), Cambridge, Mass., 1885

Bumpus, Hermon C.
The Variations and Mutations of the Introduced Sparrow, Passer Domesticus (Reprinted from *Biological Lectures, Marine Biological Laboratory*), Woods Hole, Mass., 1896-1897

Chapman, Frank M.
Criteria for the Determination of Subspecies in Systematic Ornithology (Reprinted from *The Auk,* Vol. XLI), Cambridge, Mass., 1924

Miller, Alden H[olmes]
Speciation in the Avian Genus Junco (Reprinted from *University of California Publications in Zoology,* Vol. XLIV), Berkeley, Calif., 1941

Mayr, Ernst
Criteria of Subspecies, Species and Genera in Ornithology (Reprinted from *Annals of the New York Academy of Sciences,* Vol. XLIV), New York, 1943

Mayr, Ernst
History of the North American Bird Fauna (Reprinted from *The Wilson Bulletin,* Vol. LVIII), New York, 1946

Mayr, Ernst
Speciation in Birds (Reprinted from *Proceedings of the Xth International Ornithological Conference, Uppsala, 1950*), Uppsala, Sweden, 1951

Sumner, Francis B.
Some Effects of External Conditions Upon the White Mouse (Reprinted from *Journal of Experimental Zoology,* Vol. VII), Baltimore and Philadelphia, 1909

Sumner, F[rancis] B.
The Role of Isolation in the Formation of a Narrowly Localized Race of Deer Mice (Peromyscus) (Reprinted from *The American Naturalist,* Vol. LI), New York, 1917

Sumner, F[rancis] B.
Genetic, Distributional, and Evolutionary Studies of the Subspecies of Deer-Mice (Peromyscus) (Reprinted from *Bibliographia Genetica,* Vol. IX), The Hague, Netherlands, 1932

Hall, E. Raymond
Criteria for Vertebrate Subspecies, Species, and Genera: The Mammals (Reprinted from *Annals of the New York Academy of Sciences,* Vol. XLIV), New York, 1943

INTRODUCTION

Few problems in the natural sciences have proved to be as complex, vexatious, and seemingly without solution as that of characterizing the precise nature of what is meant by the term "species." Efforts have been made to produce a definition acceptable to a majority of systematists, taxonomists, and field workers, but the question is far from resolved, even today. Some authorities, such as Richard Blackwelder, are pessimistic about finding an answer. "The more recent discussions [on this subject]," he has written, "have been as barren as the old. It would seem that this was inevitable, because the arguers have never settled the preliminary problem of what it is the argument is designed to settle. A major symposium . . . on the Species problem did not even attempt to define the problem or the issues."[1]

Historically, species definitions have fallen into three broad categories. The first is the essentialist concept. This theory holds that animate and inanimate differences in nature mirror what Ernst Mayr terms "a limited number of universals which are fixed, unchanging, and separated by well-defined discontinuities from other universals."[2] Beginning with Aristotle, logical method and the procedures of classification have been closely related. In more recent times, the ideas of John Ray and Linnaeus, in whose works the degree of morphological difference between forms was emphasized, dominated taxonomic work down to the early years of the twentieth century.

Proponents of the nominalist species concept argued that species did not exist in nature, but that they had been invented merely as a means of conveniently grouping large numbers of individuals, which were all that "Nature produce[d]." This was the view of Buffon and his followers, but enjoyed little support from American biologists, and is not considered in this anthology.

The third approach, referred to by many authorities as the "biological species concept," is more comprehensive than, and incorporates, the old morphological species concept. It has been growing steadily in importance since the emergence of the new science of genetics at the turn of the century. Here the species is seen as an animal or plant group reproductively isolated from others. Shared genetic properties bind individuals "derived from the gene pool of a given species." Ecological, geographical and other factors also play their role.[3]

George Gaylord Simpson has defined systematics as "the scientific study of the kinds and diversity of organisms and of any and all relationships among them," while taxonomy is described as "the theoretical study of classification, including its bases, principles, procedures, and rules." Classification involves the practical application of taxonomic principles, that is, "the ordering of animals into groups . . . on the basis of their relationships [i.e.] associations by continuity, similarity, or both."[4]

The effects of geography upon speciation were considered in the 1850's by both Charles Darwin and his young contemporary, Alfred Russel Wallace, who arrived independently at the theory of natural selection. Wallace, however, carried his biogeographic investigations to greater lengths. In the *Geographic Distribution of Animals* (1876) and later in *Island Life* (1880), Wallace argued that "variation" and "struggle for existence" were tied in with changes in climate and geography. If these changes had been extreme, he postulated, new species would be formed within a matter of centuries or possibly within several generations. Moritz Wagner had earlier put this viewpoint into theoretical form, stating that individual members of a species must move beyond their normal geographic range and isolate themselves for a lengthy period. Only in this fashion could "the formation of a real variety" be accomplished. Both Wagner and Darwin, however, failed to emphasize the distinction between the development of differences within a racial group and those which appeared among the descendants of a group of escapes in isolation. David Starr Jordan, the American ichthyologist, commented upon this point in 1898:

> "Distinctness is in direct proportion to isolation . . . The degree of resemblance among individuals is in direct proportion to the freedom of their movements, and variations within what we call specific limits is again proportionate to the barriers which prevent equal and perfect diffusion . . . Local peculiarities disappear with wide association, and are intensified when individuals of similar peculiarities are kept together."[5]

While evolution was an accepted doctrine in the United States by the 1870's, most of its leading proponents were Neo-Lamarckians, rather than Darwinians. This group placed emphasis upon environmental action affecting organisms, rather than the struggle for existence, and suggested that animals inherited certain acquired characteristics developed by previous generations. These included the increased or decreased use of certain organs and other "intentional" adaptations, as for example longer necks in giraffes reaching for leaves atop tall trees. The great contributions made by American paleontologists (whose work is covered in another volume in this Arno Press series) were partially responsible for the development of this school of thought, since their many discoveries of fossil forms suggested continuing skeletal modifications through successive generations. The early writings of Joel Asaph Allen reprinted elsewhere in this series are reflective of the Neo-Lamarckian viewpoint. Another variant of this school was that put forward by Edward Drinker Cope, a leader and founder of this peculiarly American approach to evolutionary thought, and an outstanding paleontologist. In his *The Origin of the Fittest* (1887) and *The Primary Factors of Organic Evolution* (1896) both reprinted elsewhere in this series, Cope suggested that some form of mental effort, coupled with "kinetogenesis" (development by motion) were additional factors to be considered. Neo-Lamarckians insisted that evolutionary development followed a definite pattern, rather than a random course, as the Darwinians argued.

American natural scientists also considered other theoretical approaches to evolution during the late years of the 19th century among them that of Wilhelm Roux (1850-1924), a German embryologist and exponent of developmental mechanics. Roux argued that development was a two-phase process, beginning in the embryonic stage. There, the developing structure of an animal determined its function. During the second, post-natal, stage, that of *functional development,* the action, or function, of the organs and other parts of an animal affected its structure. In the developing fertile egg cell, Roux suggested, there would be found a complicated machine-like construction. When the developing cell divided, this structure would also divide into its constituent, though still complex parts. In an experiment with a fertilized frog's egg which had just divided into two cells, Roux destroyed one half, and the other developed into half an embryo. Subsequent research in both Europe and America, however, proved Roux's theories favoring preformation untenable, while Neo-Lamarckianism was forced to give way in the face of growing evidence from the new field of genetics.[6]

Several European authorities, working independently, rediscovered the laws of heredity previously worked out by the Austrian monk Gregor Mendel. William Bateson (1861-1926), on the faculty of Cambridge University, published *Materials for the Study of Variation* (1894), in which he attempted to account for the causes of discontinuous variation found in his laboratory experiments. Though he attacked the older morphological approach, it was clear that he was not yet free of it. Not until the uncovering of Mendel's papers on heredity in 1900 by Hugo De Vries (1848-1935) and others did the significance of Bateson's laboratory results and the importance of heredity finally become clear. While De Vries, working primarily with plants, mistakenly tried to demonstrate that mutational jumps could be made within a single generation, Bateson and others took a more cautious approach. Bateson came up with the guiding principles of genetics and gave the new science its name (1906). He and others experimented largely with birds. The Frenchman Lucien Cuenot (1866-1951) extended Mendelian thinking to mammals. A. D. Darbishire (1879-1915) an Englishman, and W. E. Castle (1867-1962), an American, carried on extensive laboratory experiments with mice, guinea pigs, and rats. The American, Francis Sumner (1874-1945), ran the first extensive field studies of the role of genetics in mammals. Genetics, however, did not gain wide acceptance without strong initial opposition from conservative proponents of older viewpoints.[7]

A number of authorities, such as Mayr and Blackwelder, have emphasized that systematists are often confused by their unwillingness or inability to understand the difference between the species as a biological category and the species as a taxon. The former is defined in terms of the "reproductive relationship" of one population to another. The latter involves the attempt by systematists to place particular specimens in the proper categories. Opinions differ as to whether biological and taxonomic species coincide reasonably closely. Mayr has also pointed out that one must bear in mind that there are:

> "many populations in nature that have progressed only part of the way toward species status. They may have acquired some of the attributes of distinct species and lack others. One or another of the three most characteristic properties of species — reproductive isolation, ecological difference, and morphological distinguishability — is in such cases only incompletely developed."[8]

This book makes no attempt to resolve differences of opinion between proponents of various species concepts, but does present papers dealing with various aspects of the subject which have interested biologists during the last century. When examined in conjunction with another volume in this Arno Press series, *The Species Problem,* edited by Ernst Mayr, useful light will be shed on both new and older approaches to the question. The emphasis here is upon American and Canadian writers and practitioners, though several articles by one European, Karl Jordan, whose work had great influence upon certain of his American colleagues, is also included.

The selections in this book are divided into four groups. The first consists of two articles, one by Theodore Gill (1837-1914) and the other by Ernst Mayr (1904-), which summarize the state of systematics as seen by adherents of old and new approaches to the subject. Gill's article, written in 1907, appeared at a time when the theoretical and practical aspects of genetics had not yet made its influence felt. Mayr's piece, on the other hand, examines the field at the end of the 1950's, by which time the new systematics had generally been accepted in its essentials. Gill was a zoologist on the staffs of George Washington University and the Smithsonian. At home in many fields, he was an authority on fish and mammals. Mayr was on the staff of the American Museum of Natural History for more than two decades, and has since 1953 been a member of the Harvard faculty, where he has been Alexander Agassiz Professor of Zoology and Director of the Museum of Comparative Zoology. He is an authority on ornithology and systematics.

The second group of articles ("c" through "g") deals in general with the role of evolution in systematics. John T. Gulick (1832-1923), a native of Hawaii and a graduate of Williams College, was a missionary to China and Japan, as well as a working naturalist. His is one of the early papers, perhaps the first, in which population differences were ascribed to stochastic processes, rather than selection. James Mark Baldwin (1861-1934), a psychologist, taught at Toronto, Princeton, and Johns Hopkins Universities, as well as the National University of Mexico. His article is a statement of the Baldwin Principle, or Baldwin Effect. This holds that with suitable modifications, the phenotype (the set of characteristics an organism manifests, as distinguished from the genes it possesses) may remain in "a favorable environment until selection has achieved [its] genetic fixation."

Karl Jordan (1861-1959), born in Germany, was an authority on the morphology and anatomy of the *Physopoda* (thrips), mechanical selection, and the distinctions between geographical and non-geographical variation. After teaching for a short period at the Academy of Forestry in Münden and the Agricultural College at Hildesheim, Germany, he was for nearly half a century Curator of Entomology and later Director of the Zoological Museum at Tring, England. His paper is a restatement of the process of geographic speciation. Alpheus Hyatt (1838-1902) was a zoologist and paleontologist at the Museum of Comparative Zoology at Harvard. Sewall

Wright (1889-   ), a geneticist, was for many years on the faculties of the Universities of Chicago and Wisconsin. His is a seminal founding paper in population genetics in which the diversity of evolutionary processes, including genetic drift, were stressed.

The third section ("h" through "n") contains articles dealing with specific and subspecific criteria in systematics. David Starr Jordan (1851-1931) was a physician, educator and naturalist who taught at Cornell and Indiana Universities and was President of Indiana and later Stanford. An authority on vertebrate zoology, he was in charge of U.S. fur seal and salmon investigations for the federal government before and after the turn of the century. Percy Algernon Taverner (1875-1947), originally trained as an architect, was a Canadian ornithologist on the staff of the Ornithological Division of the National Museum of Canada, and for some time its chief. He developed the first system of bird banding in North America, later widely copied in the U.S. and elsewhere. Author of several standard works on the birds of Canada, he was interested in avian systematics and distribution.

Henry Fairfield Osborn (1857-1935) had a notable career as Professor of Zoology at Princeton, Curator of Paleontology and President of the Board of Trustees at the American Museum of Natural History, and Dean of the Faculty of Pure Science at Columbia University. Osborn's paper refuted the De Vriesian theory that single mutations gave rise to new species, and stressed the importance of small genetic changes. Alfred C. Kinsey (1894-1958) was Professor of Zoology at the University of Indiana, an entomologist who was an authority on gall wasps. From 1942 until his death, he was concerned with human sexual behavior. George Gaylord Simpson (1902-   ), a paleozoologist, mammalogist, and systematist, has been on the staffs of the American Museum of Natural History, the Museum of Comparative Zoology at

Harvard, and the University of Arizona.

The fourth group of selections ("o" through "u") deal with avian systematics. Ernest Thompson Seton (1860-1945) was English-born, but spent his adult life in Canada and the United States, where he became a noted writer and illustrator on animal subjects. In the article reprinted here, he argued that while scientific nomenclature was properly the province of taxonomists, vernacular names were within the province of the people. Such names, he added, should have a uniqueness "in accord with the genius of our language." Hermon C. Bumpus (1862-1943), was professor of Zoology at Brown University, a member of the staff of the American Museum of Natural History, and President of Tufts University. His specialties included invertebrate zoology, ornithology, and entomology. His article on the Introduced [English] Sparrow provided evidence for the efficacy of stabilizing selection.

Frank M. Chapman (1864-1945) was for well over half a century a member of the Department of Ornithology at the American Museum of Natural History, and Joel Asaph Allen's successor as its curator. One of the leading American ornithologists of his generation, he was an authority on the birds of the Western Hemisphere. Alden Holmes Miller (1906-1965), who at his death was Director of the Museum of Vertebrate Zoology at the University of California, Berkeley, was an ornithologist of distinction. Three articles by Ernst Mayr dealing with avian speciation and relating this subject to distributional factors concludes this section.

With the final four selections, we turn to mammalian systematics. Francis B. Sumner (1874-1945) was affiliated with the U.S. Bureau of Fisheries and subsequently held a professorship at the Scripps Institute of Oceanography. A systematist, he also made important contributions in ichthyology, marine ecology, and evolution. His studies of *Peromyscus* during the first three decades of this century were the first to work out genetic principles for mammals in the field, and established which ecological factors controlled speciation in this genus. E. Raymond Hall (1902-   ) is Summerfield Distinguished Professor Emeritus of Zoology and former Director of the Museum of Natural History at the University of Kansas. He is the author, with Keith R. Kelson, of *Mammals of North America*. His article summarizes criteria used for determining speciation in mammals, and it may be useful to compare his conclusions with those reached by Simpson for zoology and paleontology (item "m") and Mayr for birds (item "s"), all of which originally appeared together in the *Annals of the New York Academy of Science* (1943).

April, 1974   Keir B. Sterling
                                                   Tarrytown, N. Y.

---

(1) R. E. Blackwelder, *Taxonomy*, New York, John Wiley, 1968, p. 366.
(2) See Mayr's discussion in "Illiger and the Biological Species Concept," *Journal of the History of Biology*, Fall, 1968, vol. I, no. 2, pp. 163-164.
(3) Ibid. See also Julian Huxley, ed., *The New Systematics*, London, Oxford, University Press, 1940.
(4) G. G. Simpson, *Principles of Animal Taxonomy*, New York, Columbia University Press, 1961, pp. 6-11.
(5) D. S. Jordan, *Footnotes to Evolution*, New York, D. Appleton, 1913, p. 195. I have altered the order of Jordan's remarks here.
(6) See Wilhelm Roux, "The Problems, Methods, and Scope of Developmental Mechanics," (trans. by W. M. Wheeler) *Lectures, Marine Biology Laboratory, Woods Hole, for 1894*, Woods Hole, Mass., 1895, pp. 149-190.
(7) L. C. Dunn, *A Short History of Genetics*, New York, Mc Graw Hill, 1965, chs. 6 and 8; Hugo De Vries, *Species and Varieties, their origin by Mutation*, 2nd ed., Chicago, Open Court, 1906.
(8) Ernst Mayr, *Animal Species and Evolution*, Cambridge, Harvard University Press, 1963, pp. 21-22,24; Blackwelder, *Taxonomy*, pp. 166,168.

SYSTEMATIC ZOOLOGY: ITS PROGRESS AND PURPOSE

BY

THEODORE GILL

FROM THE SMITHSONIAN REPORT FOR 1907, PAGES 449–472
(WITH PLATES I–XIV)

(No. 1842)

WASHINGTON
GOVERNMENT PRINTING OFFICE
1908

JOHN RAY, 1627-1705.

SYSTEMATIC ZOOLOGY: ITS PROGRESS AND PURPOSE.[a]

By Theodore Gill.

It is most fitting that in this year, when the scientific world is commemorating the natal centenaries of two naturalists who have been regarded as the chief systematists of their times, consideration should be given to the subject and object of their old pursuits. Carl Linné, whose bicentenary has been celebrated, was the man who first provided an elaborate code of laws for the nomenclature of all the kingdoms of nature and set an example to others by provision of concise and apt diagnoses of the groups and species he recognized. Louis Agassiz, who was born during the centenary year of Linné, gave a grand impulse to the study of nature in his adopted country, raised it in popular esteem, taught new methods of work, and directed to new lines of investigation.

Of all the students of nature from the time of Aristotle to the century of Linné, none requires present notice as a systematic zoologist except John Ray, who was the true scientific father of the Swede. Born in 1627, he flourished in England during the last quarter of the seventeenth century, and died only two years before the birth of Linné.

JOHN RAY.

It was long ago truly affirmed by Edwin Lankester that "Ray has been pronounced by Cuvier to be the first true systematist of the animal kingdom, and the principal guide of Linné in the department of nature."[b] He, indeed, made a pathway in the zoological field which Linné was glad to follow, and to some extent he anticipated the brightest thoughts of the great Swede. He, for example, in a dichotomous systematic table of the animal kingdom,[c] first combined the lunged fish-like aquatic and hairy quadruped viviparous animals in a special category (Vivipara) in contrast with all the other ver-

---

[a] Address before the Section of Systematic Zoology, Seventh International Zoological Congress, August 20, 1907.—Reprinted from Science, Oct. 18, 1907, with verbal modifications and additional notes.

[b] Lankester, Edwin, "The Correspondence of John Ray," 1848, p. 485.

[c] Ray, John, "Synopsis Methodica Animalium Quadrupedum et Serpentini Generis," 1693, p. 53.

tebrates, leaving to Linné only the privilege of giving a name to the class. He recognized a group of lung-bearing animals distinguished by a heart with a single ventricle, including quadrupeds and serpents, and thus appreciated better than Linné the class which the latter named Amphibia. He likewise gave the anatomical characters, based on the heart, blood, and lungs, which Linné used for his classes.[a]

THE BEGINNINGS OF SYSTEMATIC ZOOLOGY.

Systematic zoology is a vast subject, and any address devoted to it must necessarily be very partial. It need only be partial for such an assemblage of masters in zoology as I have the great honor to address, and I shall confine the present discourse to a review of some of the elements which have made systematic zoology what it now is. I will venture, too, to submit reasons why we may have to take a somewhat different view of the achievements of some men than did our early predecessors. If in doing so I may appear to be dogmatic, I entreat you in advance to insert all the " ifs " and " I thinks " and " perhaps " that you may deem to be necessary. For the present purpose, the work of two who exercised, each for a considerable time, a paramount influence on opinion and procedure deserves notice, especially because there has been much misapprehension respecting their benefits to natural science. The two were Carl Linné and Georges Cuvier; the one exercised dictatorship from the middle of the eighteenth century till some time after its close; the other was almost

---

[a] The " Synopsis Methodica Animalium Quadrupedum et Serpentini Generis " of Ray is very scarce, and the account of his views given in various works misleading; therefore his arrangement of the Animal Kingdom, so far as the Vertebrates are involved, is here reproduced (from p. 53):

Animalium Tabula generalis.

Animalia sunt vel
- *Sanguinea*, eáque vel
  - *Pulmone respirantia*, corde ventriculis prædito,
    - *Duobus*
      - *Vivipara* [=MAMMALIA Linn.]
        - *Aquatica*; cetaceum genus.
        - *Terrestria*, Quadrupedia, vel, ut *Manati* etiam complectantur, pilosa. Animalia hujus generis amphibia terrestribus annumeramus.
      - *Ovipara*, Aves [=AVES Linn.]
    - *Unico*, Quadrupedia vivipara & Serpentes. [=AMPHIBIA pp. Linn.]
  - *Branchiis respirantia*, Pisces sanguinei præter Cetaceos omnes. [=PISCES and AMPHIBIA NANTES Linn.]
- *Exanguia*. [=INVERTEBRATA]

The arrangement of the Invertebrates is not better (nor worse) than that of Linné; that of the Vertebrates is better. Furthermore, Ray segregated the Vertebrates (as Sanguinea) from the Invertebrates (Exanguia), which wise arrangement Linné did not adopt.

Carolus Linnæus (Carl von Linné), 1707-1778.

equally dominant from the first quarter of the last century to well into the third quarter. No other men approached either of these two in influence on the work of contemporaries or successors. The evil features, as well as the good, were transmitted to and adopted by later authors. Therefore, a notice of those features may assist us to a correct judgment of the history of our subject, and may help to show why the disciples of the great Swede, as well as the great Frenchman, complicated many problems they investigated. Sufficient time has elapsed to enable us to judge knowingly and impartially.

CARL VON LINNÉ.

Linné needs no present eulogy, for this year his praises have resounded over the whole world. Born just two centuries ago (1707), he published the first edition of the "Systema Naturæ" in 1735, and his last (twelfth) in 1766. The various editions mark to some extent the steps of man's progress in the knowledge of nature during the time limited by the respective dates.

Linné's industry was great, his sympathies widespread, and his method in large part good. Compare the "Systema Naturæ" and other publications of Linné with works published by earlier authors, and the reason for the active appreciation and esteem which greeted his work will be obvious. The typographical dress and the clearness of expression left no doubt as to what the author meant, and enabled the student to readily grasp his intentions. His boldness in giving expression to new ideas insured success when they deserved it. Although Ray had already recognized four of the great groups or classes of vertebrates, he had not named two of them, and there were vernacular terms only for the birds and fishes. Linné, for the first time, applied names to the other groups, and admirable ones they were. Mammalia and Amphibia were the coinage of Linné and are still retained; Mammalia or mammals by all; Amphibia or amphibians by the majority for one of the classes now adopted.

A great advance, too—an inspiration of genius, indeed—was the segregation of the animals combined under the class of mammals. Popular prejudice was long universal and is still largely against the idea involved. Sacred writ and classical poetry were against it. It seemed quite unnatural to separate aquatic whales from the fishes which they resembled so much in form and associate them with terrestrial hairy quadrupeds. How difficult it was to accustom one's self to the idea is hard for naturalists of the present day to appreciate. Linné himself was not reconciled to the idea till 1758, although Ray had more than hinted at it over three-score years before. At last, however, in no uncertain terms, he promulgated it. It was a triumph of science over popular impressions; of anatomical consideration over superficial views.

But mingled with the great benefactions were many views which long influenced naturalists, but which modern zoology has overthrown.

LINNÆAN CLASSES.

After the tentative arrangements published in the original first, second, and sixth editions of the "Systema," Linné thoroughly revised his work, and first consistently applied the binomial method of nomenclature to all species in the tenth edition, published in 1758. Six classes were admitted with equal rank, no category being recognized between the class and kingdom. The classes were the Mammalia or mammals, Aves or birds, Amphibia, Pisces, Insecta, and Vermes. The first four of these classes correspond mainly to the Aves and nameless groups of Ray.

During the Linnæan period of activity the invertebrates were little understood, and his treatment of that enormous host, referred to his two classes Insecta and Vermes, contrasts rather than compares with that at the present time. Naturally, the vertebrates were much better comprehended, and all such then known, with a single exception, were distributed among four classes just named, the Mammalia, Aves, Amphibia, and Pisces. The solitary exception of exclusion of a true vertebrate from its fellows was the reference of the genus *Myxine* to the Vermes, next to *Teredo*, the shipworm. The first two classes were adopted with the same limits they now have, but the Amphibia and Pisces were constituted in a truly remarkable manner. The class of Amphibia was a creation of Linné, and was simply contrasted with his Pisces by having a lung of some kind or other ("*pulmone arbitrario*"), while the Pisces have exposed branchiæ ("*branchiis externis*"). The Amphibia, thus defined, were made to include as orders: (1) Reptiles, or Reptilia, having feet; (2) Serpentes, footless, and (3) Nantes, having fins.

Under the Nantes were first grouped the lampreys, the selachians, the anglers (*Lophius*), and the sturgeons (*Acipenser*). In the twelfth edition were added *Cyclopterus*, *Balistes*, *Ostracion*, *Tetrodon*, *Diodon*, *Centriscus*, *Syngnathus*, and *Pegasus*. The Nantes were added to the Amphibia partly because of the assumption that the branchial pouches of the lampreys and the selachians were lungs and partly on the authority of Dr. Alexander Garden, of Charleston, S. C., who mistook the peculiar transversely expanded and partly double air-bladder of *Diodon* for a lung. With such errors of observation as a basis, Linné apparently assumed that all the associated genera also had lungs. Gmelin, in his edition of the "Systema Naturæ" (generally called the thirteenth), corrected this error, and returned all the Nantes to the class of Pisces, thus reverting to the older view of Linné himself. The Pisces of Linné included only the genera left after the exclusion of those just named and also of *Myxine*, which last was referred to the class of Vermes between the leeches (*Hirudo*) and the shipworms (*Teredo*).

LINNÆAN GENERA.

The genera of Linné were intended and thought by him to be natural,[a] and natural groups some of the so-called genera were, but present opinion assigns to most of them a very different valuation from that given in the " System Naturæ." Some of the genera of Invertebrates were extremely comprehensive. For example, *Asterias* included all the members of the modern classes of Stelleroidea or Asteroidea and Ophiuroidea; *Echinus* was coequal with the Echinoidea; *Cancer*, *Scorpio*, *Aranea*, *Scolopendra*, and *Julus* were essentially coextensive with orders or even higher groups of the zoologists of the present time. Others were so heterogeneous that they can not be compared with modern groups. Thus *Holothuria*, in the last edition of the " Systema," was made to include four holothurians in the modern sense, a worm, a physaliid, and three tunicates; in other terms, the so-called genus included representatives of four different classes and even branches of the animal kingdom.

It has been stated by various writers that the genera of Linné were essentially coequal with the families of modern authors, but, as has been indicated, such is by no means the case. Other striking exceptions to the generalization may be shown.

Not a few of the genera of Vertebrates, although not of the superlative rank as several of the Invertebrates, were equivalent to orders of modern zoology; such were, in the main, *Simia*, *Testudo*, *Vespertilio*, and *Rana*. *Simia* included all the anthropoid Primates except man; *Vespertilio* was equivalent to the order Chiroptera less the genus *Noctilio;* *Testudo* was exactly equal to the order Testudinata or Chelonia; *Rana* to the order Salientia or Anura. A number of other genera of one or few species known to Linné were also of ordinal or subordinal value.

In striking contrast with the range of variation of such genera were others, of which several, well represented in northern waters, may be taken as examples. *Scorpæna* was distinguished simply because it had skinny tags on the head;[b] *Labrus* because it had free membranous extensions behind the dorsal spines;[c] and *Cobitis* be-

---

[a] Classis et ordo est sapientiæ, genus et species Naturæ opus.—Linn. Syst. Nat., I, 13.

[b] *Scorpæna*. Caput cirris adspersum.

[c] *Labrus*. Pinna dorsalis ramento post spinas notata.

cause it had the caudal peduncle of regular height[a] and scarcely constricted as usual in fishes. These characters are of such slight systematic importance that they have not been used in the diagnoses of the genera by modern ichthyologists. Further, use of them misled even Linné as well as his successors. Some of the consequences may be noticed.

The close affinity of the "Norway haddock" or Swedish Kungsfisk or Rödfisk (*Sebastes marinus*) to the typical *Scorpæna* was unperceived and that species referred to *Perca* and even confounded with a *Serranus*.

The typical *Labri* of the northern seas do, indeed, have filiform processes of the fin membrane behind the dorsal spines, but most of the species referred by Linné to *Labrus* do not, and among them is a common sunfish (*auritus* = *Lepomis auritus*) of America.

The genus *Cobitis* was made to include Cyprinodonts of the genera *Anableps* and *Fundulus*, and thus were associated fishes differentiated from the Loaches by characters of immeasurably more importance than the trivial one which was the sole cause of their juxtaposition.

Another conspicuous instance of a trivial character used as generic, and contrasting with very important differentials of species included under the same genus, is furnished by *Esox*. The essential Linnæan diagnostic character is the protrusion of the lower jaw.[b] Nine species were referred to the genus which represent no less than eight distinct and, mostly, widely separated families of modern systematists.[c] Several of the species do not have the prominent lower jaw, and one of them (*Lepisosteus osseus* of modern ichthyology) is especially distinguished by Linné himself on account of the shorter lower jaw.[d]

But the most marked cases of insignificance of characters used to differentiate by the side of those serving for combination are found in the class Amphibia.

The genus *Lacerta* is made to include all but one of the pedate Lizards, and the Crocodilians as well as the salamanders, but the "dragons," or Agamoid lizards with expansible ribs, are set apart in an independent genus.[e]

---

[a] *Cobitis*. Corpus vix ad caudam angustatum.

[b] *Esox*. Mandibula inferior longior, punctata. Syst. Nat., '58; '66, 424.

[c] The species are (1) *Sphyræna* (Sphyrænidæ), (2) *osseus* (Lepisosteidæ), (3) *Vulpes* (Albulidæ), (4) *Synodus* (Synodontidæ), (5) *lucius* (Luciidæ), (6) *belone* (Esocidæ), (7) *hepsetus* and (8) *brasiliensis* (Exocœtidæ), and (9) *gymnocephalus* (Chirocentridæ). Syst. Nat., '66, 513–517.

[d] Mandibula inferior brevior. Syst. Nat., '66, 516.

[e] *Lacerta*. "Corpus (Testa Alisve) nudum, caudatum" contrasting with *Draco*. "Corpus Alis volatile." Syst. Nat., '66, 349.

The genus *Coluber* was intended to embrace all the snakes, except those with a rattle or undivided abdominal and caudal scutes,[a] and hence the vipers and copperheads, so very closely related to the rattlesnakes, were combined with ordinary snakes instead of with their true relations.[b]

Many of the genera of Linné, in fact, were very incongruous, and the great Swede not infrequently failed to interpret and apply their characters in the allocation of species. A few cases furnished by common European or American fishes will illustrate what is meant.

Specimens of the common gunnell or butterfish were received by Linné at different times and once referred to his genus *Ophidion* and at another time to the genus *Blennius*, and the same species stands under both names in the last two editions of his " Systema."

The common toadfish of the Americans (*Opsanus tau*) was placed in the genus *Gadus* (*tau*) and a nearly related species of the Indian Ocean was referred to the genus *Cottus* (*grunniens*).

The common ten-pounder of the American coast served as the type and only species of the genus *Elops*, and also as a second species of the genus *Argentina*, although the characters given were in decided discord with those used for the latter genus, and in perfect harmony with those employed for the distinction of the former genus. Indeed, it might be properly assumed that the ascription of the *Argentina carolina* to *Argentina* was simply a matter of misplacement of a manuscript leaf, and such it may be even now considered, although the error is continued in the twelfth edition, having escaped the notice of Linné.

LINNÆAN NOMENCLATURE.

The code of nomenclature devised by Linné was in many respects admirable, but he did not provide sufficiently for the principle of priority in nomenclature. He set the example of changing a name given by himself or by others, when he thought a better one could be substituted; he also felt at liberty to change the intent of a genus. A few examples of many cases may illustrate.

In 1756 the name *Salacia* was given to the Portuguese man-of-war; in 1758 the name *Holothuria* was substituted; in 1766 the latter name was retained, but with a very different diagnosis, and for the first time four holothurians in the modern sense of the word were introduced.

In 1756 the names *Cenchris* and *Crotalophorus* were used for genera, two years later renamed *Boa* and *Crotalus*. In 1756 Artedi's

---

[a] *Coluber.* " Scuta abdominalia: squamæ caudales " contrasting with " *Crotalus.* Scuta abdominalia caudaliaque cum crepitaculo " and " *Boa.* Scuta abdominalia caudaliaque absque crepitaculo." Syst. Nat., '66, 349.

[b] As an example of *Coluber* a figure (tab. 3, fig. 2) of a snake with venom fangs was given.

name, *Catodon*, was retained for the sperm whale, and Artedi's *Physeter* mainly for the killers (*Orca*); but in 1758 *Physeter* was taken up for the sperm whale, for which it has been retained ever since, except by a very few naturalists.

In 1756 and 1758 *Ophidion* was used for an acanthopterygian jugular fish—the common northern butterfish, or gunnell, now generally called *Pholis*—but in 1766, under the guise of *Ophidium*, it was transferred to the Apodes and primarily used for the soft-finned (supposedly) apodal type, which is still known as the genus *Ophidium*.

In 1756 and 1758 *Trichechus* was used for the manatee alone, while the walrus was correctly associated with the seals, but in 1766 the very retrograde step was taken of associating the walrus with the manatee and retaining for the two the name *Trichechus*. Many naturalists persist to the present day in keeping the name for the walrus alone.

The example thus set by the master was naturally followed by his disciples. Many felt at liberty to change names and range of genera as they thought best and great confusion resulted, which has continued more or less down to this year of grace, 1907.

Many of the evils which have been the consequence could have been prevented or rectified if the British Association for the Advancement of Science had been logical in the code (often admirable) which it published in 1842. Instead, however, of accepting the edition of the "System Naturæ" (tenth) in which Linné first introduced the binomial nomenclature as the starting point, they preferred homage to an individual rather than truth to a principle, and insisted on the twelfth edition as the initial volume of zoological nomenclature.[a] The unfortunate consequences have been manifold. Such consequences are the natural outcome of illogical and ill-considered action and must always sooner or later follow. After these many years almost all naturalists have acceded to the adoption of the tenth edition.

---

[a] The addition of some genera and many species in the twelfth edition marked an advance in that respect of Linné's knowledge, but otherwise no firmer grasp of the materials on hand became manifest. On the contrary, one familiar with the species can scarcely fail to recognize an increase of a tendency to impatience in dealing with details and not seldom a snap judgment in the allocation of species in the genera. Indeed, under the circumstances, it would have been better if the last edition had never been published. No one who has not critically examined the Systema can have an idea of the extent of discrepancy between the generic diagnoses and contents, the duplication of species under different genera, the mistakes of synonymy, and other faults. It has been affirmed that Strickland, the chief formulator of the B. A. Code of 1842, had preferred the tenth edition, but was overruled by his less informed associates of the committee on nomenclature.

If the vertebrates were so much misunderstood by Linné, it may naturally be supposed that the invertebrates were equally or still less understood. Only one interesting case, however, can be referred to. In the ninth edition of the "Systema Naturæ" Linné had a monotypic genus *Salacia* (p. 79) with a species named *Physalis* which was evidently a *Physalia* as long understood. In the tenth edition the name *Holothuria* was substituted for *Salacia* and no holothurians in the modern sense were recognized. In the twelfth edition all the species of the former edition were retained, but the diagnosis was altered and four holothurians of recent authors were added, and thus animals of different subkingdoms or branches were confounded in the genus. Now, if we accept the tenth edition of the "Systema" as the starting of our nomenclature, obviously *Holothuria* can not be used as it has been for these many years, and it must be revived in place of *Physalia*, notwithstanding the laments of those who are distressed by such a change. The echinoderms now called holothurians must be renamed. We can imagine the clamor that will arise when some one attempts the change.[a]

Another fault of less moment—indeed a matter of taste chiefly—was committed by Linné. Very numerous names of plants and animals occur in the writings of various ancient authors and were mostly unidentifiable in the time of Linné. He drew upon this store with utter disregard of the consequences for names of new genera. Most of the ancient names can now be identified and associated with the species to which they were of old applied, and the incongruity of the old and new usage is striking. For example, *Dasypus*, a Greek name of the hare, was perverted to the armadillos; *Trochilus*, a name of an Egyptian plover, was misused for the humming birds; *Amia*, a name for a tunny, was transferred to the bowfin of North America. There was not the slightest justification for such perversion of the names in analogy or fitness of any kind; there was no real excuse for it. At the commencement of Linné's career (1737), the learned Professor Dillenius, of Oxford, strongly protested against such misusage for plant genera, but the sinner persisted in the practice till the end. Naturally his scholars and later nomenclators followed the bad example, and systematic zoology is consequently burdened with an immense number of the grossest and most misleading misapplications of ancient names revolting to the classicist and historian alike.

---

[a] After undisturbed possession of the name for nearly a century and a half, two naturalists independently, in the same month (August, 1907), challenged the right of the Holothurians to the name *Holothuria*, and contended that the typical holothurians of the moderns should be renamed—*Bohadschia* for the genus and *Bohadschiidæ* for the family. T. Gill published his remarks in Science for August 9 (p. 185) and F. Poche in the Zoologischer Anzeiger for August 20 (p. 106).

The influence of Linné continued to be felt and his system to be adopted until a new century had for some time run its course. Meanwhile, in France, a great zoologist was developing a new system which was published at length in 1817, and anew with many modifications a dozen years later (1829).

GEORGES LÉOPOLD CHRÉTIEN FRÉDÉRIC DAGOBERT CUVIER.

Georges Cuvier (born 1769) claimed [a] that before him naturalists, like Linné, distributed all the invertebrates among two classes.[b] In 1795 he published an account of memorable anatomical investigations of the invertebrates and ranged them all under six classes: Mollusks, crustaceans, insects, worms, echinoderms, and zoophytes. This was certainly a great improvement over previous systematic efforts, but from our standpoint crude in many respects. It was, however, necessarily crude, for naturalists had to learn how to look as well as to think.

Cuvier later essayed to do for the animal kingdom alone what Linné did for all the kingdoms of nature. So greatly had the number of known animals increased, however, that he did not attempt to give diagnoses of the species, but merely named them, mostly in footnotes. His superior knowledge of anatomy enabled him to institute great improvements in the system. He also first recognized the desirability of combining in major groups classes concerning which a number of general propositions could be postulated.

It was in 1812 that Cuvier presented to the Academy of Sciences [c] his celebrated memoir on a new association of the classes of the animal kingdom, proposing a special category which he called branch (embranchment), and marshaling the classes recognized by him under four primary groups: (1) the Vertebrates or Animaux vertébrés; (2) the Mollusks or Animaux mollusques; (3) the Articulates or Animaux articulés, and (4) the Radiates or Animaux rayonnés. These were adopted in the "Règne Animal." In the first (1817) edition, as in the second (1829–1830), nineteen classes were recognized, and in the latter too little consideration was given to the numerous propositions for the improvement of the system that had been suggested and urged meanwhile.

It has been generally assumed that Cuvier's work was fully up to the high mark of the times of publication, and for many years the classification which he gave was accepted by the majority of naturalists as the standard of right. To such extent was this the case that his classification of fishes and the families then defined was retained to at least the penultimate decade of the last century by the first

---

[a] Règne Animal, 1817, I, 61.
[b] Scopoli and Storr admitted more classes.
[c] Ann. Mus. Hist. Nat., Paris, 1812, 19, 73–84.

GEORGES CUVIER, 1769-1832.

ichthyologists of France. Nevertheless the work was quite backward in some respects and exercised a retardative influence in that the preeminent regard in which the great Frenchman was held and the proclivity to follow a leader kept many from paying any attention to superior work emanating from Cuvier's contemporaries.

It is by no means always the naturalist who enjoys the greatest reputation for the time being that anticipates future conclusions. A Frenchman who held a small place in the world's regard in comparison with Cuvier advanced far ahead of him in certain ideas. Henri Marie Ducrotay de Blainville (1777–1850) was the man. When Cuvier (1817) associated the marsupials in the same order as the true carnivores and the monotremes with the edentates, Blainville (1816) contrasted the marsupials and monotremes as Implacentals ("Didelphes") against the ordinary Placentals ("Monodelphes"). While later (1829) Cuvier still approximated the marsupials to the carnivores, but in a distinct order between the carnivores and the rodents, and still retained the monotremes as a tribe of the edentates, Blainville (1834) recognized the marsupials and monotremes as distinct subclasses of mammals and had proposed the names Monodelphes, Didelphes and Ornithodelphes, still largely used by the most advanced of modern theorologists.

Against the action of Cuvier in ranging all the hoofed mammals in two orders, the pachyderms (including the elephants) and the ruminants, may be cited the philosophical ideas of Blainville (1816), who combined the same in two very different orders, the Ongulogrades and the Gravigrades (elephants), and distributed the normal Ongulogrades under two groups, those with unpaired hoofs (Imparidigitates) and those with paired hoofs (Paridigitates), thus anticipating the classification of Owen and recent naturalists by very many years.[a]

Cuvier's treatment of the amphibia of Linné equally contrasted with Blainville's. As late as 1829 the great French naturalist still treated the batrachians as a mere order of reptiles of a single family, and the crocodilians as a simple family of Saurians. On the other hand, as early as 1816 Blainville had given subclass rank to the naked amphibians with four orders, and also ordinal rank to the crocodilians, and a little later (1822) he raised the subclasses to class rank. Still more, Blainville early (1816) recognized that the so-called naked serpents

---

[a] A more familiar instance of difference between Cuvier and Blainville is that involving the systematic relation of the aye-aye (*Cheiromys* or *Daubentonia*).. Cuvier, to the end of his life, referred it to the Rodents and, in the last edition of the Règne Animal, interposed it between the Flying-Squirrels (*Pteromys*) and Marmots (*Arctomys*). Blainville, on the contrary, as early as 1816, associated it with the Lemurs, to which it is now universally conceded to be most nearly related. The evidence is very conclusive. Was Cuvier unable to appreciate its significance or was he too opinionated to recant a determination once formed?

were true amphibians and gave satisfactory reasons for his assumption, though to the last Cuvier (1829) considered them to be merely a family of the ophidians. As Blainville claimed, he based his classification on anatomical facts.[a]

A pupil of Blainville, Ferdinand L'Herminier of the island of Guadeloupe, at the instance and following the lead of his master (1827), undertook the comparative study of the sternal apparatus of birds and thereby discovered a key to the natural relationship of many types which anticipated by many years the views now current. For instance, L'Herminier first correctly appreciated the differences of the ostriches and penguins from other birds, the difference between the passerines and swifts, the homogeneity of the former, and the affinity of the humming birds and the swifts. Meanwhile Cuvier, like Linné, was content to accept as the basis for his primary classification of birds, superficial modifications of the bill and feet (toes and nails) which led to many unnatural associations as well as separations, but which nevertheless have been persisted in even to our own day by many ornithologists.

Now what could have been the underlying idea which hindered the foremost comparative anatomist of his age from the recognition of what are now considered to be elementary truths and what enabled Blainville to forge so far ahead? Cuvier manifestly allowed himself to be influenced by the sentiment prevalent in his time, that systematic zoology and comparative anatomy were different provinces. It may, indeed, seem strange to make the charge against the preeminent anatomist, that he failed because he neglected anatomy, but it must become evident to all who carefully analyze his zoological works that such neglect was his prime fault. He, in fact, treated zoology and anatomy as distinct disciplines, or, in other words, he acted on the principle that animals should be considered independently from two points of view, the superficial, for those facts easily observed, and the profound, or anatomical characters. Blainville, on the contrary, almost from the first, considered animals in their entirety and would estimate their relations by a view of the entire organization.[b] Yet

---

[a] " Ses bases sont anatomiques et surtout tirées de la consideration du crâne." Bull. Sci. Soc. Philom., 1816, p. 111.

[b] The comparison instituted between Cuvier and Blainville is more than just to the former. Cuvier was not only eight years older than Blainville but longer and better established in scientific circles; he had also more control of scientific material and laboratories; he must also have known the anatomical facts as well as Blainville. The difference between the two, therefore, resulted from the manner in which they used the facilities at hand and the intellectual powers they applied to the consideration of the problems involved. While sometimes Cuvier more nearly anticipated conclusions now adopted, Blainville did so much more frequently. If, then, modern biologists are right, the man who approached nearest to them must be regarded pro tanto as the superior.

Henri de Blainville, 1777-1850.

PIERRE LATREILLE, 1762-1833.

the sentiment then prevalent was reflected by one who enjoyed a high reputation for a time as a " philosophical zoologist "—William Swainson. In "A Treatise on the Geography and Classification of Animals" (1836, p. 173), the author complained that " Cuvier rested his distinctions * * * upon characters which, however good, are not always comprehensible, except to the anatomist. The utility of his system, for general use, is consequently much diminished, and it gives the student an impression (certainly an erroneous one) that the internal, and not the external, structure of an animal alone decides its place in nature." It was long before such a mischievous opinion was discarded.

Cuvier was regarded almost universally by his contemporaries, and long afterwards, in the words of his intellectual successor, Louis Agassiz, as " the greatest zoologist of all time." [a] In view of the facts already cited and innumerable others that could be added, however, the contemporary verdict must be somewhat modified. Cuvier was a very great man of most impressive personality, wide versatility, extraordinary industry, vast knowledge of zoological and anatomical details, an excellent historian, a useful critic, and of good judgment in affairs generally, but, although a greater all-round man, as a systematic zoologist he was not the equal of a couple of his French contemporaries, Blainville and Latreille. We have either to admit this conclusion or confess that our now universally admitted views are wrong. Nevertheless, Cuvier's work was of great importance, and he first brought to the aid of systematic zoology the new science of vertebrate paleontology.

CUVIER AND PALEONTOLOGY.

The animals, and especially the vertebrates, of past ages were practically unknown to the early zoologists, and when they had large collections, as did Volta of the fishes of Mount Bolca, they identified them with modern species, or, with Scheuchzer, might consider a giant salamander as a man witness of the deluge—" Homo diluvii testis! " It was not until Cuvier, with superior knowledge of skeletal details, examined numerous bones unearthed from the Tertiary beds about Paris, that the complete distinction of animals of ancient formations from living species was recognized. Then was afforded the first glimpse of extinct faunas destined to far outnumber the existing one, but so imperfect was the great paleontologist's foresight of what lay in store for the future that he enunciated a dogma which was long accepted as sacrosanct; he called it the law of correlation of structure.

---

[a] Agassiz, " Essay on Classification," p. 286.

A striking and even amusing example of its exposition and its failure I have previously drawn attention to.

Professor Huxley, in his excellent "Introduction to the Classification of Animals" (published in 1869), in his first chapter, "On Classification in General," concluded a consideration of Cuvier's law of the correlation of structure with the following paragraphs:

> Cuvier, the more servile of whose imitators are fond of citing his mistaken doctrines as to the nature of the methods of paleontology against the conclusions of logic and of common sense, has put this so strongly that I can not refrain from quoting his words.[a]
>
> But I doubt if anyone would have divined, if untaught by observation, that all ruminants have the foot cleft, and that they alone have it. I doubt if anyone would have divined that there are frontal horns only in this class; that those among them which have sharp canines for the most part lack horns.
>
> However, since these relations are constant, they must have some sufficient cause; but since we are ignorant of it, we must make good the defect of the theory by means of observation; it enables us to establish empirical laws, which become almost as certain as rational laws, when they rest on sufficiently repeated observations; so that now, whoso sees merely the print of a cleft [fourchu] foot may conclude that the animal which left this impression ruminated, and this conclusion is as certain as any other in physics or morals. This footprint alone, then, yields to him who observes it, the form of the teeth, the form of the jaws, the form of the vertebræ, the form of all the bones of the legs, of the thighs, of the shoulders, and of the pelvis of the animal which has passed by; it is a surer mark than all those of Zadig.

The first perusal of these remarks would occasion surprise to some and immediately induce a second, more careful reading to ascertain whether they had not been misunderstood. Men much inferior in capacity to Cuvier or Huxley may at once recall living exceptions to the positive statements as to the coordination of the "foot cleft" with the other characteristics specified. One of the most common of domesticated animals—the hog—may come up before the "mind's eye," if not the actual eye at the moment, to refute any such correlation as was claimed. Nevertheless, notwithstanding the fierce controversial literature centered on Huxley, I have never seen an allusion to the lapse. And yet everyone will admit that the hog has the "foot cleft" just as any ruminant, but the "form of the teeth" and the form of some vertebræ are quite different from those of the ruminants and, of course, the multiple stomach and adaptation for rumination do not exist in the hog. That any one mammalogist should make such a slip is not very surprising, but that a second equally learned should follow in his steps is a singular psychological curiosity. To make the case clearer to those not well acquainted with mammals, I may add that *because the feet are cleft in the same manner* in the hogs as in

---

[a] "Ossemens fossiles," ed. 4me, tome 1r, p. 184.

RICHARD OWEN, 1804-1892.

JŌHANNES MÜLLER, 1801-1858.

the ruminants,[a] both groups have long been associated in the same order under the name Paridigitates or Artiodactyles, contrasting with another (comprising the tapirs, rhinocerotids and horses) called Imparidigitates or Perissodactyles.[b]

I need scarcely add that the law of correlation applied by Cuvier to the structures of ruminants entirely fails in the case of many extinct mammals discovered since Cuvier's days. Zadig would have been completely nonplussed if he could have seen the imprint of an Agriochœrid, a Uintatheriid, a Menodontid, or a Chalicotheriid.

The value of this law was long insisted upon by many. Some of the best anatomists, as Blainville, protested against its universality, but one who ranked with Cuvier in skill and knowledge of anatomy, Richard Owen, long upheld Cuvier's view. "You may be aware," he wrote in 1843, "that M. De Blainville contends that the ground— viz, a single bone or articular facet of a bone—on which Cuvier deemed it possible to reconstruct the entire animal, is inadequate to that end. In this opinion I do not coincide."[c] The many mistakes Owen made in attempting to apply the principle proves how well Blainville's contrary opinion was justified.

The numberless remains of past animals, exhumed from the many formations which the animals themselves distinguished, have entailed constant revisions of systems resulting from clearer comprehension of the development of the animal kingdom. Such revision, too, must continue for many generations yet to come.

CUVIER AND ANATOMY.

The failure to sufficiently apply anatomy to systematic zoology was especially exemplified in the treatment of the fishes which absorbed so much of Cuvier's attention in later years. He, as well as his associate, gave accounts of the visceral anatomy and was led— often misled—to conclusions respecting relations by his dissections,

---

[a] The only essential difference between the feet of hogs and ordinary ruminants is of degree in the development of the lateral hooflets. There is every gradation among the Artiodactyles, recent and extinct, between forms having the lateral hoofs aborted and those with all developed and accumbent on the ground, as in the Hippopotamus.

[b] Huxley had previously, in 1856, in an article "On the method of Palæontology" (Annals and Magazine of Natural History, 2d series, vol. 18, p. 49), called attention to the oversight of Cuvier; he quoted, in French, the passage here rendered in English, and added: "I confess that, considering the Pig has a cloven foot, and does not ruminate, the last assertion appears to me to be a little strong. But my object is not to criticise Cuvier," etc. Apparently he had forgotten the facts, however, when he wrote the Introduction referred to.

[c] Owen, Amer. Journ. Sci. and Arts, XLV, 1843, 185.

but he failed to receive enlightenment by examination of the numerous skeletons he had made. Those skeletons, pregnant with significance for the future, had no meaning for Cuvier; he never learned how to utilize them for the fishes as he did those of the mammals. His colleague and successor, Valenciennes, in the great "Histoire Naturelle des Poissons," was equally unappreciative of the importance of comparative osteology for comprehension of the mutual relations of the groups of fishes.

CUVIER'S SUCCESSORS.

The same defect in method or logic that characterized Cuvier's work was manifested by his great English successor in range of knowledge of comparative anatomy, Richard Owen. His families, for the most part, were the artificial assemblages brought together by zoologists on account of superficial characters and too often without rigorous attention to the applicability of the characters assigned. Much better was the work of the greatest naturalist of all, Johannes Müller, who advanced our knowledge of the systematic relations of all classes of vertebrates as well as invertebrates. But all were unable to free themselves from the incubus of the popular idea that all branchiferous vertebrates formed a unit to be compared with birds and mammals. Several propositions to segregate, as classes, Amphioxus and the chondropterygians had been made, and Louis Agassiz deserves the credit of claiming class value for the myzonts or marsipobranchs as well as the selachians. But it was left to Ernst Haeckel, a pupil of Müller, still happily living, to divest himself entirely of ancient prejudices and appreciate the interrelationship of the primary sections of the vertebrate branch. He for the first time (1866) set apart the amphioxids in a group opposed to all other vertebrates, then docked off the marsipobranchs from all the rest, and collected the classes generally recognized in essentially the same manner as is now prevalent. We may differ from Haeckel as to his classes of fishes and dipnoans, but his correctness in the action just noticed will be conceded by most, if not all, systematic zoologists to-day.

EMBRYOLOGY.

While Cuvier was still flourishing, a school of investigators into the developmental changes of the individual in different classes, and among them the vertebrates, was accumulating new material which should be of use to the systematic zoologist. Chief of these was Karl Ernst von Baer. In various memoirs (1826 et seq.) he subjected the major classification of animals to a critical review from an embryological point of view, recognized, with Cuvier, the existence of four distinct plans which he called types and characterized them in em-

Louis Agassiz, 1807-1873.

ERNST HAECKEL, 1834-

bryological terms—*Evolutio radiata*, *Evolutio contorta* (mollusks), *Evolutio gemina* (articulates) and *Evolutio bigemina* (vertebrates). The last were successively differentiated on account of the embryonic changes from the fishes to the mammals. "These Beiträge," Louis Agassiz justly affirmed, "and the papers in which Cuvier characterized for the first time the four great types of the animal kingdom, are among the most important contributions to general zoology ever published."

One of the most notable results, so far as systematic zoology was involved, was the deduction forced on Kowalevsky by his investigation of the embryology of tunicates, that those animals, long associated with acephalous mollusks, were really degenerate and specialized protovertebrates. This view early won general acceptance.

While embryology was very successfully used for the elucidation of systematic zoology its facts were often misunderstood and perverted. For instance, the cetaceans were regarded as low because they had a primitive fish-like form, although it must be obvious to all logical zoologists of the present time that they are derived from a quadruped stock; snakes have been also regarded as inferior in the scale because no legs were developed, although it would be now conceded by every instructed herpetologist that they are descendants of footed or lizard-like reptiles. *Ammocœtes* was considered as higher than *Petromyzon* "inasmuch as the division of the lips indicates a tendency toward a formation of a distinct upper and lower jaw," but we now know that *Ammocœtes* is the larval form of *Petromyzon*. Still more pertinent examples might be adduced without number for the inferior systematic grades, orders, families, genera, species, etc. The words high and low were used when generalized and specialized were really meant and those words, pregnant with mischief, often led their users astray as well as the students to which they were addressed.

PHILOSOPHICAL ZOOLOGY.

As knowledge of the various animal groups increased and countless new species were piling up, yearning arose to discover principles underlying the enormous mass of accumulating details, and the excogitations of various naturalists resulted in some curious speculation and expression in classificatory form. They called their outpourings philosophy or philosophical zoology, and philosophers they were called by others.

Some of the philosophers grouped animals according to supposed degrees of nervous sensibility;[a] some according to the relations of

---

[a] Lamarck (1812) contended for three categories of animals: (1) Apathetic animals and (2) sensitive animals among the invertebrates, and (3) intelligent animals, equivalent to the vertebrates.

parts to a center or an axis;[a] some under groups supposed to correspond with different systems of the body, as the alimentary, the vascular, the respiratory, the skeletal and the muscular,[b] and some would accord to each of the senses definite groups.[c]

Equally, if not more extravagant, views were entertained by many naturalists that creative power delighted in the symmetry of numbers and in circular arrangements.[d] It was contended that all groups of animals represented analogous groups in successively diminishing circles; that in a perfect system there were a definite number of subkingdoms, an equal number of classes in each subkingdom, of orders in each class, of suborders, of families, of genera, of subgenera, etc. Some maintained that three was the regnant number, others upheld four, others seven, but the most numerous and influential school con-

---

[a] Blainville (1816) proposed to divide the animal kingdom into three subkingdoms: (1) The Artiomorphes, having a bilateral form, (2) the Actinomorphes, having a radiate form, and (3) the Heteromorphes (mainly sponges and protozoans), having an irregular form.

[b] Oken (1802–1847) gave expression to his varying views in several differing classifications. In one scheme (El. Physiophilosophy, 1847, 511 et seq.) he claimed that there were five "circles" corresponding with the "animal systems:" (1) Intestinal animals (Protozoa and Radiates); (2) Vascular, sexual animals (Mollusks); (3) Respiratory, cutaneous animals (Articulates); (4) Sarcose animals (Vertebrates except mammals), and (5) Aistheseozoa, or animals "with all * * * organs of sense perfectly developed" (mammals).

[c] Oken maintained (1802–1847) "that the animal classes are virtually nothing else than a representation of the sense organs, and that they must be arranged in accordance with them. Thus, strictly speaking, there are only five animal classes: Dermatozoa (skin or touch animals), or the Invertebrata; Glossozoa (tongue animals), or the fishes * * *; Rhinozoa (nose animals), or the reptiles * * *; Otozoa (ear animals), or the birds; Ophthalmozoa (eye animals), or the Thricozoa (mammals) * * *. But since all vegetative systems are subordinate to the tegument or general sense of feeling, the Dermatozoa divide into just as many or corresponding divisions, which on account of the quantity of their contents, may be for the sake of convenience also termed classes."—Oken, El. Physiophilosophy, 1847, p. xi. For the many other assumptions on similar and divergent lines the reader must refer to the "Elements of Physiophilosophy" (1847).

[d] The style of argumentation used by the number-philosophers had long before been employed by Sizzi, a contemporary and antagonist of Galileo, who proved, to his own satisfaction, that there could be no more than seven planets. The inconsequentiality is remarkable. "There are seven windows given to animals in the domicile of the head, through which the air is admitted to the tabernacle of the body, to enlighten, to warm, and to nourish it; which windows are the principal parts of the microcosm, or little world—two nostrils, two eyes, two ears, and one mouth. So in the heavens, as in a macrocosm, or great world, there are two favorable stars, Jupiter and Venus; two unpropitious, Mars and Saturn; two luminaries, the Sun and the Moon; and Mercury alone, undecided and indifferent. From which, and from many other phenomena of nature, such as the seven metals, etc., which it were tedious to enumerate, we gather that the number of planets is necessarily seven." More follows of like tenor.

KARL VON BAER, 1792-1876.

JEAN LAMARCK, 1744-1829.

tended for five. Exactly what the philosophers thought they meant, or what strange visions they may have conjured up may never be known. But for a time (1822–1842) the school of quinarians, as they were called, claimed most of the naturalists of Britain. The most zealous of the school (William Swainson) was especially displeased with the developmental hypothesis of Lamarck and characterized the "speculations" of the great Frenchman "not merely as fanciful, but absolutely absurd."

But it was the much-contemned hypothesis of descent with modifications that was destined at last to relieve biological science of the wild and irrational speculations and classifications of the nature-philosophers, physiophilosophers, circularians, quinarians, trinarians, septenarians, and their like that flourished during the first half of the past century.

DEVELOPMENT THEORY.

Although there had been previous indications of belief that transmutation of species might have been a cause for the diversity of animal life, Jean Baptiste Pierre Antoine de Monet de Lamarck (1809) first framed a hypothesis that had a logical basis, although weakened by unproved postulates. In view of those weaknesses, it was easy to bring forth many facts that seemed to militate unanswerably against it, and such were well put forward by Cuvier; as the hypothesis, too, was very unpopular, it was for a long time stifled. In the meanwhile geological and paleontological investigation, comparative morphology, physiology, embryology, and zoogeography, as well as systematic zoology, were revealing innumerable facts that pointed all in the same direction and were only explicable collectively by the assumption that they were the result of original community of origin and subsequent deviation by gradual changes from time to time. The facts were at length collocated with extreme skill by Charles Darwin (1859) and a rational explanation of their evolution by means of natural selection made the new development theory acceptable to well-informed naturalists and logical thinkers generally.

SEQUENCE OF GROUPS.

It had been almost the universal custom from olden time, as well as during the Linnæan era, to commence the enumeration or catalogues of animals with the forms exhibiting most analogy with man and consequently the highest in the scale of organic nature. As long as species were assumed to be individually created this was perhaps the most natural course, and at least had the advantage of proceeding from the comparatively known to the almost unknown. A significant and noteworthy exception to this mode of procedure among the old naturalists was afforded by Lamarck (1809 et seq.), the precursor in this respect, as well as in recognition of descent, of the modern school.

When it became generally recognized that there had been always a progression and development from antecedent forms, naturally there was a change in the manner of exposition of a series, and the lowest forms were taken as the initial ones and followed by those successively higher in the scale of beings. Even when old prejudices were administered to and the highest animals put first in a work, it was often done in a reversed series; that is, after the supposed natural ascensive series had been determined on, that series was simply reversed in order that the highest should be the first and the lowest the last. Many of our text-books of zoology still have this characteristic, but are being rapidly replaced by those exhibiting the phyletic series.

HISTOLOGY.

One of the most noteworthy modifications of systematic zoology was the fruit of histological research. In 1839 Theodor Schwann, incited by the brilliant results of Matthias Jacob Schleiden's researches (1838) in vegetal histology, and at the suggestion of Johannes Müller, undertook investigations which led him to consider that the animal frame was built up from innumerable cells variously modified to form the different systems and organs of which it is composed. Ultimately the animals thus developed were segregated by Ernst Haeckel, and the animal kingdom was limited to them, while the simple unicellular animals which had been already designated as Protozoa were associated with unicellular plants under the general term Protista. One of the prominent features of this idea was accepted by Thomas Henry Huxley (1874) with, however, the very important modification of retaining the old conception, the animal kingdom, and keeping the name Protozoa as the collective name of the unicellular animals while taking a suggested name of Haeckel's (Metazoa) for the multicellular animals.

GRADUAL DELIMITATION OF GENERA.

As has been already noted, the animal genera of Linné were mostly extremely comprehensive, answering, when natural groups, to families, superfamilies, and even orders or classes of modern naturalists. Such contrast, however, with others of the Linnæan genera, and when this fact became recognized and it was discovered that the large genera embraced types exhibiting many differences in detail, the latter were subdivided; early in the past century, at first owing especially to French and German naturalists, the subdivision of old genera on approximately present lines was commenced and applied at

CHARLES DARWIN, 1809-1882.

THEODOR SCHWANN, 1810-1882.

different times to various classes. It is noteworthy that in some instances the authors of the new genera quite abruptly changed their minds regarding the nature of such groups. For example, Lacépède, in 1798, in the closing lecture of his course at the Museum of Natural History, recognized only 51 genera of mammals, but a few months later (in 1799), in a " tableau," admitted and defined 84 genera.

It seems to be generally supposed that there has been an uninterrupted tendency among zoologists to refinement and increase of number of genera to the present time, but such is by no means the case. Half a century ago and more some ornithologists subdivided old genera and made new ones to an extent to which none of the present time is prepared to go. For example, Charles Bonaparte, Prince of Canino, required eleven genera of gulls to include those now congregated in one. About the same time, some herpetologists were equally radical. Leopold J. F. J. Fitzinger, in 1843, distributed species which are now combined by all in the genus *Anolis* among no less than fifteen genera. The genus *Bufo*, as now understood, was split by some herpetologists into a dozen or more. These are only samples of numberless analogous cases.

THE OLD AND THE NEW.

A comparison of systematic zoology at its dawn with that of the present time is rather a contrast of different themes.

The old naturalists believed that all species of animals were created as such by a divine fiat; the modern consider that all animals are derivatives from former ones and that their differences have been acquired during descent and development.

The Linnæans based their systems on superficial characteristics, and the moderns take into consideration the entire animal.

The early systematists assumed that characters drawn from structures or parts most useful to the animals were the best guides to the relationship of the animals; the latest ones have learned to distrust the evidential value of similarity of structures unaccompanied by similarity of all parts. The former were guided mainly by physiological characters; the latter take morphological ones.

The Linnæans confined their generalizations to few categories— genera, orders, classes; the moderns exhibit the manifold modifications and coordinations of all structural parts in many categories— genera, subfamilies, families, superfamilies and various higher groups.

The old naturalists believed more or less in the existence of a regular chain of beings from high to low; the new ones recognize the boundless ramifications of all animal stocks.

The elders assumed certain forms as highest and ranged their series from high to low; the sons commence their series with the most generalized types and progress from the generalized to the more specialized.

PROSPECTS AND NEEDS.

In numerous old systematic and descriptive works—but in many cases not very old—the skeleton and other anatomical details were noticed in connection with the species described, but not seldom some of those details, if rightly interpreted, would be in contravention of the classification adopted. In fact, the anatomy was to all intents and purposes treated as an offering of curious but useless information. Such conceptions, happily, are mainly—but not entirely—of the past, and we may live to welcome the day when every animal will be treated as whole. Systematic zoology will then be regarded as the expression of our knowledge of the entire structure and as an attempted equation of the results obtained by investigations of all kinds. In fact, systematic zoology is simply an attempt to estimate the relative importance of all structural details and to correlate them so that their relative values shall become most evident. It is the scientific outcome of all anatomical or morphological knowledge and the aim is to arrange the animal groups in such a manner as to show best their genetic relations and the successive steps of divergence from more or less generalized stocks.

One consummation devoutly to be wished for is general acceptance of a standard for comparison and the use of terms with as nearly equal values as the circumstances permit. There is a great difference in the use of taxonomic names for the different classes of the animal kingdom. The difference is especially great between usage for the birds and that for the fishes. For the former class, genera, families and orders are based on characters of a very trivial kind. For example, the family of Turdidæ, or thrushes, relieved of formal verbiage, has been distinguished from neighboring families solely because the young have spots on the breast, but even this distinction is now known to fail in some instances. Extremely few, if any, of the families of oscine birds are based on characters of a kind which would be regarded as of family value in other classes of vertebrates. On the other hand, many of the families and genera of fishes are made by some excellent authorities to include types separated by striking peculiarities of the skeleton as well as the exterior. The mammals are a class whose treatment has been mostly intermediate between that for the birds and that for the fishes. Its divisions, inferior as well as comprehensive, have been founded on anatomical characters to a greater extent than for any other class. Its students are numer-

THOMAS HENRY HUXLEY, 1825-1895.

ous and qualified. Mammalogy might therefore well be accepted as a standard for taxonomy, and the groups adopted for it be imitated as nearly as the differing conditions will admit. The families of birds would then be much reduced in number and those of fishes increased. All the active herpetologists and ichthyologists of the United States have subordinated their own beliefs and ideas as to what would have been most desirable, to a greater or less extent, to approximate the desirable reduction of the terms admitted by them to a standard uniform with that adopted by mammalogists. If others would likewise sacrifice their own predilections, the lamentable inequality of usage now prevalent would be much less; such congruity would be to the great advantage of comparative taxonomy.

In these days of extreme specialization one of the greatest needs in our universities is a professor of systematic zoology with whom conference may be held as to the propriety of any systematic modification resulting from special investigation of the anatomy of any organ or part, or of any group of animals. Such conference might prevent the publication of many propositions due to exclusive consideration of an isolated subject. Perhaps the designation of systematic morphology might better indicate the nature of the suggested course. The consummation, however, it must be admitted, is more desirable than probable.

I have intentionally refrained from any consideration of the work of living zoologists. If I had undertaken this, the task of selection would have been very difficult, and at any rate the time demanded for proper consideration would have been much more than that requisite for the reminder of past discoveries. The progress of systematic zoology during recent years has been in accelerated ratio, and not a few of those whose achievements have helped to put zoology at its present level are in Boston to-day. It is from the summit of the elevation they have enabled us to reach that we look back to the deeds of old masters and can determine, better than their contemporaries or immediate successors, their relative merits.

[NOTE.—The name "Linné" has been used because it was the one that the author assumed in the last (12th) edition of his great work. The title page has "CAROLI A LINNÉ, * * * Systema Naturæ," etc. After he was ennobled (1761) he dropped the Latin form and resumed the vernacular with the addition of *a* or *von*.]

Where Are We?

ERNST MAYR

Museum of Comparative Zoology, Harvard University, Cambridge, Mass.

It would seem altogether fitting that this year's symposium be devoted to a consideration of the evolutionary theory. We are celebrating the one hundredth anniversary of the publication of Darwin's *Origin of Species* and the occasion of such a centenary is the proper moment to review the past and consider the future. It is almost universally recognized that no other theory has had as vast an impact on biology as the theory of evolution, emphasizing as it does that no organic being can be fully understood except by considering its history. Whether we are dealing with structures, functions or habits, individually or in their interactions, their understanding is one-sided and partial without a consideration of the historic processes that brought them into being.

The history of the first one hundred years since the publication of the *Origin of Species* is a fascinating one. With its many controversies, its false starts and converging pathways toward a solution of the open problems, one might say that we have completed a full circle and that we are closer to Darwin now and to Darwin's original concepts than we have been at any time during the intervening periods. The science of genetics has perhaps been responsible for the greatest deviation from the original Darwinism, in the theories of the early Mendelians, but it has likewise been responsible for a return to the original Darwinism of 1859 from which Darwin himself had deviated in the later editions of his work. Let us look at this history a little more closely. The non-biologist often overlooks that there is more to the evolutionary theory than the problem of the reality of common descent. That the living kinds of animals and plants were descended from common ancestors was accepted by the vast majority of biologists within two decades after 1859. By about 1890 essential agreement had been reached among zoologists as to the major outlines of the phylogeny of animals (very little progress in this area having been made since then). What then were the still open problems of evolutionary biology? Actually there were big ones and little ones.

The most serious omission in Darwin's theory was undoubtedly, that it did not account for the raw "material of evolution." Darwin was fully aware of the enormous store of genetic variability but had only vague notions concerning its source. Like nearly all of his contemporaries, except Mendel, he believed in blending inheritance and had to account for the rapid depletion of variability which is the inevitable result of blending. When Jenkins (1869) brought this to Darwin's attention, he embraced Lamarckism apparently as the only possible escape from this dilemma. By the time Darwin died, the thinking on inheritance was one of the most backward areas in biology.

The changes that followed were dramatic and by necessity of great importance for the evolutionary theory. I would like to take this opportunity to give a short outline of the changes of the genetic theory so far as it concerns evolution. Essentially one may distinguish three periods:

The Mendelian Period

When Weismann demonstrated the internal contradictions and improbabilities of the Lamarckian theory and emphasized the separation of soma and germ plasm, he created a new intellectual climate for genetic thinking which gave the long overlooked work of Mendel new meaning. It is not surprising that, as a consequence, after thirty-six years of neglect, Mendel was rediscovered simultaneously by three different authors. The school of genetics which dominated biology following the rediscovery of Mendel's laws, is frequently referred to as Mendelism. Those of its representatives who were interested in evolution, particularly De Vries and Bateson, were unfortunately typological saltationists and proclaimed that the new science of genetics required a saltationist interpretation of evolution. As De Vries (1906) expressed it: "The theory of mutation assumes that new species and varieties are produced from existing forms by certain leaps." But even to the more moderate Mendelians an organism was at the mercy of his mutations. To Morgan, as late as 1932, evolution was "due to occasional lucky mutants which happened to be useful rather than harmful. As one of Morgan's disciples, I held fairly similar views at that time." (Dobzhansky, 1959, p. 254). As a consequence, the period from 1900 to about 1920 saw a sharp cleavage, an almost bridgeless gap, between the evolution-minded naturalists on one hand and the experimental geneticists on the other hand. Reconciliation became possible when it was realized that not all genetic changes were spectacular mutations but that most mutations were small and inconspicuous (Baur, East, etc.) and that indeed there is no difference between mutations and the the so-called small variations which Darwin and the naturalists had considered as the principal material of evolution. When it was finally demonstrated by Schmidt (1917), Sumner (1924) and Goldschmidt (1920) that the genetic differences between subspecies, the incipient species of the

Darwinians, were genetic in nature (contrary to the claims of the mutationists) and indicative of a genetic basis for quantitative characters, all reason for disagreement between geneticists and evolutionists had been removed. The stage was set for a new genetic interpretation of evolution, for a second phase of evolutionary genetics.

Classical Population Genetics

Population genetics clearly has two separate roots. One of them, indicated by the names Fisher, Wright and Haldane, is mathematical and its contribution to the development of population genetics has been widely emphasized. The other one, indicated by such names as Sumner, Timofeeff-Ressovsky and Dobzhansky, had its roots in population systematics and is usually ignored. We will come back to the respective contributions of the two schools.

The emphasis in early population genetics was on the frequency of genes and on the control of this frequency by mutation, selection, and random events. Each gene was essentially treated as an independent unit favored or discriminated against by various causal factors. In order to permit mathematical treatment, numerous simplifying assumptions had to be made, such as that of an absolute selective value of a given gene. The great contribution of this period was that it restored the prestige of natural selection, which had been rather low among the geneticists active in the early decades of the century, and that it prepared the ground for a treatment of quantitative characters. Yet, this period was one of gross oversimplification. Evolutionary change was essentially presented as an input or output of genes, as the adding of certain beans to a beanbag and the withdrawing of others. This period of "beanbag genetics" was a necessary step in the development of our thinking, yet its shortcomings became obvious as a result of the work of the experimental population geneticists, the animal and plant breeders, and the population systematists, which ushered in a third era of evolutionary genetics.

The Newer Population Genetics

The next advance was characterized by an increasing emphasis on the interaction of genes. Not only individuals but even populations were no longer described atomistically as aggregates of independent genes in various frequencies, but as integrated, coadapted complexes. A gene is no longer considered to have one absolute selective value, but rather a wide range of potential values that may extend from lethality to high selective superiority, depending on genetic background and on the constellation of environmental factors. I have referred to this new mode of thinking as the genetic "theory of relativity" (Mayr, 1955b). Dobzhansky's "balance theory" of genetic variation is one of its aspects. The thinking of this newer population genetics is in considerable contrast to that of the classical population genetics and even more so to that of early Mendelism. This change of view is not always realized, even by professional geneticists or by those evolutionists who have no contact with population genetics. Numerous new problems result from it, some of which we shall presently discuss in more detail.

Again, we should try to place this new development into a historical perspective and determine its relationship to the preceding period. There is no doubt that the classical period of population genetics was dominated by the mathematical analyses and models of Fisher (1930), Wright (1931) and Haldane (1932).

These authors, although sometimes disagreeing with each other in detail or emphasis, have worked out an impressive mathematical theory of genetical variation and evolutionary change. But what, precisely, has been the contribution of this mathematical school to the evolutionary theory, if I may be permitted to ask such a provocative question?

Some of the younger evolutionists, perhaps not too well acquainted with the earlier literature, have ascribed to this school many of the major components of the modern synthetic theory of evolution. Others, like Waddington (1957), have questioned the magnitude of its influence.

The maintenance of genetic variability in populations, expressed in the Hardy-Weinberg formula, was known since 1908, and Gulick and other nineteenth century authors had proposed theories of genetic drift. These earlier concepts antedate classical population genetics by a wide margin. Where the mathematical theory made concrete contributions, as in Fisher's theory of balanced polymorphism, the mathematics is of the simplest kind. Observing all this, Waddington asks quite rightly: "What then gave the mathematical theory its undoubtedly immense importance and prestige?" It seems to me that the main importance of the mathematical theory was that it gave mathematical rigor to qualitative statements long previously made. It was important to realize and to demonstrate mathematically how slight a selective advantage could lead to the spread of a gene in a population. Perhaps the main service of the mathematical theory was that in a subtle way it changed the mode of thinking about genetic factors and genetic events in evolution without necessarily making any startlingly novel contributions. However, I should perhaps leave it to Fisher, Wright, and Haldane to point out themselves what they consider their major contributions. That much seems certain that the interpretation in Mendelian terms of the inheritance of quantitative characters by Mather and the development of the more sophisticated modern views on the interaction of genetic factors, on coadaptation, and on genetic homeostasis would have been impossible without the foundation laid

The Contributions of Genetics

It seems to me that we have now reached a stage in the development of the evolutionary theory where one can look back without passion or prejudice and determine the respective contributions of experimental genetics, of systematics, of paleontology, of developmental biology, and of other branches of biology, to the synthetic theory of evolution. One should do this not for purely historical reasons or to establish priorities for prestige reasons, but because the planning of future work will be helped by a clear recognition of the potential contributions that can be made by the various collaborating branches of biology. And in view of the fact that this symposium is devoted to genetics and 20th century Darwinism, I shall single out the specific contributions which genetics has made to the synthetic theory of evolution. In view of the erroneous statements found in much of the current literature I shall try to discuss the history of some of the concepts which together form the synthetic theory. Such an emphasis on history may be a wholesome counterweight against the exceedingly unhistorical attitude of our current age.

Mutation

The idea of an ever-continuing origin of new genetic variation goes back to folklore (Zirkle, 1946). It was, as stated above, a cornerstone of Darwin's theory. The version of early genetics, the peculiar mutationism of De Vries, was an essentially negative contribution which retarded a real understanding of mutation for several decades. That generation of evolutionists, who associated with the term "mutation" what De Vries had meant by it, had to die off before the term could be reintroduced into the evolutionary theory in the broader meaning of later genetics. But genetics has given up not only the saltationist interpretation of mutationism but also its faith in "mutation pressure" as the driving force of evolutionary change, a concept widespread in the 1920's and 1930's. In both respects modern genetics has returned to the thinking of Darwin and the naturalists.

Population

The claim has been made by some population geneticists that population thinking and its application to the evolutionary theory is a contribution of genetics. This overlooks that population thinking was already strongly apparent in Darwin's own work, in particular his application of Malthus' thinking to the theory of natural selection. Population thinking was even more widespread and more concrete in the second half of the nineteenth century. Systematists, following the example set by Baird, Schlegel, and others became increasingly interested in the analysis of local populations. This reached its climax in the work of Heincke (1878) and his school on populations of marine fishes. At the same period population thinking was widespread among students of birds, butterflies, beetles, and snails. Indeed, it was apparent in the writings of almost any progressive animal taxonomist in the period between 1890 and 1930. The early population geneticists (except the mathematicians) had all either started as naturalists (Dobzhansky, Sumner) or had been in close contact with taxonomists, as was Goldschmidt at the Munich museum when he initiated his study of *Lymantria* populations. It is of more than historical significance that population thinking came into genetics from systematics and not the reverse (Mayr, 1955a).

Natural Selection

No one has claimed that the theory of natural selection is an invention of the geneticist. Yet the current generation is not fully aware how antiselectionist virtually all the early Mendelians were. Natural selection survived during that period in the writings of systematists like K. Jordan rather than in those of the mutationists. It is, of course, entirely consistent that he who believes that mutations make new species and that mutation pressure directs the course of evolution does not have any use for natural selection.

Genetic Drift

The idea that random processes affect the genetic contents of local populations and thus control their phenotypes preceded the coining of the term "genetic drift" by many decades, if not generations, as correctly pointed out by Wright (1951). It was the principal basis of Gulick's interpretation of the variation of *Achatinella* in the Hawaiian Islands but similar thoughts had been expressed by even earlier authors as well as later ones (e.g. Lloyd 1912, Hagedorn 1917). What Wright did was to separate clearly the respective contributions which mutation pressure, accidents of sampling, and natural selection make to gene frequencies in populations of various sizes.

Isolating Mechanisms

The term itself, coined in 1936 by Dobzhansky is recent. That the interbreeding of species is prevented by various factors in addition to sterility was, however, known many decades earlier. One finds, for instance, discussions of these mechanisms in the writings of Darwin, Seebohm, K. Jordan, and particularly Du Rietz (1930).

Geographic Variation

The fact of geographic variation was, of course, known long before Darwin. That the differences between geographic races have a similar genetic basis as the differences between individuals within a population was established by Schmidt (1917)

Sumner (1924), and Goldschmidt (1920) and was one of the main reasons for the downfall of mutationism on one hand and of Lamarckism on the other hand. This finding permitted a selectionist interpretation of slight differences among local populations that were obviously caused by differences in the environment. A Lamarckian interpretation of geographic variation was inevitable, as long as one accepted the theories of De Vries and Bateson.

I have presented this analysis not as an attempt to depreciate the contributions of genetics but rather by eliminating false claims to permit a sharper focussing on the unique contribution of genetics. As time goes by, it becomes increasingly evident that nothing compares in significance with the demonstration of the particulate nature of the genetic material. This recognition is of such overwhelming importance that it overshadows everything else. The new interpretation of natural selection, the understanding of the meaning of recombination, the new insight in the relation of gene and character, the interpretation of quantitative inheritance, and many other aspects of the genetic theory, all are ultimately inevitable consequences of the Mendelian theory.

The second reason for the great importance of genetics is that it has provided, time after time, a causal explanation for purely empirical generalizations established by systematists and other naturalists. This is true for almost every major evolutionary theory or concept. Let me illustrate it with only one or two examples.

Let us take, for instance, the theory of geographic speciation. One after the other of the reputed cases of sympatric speciation had to be reinterpreted during the past 70 years and eventually it was found in all well analyzed cases of speciation in animals that a period of geographic isolation had to be included in the process. This was a purely empirical finding for which no immediate reason was apparent. This need for geographic isolation was finally explained by population genetics which showed that a gene pool is such a well integrated and coadapted system that it cannot be divided into two equally well integrated gene pools without a period of geographic isolation of considerable duration. The purely extrinsic separation permits the independent occurrence of two separate integration processes which eventually lead to mutual reproductive isolation.

Or, let us look at the problem of the relation between embryology and evolution which has been a source of endless disputes since the 18th century. Most pre-Darwinian evolutionists and many of Darwin's contemporaries considered the development of the individual and phylogenetic change merely as two aspects of a single phenomenon. Stating this problem in terms of genetic information has greatly clarified the issue. The genetic material, handed down from generation to generation, can be considered as coded norms of reaction.

Individual development is the uncoding of the information made available through fertilization. Evolution, on the other hand, is the generating of ever-new codes of information, of ever-new norms of reaction. Stated in these terms it is at once evident that the superficial similarity between individual and phylogenetic development is entirely spurious. One is a functional, the other a historical phenomenon.

It seems to me that the basic problem of Lamarckism, the inheritance of acquired characters, can be likewise stated far more clearly in terms of codes of information than it has been previously possible. If one wanted to demonstrate the occurrence of such inheritance, one would have to prove that the developmental pathway is not a one way street and, as Weismann emphasized some 70 years ago, one would have to prove that the information from the peripheral end organs is fed back to the germ cells in the gonads and is used to recode the inherited information in an adaptively improved manner. I hardly need to dwell on the extraordinary improbability of such an occurrence as far as the higher organisms are concerned. Whether or not the situation is different in microorganisms is perhaps still an open question. There is at least a theoretical possibility, for such an interaction owing to the shortness of the pathway between the gene and the component of the phenotype which it controls.

I have come to the end of my short historical survey of the relation between genetics and the other branches of evolutionary science. It seems evident that there is a happy symbiosis among these various fields. The naturalist has access to a vast store of observational evidence on which he bases various empirical generalizations. It is the role of the geneticist to interpret these generalizations in terms of the genetic material and to test his conclusions by experiment. I can foresee no reason for a change in this historically established pattern of co-operation. The best evidence for its success is the modern synthetic theory of evolution.

I have entitled my discussion: "Where are we?" One of the methods of fixing one's position is to look back and try to reconstruct the steps taken to reach the present position. This we have done in our historical survey. A second method is a careful scrutiny of the surroundings of our position. What do we mean by 20th century Darwinism, and what do we mean by the synthetic theory of evolution? I think its essence can be characterized by two postulates: (1) that all the events that lead to the production of new genotypes, such as mutation, recombination, and fertilization are essentially random and not in any way whatsoever finalistic, and (2) that the order in the organic world, manifested in the numerous adaptations of organisms to the physical and biotic environment, is due to the ordering effects of natural selection.

Nothing has been discovered in the decades

since these principles were first clearly stated that is in any way in conflict with these basic assumptions. Yet we must realize that this is only a beginning. It has been claimed again and again in the last 30 years that evolutionary biology is exhausted as a field of research because the synthetic evolutionary theory has supplied all the answers. Let me emphasize, in a variation of Mark Twain's saying, that the news of the death of evolutionary biology is greatly exaggerated. Just how many unsolved problems there are becomes apparent as soon as one takes up individual areas and attempts to delimit the known from the unknown. Let me try to do this for the field of evolutionary genetics. Where are the pathways to the unknown?

NATURAL SELECTION

Let us begin with some problems relating to selection. In an immensely stimulating and thoughtful recent essay on this topic, Lerner (1959) has stated correctly: "What we have learned so far about natural selection is obviously only the beginning. What remains to be learned is immeasureably more." The power of natural selection is no longer questioned by any serious evolutionist. It is understood that it is a statistical, a populational phenomenon, the essence of which consists in differential reproduction. And as soon as this is fully understood it is evident that selection can be a creative process. Yet in spite of all these advances numerous unsolved problems remain. Let me single out only four aspects of natural selection which raise doubts in my troubled mind.

1. The selection of genes versus the selection of phenotypes. Haldane (1957) has recently pointed out that selection places a considerable strain upon populations. Too rapid a rate of simultaneous selection against too many genes might eliminate the entire population. By these considerations, Haldane has called attention to a neglected aspect of selection. But is this really a serious problem? Is this not a situation where the new genetic theory of relativity should be applied? Haldane himself emphasizes correctly that we are dealing with relative values and that phenotypes are exposed to selection, not genes. Those individuals which have the greatest number of minus genes have the lowest reproductive potential. Those with the greatest number of plus genes have, quite obviously, the highest reproductive potential. Natural selection thus can deal simultaneously with a great number of variable loci. The "goodness" or "badness" of a gene will be relative depending on the fitness of the averages of all the individuals of the population. And as the population density drops, even rather poor phenotypes may survive by not being exposed to competition from better ones. Total reproductive potential is another factor introducing a relative element. This is something we have to keep in mind when we compare *Drosophila* with man. If *Drosophila melanogaster* can withstand the impact of 300 R per generation because among the 500 fertilized eggs laid by a female, one will be a zygote with a combination of genes which has been compensated for radiation damage, it does not mean that man would be able to do likewise, even if he did step up his reproductive rate to a dozen children per family. To repeat once more, I feel that Haldane has called attention to an important problem when he asked the question: "How severe a selection pressure can a population endure at any one time?" But I am not sure that his answer is necessarily correct. And with this doubt we come to a second problem.

2. The measurement of fitness. Fitness, as we have seen, is a highly relative matter. When comparing the fitness of different genotypes, it is of crucial importance to find an objective yardstick. Fitness, as R. A. Fisher pointed out long ago, is best defined operationally as the relative contribution to the gene pool of the next generation (disregarding accidents of sampling). But how to measure it comparatively? It has become routine in population genetics to measure the fitness of whole chromosomes in homozygous condition as compared to the same chromosome in heterozygous condition paired with a balanced lethal. The chromosome then, on the basis of its viability in homozygous condition, is labelled lethal, semilethal, subvital, or viable, as the case may be. Frankly, I think this is utter nonsense, as I already pointed out in 1955. Homozygous chromosomes do not occur in nature in sexual species neither do balanced lethal tester chromosomes derived from alien populations. "To compare the relative performance of two combinations which do not occur in nature and utilize it as an index of viability for a chromosome does not appear to be particularly meaningful technique," as I put it very mildly (Mayr 1955b). Dobzhansky and Wallace have shown that there is very little relation between the fitness of a given chromosome in homozygous or in heterozygous condition. The mere fact that 20–40% of all chromosomes found in wild species of *Drosophila* are lethal when homozygous should be sufficient to indicate the inappropriateness of this terminology. According to the classical hypothesis of the genetic load of population, lethality of the homozygous chromosomes might be due to deficiencies or other lethal genes. Dobzhansky's work on synthetic lethals shows that this is not necessarily the case. It is becoming more evident from day to day, as postulated by the "balance theory" (Dobzhansky, 1955, 1959), that it is the interaction of genes which determines fitness. And in outbreeding organisms, in which homozygous chromosomes do not occur, selection is for high fitness in heterozygous condition. Since every chromosome in a sexual species is unique and every diploid combination is unique, it would seem futile to determine its fitness in any

way that is meaningful. Whether we like this conclusion or not, it seems to me that there is no way of measuring the fitness of chromosomes.

And, to carry this argument one step further, what is true for chromosomes is true for individual genes. The mathematical geneticists assign in their formulae an absolute selective value S or W to a given gene, let us say gene a. Whatever usefulness such a value W may have is at once destroyed by the cautious qualification that W is to be taken as the average of all the selective values which the genotype may have in different physical and biotic environments. It is obvious then that W instead of being an absolute and uniform value, is an exceedingly relative and heterogeneous one. Even more devastating for the usefulness of the value W is the fact that fitness of the gene a is determined not only by the physical and biotic environment in which the phenotype is placed, but by the genetic environment as well in which the gene a is placed. Depending on the genetic background the same gene may be "good" or "bad," indeed in an extreme case a normally valuable gene may well be near lethal.

If there is any validity in this argument, we may have to revise many of our ideas. For instance, is the old argument really true, that virtually all mutations are deleterious, because the same loci, having mutated thousands of times in their history must have reached the optimal genes since all inferior genes, except concealed recessives, must have long since been eliminated by natural selection. This argument makes two assumptions, both of which are demonstrably wrong, first, that the selective value of a gene is absolute, second, that the totality of environments (the physical, the biotic, and the genetic) is constant. As soon as we admit the continuous changes of the three environments, we must admit that many current mutations are far more "hopeful" than would be expected on the basis of the orthodox considerations.

But I would like to carry this argument still one step further. All evolution is a sequence of unique events. This is true not only for selective values in different genotypes and different environments, it is likewise true for segregation, crossing over, and whatever other events determine evolutionary success and evolutionary change. If we set up 10 parallel selection experiments, starting each from the same, apparently identical foundation stock, we are liable to get in the end 10 somewhat different answers. The more I study evolution the more I am impressed by the uniqueness, by the unpredictability, and by the unrepeatability of evolutionary events. Let me end this discussion with the provocative question: "Is it not perhaps a basic error of methodology to apply such a generalizing technique as mathematics to a field of unique events, as is organic evolution?"

3. The population as a unit of selection. Haldane (1932) again was the first to emphasize the importance of the population-as-a-whole, as a unit of selection. There are a number of attributes of species and populations which are not of any particular selective advantage to any single individual in a population but which are of great advantage to the population as a whole. This includes such factors as rate of mutation, degree of out-breeding, sex ratio, etc. It is easy to propose models by which the genetic basis of such factors can be changed. Yet such models must make certain unrealistic assumptions, such as the total isolation of each population and the total extinction of the less well adapted populations. Normal distribution patterns make these assumptions unlikely. Here again we need more thinking and perhaps more experimentation, fully taking into account that most natural populations are open systems.

4. Reproductive success. Natural selection owes its universal success to the fact that it applies no rigid artificial criteria. Its only criterion is that of success. Fitness consequently, is quite correctly defined as the contribution to the gene pool of the next generation. And yet this eminently practical principle has one Achilles heel: it also rewards pure reproductive superiority. A male bird of paradise who adds to the gaudiness of his plumage is rewarded in the same manner as the individual who invents a slight improvement in physiology which benefits the species as a whole, and possibly all descendant species. More and more cases are being discovered where genes have become established in a species not because they add in any way to the adaptiveness of the species (in fact they usually do exactly the opposite) but because natural selection is defenseless against them. Here is clearly a situation where the two terms "of high selective value" and "of high adaptive value" are not congruent. I strongly suspect that such selection for mere reproductive success may have frequently played a role in the extinction of species. And such extinction may be very rapid. Nearly all the cases where such genes were found in living populations are cases where these genes were utilized for balanced polymorphism giving to the heterozygote genuine adaptive superiority.

The problem is of more than academic interest to man and has worried eugenicists for generations. It does not require complex mathematics to figure out what would happen if the improvident moron or low I.Q. regularly had a dozen children and the prudent, superior citizen only two. I said deliberately "if" and I am not trying to create an alarmist situation. Yet one can not afford to ignore this Achilles heel of natural selection.

Here then are four aspects of natural selection which pose unsolved problems. Our chairman, Dr. I. M. Lerner, who has recently published a book on selection, could, no doubt, easily pose four or a dozen additional problems. I could continue and show the same not only for selection but also for other major aspects of evolution, such as mutation, recombination, and errors of sampling. However, for the sake of variety, I prefer a somewhat different approach. Let me attempt, instead, to express some of the unsolved evolutionary problems in terms of the different levels of integration. Let me single out those of the chromosome, the individual, the population, the species, and the phyletic line. Each of these levels has its own evolutionary problems although nearly all of them involve mutation, recombination, selection, and errors of sampling.

GENE AND CHROMOSOME

I do not need to stress to an audience including so many distinguished geneticists how many unsolved problems the words gene and chromosome bring to our mind. Some of these are essentially technical and functional, but most of them have a very definite evolutionary import. Every textbook of cytology discusses the selective advantages of the organization of genes into chromosomes. Foremost is that such organization permits regularity of segregation and a rigid control over recombination.

Classical genetics was a rebellion against the concept of blending inheritance and when studying the independent assortment of genes, according to Mendel's second law, it unconsciously overemphasized the independence of genes. Now, with the theory of particulate inheritance universally accepted, we can afford to stress the mutual dependence of genes. This occurs on many levels. The interaction of alleles was already known to Mendel and acknowledged in the terms dominant and recessive, homozygote and heterozygote. The precise nature of this interaction is still largely a mystery. Why heterozygosity produces heterosis in some cases and a lowering of fitness in others has been discussed in books (e.g., Lerner 1954) and symposia (Cold Spring Harbor 1955) without the emergence of any definite broad generalizations. Ultimately the solution of this problem depends on the prior solution of the problem of the gene and of mutation. The usual operational definition of mutation may be satisfactory for the purposes of classical genetics but the evolutionist wonders whether there are not different classes of mutations, which are qualitatively quite different from each other. Are mutations that produce isoalleles of a different class of phenomena from highly deleterious mutations, even if both should be changes in the code of information? We do not know, and one can present evidence both for and against such an assumption. And this is only one of our uncertainties. The problem of alleles versus mutational sites (Benzer 1957, Pontecorvo 1958) is another one. Yet in all these cases, we have interaction of genes at one genetic locus (in the broad sense of the word). Genes, however, also interact at several other levels. The physiological geneticist when studying the genic control of development usually studies such interaction at the end organ, at the phenotypic periphery. The problems at this level have been widely discussed (Kühn 1955, Waddington 1956). But there is an intermediate level which is by far more interesting and by far more puzzling. I am speaking of the set of interactions of genes in the chromosomes. The importance of epistatic interactions is becoming daily more apparent. In part, they concern adjacent loci and sites, as in the classical position effect, and in the conveyor-belt-like physiological interaction of adjacent loci in microorganisms, such as *Salmonella* or *Aspergillus*. The direct influence of adjacent parts of a chromosome on each other is, however, by no means limited to lower organisms. The linear sequence of sites at the *bithorax* locus in *Drosophila melanogaster* is, according to Lewis (1951), reflected in a linear sequence of the phenotypic "sites." Levitan's (1958) findings on the differences between *cis* and *trans* positions of certain inversions in *Drosophila robusta* are further evidence for the phenotypic effect of the interaction of genes in the chromosome. I have mentioned above other evidence for such interactions (e.g. synthetic lethals) which Dobzhansky has discussed in recent papers (1955, 1959) and will no doubt do so tomorrow. These epistatic interactions create not only conceptual difficulties, but also purely practical terminological ones. As I pointed out in 1955, we have an elaborate terminology for interactions of homologous loci or sites, such as dominant, recessive, heterotic, overdominant, etc. Yet we have no equivalent terminology for epistatic interactions. Where such interactions affect merely the structural phenotype, perhaps no terminology is needed. In this case it may be sufficient to state that the particular aspect of the phenotype has a polygenic basis. However where such interactions primarily affect viability or general fitness, as in loci that are "in internal balance with each other," we have a situation which in many ways corresponds to heterosis and yet which cannot be included with "heterosis" according to the conventional definition. The term "epistatic" has been expanded from its original narrow meaning to cover all such cases of interaction of loci but it makes no distinction between interactions that create adaptive peaks and those that do not.

We feel that the patterning of chromosomes is of far greater importance than hitherto realized, yet we have not yet found the tools permitting us to analyze this phenomenon. What happens takes place on the submicroscopic level, its principal manifestations being physiological effects. Although we must be extremely open-minded in this area, there is no evidence whatsoever for the occurrence of systemic mutations in the sense of Goldschmidt (1940). Yet, there may be different types of chromosomal patterning in different species, in open and closed populations, among inbreeders and outbreeders. Our ignorance, at present, is complete.

The Phenotype of the Individual

What is exposed to natural selection is not the individual gene nor the genotype but rather the phenotype, the product of the interactions of all genes with each other and with the environment. The central position of physiological genetics in the evolutionary theory and in the evolutionary happenings, is not yet as much appreciated as it deserves. To say that all quantitative characters are the product of polygenic inheritance is merely a different way of saying that they are the product of an interaction of genes. And quantitative characters play a decisive role in contributing to fitness (Cold Spring Harbor Symposium 1955, Lerner 1954). The evolutionary problems connected with developmental physiology are numerous. Let me single out only one or two for more detailed discussion.

What is the relation of modifiers, polygenes and pleiotropy? The understanding of the interaction of genes still suffers from the typological thinking of earlier periods. A character was thought to be the product of a gene and if it turned out that additional genes affected the expression of the character they were labelled modifying genes. If it turned out that a character was the product of a considerable number of genes, all contributing to the phenotype of this character, they were labelled polygenes. Curiously enough, many authors have treated such polygenes as ad hoc mechanisms which have no other functions in the organism than to modify a given character. Goldschmidt (1953, 1954) and other adherents of such a scheme argued that it would require an impossibly large number of modifying genes. It is curious that Goldschmidt, who was so well aware of the pleiotropy of genes, was quite unable to appreciate the probability that many polygenes are nothing but the pleiotropic manifestations of genes that have other major functions. That this is correct is indicated by every major new mutation which immediately encounters in its gene pool a high number of genes which affect its dominance, its penetrance, or its expressivity. It would be absurd to assume that every gene pool contains a large reservoir of "reserve genes" which have no other function but to wait for the occurrence of major mutations and then mold them adaptively into the phenotype. I would be reluctant to state these obvious facts were it not for several recently published statements which still ignore this obvious interpretation. Let us repeat then once more, that the high degree of pleiotropy of most genes provides for abundant material for polygenic effects. Consequently, there is no reason to believe in the existence of two sharply separated classes of genes, oligogenes and polygenes. The same gene may behave as an oligogene for one character and as a polygene for another one.

Phenotypic anticipation. It is a fact well known to naturalists and animal and plant breeders that unexpected or otherwise rare phenotypic traits may be revealed in certain individuals of a population if exposed to exceptional conditions or to environmental shocks. A well-known, recently studied case is the condition of cross veinlessness in *Drosophila* which can be revealed by temperature shocks (Waddington 1953, 1957). One aspect of this phenomenon which is not nearly sufficiently stressed is that only a certain percentage of the individuals of a population will show this phenotypic response. Whether or not an individual responds in this manner is not a matter of accident. Rather, the treatment reveals those individuals which have a genotype with many of the prerequisite factors for the exceptional phenotype. Cross veinlessness is a highly polygenic character and each of the individuals that responds to the treatment with the cross vein phenotype will have a different combination of genes. Selecting such individuals and crossing them, continuously selecting for an increased penetrance, will lead to a gradual accumulation of such polygenes until finally the phenotype will appear as the result of normal development without need for an environmental shock. Such experiments have been made from the beginnings of genetics as for instance Bateson's experiment on the prevention of flowering in the sugar beet (Hall 1928). (From Haldane 1932). We are here simply dealing with a threshhold phenomenon where numerous genes contribute to a certain phenotype but where the potentiality for it will not be pushed above the visible threshhold until a sufficient number of genes have accumulated in the genotype. I entirely agree with this interpretation as presented by Milkman (unpublished thesis, 1955) and Stern (1958). It seems to me that these interpretations are clearer and simpler than Waddington's interpretation. Experiments on the "revealed genotype" shed new light on the so-called Baldwin effect and indicate the possibility that a modification of the phenotype is of evolutionary significance only if it is due to concealed polygenes and will lead to their accumulation in the gene pool.

Population

The next higher level is that of the population. That there are still many unsolved evolutionary problems on this level is true in spite of the many advances of systematics and population genetics. Let me single out just a few of such unsolved problems.

What gives the local gene pool its cohesion and coadaptation? The work of Dobzhansky and others has revealed that there is a harmony among the genes which together make up the local gene pool. Through natural selection such genes are accumulated as produce maximal viability in their allelic and epistatic interactions. This very general statement is about all we can say about the "coadaptation" or "internal balance" of the gene pool of the local population. How it operates is still largely obscure because adequate tools have not yet been found to determine by what possible mechanisms such coadaptation, such internal balance, is maintained. Sometimes it is done by the development of super genes, such as inversions but there are other mechanisms as well. In par ticular, it includes all the factors which Dobzhansky records under the collective heading "balance hypothesis." Perhaps the system owes its harmony to the multiplicity of interactions of its components, which would automatically resist any disturbance. This would be Lerner's genetic homeostasis, although I would give, in such a homeostasis, as much weight to epistatic interactions as to allelic ones. Many otherwise puzzling evolutionary phenomena may find their explanation in epistatic inertia. Let us take for example the conservative nature of the frequencies of blood group genes in human races. Even though individual blood group genes and genotypes differ in their selective values under different conditions quite drastically from each other (particularly at the ABO locus), the relative frequency of the genes remains relatively stationary in a given population apparently owing to various balance mechanisms (allelic and epistatic).

I will make only one other comment about polymorphism. It has always puzzled me why phenotypic polymorphism is so widespread in certain groups of animals, as spotting patterns in coccinellid beetles and banding patterns in gastropods, while other families are singularly uniform. Indeed, even among closely related species one may be highly polymorph, the other completely uniform. The only answer, we can give at the present time, is the very lame one that different organisms meet the challenge of the environment in different ways.

Work on the genetics of populations has revealed a far more serious population problem. It appears possible, if not probable, that the majority of the findings made on experimental populations can not be automatically applied to natural populations. Why?

Closed and Open Populations

The experimental geneticist works by necessity with closed populations. Except for the rare occurrence of mutations, genetic changes in such populations are due to recombination, selection and errors of sampling. The genetic input is negligible. It is only within recent years that we have begun to realize that these closed populations are not natural populations, not even the large populations of the population cages. Inbreeding in such populations is far greater than in natural populations, except in species that have special mechanisms to secure inbreeding. In a natural population, which is a wide-open population, perhaps 40 per cent or more of the members of an effective local breeding population are immigrants from the outside. The gene contents of these immigrants is, of course, largely the same as that of the local population. Yet, they provide an input of new genetic factors into the population which is perhaps 100 or 1000 times greater than the input caused by mutation. No one has as yet analyzed all the consequences of this situation. I have pointed out (Mayr 1954) that those kinds of genes are likely to be favored in closed populations which are superior as homozygotes or in a limited number of combinations. On the other hand those genes will be favored in open populations which can produce viable phenotypes in a vast assortment of different genetic combinations: "jack-of-all-trades genes." I would like to go one step further. Lerner (1954) and many other students of experimental populations or of domesticated animals and cultivated plants have emphasized the great importance of heterozygosity. The widespread occurrence of balanced polymorphism indicates that heterozygosity is important also in open populations. Yet, there are numerous observations which indicate that the type of heterozygosity which produces heterosis is a special case and is often the product of careful selection. Epistatic interactions may contribute more to superior fitness in open populations than allelic overdominance. The relative importance of this interaction among loci is proven by the breakdown of viability in the F_2 of interpopulation crosses.

There is every reason to believe that the internal genetic structure of an open population is quite different from that of a closed one. The relative importance of most genetic processes such as mutation, random fixation, inbreeding, heterosis, and others will be entirely different in the two types of populations.

We must keep this in mind when we consider the role of mutation in natural populations. Buzzati-Traverso (1954b) has pointed out that the genetic variability of closed populations is not inexhaustible, particularly if selection pressure is high, and that a mild amount of induction of mutations by X ray may speed up evolutionary change (see also Scossiroli 1954). The work of Wallace (1958) with largely homozygous strains of *Drosophila melanogaster* indicates likewise that in closed homozygous populations radiation-induced mutation may be beneficial by increasing heterozygosity. Let me emphasize that Wallace has not made the mistake of applying these findings to open populations. Any open population has an input of genetic novelties which may well be of the same order of magnitude as that which is induced by radiation or even greater. And yet, in the case of the open population these genetic novelties have been pretested in other populations so that all the more generally deleterious or lethal ones had already been eliminated. In open populations there is a high level of competition, in every generation, between the indigenous and the immigrant heterozygous combinations. The very imbalance of such a system makes it highly adaptive under the fluctuating environmental conditions of nature. The inbreeding closed laboratory population, living under uniform environmental conditions is an altogether different system. We must keep this in mind when we discuss radiation damage in natural populations and in man. There are few natural populations of wild animals that have a higher rate of outbreeding than civilized man. It would be altogether unscientific to apply the findings on the effects of radiation in closed laboratory populations to a species with the wide-open population structure of modern man.

I might add that the peculiar characteristics of the open populations are not adequately coped with by the mathematical geneticist, in spite of all their efforts, particularly those of Wright, to allow for the immigration factor. The mode of interaction of genes is a far more important aspect of population structure than mere gene frequency. And the immense quantitative and qualitative difference between genetic input due to mutation and that due to immigration is rarely ever mentioned. How can one appreciate the genetic dynamics of a local population in a normal outbreeding species if one does not emphasize that the genetic input through immigration may be one hundred times as large as that through mutation?

There are still other aspects of the genetics of local populations which the mathematical formulations can not represent adequately. The normal population continuum of a species consists of remarkably well adapted local populations which nevertheless are part of a continuous system of populations which actively exchange genes with each other. Local demes (the deme is here defined in the narrowest sense as the local interbreeding population) may consist for instance of 50–200 individuals among which 40–50% are immigrants. These immigrants are largely derived from neighboring demes but with an important fraction consisting of long distance immigrants. The highly skewed character of the dispersal curve is unquestionable (Bateman 1950). As I pointed out elsewhere (Mayr 1954), such a system virtually demands that there are different kinds of genes. With a number of isoalleles almost certainly present at many loci, there will be a strong premium not on the genes which in certain combinations can produce the optimal genotype but rather on genes which can produce a passable genotype in the greatest number of possible combinations.

The degree of "openness" of a population will thus determine what kinds of genes are favored by selection. That the same gene may have different selective values in populations of different sizes is merely another manifestation of the genetic "theory of relativity".

The Species

The problems we have encountered on the level of the population are compounded on the level of the species. Nothing is less suitable as a foundation for a study of species than a typological approach. Recent work on various species of *Drosophila* has shown how much structure a species has and how little it is justified to generalize from the genetics of one population to the genetics of another one, and from one species to another. The really difficult part to keep in mind at all times is that every population has individuality and yet that all populations are held together by gene exchange. And one suspects that the character of a species as a whole is determined by the populations near the ecological center of the species which are the most prosperous, have the highest population excess, and are therefore the source of the greatest amount of gene flow. From this center outward to the species border, every population, to an increasing extent, has to make a compromise between coping, on one hand, with the increasingly difficult conditions of the local environment and, on the other, with the inflow of unsuitable genes from the center of the species range. It may appear to be merely a formal difference whether one describes a species in terms of a gradient from the southernmost to the northernmost or from the most easterly to the most westerly population or, rather, as a series of more or less concentric circles around the center of the species. Yet, the second alternative may actually describe the situation far more accurately. This is perhaps most apparent where a species contracts at one periphery and expands at the opposite one. This leads to all sorts of distributional eccentricities which are otherwise not easily understood. The high frequency of gene arrangements at the southern periphery of the range of *D. subobscura* may be an illustration of this phenomenon.

Peripheral Populations

The studies of systematists as well as those of population geneticists indicate that the peripheral populations of a species, and particularly the peripherally isolated populations, may be quite exceptional in their genetic structure. What are their special characteristics? To begin with, they live near the border of the species range, in other words, under environmental conditions that are marginal for the species. Selection will be unusually severe. Gene flow is much reduced as compared to the more central populations of the species, and so far as it exists it is a one way flow from the interior of the species range toward the periphery because population surplus in the peripheral populations is negligible. As I pointed out earlier (Mayr 1954) this one-sided genetic input may well be responsible for the inability of these peripheral populations to become truly adapted to the local conditions. The simultaneous need for retaining coadaptation with the gene pool of the species as a whole and for being able to assimilate the immigrants from the more central parts of the species range, limits the opportunities for building up novel gene combinations that would permit an expansion beyond the present species range.

This limitation is not true for the peripherally isolated populations and as a result one usually finds rather deviating phenotypes in peripherally isolated populations. There are indications, although at the present time no real proof, that such populations are important in ecological shifts. Let us take for instance the problem of the shift in monophagous insects from one species of food plant to another. Most so-called monophagous insects are not entirely restricted to a single species of food plant but may be found occasionally on an accessory host. It is quite possible that such an accessory host may become the main host under the special environmental conditions of the peripherally isolated population. Protection from the gene flow of the parental species will permit not only the accumulation of genes leading to a better adaptation to this new host but also of isolating mechanisms against the parental populations. If this peripherally isolated population subsequently re-establishes contact with the parental population it will have not only a set of isolating mechanisms against the parental species but also a different food niche and will thus be able to coexist sympatrically with the parental species.

We are still far from understanding the forces that are particularly active in peripheral populations and the responses of such populations. Carson (1955) found virtual homozygosity for gene arrangements in peripheral populations of *Drosophila robusta*, while Stumm-Zollinger and Goldschmidt (1959) found that in *Drosophila subobscura* much heterozygosity is retained in the peripheral populations of Israel. However, even these populations had a higher index of free recombination than some more centrally located populations of this species. In the work of Dowdeswell and Ford (1957) it is likewise apparent that peripherally isolated populations have special properties. While the butterfly *Maniola jurtina* is phenotypically uniform over the greater part of southern England and shows comparatively little difference among the larger islands of the Scilly Islands, it shows much difference between the smaller islands or between isolated populations on larger islands. It appears that the phenotypic control of the characters in these small isolated populations is less influenced by the phenotype of the main population than would be the case if they were contiguous with it. These population differences have not yet been analyzed genetically but protection against gene flow from the large populations permits undoubtedly in all these cases a more immediate adaptation (through selection) to the special habitat conditions of the isolated areas. To what extent the diversity in gene arrangements found in the central part of the range of some species of *Drosophila* is an indication of "ecological well being" which permits the species to make experiments in niche occupation, or a method to reduce free recombination in order to protect favorable gene combination against break-up through gene flow, is still an open question. The effects of high levels of gene flow in the central part of the range of species, should not be under estimated. Except for a reduction of recombination, the only other protection a gene pool has against the disturbing effects of gene immigration, is the accumulation of genes that have, as the breeder would say, high combining powers.

The size of the local deme and the amount of gene exchange with its neighbors are thus very important. But so are numerous other factors that affect the population structure of species and the genetic system of a population. In a way every species is unique in its system and yet, certain patterns seem to emerge. Dobzhansky (1959) showed for instance for *Drosophila*, that the two common, widespread species *D. willistoni* and *D. pseudoobscura* differ in various aspects from the rare, more specialized species *D. prosaltans*. For plants Grant (1958) pointed out that the genera *Clarkia* and *Gilia*, which live in the same environment, show numerous similarities in population structure and breeding system in spite of being as little related as two genera of dicotyledons can be. Such differences and similarities can be established only through the most careful comparative work. In this respect it seems to me that the botanists are far ahead of the zoologists. Stebbins (1950), Grant (1957-58), Lewis (1955), Clausen (1951), and others have pointed out the differences among genera and families of plants with respect to species structure, frequency of polyploidy, incidence of hybridization, phenotypic stability, factors controlling the degree of outbreeding, and so forth. Some progress has been made comparing plants as a whole with animals as a whole, but such comparisons are somewhat misleading because insects and birds are treated as if they were the typical representatives of all animals. To be sure, much recent work (Hubbell 1956, Brown 1958) indicates that species structure and mechanisms of speciation are indeed essentially the same in insects as they are in birds but insects are by no means fully representative of the other invertebrates. Unfortunately our knowledge of species structure and mechanisms of speciation of the lower invertebrates is still virtually zero. A notable exception is the work of Sonneborn and associates on ciliates (Sonneborn 1957). In this he shows how numerous aspects of the biology of ciliates fall into a meaningful pattern if correlated with the degree of inbreeding or outbreeding of the particular biological species. How different the population structure of a species may be in an occasional species of invertebrates is indicated by the West Indian snails of the genus *Cerion*. (Mayr and Rosen 1956).

To me the most impressive aspect of such diversity is that it shows the multiplicity of pathways by which a high level of adaptation can be attained. There has been much argument in the recent literature on the question whether and how one can measure the adaptedness of a species. In such arguments one should not forget that there are very different ways of achieving adaptation.

Looking at numerous species of animals I can distinguish three particularly common roads to success:

1. *The narrow specialist.* This type is particularly common among parasites and host-specific insects. The universe of such a species is narrow and confined. Yet, within its universe, it reigns supreme. An insect that lives as a leaf miner in the leaves of a birch tree or a fluke that lives in the liver of a particular species of vertebrates, may be immensely successful and build up enormous populations. Yet, its fate is closely tied to that of its host and the balance between the parasite and the host is a precarious one which at any time can lead to the extermination of either the parasite or the host (with parasite). Nevertheless the advantages of becoming the supreme specialist of the narrow niche are so tempting that an extraordinarily high percentage of all animals succumbs to this temptation. It is probable that more than fifty per cent of all the species of animals should be classed as narrow specialists.

2. *The successful universalist.* Man, of course, is the outstanding example of this category. It contains species which can cope with many climates and with many ecological niches. There is, of course, no single species which is equally well adapted to all climates and to all habitats and it may be a matter of definition what to include in this category. Yet, by and large, we will have no difficulties in pointing out certain species in almost any group of organisms which lack conspicuous specializations and appear to be successful in several niches, species which the ecologist would class as "tolerant" or "euryecous." The geneticist would suspect that such species are rich in concealed genetic variability, that they are highly heterozygous, that they may tend to chromosomal polymorphism, that they have much gene flow within the system and highly developed genetic homeostasis. Yet by all these mechanisms such species become so successful at coping with a changing environment that they tend to become conservative from the evolutionary viewpoint.

3. *The opportunist.* The third type of species is characterized, up to a point, by a combination of the two extremes previously stated. These are species with considerable geographic variability, with a successful central universalist aggregate of populations, and yet with an ability to form peripheral geographic isolates. Each of the geographic isolates is able to become specially adapted to its local environment, making the best of the local opportunities. Some of the geographic isolates are sufficiently withdrawn from the gene flow of the species to be able to find their own equilibrium without having to compensate incessantly for the disturbing effects of the immigrant genes. There is much evidence to indicate that this third type of species structure is the most important for long-range evolution. Such opportunism may be expensive but it is worth the price. Of 50 attempts to find a new ecological niche, 49 may be unsuccessful either because the new population is too small or because the new niche is not sufficiently important. Yet even if only 1 out of 50 is successful, indeed if only 1 out of 250 is successful, it would lead to a continued expansion of the total evolutionary universe. It seems to me, at least as far as animals are concerned, that much of the known evolutionary progress has been achieved by this type of species. The chance that such species are well represented in the fossil record is small and some of our difficulties in reconciling the geological record with our concepts of evolution, as based on the phenomena of genetics, may be due to the probabilities of fossil preservation. The conservative successful universalist certainly has a far greater chance to be preserved in the fossil record than the geographical isolate which switches successfully into a new niche.

The Higher Category

With a consideration of the fossil record we have reached the highest level of integration which I want to discuss today. You may ask, is it legitimate to include the higher systematic categories in a discussion of genetics and evolution? Is this not nonsense, since higher categories can not be crossed successfully, and since all genetic analysis depends on the analysis of fertile crosses? This in the past has been the viewpoint of many paleontologists, who then unfortunately proceeded to postulate evolutionary mechanisms that were in clear conflict with the genetic evidence. Refuting these endeavors Rensch and Simpson have shown that all we know about evolutionary trends and the origin and evolution of the higher categories is fully consistent with the modern genetic theory. It seems to me, however, that the students of macroevolutionary phenomena have not gone nearly far enough in exploiting the findings of current genetics. Let me illustrate this with one example. Systematists (including paleontologists) still tend to correlate too closely the amount of phenotypic difference with the amount of genetic difference. The category of sibling species proves that such an equation neglects to allow for manifestations of developmental homeostasis. Sibling species are perfectly normal species, genetically speaking, which are characterized, however, by the slightness, if not complete absence of morphological differences from each other. The occurrence of such species proves that a major reconstruction of the genotype can take place without any visible effect in the phenotype. The capacity of such species for developmental homeostasis must be so great that any unbalancing genetic change is at once compensated somewhere along the developmental pathway. The frequency and wide distribution of isoalleles which were found at almost any locus where they were looked for with special methods (for instance, most recently at the White locus, Green 1959), is further evidence for the enormous strength of these homeostatic devices.

The wide occurrence of these stabilizing developmental mechanisms must be taken into consideration in the interpretation of some of the phenomena that are so puzzling to the paleontologist. One of them is the occurrence of stable types, or as stated in terms of evolutionary rates, the occurrence of bradytelic evolution (Simpson, 1944). Two explanations, both somewhat dubious, are usually cited to account for such types. One is that they occur in stable habitats and that selection itself is the conservative factor. The fact that all the associates of these species at former geological periods have since become extinct or changed drastically deprives this explanation of probability. The other hypothesis is that the lack of change is due to an enormous stability of the genetic material in these species resulting in an incredibly low rate of mutation. This postulate does not explain why the replication of the genetic material should be so nearly errorless in a few exceptional types of organisms when there are rather well established mutational rates of definite magnitudes in the vast majority of organisms. To me, at least, it would seem simpler to assume that the rates of mutation in these old types are of the same order of magnitude as in other organisms, but that developmental homeostasis is so highly developed as to prevent completely the phenotypic penetrance of the concealed genetic changes.

One might utilize the same phenomenon of developmental homeostasis to explain a diametrically opposite evolutionary phenomenon, the sudden, sometimes almost explosive, breaking up of morphological types. Paleontologists have described numerous instances where a well-known fossil type, usually a marine invertebrate, suddenly blossoms out into numerous, more or less different subtypes. Such evolutionary explosions may take place after the parental type had been morphologically stable for periods of 20 to 40 million years. None of the explanations that are usually advanced, such as bursts of mutation or geological cataclysms, is particularly convincing. It would seem more likely that some special event such as a sudden shift in ecological balance or the rare occurrence of a successful hybridization has led to the upsetting of the developmental homeostasis and has permitted new selective forces to act upon the newly available phenotypic variation. The history of domestic animals offers abundant proof for the emergence of rich stores of phenotypic variation as soon as the developmental and genetic homeostasis is broken down owing to close inbreeding, extreme selection, or both. Why not apply this finding to the phenomenon of explosive evolution?

Conclusion

In concluding my remarks I would like to make two observations. The first is that in spite of the almost universal acceptance of the synthetic theory of evolution, we are still far from fully understanding almost any of the more specific problems of evolution. I have tried to demonstrate this for every level from the chromosome up to the species and evolutionary line. There is still a vast and wide open frontier. And I would like to make a second point.

We live in an age that places great value on molecular biology. Let me emphasize the equal importance of evolutionary biology. The very survival of man on this globe may depend on a correct understanding of the evolutionary forces and their application to man. The meaning of race, of the impact of mutation, whether spontaneous or radiation-induced, of hybridization, of competition,—all these evolutionary phenomena are of the utmost importance for the human species. Fortunately the large number of biologists who continue to cultivate the evolutionary vineyard is an indication of how many biologists realize this: we must acquire an understanding of the operation of the various factors of evolution for the sake not only of understanding our universe, but indeed very directly for the sake of the future of man.

References

BATEMAN, A. J. 1950. Is gene dispersion normal? Heredity, *4:* 353-363.
BENZER, SEYMOUR. 1957. The Elementary Units of Heredity. In: *The Chemical Basis of Heredity.* (W. D. McElroy and B. Glass, eds.): 70-93, Baltimore: Johns Hopkins Press.
BROWN, W. J. 1958. Sibling Species in the Chrysomelidae. Proc. 10th Int. Cong. Entomology 1: 103-110.
BUZZATI-TRAVERSO, A. A. 1954a. Conclusions and perspectives. Symp. on Genetics of Pop. Structure. IUBS Publ. B15: 126-138.
——— 1954b. On the role of mutation rate in evolution. Proc. 9th Int. Cong. Genetics *1:* 450-462.
CARSON, H. L. 1955. The genetic characteristics of marginal populations of *Drosophila.* Cold Spr. Harb. Symp. Quant. Biol. *20:* 276-287.
CLAUSEN, JENS C. 1951. *Stages in the Evolution of Plant Species.* New York: Columbia Univ. Press.
DE VRIES, HUGO. 1906. *Species and Varieties, Their Origin by Mutation.* (Lectures delivered at Univ. of Calif.) D. T. MacDougal, ed., 2nd edition, Chicago.
DOBZHANSKY, TH. 1955. A review of some fundamental concepts and problems of population genetics. Cold Spr. Harb. Symp. Quant. Biol. *20:* 1-15.
——— 1959. Variation and evolution. Proc. Amer. Phil. Soc., *103* (2): 252-263.
DOWDESWELL, W. H., FORD, E. B. and MCWHIRTER, K. G. 1957. Further studies on isolation in the butterfly *Maniola jurtina* L. Heredity *11* (1): 51-65.
DU RIETZ, G. E. 1930. The fundamental units of botanical taxonomy. Svensk. Bot. Tidsskr. *24:* 333-428.
EISELEY, LOREN. 1958. *Darwin's Century, Evolution and the Men Who Discovered it.* New York: Doubleday and Co.
FISHER, R. A. 1930. *The Genetical Theory of Natural Selection.* Oxford: Clarendon Press.
GRANT, VERNE. 1957. Plant species in theory and practice. In: *The Species Problem* (ed., Ernst Mayr), Symp. AAAS Meeting 1955, publ. 50: 39-80.
——— 1958. The regulation of recombination in plants. Cold Spr. Harb. Symp. Quant. Biol. *23:* 337-363.
GREEN, M. M. 1959. The discrimination of wild-type isoalleles at the White locus of *Drosophila melanogaster.* Proc. Nat. Acad. Sci., *45* (4): 549-553.

GOLDSCHMIDT, RICHARD H. 1934. *Lymantria*. Bibl. Genet., *11:* 1-180.
GOLDSCHMIDT, RICHARD H. 1940. *The Material Basis of Evolution*. New Haven: Yale Univ. Press.
1953. Pricking a bubble. Evolution, *7:* 264-269.
1954. Presidential Address. Different Philosophies of Genetics. Proc. 9th Int. Cong. Genetics, 1: 83-99.
HALDANE, J. B. S. 1932. *The Causes of Evolution*. London: Longmans, Green.
1957. The Estimation of Viabilities. Journ. Genetics, *54* (2): 294-296.
HAGEDOORN, A. C. and A. L. 1917. Rate and evolution. Amer. Nat., *51:* 385.
HALL, A. L. 1928. Bateson's experiment on bolting in sugar beet and mangolds. Journ. of Genetics, *20:* 219-231.
HEINKE, F. 1878-1882. *Die Varietäten der Herings*. 2 Teile, Berlin.
HUBBELL, T. H. 1956. Some aspects of geographic variation in insects. Ann. Rev. Entomology, *1:* 71-88.
KÜHN, ALFRED. 1955. *Vorlesungen über Entwicklungsphysiologie*. Berlin.
LERNER, I. M. 1954. *Genetic Homeostasis*. Oliver and Boyd.
1959. The concept of natural selection: a centennial view. Proc. Amer. Phil. Soc., *103* (2): 173-182.
LEVITAN, MAX. 1958. Non-random associations of inversions. Cold Spring Harb. Symp. Quant. Biol. *23:* 251-268.
LEWIS, E. B. 1951. Pseudo-allelism and gene evolution. Cold Spring Harb. Symp. Quant. Biol., *16:* 159-174.
LEWIS, HARLAN L. 1955. Specific and infraspecific categories in plants. Biological Systematics, 16th Ann. Biol. Colloquium, Oregon State College, 13-20.
LLOYD, R. E. 1912. *The Growth of Groups in the Animal Kingdom*. London.
MAYR, ERNST. 1954. Change of genetic environment and evolution. From: *Evolution as a Process*, 157-180. (J. Huxley, ed.), Allen and Unwin.
1955a. Karl Jordan's contribution to current concepts in systematics and evolution. Trans. Royal Ent. Soc. Lond., *107:* 45-66.
1955b. Integration of genotypes: synthesis. Cold Spring Harb. Symp. Quant. Biol. *20:* 327-333.

MAYR, ERNST, and ROSEN, C. B. 1956. Geographic variation and hybridization in populations of Bahama snails (*Cerion*). Amer. Mus. Novitates, no. 1806.
PONTECORVO, G. 1958. *Trends in Genetic Analysis*. New York: Columbia University Press.
SCOSSIROLI, R. E. 1954. Effectiveness of artificial selection under irradiation of plateaued populations of *Drosophila melanogaster*. Symp. of Pop. Structure, pp. 43-66. Ser. B, 15; Naples: Int. Union Biol. Sci.
SCHMIDT, J. 1917. Statistical investigations with *Zoarces viviparus*. Jour. of Gen. *7:* 105-118.
SIMPSON, G. G. 1944. *Tempo and Mode in Evolution*. New York: Columbia University Press.
SONNEBORN, T. M. 1956. Breeding systems, reproductive methods, and species problems in Protozoa. In: *The Species Problem* Ernst Mayr, ed. Symp. AAAS Meeting 1955, publ. 50, 155-324.
STEBBINS, G. L., JR. 1950. *Variation and Evolution in Plants*. New York: Columbia University Press.
STERN, CURT. 1958. Selection for subthreshold differences and the origin of pseudoexogenous adaptations. Amer. Nat. *93:* 313-316.
STUMM-ZILLINGER, E., and GOLDSCHMIDT, E. 1959. Geographical differentiation of inversions systems in *Drosophila subobscura*. Evolution *13:* (1): 89-98.
SUMMER, F. B. 1924. The stability of subspecific characters under changed conditions of environment Amer. Nat. *58:* 481-505.
WADDINGTON, C. H. 1953. Genetic assimilation of an acquired character. Evolution *7:* 118-126.
1956. *The Principles of Embryology*. London: Allen and Unwin.
1957. *The Strategy of the Genes*. London: Allen and Unwin.
WALLACE, BRUCE. 1958. The average effect of radiation induced mutations on viability in *Drosophila melanogaster*. Evolution *12* (4): 532-552.
WRIGHT, SEWALL. 1931. Evolution in Mendelian populations. Genetics *16:* 97-159.
1951. Fisher and Ford on the "Sewall Wright Effect" Amer. Scientist *39:* (3): 452-459.
ZIRKLE, C. 1946. The early history of the idea of the inheritance of acquired characters and pangenesis Trans. Amer. Phil. Soc. *35* (2): 91-151.

On Diversity of Evolution under one set of External Conditions.
By Rev. John T. Gulick.

[Read November 21, 1872.]

The terms "Natural Selection" and "Survival of the Fittest" present different phases of a law which can act only where there is variation. The words in which the law is expressed imply that there are variations which may be accumulated in different proportions according to the differing demands of external conditions.

What, then, is the effect of these variations when the external conditions remain the same? Or can it be shown that there is no change in organisms that is not the result of change in external conditions? Again, if the initiation of change in the organism is through change in the "Environment," by what law is the cessation of change determined? If change continues in the organism long after the essential conditions of the "Environment" have become stationary, how do we know that it is not perpetual? Does the change, whether transitory or continuous, expend itself in producing from each species placed in the new "Environment" just one new species completely fitted to the conditions? or may it produce from one stock many that are equally fitted? If the latter, what is the law or condition that determines their number, their affinities, and the size and position of their respective areas, as related to each other and to the whole available area?

Facts throwing Light on the Subject.

I believe that in the relations of species to each other as distributed in nature, we shall find light on the subject. I call attention at this time to the variation and distribution of terrestrial mollusks, more especially those found on the Sandwich Islands; but similar facts are not wanting elsewhere.

The land-shells of the Sandwich Islands not only differ in species from those of other countries, but they belong, for the most part, to a group of genera found nowhere else. These are the *Achatinellinæ*, of which there are seven arboreal genera (*Achatinella, Bulimella, Helicterella, Laminella, Partulina, Newcombia*, and *Auriculella*), and three ground-genera (*Carelia, Amastra*, and *Leptachatina*).

Some of these genera are confined, in their distribution, to a single island. The average range of each species is five or six miles, while some are restricted to but one or two square miles, and only a very few have the range of a whole island.

The forest-region that covers one of the mountain-ranges of Oahu is about forty miles in length and five or six miles in breadth. This small territory furnishes about 175 species, represented by 700 or 800 varieties. The fall of rain on the northeast side of the mountain is somewhat heavier than on the opposite side, and the higher ridges of the mountains are cooler than the valleys; but the valleys on one side of the range have a climate the same in every respect. The vegetation in the valleys differs somewhat from that on the ridges; but the vegetation of the different valleys is much the same; the birds, insects, and larger animals are the same. Though, as far as we can observe, the conditions are the same in the valleys on one side of the range, each has a molluscan fauna differing in some degree from that of any other. We frequently find a genus represented in several successive valleys by allied species, sometimes feeding on the same, sometimes on different plants. In every such case, the valleys that are nearest to each other furnish the most nearly allied forms; and a full set of the varieties of each species presents a minute gradation of forms between the more divergent types found in the more widely separated localities.

No theory is satisfactory that does not account, 1st, for their being distributed according to their affinities in adjoining areas more or less distinctly defined, and, 2nd, for their being restricted to very small areas.

External Conditions not the Cause.

I think the evolution of these different forms cannot be attributed to difference in their external conditions :—

1st. Because in different valleys, on the same side of the mountain, where food, climate, and enemies are the same, there is still a difference in the species.

2nd. Because we find no greater difference in the species when we pass from the more rainy to the drier side, than when we compare the forms from valleys on the same side of the mountain, separated by an equal distance.

3rd. Because if, failing to find a reason in the more manifest conditions, we attribute the difference in the species to occult influences, such as magnetic currents, we must suppose that there are important differences in these hidden conditions for each successive mile, and that their power at the Sandwich Islands is a thousand times greater than in most countries.

Separation and Variation Correlative Factors in the Evolution of Species.

If we would account for the difference and for the limited distribution of these allied forms on the hypothesis of Evolution from one original species, it seems to me necessary to suppose two conditions, both of which relate to the state of the species—namely, Separation and Variation. I regard Separation as a condition of the species and not of surrounding nature, because it is a state of division in the stock which does not necessarily imply any external barriers, or even the occupation of separate districts. This may be illustrated by the separation between the castes of India or between different genera occupying the same locality.

To state the conditions more fully:—

1st. We must suppose that they possess or have possessed an inherent tendency to variation, so strong that all that is necessary to secure a divergence of types in the descendants of one stock is to prevent, through a series of generations, their intermingling with each other to any great degree. This supposition is not at variance, but rather in accordance, with facts that are observed in analogous cases in the history of man and of domestic animals of one original stock, that are kept entirely apart. But this condition alone would not be enough to account for the species of *Achatinellinæ* being confined to areas so much smaller than usual; for if this tendency has produced such results in the distribution of one family, why does it not in all?

Migration and Variation opposing Factors in the Limitation of Areas.

2nd. To account, therefore, for the small areas, we must further suppose that, as compared with other families, there is a disproportion between the tendency to variation and the tendency and opportunities to migrate. Either the tendency to variation in this family is very much greater than usual, or their tendency to migrate is weaker and their opportunities fewer than usual. According to *à priori* reasoning, the areas occupied must vary directly as the tendency, power, and opportunities for migrating, but inversely as the tendency to variation.

If the amount of migration is greatly expanded in proportion to the tendency to variation, the areas must be expanded; if, on the other hand, the tendency to variation is expanded as compared with the amount and extent of migration, the areas occupied by the different species must be correspondingly contracted.

If the power of migrating and the opportunities for being transported are very limited in any family of creatures, we may expect that the areas occupied by the different species and varieties of that family will be more restricted than the areas occupied by the species of other families that have greater opportunities for migrating but the same tendency to variation. When we find that in Europe and North America nearly every species of *Helix* occupies an area many thousand times as large as the area occupied by any *Achatinella*, we naturally ask whether the difference can be accounted for by circumstances that limit the dispersion of the latter, or whether the results are to be attributed to a stronger tendency to variation. It is evident that to the forest species, that live on trees found chiefly in the valleys, the mountain-ridges separating the valleys must be partial barriers; but the valleys cannot be barriers to the species occupying the ridges, for the ridges rising between the valleys are all spurs from the one central range that forms the backbone of the island. In accordance with these facts we find that the distances over which the ridge species are distributed are usually somewhat greater than those reached by the valley species. But even the ridge species are limited in their distribution to very small areas. Few have a range of territory more than six or eight miles in length and three or four miles in breadth; and many are restricted to half that area. Though some of the groups of species are found both in the valleys and on the ridges, so that no barriers intervene to break the continuity of their intercourse, we still find them distributed over small areas, and these areas again divided amongst subordinate varieties. The streams that flow through these valleys cannot serve in carrying the shells from one valley to another; but the separation from this cause can be no greater than that which is experienced by mollusks inhabiting mountain-valleys in other countries. It therefore appears that the limited range of the species of this family receives but slight explanation from the nature of the country. Neither can we suppose that the power of locomotion in this family is so immeasurably below that possessed by the *Helices* of Europe and America, and by the *Achatinæ* of Africa, as to account for the excessive disproportion in the areas occupied, as well as in the amount of divergence between the types found in any locality and those found at given distances. In Africa some of the species of *Achatina* have a range of more than a thousand miles, while on the island of Oahu the most widely diffused species of the arboreal genus *Achatinella* is restricted to about ten miles, and the utmost limit gained by any species of the ground-genus *Amastra* is about twenty miles. Again, the difference of type is quite as great between the species of *Achatinella* found in the mountains near the eastern end of Oahu and those found forty miles distant, on the other end of the same range of mountains, as the difference between the species of *Achatina* found in Sierra Leone and those in the region of Port Natal, nearly four thousand miles distant.

The birds that prey upon these snails are probably few; but the forests are populous with fruit- and nectar-feeding birds, that might be supposed to give as effectual means of transportation as could be given by any. The number of species represented by these birds is no doubt less than would in most cases be found in an equal extent of continental forest; but the number of individuals is probably greater than the average number inhabiting equal areas in other parts of the world.

If we find no reason for attributing the small areas occupied by these species to deficient means of transportation, may we not believe that rapidity of variation has had influence in determining the result?

Stability of Type as affected by Cultivation.

It is known that there is a great difference in the stability of type in different species of plants and animals that have been subjected to cultivation. One produces striking varieties in a single generation; another requires careful selection of certain characters for many generations before well-marked varieties can be secured. We also know that continued cultivation will, in many instances, break down the stability of type in a species that, in the first place, adhered with great persistency to one form. It often happens that when the stability has once been disturbed, a wide range of variation may afterwards be obtained with comparative rapidity.

Is it not possible that similar changes may sometimes take place in species in their wild state? Two important elements of the cultivation which tends to develop varieties are the removal of competitors and enemies, and the abundant supply of nourishment; but both these conditions may sometimes be furnished by nature without the intervention of man.

The Natural Selection that prevents Variation.

The more severe the competition the more rigidly does Natural Selection adhere to the one form that is best suited to meet that competition, or, according to the language in which Professor Owen has stated the doctrine, the more certainly does the "Battle of Life" extinguish all variations from that one form. When a species is subjected to severe competition of the same kind for countless generations, we may well believe that it gains a stability of type that is not found in one that has during the same time been, either comparatively free from competition, or under the influence of a succession of different competitors and enemies*.

Stability of Type in Island Fauna may be impaired:—

1st. By Freedom from the Competition that limits Variation.— We can see that when animal life commences upon an island where vegetation has already become abundant, the first species that appears on the arena, unless immediately followed by other creatures capable of being either friends or foes, will enjoy for a time complete freedom from competition. If the vegetation is suited, it will also have an abundance of food. Under these circumstances every variation that occurs, unless decidedly malformed, will have a chance of living and exerting an influence upon the final result.

*2nd. By Competition accelerating Variation.—*If the introduction of competitive animals is long delayed, the first struggle for life will occur between the members of the one stock. But competition of this kind does not tend to prevent variation, but rather to accelerate it, by driving portions of the race into new spheres. Supposing the animals first inhabiting the island to be a species of arboreal mollusks, there would soon be an excess of occupants on the trees best suited to them in the region where they first appeared. The portion of the population that would survive this exigency would, in the first place, be those that found sustenance on trees of other kinds. Some of these would either themselves, or through their descendants, reach localities where the trees are again found on which the stock commenced its career. Those that, in this way, returned to the original trees, would have acquired some new tendencies to variation through the ordeal through which they had passed; and those that remained upon the other kinds of trees would rapidly develop new characters: in either case, there would be no outside competition limiting them to one definite form. New forms of variation would

---

* The only terrestrial mollusks with which the *Achatinellinæ* have to compete are a few *Helices* much inferior in size, and not arboreal in their habits.

have an opportunity of being preserved. New shades of colour, for example, would not expose the owners to the attacks of enemies. Variations of shape, if not inconsistent with the pursuit of food, would be no disadvantage.

3rd. By continual Change in the Character of the Natural Selection.—Still further, we can see that when competition arises from the *gradual* introduction of animals, either friendly or hurtful to the first occupants, the character of the Natural Selection, to which they would thus be subjected would be continually changing; no one set of characters would have constant advantage through a long series of successive generations.

In these ways the persistence of form might be impaired, and the variability which we may believe exists in some degree in all organisms might be greatly increased beyond what is usually found. This tendency to comparatively rapid variation having been established, the evolution of species would be correspondingly rapid, and the areas of each proportionately limited.

Imaginary Case, illustrating Evolution without change in the External Conditions.

If a bird should carry a leaf bearing two individuals of some species and drop it a mile beyond the limits already reached by others of that species, they might there find the same trees to which they were accustomed, and multiply for some tens of years before the first scattering individuals from the slowly advancing wave of migration would reach them. They might, by this time, have increased to many thousands; and having been entirely separated from the original stock for a considerable number of generations, with a preexisting tendency to rapid variation, a certain variety of form and colour might have partially established itself amongst them. The arrival of a few individuals representing the old stock would, amongst the multitudes of the new variety, have no influence in bringing back the succeeding generations to the original form. The new characters would become from year to year more distinctly set. Owing to an intervening ridge acting as a partial barrier, the number of individuals of the original stock coming amongst them might be always restricted; and even if no such barrier existed, the individuals arriving from abroad could never be more than a very small number compared with those produced on the spot and possessing the local characteristics.

Changes produced by the Introduction of Enemies.

At this point one other inquiry naturally arises:—If the multitude of varieties and the restricted distribution of both varieties and species is in any degree due to freedom from severe competition, what would be the effect if, by degrees, many birds and insects, hostile to these snails, should find their way to the Sandwich Islands and become numerous in those mountain-regions? One of the first effects would naturally be the disappearance of many varieties and species by which the different forms of each genus are now so minutely gradationed together. Certain protective colours would be made to prevail, to the partial exclusion of some of the brilliant contrasts of colour. The same enemies being found in all the valleys of an island, the forms that proved to be best fitted to survive in one valley would have the advantage everywhere, and therefore gradually spread from valley to valley. The distribution of species and their separation from each other by distinct forms would thus become similar to what is found in the case of continental species.

The destruction of forests by the introduction of cattle and goats is now causing the extinction of some of the species.

Recapitulation and Conclusion.

A comparison of the distribution of island mollusks with the widely contrasted distribution of continental species, leads me to believe that the evolution of many different species may take place without any difference in the food, climate, or enemies that surround them. The rapidity of evolution or the time within which a certain amount of change is effected must depend upon the average amount of change in one direction in a single genetation, and the rapidity of succession in the generations. Ten thousand years would make but little difference in a species of cedar, in which the life of a single tree might count a third of that period. But in the case of some species of insects the same period might cover ten thousand generations; and though the change in each generation might be as imperceptible as in the cedar, the aggregate of change for the whole period might be very apparent.

We must also bear in mind, the Natural Selection arising from severe competition with species that have a wide range tends to prevent variation and give a wider diffusion to forms that would otherwise be limited in their range and variable in their type. Natural Selection is as efficient in producing permanence of type in some cases as in accelerating variations in other cases.

If we suppose separation without a difference of external circumstances is a condition sufficient to ensure variation, it renders intelligible the fact that, in nearly allied forms on the same island, the degree of divergence in type is in proportion to the distance in space by which they are separated. The difference between two miles and ten miles makes no change in climate; but it is easy to believe that it is the measure of a corresponding difference in the time of separation. In forms that differ more essentially, the separation may have been as complete and as long-continued in the case of those which now inhabit one valley as in the case of those which are separated by the length of an island. When a wide degree of divergence has been established, hybridation would be precluded. We accordingly find that the difference between species of different genera or subgenera is in most instances equally great whether we take for comparison those from the same or from different valleys.

If, on the other hand, we suppose that a difference in the external conditions is necessary to the evolution of distinct forms, these and other similar facts remain unexplained.

A NEW FACTOR IN EVOLUTION.

By J. Mark Baldwin.

In several recent publications I have developed, from different points of view, some considerations which tend to bring out a certain influence at work in organic evolution which I venture to call "a new factor." I give below a list of references[1] to these publications and shall refer to them by number as this paper proceeds. The object of the present paper is to

[1] References:

(1). *Imitation: a Chapter in the Natural History of Consciousness*, Mind (London), Jan., 1894. Citations from earlier papers will be found in this article and in the next reference.

(2). *Mental Development in the Child and the Race* (1st. ed., April, 1895; 2nd. ed., Oct., 1895; Macmillan & Co. The present paper expands an additional chapter (Chap. XVII) added in the German and French editions and to be incorporated in the third English edition.

(3). *Consciousness and Evolution*, Science, N. Y., August, 23, 1895; reprinted printed in the American Naturalist, April, 1896.

(4). *Heredity and Instinct* (1), Science, March 20, 1896. Discussion before N. Y. Acad. of Sci., Jan. 31, 1896.

(5). *Heredity and Instinct* (II), Science, April 10, 1896.

(6). *Physical and Social Heredity*, Amer. Naturalist, May, 1896.

(7). *Consciousness and Evolution*, Psychol. Review, May, 1896. Discussion before Amer. Psychol. Association, Dec. 28, 1895.

gather into one sketch an outline of the view of the process of development which these different publications have hinged upon.

The problems involved in a theory of organic development may be gathered up under three great heads: Ontogeny, Phylogeny, Heredity. The general consideration, the "factor" which I propose to bring out, is operative in the first instance, in the field of *Ontogeny*; I shall consequently speak first of the problem of Ontogeny, then of that of Phylogeny, in so far as the topic dealt with makes it necessary, then of that of Heredity, under the same limitation, and finally, give some definitions and conclusions.

I.

Ontogeny: " Organic Selection " (see ref. 2, chap. vii).—The series of facts which investigation in this field has to deal with are those of the individual creature's development; and two sorts of facts may be distinguished from the point of view of the *functions which an organism performs in the course of his life history*. There is, in the first place, the development of his heredity impulse, the unfolding of his heredity in the forms and functions which characterize his kind, together with the congenital variations which characterize the particular individual—the phylogenetic variations, which are constitutional to him; and there is, in the second place, the series of functions, acts, etc., *which he learns to do himself in the course of his life*. All of these latter, the *special modifications which an organism undergoes during its ontogeny*, thrown together, have been called "acquired characters," and we may use that expression or adopt one recently suggested by Osborn,[2] "ontogenic variations" (except that I should prefer the form "ontogenetic variations"), if the word variations seems appropriate at all.

[2] Reported in *Science*, April 3rd.; also used by him before N. Y. Acad. of Sci., April 13th. There is some confusion between the two terminations "genic" and "genetic." I think the proper distinction is that which reserves the former, "genic," for application in cases in which the word to which it is affixed qualifies a term used *actively*, while the other, "genetic" conveys similarly a *passive* signification; thus agencies, causes, influences, etc., and "ontogenic phylogenic, etc.," while effects, consequences, etc, and "ontogenetic, phylogenetic, etc."

Assuming that there are such new or modified functions, in the first instance, and such "acquired characters," arising by the law of "use and disuse" from these new functions, our farther question is about them. And the question is this: How does an organism come to be modified during its life history?

In answer to this question we find that there are three different sorts of ontogenic agencies which should be distinguished—each of which works to produce ontogenetic modifications, adaptations, or variations. These are: first, the physical agencies and influences in the environment which work upon the organism to produce modifications of its form and functions. They include all chemical agents, strains, contacts, hindrances to growth, temperature changes, etc. As far as these forces work changes in the organism, the changes may be considered largely "fortuitous" or accidental. Considering the forces which produce them I propose to call them "physico-genetic." Spencer's theory of ontogenetic development rests largely upon the occurrence of lucky movements brought out by such accidental influences. Second, there is a class of modifications which arise from the spontaneous activities of the organism itself in the carrying out of its normal congenital functions. These variations and adaptations are seen in a remarkable way in plants, in unicellular creatures, in very young children. There seems to be a readiness and capacity on the part of the organism to "rise to the occasion," as it were, and make gain out of the circumstances of its life. The facts have been put in evidence (for plants) by Henslow, Pfeffer, Sachs; (for micro-organisms) by Binet, Bunge; (in human pathology) by Bernheim, Janet; (in children) by Baldwin (ref. 2, chap. vi.) (See citations in ref. 2, chap. ix, and in Orr, *Theory of Development*, chap. iv). These changes I propose to call "neuro-genetic," laying emphasis on what is called by Romanes, Morgan and others, the "selective property" of the nervous system, and of life generally. Third, there is the great series of adaptations secured by conscious agency, which we may throw together as "psycho-genetic." The processes involved here are all classed broadly under the term "intelligent," i. e., imitation, gregarious influences, maternal instruction, the lessons of pleasure and pain, and of experience generally, and reasoning from means to ends, etc.

We reach, therefore, the following scheme:

Ontogenetic Modifications.	*Ontogenic Agencies.*
1. Physico-genetic.	1. Mechanical.
2. Neuro-genetic.	2. Nervous.
3. Psycho-genetic.	3. Intelligent.
	Imitation.
	Pleasure and pain.
	Reasoning.

Now it is evident that there are two very distinct questions which come up as soon as we admit modifications of function and of structure in ontogenetic development: first, there is the question as to how these modifications can come to be adaptive in the life of the individual creature. Or in other words: What is the method of the individual's growth and adaptation as shown in the well known law of "use and disuse?" Looked at functionally, we see that the organism manages somehow to accommodate itself to conditions which are favorable, to repeat movements which are adaptive, and so to grow by the principle of use. This involves some sort of selection, from the actual ontogenetic variations, of certain ones—certain functions, etc. Certain other possible and actual functions and structures decay from disuse. Whatever the method of doing this may be, we may simply, at this point, claim the law of use and disuse, as applicable in ontogenetic development, and apply the phrase, "Organic Selection," to the organism's behavior in acquiring new modes or modifications of adaptive function with its influence of structure. The question of the method of "Organic Selection" is taken up below (IV); here, I may repeat, we simply assume what every one admits in some form, that such adaptations of function—"accommodations" the psychologist calls them, the processes of learning new movements, etc.—*do occur*. We then reach another question, second; what place these adaptations have in the general theory of development.

Effects of Organic Selection.—First, we may note the results of this principle in the creature's own private life.

1. *By securing adaptations, accommodations, in special circumstances the creature is kept alive* (ref. 2, 1st ed., pp. 172 ff.). This is true in all the three spheres of ontogenetic variation distinguished in the table above. The creatures which can stand the "storm and stress" of the physical influences of the environment, and of the changes which occur in the environment, *by undergoing modifications of their congenital functions or of the structures which they get congenitally—these creatures will live; while those which cannot, will not.* In the sphere of neurogenetic variations we find a superb series of adaptations by lower as well as higher organisms during the course of ontogenetic development (ref. 2, chap. ix). And in the highest sphere, that of intelligence (including the phenomena of consciousness of all kinds, experience of pleasure and pain, imitation, etc.), we find individual accommodations on the tremendous scale which culminates in the skilful performances of human volition, invention, etc. The progress of the child in all the learning processes which lead him on to be a man, just illustrates this higher form of ontogenetic adaptation (ref. 2, chap. x–xiii).

All these instances are associated in the higher organisms, and all of them unite to *keep the creature alive*.

2. By this means *those congenital or phylogenetic variations are kept in existence, which lend themselves to intelligent, imitative, adaptive, and mechanical modification during the lifetime of the creatures which have them.* Other congenital variations are not thus kept in existence. So there arises a more or less widespread series of *determinate variations in each generation's ontogenesis* (ref. 3, 4, 5).[3]

[3] "It is necessary to consider further how certain reactions of one single organism can be selected so as to adapt the organism better and give it a life history. Let us at the outset call this process "Organic Selection" in contrast with the Natural Selection of whole organisms. . . . If this (natural selection) worked alone, every change in the environment would weed out all life except those organisms which by accidental variation reacted already in the way demanded by the changed conditions—in every case new organisms showing variations, not, in any case, new elements of life-history in the old organisms. In order to the latter we would have to conceive . . . some modification of the old reactions in an organism through the influence of new conditions. . . . We are, accordingly, left to the view that the new stimulations brought by changes in the environment

The further applications of the principle lead us over into the field of our second question, i. e., phylogeny.

II.

Phylogeny: Physical Heredity.—The question of phylogenetic development considered apart, in so far as may be, from that of heredity, is the question as to what the factors really are which show themselves in evolutionary progress from generation to generation. The most important series of facts recently brought to light are those which show what is called "determinate variation" from one generation to another. This has been insisted on by the paleontologists. Of the two current theories of heredity, only one, Neo-Lamarkism—by means of its principle of the inheritance of acquired characters—has been able to account for this fact of determinate phylogenetic change. Weismann admits the inadequacy of the principle of natural selection, as operative on rival organisms, to explain variations when they are wanted or, as he puts it, "the right variations in the right place" (*Monist*, Jan., '96).

I have argued, however, in detail that the assumption of determinate variations of function in ontogenesis, under the principle of neurogenetic and psychogenetic adaptation, does away with the need of appealing to the Lamarkian factor. In the case i. g., of instincts, "if we do not assume consciousness, then natural selection is inadequate; but if we do assume consciousness, then the inheritance of acquired characters is unnecessary" (ref. 5).

"The intelligence which is appealed to, to take the place of instinct and to give rise to it, uses just these partial variations which tend in the direction of the instinct; so the intelligence *supplements* such partial co-ordinations, makes them functional, and *so keeps the creature alive*. In the phrase of Prof.

themselves modify the reactions of an organism. . . . The facts show that individual organisms do acquire new adaptations in their lifetime, and that is our first problem. If in solving it we find a principle which may also serve as a principle of race-development, then we may possibly use it against the 'all sufficiency of natural selection' or in its support" (ref. 2, 1st. ed., pp. 175-6.)

Lloyd Morgan, this prevents the 'incidence of natural selection.' So the supposition that intelligence is operative turns out to be just the supposition which makes use-inheritance unnecessary. Thus kept alive, the species has all the time necessary to perfect the variations required by a complete instinct. And when we bear in mind that the variation required is not on the muscular side to any great extent, but in the central brain connections, and is a slight variation for functional purposes at the best, the hypothesis of use-inheritance becomes not only unnecessary, but to my mind quite superfluous" (ref. 4, p. 439). And for adaptations generally, "the most plastic individuals will be preserved to do the advantageous things for which their variations show them to be the most fit, and the next generation will show an emphasis of just this direction in its variations" (ref. 3, p. 221).

We get, therefore, from Organic Selection, certain results in the sphere of phylogeny:

1. *This principle secures by survival certain lines of determinate phylogenetic variation in the directions of the determinate ontogenetic adaptations of the earlier generation.* The variations which were utilized for ontogenetic adaptation in the earlier generation, being thus kept in existence, are utilized more widely in the subsequent generation (ref. 3, 4). "Congenital variations, on the one hand, are kept alive and made effective by their use for adaptations in the life of the individual; and, on the other hand, adaptations become congenital by further progress and refinement of variation in the same lines of function as those which their acquisition by the individual called into play. But there is no need in either case to assume the Lamarkian factor" (ref. 3). And in cases of conscious adaptation: "We reach a point of view which gives to organic evolution a sort of intelligent direction after all; for of all the variations tending in the direction of an adaptation, but inadequate to its complete performance, *only those will be supplemented and kept alive which the intelligence ratifies and uses.* The principle of 'selective value' applies to the others or to some of them. So natural selection kills off the others; and the *future*

development at each stage of a species' development must be in the directions thus ratified by intelligence. So also with imitation. Only those imitative actions of a creature which are useful to him will survive in the species, for in so far as he imitates actions which are injurious he will aid natural selection in killing himself off. So intelligence, and the imitation which copies it, will set the direction of the development of the complex instincts even on the Neo-Darwinian theory; and in this sense we may say that consciousness is a ' factor' " (ref. 4).

2. *The mean of phylogenetic variation being thus made more determinate, further phylogenetic variations follow about this mean, and these variations are again utilized by Organic Selection for ontogenetic adaptation.* So there is continual phylogenetic progress in the directions set by ontogenetic adaptation (ref. 3, 4, 5). " The intelligence supplements slight co-adaptations and so gives them selective value; but it does not keep them from getting farther selective value as instincts, reflexes, etc., by farther variation " (ref. 5). " The imitative function, by using muscular co-ordinations, supplements them, secures adaptations, keeps the creature alive, prevents the 'incidence of natural selection,' and so gives the species all the time necessary to get the variations required for the full instinctive performance of the function " (ref. 4). But, " Conscious imitation, while it prevents the incidence of natural selection, as has been seen, and so keeps alive the creatures which have no instincts for the performance of the actions required, nevertheless does not subserve the utilities which the special instincts do, nor prevent them from having the selective value of which Romanes speaks. Accordingly, on the more general definition of intelligence, which includes in it all conscious imitation, use of maternal instruction, and that sort of thing—no less than on the more special definition—we still find the principal of natural selection operative " (ref. 5).

3. *This completely disposes of the Lamarkian factor as far as two lines of evidence for it are concerned.* First, the evidence drawn from function, " use and disuse," is discredited; since by " organic selection," the reappearance, in subsequent generations, of the variations first secured in ontogenesis is ac-

counted for without the inheritance of acquired characters. So also the evidence drawn from paleontology which cites progressive variations resting on functional use and disuse. Second, the evidence drawn from the facts of "determinate variations;" since by this principle we have the preservation of such variations in phylogeny without the inheritance of acquired characters.

4. *But this is not Preformism in the old sense; since the adpatations made in ontogenetic development which "set" the direction of evolution are novelties of function in whole or part* (although they utilize congenital variations of structure). And it is only by the exercise of these novel functions that the creatures are kept alive to propagate and thus produce further variations of structure which may in time make the whole function, with its adequate structure, congenital. Romanes' argument from "partial co-adaptations" and "selective value," seem to hold in the case of reflex and instinctive functions (ref. 4, 5), as against the old preformist or Weismannist view, although the operation of Organic Selection, as now explained, renders them ineffective when urged in support of Lamarkism. "We may imagine creatures, whose hands were used for holding only with the thumb and fingers on the same side of the object held, to have first discovered, under stress of circumstances and with variations which permitted the further adaptation, how to make use of the thumb for grasping opposite to the fingers, as we now do. Then let us suppose that this proved of such utility that all the young that did not do it were killed off; the next generation following would be plastic, intelligent, or imitative, enough to do it also. They would use the same co-ordinations and prevent natural selection getting its operation on them; and so instinctive 'thumb-grasping' might be waited for indefinitely by the species and then be got as an instinct altogether apart from use-inheritance" (ref. 4). "I have cited 'thumb-grasping' because we can see in the child the anticipation, by intelligence and imitation, of the use of the thumb for the adaptation which the Simian probably gets entirely by instinct, and which I think an isolated and weak-minded child, say, would also come to do by instinct'" (ref. 4).

5. It seems to me also—though I hardly dare venture into a field belonging so strictly to the technical biologist—that *this principle might not only explain many cases of widespread "determinate variations" appearing suddenly, let us say, in fossil deposits, but the fact that variations seem often to be "discontinuous."* Suppose, for example, certain animals, varying, in respect to a certain quality, from a to n about a mean x. The mean x would be the case most likely to be preserved in fossil form (seeing that there are vastly more of them). Now suppose a sweeping change in the environment, in such a way that only the variations lying near the extreme n can accommodate to it and live to reproduce. The next generation would then show variations about the mean n. And the chances of fossils from this generation, and the subsequent ones, would be of creatures approximating n. Here would be a great discontinuity in the chain and also a widespread prevalence of these variations in a set direction. This seems especially evident when we consider that the paleontologist does not deal with successive generations, but with widely remote periods, and the smallest lapse of time which he can take cognizance of is long enough to give the new mean of variation, n, a lot of generations in which to multiply and deposit its representative fossils. Of course, this would be only the action of natural selection upon "preformed" variations in those cases which did not involve positive changes, in structure and function, *acquired in ontogenesis;* but in so far as such ontogenetic adaptations were actually there, the extent of difference of the n mean from the x mean would be greater, and hence the resources of explanation, both of the sudden prevalence of the new type and of its discontinuity from the earlier, would be much increased. This additional resource, then, is due to the "Organic Selection" factor.

We seem to be able also to utilize all the evidence usually cited for the functional origin of specific characters and groupings of characters. So far as the Lamarkians have a strong case here, it remains as strong if Organic Selection be substituted for the "inheritance of acquired characters." This is especially true where intelligent and imitative adaptations are involved, as in the case of instinct. This "may give the reason, e. g., that instincts are so often coterminous with the limits of species. Similar structures find the similar uses for their intelligence, and they also find the same imitative actions to be to their advantage. So the interaction of these conscious factors with natural selection brings it about that the structural definition which represents species, and the functional definition which represents instinct, largely keep to the same lines" (ref. 5).

6. It seems proper, therefore, to call the influence of Organic Selection "a new factor;" for it gives a method of deriving the determinate gains of phylogeny from the adaptations of ontogeny without holding to either of the two current theories. *The ontogenetic adaptations are really new, not performed; and they are really reproduced in succeeding generations, although not physically inherited.*

(*To be continued.*)

A NEW FACTOR IN EVOLUTION.

By J. Mark Baldwin.

(*Continued from page 451*).

III.

Social Heredity.—There follows also another resource in the matter of development. In all the higher reaches of development we find certain co-operative or "social" processes which directly supplement or add to the individual's private adaptations. In the lower forms it is called gregariousnes, in man sociality, and in the lowest creatures (except plants) there are suggestions of a sort of imitative and responsive action between creatures of the same species and in the same habitat. In all these cases it is evident that other living creatures constitute part of the environment of each, and many neuro-genetic and psycho-genetic accommodations have reference to or involve these other creatures. It is here that the principle of imitation gets tremendous significance; intelligence and volition, also, later on; and in human affairs it becomes social co-operation. Now it is evident that when young creatures have these imitative, intelligent, or quasi-social tendencies to any extent, they are able to pick up *for themselves*, by imitation, instruction, experience generally, the functions which their parents and other creatures perform in their presence. This then is a form of ontogenetic adaptation; it keeps these creatures alive, and so produces determinate variations in the way explained above. It is, therefore, a special, and from its wide range, an extremely important instance of the general principle of Organic Selection.

But it has a farther value. *It keeps alive a series of functions which either are not yet, or never do become, congenital at all.* It is a means of extra-organic transmission from generation to generation. It is really a form of heredity because (1) *it is a handing down of physical functions*; while it is not physical heredity. It is entitled to be called heredity for the further reason (2) that *it directly influences physical heredity in the way mentioned*, i. e., it keeps alive variations, thus sets the direction of ontogenetic adaptation, thereby influences the direction of the available congenital variations of the next generation, and so determines phylogenetic development. I have accordingly called it "Social Heredity" (ref. 2, chap. xii; ref. 3).

In "Social Heredity," therefore, we have a more or less conservative, progressive, ontogenic atmosphere of which we may make certain remarks as follows:—

(1) *It secures adaptations of individuals all through the animal world.* "Instead of limiting this influence to human life, we have to extend it to all the gregarious animals, to all the creatures that have any ability to imitate, and finally to all animals who have consciousness sufficient to enable them to make adaptations of their own: for such creatures will have children that can do the same, and it is unnecessary to say that the children must inherit what their fathers did by intelligence, when they can do the same things by intelligence" (ref. 6).

(2) *It tends to set the direction of phylogenetic progress* by Organic Selection, Sexual Selection, etc., *i. e.*, it tends not only

to give the young the adaptations which the adults already have, but also *to produce adaptations which depend upon social coöperation; thus variations in the direction of sociality are selected and made determinate.* "When we remember that the permanence of a habit learned by one individual is largely conditioned by the learning of the same habits by others (notably of the opposite sex) in the same environment, we see that an enormous premium must have been put on variations of a social kind—those which brought different individuals into some kind of joint action or coöperation. Wherever this appeared, not only would habits be maintained, but new variations, having all the force of double hereditary tendency, might also be expected" (ref. 3). Why is it, for example, that a race of Mulattoes does not arise faster, and possess our Southern States? Is it not just the social repugnance to black-white marriages? Remove or reverse *this influence of education, imitation, etc.*, and the result *on phylogeny* would show in our faces, and even appear in our fossils when they are dug up long hence by the paleontologist of the succeeding aeons!

(3) *In man it becomes the law of social evolution.* "Weismann and others have shown that the influence of animal intercourse, seen in maternal instruction, imitation, gregarious coöperation, etc., is very important. Wallace dwells upon the actual facts which illustrate the 'imitative factor,' as we may call it, in the personal development of young animals. I have recently argued that Spencer and others are in error in holding that social progress demands use-inheritance; since the socially-acquired actions of a species, notably man, are socially handed down, giving a sort of 'social heredity' which supplements natural heredity" (ref. 4). The social "sport," the genius, is very often the controlling factor in social evolution. He not only sets the direction of future progress, but he may actually lift society at a bound up to a new standard of attainment (ref. 6). "So strong does the case seem for the Social Heredity view in this matter of intellectual and moral progress that I may suggest an hypothesis which may not stand in court, but which I find interesting. May not the rise of social

life be justified from the point of view of a second utility in addition to that of its utility in the struggle for existence as ordinarily understood, the second utility, *i.e.*, of giving to each generation the attainments of the past which natural inheritance is inadequate to transmit. When social life begins, we find the beginning of the artificial selection of the unfit; and this negative principle begins to work directly in the teeth of progress, as many writers on social themes have recently made clear. This being the case, some other resource is necessary besides natural inheritance. On my hypothesis it is found in the common or social standards of attainment which the individual is fitted to grow up to and to which he is compelled to submit. This secures progress in two ways: First, by making the individual learn what the race has learned, thus preventing social retrogression, in any case; and second, by putting a direct premium on variations which are socially available " (ref. 3).

4. The two ways of securing development in determinate directions—the purely extra-organic way of Social Heredity, and the way by which Organic Selection in general (both by social and by other ontogenetic adaptations) secures the fixing of phylogenetic variations, as described above—seem to run parallel. Their conjoint influence is seen most interestingly ingly in the complex instincts (ref. 4, 5). We find in some instincts completely reflex or congenital functions which are accounted for by Organic Selection. In other instincts we find only partial coördinations ready given by heredity, and the creature actually depending upon some conscious resource (imitation, instruction, etc.) to bring the instinct into actual operation. But as we come up in the line of phylogenetic development, both processes may be present *for the same function;* the intelligence of the creature may lead him to do consciously what he also does instinctively. In these cases the additional utility gained by the double performance accounts for the duplication. It has arisen either (1) by the accumulation of congenital variations in creatures which already performed the action (by ontogenetic adaptation and handed it down socially), or (2) the reverse. In the animals, the social

transmission seems to be mainly useful as enabling a species to get instincts slowly in determinate directions, by keeping off the operation of natural selection. Social Heredity is then the lesser factor; it serves Biological Heredity. But in man, the reverse. Social transmission is the important factor, and the congenital equipment of instincts is actually broken up in order to allow the plasticity which the human being's social learning requires him to have. So in all cases both factors are present, but in a sort of inverse ratio to each other. In the words of Preyer, "the more kinds of co-ordinated movement an animal brings into the world, the fewer is he able to learn afterwards." The child is the animal which inherits the smallest number of congenital co-ordinations, but he is the one that learns the greatest number (ref. 2, p. 297).

"It is very probable, as far as the early life of the child may be taken as indicating the factors of evolution, that the main function of consciousness is to enable him to learn things which natural heredity fails to transmit; and with the child the fact that consciousness is the essential means of all his learning is correlated with the other fact that the child is the very creature for which natural heredity gives few independent functions. It is in this field only that I venture to speak with assurance; but the same point of view has been reached by Weismann and others on the purely biological side. The instinctive equipment of the lower animals is replaced by the plasticity for learning by consciousness. So it seems to me that the evidence points to some inverse ratio between the importance of consciousness as factor in development and the need of inheritance of acquired characters as factor in development" (ref. 7).

"Under this general conception we may bring the biological phenomena of infancy, with all their evolutionary significance: the great plasticity of the mammal infant as opposed to the highly developed instinctive equipment of other young; the maternal care, instruction and example during the period of dependence, and the very gradual attainment of the activities of self-maintenance in conditions in which social activities are absolutely essential. All this stock of the development theory is available to confirm this view" (Ref. 3).

But these two influences furnish a double resort against Neo-Lamarkism. And I do not see anything in the way of considering the fact of Organic Selection, from which both these resources spring, as being a sufficient supplement to the principle of natural selection. The relation which it bears to natural selection, however, is a matter of further remark below (V).

"We may say, therefore, that there are two great kinds of influence, each in a sense hereditary; there is *natural heredity* by which variations are congenitally transmitted with original endowment, and there is '*social heredity*' by which functions socially acquired (*i. e.*, imitatively, covering all the conscious acquisitions made through intercourse with other animals) are also socially transmitted. The one is phylogenetic; the other ontogenetic. But these two lines of hereditary influence are not separate nor uninfluential on each other. Congenital variations, on the one hand, are kept alive and made effective by their conscious use for intelligent and imitative adaptations in the life of the individual; and, on the other hand, intelligent and imitative adaptations become congenital by further progress and refinement of variation in the same lines of function as those which their acquisition by the individual called into play. But there is no need in either case to assume the Lamarkian factor" (ref. 4).

"The only hindrance that I see to the child's learning everything that his life in society requires would be just the thing that the advocates of Lamarkism argue for—the inheritance of acquired characters. For such inheritance would tend so to bind up the child's nervous substance in fixed forms that he would have less or possibly no unstable substance left to learn anything with. So, in fact, it is with the animals in which instinct is largely developed; they have no power to learn anything new, just because their nervous systems are not in the mobile condition represented by high consciousness. They have instinct and little else" (ref. 3).

IV.

The Process of Organic Selection.—So far we have been dealing exclusively with facts. By recognizing certain facts we have reached a view which considers ontogenetic selection an important factor in development. Without prejudicing the statement of fact at all we may enquire into the actual working of the organism is making its organic selections or adaptations. The question is simply this: how does the organism secure, from the multitude of possible ontogenetic changes which it might and does undergo, those which are adaptive? As a matter of fact, all personal growth, all motor acquisitions made by the individual, show that it succeeds in doing this; the further question is, how? Before taking this up, I must repeat with emphasis that the position taken in the foregoing pages, which simply makes the fact of ontogenetic adaptation a factor in development, is not involved in the solution of the further question as to how the adaptations are secured. But from the answer to this latter question we may get further light of the interpretation of the facts themselves. So we come to ask how Organic Selection actually operates in the case of a particular adaptation of a particular creature (ref. 1; ref. 2, chap. vii, xiii; ref. 6, and 7).

I hold that the organism has a way of doing this which is peculiarly its own. The point is elaborated at such great length in the book referred to (ref. 2) that I need not repeat details here. The summary in this journal (ref. 6) may have been seen by its readers. There is a fact of physiology which, taken together with the facts of psychology, serves to indicate the method of the adaptations or accommodations of the individual organism. The general fact is that the organism concentrates it energies upon the locality stimulated, for the continuation of the conditions, movements, stimulations which are vitally beneficial, and for the cessation of the conditions, movements, stimulations, which are vitally depressing and harmful. In the case of beneficial conditions we find a general *increase of movement, an excess discharge of the energies of movement in the channels already open and habitual; and with this, on the psychological side, pleasurable consciousness and attention.* Attention to a member is accompanied by increased vasomotor activity, with higher muscular power, and a *general dynamogenic heightening in that member.* "The thought of a movement tends to discharge motor energy into the channels as near as may be to those necessary for that movement" (ref. 3). By this organic concentration and excess of movement many combinations and variations are rendered possible, from which the advantageous and adaptive movements may be selected for their utility. These then give renewed pleasure, excite pleasurable associations, and again stimulate the attention, and *by these influences the adaptive movements thus struck are selected and held as permanent acquisitions.* This form of concentration of energy upon stimulated localities, with the resulting renewal by movements of conditions that are pleasure-giving and beneficial, and the subsequent repetitions of the movements, is called the " circular reaction."[4] (ref. 1, 2). It is the selective property which Romanes pointed out as characterizing and differentiating life. It characterizes the responses of the organism, however low in the scale, to all stimulations—even those of a mechanical and chemical (physico-genic) nature. Pfeffer has shown such a determination of energy toward the parts stimulated even in plants. And in the higher animals it finds itself exactly reproduced in the nervous reaction seen in imitation and—through processes of association, substitution, etc.—in all the higher mental acts of intelligence and volition. These are developed phylogenetically as variations whose direction is constantly determined, by this form of adaptation in ontogenesis. If this be true—and the biological facts seem fully to confirm it—this is the adaptive process in all life, and this process is that with which the development of mental life has been associated.

It follows, accordingly, that the three forms of ontogenetic adaptation distinguished above—physico-genetic, neuro-genetic, psycho-genetic—all involve the sort of response on the part of the organism seen in this circular reaction with excess discharge; and we reach one general law of ontogenetic adaptation and of Organic Selection. " The accommodation of an organism to a new stimulation is secured—not by the selection of this stimulation beforehand (nor of the necessary move-

[4] With the opposite (withdrawing, depressive affects) in injurious and painful conditions.

ments)—but by the reinstatement of it by a discharge of the energies of the organism, concentrated as far as may be for the excessive stimulation of the organs (muscles, etc.) most nearly fitted by former habit to get this stimulation again (in which the " stimulation " stands for the condition favorable to adaptation). After several trials the child (for example) gets the adaptation aimed at more and more perfectly, and the accompanying excessive and useless movements fall away. This is the kind of selection that intelligence does in its acquisition of new movements" (ref. 2, p. 179; ref. 6).

Accordingly, *all ontogenetic adaptations are neurogenetic.*[5] The general law of "motor excess" is one of *overproduction;* from movements thus overproduced, adaptations survive; these adaptations set the determinate direction of ontogenesis; and by their survival the same determination of direction is set in phylogenesis also.

The following quotation from an earlier paper (ref. 7) will show some of the bearings of this position:

"That there is some general principle running through all the adaptations of movement which the individual creature makes is indicated by the very unity of the organism itself. The principle of Habit must be recognized in some general way which will allow the organism to do new things without utterly undoing what it has already acquired. This means that old habits must be substantially preserved *in the new functions;* that all new functions must be reached by gradual modifications. And we will all go further and say, I think, that the only way that these modifications can be got at all is through some sort of interaction of the organism with its environment. Now, as soon as we ask how the stimulations of the environment can produce new adaptive movements, we have the answer of Spencer and Bain—an answer directly confirmed, I think, without question, by the study both of the child and of the adult—i. e., by the selection of fit movements from excessively produced movements, that is, from *movement variations.* So granting this, we now have the further question:

[5] Barring, of course, those violent compelling physical influences under the action of which the organism is quite helpless.

How do these movement variations come to be produced *when and where they are needed?*[6] And with it, the question: How does the organism *keep those movements going* which are thus selected, and *suppress* those which are not selected?

"Now these two questions are the ones which the biologists fail to answer. But the force of the facts leads to the hypotheses of "conscious force," "self-development" of Henslow and "directive tendency" of the American school—all aspects of the new Vitalism which just these questions and the facts which they rest upon are now forcing to the front. Have we anything definite, drawn from the study of the individual on the psychological side, to substitute for these confessedly vague biological phrases? Spencer gave an answer in a general way long ago to the *second* of these questions, by saying that in consciousness the function of pleasure and pain is just to keep some actions or movements going and to suppress others.

"But as soon as we enquire more closely into the actual working of pleasure and pain reactions, we find an answer suggested to the *first* question also, *i. e.*, the question as to how the organism comes to make the kind and sort of movements which the environment calls for—the *movement variations when and where they are required.* The pleasure or pain produced by a stimulus—and by a movement also, for the utility of movement is always that it secures stimulation of this sort or that —does not lead to diffused, neutral, and characterless movements, as Spencer and Bain suppose; this is disputed no less by the infant's movements than by the actions of unicellular creatures. There are characteristic differences in vital move-

[6] This is just the question that Weismann seeks to answer (in respect to the supply of variations in forms which the paleontologists require), with his doctrine of 'Germinal Selection' (*Monist*, Jan., 1896). Why are not such applications of the principle of natural selection to variations *in the parts and functions of the single organism* just as reasonable and legitimate as it is to variations in separate organisms? As against "germinal selection," however, I may say, that in the cases in which ontogenetic adaptation sets the direction of survival of phylogenetic variations (as held in this paper) the hypothesis of germinal selection is in so far unnecessary. This view finds the operation of selection *on functions in ontogeny* the means of securing "variations when and where they are wanted;" while Weismann supposes competing germinal units.

ments wherever we find them. Even if Mr. Spencer's undifferentiated protoplasmic movements had existed, natural selection would very soon have put an end to it. There is a characteristic antithesis in vital movements always. Healthy, overflowing, outreaching, expansive, vital effects are associated with pleasure; and the contrary, the withdrawing, depressive, contractive, decreasing, vital effects are associated with pain. This is exactly the state of things which the theory of selection of movements from overproduced movements requires, *i. e.*, that increased vitality, represented by pleasure, should give the excess movements, from which new adaptations are selected; and that decreased vitality represented by pain should do the reverse, *i. e.*, draw off energy and suppress movement.[7]

"If, therefore, we say that here is a type of reaction which all vitality shows, we may give it a general descriptive name, *i. e.*, the "Circular Reaction," in that its significance for evolution is that it is not a random response in movement to all stimulations alike, but that it distinguishes in its very form and amount between stimulations which are vitally good and those which are vitally bad, tending to retain the good stimulations and to draw away from and so suppress the bad. The term 'circular' is used to emphasize the way such a reaction tends to keep itself going, over and over, by reproducing the conditions of its own stimulation. It represents habit, since

[7] It is probable that the origin of this antithesis is to be found in the waxing and waning of the nutritive processes. "We find that if by an organism we mean a thing merely of contractility or irritability, whose round of movements is kept up by some kind of nutritive process supplied by the environment—absorption, chemical action of atmospheric oxygen, etc.—and whose existence is threatened by dangers of contact and what not, the first thing to do is to secure a regular supply to the nutritive processes, and to avoid these contacts. But the organism can do nothing but move, as a whole or in some of its parts. So then if one of such creatures is to be fitter than another to survive, it must be the creature which by its movements secures more nutritive processes and avoids more dangerous contacts. But movements toward the source of stimulation keep hold on the stimulation, and movements away from contacts break the contacts, that is all. Nature selects these organisms; how could she do otherwise? . . . . We only have to suppose, then, that the nutritive growth processes are by natural selection drained off in organic expansions, to get the division in movements which represents this earliest bifurcate adaptation." (Ref. 2, p. 201).

it tends to keep up old movements; but it secures new adaptations, since it provides for the overproduction of movement variations for the operation of selection. This kind of selection, since it requires the direct coöperation of the organism itself, I have called 'Organic Selection.'"

The advantages of this view seem to be somewhat as follows:

1. It gives a method of the individual's adaptations of function which is *one in principle with the law of overproduction and survival now so well established in the case of competing organisms.*

2. It reduces nervous and mental evolution to strictly parallel terms. The intelligent use of phylogenetic variations for functional purposes in the way indicated, puts a premium on variations which can be so used; and thus sets phylogenetic progress *in directions of constantly improved mental endowment.* The circular reaction which is the method of intelligent adaptations is liable to variation in a series of complex ways which represent phylogenetically the development of the mental functions known as memory, imagination, conception, thought, etc. We thus reach a phylogeny of mind which proceeds in the direction set by the ontogeny of mind,[8] just as on the organic side the phylogeny of the organism gets its determinate direction from the organism's ontogenetic adaptations. And since it is the one principle of Organic Selection working by *the same functions* to set the direction of both phylogenies, the physical and the mental, the two developments are not two, but one. Evolution is, therefore, not more biological than psychological (ref. 2, chap. x, xi, and especially pp. 383–388).

3. It secures the relation of structure to function required by the principle of "use and disuse" in ontogeny.

4. The only alternative theory of the adaptations of the individual are those of "pure chance," on the one hand, and a "creative act" of consciousness, or the other hand. Pure chance is refuted by all the facts which show that the organism does not wait for chance, but goes right out and effects new adaptations to its environment. Furthermore, ontogenetic

[8] Prof. C. S. Minot suggests to me that the terms "ontopsychic" and "phylopsychic" might be convenient to mark this distinction.

adaptations are determinate; they proceed in definite progressive lines. A short study of the child will disabuse any man, I think, of the "pure chance" theory. But the other theory which holds that consciousness makes adaptations and changes structures directly by its *fiat*, is contradicted by the psychology of voluntary movement (ref. 4, 6, 7). Consciousness can bring about no movement without having first an adequate experience of that movement to serve on occasion as a stimulus to the innervation of the appropriate motor centers. "This point is no longer subject to dispute; for pathological cases show that unless some adequate idea of a former movement made by the same muscles, or by association some other idea which stands for it, can be brought up in mind the intelligence is helpless. Not only can it not make new movements; it can not even repeat old habitual movements. So we may say that intelligent adaptation does not create coördinations; it only makes functional use of coördinations which were alternatively present already in the creature's equipment. Interpreting this in terms of congenital variations, we may say that the variations which the intelligence uses are alternative possibilities of muscular movement" (ref. 4). So the only possible way that a really new movement can be made is *by making the movements already possible so excessively and with so many varieties of combination, etc., that new adaptations may occur.*

5. The problem seems to me to duplicate the conditions which led Darwin to the principle of natural selection. The alternatives before Darwin were "pure chance" or "special creation." The law of "overproduction with survival of the fittest" came as the solution. So in this case. Let us take an example. Every child has to learn how to write. If he depended upon chance movements of his hands he would never learn how to write. But on the other hand, he can not write simply by willing to do so; he might will forever without effecting a "special creation" of muscular movement. What he actually does is to *use his hand in a great many possible ways as near as he can to the way required;* and from these excessively produced movements, and after excessively varied and numerous trials, he gradually selects and fixes the slight successes made

in the direction of correct writing. It is a long and most laborious accumulation of slight Organic Selections from overproduced movements (ref. for handwriting in detail, 2, chap. v; also 2, pp. 373, ff.).

6. The only resort left to the theory that consciousness is some sort of an *actus purus* is to hold that it *directs* brain energies or selects between possible alternatives of movement; but besides the objection that it is as hard to direct movement as it is to make it (for nothing short of a force could release or direct brain energies), we find nothing of the kind necessary. The attention is what determines the particular movement in developed organisms, and the attention is no longer considered an *actus purus* with no brain process accompanying it. The attention is a function of memories, movements, organic experiences. We do not attend to a thing because we have already selected it, or because the attention selects it; but *we select it because we—consciousness and organism—are attending to it.* "It is clear that this doctrine of selection as applied to muscular movement does away with all necessity for holding that consciousness even directs brain energy. The need of such direction seems to me to be as artificial as Darwin showed the need of special creation to be for the teleological adaptations of the different species. This need done away, in this case of supposed directive agency as in that, the question of the relation of consciousness to the brain becomes a metaphysical one, just as that of teleology in nature became a metaphysical one; and it is not to much profit that science meddles with it. And biological as well as psychological science should be glad that it is so, should it not?" (ref. 6; and on the metaphysical question, ref. 7).

V.

A word on the relation of this principle of Organic Selection to Natural Selection. Natural Selection is too often treated as a positive agency. It is not a positive agency; it is entirely negative. It is simply a statement of what occurs when an organism does not have the qualifications necessary to enable it to survive in given conditions of life; it does not in any way define positively the qualifications which do enable other organisms to survive. Assuming the principle of Natural Selection in any case, and saying that, according to it, if an organism do not have the necessary qualifications it will be killed off, it still remains in that instance to find what the qualifications are which this organism is to have if it is to be kept alive. So we may say that *the means of survival is always an additional question* to the negative statement of the operation of natural selection.

This latter question, of course, the theory of variations aims to answer. The positive qualifications which the organism has arise as congenital variations of a kind which enable the organism to cope with the conditions of life. This is the positive side of Darwinism, as the principle of Natural Selection is the negative side.

Now it is in relation to the theory of variations, and not in relation to that of natural selection, that Organic Selection has its main force. Organic Selection presents *a new qualification of a positive kind* which enables the organism to meet its environment and cope with it, while natural selection remains exactly what it was, the negative law that if the organism does not succeed in living, then it dies, and as such a qualification on the part of the organism, Organic Selection presents several interesting features.

1. If we hold, as has been argued above, that the method of Organic Selection is always the same (that is, that it has a natural method), being always accomplished by a certain typical sort of nervous process (*i. e.*, being always neuro-genetic), then we may ask whether that form of nervous process—and the consciousness which goes with it—may not be a variation appearing early in the phylogenetic series. I have argued elsewhere (ref. 2, pp. 200 ff. and 208 ff.) that this is the most probable view. Organisms that did not have some form of selective response to what was beneficial, as opposed to what was damaging in the environment, could not have developed very far; and as soon as such a variation did appear it would have immediate preëminence. So we have to say either that selective nervous property, with consciousness, is a variation,

or that it is a fundamental endowment of life and part of its final mystery. " The intelligence holds a remarkable place. It is itself, as we have seen, a congenital variation; but it is also the great agent of the individual's personal adaptation both to the physical and to the social environment " (ref. 4).

" The former (instinct) represents a tendency to brain variation in the direction of fixed connections between certain sense-centers and certain groups of coördinated muscles. This tendency is embodied in the white matter and the lower brain centers. The other (intelligence) represents a tendency to variation in the direction of alternative possibilities of connection of the brain centers with the same or similar coördinated muscular groups. This tendency is embodied in the cortex of the hemispheres" (ref. 4).

2. But however that may be, whether ontogenetic adaptation by selective reaction and consciousness be considered a variation or a final aspect of life, it is a *life-qualification of a very extraordinary kind*. It opens a new sphere for the application of the negative principle of natural selection upon organisms, *i. e.*, with reference to *what they can do*, rather than to what they are; to the new use they make of their congenital functions, rather than to the mere possession of the functions (ref. 2, pp. 202 f.). A premium is set on congenital plasticity and adaptability of function rather than on congenital fixity of function; and this adaptability reaches its highest in the intelligence.

3. It opens another field also for the operation of natural selection—still viewed as a negative principle—through the survival of particular overproduced and modified reactions of the organism, by which the determination of the organism's own growth and life-history is secured. If the young chick imitated the old duck instead of the old hen, it would perish; it can only learn those new things which its present equipment will permit—not swimming. So the chick's own possible actions and adaptations in ontogeny have to be selected. We have seen how it may be done by a certain competition of functions with survival of the fit. But this is an application of natural selection. I do not see how Henslow, for example, can get the so-called "self-adaptations"—apart from " special

creation "—which justify an attack on natural selection. Even plants must grow in determinate or " select ". directions in order to live.

4. So we may say, finally, that Organic Selection, while itself probably a congenital variation (or original endowment) works to secure new qualifications for the creature's survival; and its very working proceeds by securing a new application of the principle of natural selection to the possible modifications which the organism is capable of undergoing. Romanes says: " it is impossible that heredity can have provided in advance for innovations upon or alterations in its own machinery during the lifetime of a particular individual." To this we are obliged to reply in summing up—as I have done before (ref. 2, p. 220)—we reach "just the state of things which Romanes declares impossible—heredity providing for the modification of its own machinery. Heredity not only leaves the future free for modifications, it also provides a method of life in the operation of which modifications are bound to come."

VI.

The Matter of Terminology.—I anticipate criticism from the fact that several new terms have been used in this paper. Indeed one or two of these terms have already been criticised. I think, however, that novelty in terms is better than ambiguity in meanings. And in each case the new term is intended to mark off a real meaning which no current term seems to express. Taking these terms in turn and attempting to define them, as I have used them, it will be seen whether in each case the special term is justified; if not, I shall be only two glad to abandon it.

Organic Selection.—The process of ontogenetic adaptation considered as keeping single organisms alive and so securing determinate lines of variation in subsequent generations. Organic Selection is, therefore, a general principle of development which is a direct substitute for the Lamarkian factor in most, if not in all instances. If it is really a new factor, then it deserves a new name, however contracted its sphere of application may finally turn out to be. The use of the word

"Organic" in the phrase was suggested from the fact that the organism itself coöperates in the formation of the adaptations which are effected, and also from the fact that, in the results, the organism is itself selected; since those organisms which do not secure the adaptations fall by the principle of natural selection. And the word "Selection" used in the phrase is appropriate for just the same two reasons.

Social Heredity.—The acquisition of functions from the social environment, also considered as a method of determining phylogenetic variations. It is a form of Organic Selection but it deserves a special name because of its special way of operation. It is really heredity, since it influences the direction of phylogenetic variation by keeping socially adaptive creatures alive while others which do not adapt themselves in this way are cut off. It is also heredity since it is a continuous influence from generation to generation. Animals may be kept alive let us say in a given environment by social coöperation only; these transmit this social type of variation to posterity; *thus social adaptation sets the direction of physical phylogeny and physical heredity is determined in part by this factor.* Furthermore the process is all the while, from generation to generation, aided by the continuous chain of extra-organic or purely social transmissions. Here are adequate reasons for marking off this influence with a name.

The other terms I do not care so much about. "Physicogenetic," "neuro-genetic," "psycho-genetic," and their correlatives in "genic," seem to me to be convenient terms to mark distinctions which would involve long sentences without them, besides being self-explanatory. The phrase "circular reaction" has now been welcomed as appropriate by psychologists. "Accommodation" is also current among psychologists as meaning single functional adaptations, especially on the part of consciousness; the biological word "adaptation" refers more, perhaps, to racial or general functions. As between them, however, it does not much matter.[9]

[9] I have already noted in print (ref. 4 and 6) that Prof. Lloyd Morgan and Prof. H. F. Osborn have reached conclusions similar to my main one on Organic Selection. I do not know whether they approve of this name for the "factor;" but as I suggested it in the first edition of my book (April, 1895) and used it earlier, I venture to hope that it may be approved by the biologist.

ON MECHANICAL SELECTION AND OTHER PROBLEMS.
By KARL JORDAN, Dr. Phil.

ON MECHANICAL SELECTION AND OTHER PROBLEMS.
By KARL JORDAN, Dr. Phil.

THE peculiar kind of variation I have to deal with in this paper concerns some accessory organs of the reproductive system of a group of insects, and bears closely upon Romanes's theory of Physiological Selection. It is *a priori* evident that the demonstration of the occurrence of an extensive variability in any part of those organs which are the most important for the preservation of animated nature must be of far-reaching consequence in respect to the origin of species. As the mere statement of the facts would be unintelligible to the general reader who has no special knowledge of the animals in question, I shall endeavour to interpret the facts. However, before I begin to give the details, it is necessary to come to an understanding about some questions of a general character, and this necessitates my entering, rather reluctantly, upon a ground which nowadays has so often been traversed, with and without success: I should not do so, if the facts I bring forth, and the conclusions I have to derive from them, did not put me in contraposition to a good many naturalists. Being surrounded by cabinets full of specimens, I shall, during this excursion on a theoretical ground, always be reminded that possibilities arrived at by general reasoning are often impossibilities in nature, in so far as what is *a priori* possible or even probable may be found not to occur in nature.

I.—INTRODUCTORY NOTES.

Everybody who is a little acquainted with the diagnostic works on Zoology or Botany will know sufficiently that a continuous question of contest amongst us species-makers is whether a given form of animal or plant is a "distinct species" or not; and he will also have become aware that in many cases the contending parties do not come to an understanding, because, though using the same term "species," the mutual conception of that term is widely different. And this is so not only amongst us species-makers, but we meet with the phenomenon also in the essays of a more philosophical kind which bear upon the theory of evolution. It would almost appear, in fact, as if a "species" is that which a respective author chooses to consider a "species." It is not necessary to give any illustration taken from systematic works, first, since we do not think it does much harm to the value of purely diagnostic articles whether the term "species" is always applied in the same sense by the same or various authors; secondly, because illustrations can be found in any volume containing descriptions of varieties and species. In natural philosophy, however, so far as it endeavours to explain by theories the diversities in animated nature, it is all important that an author has a fixed idea of that diversity which he in his writings calls a "species," and therefore we will give here an illustration of the before-mentioned phenomenon taken from this side of our science.

In a short article in which he claims priority over Romanes with regard to physiological selection, Dahl* proceeds to say that a separation of the varieties of one species into more species within the same locality is not possible, if the various

* *Zool. Anzeiger* 1889. p. 262.

varieties did not possess, besides the (morphological) distinguishing characters, either antipathy against mutual intercrossing, or mutual sterility, or both qualities; and he gives the following illustration :—

"The caterpillars of the two closely allied species of butterflies *Gonopteryx rhamni* and *G. cleopatra* became adapted to different species of plants, the one to *Rhamnus frangula* [and *cathartica*], the other to *Rhamnus alpina*, i.e. the apparatus of digestion became so modified that those species of plants could be made use of [physiologically] in the best possible manner. With the condition of the organs of digestion was associated, perhaps indirectly, a somewhat different yellow colour in the imago-state, and here the preference could begin to act. . . . Against a thousand butterflies adapted to *Rhamnus alpina* there occurred perhaps two which were prepossessed in favour of a deeper yellow colour, two which preferred a lighter yellow tint, while those thousand were indifferent [to the deeper or lighter tint of yellow]. The first-named [two] specimens copulated with individuals adapted to *Rhamnus alpina*, and gave birth only to fully fertile offspring. The second group [two specimens] copulated with light-coloured individuals adapted to *Rhamnus frangula*, and gave birth to non-fertile intergraduates. Of the thousand specimens one half copulated with specimens adapted to *Rhamnus alpina*, the other half with specimens adapted to *Rhamnus frangula*; of the two halves, therefore, only the first gave birth to fully fertile offspring. . . . It is easily to be seen that the number of those individuals which were prepossessed in favour of closely allied specimens [in colour] would grow very quickly, and that the number of the others must very soon become reduced, even if this number at the beginning [of the selection] is still more preponderating [than in the above illustration]."

As the author expressly states that the divarication of species at the same locality is not possible if together with the "*development of the distinguishing characters* there did not take place *a development of dislike*," etc., we must take it as the meaning of the illustration that *Gonopteryx rhamni* and *G. cleopatra* have developed from a common ancestor into two species by means of psychological selection. Or, to put it figuratively, the species A developed into two varieties, A^1 and A^2, in consequence of the caterpillars becoming changed by the influence of the different food-plants; then psychological selection on the part of some females set in and modified those two varieties more and more, so that the varieties A^1 and A^2 became in the course of time two "species" B and C. Which characters does the author attribute to the two varieties A^1 and A^2 at the time when selection commenced to act? There were four distinguishing characters: (1) A^1 and A^2 were different in the chemistry of their body; (2) they were so different in colour that the *females* could perceive the difference; (3) A^1 produced only pale specimens, and A^2 only deeper-coloured specimens, i.e. they bred true; (4) the cross-products were not fit to propagate. Now, if forms so widely distinguished morphologically and physiologically as A^1 and A^2 are not "species," then we fear there are no species at all. The author apparently confounds the transmutation of one species into one other and the separation of a species into two or more. The illustration does not show how by means of psychological selection (or any other modifying factor) *G. cleopatra* and *G. rhamni* have in the same locality developed from one common ancestor into two species, but shows how the two "species" *cleopatra* and *rhamni* might become the one a paler and paler, the other a deeper and deeper yellow "species." At the time when selection set in in Dahl's illustration there were already two "species," and hence the specific distinctness of the two is not the outcome of psychological selection.

In order to avoid similar confusion we have to come to a conclusion which definition of the term "species" as opposed to the lower degrees of variation we will accept, before we discuss the variability of the genital armature within the limits of a "species."

As most of us pretend to be evolutionists, let us take it as an axiom, which we need not discuss here any more, that the great divergency exhibited in animated nature is the result of the development of the various forms of animals and plants from a common ancestor. Before Darwin brought forward his theory of evolution naturalists had to solve one question: which are the differences found to exist among the various forms of animals and plants? The question which is put to naturalists nowadays is, however, twofold: (1) which kinds of divergency do we find to exist in nature? (2) how has this divergency come about?

It seems to us to be a rather general assumption amongst naturalists, no matter whether they treat animated nature from the point of view of a philosopher, or whether they work with the individual specimens, that the variation of the individuals which belong to a complex called species is such that, with the exception of marked di- and polymorphism, we can draw up a series of specimens which form a continuous chain from one extreme variety to the other, the differences between the adjoining links being extremely slight. And it is almost natural that the assumption should be so general. The question which governs zoological work is one of specific differences on the side of us systematists, and one of the causes of the specific and non-specific characters on the side of the biologists, and over the consideration of "characters" it has been lost sight of that speaking of a specific character and the variability of characters means, in fact, speaking of *abstracta*, while our work must be based upon *concreta*, upon the individual specimens. The variation of an organ may be continuous; in a series of individuals a certain organ may even be constant; but that does by no means imply that the individuals as opposed to one of their organs form a continuous chain. An individual has many "characters," and these do not vary in the same manner and degree in the various specimens; some individuals may be almost identical in one or some characters, while they are widely different in other respects, and this will at once become manifest to everybody who actually tries to put together a "continuous" series of specimens. An illustration will bring the fact more closely to mind. A, B, C, D may represent individuals; a, b, c, three characters; a^1, a^2, a^3, a^4, etc., b^1, b^2, etc., c^1, c^2, c^3, etc., may be minute degrees of development of the characters. Now, if we arrange the individuals according to the gradual development of the character a, and thus have a continuous series in respect to this character, the chain of individuals will nevertheless be discontinuous, as in the following diagram :—

$$\begin{array}{cccc} A & B & C & D \ldots \\ a^1 & a^2 & a^3 & a^4 \ldots \\ b^{10} & b^1 & b^6 & b^{15} \ldots \\ c^1 & c^3 & c^7 & c^3 \ldots \\ \vdots & \vdots & \vdots & \vdots \end{array}$$

If we further arrange the series according to the continuity of character b or c, not only will the individuals $A, B, C \ldots$ stand in other places within the new series, but also new individuals must be introduced to make the chain again continuous. Hence it will be sufficiently clear that, notwithstanding the variation of every organ of a species may be continuous, the individuals each being a sum of organs form a discontinuous series. We can express this phenomenon in other words as the law of *independent variation* of organs as opposed to the *correlative variation* of organs.

There are many species in which not only the individuals form discontinuous series, but which exhibit also one or more discontinuously variable characters; and it is especially the appearance of such discontinuity which Bateson so amply illustrates in *Materials for the Study of Variation* (London, 1894). In this case the specimens belonging to a species can be arranged into groups according to a discontinuous character, each group containing individuals in which that character is continuous or nearly so, while from one group to the other that character exhibits a more or less wide gap. In diagnostic zoology the groups of individuals of polymorphic animals have often been mistaken for distinct species. The study of independent discontinuous variation of two or more organs can, however, often help us to distinguish polymorphism from specific distinctness. In the following diagram the specimens are first arranged into groups according to character a, and then into groups according to character b:—

I.	A	B	C	D	E	F
	a^1	a^2	a^3	a^7	a^8	a^9
	b^8	b^9	b^1	b^2	b^3	b^7
	c^{14}	c^7	c^6	c^5	c^9	c^{12}
	⋮	⋮	⋮	⋮	⋮	⋮
II.	C	D	E	A	H	I
	b^1	b^2	b^3	b^8	b^9	b^{10}
	a^3	a^7	a^8	a^1	a^{11}	a^{12}
	c^6	c^5	c^9	c^{14}	c^2	c^3
	⋮	⋮	⋮	⋮	⋮	⋮

In the first arrangement (I.) $A B C$ form one, $D E F$ another group; in the second arrangement $C D E$ belong to one, and $A H I$ to another. In each case the individuals $A B C D E F H I$ form two groups not connected by intergradations, each group, therefore, being conformable to what we species-makers call "species" as a rule; but in the first arrangement the specimen C belongs together with A and B to a "species," while in the second arrangement it stands together with D and E, which, according to arrangement I., are specifically different from A, B, and C. The danger of arriving at erroneous results by taking into account only one organ is so evident and so well known that one must wonder how it is possible that nevertheless a classification—for example, of *Lepidoptera*—based upon wing-markings alone, or upon neuration, can be expected to be an exact expression of the blood-relationship of the classified forms.

Which characters are correlated and which independent in respect to variation can only be found out by the study of individual variation; where this study is neglected correlated characters are often taken to be independent diagnostic characters. The length of the wing and the breadth of the wing-bands are often correlated in *Lepidoptera*, and so is in many cases the lesser or greater concavity of the outer margin of the forewing of butterflies depending on the length of the wing; to say in such cases that a certain form of butterfly differs from another form in the wing being longer, externally more concave and having narrower bands, would amount to the same as if in the case of two black forms with yellow spots we should say that the one is differentiated from the other by the yellow spots being restricted and by the black ground-colour being extended.

The independent variability of organs found to exist among the individuals derived from one *female*, or among the individuals flying in the same locality—in short, among specimens about the specific identity of which there is no doubt—leads us to the conclusion that the same organs of the same species vary also independently in other localities. The distinguishing characters of a geographical race thus can be shown to be independent of one another, and that the association of distinguishing characters is not correlation of these characters, which is of the greatest importance, if we come to consider the causes of the divergence exhibited by such races, the degree of relationship of the races, and the probable history of their geographical distribution. We mention in this place only that, as the wing-form and wing-pattern of the *Papilios* vary independently of the copulatory organs, we shall find in the body of this paper ample illustrations of those three phylogenetical and biological phenomena. The first we have to do is in every case to study the individual variation in order to know which characters are correlated and which not in respect to variation, then to study the characters of the same organs of the geographical representatives of the same species, and finally to compare the results we arrive at, if we draw from each single independent character conclusions as to subspecific and specific distinctness, the influence of the evolutionistic factors, and the history of this influence.

The occasional appearance of individuals, mostly single or in very small numbers, amongst the normal specimens, differing widely from the latter in one or more points, we have to classify as a peculiar kind of di- and polymorphism. The so-called sports and monstrosities, between which terms there is no line of delimitation, as there is also no such line between normal and abnormal varieties, belong to this kind of discontinuous variation, and it is a rather general assumption that the characters of such sports and monstrosities get swamped away by intercrossing. Is this really true in the verbal sense? If swamped away means "not appearing in the direct offspring" of that abnormal specimen, it might often be correct; but if it means disappearing for ever, never reappearing amongst the individuals constituting the species, it is not true. Many sports and even monstrosities are found again and again, as every collector knows; and when an exceptionally abnormal variety has been found once, there is no reason for the assumption that it will not be found a second time. In fact, the collectors, who work with individuals, expect and hope to come across such a variety themselves. If, for example, somebody had published this year that from one of his pupae of *Acherontia atropos* (death-head moth) a specimen with entirely ochreous wings had emerged, we are quite sure that all those who are in possession of a considerable number of pupae of this insect would look forward to the emergence of the imagines with the eager expectation to find such an individual among them.*

The repeated occurrence of the same abnormal variety is, to our mind, of the greatest importance, as the repetition of the phenomenon is a proof that the variety, how abnormal and rare it ever may be, is not entirely swamped away from the species, and that the constitution of the individuals of the respective species, taken as a whole, is such that under certain circumstances, whichever they may be, the variety will be produced, or even must be produced. It is, therefore, quite intelligible that, when those certain circumstances become more frequent, the variety which we now call abnormal might become normal. From this point of view, the record of unusual varieties, including monstrosities, is by no means so unimportant as it appears to be to the systematist who recognises only "distinct" species. On the contrary, we

---

* For illustrations see Bateson, *Materials for t Study of Variation*, London, 1894.

maintain with Eimer that abnormal varieties show distinctly the directions in which a species is able to develop. We give an illustration. For certain reasons which we cannot explain here, as they concern details of structure, we must consider *Papilio sarpedon* to be derived from a form which had green spots within the cell of the forewing and a submarginal row of green markings, and that *P. sarpedon* is now on the way to lose the green markings altogether. The specimens of *P. sarpedon* from the Solomon Islands have preserved some of the additional spots, while in the forms from other localities the forewing has only a median macular band; in South India and Ceylon the first spot of this band disappears very often, and in China the band of the hindwing is often reduced to a costal patch. Now, Mr. de Nicéville figures a "sport" from Sumatra [*] in which the median band of both wings has disappeared, except a small spot, and which therefore has an almost entirely black upperside. The specimen is now in the Tring Museum, which contains also another individual of *P. sarpedon* (from Sikkim) that has the median band of the forewing dusted over with black scales, which conceal partly the green ground-colour of the wing-membrane. The direction of development indicated by these two "sports" coincides with that which we are led by structural characters to consider the real direction of development of *P. sarpedon*, and the importance of this coincidence is obvious.

Apart from polymorphism among the individuals in the same locality, we observe very commonly a polymorphism in which the various forms are separated geographically. This geographical polymorphism we shall have to deal with later on, and therefore we restrict ourselves here to the remark that the various geographical varieties are, just like the forms of ordinary polymorphism, often not connected by intergradations, and thus are likely to mislead us species-makers to treat them as "distinct species"; in the latter respect geographical and non-geographical polymorphism have much in common, because they are homologous phenomena, as we shall soon see; and we may add that geographical separation of different forms cannot be an *a priori* criterion of specific distinctness, though this has often enough been alleged.

The individual variation including non-geographical and geographical polymorphism, as dealt with in the preceding lines, concerns especially the individuals which exist as contemporaries; now let us briefly consider the historical side of the variation. According to the theory of evolution, the descendants have in the course of time (gradually or *per saltum*, in respect to the distinguishing characters as well as to time) become dissimilar to their ancestors; we assume generally that a long period was necessary to bring a considerable change in an animal or plant about. A period of one or two hundred years is by many systematists considered quite insufficient to alter in nature a form of animal or plant so far that the change is obvious, or that any transformation has taken place at all within the same locality. Though we accept the theory of the transmutation of the species in the course of time as the base of scientific work in Natural History, we nevertheless *identify* the forms found in our days with the forms which our forefathers in science had before them; without hesitation we treat the forms, whether they be species or variety, captured in 1896, as identical with those which Linné received a hundred and fifty years ago, or which Merian or Seba obtained about two hundred years back. And when we observe a difference between our recent specimens and the old pictures, differences which are by no means always explainable by the assumption of incorrectness of the drawings,

---

[*] *Journ. Bombay N. H. Soc.* 1893. p. 54. t. L. f. 11 (♂).

the very last we think of in systematic work is that the form we now receive from a particular locality has actually changed during the last fifty or two hundred years. Of course, our knowledge of a single specimen of a certain form preserved from the collections made during the last century, and our knowledge of an old figure of that form, does not enable us to draw any conclusions as to the extent of variation of that form at the time of its capture, nor does such poor knowledge admit any conclusion as to a probable transformation. The evidence is gone, and that is very much to be regretted.* But should not the want of evidence from the past be a hint to us to procure more evidence for the scientists who come after us? It is all important to prove that the transmutation of species upon which the theory of evolution is based actually takes place; and it would be a good object for small provincial museums to preserve long series of specimens, with exact dates, at least of the species of those families of the respective district which are known to be easily influenced by the transmutating factors, families which moreover are easily preserved, such as butterflies and moths, wingless beetles, and landshells; museums with such collections would not be simply a sight for the public, but they would be of scientific value, and their scientific success would be far superior to that which they can obtain by gathering sparse material from all countries of the globe. There would certainly be no want of voluntary help from the side of private collectors, in civilised countries. The study of the local fauna and flora and their variation carried out in this sense, though many systematists and species-makers (which terms are not always synonymous) look down upon the study of aberrations as being unscientific, stands far above the mere descriptions of new species, which mostly do not help the least to solve the all-governing questions of evolution, but add simply some more "species" to the hundreds of thousands of "species" already made known.

However, we are not entirely left without evidence that transformation is actually going on under our eyes. We know from many plants † and a small number of animals that geographical races, when reared under conditions other than those of the country where the race lives, change in characters; we refer to the Porto Santo rabbits,‡ and *Polyommatus phlaeas*,§ and *Pieris brassicae*. ‖ And a striking

---

* Since the above was written I have received Oberthür's *Études d'Entomologie* XX. 1896. Here Oberthür tries to demonstrate from copies of old figures of *Lepidoptera* and figures of recently caught specimens that no change has taken place in these insects since Linné's time. The differences, however, which are exhibited by these old drawings and the figures of the recent specimens are not only so conspicuous that identity is entirely out of the question, but the differences are such that if they actually were found in contemporary specimens, many prominent Lepidopterists, including Mr. Ch. Oberthür himself, would treat these differences as being of "specific" value. In fact, Clerck's figure of *Papilio deiphobus* resembles much more that insect which Felder separated as a distinct species (*Papilio deiphontes*) than it does Oberthür's figure of a specimen of *deiphobus* caught in 1893. Oberthür must have been led by that kind of reasoning indicated in the above text, else he could not have spoken of a "proof" that the species in question have not undergone "the least modification." This kind of reasoning is: first the figures of Clerck are mentally corrected according to the characters of recent specimens, then the characters thus corrected are pronounced to be identical with those of recent specimens.

I by no means will say that the particular species in question have been transformed since Linné's time; but I maintain that if there is any evidence as to transformation or non-transformation, the evidence is certainly in favour of transformation.

† Kerner, *Gute und schlechte Arten*, gives many illustrations of the transformation of plants.

‡ Darwin, *Variation of Animals and Plants under Domestication*, 2nd edition, London, 1888. p. 119.

§ Weismann, *Studies in the Theory of Descent*. From pupae of the southern form of *Polyommatus phlaeas* brought from Italy to Germany imagines emerged which were intermediate between the Italian and German forms.

‖ Scudder, *Butterflies of New England* II. 1889. p. 1175, gives a detailed account of the gradual spread of *Pieris rapae*, which butterfly was first noticed in Canada in 1860, and describes a new variety (*novangliae*) into which *rapae* has developed.

illustration of the fact that in a certain district a transformation of a species takes place has been published during recent years from different sides. A moth of the family Geometridae, *Amphidasis betularius*, has a black and a white form (with intergradations): the black form (*doubledayarius*), which for a long time was known to occur in Great Britain, was very rarely found in North-Western Germany; during the last ten to twenty years, however, the black form has become more frequent, and is at the Lower Rhine now nearly as common as the white form. Without trying to give an explanation of this phenomenon, which speaks for itself, it is of interest to note that in countries with a rational cultivation of the soil the changes in the immediate neighbourhood of the animals are very varied: the disappearance of swamps, planting of new forests on barren hills, artificial watering of pasture land, and so on, will not only destroy many forms, and give opportunity for other forms coming into the district; but these changes will most probably also have a certain amount of influence on those animals and plants which remain residents, at least on those forms which, like butterflies and plants, are very sensible to a change in the biological factors; and it would be highly interesting to have correct data whether perhaps a visible transformation of some animals and plants would be effected in a shorter time in a country with an intensive cultivation of the soil (like France) than in a country (like England) in which agriculture is more stagnant. Certainly we will not maintain that all the forms of animated nature have changed during the last hundred and fifty years, but will only draw attention to the historical side of the question, which cannot be entirely neglected. Palaeontology teaches us that there are forms which for a very long time have not changed, at least not in those parts which have been preserved; while, on the other hand, we know from palaeontology that in most cases a change must have taken place.

At subsequent geological epochs the earth has been inhabited by different faunas and floras; that we know. If we had now before us all the specimens of animals and plants which ever have existed, and tried to group them as we do with the specimens in our collections, what would be the result? The gaps between the various forms would all be filled up by intergradations;* the Gorilla would be as different from the Giraffe as he is now; but tracing both animals back to their common ancestral form, we could draw up a series of individuals which would perfectly connect the two animals, and which would not allow us to draw a line of division; Gorilla and Giraffe thus would appear to be merely the extremes of the series. If, as we mostly practically do, the "species" are based only upon morphological differences, all the animals and plants would belong to one variable species. Any definition, therefore, based solely upon morphological differences must necessarily be a failure, and it follows also that the allegation that a certain form is a "distinct species" because there are no intergradations between it and the allied forms is in discordance with the theory of evolution, according to which the intergradations did occur, but are perhaps now extinct.

If we trace the line of ancestrals of two given types back to the form from which both have descended, we have for each type a separate line until we come to the primordial type in which both lines combine. The line of ancestors of every species comprises the different steps of development from the primordial form down to the recent form, and represents the historical polymorphism of the species, in opposition to contemporary polymorphism. Are the differences between the steps of development "specific" differences? *Hipparion* and *Equus*, though assumed to stand in

* Romanes, *Darwin and After Darwin* II. London, 1895. p. 282.

the relationship of ancestor and descendant, are treated as members of different families; if it is right to keep ancestor and descendant, in spite of the intergradations which have existed according to the theory of evolution, in separate families in this case, why do we not do it in the case of the horses which our forefathers rode and those which live now? If it is the presence of morphological difference which leads us to split up in the one case, and the absence of such difference in the other case to unite, why then are *Distomum*, *Redia*, and *Cercaria*; *Rhabdonema* and *Rhabditis*; *Vanessa levana* and its offspring *prorsa*, the same species? Morphological difference alone is not a criterion of specific distinctness.

Besides the morphological differences among animals we observe a mental or psychological difference. It has often been noticed that in mixed flocks of sheep* or cattle the individuals belonging to the same race often keep separate from the rest, and it is also well known that in the state of nature strangers are driven away or even killed. Though psychological selection, as manifested in these cases, is of importance to accelerate the transformation of a variety and to fix the varietal characters, we believe that among domesticated animals it is very often not racial community that keeps the individuals together, but the circumstance of being accustomed to one another. In some districts of Germany all the geese of a village are driven to pasture in one large flock; when on the pasture ground, the individuals belonging to each house keep together, even if they are not the offspring of the same parents, but are brought together from different villages as goslings: this keeping together is an expression not of community of characters, nor of community of descent, but of community of the stable.

For us systematists that kind of psychological variation is of more interest which is the immediate outcome of morphological differences in the organs of sense and in the organs which are destined to affect the senses. The variability of the organs of sense among higher animals, and the difference of discriminating power among the individuals of the same species and race, are facts so well known that it is sufficient to remind the reader of the variability in the eyesight of men, or of the difference between dogs in regard to the power of smell. Among lower animals the senses are often very differently developed in the various families; in insects the power of discriminating form and colour seems generally to be rather weak, while in some families the organ of smell is highly developed. Carrion- and dung-beetles are able not only to smell the carrion or dung from a great distance, but, what is more important, they distinguish between the scent of carrion and that of dung. The sexes of *Lepidoptera* and *Coleoptera* are brought together by differences in the scent of the sexual scent-glands of the various forms; *females* of certain moths attract great numbers of *males*, even if the *female* is kept in a box with holes, and there is not the least doubt that the *males* of the various species of a genus follow the scent of the respective *females* and not that of allied species. As individuals of one species sometimes come to a *female* of another species, and as, further, dung-beetles are occasionally attracted by carrion, we must conclude that the specimens of dung-beetles really perceive the scent of dung *and* carrion, and that the *males* of insects not only perceive the scent of the *females* of their own species, but also that of other species. The phenomenon that dung-beetles come to dung, and that the *males* of a given species of insect follow the scent of the *females* of their species, cannot be explained by the assumption that the organ of smell of the specimens is so constructed that the insect is not able to perceive any scent but that particular one. The specimens of allied species, therefore,

* Darwin, *Variation of Animals*, etc., 2nd Ed. London, 1888. p. 145.

discriminate between the various scents, and follow that scent which, to put it more personally, incites them most. We have here a psychological selection. Now, if a variation takes place in the organ of smell and in the scent-producing glands, it seems to us evident that the effect will be such that the varietal individuals do not follow the same scent as that to which the normal specimens give preference, but select a varietal scent. Careful observations about psychological variation are extremely scarce; but we can give a beautiful illustration on the authority of Professor Standfuss,* who says that at Zürich *males* of *Callimorpha dominula* came very rarely to *females* of the Italian variety *persona*, while they were attracted in great numbers by the *females* of *dominula*. We see here that morphological and psychological variation can go hand in hand. If the power of discriminating scent is so highly developed throughout the order of *Lepidoptera* as in the case of *Callimorpha dominula*, we have a ready explanation of the phenomenon that the scent-organs, compared with the colour and pattern of the wings, are so constant in *Lepidoptera*. It is readily conceivable that the relative constancy of the scent-organs of *Lepidoptera* is a consequence of psychological selection based upon the difference in the organ of smell and in the scent-glands. And if we must admit the probability of the influence of this kind of mental selection, we are justified in concluding that psychological selection takes place in all animals which have one or the other organ of sense highly developed. As, however, with the various organs of sense only certain kinds of characters are perceivable, it is obvious that in different groups of animals psychological selection will affect different sets of characters according to which organ of sense has a high power of discrimination. Thus it seems to us intelligible that there exists such a strong contrast between *Lepidoptera* and birds in respect to the constancy of colour and markings. The great variability in colour in the order of *Lepidoptera* and the relatively slight variability † in scent-organs can be accounted for by the presence of selection as to scent-organs and the absence of selection as to colour,‡ while the surprisingly great constancy even in the shade of colour among birds may largely be due to the sharp eyesight of these animals rendering them capable of distinguishing between shades of colour, and hence inducing them to associate with specimens of their own colour and to drive differently coloured individuals away.

The effect of the variation of psychical qualities as dependent on the variability of the organs of sense can, like the variation of morphological characters, be classified in individual, geographical, and historical polymorphism. The considera-

* Standfuss, *Handbuch für Schmetterlingssammler*, Jena, 1896. p. 107.

† The scent-organs of *Lepidoptera*, especially those of the *males*, are not constant in every case; there are even species in which they are very variable; and the colour of birds is also not constant in every species.

‡ Eimer, *Artbildung und Verwandtschaft bei Schmetterlingen*, tries to "prove" on the ground of his studies on the wing-pattern of Papilios that the transmutation of animated nature takes place without natural selection. But even if this be true as regards a direct effect of selection on the wing-pattern, it is conceivable that psychological selection, which is a part of natural selection, can have, besides the direct influence on "structural" characters, an indirect influence on the distribution of colour on the wings Eimer is in so far quite right as he says with Darwin (not against Darwin) that incipient varieties cannot have been originated by natural selection; where a "selection" takes place, there must already be a difference amongst the individuals. As Eimer expressly says that natural selection cannot produce species but only preserve species, and as, further, Eimer actually does not show how species (conforming to Eimer's definition of species) but only how varieties originate, I cannot perceive why we should not attribute to natural selection the preservation of *varietal* characters. Variation is, according to Darwin, not the outcome of natural selection, but is the premiss of natural selection; varietal characters can be preserved by selection and increased by survival of the fittest. I do not perceive any great contrast between Darwin and Eimer in respect to these points.

tion of these three kinds of polymorphism in the previous lines led us to the conclusion that morphological differences of any kind and degree are not decisive criteria as to specific distinctness; the systematist actually sinks his species in spite of distinguishing characters as soon as it is proved that the morphologically different forms appear among the offspring of the same *female*. The most general case of bodily difference which is not regarded as being specific is the difference between *males* and *females*; notwithstanding the great dissimilarity which the sexes so often exhibit, not only in the reproductive organs, but also in other morphological characters, the systematist puts *male* and *female* together in one species, and hence makes at once the concession that his term "species" is not a purely morphological one, but that the higher criterion of the term is of a physiological kind.

Although morphological identity means also specific identity, the inverse that specific identity means morphological identity is not correct. The question, therefore, is now, which physiological divergence we will take as the real criterion of specific distinctness.

We have seen above that the line of ancestors of a given type can be divided into a recent portion which is independent of the lines of ancestors of all other types, and into a remote portion in which the lines of several types are combined. The specific difference which now keeps allied types separate was absent from them in remote times; what now is specifically different was formerly specifically identical. Hence a definition of "species," *i.e.* a definition of what makes two types specifically different, has to exclude any relation to the ancestral forms of the given types, but has to take into consideration the contemporary types and their descendants. For the sake of argument let us assume all types of animals and plants were monogamic, so that every individual would produce offspring without copulation with another individual. The question as to the characters of the descendants would be twofold: first, the descendants of a type (taken as a whole) become under certain conditions so changed that the gap which separates them from the descendants of another type is entirely filled up by intergradations, or that the descendants of both types entirely fuse together in characters; or, second, the sum of the descendants of each type remains under all conditions separated by a morphological gap from all other sums of descendants, whether the characters change or not. We have already referred to the transplantation of plants and animals by which it has been proved that forms which were unhesitatingly considered specifically different became identical. The divergency in the development of forms has in these cases been annihilated, and we must conclude that similar divergencies will in nature also be annihilated when the necessary conditions arise. If, therefore, *all* the various types in animated nature were different only to that degree, it follows that under favourable circumstances all these different forms would fuse together to one single type. Divergency would change into convergency and identity, and there would be no question as to species. As we, however, observe that fusing together is restricted to the nearest allied types, and that with the greater divergency of allied forms the possibility of fusing decreases until the forms remain separate under any condition, the term "species" as an expression of the divergency in nature must be an expression of that divergency which, though starting from identity in the ancestral forms, will never again develop into identity.

In the case of sexual propagation the question is more intricate in consequence of the intercrossing which takes place between different types. The question as to the lines of descendants is threefold: the line of descendants of a given type fuses completely together with that of one or more other types under certain conditions, and the descendants are qualified for propagation; secondly, the line of descendants remains separate under every condition; and thirdly, the lines of descendants are partly fused by the appearance of cross-products, which, however, are not fully qualified for propagation. The two first points are the same as those mentioned under monogamic propagation. The third case is of little consequence; the cross-products, though often obscuring the fact of the independence of the descendants of two given types, have really no influence upon the lines of descendants, as the offspring of the cross-products soon become extinct, and therefore do not affect the characters of those descendants of the present types which are not cross-products. In asexual and in sexual propagation we observe that the lines of descendants of the various types exhibit this contrast, that they are either capable of fusion or not capable of fusion: in the first case divergent development changes under certain conditions into convergent development, and ends in identity; in the second case the development ends under every condition in divergency.

In order to see clearly which kind of divergency in animated nature we shall have to term *specific*, we will shortly recapitulate those points which have to be taken into consideration.

1. The presence of morphological distinguishing characters is not a final criterion of specific distinctness; a definition of species based solely upon such differences not only would not take into account individual, geographical, and historical polymorphism, but would, if consequently applied, make every individual specifically distinct, as we have seen that the sum of the characters of every individual is different from the sum of the characters of every other individual.

2. Though according to the theory of evolution every species is the outcome of the transmutation of another, ancestral species, we have only morphological characters to distinguish ancestral and descendant species by. Therefore, considering what is said under 1, a final criterion whether the different types which form the direct line of ancestors of a given type are " specifically " different is wanting; palaeontology provides us with morphologically different specimens, which can never be proved to have been specifically distinct. And as, further, allied species have at a former period not been specifically different, a definition of the term " species " based upon evolution must leave the line of extinct ancestors altogether out of consideration.

3. As the theory of evolution further implies that a species can in the course of time develop into one or more descendant species, the term " species " rigorously applied must be restricted to contemporary individuals. Hence the definition of the term " species " as designating a certain kind, and always the same kind, of diversity throughout animated nature, has to be arrived at by comparing the divergency of coexistent types.

4. The theory of creation explains the diversity in animated nature by assuming that every species from the time of its special creation to its extinction is a unit separate from every other unit; allied species have never been and will never be the same. The theory of evolution, abandoning special creation of each species, puts in its place divergent development from a common form, but to explain the actually existing great discrepancy in nature must assume, like the theory of creation, that when a certain degree of divergency is attained the form of animal or plant exhibiting this divergency can never become one with any other form. This degree of divergency stands in the same contraposition to all lower degrees of

development, which allow a fusion of the respective forms, as the negative does to the affirmative, and therefore is different from the lower degrees of development not only as to degree, but also as to kind.

5. From the fourth point it follows that if specific difference means a difference of kind, not only of degree, the chief criterion of specific distinctness of a given form of animal or plant is the impossibility of fusion with other forms.

Hence we have to accept the following definition of the term "species":—

A species is a group of individuals which is differentiated from all other contemporary groups by one or more characters, and of which the descendants which are fully qualified for propagation form again under all conditions of life one or more groups of individuals differentiated from the descendants of all other groups by one or more characters.

The reasons why we do not accept one of the definitions of species as given by other authors are best stated by shortly discussing the definitions of Eimer, Romanes, and Wallace.

I. Eimer's definition * is a purely physiological one :—

"Species are groups of individuals which are so modified that sexual intercourse does no longer take place between them and other groups, or that successful intercourse is not unlimitedly possible."

Apart from the criteria of the term "species" in this definition being negative, we cannot accept the definition simply because it is not a definition in consequence of its allowing us a *choice* between two different criteria. Let us discuss these criteria separately :—

1. "Species are groups of individuals which are so modified that sexual intercourse does no longer take place."

According to this definition the prevention of intercrossing is a consequence of the modification of the groups of individuals. From several sides attention has been drawn to the occurrence of psychological selection among the individuals of the same species, especially among specimens belonging to varieties of domesticated animals. As in these cases intercrossing does not take place on account of a dislike of the respective varieties, *i.e.* as the prevention of intercrossing is a consequence of the modification of a species into varieties, these varieties would conform to Eimer's definition, and therefore would have to be considered as distinct species.

2. "Species are groups of individuals which are so modified that successful sexual intercourse is not unlimitedly possible."

In the second part of his studies on Papilios † Eimer claims priority over Romanes as to the principal idea upon which Romanes based his theory of Physiological Selection. The chief premiss of this theory is the occurrence of such a variation among the individuals of the same species that a group of specimens does not only not cross with the rest of the individuals on account of dislike, but is infertile or restrictedly fertile with them, whereas the specimens of that group are *inter se* completely fertile. If such a physiological variation occurs, and there is no reason why it should not occur, that group of individuals would constitute a species according to Eimer's definition, in spite of the absence of morphological distinguishing characters, and in spite of the offspring of the physiological variety belonging partly to the

---

* Eimer, *Artbildung und Verwandtschaft bei Schmetterlingen.* 1889. p. 16. Romanes, in *Darwin and After Darwin*, 1895. p. 229, says that the "purely physiological definition is not nowadays entertained by any naturalist."

† Eimer, *Artbildung und Verwandtschaft bei Schmetterlingen* II. Jena, 1895. p. 14.

normal form of the species. Even if the physiological difference is accompanied by a morphological difference, the variety cannot be considered a species as long as some of its offspring are of the normal form; A and B are not specifically distinct as long as A produces B, and B produces A.

II. Romanes's definition* is as follows:—

"A group of individuals which, however many characters they share with other individuals, agree in presenting one or more characters of a peculiar and hereditary kind, with some certain degree of distinctness."

This definition, if the words "and hereditary" were left out, would be based solely upon the presence of morphological difference. "Evolutionists," says Romanes, "have more and more grown to lay stress on the hereditary character of such peculiarities as they select for diagnostic features of specific distinctness. Indeed it is not too much to say that, at the present time, evolutionists in general recognise this character as, theoretically, indispensable to the constitution of a species." † Our objection against the word "hereditary," which does not "supply exactly that objective and rigid criterion of specific distinctness" which we necessarily require, is twofold:—

1. There are characters of an hereditary kind which are not specific. Indeed many of those peculiarities of individual varieties which regularly appear in every brood are hereditary, but do not make the individuals which exhibit them specifically distinct. Among the individuals of *Papilio sarpedon*,‡ to give an illustration, appear specimens (in New Britain regularly) which have one or two additional green spots on the forewing; in specimens from the Solomon Islands additional green spots are always present; as for certain reasons we consider *Papilio sarpedon* to be the descendant of a species with a greater number of markings than *sarpedon* now has, the presence of such additional markings we have to explain by the assumption that these markings are inherited from the ancestors. We observe here restricted inheritance of a peculiar character in the individuals from various parts of the Indo-Australian regions, and constant inheritance of that character on the Solomon Islands. Notwithstanding that it is an hereditary character which distinguishes the respective individuals, these specimens are not specifically distinct from those individuals which are devoid of that character.

2. Specific characters are not unrestrictedly hereditary. It is a well-known fact that the various forms of some polymorphic species of butterflies are so distributed that the species is polymorphic in one, monomorphic in other districts. *Papilio aegeus* has a number of different *female*-forms in New Guinea, the Aru and Key Islands, while in Australia there is only one kind of *female*. *Papilio clytia* is remarkably dimorphic in India in both sexes—the varieties are called *clytia*-form and

* Romanes, *Darwin and After Darwin* II. 1895. p. 231.
† Ibid.
‡ In this article I have given the names of species and varieties without adding the "author" for two reasons. (1) The original meaning of a name has mostly changed, its application having become restricted or extended. The reference to the original author, therefore, would give us quite an erroneous idea about the extent in which the name is used in this article. Various writers apply the same name with the same original author to different forms of *Lepidoptera*; hence it is necessary, in order to avoid confusion, to say precisely in which sense a name is here made use of. (2) As it is the contents of a name and not the name I have to deal with, the mere annexation of an author's name to the specific or varietal name would not convey to the reader any idea about the meaning of the name; it is, therefore, necessary to give a reference to a book where the respective form has been figured or sufficiently described. Instead of giving the reference behind each name, I annex to this article, for the sake of simplicity, an alphabetical list of the species and varieties with the necessary references.

dissimilis-form—while on the Andaman Islands only a *dissimilis*-form and on the Philippines only a *clytia*-form occurs; both the Andaman and the Philippine monomorphic insects are not specifically distinct from the Indian dimorphic insect. The nearest ally of *Papilio clytia* is the very variable *Papilio paradoxus*.* If we assume that the Andaman and Philippine insects are the descendants of the Indian dimorphic *clytia*, or the latter the descendant of the former, or all three the descendants of a common ancestor, it is evident that the characters by which formerly (before the separation into three geographical races) the species was distinguished from *paradoxus* are not the same as those by which all forms from the three localities together (which forms now constitute the species) are at present distinguished from *paradoxus*.

We may add a few more illustrations from other well-known butterflies. The large Papuan butterflies usually called *Ornithoptera* are in the male sex of an orange, green, or blue colour. For the sake of argument let us assume that the green *priamus*, orange *croesus* and *lydius*, and blue *urvillianus* were derived from a common ancestor which had an orange male. The nearest ally is *O. tithonus*, which has a green male. In the diagnosis of the green ancestral forms of *tithonus* the difference in colour between it and the orange ally would have been a specific difference; the recent *tithonus* cannot be differentiated from *priamus* by the shade of colour, as both insects are green. This one specific character has not remained a specific character in *tithonus*, though perhaps the ancestral *tithonus* and the recent one are identical in colour, and though the character itself, therefore, has proved to be hereditary; on the other side, that one distinguishing character of the orange ancestor of *priamus*, *croesus*, *lydius*, and *urvillianus* has been inherited only by *lydius* and *croesus*, not by *priamus* and *urvillianus*, and therefore has proved to be only restrictedly hereditary.

One of the characters by which *Papilio eurypylus* is distinguished, for instance, from *P. sarpedon*, is the position and extent of the red costal mark on the underside of the hindwing. This mark is in *eurypylus* a small red spot before the costal nervure, while in *sarpedon* the spot is extended beyond the costal to the subcostal nervure. Quite recently Mr. Walter Rothschild received a series of specimens of *P. eurypylus* from the Kei Islands,† among which is one that has the red mark extended to the subcostal vein. The extent of this spot, which until now was a specific character of *eurypylus*, is no longer of specific value, and cannot serve to distinguish *eurypylus* from *sarpedon* or *macfarlanei*: in respect to the latter two species the form of the spot is hereditary, but no longer specific; in respect to *eurypylus* it is not unrestrictedly hereditary, and also no longer of specific value. Now, if it should happen that all specimens of *eurypylus* acquired the extended red mark and became at the same time in other characters still more different from *sarpedon* and *macfarlanei*, would then *eurypylus*, *sarpedon*, and *macfarlanei* of our days not be specifically different, because their offspring exhibited other distinguishing characters?

The objection against the use of the word "hereditary" in the definition of the term "species" which we have here raised is based immediately upon the assumption of the transmutation of species and their components (varieties). The species A and B, to put it figuratively, may have as specific characters a and b respectively; in the

---

* *Papilio echidna* from the lesser Sunda Islands is a *dissimilis*-form, but is perhaps specifically distinct; I do not take this insect into account here.

† Collected by Captain Cayley Webster. The importance of the record of individual aberrations is here again evident.

course of time, which need not be long, the species A and B develop into A^1 and B^1, with the characters a^1 and b^1 respectively; this change in characters may perhaps be due to the respective areas occupied by the species being suddenly extended or restricted. The characters a and b are, therefore, not inherited by A^1 and B^1; but A and B (and A^1 and B^1) are nevertheless specifically distinct.

III. Wallace's definition of the term "species"* is a combination of definitions I. and II., with the addition that the specific characters are of an *adaptive* kind. As the objections raised under I. and II. apply also to Wallace's definition, we can restrict our remarks to a short discussion of that latter point. As a species is not only opposed to every other species, but is also to be distinguished from variety, the definition of the term "species" must be a guide for the distinction of species from variety. The kind of characters, therefore, to be mentioned in the definition must exclusively be distinctive of species; a certain quality alleged to be required to make a group of individuals specifically distinct must not be a quality that distinguishes variety from variety. The question is, therefore, are there varieties (as opposed to species) the distinguishing characters of which are adaptive? The theory of Natural Selection, so much supported by Wallace, gives an affirmative answer. All the varieties which are selected as the fittest are varieties with special adaptive characters. If we accept Natural Selection as a factor in evolution, we have consequently to concede that both species and varieties are "adapted to slightly different conditions of life." Hence it is evident that Wallace's definition of the term "species" includes a quality which is not exclusively specific, but applies also to the term "variety," and that we have to cancel altogether the restriction that specific characters are of an adaptive kind.

The principal objection here raised against those three definitions, which may be taken as fairly representing the various views of modern authors, is that the definitions, even when accepted as giving the general distinction between any two species, do not furnish us with a general criterion between species and varieties. This sounds baroque, but is a fact. There are varieties (as opposed to "species") which do no longer have sexual intercourse with the other individuals of the same species, and we must also assume that sometimes such an intercourse is not possible; there are varieties which exhibit hereditary distinguishing characters; and there are also varieties with adaptive distinguishing characters. The consequence of accepting a definition of the term "species" which does not exclude every kind of variety (*varietas* as opposed to *species*) leads naturally to the conclusion that there is no real distinction between "species" and "variety," and that it is purely conventional whether we call a form species or variety, an opinion by no means rarely met with even amongst us species-makers. For example, Butler says: † "For some years past I have held the view that what is generally understood by the term "species" (that is to say, a well-defined, distinct, and constant type, having no near allies) is non-existent in *Lepidoptera*, and that the nearest approach to it in this order is a constant, though but slightly differing, race or local form—that genera, in fact, consist wholly of a gradational series of such forms."

According to our definition there is, however, a real distinction between the

---

* *Darwinism* 1889. p. 167.

† *Ann. Mag. N. H.* (5). XIX. p. 103. It is easily perceivable from Dr. Butler's work on *Lepidoptera*, for instance from his revision of a group of butterflies called *Euploeinae*, that what Butler regards above as "species" is subspecies or geographical variety, the gradational series of which constitute the "species," and that his "genus" is the species.

terms "species" and "variety," a distinction which indeed everybody silently accepts who considers the enormous numbers of different forms of animals and plants to be the outcome of divergent development, and expresses this development by the conventional figurative tree. If the specimens represented by any given portion of a branch of this tree were so constructed that under favourable circumstances they would be identical with the specimens represented by a portion of another branch, *i.e.* if any two branches could, and can, long after the common origin, merge together, then branch off again, merge together again, and so on, it would be preposterous to assume that this should never have happened. But if we thus should have to concede the possibility that the lines of ancestors of any two forms of plants or animals, say of the lion and the giraffe, were such that they first became widely divergent, then identical, then again widely divergent, not only the figurative tree, but also the kind of evolution it is meant to illustrate, would be pure nonsense. Therefore we take it that we are actually agreed upon that part of our definition which says that from a certain point a branch of the tree *cannot* merge together again with any other branch; now, if we call every form which has reached this degree of development *specifically* distinct, we have an absolute distinction between species and the lower degrees of development.

The question of specific distinctness or non-distinctness is therefore twofold: first one of morphological, and second one of physiological difference. As the systematist is practically not able to test by experiment the presence of the second distinction, it is obvious that he never can *prove* with *certainty* from the specimens alone whether the distinguishing morphological characters they exhibit are of specific value or not. However, we are able to arrive at a probably correct conclusion without testing in each case the specific distinctness, if we take into account the way in which divarication of species comes about, and if we further compare the characters of such forms as have been tested to be specifically distinct.

For our present purposes it is quite irrelevant whether the causes of the transmutation and divarication of species are those factors which are maintained by the Neo-Darwinians to be the sole agents, or those which the Neo-Lamarckians consider to be alone effective. Hence we shall abstain from any discussion of the much contested final causes of divergent development, and shall simply ask, which is the way that leads from variability of a species to divarication of this species into more species? Our purpose allows us to simplify the question still more and to restrict the discussion to the two points: first, can a species develop into two (or more) species without isolation? secondly, can isolation as such transmute one species into two (or more)?

The most extreme kind of variability of a species logically possible is that in which the varieties composing the species are not only morphologically but also physiologically different. As upon the occurrence of such variation Romanes's theory of Physiological Selection, which we have had to allude to several times, is founded, we may be allowed to annex our notes to a short discussion of this theory as far as it bears upon our particular question.

In order to explain the infertility or restricted fertility between different species Romanes assumes that the divarication of one species into more species has something to do with the occurrence of such a variation that some individuals of a given species are not fertile with the rest of the species, but are fully fertile *inter se*. That variety, though living in the same district as the normal form of the species, will develop divergently, according to the theory, and give rise to a new species, as it is physiologically so separated from the normal individuals that intercrossing is excluded. Is this conclusion correct?

We have seen above that aberrational individuals occur again and again among the normal specimens; if therefore a physiological variety such as Romanes's theory demands occurs once, a species is able to produce it again and again, as long as the circumstances under which the species lives do not change. But to make the case as favourable as possible we will assume that the normal individuals, which may be designated as N, gave birth to the variety V only once, so that if all the specimens of the variety were killed the variety would never reappear. Now, the offspring of the variety V will in the first and following broods belong either all to the variety V, or partly to the variety V and partly to the normal form N. The first assumption nobody can admit to occur, as this would mean that V is a species suddenly branched off from N, and as further we should have, instead of the explanation of the phenomenon of sterility between species which it is the aim of the theory to give, merely the statement that species V has sprung up among the individuals of species N with which it is infertile. Hence we have to do only with the second case, that part of the offspring of V, at least part of the first brood after the origin of V, belong to N; these offspring of V we will designate Nv in order to indicate that, though they are normal, they are the immediate descendants of parents with varietal characters. Nv cross *inter se* and with N, not with V; which characters will the offspring of Nv have, which itself, we repeat it, is the descendant of V? The offspring of Nv could be either all identical with Nv ($=N$), or partly identical with Nv and partly with V. The first alternative means that each specimen of V produces two groups of individuals, Vv and Nv, of which one group (Nv) never will give rise to a form similar to its parent-form V, while the other group (Vv) produces both the parent- and grand-parent-form. Although the assumption here made, that some of the offspring of one specimen breed perfectly true (Nv), while the other offspring do not breed true, is quite at variance with our experience, and therefore not acceptable to any naturalist, we will nevertheless accept the assumption for the sake of argument. As V breeds in and in, according to that assumption, it is argued that in the course of time V will become so modified that it will also breed true like N and then be specifically distinct. Apart from Nv and V here being already two species from the beginning, one (Nv) breeding true, the other (V) producing *per saltum* in every generation some specimens of Nv, an illustration will show at once that before the above-mentioned modification is effected the variety V will altogether be swamped away. Let us assume (1) that in an isolated district two hundred specimens of a certain species (one hundred of each sex) could find subsistence; (2) that each *female* would produce ten *females* and ten *males*; (3) that at a given time one-tenth of the specimens conformed to the variety V, *i.e.* were sterile with the other nine-tenths (N), but fertile *inter se*; and (4) that 80 per cent. of the offspring of V belonged to V, the other 20 per cent. to N, which breeds true,—for convenience we shall take into account only the *females*,—the numbers of the forms N and V in the successive broods would be as follows :—

	i.	ii.	iii.	iv.	....
$N♀$	90	92	94	95	....
$V♀$	10	8	6	5	....

We see that the number of specimens of V diminishes, and that extinction of

V must soon be the result, if at the first appearance of V the number of individuals of this variety was not very great. With every successive brood the percentage of varietal specimens produced by V will become higher, and to make up for this we have assumed that the very first specimens of V produced already 80 per cent. of the varietal form.

There remains now only the second alternative, that Nv as the descendant of V produces both V and N, which agrees with what we know of the propagation of varieties which occur among normal specimens, and hence is the only acceptable alternative. The offspring Vn of Nv crossing with V will bring the blood of Nv and N, which interbreed, into V, just as Nv (as the offspring of V) brings the blood of V into N, and this would go on as long as V and N exist together in the same locality. Though N and V are mutually sterile, the blood of N comes into V by means of Vn, and the blood of V into N by means of Nv; this *indirect intercrossing* will completely annihilate the effect of the assumed mutual sterility of N and V. The following diagram will serve to illustrate these lines:—

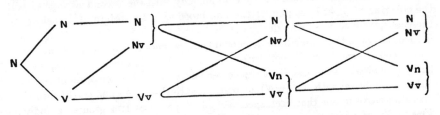

It can easily be shown that after a certain time N and V will occur in equal numbers.*

The physiological selection will, therefore, in no case result in divarication of a species into two, but the outcome of the physiological variation will be either dimorphism of the species, when both the normal and the varietal form are equally favoured in respect to the circumstances of life, or extinction of that form which is the least favoured. If however the most favourable kind of variation does not lead to the origin of a new species beside the parent one, no other variation will lead to this end. Hence we must conclude that a divarication of a species into two or more species cannot come about so long as the divergent varieties live so together that a direct or indirect intercrossing is not prevented.

Having thus disposed of the possibility of the divarication of species without the help of some kind of local separation, we have to consider the other question: whether local separation as such can be able to give origin to a variety and to transform a variety into a species. The theory of isolation as promoted by Wagner says that the peculiar characters of some isolated specimens will by breeding in and in finally be transmitted to all the descendants of those specimens, and their degree of divergency become in the course of time so much higher that these descendants represent a new species.

Experiments teach us that aberrant specimens of a species occurring amongst the normal specimens produce, when crossed together, offspring which partly are of the normal, partly of the varietal form of the species; from black specimens of the moths *Amphidasis betularius* and *Liparis monacha* are obtained both black and white individuals. To make the circumstances most favourable for the eventual

* Murphy, *Habit and Intelligence*, London, 1879. p. 241.

divergent development of the variety we will assume (1) that a number of varietal specimens are *completely* isolated from the rest of the species; (2) that 80 per cent. of the offspring of these specimens belong to the varietal form. The specimens now have to propagate under the further premiss (3) that the normal and the varietal forms exist under exactly the same conditions of life, so that every other transmuting factor besides mechanical isolation is excluded. By mechanical isolation we understand a separation of the animals or plants in question by a mechanical barrier, so that an intercrossing with the original stock is prevented; experimentally the case could be demonstrated by rearing wingless animals side by side, but separated by an adequate fence. Under the above premisses ten *females* of a variety, each producing twenty *females*, kept in an enclosure representing the isolated locality, would give birth to a hundred and sixty *females* of the varietal (V) and forty of the normal form (N). If the latter produce also each twenty *females*, of which 80 per cent. might be taken as normal and 20 per cent. as varietal, and if the locality is fit to provide food for a thousand *females* (and a thousand *males*, which are not taken into account), the numbers of both forms would in the succeeding broods be as follows:—

	i.	ii.	iii.	iv.	v.
N	—	40	320	392	435
V	10	160	680	608	565

The result is here again that after a small number of broods both the varietal and normal forms will exist in equal numbers in the isolated district. The varietal form can never become the sole inhabitant of that district unless the circumstances of life are such that the normal form is less favoured by them, *i.e.* unless there is some transmuting factor active besides isolation. Isolation as such is not an active factor which produces a character, but is a factor which merely preserves a character produced by some other factor; isolation has, therefore, no direct effect. The reason that the effect of isolation has by many authors been so much overestimated is so obvious that we scarcely need mention it: the differences exhibited by geographically isolated forms of the same species are often attributed to the direct effect of isolation, because isolation and morphological difference were seen to be always associated, while no other transmuting factor seemed to be obvious to those who were unacquainted with the experiments made in this direction. Apart from experiments, there are many geographical races which, on closer examination, show at once that their characters cannot be the outcome of the isolation of some ancestral specimens which accidentally exhibited the respective distinguishing character. Wallace[*] was the first to draw attention to a peculiarity common to a great many species of butterflies on the island of Celebes: these species or varieties have much longer and more falcate forewings than the races from the other islands of the Indo-Australian region. On the island of Sumba, which lies south of Flores, that character is also found in some species. The Chinese races of butterflies and moths have generally the black colour more extended than the respective Indian races. The butterflies and moths of Sumatra and Borneo[†] are mostly much darker than the races of the same species from Malacca and Java. The Queenslandian races are often pale, those from the Kei Islands have the markings often restricted, and so on.

[*] *Proc. Linn. Soc. Lond.* XXV. p. 18.
[†] Hagen, *Iris* 1894. p. 17.

Characters like these, common to a multitude of racial forms living in one isolated district, cannot be accounted for by a direct effect of isolation: it would be almost ridiculous to assume that the first specimens of a great number of species which came to Celebes had all long and falcate wings, while the specimens dispersed over the neighbouring groups of islands had short wings; or that the first specimens which came to Sumatra and Borneo were dark, while the individuals of the same species which migrated to Java were in so many cases less black.

We now have seen that the geographical isolation of aberrant specimens has not been and is not the means of the divarication of species, and that the effect of the transmuting factors acting upon the specimens of a species within the same locality is at the highest marked polymorphism; therefore there is only one way possible by which the divarication of a species into two or more can come about—that is, the combination of isolation and transmuting factors. *The isolation of one or more* (Neo-Darwinian and Neo-Lamarckian) *factors* is the means by which the specimens of a species which are subjected to these isolated factors, whichever they may be, become different from those specimens which stand under other influences, no matter whether the first specimens which became isolated as to the transmuting factors were normal or aberrational. This assumption corresponds completely with the result of experiments, and explains all the peculiarities in the characters of geographical races and representative species. And we shall see in the third part of this paper that there are instances in which the geographical isolation can be very incomplete, and in which, nevertheless, the divergent development will lead to specific distinctness of the biologically isolated specimens.

The geographical races thus produced we must assume to be first inconstant, to become more and more constant and divergent by the incessant influence of the transmuting factors, and to develop finally into a form which is so modified that it never will fuse either with the parent-form or the sister-forms, and that it therefore agrees with the definition of the term "species."

As this kind of divarication of species is the only possible * one, and hence geographical polymorphism of a species the beginning of the ramification into more species, the study of localised varieties is of the greatest importance in respect to the theory of evolution; the study of geographical races, or subspecies, or incipient species, is a study of the origin of species. The meaning of the term "subspecies,"† nowadays generally applied to geographical or localised forms, is evolutionistic, and, in fact, the only evolutionistic idea which has penetrated into that work of systematists which is purely diagnostic. Every scientist who pretends to be an evolutionist must perceive the importance of subspecies. Whoever persistently ignores the existence of subspecific characters ought to have the courage which I admire in Charles Oberthür—great courage it certainly requires to defend a standpoint against the bulk of naturalists—to define the species as a created entity.

Eimer, *Artbildung und Verwandtschaft bei Schmetterlingen*, gives beautiful examples of the various degrees of divergency of localised varieties. Whether only one or a few specimens exhibit in a given locality a character not found elsewhere; whether a greater number of individuals in a certain district are characterised by a

---

* It is scarcely necessary to add that the area to which a certain transmuting factor is restricted need not be a political or physiographical district.

† This term had already been applied to geographical races before the appearance of Darwin's *Origin of Species*.

peculiarity not met with, or rarely met with, in other districts; whether all the localised specimens are different from the rest of the species; whether a localised variety is or is not connected with the other varieties by intergraduate specimens,— in every case the presence of a localised peculiarity indicates that the individuals inhabiting the respective locality are on the way to develop divergently in consequence of some biological peculiarity of the locality. All these degrees of divergency are distinguished from that higher degree which we have taken as the criterion of specific distinctness by not conforming to the physiological part of our definition of species. By experiments it has been proved * that geographical forms lose their distinguishing characters and fuse together with other forms of the same species. Therefore, if all the coexistent specimens of a species were at our disposal, the definition of the term "subspecies" would be as follows:—

A subspecies is a localised group of individuals of a species the mean of the characters of which is different from the mean of the characters of all the other localised groups, and which will, under favourable circumstances, fuse together with other groups.

However, the material contained in collections is, compared with the actual number of specimens existing of each variety, extremely meagre, though nowadays systematists comprehend more and more that a few specimens of each species are insufficient for a serious study, and hence try to bring together long series from every locality. The conclusions, in respect to variation, which we draw from the inadequate material we have to work with, must necessarily often be erroneous. If, for example, our series shows a variation of a character (expressed in numbers) between twenty and fifty, there may, in fact, exist individuals which stand outside these limits. Rare varietal specimens, which hitherto have been found only in a certain locality, may very well occur elsewhere; a certain variety may appear to us more common in one locality than in another, and hence the mean of the characters in the first locality to be different from the mean in other localities, because a collector paid more attention to varietal specimens in the first locality. This imperfectness of our knowledge we have to take into account, and we must, therefore, restrict the application of the term "subspecies" in order to avoid deception as far as possible.

The above definition has not had regard to the degree of divergency attained by the localised form. Now, we ask, which then is the lower limit of application of the term "subspecies"? The diversity which the sexes exhibit in respect to localised variation gives us the answer. We know a good many cases in which the *males* are in various districts not distinguishable, while the *females* are very different, and cases in which the variation takes place in the *male* sex and not in the *female* sex. For illustration we refer to the following insects: *Papilio semperi*, from the Philippine Islands, varies in the *female* sex according to locality, while the *males* from the various islands, in spite of individual variability, are not distinguishable; *Papilio oenomaus*, from Timor and Wetter, is on these two islands the same in the *male* sex, while the *females* are conspicuously different; and so it is with *Papilio phestus*, from New Britain, New Ireland, and the Solomon Islands. If we apply in these cases, the importance of which we shall soon endeavour to show, the term "subspecies," we have a rule which can guide us in all other cases—namely, as the numbers of specimens of each sex can be taken as being (roughly) equal, we shall have to use the term "subspecies" when a localised variation is such that about half

* Kerner *Gute und Schlechte Arten*; Weismann *Studien zur Descendenztheorie*.

of the individuals belong to the varietal form. All lower degrees of localised variation may be termed "localised aberration" (ab. loc. = *aberratio alicuius loci*).*

We have already referred to some observations which show that localised divergent development is going on under our eyes. There are certainly species which are at present stationary, and perhaps have been so for a long time; but so much is certain that nearly all the species which have a wider distribution (except a number of "globe-trotters") exhibit some kind of local variation, and that, therefore, since local variation is the beginning of the divarication of a species into more species, the more widely distributed species are at the present period actually in a state of divergent development. We have examined a great many *Lepidoptera*, both butterflies and moths, in regard to this question, and find that there are very few species which are not split up into geographical races, although the differences between the subspecies are often extremely minute. The degree of divergency depends especially on the sensibility of the species in respect to the transmuting factors, and on the degree of isolation and intensity of the latter, as well as on the degree of geographical isolation. Wingless animals, and plants without means of dispersal, are generally more easily affected and on smaller areas than animals and plants with good means of dispersal. Wingless beetles, for example, such as *Carabus*, vary enormously according to locality; in the Alps, for instance, there are a great many subspecies of *Carabus* each confined often to one mountain.

The number of subspecies into which the Indo-Australian *Papilios* have developed is very great,† and, when studying these insects, we were surprised to find that, in opposition to the general view, not the *males* but the *females* appear to be the first affected by localised transmuting factors. In *all* cases, without exception, where the distinguishing characters of a subspecies are found only in one sex, it is the *female* and not the *male* which exhibits them; and further, in those subspecies which are obviously different in all specimens of one sex, slightly or even only occasionally different in the other sex, it is again the *female* that is the more aberrant sex. If we further take into account the *local aberrations* as far as they constantly and commonly appear among the normal specimens, we have thirty-six cases among the Indo-Australian Papilios in which the localised variation is entirely or almost entirely restricted to the *female* sex, while there is not a single subspecies which is in the *male* much more different from the allies than in the *female*. In seasonal forms of Papilios the *females* again exhibit a greater amount of divergency than the *males*, a phenomenon which is strikingly illustrated by the Japanese *Papilio machaon hippocrates*. The variability is in the *females* of the Indo-Australian Papilios altogether greater than in the *males*, or, to express it biologically, the *females* are more easily affected by the causes of variation than the other sex. If localised variation is the beginning of the divarication of a species into more species, and we have seen that this is the only possible way by which divarication can come about, those phenomena, which relate especially to pattern, admit no other explanation than that, at least in all cases where the localised variation is restricted, or nearly so, to the *females*, the transmutation of the species begins in the *female* sex, and that, therefore, the *female* is in advance of the *male* in respect of the development into new species. Eimer and Fickert in their studies on the Papilios come to the opposite result; and that is, we think, due to their assumption that the original pattern of the wings of Papilios consisted of "longitudinal" bands

---

* See this journal, 1895. p. 180.
† See Rothschild, NOV. ZOOL. 1895. p. 463.

standing at right angles to the veins; if Eimer had assumed that the original pattern of the wings consisted of "longitudinal" bands running in the direction of the veins,* he would have arrived at a "preponderance" of the *female* sex.

The degree of divergency is, in many subspecies, so minute that the peculiarity would escape even a skilled eye but for a carefully working systematist having drawn attention to it, and having fixed, so to say, the minute peculiar character by naming the subspecies. Romanes has severally referred to minute specific characters in order to confute Wallace's opinion that all specific characters are useful; as specific characters are only higher degrees of development of subspecific characters, the question whether subspecific characters have originated and are accumulated by Natural Selection is no less important than the same question in respect to specific characters. Dixon † and Allen ‡ give ample illustrations of the question as regards birds. Of *Lepidoptera* we mention, out of hundreds of cases, only two: *Papilio nomius* has a western subspecies (Ceylon to Assam) and an eastern subspecies (Burma, Tenasserim, Tonkin, Hainan) which are *constantly* differentiated by one minute character, namely a brown line situate in the eastern form on the praecostal vein on the underside of the hindwing, which short and thin line is absent from the specimens of the western subspecies; that this distinguishing character is indeed minute will be admitted if we add that, though it is the only constant difference of the two subspecies we can find, it has never been mentioned by any specialist until 1895, in Mr. Walter Rothschild's Revision of the Eastern Papilios.§ *Papilio agamemnon argynnus* from the Kei Islands and *P. agamemnon neopommeranius* from Neu-Pommern differ from all the other subspecies of *agamemnon* in the hindwings being above nearly devoid of markings, which renders the two subspecies extremely similar; but there is one constant character, which can easily be perceived with the help of a lens, that distinguishes *neopommeranius* and *argynnus*: in *neopommeranius* the spots of the median row of the forewing beneath are scaled all over, while in *argynnus* the outer portion of each spot is scaleless. Though nobody can very well entertain the opinion that such differences are due to the direct action of Natural Selection, one can evade the weight of the minute distinguishing characters by the assumption that these minute characters are correlated to some other, unknown, character which is of a useful kind. *Lepidoptera*, however, furnish us with the means to repudiate this evasive answer. All those species which are said to be mimetic, and which have been quoted again and again as excellent illustrations of the marvellous effect of Natural Selection, have certain characters of colour or form which are attributed to the direct (not indirect) influence of Natural Selection. Now, if such a character varies geographically in the mimetic species and at the same locality in the imitated species by minute degrees, the minute difference between the geographical forms of the mimetic species ought logically also to be attributed to the direct action of Natural Selection. Such cases are, however, not rare among insects. We refer, for the sake of illustration, to one of the most striking examples of mimetic adaptation. *Papilio caunus* of the Malayan region has a style of marking quite unusual for a *Papilio*, and resembles another, nauseous, butterfly, *Euploea rhadamanthus*, to a surprising

---

* According to Eimer, the costae and rows of punctures on the elytra of beetles are transverse, while the bands (of *Cerambycidae, Cantharidae*, etc.) which stand at right angles to those rows, and which are often continuations of the transverse bands of the sterna and abdomen, must be called longitudinal.

† *Evolution without Natural Selection*, London, 1885.

‡ *On the Mammals and Winter Birds of East Florida*, in *Bull. Mus. Comp. Zool.* II. 1871.

§ Nov. Zool. 1895. p. 422.

degree. The three Malayan subspecies of *Papilio caunus*, inhabiting the first Malacca and Sumatra, the second Borneo, and the third Java, differ from one another by closer examination in the size of the white markings; the differences are, however, so slight that they do not affect the general aspect of the specimens; in fact, the Java *caunus* resembles the Bornean *rhadamanthus* just as much as the Bornean *caunus* does. The difference between the three *caunus*-forms are certainly not such that if the differences were altogether absent the specimens would be less protected; there are, to be sure, very few cases of mimicry in which the resemblance of the mimetic and imitated species is greater than that of any of the three *caunus*-forms with any of the three *rhadamanthus*-forms. If, therefore, mimicry is of value to the imitating species in all cases where the resemblance is of a much more superficial kind, we cannot see why it was necessary at all to have the white markings of the Borneo *caunus*, compared with the Malacca and Java *caunus*, a little reduced. Further, though the resemblance between the species in question is very great, there are still differences in markings between the mimetic and the imitating species in each locality which are greater than the differences between the three *caunus*-forms. It certainly requires a great deal of faith in the omnipotency of Natural Selection to believe that the slight reduction of the white markings in the Borneo form of *Papilio caunus* is due to a survival of those specimens of *caunus* in which the white markings were a little smaller than usual; the enemies of *caunus* to which we attribute the execution of the selection then must have been possessed of a much keener power of discrimination of markings than the entomologists who, until the appearance of Mr. Walter Rothschild's Revision of the Eastern Papilios, treated the Bornean *caunus* as identical with the Java *caunus* as figured by Westwood! If we, however, admit that the slight distinguishing characters of the three Malayan *caunus*-forms cannot possibly be due to the action of Natural Selection, but must be the effect of some other transmuting factor, it is evident that also in all other cases of minute distinguishing characters we need not refer to Natural Selection as the cause of the minute difference. The presence of minute distinguishing characters allows, therefore, a restriction of the possible causes of the divergency of the respective forms; and, as we thus have to admit the importance of insignificant distinctions in respect to evolution, it will be obvious that the importance increases as the degree of distinctness decreases.

We have said above that we take as the lower limit of the application of the term "subspecies" such cases in which about half of the individuals are characterised by some peculiarity : which is the upper limit? or, when have we to begin to call a form specifically distinct? According to our definitions of the terms "species" and "subspecies," the distinction between subspecies and species is a biological one, the presence of which, as mentioned on p. 442, we systematists are not able to directly prove or disprove from the material we are working with. As we know now from experiments and careful field observations that morphologically very different forms, connected or not by intergradations, can, in spite of the conspicuous differences, be one species (individual and seasonal polymorphism, heterogenesis, etc.), it is *a priori* evident that also geographically separated forms, in spite of their being morphologically distinct and in spite of their not being connected with one another by intergradations, can very well be subspecies of one species, *i.e.* can under favourable circumstances fuse into one form. The actual proof of specific distinctness the systematist as such cannot bring; we species-makers do, in fact, not pretend, at least many of us do not, that in every case the form which we pronounce to be a species is really a species ; we work, or ought to work, with the mental reservation that the specific distinctness of our *species novae* deduced from morphological differences will be corroborated by biology (in the widest sense). The work a systematist has to do is twofold : above all, he is a registrar of facts observed upon the body of individuals, and secondly he has to draw conclusions from those facts.

All our knowledge of nature is based upon the knowledge of single phenomena. In Natural History the base of our knowledge is the individual : the characters of individuals and sums of individuals are the A B C of this science ; they are to the naturalist what words are to the philologer. To render the characters clear is the first task to be solved ; before this task is completed we are not able to draw correct conclusions. Although nowadays the recorder of facts, the diagnosticist, does not rank high in science, every theory in Natural History depends especially on the correctness of the facts furnished by the diagnosticist ; when that record lacks correctness the theory based upon it must break down. As an excellent illustration of this we may regard Weismann's theory[*] of "Phyletic Parallelism in Metamorphic Species," as far as it asserts the existence of an incongruity in the classification of *Lepidoptera* based on larval or based on imaginal characters. The "*Rhopalocera*" are by no means a sharply defined group in the imago state ; neither the erect position of the wing of the resting butterfly and the colour of the wings, nor the clubbed antennae, are characters applying to all "*Rhopalocera*" and exclusively to *Rhopalocera*; and as there is no single character by which all the *Rhopalocera* are distinguished from all other *Lepidoptera*, we can also not expect that the larvae of *Rhopalocera* form a sharply defined group distinguished as a whole from all other Lepidopterous larvae. The apparent incongruity in the classification according to the larval or imaginal state of "*Bombycidae*" and "*Notodontidae*" again is not due to an incongruent development of larvae and imagines, but to the fact that Lepidopterists have placed in these (and other) families the most heterogeneous things in consequence of an entirely inadequate knowledge of the forms classified.

We learn from this illustration first that diagnostic work is the true basis of evolutionistic theories and hence of the highest importance, and secondly that the record of facts must be exact. Huxley says that the record of facts is not scientific if the facts do not permit of the drawing of general conclusions. In the above case the blame is much more on the side of the systematists who gave the clubbed antennae as distinguishing character of butterflies, than on the side of Weismann who accepted this statement as correct. If, therefore, diagnostic work is intended to meet the claim of furnishing facts from which general conclusions as to evolution (classification, variation, etc.) can safely be drawn, or if a diagnosticist claims to have his work regarded as scientific, it must be well distinguished between the description of the characters of individuals and the statement of an opinion deduced by the diagnosticist from the characters of the individuals ; the record of the characters of individuals, or the statement of facts, ought to precede the statement of the personal conclusion, which perhaps is entirely wrong. In the case of species and lower degrees of divergency diagnosticists mostly lose sight of this ; when we describe a number of individuals as belonging to a new species we present very often to the reader, not the characters of the specimens, but a ready-made conclusion which asserts (1) that the specimens are specifically identical, and (2) that the species varies in colour, markings, structure, and size to such and such an extent. The specific

---

[*] Weismann, *Studies in the Theory of Descent*, London, 1882. II. p. 432.

identity and the specific distinctness of the specimens are certainly our deduction, and the variation of the species thus erected is also ours. It is not rarely that one meets with diagnoses of species which give the average of some character of the individuals—for example, the average size—which perhaps is not found in any of the specimens measured, or if found may occur very rarely compared with the greater abundance of large and small individuals.* Most deceptive are those diagnoses which contain statements like these: "Colour brown to black; size 50 to 60 mm.; *habitat* India to New Guinea." In such diagnoses the facts are veiled, and we are easily deceived by taking the diagnosis as being the record of facts, while it is a mere statement of an opinion. The erroneous view expressed by Romanes † that geographical races are less abundant among animals than among plants, and Pagenstecher's view ‡ that moths do not vary to any extent according to locality, are the consequence of such deceptive statements on the side of the diagnosticists. The description of a species or variety, therefore, ought to be a pure statement of facts; as said above, the facts which the diagnosticist deals with are the characters of individuals; a pure statement of facts, with the exclusion of any statement arrived at by reasoning, we should have when the characters of the different individuals were recorded in such a way that from the description it would be plainly visible which characters belong to each single individual. When this is done, the statement of our opinion as to the specific identity of the specimens, the variation and distribution of the species, etc., cannot affect the facts, and, therefore, cannot do much harm, even if our opinion should be wrong. Hence we take it that the description of a species or variety being intended to be a statement of facts, not of conclusions, ought to be the description of one individual to the characters of which the different characters of other individuals are so annexed that a mistake as to which individual a respective character belongs cannot occur. That specimen round which the others are grouped in the description is the *type-specimen* of the description, and as the description is the description of the species or variety (as far as the individuals of the species or variety are known at the time), in the same sense as a figure of an individual is meant to represent the species or variety, that specimen is correctly called *type-specimen* of the species or variety respectively.

Besides the pure record of morphological facts, the diagnosticist has to draw inferences from the facts; and as the recorder of facts ought to know the facts best, the conclusions the diagnosticist arrives at ought to be generally correct if the method of reasoning is correct. The inferences which concern us here are such as to the specific or non-specific distinctness of groups of individuals, and hence we shall restrict our discussion to this kind of conclusions.

If we received a bird of Paradise with conspicuous ornamental feathers, even if the species were quite unknown to us, we should at once pronounce the specimen to be a *male*, though we know nothing about its having been a physiological *male*; and if the quills of the ornamental feathers were surrounded by a horny sheath, we should conclude that the individual was not yet in full plumage. As Dr. Martin has succeeded in breeding tailed and tailless *females* of *Papilio memnon* from the eggs of one *female* in Sumatra, we must conclude that also in other districts where the two *female* forms occur both can be produced by each of the two. What is found to be true in a number of cases we are bound to conclude to be true in all

---

* See Bateson & Brindley, *P. Z. S.* 1892. pp. 585 ff.
† *Darwin and After Darwin*, London, II. 1895. p. 209.
‡ *Jahrb. Ver. Nass.* 1896. p. 158.

similar cases. This inductive method of reasoning may often lead to wrong inferences, as the correctness of the latter depends first on the premiss that there are cases which are really proved to be true, and secondly that the cases which we believe to belong to the same category as those proved are really similar cases.

If we apply this to our question as to specific or non-specific distinctness, it is evident that the conclusion of the diagnosticist can be correct only under the condition that the specific or non-specific distinctness of some forms is proved by experiment, and that he is so well acquainted with the morphology of the forms in question that he can with great probability of correctness decide whether the required similarity is actual or superficial. If in a given group of forms the specific distinctness of any form is not proved, we have to resort to a proved case in an allied group of forms; of course, the more dissimilar the forms referred to for comparison are, the more it becomes probable that our inference is not correct. In most cases it is, therefore, circumstantial evidence we have to judge from, and, as many an innocent man has been condemned by a competent judge on the ground of circumstantial evidence, we cannot very well expect to be always right in our judgment of the specific value of the differences of forms. Though the special evidence furnished by morphology and biology is to be carefully considered in every single case, there are nevertheless some general arguments which apply to a multitude of cases. The question as to specific identity or non-identity concerns first forms which occupy the same area, or whose areas overlap, and secondly forms which inhabit localities separated from one another by districts that are not inhabited, or not inhabitable, by them.

We have above tried to show that a species can develop into more species only with the help of isolation of the varietal forms. If, therefore, two allied species are found to inhabit the same district, no matter whether the areas are totally or only partly the same, it is obvious that at a former period, when the species in question were not yet so far advanced in divergent development, they must have occupied separate areas. From the fact of cohabitation (in a wide sense) the further inference must necessarily be drawn that the possibility of cohabitation without fusion is due to the forms having become so divergent that they are indifferent to one another. The time which has elapsed since the two forms now living together became specifically different must therefore be much greater than that elapsed since the formation of the geographical representatives of those two species. If comparative anatomy and morphology are of any value as to the judgment of the phylogeny of species, the morphological differences between a species and an ally which branched off at an early period must be greater than the differences between the same species and its younger geographical representative species, and still greater than the differences between the geographical forms of the species. If in a given case we have to decide whether A and B, which live together, are two different species, or two forms of one species, the morphological characters of A compared with those of B and the geographical representatives of B will have to guide us in our judgment. There are three possibilities resulting from the comparison. First, the morphological differences between A and B are greater than those between B and its representatives; in this case A and B must be considered specifically distinct, until experiment proves the reverse. Secondly, the morphological differences between A and B are less great than those between B and any one of its representatives; in this case A and B are specifically identical. Thirdly, the differences between A and B are equal in morphological value to those between B and any one of its representatives;

in this case A has to be put in the same relation to B in which that respective representative stands to B, i.e. it must be considered either as specifically different or as specifically identical with B, according to the specific distinctness or non-distinctness of that representative.

The same kind of evidence we may employ when we have to come to a decision as to the specific distinctness of geographically separated forms which are not connected by intergradations. But when that evidence is not conclusive enough, we may have recourse to the evidence furnished by the variation of the forms. We must accept as a general law that forms which are connected by all intergradations, or forms which overlap in characters, are specifically identical; geographical form, agreeing with this law are, therefore, to be accepted as specifically non-distinct. If we now compare the various organs of the species in respect to the effect which the causes of variation have upon them, we shall find that a number of characters are easily affected and show a variation between wide limits, while other characters remain comparatively constant. Organ a varies, for example, in species A, as far as we know at the time, from ten to a hundred, while organ b varies only from thirty to thirty-five. Now if it is proved that in a number of allied species a similar difference in respect to the variability of the organs a and b takes place, we can with great probability of correctness conclude that a form B similar to A is specifically distinct from A if the character of the organ b is far outside the range of variation observed in A, and, on the other hand, that B is a form of the species A if the character of the organ b comes within the limits of variation observed in A, no matter whether A and B are very similar or very dissimilar in respect to the variable character of the organ a.

As long as the special evidence does not force us to conclude otherwise, the diagnosticist has to go by the following two general rules:—

1. If is found that A and B stand in a certain relation to one another (sexes, aberrations, seasonal forms, subspecies, species), and that the allied forms C and D differ from one another in a similar way as A does from B, C and D have to be put into the same relation to each other in which A stands to B.

Illustration.—We know by breeding experiments that in Japan the spring brood of *Papilio sarpedon* is smaller and has a wider band than the summer broods; in North India we find an insect very similar to the Japanese one, and observe that in April and May a form flies which is small and has a wide band, and that later in the year all the specimens belong to another form which is somewhat larger and has a narrower band; as in Japan the smaller and the larger forms are proved to be seasonal forms of one species, we are logically bound to regard also the smaller and the larger forms in India as belonging to the spring and summer broods respectively of one species. A good number of allied *Papilios* show in India the same phenomenon; specimens collected during the first half of the year are smaller and have wider bands than the specimens collected later on; though it has not been proved by rearing that we have here actually to do with spring and summer forms, it would be illogical to regard the spring and summer specimens as specifically distinct.

2. If it is found that A and B stand in a certain relation to one another, which relation is either proved by experiment or arrived at by general reasoning, a specimen or specimens differing from A and B in a similar way as A does from B have to be considered as a third form C standing in the same relation to A and B as A does to B.

Illustration.—The islands of Sambawa, Alor, Wetter, Timor, Letti, the Tenimber Islands, North Australia, and the New Hebrides, are each inhabited by a subspecies of *Papilio canopus*; the subspecies differ from one another in the shape of the hindwing, and in the extent, presence, and partial absence of the wing-markings. Lately Mr. Rothschild received a specimen of *Papilio* from the island of Sumba differing from the Sambawa and the Alor forms of *canopus* in a similar way as these do from one another and from the Timor form, namely in the partial absence and in the extent of the markings. The only logically possible way, accepted by Mr. Rothschild, was to treat the Sumba specimen also as a form of *canopus*.

When the evidence leads to the conclusion that the differences exhibited by a number of specimens, or forms represented by specimens, are not specific, it is self-evident that the various forms belong to one species. This species, then, consists of a number of different varieties, every single individual of which, however aberrant it may be, represents the species, and every single peculiarity of any individual is a peculiarity of the species; all the specimens of all the various forms taken together are "the species" as opposed to every other species. The diagnosis of this species, which must not be confounded with the description of the species, is therefore a diagnosis of a sum of varieties; and as a diagnosis is analogous to the definition of a term, it must contain all the distinguishing characters common to all the specimens, and hence must apply to each single specimen. Besides the specific distinguishing characters each variety has one or more characters of its own which form the diagnosis of each respective variety. The discovery of a new variety, which was hitherto unknown on account of the incompleteness of our knowledge, or which has sprung up in consequence of the area of the species having recently become extended (*Pieris rapae* ab. loc. *novangliae*, for example), may necessitate an alteration of the diagnosis of the species to which the new variety belongs. As the forms diagnosticated for the sake of convenience are fixed by a name given to each of them, it is obvious that the only way logically possible to name a species and its subordinate components is to give a name of its own to the species, one to each subspecies as subordinate to the species, and one to each individual aberration as subordinate to the subspecies. Diagrammatically it can be illustrated thus:—

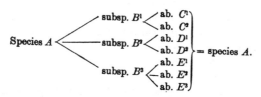

Every individual forms together with other individuals a group characterised by a peculiarity not met with in the rest of the individuals, which, therefore, form another group (or more), termed here *individual aberration = ab*. A number of individual aberrations are the components of a subspecies, and a number of subspecies the components of a species. Each specimen will require in this case a specific, a subspecific, and an aberrational name. In many cases, however, the nomenclature will become much simplified by aberrational names being unnecessary, because the individuals do not vary to such an extent within a subspecies that aberrational names are required; the simplification will be still greater when, besides the aberrational names, the subspecific names are not required. In order to avoid grave mistakes it is necessary that we insert before the aberrational name some sign to indicate that the name is meant for an aberration. For, in species which do not vary according to locality we often have important individual variation, and therefore have to designate individuals with aberrant characters, besides by the generic name, by a specific and an aberrational name only, thus : *Papilio gambrisius* ab. *abbreviatus*; *P. gambrisius abbreviatus* would have an entirely different and erroneous meaning. The abbreviations employed as a sign may be *ab.* for the usual individual aberration, *ab. loc.* for localised individual aberration, ♂-*ab.* and ♀-*ab.* for aberrations occurring only in one sex, ♀-*f.* for a constantly appearing form of polymorphic species, ♀-*f. loc.* when the form is localised, *gen. vern.* and *gen. aest.* for seasonal forms. In this way the various kinds of individual variation can be distinguished by the special sign employed, which would not be possible if the aberrational name were simply annexed to the subspecific or specific name.

Since the diagnosticist when describing a form very often does not know whether this form will ultimately turn out to be cospecific with other forms, or whether it is actually specifically distinct, and as, further, a great many forms have been diagnosticated as species which now are known to be subspecies (and the reverse), the question arises how the above system of nomenclature must be carried out. For the sake of simplicity we shall take into account solely a species with its subspecies ; then we have the following possible cases :—

1. The first diagnosis and description are so general that they apply very well to a certain species A, but do not give any character from which we could see which one of the subspecies (B^1, B^2, B^3) of A the author has had before him. In this case the name given by the author must be kept for the species A, and each subspecies requires another name.

2. The first diagnosis and description apply to two or more forms (but not to all) which are now known to be subspecies of a certain species, and are so general that we do not know whether the author had one or more forms before him. In this case again the first name must be employed for the species, and each subspecies requires another name.

3. The diagnosis applies not to one entire species A, but to one particular subspecies B^1 of A ; the other subspecies B^2 and B^3 of A either were not known to the respective author, or their specific identity with B^1 was not recognised by him. B^1 may be the first described of the three forms. Which name must be used for the species A ? Illustration : Linné described the Amboina form of a beautiful insect under the name of *Papilio priamus* ; we know now that this Amboina form is a subspecies of a species which ranges over nearly the whole of the Papuan region and has developed into several subspecies. Linné's name of *priamus* was given, not to the entire species, but to one particular subspecies, and there is not the least doubt that this name must be kept to designate that particular subspecies. Now, how have we to call the entire species ? A short consideration of what a name is meant for and how systematists employ a name will give a satisfactory answer. A diagnosticist describes a species x from a number of individuals ; further researches show that the characters in the original description apply only to a certain number of specimens ; aberrant specimens are found, and the result is that the original description of the species has to be largely modified ; but, in spite of this, the original name is kept for the species. Illustration : Linné's description of *Papilio podalirius* does not apply to certain aberrations which occasionally occur among the

normal specimens; nevertheless we include these aberrations under Linné's name, and thus extend the meaning of the name.

As the number of specimens at the disposal of the author is always comparatively small, a name will, in consequence of further research, always cover a larger field than it did when first applied. If we keep this extension of the meaning of a name in view, it is obvious that the name of a certain form has to include all subsequently discovered forms which are specifically identical with the first form. Just as the name of *Papilio podalirius* comprises the so-called normal as well as the aberrational specimens, the name of *Papilio priamus* comprises the particular subspecies *priamus* described by Linné as well as all the more recently discovered forms called *poseidon, euphorion, richmondius*, etc., as the following diagram shows:—

$$\text{Papilio priamus} \begin{cases} \text{priamus,} \\ \text{poseidon,} \\ \text{euphorion,} \\ \text{etc.} \end{cases}$$

That is to say, the first name given to any member of a species is to be taken as the name of the entire species. The consequence is that the name of the species must be repeated when the respective component to which it originally was given is to be designated. Thus it might very often happen that a particular individual aberration had to be called after this pattern: *Papilio polytes polytes* ♀-f. *polytes*. The meaning of this name is exclusive and hence precise, and that is the highest praise we can give to a name: ♀-f. *polytes* shows that the *female* sex of the subspecies *polytes* is polymorphic, ♀-f. *polytes* being co-ordinate to one or some other aberrations of that sex (♀-f. *cyrus*, ♀-f. *romulus*); *polytes* ♀-f. *polytes* means that the particular *female* form was the first described; *polytes polytes* has again the meaning that the particular subspecies was the first described of all those which belong to the species *polytes*.

A few illustrations will more especially show the convenience of this method of nomenclature. Boucard described one of those beautiful Central-American beetles which belong to the genus *Plusiotis* under the name of *aurora*; the specimen has remained unique as far as we know, while many individuals have afterwards been found which, though specifically identical with the first-described specimen, differ from the latter very conspicuously in colour, being green instead of aurora-colour. The aurora-coloured individual is apparently a so-called accidental aberration, while the green individuals are the normal (or morphologically typical) ones. According to the old style of nomenclature the two forms would have to stand as *Plusiotis aurora* (accidental aberration) and *Plusiotis aurora* ab. *chrysopedila* (normal form). How absurd this kind of nomenclature is will easily be understood if we take, instead of these beetles, an albinistic specimen and normal individuals of a mammal or bird. Our method treats both forms as forms of one species, *Plusiotis aurora* ab. *aurora* and *Pl. aurora* ab. *chrysopedila*, the species *aurora* thus being composed of a normal form (ab. *chrysopedila*) and an aberrant form (ab. *aurora*).

A European moth of the genus *Hepialus* has developed into two subspecies, one with ♂ and ♀ nearly the same in colour (*hethlandicus*), and the other with ♂ and ♀ very different in colour (*humuli*); the first is said to be phylogenetically the older form, and therefore represents morphologically the typical one of the two; the first described, however, is the sexually dimorphic form *humuli*, and therefore the typical one in a nomenclatorial sense. Morphology and nomenclature come into contest if we employ the old style of naming the forms, while the contest is entirely avoided by accepting our method, according to which the species *Hepialus humuli* comprises two forms, *Hepialus humuli humuli* and *Hepialus humuli hethlandicus*.

A species of *Pieris* was described by Linné in 1758 as *Pieris napi*, while the Alpine and boreal variety of it, which has a different appearance, received in 1808 the name of *Pieris bryoniae*. From the experiments with this insect carried out by Weismann and others, the inference has been drawn * that *bryoniae* is phylogenetically the older one of the two forms, and that, therefore, the species ought to bear the younger name of *bryoniae* instead of the older name of *napi*. As the meaning of *Pieris napi* var. *bryoniae*, which is the name of the Alpine and boreal butterfly according to the old style of nomenclature, is that *bryoniae* is a variety originated in consequence of the variation of *napi*, an alteration is indeed necessary if the above interpretation of the experiments is correct, and thus evolutionists would have to play havoc with the names of all those numerous species of which a younger form happens to be described first. We have, however, endeavoured to show † that *the species is represented neither by the white form napi, nor by the darker form bryoniae, but is composed of napi and bryoniae; the species is not congruent with the ancestral form of the recent forms, but is congruent with the sum of the recent forms*, and its name is, therefore, independent of the name of that form which is supposed to be phylogenetically the oldest of the component forms. According to our method of nomenclature the name of the *species* in question would be *Pieris napi*, the name of the Alpine and boreal form *P. napi bryoniae*, and that of the form inhabiting the rest of Central and Northern Europe *P. napi napi*. If in theoretical treatises it is necessary to distinguish nomenclatorially the oldest from the younger forms of a species, it could be done by adding (f. prim.) = *forma primigenia*, or some such sign, to the name—*P. napi bryoniae* (f. prim.).

The various points in these introductory notes have been very cursorily dealt with; but we are in hopes that the remarks, in spite of their shortness, will serve to explain our interpretation of the facts of variation we are now going to bring before the reader.

II.—THE VARIATION OF THE GENITAL ARMATURE OF CERTAIN PAPILIOS.

The prehensile organs situated round the orifice of the sexual system of insects have for about fifty years been made use of for diagnostic purposes, and it was, and is, a general belief that the genital armature is of such great constancy in every species that peculiarities exhibited by certain individuals in these internal ectodermal organs, and not found in other individuals which otherwise are very slightly different from those, are of specific value. As we have shown in the introduction that every individual has its individual peculiarities, a slight distinguishing character of an individual, besides the sexual armature, can always be found, and therefore the above opinion leads practically to the assertion that a specimen with some kind of peculiarity in the sexual armature is specifically distinct from the specimens which do not have that peculiarity. On the other hand,

---

* Weismann, *Studies in the Theory of Descent* I. London, 1882. pp. 61 ff.
† Compare also Lorenz, *Sitz.-Ber. zool. bot. Ges. Wien* 1892. p. 17; Jordan, Nov. Zool. 1895. p. 182; Hartert, *Ibis* 1896. p. 363.

comparatively very few authors * have given expression to the opinion, not only that there is a certain amount of variability found in those organs, but that one can by no means rely upon them in the judgment of specific distinctness or non-distinctness. During our researches on the Eastern Papilios † we came across some striking cases of variability of the copulatory organs which made it evident to us that the above assertion of an extensive variability was not the outcome of superficial research, and this induced the Honourable Walter Rothschild to charge me with investigations in the matter. As Mr. Rothschild knew from the study of the external characters of the Eastern Papilios that a decision about the specific distinctness of these variable insects with some certain degree of correctness could not be arrived at unless one had ascertained with some probability of correctness the limits of individual variation of each form (no matter whether the form was described as species or variety), and hence in order to come as nearly as possible to the knowledge of the limits of variation it was necessary to compare a great many specimens, he liberally put the long series of individuals of his collection at my disposal, and to this the results of our investigations are largely due. It seemed to us that in the first place the aim of our researches had to be to ascertain whether the alleged constancy of the genital armature was, at least in most species, real, especially as compared with the distinguishing characters derived from the wing-pattern. A little consideration, however, showed us that this was scarcely necessary. First, if we accept the statement that every specifically distinct form is to some degree different from the allied forms in the genital armature as being true, it by no means follows that the inverse is correct, namely that forms presenting in the sexual organs some differences from the allied forms are specifically distinct. Hence the proof of the variability of the organs in question would not imply that these organs are useless for diagnostic purposes, though their taxonomic value would certainly be lessened. Secondly, if two or more allied species are different in the sexual organs we have to conclude from the theory of evolution that the present differences are the outcome of divergent development of the allied species from a common ancestral species which itself had the sexual organs either different from all its descendant species, or from all but one ; if we concede this, and all followers of Darwin have to do so, it is self-evident that the ancestral species must have been variable in the sexual armature. As, therefore, in the ancestors of our present species the genital armature must be assumed to have been so variable that the variation could lead to specific separation, we cannot but assume *a priori* that in all the species of the present which are in the state of diverging into varieties the genital organs must exhibit not only some variability, but also variation to such an extent that the genital characters of a certain variety could be increased by the factors of evolution and ultimately be transformed into specific characters, unless one tries to avoid these consequences either by abandoning evolution altogether, which would be acceptable, or by maintaining that evolution is not going on during the present epoch, which would be ridiculous.

This consideration made it pretty clear to us that the more important part of our investigations would have to be, not to prove the occurrence of variation of the organs in question, but to ascertain the kind of variation, especially to accumulate such facts from which could be seen whether there is individual,

* Perez, *Ann. Soc. Ent. Fr.* p. 74 (1894) ; Edwards, *Canad. Ent.* p. 56 (1894) ; Kolbe, *Ent. Nachr* p. 133 (1887).
† Nov. Zool. p. 269 (1895).

seasonal, and geographical polymorphism in the sexual armature. And if such variation be found, the next task would be to compare this variation with that of the external organs, especially with that of the wings, in order to see, first, whether the genital organs and the pattern of the wings were independent of one another in respect to their variation; secondly, whether, in spite of this independence, there are certain kinds of varieties which are characterised by more or less constant peculiarities in the wing-pattern associated with, but not correlated to, peculiarities in the genital armature. As such varieties could easily be mistaken (in fact, have largely been mistaken) for distinct species, and hence would have the appearance of incipient species, and as, further, the divarication of a species can come about only by means of branching into subspecies, we could *a priori* expect to find such a combination of characters in geographical races or subspecies.

As the present paper stands in close connection with the classificatory investigation on the Papilios we are carrying on, the researches here demonstrated are restricted to that group of insects; and this we deem the more necessary, because the correctness of the results of such work depends to a great extent on the full acquaintance with the various forms dealt with. Though in a monograph of the Papilios all the forms of all the species must be taken into consideration, we have abstained from treating upon all the Palaearctic and Indo-Australian Papilios in this paper for the good reason that, as in every form at least all the more prominent varieties in the genital armature have to be described and figured, the detail of the paper would be so immense that in consequence of the great amount of detail the single facts of variation would be much obscured. Hence we have thought it best to demonstrate the variation of the genital armature on a small number of species which have been so selected that they very well illustrate, first, the amount of variation; secondly, the kind of variation; and thirdly, both the amount and kind of variation within several, morphologically very different, groups of *Papilio*.

As far as we know, systematists have, as regards *Lepidoptera*, only made use of the *male* genital armature for diagnostic purposes. Salvin * mentions the presence of a kind of armature at the orifice of the vagina, but has not succeeded, in consequence of an inadequate method of preparation, in bringing it forward for the purpose of classification. We first came across the vaginal armature when we studied the morphology of the abdomen of the Papilios with a view of discovering characters which could help us in coming to a decision about the extent of the genera into which the Papilios must be classified, and soon found out, on the one side, that the morphology of the abdomen of the *females*, including the vaginal armature, presents excellent generic characters to the systematist (compare Pl. XIX., f. 181, 182), and, on the other side, that the detail of the structure was of the highest taxonomic value as to the delimitation of species. Therefore we have selected a few of the species examined to illustrate the peculiar structure and the variation of the vaginal armature, and the form of the eighth abdominal segment.

A. Male Genital Armature.

The clasping apparatus of the *male* comprises three organs: (1) a dorsal hook called by Gosse *uncus*; (2) the lateral *valves* or *claspers*, bearing on the inner side ridges, teeth, and hooks called *harpe*; and (3) the *scaphium*, which is situated

* *Biol. Centrali-Amer., Rhop.* II. p. 189 (1890).

immediately below the anus and above the penis, and becomes visible when the valves are removed. The homology of these organs has been explained by Dr. Peytoureau in his work entitled *Contribution à l'étude de la Morphologie de l'Armure génitale des Insectes*,* and we therefore refer the reader to that work ; it is here sufficient to mention that Dr. Peytoureau comes to the result that the *valves* are lateral wings to the *ninth* segment, which itself becomes visible only by dissection, and that the dorsal *uncus* and the ventral *scaphium*, between which the *anus* is found, represent the *anal* or *tenth* segment ; the penis, therefore, has its position between the *ninth* and *tenth* segments.

The *uncus, scaphium*, and the *valve* with the *harpe* are of classificatory value, and if we intended to explain here the complete morphology of the genital armature of the Papilios we certainly should have to take all three organs as well as the *penis* into account ; merely for the sake of simplifying matters we have restricted our notes almost to the *valve* and *harpe*.

The harpe of the Papilios is a fold of the inner sheath of the valve partly raised to ridges, teeth, hooks, rod-like processes, etc., which are sometimes of rather a complicated structure. We have generally given a figure of the harpe and valve as they appear when viewed with the eye perpendicularly above the plane of the valve, while the figures representing the harpe or parts of it are so drawn that the planes of harpe and paper are the same. Very much depends on the position in which the eye is to the harpe, if the comparative study of these organs is to be of any use ; a curved or twisted ridge or process appears very different when viewed under a different angle ; and hence we have endeavoured to represent the same organ of the various species and subspecies in the same position, so that a comparison of the figures gives an exact idea of the differences in the organs.

1. **Papilio machaon** ; † f. 39 to 45.

This species occurs nearly all over the Palaearctic Region, inclusive of China and Japan, and is found also on the Indian side of the Himalayas at higher elevations, as well as in the Nearctic Region. The lines of delimitation between the Old World forms of *P. machaon* are very difficult to draw, in fact cannot be drawn, as the forms overlap in characters. The most remarkable varieties are the summer brood of the Japanese *machaon*, and the subspecies from the interior of Sikkim and the higher parts of Western China. As we shall have to describe and figure the sexual armature of the various forms of *P. machaon* in another paper, we introduce the species here merely because it is the only British representative of the entire group, and therefore will enable the British entomologist to verify our observations.

The *valve* of *P. machaon* is of a triangular shape, and, though somewhat variable in outline according to the individual specimens, does not present any obvious differences in the specimens of the different subspecies.

The *harpe* is a longitudinal fold lying along the ventral margin of the valve ; it is distinctly raised and leans somewhat over dorsally. The basal half or so is rounded and simple, whereas the apical half is compressed, with the upper free edge denticulate, so that it resembles the blade of a saw (f. 39—43, ventral view). The

---

* *Revue Biologique du Nord* VII. 1895. The author gives on pp. 13 to 50 a list of works dealing with the abdomen of insects.

† See note on p. 439.

basal, rod-like, portion either immediately runs out into the saw (f. 43), or has to curve a little dorsally to join the saw, the latter being a little more dorsal than the former; this variation is independent of locality.

a. *P. machaon flavus* from Great Britain.

The denticulate portion of the harpe is generally longer than in *machaon machaon* from Germany and in *machaon sphyrus* from South Europe and Asia Minor. F. 39, 40, and 41 are taken from three British specimens, and represent the amount of variation found by us in our series. The proportion of the length of the saw to the entire harpe is in the three specimens 19 : 30, 15 : 30, and 13 : 30 ; the variation in the length of the denticulate portion amounts therefore to almost 50 per cent. of the length of the saw figured in f. 141.

The *uncus* of the British specimens (f. 44, dorsal view) is nearly always slenderer than that of the Continental individuals (f. 45), and agrees very well with that of the Japanese subspecies (spring and summer broods).

b. *P. machaon machaon*.

In f. 45 the *uncus* of an individual from Switzerland is represented to show the divergence from f. 44.

c. *P. machaon sphyrus*.

The harpe of this southern subspecies as well as the harpe of *machaon machaon* agrees on the whole with f. 40, but in some specimens from Asia Minor and Palestine the denticulate portion is remarkably short. F. 42 represents an extreme, the proportion of saw to entire harpe being 11 : 30 ; the saw is in this individual from Syria more than 70 per cent. shorter than in the British specimen represented by f. 39, an amount of variation which is higher than we anticipated. The harpe of a Palestine specimen, as drawn in f. 43, is abnormal in so far as the saw does rise gradually, not abruptly.

d. *P. machaon hippocrates* from Japan.

Besides the *uncus* mentioned before, we do not see any difference between the genital armature of this remarkable form and the European *machaon*. The spring and summer broods, though so conspicuously different in size and pattern, also do not exhibit, to our knowledge, any distinguishing character in the organs in question.

As the swallow-tails found in the Wicken Fens, near Cambridge, are doubtless one species, the variation of the species in the length of the denticulate portion of the harpe, or, in other words, of the prehensile part of the harpe, amounts to nearly 50 per cent. As further the Syrian specimens do not differ in the wing-pattern so much from British specimens as in Germany and in Syria the individuals of the first brood very often differ from those of the second brood, and as a line of separation between the characters of British and Syrian specimens is altogether absent, we have also to admit that f. 39 and 42 are taken from the same species, *i.e.* that the variation of the prehensile organ amounts to 73 per cent.

Our conclusions are : (1) The genital armature of *machaon* does not afford any characters by which the various geographical races, distinguished especially by differences in the wing-markings, can be constantly differentiated from one another. The harpe of the British form has, however, on the whole, the longest prehensile part, while the specimens from Syria and Palestine have that denticulated portion on an average shorter than specimens from England, Central Europe, Sikkim, China, and Japan; the *uncus* is thinner in nearly all British and Japanese individuals than in most individuals from the interjacent countries.

(2) The spring and summer broods are not different in the genital armature.

(3) The prehensile portion of the harpe varies considerably in length; the amount of variation is 73 per cent.

2. Papilio aegeus; f. 1 to 11.

According to the wing-pattern the species comprises a subspecies inhabiting New Guinea and Aru (*aegeus ormenus*), another found in Australia (*aegeus aegeus*), a third inhabiting the Kei Islands (*aegeus keianus*), a fourth found on the Banda Islands (*aegeus adrastus*), and a fifth is met with in New Britain (*aegeus bismarckianus*).

Of these we could dissect very long series of the first two forms, while the *males* of the three other subspecies we have not examined for want of sufficient material.

P. aegeus ormenus is in both sexes a very variable insect as regards colour, while the Australian form *aegeus aegeus* is rather constant in that respect.

The *valve* (f. 1, seen from above) is of the usual triangular shape; its ventri-apical angle is sometimes more, sometimes less rounded, regardless of locality as well as of wing-pattern.

The *harpe* lies as in *P. machaon* along the ventral edge of the valve, extending from the base to the apex; it is a rather thin blade with the free upper edge sharp, but not conspicuously dentate, bearing only a fine denticulation at the projecting portions, and leans over dorsally. Before the middle it widens out triangularly to form the *submedian projection*, and its free apical portion is raised above the level of the margin of the valve and forms the *apical projection*. The length and form of these two projections vary individually and subspecifically.

a. *P. aegeus ormenus*; f. 1 to 6.

According to the development of the subapical white band of the forewing the *males* belong to three varietal forms: ♂-ab. *ormenus*, with the band complete on upper- and underside; ♂-ab. *pandion*, with the spots of the band partly obliterated; ♂-ab. *othello*, with the band absent. The three forms occur together in the same locality and are connected by all intergradations. The variation of the sexual armature is entirely independent of that of the variation of the band; individuals of ♂-ab. *othello* are different *inter se*, while some of them agree with certain specimens of ♂-ab. *pandion* or ♂-ab. *ormenus*, and so it is with the latter aberrations. The following remarks, therefore, refer to every form of the *male*.

F. 1 is taken from an individual from Dutch New Guinea (coast near the Arfak Mountains); in this perpendicular view the apical projection of the harpe, being more erect than the shorter submedian one, appears to be short. The harpe of the same individual is represented in f. 2; the submedian projection is broad and triangular, and differs obviously from the same projection of f. 3 to 5, which are taken from individuals from the same locality and represent very well the amount of variation we have found in *aegeus ormenus*. In f. 5 the submedian projection is very small, in f. 4 very high, in f. 3 and 4 much slenderer than in f. 2. A specimen (f. 6) from Finschhafen, German New Guinea, has the projection as broad as it is in f. 2, and nearly as high as it is in f. 4. The usual form of the projection met with in most examples from all parts of New Guinea is that of f. 2. The length of the projection varies about from 1 : 2.

The free apical projection of the harpe varies in a similar way in outline and height, as will be seen by comparing f. 2 to 6 ; the extremes which came under our notice are shown in f. 4 and 5, the length varying about from 2 : 3. The normal size of the apical projection is that of f. 2.

The specimens from British New Guinea, the D'Entrecasteaux Islands, and Woodlark Island have in the harpes no character by which we could distinguish them from the individuals from Northern New Guinea.

b. *P. aegeus aegeus* from Queensland and New South Wales ; f. 7 to 11.

Though the Australian form of *aegeus* is in respect to the wing-pattern very constant as compared with *aegeus ormenus*, the variation in the genital armature is just as great as in the New Guinean subspecies. The *male* corresponds in the pattern of the forewing to *aegeus ormenus* ♂-ab. *ormenus*.

The commonest forms of the submedian projection met with are shown in f. 8 and 11; on the whole, the basal edge of the projection is more vertical than in *aegeus ormenus*, but this does not apply to every specimen ; the character is especially often obvious in the individuals from New South Wales and Southern Queensland. The apical projection is in a few examples a little higher than in *aegeus ormenus*. F. 7 and 8 are taken from two specimens from Cairns, North Queensland ; f. 9 represents an individual from Cedar Bay, thirty miles south of Cooktown ; f. 10, with an abnormally high and slender submedian projection, is taken from a Queensland individual without exact locality, while f. 11 represents a New South Wales individual. The variation in the length of the submedian projection is in this selected series not so great as in f. 2 to 6, as we did not find an individual in which the submedian projection was as feebly developed as in f. 5.

The importance of the differences exhibited in f. 2 to 11 will at once become obvious when we compare the harpes of the two nearest allied species.

3. Papilio inopinatus ; f. 12.

P. aegeus is on the Tenimber Islands represented by an insect which is comparatively very constant in external characters, and is in colour and pattern always separated from *P. aegeus* by a wide gap. Though the absence of intermediate specimens is not a proof that the Tenimber insect named *inopinatus* is specifically distinct from *P. aegeus*, we have to treat *inopinatus* as a species for the following reasons : the external differences between *inopinatus* and *aegeus* are greater than, or as great as, the differences between the relative forms which are regarded as distinct species ; if *lowi* and *mayo* are kept separate from *memnon—rumanzovius, deiphontes*, and *deipylus* as distinct from *deiphobus—gambrisius* specifically separate from *aegeus*, then *inopinatus* is likewise to be treated as a distinct species. Further, the variation within the *male* sex of *P. aegeus* from New Guinea, Australia, Aru, Kei, and Banda Islands takes place between such limits that the difference between the extremes is not so great as that between *aegeus* and *inopinatus* ; the same applies to the variation of that *female* form of *ormenus* which corresponds to the *female* of *inopinatus*.

Thus we think it fairly safe to consider *inopinatus* specifically distinct. The evidence is to some extent corroborated by the difference exhibited by the harpe. We have examined only a few individuals, all of which have the median thin portion of the harpe (f. 12) longer than it is in *aegeus*, especially than we found it to be in *aegeus ormenus* ; the submedian projection is broad and low, and the apical projection is likewise shorter than in *aegeus*. A comparison of f. 2 to 11 with 12 evidently shows, however, that the distinguishing points in the harpe of *inopinatus* are of much less weight in the judgment of the specific distinctness of the insect than the external features ; the difference between f. 4 and 5 and f. 8 and 11 is far greater than that between f. 12 (*inopinatus*) and 8 or 3 (*aegeus*).

4. Papilio tydeus ; f. 13 to 16.

This is the representative species of *aegeus* on the Northern Moluccas ; the same reasons which induce us to treat *inopinatus* as specifically distinct apply also to this insect. The external differences from *aegeus* are in *tydeus* not quite so great as in *inopinatus* ; in opposition to this the harpe of *tydeus* differs much more widely from that of *aegeus*. The clasper or valve of *tydeus* (f. 16) is larger than that of *aegeus*, and the harpe therefore longer. The submedian and apical projections of the harpe are more bent over dorsally, as will be seen by comparing f. 1 and 16. The submedian projection is broad and high, and the apical one is conspicuously higher and more erect than in the allies.

We have examined four individuals from Halmaheira and three from Batjan ; there is no localised difference in the harpe of *tydeus*. F. 13, 14, and 16 are taken from Halmaheira individuals ; f. 15 represents the harpe of one of the Batjan specimens. The variation in the form and length of the projections is obvious ; the apical projection in f. 15 is almost half as broad again as that in f. 14.

The facts here illustrated are as follows :—

(1) The variation of the pattern of the wings of *P. aegeus ormenus* is entirely independent of the variation of the harpe.

(2) The difference in the harpes of *aegeus ormenus* and *aegeus aegeus* is very slight and applies only to scarcely half of the number of specimens examined.

(3) *P. inopinatus*, though in external characters very different from *aegeus*, exhibits in the harpe only a slight, but according to the specimens examined rather constant, difference from *aegeus*.

(4) *P. tydeus*, which is in colour and pattern separated from *aegeus* by a less wide gap, has the harpe in all the specimens examined conspicuously different from that of *aegeus* and *inopinatus*.

(5) If we take the length of the harpe = 100, the projections measured from the plane of the valve to the tip of the projections vary in length as follows :—

P. aegeus : submedian projection from 18 to 33 ; apical projection from 33 to 45.
P. inopinatus   „   „   „   18   ;   „   „   „   32.
P. tydeus   „   „   „   30 to 33 ; „   „   „   51 to 63.

Or, in *aegeus* the variation of the submedian projection amounts to 56 per cent. of the length of longest submedian projection observed, while the variation of the apical projection is 27 per cent.

(6) *P. aegeus* and *tydeus* differ somewhat in the size and outline of the valve.

5. Papilio polytes; f. 17 to 37.

The *male* sex of this insect is in the pattern of the wing not very variable, while the *female* exhibits a very great amount of individual and local variation. The species ranges in a good number of subspecies over the Indian and Malayan Subregions, and goes as far as the Moluccas; in New Guinea, the Aru Islands, Queensland, and the islands farther east it is represented by *P. ambrax* and *P. phestus* respectively.

The *valve* is more or less triangular (f. 17 and 31), and varies individually. The *harpe* has the same position as in *P. aegeus*; it is a ventral longitudinal fold; the basal half (or so) is stick-like, while the apical portion is abruptly raised into a thin blade, which leans over dorsally so that its ventral surface is visible when the valve is viewed with the eye above the plane of the valve (as in f. 17 and 31); the harpe thus has the appearance of a hatchet; the upper free edge of the blade is very finely denticulate; the tip of the blade projects free for a short distance, and is somewhat curved dorsally, often forming a blunt hook. The outline of the harpe varies according to locality and to the individual specimen.

a. *P. polytes polytes* from N.W. India to Malacca, Natuna Islands, Tonkin; f. 17 to 26.

The blade of the harpe is highest near its basal (subperpendicular) edge, where it is slightly angulate; a second, more distinct, angle is formed just before the edge slopes down towards the apex of the harpe. The apex is scarcely produced or very slightly so. The degree of variation found by us in the specimens from Continental India is represented by f. 18 to 21.

The form of the harpe as shown in f. 18 is that found in most specimens; the figure is taken from a Kumaon individual (caught in June 1893). The blade of the harpe of the individual from Bankipore (captured March 20th, 1893) (f. 19) is much less steep basally than usual, its dorsal edge being much reduced in length; the two angles are conspicuous. The two Burmese (Bassein) examples from which f. 20 and 21 are drawn are especially remarkable for the development of the free apical projection.

In external features *P. polytes* from the Natuna Islands (between Malacca and Borneo) forms a transition from *polytes polytes* to *polytes theseus*. From a long series of individuals examined (captured by A. Everett in September and October 1893) the three most different harpes are here figured (f. 24, 25, 26); the blade agrees very well in shape with that of Indian individuals. The thin carina running from the upper edge of the handle of the harpe along the base of the blade varies from being absent to being well marked.

b. *P. polytes borealis* from China.

All the specimens examined agree in the harpes with *polytes polytes*. The individuals of the interesting variety *P. polytes borealis* ♂-ab. *thibetanus*, in which the white discal markings of the hindwing are partly obliterated, also do not exhibit any peculiarity in the genital armature.

As said in Mr. Rothschild's Revision of the Eastern Papilios, the specimens of *P. polytes* from the Loo Choo Islands (south of Japan) stand intermediate in pattern between *P. polytes borealis* and the Malayan *P. polytes theseus*. The harpe of the Loo Choo *polytes* is in so far remarkable that it differs from the harpe of *polytes*

polytes and *polytes theseus* in being (in individuals of equal size) larger (f. 22 and 23); the angles of the blade are less prominent than in *polytes polytes*, by which character it leads over to *P. polytes theseus* (f. 28).

c. *P. polytes nikobarus* from the Nicobars and Andamans; f. 27.

We have now received more material from these islands, and find that *polytes nikobarus* can very well be kept separate from *polytes polytes*, as the greater proportion of the specimens from the Andaman and Nicobar Islands are somewhat different from *polytes polytes*. The harpe confirms this opinion; the outline of its blade is more rounded (f. 27), not exhibiting the two angles found in *polytes polytes* (f. 18), and resembles somewhat that of *P. polytes theseus* from Java (f. 29), except in the apex being obviously truncate.

d. *P. polytes theseus* from the larger and lesser Sunda Islands; f. 28 to 30.

This subspecies differs in the *male* from *P. polytes polytes* in being generally smaller and in having the tail to the hindwing more or less obliterated. The harpe of *P. polytes theseus* is distinguished from that of *polytes polytes* by being in the blade absolutely longer, less raised, much more evenly rounded, and by the apex being more pointed.

Sumatra and Borneo individuals have the blade shorter and higher than Java examples, and hence lead over to *polytes polytes*; f. 28 is taken from a small example from the Kina Balu (North Borneo). Timor specimens agree generally with the Javan individuals in the form of the harpe; f. 30, however, is aberrant in having the second angle of the edge of the blade faintly marked and the apex distinctly truncate in a ventral view.

e. *P. polytes alcindor* from Celebes and Saleyer; f. 34.

The *female* of this subspecies is very aberrant, the *male* much less so. The *male* has one character in common with *P. polytes polytes* and *P. polytes theseus*, namely the presence of blue scales on the underside of the hindwing outside the macular white band, which scales are absent from the subspecies flying on the Philippine and Moluccan Islands. So insignificant as this character is, it becomes interesting when we see that the harpe of *polytes alcindor* (f. 34) comes closer in outline to that of *polytes polytes* (compare f. 25) than to that of *polytes alphenor* from the Sulla Islands (f. 33) or from the Philippines (f. 32). We observe, however, that the tip of harpe of *polytes alcindor* (f. 34), though short, is visibly curved upwards in a similar way as in *polytes alphenor*.

f. *P. polytes perversus* from the islands of Sangir and Talaut; f. 35.

In external features this form combines to a certain extent in the *male* the characters of *polytes alcindor* (Celebes) and *polytes nicanor* (Halmaheira). The harpe comes nearest to that of *alcindor*; has the tip, however, a little more hook-shaped.

g. *P. polytes alphenor* from the Philippine Islands (inclusive of Palawan), the Sulla Islands, and the Southern Moluccas; f. 31 to 33.

The *males* from these various islands agree very well with one another in external characters; the polymorphic *females*, however, are partly different according to locality, thus showing that the insect is on the way to develop into several local races. The harpes of the *males* are, according to locality, slightly different.

In all the specimens examined the blade of the harpe is conspicuously longer and slenderer than in all the other subspecies of *P. polytes*, and reaches close to the apical margin of the valve (f. 31, taken from a Mindoro example); the narrow apical portion is especially long, and often rather strongly hook-shaped, the hook projecting above the level of the raised ventri-apical edge of the valve.

The harpe of the average example from the Philippines is represented in f. 32; it has a very different appearance from the harpe represented in f. 18 to 26, while the difference between f. 32 and 29 is less conspicuous; the variation of the Philippine examples takes place in such a direction that some individuals approach f. 35.

Of the *males* from the Sulla Islands Mr. Rothschild says that they are not exactly identical with the Philippine *alphenor* in wing-pattern, but approach a little *P. polytes perversus*. The harpes of the Sulla *males* (f. 33) are, on the contrary, still slenderer than in Philippine *alphenor*, thus indicating that a relationship in wing-pattern is not necessarily corroborated by the development of the genital armature.

From the Southern Moluccas we had unfortunately only one individual at our disposal; the harpe of this specimen is shaped nearly as in *polytes perversus* (f. 35).

In the harpes, therefore, the *males* of *alphenor* from the three localities (Philippine Islands, Sulla Islands, and Southern Moluccas) are fairly well distinguishable, though there is no distinct line of separation.

h. *P. polytes nicanor* from Halmaheira and Batjan; f. 36.

The blade of the harpe is much higher than in *polytes alphenor*, and in this respect *nicanor* comes nearest to the Indian *polytes polytes*; the apex of the harpe is produced, nearly as in *alphenor*. F. 36 is taken from a Halmaheira individual.

6. Papilio ambrax; f. 37 and 38.

The outline of the valve (f. 38) of this species varies individually. The harpe is generally formed as in f. 37, but is sometimes somewhat slenderer and at the apex less hooked.

We merely give the two figures of the valve and harpe of an *ambrax* individual (from German New Guinea) in order to enable the reader to compare them with f. 18 to 36. The differences between f. 37 (*ambrax*) and 36 or 34 (*polytes*), belonging to two species, are not so conspicuous as the divergency exhibited by various harpes within the species *polytes*.

The facts of variation observed in *P. polytes* are as follows :—

(1) The valve is variable in the individual specimens, but does not exhibit in the specimens examined any obvious variation according to locality.

(2) The conspicuous individual aberration *P. polytes borealis* ♂-ab. *thibetanus* has no peculiarity in the genital armature.

(3) The individual variation in the blade of the harpe of the individuals from India is such (f. 18 and 19) that the extremes stand further apart than many individuals of *polytes polytes* do from many individuals of *polytes theseus*.

(4) The harpe of *polytes theseus* from Java and the lesser Sunda Islands is well distinguished from that of the Indian *polytes polytes*, but there is no parting line on account of the intermediate form of the harpe of *polytes theseus* from Borneo and Sumatra.

(5) Certain Timor individuals of *polytes theseus* come in the harpe (f. 30) very near *polytes perversus* from Sangir and Talaut (f. 35), and this again is very close to *polytes alcindor* from Celebes (f. 34).

(6) The harpe of *polytes alphenor* from the Philippine and Sulla Islands is in the height of the blade nearer related to *polytes theseus* from Java and the lesser Sunda Islands than to *polytes theseus* from Borneo and *alcindor* from Celebes, which subspecies are geographically the nearest to *alphenor*; while, on the other hand, the apex of the harpe is in all the subspecies inhabiting Celebes, the Philippines, Sangir and Talaut, the Sulla Islands, and the Moluccas, somewhat curved upwards (and towards the dorsal margin of the valve), and not turned up, or rarely so, in *polytes theseus*.

(7) The subspecies most conspicuously different in the harpe is *P. polytes alphenor*, while the subspecies most conspicuously different in the shape and pattern of the wing is *P. polytes nicanor*.

7. Papilio euchenor; f. 51 to 64.

This insect is purely Papuan, being found in New Guinea, the Aru and Kei Islands, the D'Entrecasteaux Islands, Woodlark Island, and on New Britain and New Ireland; it has no near relative. Up to 1895 the specimens from these various localities had been treated as identical; Mr. Rothschild in his Revision found, however, that the individuals in his collection from the Bismarck Archipelago are in both sexes conspicuously different from the New Guinea specimens, and that, on the other side, the individuals from Aru are in the *female* sex, not in the *male*, also *constantly* different as far as the great material examined can be taken as furnishing a *proof* of a constancy of distinction. Lately, Mr. Rothschild observed, moreover, that the specimens from New Ireland are again in both sexes distinguishable from the individuals from New Britain; so that there are four well recognisable forms, to which is to be added a fifth from Woodlark Island * described as a distinct species some forty years ago, but scarcely different in the ♂ in external features from the New Guinea form. The question is now, are the five forms one species, or do they belong to more species? An answer is in this case extremely difficult to give: first, because *euchenor* stands quite isolated amongst the Indo-Australian Papilios, and thus does not allow us to compare the distinguishing characters of other forms assumed, or proved, to be specifically distinct; secondly, because the main portion of New Britain, which geographically is nearest to New Guinea and hence may perhaps be inhabited by an *euchenor* form intermediate between the New Guinea form and the New Britain form, is entomologically an entire blank, all the specimens received from New Britain being caught in the north-east of the island. The external characters of the various forms, however, allow us to set at rest the question; as *obsolescens* from Aru and Kei, and *godarti* from Woodlark, are in the *male* not always distinguishable from *euchenor* from New Guinea, these three forms have to be treated as subspecies of one species (the name of which is *euchenor*).

The two forms from the Bismarck Archipelago, *depilis* from New Britain and *novohibernicus* from New Ireland, have several conspicuous characters in common by which they are differentiated from *euchenor euchenor*, *euchenor obsolescens*, and *euchenor godarti*, while the differences between *depilis* and *novohibernicus* are

* And perhaps a sixth from the D'Entrecasteaux Islands.

quantitatively much slighter; hence *depilis* and *novohibernicus* must (on the ground of external features) be considered as nearer related to one another than to the other *euchenor* forms. As in the present paper it is our purpose to demonstrate the variation of the genital armature within the limits of a species, and as therefore we have to avoid, as far as possible, any error as to the actual specific identity of the forms included by us in the limits of a respective species, we will, merely for the sake of being on the safe side, assume that *depilis* is a species distinct from *euchenor*. The external characters by which *novohibernicus* is distinguished from *depilis* are found in every specimen of our short series ; if, therefore, the constant presence of a distinguishing character is considered sufficient to make the respective form specifically distinct (as some naturalists do), *novohibernicus* is also a distinct species. However, the distinguishing characters of *novohibernicus* amount quantitatively not to more—about the qualitative amount of these characters we know nothing—than the differences do which are observed between New Guinea individuals, differences in the extent of the yellow markings which are not thought to indicate anything else but individual variability within the same species ; consequently we must assume that the differences between the individuals from New Britain and those from New Ireland being quantitatively the same are also qualitatively the same, *i.e.* do not indicate more than divergency of individuals of the same species. Hence the characters distinguishing *depilis* and *novohibernicus* must correctly be considered as not being of specific value. The various forms in question are therefore to be grouped as follows :—

1. *P. euchenor* { *euchenor* ; New Guinea and islands near it.
godarti ; Woodlark Island.
obsolescens ; Aru and Kei Islands.

2. *P. depilis* { *depilis* ; New Britain.
novohibernicus ; New Ireland.

The genital armature of the five forms is in accordance with this division.

The *valve* of *P. euchenor* (f. 51) is very large, strongly convex outwardly, with the apical margin rounded, the ventri-apical angle not being triangularly produced as in *P. aegeus* and most other species ; it exhibits some individual variability in the outline, especially in the ventri-apical portion. The armature of the valve consists of a fold running along the ventral margin of the valve in a slightly oblique direction, turning near the apex round towards the dorsal edge of the valve, running from here as a thin fold backwards to the base, first in a directly basal, then in a dorso-ventral direction, and thus returning to the starting-point ; from the basal dorso-ventral portion an oblique fold (f. 51, *d*) starts, traverses the (concave) valve, and widening out joins the ventral longitudinal ridge. The ventral portion is raised into a ridge. armed at both ends with a process, of which the first (*c*) is here called "basal hook," the second (*b*) "ventri-apical hook"; the basal ridge leans over ventrally, so that in a view perpendicular to the valve the dorsal surface of the ridge is visible (as in f. 51); it is highest near the basal hook (f. 52); the outline is not constant. The ventri-apical hook is directed in a basi-apical direction leaning over ventrally, with the tip protruding above the elevated edge of the valve. The vertical, ventri-dorsal, portion of the fold is less high than the ventral ridge, thinner and denticulate ; it leans over apically, so that in a perpendicular view the basal surface of the ridge is visible ; at the dorsal end it is produced into a slender and very sharp hook (*a*), the "dorsal hook," which is curved in an apici-ventral direction; just underneath the hook the ridge is highest; the number of teeth is variable.

The variation according to locality affects especially the length of the longitudinal (ventral) and the ventri-dorsal ridges, their outline and armature, and the height of the oblique fold.

a. *P. euchenor euchenor*; f. 51 to 54, 59 to 61.

The variation noticed by us in the ventral ridge of New Guinea individuals is represented in views from the dorsal side in f. 52 to 54. The usual form of the ridge is given in f. 52, taken from an example from Constantinhafen, German New Guinea (f. 51 is taken from the same individual). In f. 53, representing a specimen from the same locality, the ridge is very high in the basal third and then rather suddenly diminishes in height; and in f. 54, taken from a specimen from Simbang, near Finschhafen, Huon Golfe, the abruptly raised basal portion is rather angulate.

Our specimens from the D'Entrecasteaux group of islands, east of New Guinea, as well as those from Waigeu, west of New Guinea, come in respect to the ventral ridge within the limits of variation as illustrated by f. 52, 53, 54. All the specimens have the angle formed by the sudden break in the outline of the ridge produced into the beak-like basal hook.

The vertical (ventri-dorsal) ridge is represented separate from the ventral ridge in order to be able to give the exact outline; f. 59, 60, 61, are taken from the same individuals as f. 52, 53, 54 respectively. As f. 59 to 61 are drawn from a basal view of the ridge, the ventri-apical hook (*b*) has a different appearance from that in f. 52 to 54.

The higher dorsal (left-hand side in figure) portion of the ridge is dentate; just at the highest point, or close to it, stands nearly always a stronger tooth, which in f. 59, however, is obsolete. The specimens from the D'Entrecasteaux Islands have nearly all the ridge toothed similarly to f. 61, a character which becomes more obvious in the individuals from Woodlark Island. The interesting aberration from Jobi Island, *P. euchenor euchenor* ab. *eutropius*, does not present any peculiarity in the valval armature.

b. *P. euchenor godarti*; f. 58.

We have examined three specimens of this form, which is all known to exist in collections, except Montrouzier's type-specimen which is perhaps (?!) preserved in the Paris Museum. The only distinguishing character in the valve and its armature found in all three individuals concerns the oblique fold (*d*), which is higher than in either *euchenor obsolescens* or *euchenor euchenor*; the valval cavity before and behind the fold is consequently deeper. The ventral and ventri-dorsal ridges are not constantly different from those of the New Guinea individuals; the dentition of the vertical fold is, however, in all three examples rather plentiful. The more aberrant harpe of the three is represented by f. 58 and 64; the ventral ridge (f. 58) is distinguished by the long basal and relatively short ventri-apical hook, and by the upper edge of the ridge being feebly and widely bisinuate; the ventri-dorsal ridge is multidentate.

c. *P. euchenor obsolescens*; f. 55, 56, 57, and 62, 63, 64.

In Aru individuals the ventral (longitudinal) ridge is somewhat shorter than in *euchenor euchenor*, and the vertical ridge accordingly longer, as will be seen by comparing f. 55, 56, and 62, 63, with f. 52 to 54 and 59, 60 respectively. In outline the longitudinal ridge is on the whole not different from that of New Guinea examples, but the usual form of the ridge is not that represented in f. 56, which is similar to f. 52, representing the usual form of the ridge in *euchenor euchenor*, but a form resembling f. 55. In the latter figure there is an additional tooth upon the crown of the basal dilatation of the ridge, which we have seen only in this one individual. The ventri-apical hook is in the specimens examined shorter than in *euchenor euchenor*. The vertical ridge (f. 62 and 63) shows a certain amount of variability in the dentition. The ventri-apical hook appears to be somewhat more curved towards the right-hand side than in f. 59 to 61; this is due to the hook being more erect than in *euchenor euchenor*, less leaning over ventrally and apically.

From the Kei Islands three individuals have been examined which, in external features, do not exhibit obvious differences from the Aru specimens in the Tring Museum. The harpe presents, however, in the three examples some slight distinguishing characters. The longitudinal ridge is (f. 57) still shorter than in Aru individuals, the ventri-apical hook stands still more erect to the plane of the valve and the upper edge of the ventral ridge (as will be seen both from f. 57 and f. 64), and the vertical ridge (f. 64) is provided with many strong teeth.

8. Papilio depilis; f. 65 to 71.

The *valve* is ventrally a little more rounded than in *P. euchenor*, but this character is not constant. The armature of the valve (f. 65) is, however, obviously different. A comparison of f. 51 and 65 will show that the valves with the armature are in both species closely related; there are the same folds, ridges, and hooks in *depilis* which we have found in *euchenor*, but the organs have differently developed. The ventral (longitudinal) and apical (vertical) ridges of *euchenor* stand in *depilis* both so oblique that the angle formed by them in *euchenor* has almost disappeared. The ventri-apical hook does not lean over to the ventri-apical side of the valve, but to the dorsal side, so that the point of the hook will in *depilis* meet in copulation quite a different spot in the vaginal region of the *female* than in *euchenor*. The dentate vertical ridge is much shorter (f. 70, 71), sinuate in or near the middle, with the two higher parts at the side of the sinus dentate. The oblique fold (*d*) joins the ventral ridge near the apex (f. 66 to 69), not in the middle as in f. 52 to 58, and is basally not rounded but strongly compressed.

a. *P. depilis depilis*; f. 65 to 67, and 70.

The ventral ridge is throughout its length very high; its outline is variable. In f. 66 the upper edge of the ridge is undulate; the basal angle (*e*) is without the beak-like hook found in every specimen of *euchenor*. In a second individual (f. 67) the basal hook is indicated by a very minute tooth; in the middle the ridge is triangularly dilated. In a third specimen (not figured) the ridge is again without the basal hook, and is in the middle also higher than at the basal angle, but not so triangularly dilated as in the second example.

The ventri-dorsal dentate ridge (f. 70) leans strongly over to the apical side of the valve, and hence appears less high than in *euchenor*, but there are specimens of *euchenor* which in this respect are scarcely different from *depilis*. Besides the dorsal hook (*a*) there are three longer subdorsal and two smaller subventral teeth separated by a sinus; in a second specimen these teeth are all obsolete, while in a third only two are present.

b. *P. depilis novohibernicus*; f. 68, 69, and 71.

The harpe is very slightly different from that of *depilis depilis*; the portion of the ventral ridge which is produced into the ventri-apical hook forms a slight angle with the basal portion of the ridge, being a little more curved dorsally than in *depilis depilis*; the vertical dentate ridge is shorter than in that subspecies.

In three out of four examples the ventral ridge is highest beyond the middle, as in f. 68 and 69; while in the fourth specimen the dilatation takes place before the middle in a somewhat similar way as in f. 53. The basal hook (c) is in f. 68 scarcely indicated; in f. 68 it is as strong as in many *euchenor*. The dorsal ridge is generally shaped as in f. 71, but the teeth are sometimes much feebler than in the figure.

The facts of variation illustrated by f. 51 to 71 are as follows:—

(1) In the two closely allied species *P. euchenor* and *P. depilis* the armature of the valve is built up after exactly the same plan, but in the detail of structure there are conspicuous differences.

(2) The subspecies of *euchenor* from New Guinea and that from Woodlark present in the specimens examined no constant difference except in the oblique fold (d); the subspecies from Aru is so slightly different that the distinguishing character is scarcely noticeable if one does not compare several specimens; the Kei Island individuals are more obviously different than the Aru specimens (and represent probably another local form). The two subspecies of *depilis* are, according to the seven specimens examined, slightly different in the *male* genital armature.

(3) The individual variation within each subspecies is such that the differences between the harpes of several individuals from the same place (and hence most certainly belonging to the same species) are more obvious than those of the subspecies *inter se*.

(4) The only specimen known of the aberration *P. euchenor euchenor* ab. *eutropius*, which is abnormal in the pattern of the forewing, does not present any peculiarity in the genital armature.

9. Papilio cloanthus; f. 149 to 155.

The range of this insect is rather widely interrupted, the species being found all over North India, Upper Burma, and Central and Western China, and again in the mountainous regions of N.E. Sumatra; from the mountains of Malacca, Tenasserim, and Siam *P. cloanthus* is not known. The external features of the specimens from the various localities are such that we can group the individuals according to locality in three forms: an Indian, a Chinese, and a Sumatran form. The first two are not always distinguishable in pattern, and hence are certainly not specifically distinct from one another. The Sumatran specimens, at least all individuals of our long series (forty odd examples), are constantly different in the colour and extent of the markings; the divergency from Indian specimens is, however, not very conspicuous, which will be admitted if we call to mind that de Nicéville[*] especially says that the Sumatran insect is "identical" with the Indian one. As the minuteness of the distinguishing characters of the insect is, according to what has been said in the introductory notes, *a priori* no objection to the constantly found characters being of specific value, there must be other reasons brought forward which

---

[*] *Journ. As. Soc. Beng.* p. 526. n. 607 (1896).

force us to treat the Sumatran form not as a separate species. North Indian *cloanthus* occur in several broods; the individuals of the spring and those of the summer broods are in the extent of the markings rather obviously different, and this proves that the species is in pattern easily modified by the transmuting factors; the extreme individuals of the spring brood differ in the extent of the markings more from the extreme examples of the summer brood than certain Indian individuals do from certain Sumatran ones. Further, the differences of certain Chinese specimens from Indian individuals are quantitatively greater than those between Indian and Sumatran examples. Hence it is correct to accept Mr. Rothschild's opinion * and to treat the three insects in question as three subspecies of *P. cloanthus*.

The valve (f. 149) of *P. cloanthus* is, as in all the allied species, rather small; at the apex it is deeply sinuate. The sinus divides the apical third of the valve in a smaller dorsal lobe (b) and a larger ventral lobe (a). The ventral edge of the valve is angulate (f), and from this angle to the tip of the ventral lobe densely beset with irregular rows of thin and sharp teeth. The internal sheath of the valve is raised into a distinct fold (e), which begins ventrally at the base—in f. 149 and 150 the right-hand side is the ventral side of the valve—runs for a short distance along the ventral margin of the valve, turns in a rather even curve round towards the dorsal side, forms a subdorsal tooth (c), and then traverses longitudinally the dorsal lobe, being here raised into a short dentate ridge (d), the "dorsal ridge" which stands almost perpendicular upon the plane of the lobe leaning very feebly over ventrally. In f. 149 and 150 the subdorsal tooth (c) is visible almost in its entire length, because it is bent over apically, while the dorsal ridge appears much less high than it really is.

The variation of the species according to locality relates especially to the form of the valve, the length of the subdorsal tooth, and the form of the dorsal ridge.

a. *P. cloanthus cloanthus* from Kulu to the Shan States, at higher elevations; f. 149, 151, 152.

The sinus of the valve is $\frac{3}{4}$ mm. deep. The ventri-dorsal fold (e) is slightly curved. The subdorsal tooth (c) is high, simple, and when seen from the dorsal side, as in f. 151 and 152, reaches so far that its tip appears to be above the dorsal ridge. The length of the tooth is variable: the lower extreme met with by us is represented by f. 152 (Shan States); in a second individual from the Shan States the tooth is nearly as long as in f. 151, which is taken from a Sikkim specimen and represents the usual form of the tooth. As both the Chinese and Sumatran subspecies have the tooth generally considerably shorter, the mountainous regions of Siam, Tenasserim, and Malacca on the one side, and of Upper Tonkin on the other, will most probably yield (if inhabited by *cloanthus*) individuals more often, or even constantly, intermediate between the three subspecies in respect to the length of the tooth.

The dorsal ridge (f. 151, 152) of *cloanthus cloanthus* is a little longer than high, and is in all our specimens rather strongly denticulate.

The individuals of the spring brood are not different in the valve and harpe from the individuals of the later broods.

b. *P. cloanthus clymenus* from Western and Central China; f. 153.

In the outline of the valve all the individuals examined agree with specimens

* Nov. Zool. p. 445 (1895).

of *cloanthus cloanthus*, and so they do in the form of the ventri-dorsal fold. The subdorsal tooth is in every specimen shorter than that of f. 151 (normal length of tooth of *cloanthus cloanthus*), but there occur individuals which have the tooth nearly as long as it is in f. 152. A specimen of *cloanthus clymenus* with the tooth of a length which is normal for *clymenus* is represented by f. 153. We have examined some individuals which externally are not different from Sikkim specimens, and observed the highly interesting fact that in these individuals the tooth is as short as in f. 153, while in some other examples which differ in the extent of the black colour on the wings considerably from *cloanthus cloanthus* the tooth approaches in length that of f. 152. This is a remarkable illustration of what we have said in the introduction, namely that specimens can be similar or identical in one set of characters, while in another, independent, set they are dissimilar; the Chinese individuals of *cloanthus* which in pattern are like Indian *cloanthus cloanthus* are nevertheless individuals of *cloanthus clymenus*, distinguished from *cloanthus cloanthus* by a character of the genital armature not found, to our knowledge, in any individuals from India. We shall have to refer again to this fact later on.

The dorsal ridge of the Chinese *cloanthus* is similar to that of the Indian specimens; we have not found any difference that can be pronounced constant; in many individuals the ridge is a very little longer, and the teeth are often more numerous and smaller.

c. *P. cloanthus sumatranus* from the mountainous districts of N.E. Sumatra; f. 150, 154, 155.

The sinus of the valve is much smaller than in the two preceding subspecies; the dorsal edge of the dorsal lobe is more rounded, and the ventral lobe is considerably blunter. The dentition of the ventral edge of the valve is, especially near the blunt angle (*f*), extended upon the outside of the valve to such a degree that five or six small, but strongly chitinised, teeth stand irregularly one above the other.

The ventri-dorsal fold (*e*) is straighter and more raised than in *cloanthus cloanthus* and *cloanthus clymenus*, its edge is less rounded off and, especially near the subdorsal tooth, slightly notched or faintly denticulate.

The subdorsal tooth is as short as in *cloanthus clymenus*, but at the base broader, in consequence of the fold, of which it is a process, being higher; mostly it bears two or three faint teeth at the ventral edge.

The dorsal ridge is obviously shorter than in the rest of the species, but is of the same height; the free edge is less dentate, often simply sinuate. F. 154 and 155 illustrate the degree of variation in the dorsal ridge and the subdorsal tooth noticed by us.

The principal facts of variation as illustrated by f. 149 to 155 are as follows:—

(1) The three subspecies of *P. cloanthus* agree in the valve and its armature, but exhibit some differences in the detail of structure.

(2) The Indian and the Chinese forms differ constantly in the length of the subdorsal tooth, though the extremes come very close.

(3) The Sumatran form is aberrant in the form of the valve, the ventri-dorsal ridge, the subdorsal tooth, and the dorsal ridge.

(4) In the length of the subdorsal tooth the Sumatran and Chinese forms agree with one another, while they disagree with the Indian subspecies which inhabits interjacent countries.

(5) The individual variation of the genital armature in the Chinese subspecies is entirely independent of external characters.

(6) The seasonal dimorphism in external characters obviously marked in the Indian *P. cloanthus cloanthus* does not affect the genital armature in any way, as far as we could ascertain.

10. Papilio sarpedon; f. 96 to 148.

Though some people have treated *P. sarpedon* and *P. cloanthus* as belonging to two different genera, the insects are, nevertheless, very closely allied to one another, more closely than to any other species. This does not only follow from a careful comparison of the wing-pattern of the two Papilios, but also from the structure.

The range of *P. sarpedon* comprises the whole of the Indo-Australian Region, including Japan (except the north of it). The number of subspecies into which the insect has developed is very great. As it is one of the commonest species we could examine a large number of specimens, and to this it is due that we here came across an individual which stands in the genital armature far outside the usual limits of variation of the subspecies to which it belongs.

If we select some of the extreme forms here treated as subspecies of one species, for example the Indian, Celebensian, and the Solomon Island forms, their external differences are so very conspicuous that one might easily be misled to consider these forms specifically distinct. A comparison of the representatives from all the various localities, however, convinces us that the differences in colour, pattern, shape, and size between every two nearest allied forms are very slight, and do in some forms not even apply to every specimen; and we observe further that, where the differences are constantly met with, the characters amount quantitatively not to more than the differences between the seasonal forms of Japanese *sarpedon*, or than the differences found between certain Indian examples. Hence we think it to be quite correct to accept Mr. Rothschild's statement that all the forms dealt with in the following lines are subspecies of one species.

As we now are acquainted with the more simple armature of the valve of *P. cloanthus*, that of *P. sarpedon* will be more easily understood. A comparison of f. 96 and 149 will at a glance show the great similarity in the apparatus of the two species. The valve, though differing in outline from that of *P. cloanthus*, has the same apical sinus, and the armature has nearly the same position.

The ventral lobe of the valve (f. 96, *a*) is longer, mostly broader, than the dorsal one (*b*); its ventral edge is denticulate, as in *cloanthus*, but there is only one row of teeth, and the toothed portion extends farther down towards the base. There is a good deal of variation in the shape of the lobes and the depth of the sinus, both in respect to individuals and to geographical races.

The fold (*e*) formed by the inner sheath of the valve begins ventrally at the base of the valve, as in *cloanthus*, runs in an oblique direction to somewhat beyond half-way to the apex, turns round here towards the dorsal side, forms, when having arrived at a level with the ventral margin of the dorsal lobe, a subdorsal tooth (*c*), and then is continued in a longitudinal direction to form a dorsal ridge (*d*). The homology of the organs is obvious. The fold *e* is homologous to the fold *e* in *cloanthus*, but is here less raised, takes a somewhat different course, and is at the point where it curves round towards the dorsal side often feebly toothed. The subdorsal tooth corresponds to that of *cloanthus*, but is here much less chitinised, usually broadened at the tip, and is in fact a dilatation of the fold *e* partly rolled in so as to form a half-cylinder; though this tooth is variable, we shall not refer to it under each subspecies. The dorsal ridge corresponds to the dorsal ridge of *cloanthus*, but is here of a more complicated structure. In a view from the dorsal side (f. 97) it will be seen that the dorsal ridge of f. 96 consists of a dentate high basal portion and a free rod-like apical process (*g*), the *dorso-apical process*, which is curved upwards—that means towards median plane of the abdomen—is denticulate at the tip, and protrudes beyond the tip of the dorsal lobe. At the ventral side of the dorsal ridge a longitudinal fold (f. 96, *i*, and 98, *i*) will be noticed which is continued to the apex of the dorsal lobe; that vertical portion of the fold (f. 98, *h*) which runs up to the upper edge of the ridge is of high importance, as it develops in most subspecies to a peculiar organ.

a. *P. sarpedon sarpedon*; f. 96 to 114.

This form occurs all over India (except S. India and Ceylon) to Java, the Philippine Islands, and Japan; in China it is replaced by another subspecies. There are some external characters by which the individuals from Java, Borneo, and the Philippines can be distinguished from the individuals from N.W. India and N. India, but these characters are very slight; besides, the Malayan individuals lead over to the more different forms from the lesser Sunda Islands, and hence remain best included in *sarpedon sarpedon*. The genital armature is entirely in accordance with this statement.

We have examined specimens from N.W. India, Sikkim, Assam, Burma, Shan States, Tenasserim, Cochin China, Sumatra, Nias, Java, Natuna Islands, Borneo, Palawan, Mindoro, Luzon, the Riu Kiu Islands (= Loo Choo Islands), and Japan. The specimens from all these localities agree so well with one another, apart from individual peculiarities, that we did not succeed in finding in the genital armature any character by which the specimens from one or the other place could be recognised.

The outline of the valve normally met with in *sarpedon sarpedon* is represented by f. 96, which is taken from a Kumaon individual. The sinus is about ½ mm. deep; the dorsal lobe is rounded at the apex; the ventral one is also rounded, its ventral edge feebly incurved. The dorsal ridge is in the dentition very variable; the important feature is that normally the lateral fold *h* of f. 98 is very slight and incomplete, as in f. 99, or even absent.

The variability in the form of the valve is illustrated by f. 96, 97, 113, and 114. In f. 97, taken from an example from the Shan States, the sinus is narrow, and the apex of the ventral lobe also very narrow. F. 113 and 114 represent two other Shan States specimens: one has the ventral lobe much produced, and the sinus accordingly deep; in the other the same lobe is very short and broadly rounded. In the individuals from the Malayan region the ventral lobe is often, but by no means regularly, more produced and slenderer than in the average Indian specimen; and this is not surprising, as in the various subspecies from the lesser Sunda Islands the lobes are constantly long and slender.

The variation of the dentition of the dorsal ridge is very great, as will be seen from f. 98 to 100 and 104 to 110. Sikkim specimens are represented by f. 99, 100, and 108; individuals from the Shan States by f. 105 and 106; a Malacca specimen by f. 107; two Bornean examples (from Mount Mulu) by f. 109 and 110.

Every individual specimen examined exhibited some peculiarity in the number, size, and form of the teeth of the dorsal ridge.

The lateral (ventral) fold (h) of the dorsal ridge (f. 98) is in f. 107 (Taiping, Malay Peninsula) absent, in f. 99 (Sikkim) very short, in f. 106 (Shan States) complete but very slightly raised, and is more distinct in f. 98 (Kumaon). In f. 100 (Sikkim) the fold runs up to a small tooth, which has a somewhat transverse position to the longitudinal ridge; in f. 108 the fold is raised to a low but obvious dentate ridge, which we shall call "*transverse ridge.*"

In a Mindoro specimen caught by Mr. A. Everett in December 1894 the additional transverse ridge is raised above the level of the dorsal ridge, and forms a conspicuous, broad, tooth-like prominence; in f. 111 the dorsal ridge with the transverse prominence is represented in a view vertical upon the plane of the valve; and in f. 112 we give a view of the same organ from the apical side (with the eye a little above the valve).

Still more aberrant is a specimen from the Shan States caught by Mr. Roberts in the same district where the individual was obtained from which f. 106 is taken. The increase in the size of the transverse ridge, as illustrated by f. 107, 99, 106, 98, 100, 108, 111, and 112, reaches in f. 101 to 103 the maximum. The transverse ridge h is higher, strongly dentate, and more extended than the dentate portion of the dorsal ridge d; in f. 101 the organ is seen from above, in f. 102 from the ventral side, and in f. 103 from the apical side (compare f. 111 and 112).

We have examined more than a hundred specimens of *P. sarpedon sarpedon*, and found only one that has the additional transverse ridge so extraordinarily developed; the significance of this variation is obvious if we compare the special structure of the dorsal ridge of *P. sarpedon anthedon, milon, choredon, teredon*, etc. We shall have to refer to this particular case again.

b. *P. sarpedon semifasciatus* from China.

The Chinese subspecies of *sarpedon* is in the markings not always distinguishable from *sarpedon sarpedon*, but the greater number of the individuals from Central and Western China have a very remarkable character in the band of the hindwing being more or less obliterated. Such specimens with almost entirely black hindwings are, in respect to pattern, quantitatively more different from *sarpedon sarpedon* than the individuals of any other subspecies are. Though we dissected a long series of Chinese individuals, we did not perceive any character in the genital armature by which they could be differentiated from *sarpedon sarpedon*; and this concerns the examples which are most aberrant in pattern, as well as specimens with the ordinary *sarpedon sarpedon* pattern. The fact is of high interest, as it distinctly shows that a great external discrepancy of a localised form does not necessarily imply that there is also a peculiarity in the genital armature of the form.

c. *P. sarpedon adonarensis* from Adonara and Sambawa; f. 127.

The interesting external features of this form are pointed out by Mr. Rothschild on p. 324 of this volume. In pattern it comes near the subspecies from the neighbouring islands of Sumba, Timor, and Wetter; and we were rather surprised when we found that in the armature of the valve the specimens from Sambawa and Adonara disagree with those subspecies, and agree much better with the Indo-Malayan *sarpedon sarpedon*, as the dorsal ridge is dentate and has the transverse ridge as feebly developed as it normally is in Indian examples. F. 127 represents the dorsal ridge of an Adonara individual from the ventral side; it will be noticed that the dentate part of the ridge is markedly less extended than in *sarpedon sarpedon*, and that the transverse fold *h* has not developed to a ridge, as in f. 123 (Sumba), or f. 128 (Wetter) and f. 129 (Timor). The valve of *adonarensis* stands in shape intermediate between *sarpedon sarpedon* and *sarpedon jugans*, the lobes being slenderer than in the former, and shorter and broader than in the latter.

The individuals from the island of Lombok approach in external characters and in the genital armature still more the Indo-Malayan form.

d. *P. sarpedon teredon* from South India and Ceylon; f. 115 to 120, 133, 134.

The range of this subspecies is separated from that of *sarpedon sarpedon* by a wide area where most probably the species does not occur. We have examined above thirty specimens of *P. sarpedon teredon* from South India and Ceylon, all of which are in external features and in the genital armature well distinguishable from all other forms of *sarpedon*.

The valve is much narrower and the sinus considerably deeper than in *sarpedon sarpedon*. The ventral lobe (f. 115, 133, 134) is usually slender in its apical half, and curved towards the median axis of the abdomen, thus forming almost a hook-like organ; the dorsal lobe is sometimes strongly pointed (f. 133). The inconstancy of the outline of the lobes is illustrated by f. 133 and 134, which are taken from South Indian individuals.

The ventral (longitudinal) portion of the valval fold extends farther towards the apex of the valve than in the preceding forms, and hence the ventri-dorsal portion has a more oblique direction. The dorsal ridge is much less raised than in *sarpedon sarpedon*; the denticulation is absent; only the middle portion of the ridge, which corresponds to that part of the ridge of *sarpedon sarpedon* in f. 98 and 102 which bears the transverse fold or transverse ridge respectively, is elevated; it has, combined with the transverse fold (*h* of f. 98), developed in a vertical and transverse direction to a strong tooth-like transverse ridge, which slightly leans over basally and dorsally. As in f. 115, taken from an example from Trichopolis, the transverse ridge is too inconspicuous, we give an enlarged figure of the dorsal ridge of the same specimen in the same position (f. 116), and also a figure of the organ from the ventral side (f. 117). In f. 118 to 120 the transverse ridge alone is represented from the apical side; mostly the ridge is shaped as in f. 119; its edge is usually not dentate, but there occur specimens, like that from which f. 118 (South India) is taken, which have the transverse ridge dentate; in one of the Ceylonese examples (f. 120) the ridge is considerably smaller than in f. 119.

The individuals which belong to the ab. *thermodusa* have no character in the valve and its armature that is peculiar to them.

In external features and in the genital armature *teredon* comes much closer to the forms from the lesser Sunda Islands than to *sarpedon sarpedon*, which inhabits the interjacent countries.

e. *P. sarpedon jugans* from Sumba; f. 121 to 126.

"This form combines to a certain extent the characters of *sarpedon sarpedon* and *sarpedon timorensis*," and its "genital armature resembles more that of *timorensis* than that of *sarpedon* and *adonarensis*" (Rothschild, this volume, p. 324).

The sinus of the valve (f. 121) is deep. The ventral lobe is very slender, often of almost rod-like appearance, sometimes even narrower than that of *sarpedon teredon*. The apex of the dorsal lobe is mostly rounded. The dorsal ridge is denticulate in the four specimens in the Tring Museum, but a comparison of f. 123 with f. 108 shows that the basal half of the ridge is very feebly raised; as in *teredon* (f. 117), the elevation is restricted to the median part of the ridge and the transverse ridge. F. 121, 122, 123, 125 are taken from the same individual; 122 and 123 give a view from the apical and ventral sides of the valve respectively, while f. 125 represents the median part of ridge alone in a view from above (as in f. 121). The dorsal ridge of another individual is represented from the apical side in f. 124; the transverse ridge is here much less developed than in the other specimen, in fact not more than in the Mindoro example represented by f. 112; the ridge d of the latter is, however, scarcely indicated in f. 124; the transverse ridge of f. 124 is enlarged in f. 126, which gives it in a view from above. The differences between f. 122 and 124, and 125 and 126, are very conspicuous.

f. *P. sarpedon timorensis* from Timor and Wetter; f. 128, 129.

We have only two specimens of this interesting form, one from Dili, Portuguese Timor, and the other from Wetter; the two individuals disagree somewhat with one another in external characters as well as in the form of the transverse ridge, but the differences are such that they may very well be individual and not subspecific.

The valve agrees with that of *jugans*, but the ventral lobe is less slender. The dorsal ridge, as in *teredon* from Ceylon and South India, is not denticulate; the transverse ridge is tooth-like, nearly as in *teredon*; f. 128 (Wetter) and 129 (Timor) represent the transverse ridge in a view from above; the figures may be compared with f. 125 and 126 (Sumba), 132 (Queensland), 138 (New Britain), 143 (Celebes), and 148 (Guadalcanar, Solomon Islands).

g. *P. sarpedon choredon* from Australia and New Guinea (including the islands near its coast); f. 130 to 132, 135 to 137.

The specimens from Waigeu and the northern parts of New Guinea are in external characters sometimes slightly different from ordinary individuals from Australia, and lead over to the next subspecies, which inhabits the Bismarck Archipelago.

The sinus of the valve is very deep (f. 135, Queensland). The ventral lobe is very prominent, its upper edge straight, its ventral (denticulate) edge evenly rounded; in breadth the ventral lobe is intermediate between *timorensis* and *jugans* on the one side, and *sarpedon sarpedon* on the other; in some examples the lobe is a third broader than in others. The dorsal lobe is in all examples we have seen rounded at the apex. The fold e of f. 135 is more curved than in *sarpedon sarpedon*, and takes about the same course as in *sarpedon teredon*.

The dorsal ridge divides basally in a dorsal (r) and a ventral (s) portion which correspond to the two slight folds marked r and s in f. 96. Now, in very many specimens both from Australia and New Guinea, only the ventral branch of the ridge participates in the formation of a high transverse ridge, as in f. 135 and in f. 136; the latter figure is taken from an individual from Redscar Bay, British New Guinea, and is a highly enlarged view of the transverse ridge and the adjoining parts of the dorsal ridge. In this case the dorsal ridge is rounded off and has no teeth, except at the tip of the apical rod-like processus, which is longer than in all the preceding subspecies. F. 137 represents the dorsal ridge with the transverse ridge of a Redscar Bay individual in an apical view. Another extreme in the structure of the dorsal ridge is represented by f. 130 to 132, which are taken from an example from Cairns, North Queensland. F. 130 gives a view from the apical side; the dorsal ridge d is provided with teeth, and the transverse ridge h is joined to the dorsal ridge, and is not a separate structure as in f. 136; f. 131 gives the ridges in a ventral view, while f. 132 is taken from above. The intergradations between the extremes figured here are equally abundant in Australia and New Guinea.

The individual variation of *choredon* in the direction from f. 130 to 137, from the transverse ridge forming one piece with the dorsal ridge (as in *sarpedon sarpedon, teredon, jugans*), to the other extreme where the transverse ridge stands isolated, will serve to comprehend the still more exaggerated development of the transverse ridge in some of the following subspecies.

h. *P. sarpedon imparilis* from New Britain ; f. 138.

The external differences between this subspecies and the preceding one, though slight, are prominent enough to enable us to distinguish all our New Britain individuals from *choredon*. In the genital armature *imparilis* comes again very close to *choredon*; the lobes of the valve are, however, more pointed (f. 138); the transverse ridge is smaller, stands less oblique, and leans over towards the base of the valve; the dentition of the transverse ridge is as variable as in *choredon*.

The specimens with additional spots on the forewing have no character in the valve and the armature peculiar to them.

i. *P. sarpedon impar* from New Georgia, Solomon Islands.

The *male* of this insect is unknown; the *female* is in pattern midway between the preceding and the following form.

k. *P. sarpedon isander* from Guadalcanar and Bougainville, Solomon Islands; f. 148.

Though this form is so very aberrant in markings that it has been described by Godman & Salvin as a distinct species, and has also been kept separate from *sarpedon* by Mr. Rothschild in his Revision (who, however, informs me now that, in consequence of the receipt of more specimens, he must sink it to the rank of a subspecies of *sarpedon*), the valve and its armature are only slightly different from that of *imparilis*. The ventral lobe of the valve is broader, blunter, and shorter. The transverse ridge is not separated from the dorsal ridge; it is high, mostly simple, and seldom notched or faintly dentate. F. 148 is taken from an individual from Guadalcanar.

l. *P. sarpedon dodingensis* from the Northern Moluccas (Halmaheira and Batjan); f. 139 and 140.

The sinus of the valve (f. 139) is very deep and especially narrow, being about twice as deep as broad. The ventral lobe of the valve is shaped nearly as in *imparilis*. The dorsal ridge is bifurcate as in *choredon*, but the bifurcation takes place not far from the apex of the valve; the ventral part of the ridge is raised and forms the usual transverse ridge. From f. 140, which is taken from the same specimen as f. 139 and gives a view of the dorsal ridge from the apical side, it will be observed that the dentate ridge differs essentially from the transverse ridge of f. 103, 130, 137, 142, 144, in so far as it is not homologous to the fold *h* of f. 98, but to the middle portion of the dentate dorsal ridge to which that fold is joined. In a Batjan example the ridge, which in f. 140 is tridentate, is much reduced. The apical process of the dorsal ridge is long.

m. *P. sarpedon anthedon* from the Southern Moluccas (Amboina, Ceram); f. 141 and 142.

The sinus of the valve is twice as broad as in *sarpedon dodingensis*, but not so deep as in that form. The ventral lobe of the valve varies a good deal individually, and has generally the outline of that of *choredon* or *dodingensis*; the dorsal lobe is broader than the ventral one, and rounded at the apex. The dorsal ridge has the apical (rod-like) process prolonged, as is the case in *dodingensis*. The transverse ridge is always strongly developed, and nearly always so obviously detached as in f. 142, which is taken from a Ceram individual of which f. 141 represents the valve and harpe from above.

In external features *anthedon* is much more closely allied with *dodingensis* than with *choredon*; in the valve and its armature the difference between the former is, on the contrary, much greater than that between *anthedon* and *choredon*. *Anthedon* is distinguished from *choredon* only in the free apical process of the dorsal ridge being longer, and in the transverse ridge leaning more evidently over basally, as will be seen from comparing f. 135 and 141, and 137 and 142; while it differs from *dodingensis* in having a broader dorsal lobe, a much wider sinus, and a differently shaped and differently situate transverse ridge. The relationship indicated by pattern is therefore not the same as that which we must deduce from the structure of the genital armature.

n. *P. sarpedon monticolus* from Bonthain Peak, S. Celebes; f. 146 and 147.

The discovery of a mountain form of *sarpedon* in Celebes which is in general appearance very different from the form inhabiting the lower districts of the island is highly interesting. The subspecies resembles on superficial examination Indo-Malayan individuals of *sarpedon sarpedon*; the actual affinities of the subspecies are, in respect to the external characters, as follows: In the green colour of the markings *monticolus*—it ought to be *monticola!*—agrees with *sarpedon sarpedon*, *sarpedon jugans*, *sarpedon adonarensis*, and *sarpedon timorensis*, and differs conspicuously from its compatriot *milon* and the Moluccan races. In the shape of the wings it resembles *sarpedon sarpedon* from the Sunda Islands. In the shape of the median band of the forewing it comes nearest to *sarpedon dodingensis* from the Northern Moluccas, except in the third spot being larger than the fourth, in which character it agrees with *timorensis*. The underside of the hindwing is in *monticolus*, *milon* (Celebes), and *dodingensis* (Northern Moluccas), in opposition to all other subspecies, provided with a red mark before the median cell between veins 6 and 7. And, lastly, the genital armature of *monticolus* stands intermediate in structure between that of *milon* and that of *dodingensis*, and hence is obviously different from that of *sarpedon sarpedon*.

We have to lay great stress upon the mixture of characters found in *P. sarpedon monticolus*. As the relationship to *milon* from Celebes, *sarpedon* from the Sunda Islands, and *dodingensis* from the Northern Moluccas is equally great in the characters of the wing and valve, we have no reason whatever to say that *monticolus* is derived from *sarpedon sarpedon*, or from *sarpedon dodingensis*, and not from *sarpedon milon*; and hence a conclusion, based upon the occurrence of a special form of *P. sarpedon* on Bonthain Peak, as to the probable geological history of Celebes—connection with the Sunda Islands or with the Northern Moluccas—would lack the necessary facts. We shall have to refer again to *monticolus* in the last chapter of this paper.

The valve is nearly shaped as in *sarpedon milon*, i.e. the ventral lobe is much longer and narrower than in *sarpedon sarpedon*. The dorsal ridge (f. 146, from apical side) bears a transverse dentate ridge, similar in position to that of *dodingensis* (f. 140), but differently shaped; the dentate ridge is much less extended than in ordinary examples of *milon* (f. 144). F. 146 gives a dorsal view of the ridge.

o. *P. sarpedon milon* from Celebes, the Sulla Islands, and the island of Talaut; f. 143, 144, and 145.

This long-winged and narrow-banded subspecies has the blue colour of the wing-band in common with the Moluccan races; the nearest ally in respect to the pattern of the wing is *dodingensis*, which has, like *milon* and *monticolus*, an additional red spot before the cell on the underside of the hindwing, and has the median band of both wings also obviously narrower than it is in the subspecies from Amboina and Ceram.

The sinus of the valve of *milon* (f. 143) is deep and broad; the ventral lobe of the valve is very slender and long, resembling somewhat that of *jugans* from Sumba (f. 121) and *teredon* from Ceylon and S. India (f. 133). The apical processus of the dorsal ridge is nearly as long as in the two Moluccan races. The transverse ridge has developed to a broad saw-like organ, which when seen from above sometimes almost extends to the sinus of the valve. In the ventral view of the dorsal ridge (f. 145, Celebes specimen) the denticulate portion of the dorsal ridge itself (*d*) is plainly visible; this portion of the ridge is not developed in the Moluccan races, but in *monticolus* from Bonthain Peak (f. 147). The transverse ridge (*h*) is in this specimen almost perpendicular. A view of the same organ from the apical side of the valve is given in f. 144; the ridge is larger than in any other subspecies, but there is some individual variation in the size of the ridge.

We have only one ♂ from the island of Mangiola (Sulla Islands), which, like the *female*, differs from the Celebes individuals in having a narrower median band to the wings. The valve and its armature is in this ♂ very slightly different from that of our Celebes examples; but, of course, we cannot tell from one individual whether the difference is due to individual or to local variation.

The more prominent facts of variation illustrated by f. 96 to 148 are as follows:—

(1) The genital armature varies according to locality; the Chinese subspecies agrees in the apparatus, however, with the Indo-Malayan one, in spite of very prominent external differences.

(2) The distinguishing characters in the genital armature of the subspecies are found in the valve as well as in the harpe.

(3) The individual variation within the Indo-Malayan subspecies, of which a large material has been examined, is so great that the difference between every two nearest allied subspecies is small compared with the difference exhibited by the extreme individuals of the Indo-Malayan subspecies.

(4) The Ceylonese and South Indian subspecies is in the genital armature and in the shape of the wing nearer related* to the forms inhabiting Sumba, Wetter, and Timor, than to the Indian form.

(5) The subspecies which are in colour and pattern nearly allied* (*sarpedon sarpedon* from India, *sarpedon choredon* from Australia, *sarpedon monticolus* from Celebes) need not be similar in respect to the genital armature.

(6) The spring and summer broods in North India and Japan are not different in the genital armature.

11. Papilio bathycles; f. 46 to 50.

We have selected this species out of a number of "green" Eastern Papilios for two reasons: first, because the harpe of the species has long spines and processes the variation of which is more easily demonstrated; secondly, because the three subspecies of the species (Java; Malacca, Sumatra, Borneo; North India and Burma), though in external characters not always distinguishable (at least as regards those two of them of which we have a long series), are in the harpes, according to our material, always different. We believe that the great gap between the harpe of the North Indian form and that of the Malayan form, and the smaller gap between the Malayan and the Java subspecies, are due to the circumstance that we have not examined specimens from the interjacent countries, Tenasserim and South-West Sumatra.

As the valve of *P. bathycles* does not vary to any extent, we figure only the harpe. The harpe consists of a fold extending ventrally from the base of the valve to near the apex, and from here to near the dorsal edge of the valve; the fold is apically produced into a ventral spine (a), a ventral ridge (b), and into a dorsal ridge (c), all of which extend in an apical direction; the fold between the dorsal ridge and the ventral ridge is very slightly raised.

a. *P. bathycles bathycles* from Java; f. 46.

We have examined but four individuals, the harpe of one of which is represented by f. 46. In all four specimens the ventral spine (a) is simple, without denticulation; the ventral ridge is dentate, either of the form as represented in f. 46, with a small dentate dilatation at its dorsal edge, or the place of this dilatation is occupied by some small teeth; the dorsal ridge is broad, denticulate, and has generally one larger tooth, as in f. 46; the surface of the ridge is basally limited by a faint fold.

b. *P. bathycles bathycloides* from Sumatra, Malacca, and Borneo; f. 47.

F. 47 is taken from a Bornean example. The ventral spine is either simple, or, as in figure, denticulate at the tip; the ventral ridge is longer and basally much narrower than in the Java form, being there without dilatation or denticulation; the dorsal ridge is nearly shaped as in *bathycles bathycles*, but the fold e is more obviously marked and has a different direction.

c. *P. bathycles chiron* from Sikkim, Assam, Burma; f. 48, 49, 50.

The material of this form which stood at our disposal was much larger than that of the two preceding subspecies, and consequently the amount of variability

---

* The words "allied" and "related" mean here simply "similarity," not phylogenetic relationship.

observed in *chiron* greater than the amount observed in *bathycloides* and *bathycles*. As in wing-pattern *bathycloides* and *chiron* are not always distinguishable with certainty, while *bathycles* is more easily recognisable, the dissimilarity in the harpes of *bathycloides* and *chiron* is much greater than the difference between the harpes of *bathycloides* and *bathycles*. Comparing f. 46 (*bathycles*) and 47 (*bathycloides*) with f. 48 to 50, which represent *chiron*, we observe as main character of the harpe of *chiron* that the dorsal ridge (*c*) is reduced to a small triangular tooth. Besides, the dorsal tooth *c* stands nearer the ventral ridge owing to an increase in breadth of the ventral portion of the harpe; the ventral spine has a somewhat different direction, and the ventral ridge is differently shaped.

The division of the ventral ridge of *bathycles* into two lobes (f. 46) is in f. 48, taken from a Shillong (Assam) example, more obvious; in f. 49 (Sikkim) the bifurcation is complete and the denticulation is much reduced; in an individual from the Shan States (f. 50) the ridge is divided into three teeth.

The ventral spine is either simple (f. 49), or denticulate at the tip.

The dorsal, triangular, ridge varies a good deal in size, but to our knowledge this variation is independent of locality, as is also the variation of the ventral ridge and ventral spine of *chiron*.

There often occur specimens of *chiron* in which the ochreous costal mark on the underside of the hindwing is wanting (ab. *chironides*), a character by which also *bathycloides* is distinguished from *bathycles*; in the genital armature of ab. *chironides* there is no peculiarity, and we also failed to find any distinguishing character in the genital armature of the spring specimens from Sikkim (broad-banded) as compared with the summer specimens (narrow-banded).

The facts of variation illustrated by f. 46 to 50 are as follows:—

(1) The three subspecies of *P. bathycles*, the Javan (*bathycles*), the Malayan (*bathycloides*), and the Indian (*chiron*), have the harpe built up after the same plan, but the detail of the structure furnishes in the specimens examined obvious distinguishing characters.

(2) The Malayan and the Indian forms are nearest related in pattern, while the Malayan and the Javan forms are nearest related in the structure of the harpe.

(3) The individuals of the spring and summer broods of *chiron*, and the specimens of *chiron* ab. *chiron* and *chiron* ab. *chironides*, are in the genital armature the same.

12. **Papilio aristeus**; f. 72 to 83.

This species ranges from Sikkim all over the Indo-Australian Archipelago, and is replaced on Celebes and some islands south of it by a close ally (*P. rhesus*), and on the islands of the Bismarck Archipelago by another close ally (*P. paron*). Mr. Rothschild, Nov. Zool. 1895. p. 418, distinguishes four subspecies, which, arranged according to the geographical position of the district which each of them inhabits, are as follows:—

(1) *P. aristeus anticrates* in Northern India (Sikkim and Assam).

(2) *P. aristeus hermocrates* from Upper Burma to Timor and the Philippine Islands.

(3) *P. aristeus aristeus* on the Moluccas.

(4) *P. aristeus parmatus* in Queensland, on New Guinea and Waigeu.

That these four forms are really one species there can scarcely be any doubt, for the distinguishing characters are by no means perfectly constant.

P. *aristeus, rhesus, paron, antiphates*, etc., and all the green Eastern Papilios have in neuration a prominent character in common : in all of them the first subcostal branch is *invariably* anastomosed to the costal nervure. The relationship indicated by this phenomenon is doubtless blood-relationship, not simply form-relationship, as the structure of the larvae, pupae, and the morphology of the imagines point in the same direction. The form of the ninth abdominal segment in the *male*, and especially that of the eighth in the *female*, is in several groups of Indian Papilios with that peculiarity in the neuration morphologically similar, while those segments are in *P. podalirius, ajax*, and allies, which have a superficial resemblance to *aristeus, antiphates*, etc., of quite a different form (compare f. 181, *podalirius* ♀, and 182, *sarpedon* ♀).

The peculiar form of the valve of *P. aristeus* will easily be understood if one compares with f. 72 (*aristeus anticrates*) the valve of *P. sarpedon* (f. 96).

The apical sinus which in *sarpedon* occupies the middle of the apex is in *aristeus* more dorsal and is very narrow; in a view perpendicular on the plane of the valve, as in f. 72, the sinus is concealed by a brush-like organ (*b*); but in f. 75, which represents the brush-like organ from the ventral side (*minus* the bristles), the narrow sinus between *a* and *b* is visible.

The organ *b* is homologous to the dorsal lobe of the valve of *sarpedon*, and therefore will here be called so. The ventral lobe of the valve (*a*) is very broad and rounded, or somewhat triangular. The position of the two lobes against one another can easily be imitated with the thumb (representing the dorsal lobe) and the four other fingers (representing the ventral lobe) by moving the thumb inwards.

The middle portion of the inner edge of the thick and raised ventral margin of the valve is furnished with short bristles.

The armature of the valve is also homologous to that of *sarpedon*. The usual fold of the inner sheath begins ventrally at the base of the valve, is soon curved a little dorsally, and then suddenly raised into a strongly chitinised, denticulate, ridge (*e*); from here the fold turns dorsally and becomes soon dilated into a long process (*e*), the subapical tooth of *sarpedon*, which we shall call *subapical process*; this process leans strongly over dorsally and apically, is curved, and penetrates between the dorsal and ventral lobe of the valve, so that its tip is visible when the valve is viewed from the dorsal side. The position of the subdorsal process will become clear from f. 75 and 82.

The variation according to individuals and according to locality affects the dorsal lobe (*b*), the ventral denticulate ridge (*e*), and the subdorsal process (*c*).

a. *P. aristeus anticrates* from Sikkim and Assam; f. 72, 73, 75.

The dorsal lobe of the valve is always very broad; its free apical portion is ovate. The outline is not always the same: sometimes the lobe is a little slenderer than in f. 75, which represents the average form and is taken from a Sikkim example; sometimes it is a little shorter.

The ventral ridge appears in the view from above (f. 72, Sikkim individual) almost straight; the ridge is concave dorsally, its middle portion being more ventral than the basal and apical edge. When seen from the dorsal side (f. 73) the denticulation of the ridge is more prominent; in all specimens examined the longest, most basal, tooth (right-hand side in figure) is nearly horizontal. The subdorsal process is simple, without denticulation.

There is no difference in the genital armature between the darker and the less dark specimens from Sikkim and Assam.

b. *P. aristeus hermocrates* from Burma to Timor, Borneo, Palawan, and the Philippine Islands; f. 74, 76 to 80.

We have examined individuals from the Shan States, Borneo, Palawan, the Philippine Islands, Sumba, Kalao, Wetter, and Timor. The individual variation in the pattern of the wings and in the size of the specimens is considerable, but we cannot find distinguishing characters which would necessitate a division of *hermocrates* into more subspecies; the Wetter and Timor individuals, from which islands we have only three altogether, are faintly different in the shape of the forewing, and may perhaps, on receipt of more material, be separable from *hermocrates*. An examination of the genital armature likewise did not furnish us with characters by which specimens from the various localities could be recognised, except the Shan States individuals, which are more similar in part of the genital armature to the North Indian *anticrates*.

The chief distinction in the genital armature of *hermocrates* concerns the form of the dorsal lobe, which normally is shaped as in f. 78, and the form of the ventral ridge, which has the basal tooth (f. 74) less horizontal than it is in *anticrates*.

We have two individuals from Muong Gnow, Shan States, one darker than many Philippine Islands specimens, one intermediate in the development of the bands between *anticrates* and ordinary Philippine *hermocrates*. The dorsal lobe of the dark specimen (f. 76) is scarcely different from that of *anticrates*, being much broader than it normally is in *hermocrates* (compare f. 76 to 80); in the whiter individual, which in the wing-pattern is scarcely different from dark Sikkim examples of *anticrates*, the dorsal lobe is considerably narrower (f. 77).

The series of f. 76 to 80 represents the variation of the dorsal lobe. F. 78 is taken from a specimen from Kudat, North Borneo; the lobe is still narrower than in f. 77; the next two figures (79 and 80) represent the lobe of two Palawan individuals. The specimens from Kalao, Timor, Wetter, and Sumba have the lobes as in f. 78 to 80.

The *gradual* decrease in the size of the dorsal lobe from f. 75 (*anticrates*) to f. 76 (*hermocrates*), and from there to f. 80, is evident; and it is further obvious that the difference between the extreme form of the lobe of *hermocrates* (f. 80) and the lobe of *aristeus*, as well as the difference between the other extreme of *hermocrates* (f. 76) and *anticrates* (f. 75), amounts quantitatively to much less than the difference between the extremes of *hermocrates* (f. 76 and 80).

c. *P. aristeus aristeus* from the Southern and Northern Moluccas; f. 81 to 83.

The individuals from the various islands of the Moluccas, though individually variable, exhibit neither in the wing-pattern nor in the genital armature any character confined to one or the other of the islands, or group of islands. This subspecies is in pattern most nearly related to *hermocrates*, but is easily distinguished by the ground-colour of the underside being obviously darker.

The dorsal lobe of the valve (f. 81 and 82) is in all specimens examined slenderer than in *hermocrates*; the individual variation is slight. The ventral ridge is much more extended than in the two preceding subspecies ; when seen from above, as in f. 82, it is a kind of half-ring ; f. 83 gives the same organ in a view from the dorsal side of the valve. The difference between the ridge of *aristeus* and that of *hermocrates* and *anticrates* is obvious enough without further comment. Intergradations in the form of the organ are unknown to us. F. 81 to 83 are taken from a Halmaheira individual.

The subdorsal process is denticulate, and is longer than in *hermocrates* and *anticrates*, projecting distinctly beyond the upper edge of the ventral lobe of the valve (f. 82).

d. *P. aristeus parmatus* from Queensland, New Guinea, and Waigeu.

This subspecies agrees externally so well with the Indian *P. aristeus anticrates* that Professor Eimer did not perceive the slight differences in colour and pattern which generally separate *parmatus* from *anticrates*, and which Mr. Rothschild pointed out on p. 419 of Vol. II. of this journal.

Although we know from several species that in forms which in the wing-colour and pattern are the most closely allied the genital armature is more different than in externally less closely allied forms, we were nevertheless much surprised to find that *aristeus parmatus* has, in opposition to *aristeus anticrates*, the same genital armature as *aristeus aristeus* ; we dare say exactly the same, as *we have not found a single character* by which the apparatus of *parmatus* could be distinguished from that of the Moluccan insect. This discovery is very important, as it shows evidently that, in order to understand the relation of a form, it is necessary to compare sets of entirely independent characters, and as it further proves that a form externally similar to another can be dissimilar in the genital armature, while it agrees in these organs with an externally very different form.

13. Papilio rhesus from Celebes, Saleyer, and Djampea; f. 84, 85.

Though Eimer removes this insect from *aristeus* and puts it into another group of species along with American species, such as *ajax*, all the characters by which *P. rhesus* is distinguished from *P. aristeus* are developments of the characters of *aristeus*. The only argument in favour of a relationship with *ajax* is the number of the black bands on the forewing ; in *rhesus* there is one band less than in *aristeus* ; but even this argument is not valid, as there very often occur specimens in which the usually absent band is indicated by a black spot, and as there are even examples in which this spot has developed to a distinct band. All the other characters of the wing-pattern speak against a relationship with *ajax*, and this statement is corroborated by the morphology of the *male* sexual armature and by the form of the eighth abdominal segment of the *female*. To begin with the latter, it will be sufficient to say that the eighth segment in *rhesus* is complete, as in *aristeus*, *agamemnon*, *sarpedon*, etc. (see f. 182, *sarpedon sarpedon* ; f. 185, *aristeus parmatus* ; f. 187, *macfarlanei*) ; whereas in *ajax* ♀ the eighth segment is incomplete, as in f. 181 (*podalirius*). The similarity in the armature of the valve of *rhesus* with that of *aristeus* will become evident by comparing f. 84 and 85 with 75 and 83.

The dorsal lobe of the valve of *rhesus* (f. 84) is much broader than even in the Indian *aristeus anticrates*, except the free apex, which is comparatively narrower; the subdorsal process is thin, and, as in *anticrates*, not dentate ; the ventral ridge is almost shaped as in *aristeus aristeus* and *aristeus parmatus*, but the basi-dorsal portion of the half-ring is much higher than in these subspecies; f. 84 gives a dorsal view of the ridge.

The affinities of *rhesus* are as follows: in the pattern of the wings it comes nearest to *hermocrates*, in the form of the dorsal lobe of the valve and the subdorsal process nearest to *anticrates* and *hermocrates*, in the form of the ventral ridge nearest to *aristeus* and *parmatus*.

The facts of variation illustrated by f. 72 to 85 are as follows:—

(1) *P. aristeus hermocrates* exhibits obvious variability in the shape of the dorsal lobe of the valve; the extreme specimens come very near *anticrates* and *aristeus* respectively.

(2) That Shan States individual of *hermocrates* which is most "typical" in pattern has the dorsal lobe nearly identical with *anticrates*, while a specimen from the same locality which in pattern is almost identical with certain North Indian individuals of *anticrates* has the dorsal lobe much narrower.

(3) There are no intergradations between the form of the ventral ridge of *aristeus parmatus*, *aristeus aristeus*, and *rhesus* on one side, and *aristeus anticrates* and *aristeus hermocrates* on the other.

(4) The two Eastern subspecies, *parmatus* and *aristeus*, geographically the nearest related, are identical in the genital armature and conspicuously different in pattern.

(5) The two subspecies, *anticrates* and *parmatus*, inhabiting the extreme parts of the range are in pattern closely related, and in the genital armature very different.

14. Papilio alcinous from Japan, the Loo Choo Islands, Formosa, and China; Nov. Zool. 1895. t. VI.

China is inhabited by several forms related to the Japanese *alcinous*, namely *confusus*, *impediens*, *plutonius*, and *mencius*. Of these we shall take here into consideration only the first, *confusus*, which differs externally in the ♂ from the Japanese Papilio in the body being more extended red, especially in the front of the head being clothed with red and with black hairs (while in the Japanese insect the head is all black), and in the submarginal spots of the hindwing being generally more brilliant red. The Formosa specimens agree with this Chinese form in colour, and so do the individuals from the Loo Choo Islands. The examination of the harpes of a long series of Chinese and Japanese individuals led to remarkable results which are laid down in Vol. II. of this journal. The results were as follows:—

The harpe of the black-headed Japanese insect is, in all specimens examined, very different from the harpe of the Chinese individuals, except in one individual which has the harpe like the Chinese specimens.

The red-headed Loo Choo specimens have the harpe like the black-headed Japanese specimens. We have here, therefore, a combination of the external characters of the Chinese form with the genital characters of the Japanese form.

Amongst the red-headed Chinese specimens one was found in which the harpe has the dentate ridge of the Japanese form and the free apical process of the normal Chinese form, and hence combines the characters of the harpe of the red-headed Chinese and black-headed Japanese Papilios.

The facts are briefly as follows :—

(1) Japan : Head black. Harpe with dentate ridge, without free apical process ; one specimen like (3).

(2) Loo Choo Islands : Head as in (3). Harpe as in (1).

(3) China and Formosa : Head red. Harpe without dentate ridge, with free apical process ; one specimen with dentate ridge (= 1) and with free apical process (= 3).

The only inference from these facts logically possible is that neither the red colour of part of the hairs of the head, nor the form of the harpe, is in the Papilios in question of specific value. The present case reminds one strongly of that of *P. sarpedon* as explained on p. 478.

15. Papilio aristolochiae; f. 86 to 95.

The group of Papilios to which the present species belongs has the valve much reduced ; the harpe does not vary to any extent, while the *uncus* (dorsal part of the tenth segment, according to Peytoureau) is very variable. To show the amount of variation in one of the species, we figure the uncus of three Sikkim individuals of *P. aristolochiae aristolochiae* (f. 86 to 88), of three Sambawa specimens of *P. aristolochiae austrosundanus* (f. 89 to 91), and of four Bunguran examples of *P. aristolochiae antiphus* (f. 92 to 95). The variability of the uncus within each subspecies of *aristolochiae* is so great that we fail to perceive any character in that organ which could serve to distinguish the three (in external features easily recognisable) subspecies by.

B. Female Genital Armature.

Peytoureau,[*] in his researches on the morphology of the last segments of the abdomen of the *female Lepidoptera*, comes to the conclusion that there are, as in the *male*, ten segments, the anal segment of the *female* imago, consisting of two lateral pieces comparable in form to the valves of the *male* and having the function of a protector of the anus and the orifice of the oviduct, being homologous to the ninth and tenth segments of the pupa and larva ; the orifice of the oviduct has the same position as the penis underneath the anus between the ninth and tenth segments. The variation of the anal segment is of no importance for our present purposes. The orifice of the vagina is separated from the orifice of the oviduct, and is situated between the seventh and the eighth segments; in consequence of the development of the copulatory apparatus the eighth, not the seventh, segment has undergone great changes. The form of the segment, as well as of the vaginal armature, is in a live specimen easily perceivable when one gently presses the abdomen ; in a dried-up individual the apparatus is concealed, being entirely removed into the vaginal cavity, the opening of which is closed by the seventh and eighth segments being in contact. As it was, in the first place, not the aim of our researches to find characters in the morphology of the abdomen by which the higher divisions in the system of *Lepidoptera*, such as families and subfamilies, could be distinguished, but to try whether the morphological characters of the last segments could be made use of in diagnostic work relating to the lower degrees of division, genera and species, and particularly whether within the limits of a species there was in the copulatory apparatus a variation similar in kind and extent to that of the *male* organs of copulation, it was necessary to study the organs in question *in situ*, in order to be able to

[*] See note, p. 461.

compare the position and direction and the outline of the homologous parts of the apparatus of the allied insects, and thus to be able to observe minute differences in the most nearly related forms. As in live specimens, or in individuals shortly after death, a pressure of the abdomen sufficed to bring the organs in question fully in view, it was evident that a method of preparation of dried specimens would be successful, if the intersegmental membranes of the end of the abdomen were so relaxed that the whole apparatus could be pushed out without destroying the connection between the several organs. The method of preparation we employ is quite sufficient for our purpose, and has the great advantage of being very simple. We cut off the last four segments, soak them in hot alcohol and water, remove the eggs, scales, etc., with the help of a pin and brush, press the segments gently, and leave the abdomen in alcohol and water until the segments are freely moveable; then we press the head of a pin from the inside of the abdomen against the vaginal bulb, and push it gently out, taking care that the membranes do not get torn. The eggs and the *bursa copulatrix* ought to be preserved.

While in a live specimen the membraneous and the more chitinised parts in the vaginal region are easily distinguished by their colour, in a dry individual all the parts are more or less brown, and, though the strongly chitinised organs are recognisable by their gloss, the exact limits between the membrane and the chitinised pieces which it connects are often obscured.

According to the development of the eighth segment the Eastern Papilios can be divided into two groups—such in which the eighth segment forms a complete ring without longitudinal sutures (f. 182), and such in which the ventral plate is absent. To the first group belong nearly all those Indian and African species examined in which the first subcostal nervule of the forewing anastomoses with the costa, except *P. mandarinus* and allies, and to the second group all the other Papilios examined, inclusive of *Troides = Ornithoptera*. The combination of an obvious character in neuration with a still more conspicuous particularity in the development of the eighth abdominal segment of the *females* of the species allied to *agamemnon, sarpedon, codrus, aristeus,* and *antiphates* of the Indo-Australian Region, and of *antheus, leonidas, pylades,* etc., from Africa, indicates certainly more than mere similarity in form.

The special copulatory apparatus consists of processes, ridges, tubercles, and folds near the vaginal orifice, all more or less chitinised, often dentate and hook-shaped. It is not a modification of the ventral plate of the eighth segment, but is an independent structure of the intersegmental membrane. The variety of the apparatus in the various species is startling, as a glance at f. 156 to 182 will show. The orifice of the vagina, marked v in the figures on Pl. XIX., has in the diverse groups of *Papilio* a different position; in *P. alcinous, philoxenus,* and allies, it has a ventral (*recte* basal or anterior) (f. 179), in *podalirius* and allies (f. 181) a more dorsal (*recte* apical or posterior) position.

The variability of the copulatory apparatus we explain by figures only of four species; this must suffice for the present. But to demonstrate the classificatory value of the armature we give figures of seven more species.

16. Papilio machaon; f. 156 to 159.

The figures are taken from a British individual. When examining the intersegmental membrane which bears the orifice of the vagina and the chitinised prehensile apparatus we perceive (f. 157) a somewhat lyriform or horseshoe-shaped buckle (*b*), which is a chitinised portion of the membrane itself, and does not protrude free; its ventral and lateral edges form a slightly raised ridge, as can be seen in f. 156. The ventral part of this buckle is produced into two processes (*a*), which are thin in a dorso-ventral sense, but divided longitudinally into a left and a right half, which stand at a blunt angle to one another in consequence of the middle line of the process being more dorsal than the lateral edges. At the base each process is constricted (f. 158), while the edge of the apical half is armed with teeth. In a side view the process (f. 156 and 159) appears plainly as a dilatation of the buckle *b*. The vagina *v* (f. 159) is behind the middle of the median piece of the buckle, and marked in f. 157 behind the two processes as a black spot.

The series of *females* examined has not been very great, and hence we do not yet know whether there is not some variation in the form of the armature in the various subspecies. From what we have seen we must conclude that if there is such a variation it must be very slight, as neither the European and Asiatic nor some American forms of *P. machaon* presented any obvious deviation from what we have figured on Pl. XIX. And this would be entirely in accordance with what we have found in the *male* genital armature of *machaon*, as far as a distinction between the genital armature of the various subspecies of this species goes. We can safely say that in *P. machaon* from Europe and Asia—

(1) There is no character in the genital armature of both sexes of any of the subspecies by which the greater percentage of the individuals could be distinguished;

(2) There are obvious characters in the colour and pattern of the wing and body of each subspecies by which the greater percentage of the individuals, but not every specimen, can be distinguished.

The genital armature of *P. xuthus* is entirely different from that of *machaon*.[*]

17. Papilio polytes; f. 160 and 161.

The *female* of this species is polymorphic; we have examined the three *female* forms from Ceylon and North India, the two forms from Borneo, the three from the Philippine Islands, and also a single specimen of each of the subspecies from Celebes, Sangir, Sulla Islands, Amboina, and Halmaheira.

There is a good deal of individual variation in the apparatus of *P. polytes*, but this variation is entirely independent of the colour and pattern of the wings, and hence the various forms of the *female* of each subspecies must be pronounced to be identical in the genital armature; ♀-f. *romulus*, ♀-f. *cyrus*, and ♀-f. *polytes* do not present any character in the vaginal armature by which one form is distinguished from the other, and so it is with ♀-f. *theseus* and ♀-f. *virilis*, and so on.

The intersegmental membrane (f. 160) is just underneath the eighth segment at each side thickened and corrugated, the corrugated portion having an elongate-ovate outline; round the orifice of the vagina there is a low ridge continued upwards to the hinder end of the corrugate mark; this ridge is raised into five processes, one ventral and two on each side lateral. The form of the processes is visible in f. 161, which represents the apparatus flattened out and viewed from the anal side; the ventral process is rounded, the lateral ones are sharply pointed, the lower one is the longest. The length of the lateral processes is variable. Above the vagina there is a rounded tubercle, *t*, present in all allies.

[*] This confirms Dr. Seitz's opinion that *xuthus* is not a very near relation of *machaon*. See *Soc. Ent.*, 1895. p. 130.

The various subspecies do not exhibit any obvious difference in the form of the ridge, except perhaps the forms from the Moluccas. We have examined only one individual of the two subspecies from Amboina and Halmaheira. These individuals showed some difference in the length of the lateral teeth; but as in the Indian, Bornean, and Philippine insects, of which we have examined several specimens, the teeth vary according to the individual specimens, the difference in the Halmaheira and Amboina examples can just as well be due to individual as to subspecific variation. In the *male* sex we have found that the various subspecies are in the genital armature connected by intergradations, but that nearly all the specimens of each subspecies are pretty well recognisable by the form of the harpe. In the *female* sex the difference in the vaginal armature, if present, is certainly much fainter; but we shall see later on that the difference corresponding to that in the harpes does not exist in the chitinised processes, but in the folding of the membrane, and is, when slight, not perceivable in consequence of the shrivelling of the membrane in the dry specimens.

18. Papilio ambrax; f. 162.

Though in pattern the *female* of this species comes very near that of *P. polytes nicanor*, so near indeed that entomologists like Kirsch and Snellen have mistaken it for *polytes*, there is a constant and very conspicuous difference in the vaginal armature of the two species. F. 162 represents the apparatus of a specimen from New Guinea; the ventral process *a* of f. 161 is here wanting, and this character we have found to be constant in all the twenty odd *females* of *ambrax* examined. Besides, the lower lateral tooth is generally somewhat shorter than in *polytes*. The variability of the ridge is similar to that of *polytes*. The Queenslandian subspecies does not, to our knowledge, present any distinguishing character from the New Guinean subspecies in the form of the ridge.

The representative species from the Bismarck Archipelago and the Solomon Islands we have not yet examined.

19. Papilio aegeus; f. 163 to 169, 176.

The individual variability of the vaginal armature is in this species very great, while an examination of a longer series of specimens from New Guinea (*aegeus ormenus*) and Queensland (*aegeus aegeus*) proved that the variation according to locality is very slight. The vagina is surrounded by three high processes which are dilatations of a high ridge. The position and general form of the armature is represented by the half-diagrammatic figure 165. The ventral process is the largest, and is narrowed from the middle towards the base and the apex; the apical half is dentate, divided by a median sinus into two lobes of variable length (f. 166 and 167), and is curved dorsally (f. 163); ventrally the ridge is, like the processes of *P. machaon*, longitudinally concave, and is provided with a rounded middle keel (f. 164, *m*) which does not reach the apex. The lateral process *b* is seldom simple in outline; mostly it is more or less strongly dentate (f. 166). Above the orifice of the vagina *v* there stands the supravaginal tubercle *t*. The intersegmental membrane is, moreover, raised into a vertical fold *f* (f. 163 and 164), which stands in connection with the vaginal ridge.

a. *P. aegeus ormenus*; f. 163 to 166, 176.

The *female* of this subspecies is polymorphic. As in the case of *P. polytes*, we have found no characters in the genital armature peculiar to one or the other of the *female* forms. The specimens figured are from near Dorey, Dutch New Guinea. In f. 166 the left lateral ridge is not drawn; the ventral process of this specimen is at the apex much deeper sinuate than that of f. 164, and the lateral process is much broader and multidentate, while the process is not dentate in f. 163 and 164. The anal segment and the supravaginal tubercle *t* is represented by f. 176.

b. *P. aegeus aegeus* ; f. 167 to 169.

The *female* of this subspecies is monomorphic in colour and pattern. The variability of the vaginal armature is illustrated by f. 167 to 169. The ventral process is slenderer at the base than in *aegeus ormenus*, and the lateral process has mostly on the outer side a slightly raised carina (f. 168 and 169). In f. 167 the teeth of the lateral and ventral processes, and in f. 169 those of the lateral process, are mostly obliterated; the dentition of the lateral process represented in f. 168 is restricted to the upper edge.

20. Papilio rumanzovius ; f. 170 and 178.

There are four series of species of Papilios which are closely allied to one another, the groups of *P. memnon, P. aegeus, P. deiphobus*, and *P. lampsacus*. We represent for comparison the *female* sexual armature of a species of the first three groups; the close relationship of the insects in question finds an expression in the similarity of the genital armature of *male* and *female*.

F. 170 is taken from a specimen of *P. rumanzovius* ♀-f. *rumanzovius*. The ventral process is broadest towards the base, its apex is irregularly truncate, the dentition of the process of *aegeus* (f. 166 to 169) is absent; the median keel on the ventral side of the process of *aegeus* is here separated into a number of irregular, longitudinal, slightly raised folds and wrinkles. The lateral process is less high and leans over dorsally. The supravaginal tubercle, which in *aegeus* (f. 176) is small, and near which the membrane is inobviously folded, is in *rumanzovius* large (f. 178), and the membrane between it and the anal segment (ix. + x.) is regularly and heavily folded transversely, the folds being limited laterally by a longitudinal fold.

21. Papilio memnon; f. 171 to 175, 177.

Like *P. polytes* and *P. aegeus ormenus*, this species has a polymorphic *female* in most localities; we have examined a number of tailed and tailless individuals from North India, Java, Nias, and Borneo, and come to the same conclusions as in the case of the two before-mentioned Papilios, that the subspecies are extremely slightly different in the *female* genital armature, and that the various forms of the *female* of each subspecies do not exhibit any characters in the vaginal apparatus peculiar to one or the other of the forms. The individual variability is as great, or even greater, than in *ormenus*.

The chief features by which the vaginal armature of *memnon* can be distinguished from that of *aegeus*, to which it comes rather near, are as follows: the ventral process is broader basally, and the median keel of the underside (ventral side) is widened out laterally into two ridges which rise under a small angle from the plane of the main process (f. 173, *m*). The lateral process (*b*) is higher than in *P. aegeus*, and divides very often at its dorsal edge in an outer (f. 174 and

175, *o*) and inner ridge (*u*). The outline of the ventral and lateral ridges is very variable, as f. 172 to 175 show.

The supravaginal tubercle and the anal segment are represented in f. 177; the tubercle is similar to that of *P. aegeus*, but it is surrounded by two rather heavy folds which in *aegeus* are scarcely marked.

22. Papilio alcinous; f. 179 and 180.

The vaginal orifice is more ventral than in the preceding species, and the armature is very simple. Just above the vagina the membrane is smooth and more strongly chitinised, and bears a roundish impression (f. 179, *i*). From the lower edge of the eighth segment downwards extends a broad fold, *f*, which is also more chitinised than the rest of the intersegmental membrane, especially at two points where the fold has a bulbose appearance.

We have examined but two individuals each of the Chinese *P. alcinous confusus* and the Japanese *P. alcinous alcinous*, and find some obvious distinguishing characters between the two subspecies. In the Chinese form (f. 180) the membrane underneath the orifice of the vagina is strongly folded longitudinally, in the Japanese insect the membrane is smooth there; the supravaginal impression *i* is in *confusus* small, in *alcinous* large; the lateral fold is in *confusus* extended down to the ventral side as a conspicuously raised fold, while in *alcinous* the ventral portion of the fold is only slightly raised. A specimen from the Loo Choo Islands we unfortunately could not compare.

In the so-called nauseous species of *Papilio* allied to *alcinous*, *philoxenus*, and *aristolochiae* the vaginal armature is as a rule very simple.

23. Papilio podalirius; f. 181.

We figure the end of the abdomen of this European species for two reasons: first, because it enables the reader who is not acquainted with the exotic Papilios to compare the two common European species; and secondly, because the form of the eighth segment is entirely different from that of the Asiatic *P. antiphates*, *aristeus*, etc., which bear a superficial resemblance in pattern to *P. podalirius*.

The vaginal orifice is more dorsal, or rather apical, than in the preceding and following species here dealt with; the membrane round the orifice is strongly chitinised, forming a round plate-like organ, the edge of which is raised; the middle portion of the organ gradually raises and is highest just underneath the vaginal orifice; near the ventral edge of the eighth segment the edge of the discus-like organ fuses together with a strong oblique fold which is parallel to the ventral margin of that segment. The latter itself is incomplete, there being no ventral plate to it, but the posterior lateral angles almost meet above the vagina; the segment thus stands in development intermediate between the preceding species and the species here following. In the American allies of *P. podalirius* (*P. ajax*, etc.) the eighth segment is as in *podalirius* (or nearly so), and not as in *antiphates* and *aristeus*, which have it as in f. 182 (or nearly so). In *P. leosthenes* from Australia the segment agrees on the whole also with that of *podalirius*, and has also in *P. mandarinus* no ventral plate. Hence, in the development of the eighth abdominal segment *podalirius*, *leosthenes*, *ajax*, would be more nearly related to one another than to *aristeus*, *antheus*, etc., a statement which stands in direct opposition to the kind of relation at which Professor Eimer arrives by his interpretation of the wing-pattern; while, on the other hand, the absence of a ventral connection of the lateral plates of the summit in *podalirius*, *mandarinus*, *ajax*, *leosthenes* is, to a certain extent, in favour of Eimer's opinion that there is a close relationship between the first two and between the last two species.

24. Papilio sarpedon; f. 182 to 184.

The ventral portion of the eighth segment covers the vaginal region roof-like. The intersegmental membrane bears a vertical strong fold (f), and there is between the fold and the vaginal orifice a strongly chitinised, smooth, impression (i). The margin of the orifice of the vagina is slightly raised laterally and produced ventrally into a process (a), the form of which will be seen from f. 183, which represents the organ viewed from the dorsal side. F. 182 and 183 are drawn from a Japanese individual. F. 184 is taken from a specimen of *P. sarpedon choredon* from Queensland; the process is broader at the base, and the angles at the apical sinus are more rounded; the difference between f. 183 and 184 is slight, but seems to be fairly constant. We have not found any difference between the processes of *P. sarpedon sarpedon, sarpedon semifasciatus, sarpedon teredon,* and *sarpedon milon* (!). The individual variation in the form of the process is extremely slight in the specimens examined.

25. Papilio aristeus; f. 185 and 186.

We have examined some Queenslandian *females* of *P. aristeus parmatus*, and a *female* of *P. aristeus hermocrates* from Kalao (between Celebes and Flores); both subspecies agree in the form of the eighth segment and in the genital armature, but in *hermocrates* the outline of the ventral (median) piece of the summit is indicated by a fine ridge. F. 185 and 186 are taken from a Queensland example.

The vaginal armature consists of two chitinised lappets (f. 186, a and a^1) which protrude ventrally, and of which the lateral edge runs upwards as a strong ridge, which is homologous to the vertical fold of f. 182, to the ventral margin of the eighth segment.

26. Papilio macfarlanei; f. 187 and 188.

At each side of the vagina there is a broad and strongly chitinised ridge (f), homologous to the vertical fold of f. 182, and bearing on the upperside a covering of irregularly folded small lappets. The margin of the vaginal orifice is ventrally produced into a bifurcate process (a in f. 187 and 188) which corresponds to the two lappets of fig. 186.

27. Papilio agamemnon; f. 189.

The apparatus is similar to that of *P. macfarlanei*, but the fork-like process has no stem and is connected with the folded intersegmental membrane (n) by a short bar visible in f. 189 between the two branches of the fork. We have not noticed any variability in the armature of this species.

Though our researches on the genital armature of the *female* sex of *Papilio* are quite incomplete, we are nevertheless justified in drawing some general conclusions from the facts of variation illustrated by f. 156 to 189.

(1) The prehensile apparatus consists of the armature of the orifice of the vagina, and the special folding of the intersegmental membrane.

(2) The folds of the intersegmental membrane cannot be studied from the dried-up individuals in such a manner that one perceives the minute differences between the folds of the individuals; to this purpose it is necessary to compare live specimens, or material preserved in an adequate fluid. The armature of the valve of the *male* is during copulation pressed against the intersegmental membrane of the *female*, and the spines, hooks, processes, etc., of the valve find a hold on the ridge-like folds of the intersegmental membrane ; this is plainly visible in a ♂ and ♀ of *P. memnon* in the Tring Museum which are still united to one another. The special armature of the vagina takes hold on the internal portion of the ninth segment of the *male* and on the scaphium. There is sometimes a peculiarly shaped, broad, vertical, chitinous plate underneath the scaphium, for example in the *males* of the yellow *Troides* (= *Ornithoptera*), such as *T. helenus*, to which a long and strong ventral hook in the *female* corresponds.

(3) One would expect that the variation of the harpe in the *male* is in the same species accompanied by a corresponding variation in the *female* in those parts which during copulation are in contact with the harpe ; hence the variation in the *female* genital apparatus parallel to the variation in the *male* harpe must be searched for in the special kind of folding of the intersegmental membrane, while the variation in the outline of the processes and ridges at the mouth of the vagina can have nothing to do with an adaptation to the special form of the harpe.

(4) In the only species examined in which the harpes vary very conspicuously according to locality, in *P. alcinous*, the intersegmental fold of the ♀ on which the harpe takes a hold is found to be different in the Chinese and the Japanese subspecies.

(5) Though the individual variation of the vaginal armature is often great (*P. aegeus* and *memnon*), a variation according to locality is not observed, or is slight.

(6) In species with polymorphic *females* the vaginal armature is the same in the various *females* (apart from individual variability).

III.—CONCLUSIONS.

The demonstration of the kind and the extent of the variation of the genital armature in both sexes of some species of *Papilio* which is contained in the preceding chapter, notwithstanding the fact that our researches are still incomplete and our notes on the vaginal apparatus even preliminary, enables us to compare the variation of the genital armature with that of other organs and to draw with safety some general conclusions. If we almost restrict the comparison to the colour, pattern, and shape of the wings, mentioning only the prominent characters of the respective insects, we do so in order to avoid unnecessary detail, and secondly, because the distinguishing characters of species, subspecies, and aberrations of butterflies as given in the works of Lepidopterists refer especially to the wings.

As the true basis of work in Natural History is the comparison of specimen with specimen of the lowest classificatory entity, the species, a study of the genital armature has to begin with the comparison of the armature of such individuals as are doubtless specifically the same. To determine the limits of variation of the genital armature of a species is a difficult task. Though the number of specimens of the commoner species could easily be enlarged without any great pecuniary sacrifice, the student of the variation of an internal organ, which is not visible without dissection, is at a great disadvantage compared with the student of external characters, because his series of individuals is always taken at random ; he has no security whatever that, among the hundreds of specimens which perhaps are at his disposal, there are really individuals which in the genital armature represent the extremes of that species, or come near the extremes ; while, on the other hand, the specimens conspicuously aberrant in wing-pattern, or colour, or shape, are recognised and preserved by collectors, which gradually results in an accumulation of individuals which fairly represent the various degrees, inclusive of the more extreme ones, of the variation of the species in external characters. Hence it is by no means to be wondered at that there are so many individuals known which in external characters stand far outside the *usual* limits of variation of the respective species, while the internal genital armature seemed to vary only within very narrow limits. On the contrary, it must surprise us that we nevertheless *did* succeed in finding among our specimens such individuals as exhibit in the genital organs characters so widely divergent from the normal of the respective race of *Papilio* that the individual not only forms a transition to another geographical representative, but goes in the development of the peculiar character even beyond allied forms. The specimen of *P. sarpedon sarpedon* (f. 101, 102, 103) and those of *P. alcinous* figured on Pl. VI. of Vol. II. of this journal (f. 11 and 13) are such "unusual" varieties, which perhaps are in nature not rarer than individuals that in external characters come close to, or are identical with, one of the allied forms; and we have no doubt that a continued investigation in the matter will bring to light varieties in the genital armature which in the degree of divergency are equal to so-called sports. The individual variability found in our series of British *P. machaon* to amount to 40 per cent. in the gradual increase of the length of the prehensile portion of the harpe, and in Palaearctic *machaon* to 70 per cent., is quantitatively scarcely inferior to the variation exhibited in external characters, and the same applies to the other Papilios dealt with in the preceding chapter. In the *female* genital armature we meet with an equally great variability in many of the species, especially in all those which have a more complicated armature ; the variability of the complicated apparatus, which consists of spines, hooks, dentate ridges, and so on, appears generally to be greater than that of a simpler armature, because the difference between the apparatus of two individuals is more easily perceivable in a complicated than in a simple organ, even if the difference in the latter expressed in proportional numbers is greater than in the former.

If we group the individuals of a species from one locality according to the similarity in the genital armature, and then again according to a conspicuous external character, the first series of groups and the second are not the same ; we have found again and again that specimens similar in a certain character in colour, pattern, or shape of wings have no particularity in the genital armature in common by which they are distinguished from the individuals that do not possess that external character; specimens aberrant in wing-pattern may be normal in the genital armature, and individuals abnormal in the latter may be normal in the former. The harpe of *P. aegeus ormenus* with a band on the forewing is the same as in the aberration without the band; *P. euchenor euchenor* ab. *eutropius* does not differ in the harpe from the ordinary *male* ; *P. polytes borealis* ab. *thibetanus* has a normal harpe, and so on. Still more obvious is the independence of the variation of the genital armature in the species with polymorphic *females* ; the various *females* of *memnon*, tailed and tailless, those of *polytes*, and also those of *aegeus ormenus*, though individually very variable in the vaginal armature, agree with one another (respectively) in these organs. Marked dimorphism, in which the two forms are not connected by intergradations (tailed and tailless *female* of *P. memnon*), we have not noticed in the genital armature, unless f. 171 and 173, in which the lateral ridge is not divided, and f. 174 and 175, in which the lateral ridge is divided into the two ridges *o* and *u*, represent a kind of dimorphism ; both the simple and the divided ridge occur in tailed as well as in tailless individuals of *P. memnon* ; intergradations between the two developments of the lateral ridge are unknown to us, but may occur.

Of equally great interest are those cases in which a specimen of a certain subspecies resembles another subspecies externally, but stands further away from this subspecies in the genital armature than another specimen does that externally is dissimilar to the subspecies. We recall to mind the two individuals of *P. aristeus hermocrates* from the Shan States mentioned on p. 487, and add that, if variation in pattern and variation in the sexual organs were in any way connected with one another, one would have expected that the paler individual resembling certain Sikkim specimens of *anticrates* had the dorsal lobe of the valve broader than the other individual, which in pattern is like dark Philippine examples ; as, however, just the reverse is the case, the broader dorsal lobe of the valve and the narrower and shorter black bands of *anticrates*, and the narrower dorsal lobe and more extended black bands of *hermocrates*, must be considered as independent characters.

It is possible that there are species in which a great variability, resp. constancy, of the pattern of the wing is associated with a great variability, resp. constancy, in the sexual armature, but so much is certain that in all the species we have examined there is no correlation between the directions in which the wings and the sexual armature vary. Hence we may pronounce it as a general law *that the direction of the variation of the genital armature within a species of Papilio is entirely independent of the variation of the wings.*

We need scarcely mention that this law applies also to other *Lepidoptera* ; our researches in groups other than *Papilio* are, however, so limited that we prefer for the present to express the law as above. It would be of great interest to study the variation of the genital armature of specimens which have artificially been exposed to transmuting factors, such as heat and cold, and to compare the results with those arrived at by the examination of the individuals roaming at large. Most probably the artificially produced colour-varieties will be normal in the genital armature. To this conclusion we are led by the experience gained from the examination of seasonally dimorphic species. We paid special attention to the copulatory apparatus of such species with the hope of finding in one or the other *Papilio* differences in the apparatus of the spring and summer brood (or broods), but completely failed to come across a species which, both in the wing-markings and in the sexual organs, showed seasonal dimorphism. The spring forms of the Japanese *P. machaon, P. xuthus*, and *P. sarpedon* are in the harpe and the vaginal armature the same as the differently coloured respective summer forms, and so it is with the Sikkimese species which exhibit in the wings a marked seasonal dimorphism (*P. sarpedon, cloanthus, eurypylus, bathycles*). In the case of winter and summer forms it is therefore evident that the influences which bring about a change in the wings have no apparent effect on the sexual armature. This corroborates the above statement of the independence of the variation of the sexual organs and the wing-pattern, and opens a wide field for theoretical research, inviting at the same time to a comparison of the seasonal with the geographical variation.

In the introduction we have mentioned the degrees of geographical variation observed in Papilios; the lower degrees, called above localised aberration, concern the colour, pattern, and shape of the wings, and cannot be expected to be noticeable in the genital armature without long-continued research; we have therefore to do only with the higher degrees of geographical variation of a species which is termed *subspecific* variation.

The valve, when of the usual more or less triangular outline, does not exhibit obvious subspecific variation; while the variation according to locality is conspicuous, when the valve is divided into lobes, as is the case in *P. sarpedon*, *aristeus*, and others. The variation is such that the specimens from a certain locality are generally well distinguished from the specimens of a certain other locality; but if we take individuals from all the various districts inhabited by the respective species, the lines of delimitation of the various forms become mostly obscure and disappear. North Indian specimens of *sarpedon* and individuals from Sambawa are at once distinguished by the form of the valval lobes, but examples from Lombok, Java, and Sumatra overbridge the gap; Queenslandian and Sikkim *aristeus* are in the dorsal lobe of the valve quite dissimilar, but specimens from the interjacent countries form a continuous series of intergradations between those two extremes. The armature of the valve varies in a similar way. We have cases in which the prehensile organ of part of the individuals of the various localities has some peculiarity; thus in Great Britain the prehensile portion of the harpe of a good many specimens of *P. machaon* is long, while in many individuals from Syria the dentate portion is very short; the subspecific difference in the harpe of *P. aegeus* from New Guinea and Australia is very slight and applies only to part of the specimens. In other species the differences between the individuals from various districts become more constant, so that with very few exceptions it is possible to tell from the examination of the harpe from which place a respective specimen came. While in other cases again, such as *P. bathycles*, the individuals examined were constantly different in the harpe according to locality.

The localised peculiarity in the harpe is, therefore, found in many, or in nearly all, or in all the individuals from the respective locality, and when the distinguishing character applies to all the specimens, there occur intergradations either in other places (*P. aristeus*), or intergradations are unknown (*P. bathycles*).

In the *female* sex the armature at the mouth of the vagina does not vary so obviously according to locality as the harpe of the *male* does; but when the difference in the harpes of the subspecies is very conspicuous, as in *P. alcinous alcinous* and *P. alcinous confusus*, the difference in the corresponding part of the vaginal bulb is also prominent.

If we keep in mind that the variation of the copulatory organs is independent of the variation of the wing-characters, it is to be expected that a division of a species into subspecies will result in a different number of subspecies according as we take for the basis of division solely the wing-characters, or the genital armature, or both united; and it would likewise not be surprising if in a certain case the number of subspecies inhabiting a certain region would by both divisions be the same, but the lines of division be different. Among the species we have examined the latter phenomenon did not occur; when the numbers of the subspecies were the same, the limits of division were also the same, whether the delimitation of each subspecies was carried out according to the wing-characters, or according to the apparatus of copulation. This result is most probably owing to the areas to which the various subspecies of the Papilios examined are restricted being mostly islands.

Hence we can accept it as a general rule applying to most subspecies that the distinguishing characters taken from the wings are associated with distinguishing characters in the genital armature, at least in the *male* sex. There are, however, many exceptions to this rule, in so far as a good number of subspecies characterised by some certain peculiarity in the wings have no peculiarity in the apparatus of copulation. The Chinese specimens of *P. sarpedon*, so very conspicuously aberrant in pattern, are in the valve and harpe identical with the Indian and Japanese form of *sarpedon*; the Moluccan subspecies of *P. aristeus*, though in the extent of the markings very different from the Queenslandian representative, is in the sexual organs the same; so that in an arrangement of the forms according to the development of the apparatus of copulation of the *male*, *P. aristeus aristeus* and *P. aristeus parmatus*, and *P. sarpedon sarpedon* and *P. sarpedon semifasciatus*, would have to be united. We meet, further, many subspecies which in external characters are conspicuously and almost constantly different, while there is only a slight and very inconstant character of distinction in the genital apparatus. We have found it correct in all the species examined that, if in a given species the character of distinction of a certain subspecies lies in one only of two independently varying organs, wings and apparatus of copulation, it is invariably the wing which exhibits the character by which the subspecies is distinguished. This coincides in a remarkable manner with the above observation that seasonal dimorphism does not affect the sexual armature in the case of the North Indian and Japanese Papilios, and renders it certain that the effect of (seasonally or geographically) isolated transmuting factors is first noticeable in the wing-markings, and that accordingly the transmutation of a species begins with a change in the wing-markings. Hence it is also correct to say that in an evolutionistic sense the wing-markings are less constant than the genital armature, and that consequently a difference in the latter constantly met with in the individuals of two forms of *Papilio* is much more likely to be of specific value than a difference in the wing-pattern, a conclusion which borders closely upon one of the main objects of our researches, the question as to the classificatory value of the sexual armature, which we will now briefly discuss.

A. Taxonomic Value of the Organs of Copulation.

As our researches have proved that there is a certain amount of individual variation in the sexual armature, the limits of which can, as we have seen above, only be determined by continued examination of a large material, and as we further have found that there is geographical polymorphism in the organs in question in which the localised forms are connected by intergradations (*P. sarpedon sarpedon* and *adonarensis*, *P. aristeus hermocrates* and *anticrates*), the premises of the discussion of the question which characters are and which are not of specific value are precisely the same as in the case of external characters. Referring to what we have said in the introduction about our means of recognition of specific distinctness, it will be sufficient to recall to mind the fact that the presence of a character in a certain form not met with in other forms is *a priori* not a proof of the respective form being specifically distinct from its allies, and that it is circumstantial evidence which has to guide us, in the absence of experiments, in our judgment. Now, what is the evidence we derive from the examination of the organs of copulation?

We know the larval and imaginal state of a good many Papilios, and are able to base on our knowledge of their organisation, quite apart from the characters of the organs of copulation, a classification in which the most nearly allied species are, with a certain degree of correctness, grouped together. If we now compare of a well-studied group such forms about the specific distinctness of which we do not entertain any doubt—if, for example, we examine the sexual armature of *P. memnon, mayo, oenomans, polymnestor, lampsacus, rumanzovius*, etc., which in their general organisation are all more or less closely allied to *P. memnon*, or take for examination *P. antiphates* and *androcles*, or *P. aristolochiae* and *P. polydorus*, or the various yellow species of *Troides* (= *Ornithoptera*)—the first thing we notice is that the imagines of each species exhibit, besides the distinguishing characters in colour and structure externally visible, also characters in the genital armature peculiar to the respective species. As we have not found amongst all those species the specific distinctness of which we could deduce from characters other than such of the organs of copulation a single exception to that rule, we hope we are justified in generalising the statement that every species of *Papilio* is different from every other species in the sexual armature; and this generalisation we believe the more confidently to be correct as also in other groups of *Lepidoptera* specifically distinct forms are characterised by some peculiarity in the copulatory organs, although the difference between allied species is sometimes—for example, among *Aganaidae*— very slight. If the generalisation is correct—*i.e.* if every form that is, according to our definition of the term " species," specifically distinct, which, we repeat, can only be *proved* by experiment, and we will assume the generalisation to be correct—then it follows necessarily that forms which are identical in the genital armature are also specifically identical. If we apply this conclusion to the species of *Papilio* externally polymorphic, it is evident that the genital armature of *male* and *female* is an excellent criterion of specific identity. The various varieties of the *male* of *P. aegeus ormenus* from New Guinea, of *Troides priamus poseidon* from the same country, of *P. memnon*, and so on, many of which have been described as distinct species, are thus easily demonstrated to be specifically the same. Still more important is the application to the *female* sex. The numerous species with polymorphic *females* which so often are quite unlike each other, as in the case of *P. memnon, P. aegeus ormenus, P. polytes*, and the African *P. phorcas* and *merope*, and some American species—an examination of the genital apparatus of a number of specimens will at once make it clear whether the forms in question belong certainly to one species, or whether they eventually can belong to more species. We say intentionally " a number of specimens," for the examination of one example of each form is quite inadequate, and may lead to entirely erroneous conclusions. F. 171 and 173 give an excellent illustration, the first being taken from a tailless example of *P. memnon agenor* from Sikkim, and the other from a tailed specimen from the same locality. Now, considering that f. 167 represents a species very different from *memnon*, it would by no means be preposterous to conclude, if one had to judge only from the three figures and the conspicuous external differences of the specimens, that f. 171 and 173 represented likewise specifically different Papilios. It will generally be quite sufficient to examine a number of individuals of one of the forms, and to determine thus the probable limits of variation; the vaginal apparatus of a specimen of another form, if specifically the same as the first, then will most probably come within the limits of variation of the first; if it stands outside those limits one has to examine some more specimens. It is perhaps not unnecessary to repeat that the tailed and tailless *females* of *P. memnon* have been *proved* by rearing to belong to one species.

The second point which strikes one when comparing the sexual armature of species of the same group or closely allied groups, and which is of no less great practical consequence, concerns the fact that the species which we must regard as more or less close relatives on the ground of their general organisation bear in the structure of the organs of copulation (inclusive of the modified eighth segment of the *female*) a greater resemblance to one another than to the species which stand further away in the system. Representative species, such as *P. aristeus* and *rhesus*; *aegeus, tydeus, gambrisius, inopinatus; euchenor* and *depilis*, etc., have the organs in question built up after exactly the same plan, and the differences exhibited by these representative species in the organs of copulation are often not only far less obvious than the external differences of the species, but are sometimes so slight that the degree of divergence between two species is less than the degree of divergency between the extreme individuals of one of the two species; we refer for illustration to the figures given of the harpes of *P. aegeus* and *inopinatus, polytes* and *ambrax*. In less closely related species the similarity in the sexual armature is not so great as in those representative species which are phylogenetically younger forms; the organs are more divergently developed, and the peculiar modification of one or the other part of the apparatus often obscures the actual homology, so that the organs superficially compared are very dissimilar in appearance. Nevertheless a comparative study at once shows that the superficially dissimilar apparatus are developments of the same type. In *P. sarpedon* and *cloanthus* the close relationship of the valves and harpes (f. 96 and 149) is obvious. In *aristeus* (with *rhesus*, f. 72 to 82) the organs are more strongly modified, but the homology of the single parts with those of *P. cloanthus* is not difficult to perceive; *antiphates* has the organs again similar to those of *aristeus*, and the externally very dissimilar *P. delesserti, leucothoë*, and *xenocles* resemble *aristeus* in the structure of the *male* organs to a surprising extent, the principal prehensile parts of the valve consisting in these species, as in *aristeus* and *antiphates*, of a ventral dentate ridge, a subdorsal process, and the dorsal lobe of the bipartite valve, which parts respectively are homologous to *e, c, b* of f. 72 (*aristeus*). The vaginal armature agrees in this respect with the *male* armature; here again the most closely allied species have a greater similarity in the organs in question to one another than to less nearly related species: *P. ambrax* and *polytes* (f. 161 and 160); *P. aegeus, rumanzovius,* and *memnon* (f. 166, 170, and 171); *P. sarpedon, aristeus, agamemnon,* and *macfarlanei* (f. 182 to 187) illustrate this fact sufficiently. Although it would seem to follow from these statements that the degree of blood-relationship of two species to a third could easily be made out from the degree of similarity in the form of their organs of copulation, this inference of the facts would nevertheless be hasty and most probably erroneous, if we generalised the conclusion so as to apply to every species. Our researches convince us that it is true that in every group of closely allied species of *Papilio* the characters of the three independently variable organs, wing, harpe of *male*, and the vaginal armature of *female*, are such that each of these organs of every species is morphologically closer related to the respective organ of every other species of the group than to that of any species standing outside the group.

But within the group a species could very well be nearly related to another in the pattern of the wing, while it comes in the structure of the copulatory organs of the *male* or *female* closer to a third species, so that the relationship of the species *inter se* would appear to be different according as we take as the standard of arrangement the deviation in the one or the other organ.

If, however, it is true that the species which belong to the same group (of allied species) are in the genital armature more similar to one another than to the species of other groups, we have consequently to conclude that the organs of copulation are a safe guide to determine to which group a species belongs, which means that these organs can be made use of in generic classification. The great help which the genital armature affords to the systematist is beautifully illustrated by those species which bear to each other a superficial resemblance in pattern, such as *P. clytia* and *leucothoë* or *macareus*, *P. rhesus* and *philolaus*, which on the ground of the alleged relationship in pattern have been considered allied species. Though a careful examination of the wing-markings soon convinces us that we have here and in many other Papilios to do with an analogous development in pattern, as in the case of mimetic forms, which does not indicate blood-relationship, the demonstration of the dissimilarity in the morphology of the organs of copulation of those superficially similar species will be a more convincing guide in which direction the actual relationship is to be sought for.

Having thus arrived at the two conclusions, firstly, that identity in the sexual armature means specific identity, and secondly, that close relationship in these organs points to generic identity, the question arises, whether there are in the organs of copulation of specifically distinct Papilios, as opposed to specifically non-distinct forms, characters which *a priori* could be recognised as being of specific value, and hence would enable us to draw up a general rule applying to every case by which specifically distinct forms could be distinguished from specifically identical forms. The occurrence of individual and geographical variation in the copulatory organs, the latter kind of variation always associated with an independent variation in the wing-markings, renders it alone highly improbable that a general distinction between specific and subspecific characters is possible. A comparison, however, of the degree of divergency between subspecies with the degree of divergency between closely allied species proves that the quantitative amount of divergency between specifically identical forms can be superior to the quantitative amount of divergency between allied species. *P. inopinatus* differs in the harpe (f. 12) not so much from certain examples of *P. aegeus* as some of the latter do from one another (f. 4, 5); the difference between the harpe of *P. polytes polytes* (f. 18) and *P. polytes alphenor* (f. 33) is greater than that between *P. polytes* and *P. ambrax* (f. 35, 37); the difference between *P. aristeus parmatus* and *P. aristeus anticrates* in the harpes (f. 73, 75, 81, 83) is not inferior to the difference between *P. aristeus anticrates* and *P. rhesus*, which latter is considered to be a species distinct from *aristeus* (f. 82, 84). If, however, there are such cases like these in which that difference which is the greater in quantity is the smaller in quality (in respect to specific distinctness or non-distinctness), it consequently follows that it is impossible to say *a priori* which degree of quantitative divergence in the organs of copulation is in all Papilios of specific value. Hence a peculiarity observed in the sexual armature of an individual can be an aberrational, a subspecific, or a specific peculiarity; which of the three it actually is, we can learn only from a careful weighing of all the evidence. We have said in the introduction that the evidence remains quite incomplete as long as we do not know the limits of variation of the forms in question ; as these limits, however, can be made out, to a certain degree of correctness, only by comparing many individuals, it becomes self-evident that the question whether a certain peculiar character found in the genital armature of an individual or a number of individuals indicates specific distinctness or not, can only be decided with a certain degree of correctness when the variation of the sexual armature of the allied forms is known, *i.e.* after the examination of a great material.

When we said above that the degree of relationship between a number of closely allied forms could very well appear different according as we take the sexual armature or the wing-pattern as a guide in the arrangement of the forms, we did not give illustrations, because we had to recur to the phenomenon. If it really is true that there are cases in which two forms are in colour and markings the most closely related, while the structure of the genital armature stands in opposition to this relationship, pointing in quite a different direction, a comparison of the characters of the wings and the characters of the apparatus of copulation of the most closely allied forms, which are always geographical representative forms, with the aim of deducing the different relationship as indicated by similarities in the one or in the other organ, must throw a highly interesting light upon those questions which relate to the probable history of the geographical distribution and the origin of those forms.

B. PHENOMENA IN THE VARIATION OF THE ORGANS OF COPULATION RELATING TO SOME QUESTIONS OF THE GEOGRAPHICAL DISTRIBUTION OF ANIMALS.

When we speak here of representative forms, we do so regardless of their being specifically distinct from one another or not ; whether a geographical representative has reached the degree of divergent development which we call specific, or is still a subspecies, is of no importance for the following discussion ; the important point is, that there are differences between the representative forms.

Now, when we see that some Indian and Australian representatives are very similar in colour to one another and dissimilar to the forms inhabiting the interjacent countries, does that mean that the Indian and the Australian forms have separated after the more dissimilar representatives had branched off? When we find the Celebensian mountain form of *P. sarpedon* to bear a much greater resemblance in colour to the form inhabiting the Sunda Islands than to the form found on Celebes at lower elevations, is it correct to conclude that the Celebensian mountain form is a descendant of the Sunda Island form and indicates that there was at a former period a closer connection between Celebes and the Sunda Islands ? When we observe in a great number of cases that the Javanese forms are similar to those from Malacca, while the forms from Sumatra and Borneo again are similar to one another, have we to infer from this fact that there was at one time a land-connection between Malacca and Java independent of Sumatra and Borneo? When West Africa and Madagascar are inhabited by two representatives with a narrow and interrupted band, while the East African form has a broad and uninterrupted band, can we conclude that the Madagascar form is the ancestor (or descendant) of the West African form ? When, finally, we find Central America, the Greater Antilles, the Lesser Antilles, and Venezuela inhabited each by a representative bird or Papilio, and notice that in colour the forms from the first and third and the forms from the second and fourth localities are respectively similar, are we justified in so interpreting the fact as to say the Lesser Antilles have received the bird or insect from

Central America, while the form of the Greater Antilles is an immigrant from Venezuela? We are not going to deny that in a given case applying to one or the other of those questions an affirmative answer could be in accordance with the actual history of the origin of the respective form; but we shall endeavour to explain that similarities between representative forms are capable of being explained otherwise, and that there are many cases in which the agreement of two forms in respect to a certain character *must* be explained otherwise. Let us then inquire into the facts brought to light by our researches on the Eastern Papilios. As in the body of this paper the facts we have to refer to have been more fully dealt with, it will suffice to mention them here briefly without going in for a description of the organs concerned.

Papilio alcinous alcinous from Japan is represented on the Loo Choo Islands by *P. alcinous loochooanus* and in China by *P. alcinous confusus*. There are two prominent characters by which the forms are distinguished: *alcinous* has a black head and a dentate harpe, *loochooanus* a red head and a dentate harpe, and *confusus* a red head and a non-dentate harpe. In the colour of the head, therefore, *loochooanus* agrees with *confusus*, while in the harpe it agrees with *alcinous*; hence the relationship deducible from the similarity in colour of the head is directly opposed to the relationship indicated by the harpe, and it would be equally incorrect to say that *loochooanus* is a descendant of *confusus* on account of the similarity in colour, or to conclude that it has descended from *alcinous* because the harpes are the same; there would be just as much probability of correctness, if we consider only the naked facts here adduced, in the assumption that *alcinous* is the offspring of *loochooanus*, which itself descended from *confusus*, or that the reverse expresses the phylogenetic connection between the three forms, or that the three localities were inhabited at a former period by one form which later on became differentiated into the present three Papilios.

Papilio bathycles chiron from Sikkim, Assam, and Burma, and *P. bathycles bathycloides* from Malacca, Sumatra, and Borneo, are not always distinguishable in pattern, there occurring often specimens which, according to the wing-markings, could be regarded as belonging either to the one or to the other subspecies.

The Javanese representative *P. bathycles bathycles* stands in pattern not so close to *bathycloides* as this does to *chiron*. In the harpes *bathycloides* and *bathycles* come very near each other, while *chiron* exhibits a conspicuous difference from both in the development of the dorsal process (f. 46 to 50). Here again the form inhabiting the interjacent districts is more similar in the *male* copulatory organs to the one, in colour to the other representative.

The three subspecies of *P. cloanthus* inhabiting respectively China (*clymenus*), North India and Burma (*cloanthus*), and Sumatra (*sumatranus*), when arranged according to the similarity in pattern, would stand thus: *clymenus—cloanthus—sumatranus*, whereas in an arrangement according to the development of the harpe *cloanthus* would come first: *cloanthus—clymenus—sumatranus*.

The numerous subspecies of *P. sarpedon* provide us with a number of interesting facts. The Papuan form *choredon* of *sarpedon* inhabiting Australia and New Guinea (inclusive of the islands near it) bears in colour and pattern a rather close resemblance to the specimens of the spring brood of *sarpedon sarpedon* from North India, but is distinguished from the Indian form in the shape of the valve and harpe, in which organs *choredon* is similar to *anthedon* from the Southern Moluccas. The latter, however, differs in colour from *choredon*, agreeing in that respect with *dodingensis* from the Northern Moluccas, which again has the harpe and valve different from those of *anthedon*, and is moreover distinguished from its colour-ally by the presence of an additional red spot on the underside of the hindwing. This spot is present only in *dodingensis* and the two Celebensian forms, *milon* found at lower elevations, and *monticolus* discovered recently at higher elevations on Mount Bonthain. The mountain form *monticolus* agrees in colour with *sarpedon sarpedon* from India and the Sunda Islands, also with the three representatives from the lesser Sunda Islands, comes in the shape of the band of the forewing near *dodingensis*, agrees in the size of the third spot with *timorensis*, stands in the *male* genital armature intermediate between *milon* and *dodingensis*, and hence differs in this respect obviously from *sarpedon*, has the before-mentioned red spot like *milon* and *dodingensis*, and differs from *milon* conspicuously in the shape of the wings, in which it agrees with *sarpedon* from the greater Sunda Islands. Would it really be possible to infer from such characters which contradict each other, pointing each in a certain direction of its own, the history of the origin of *choredon, dodingensis, monticolus,* etc.? Does not such a mixture of relations in opposite directions rather indicate that the similarities and dissimilarities found in representative forms are no signs of a closer or less close phylogenetic connection, and that we have to search for another explanation of the phenomena which will meet the contradiction of the characters?

Papilio aristeus anticrates from North India resembles the Queenslandian representative *parmatus* so much in pattern that Professor Eimer in his somewhat superficial work on the Papilios allied to *podalirius* ("Schwalbenschwänze") did not perceive the slight differences; whereas in the development of the valve and harpe *anticrates* and *parmatus* are opposite extremes. The forms inhabiting the interjacent countries are all much darker, and therefore have a closer resemblance to each other than to *anticrates* and *parmatus* (apart from some intergraduate examples). Hence the forms geographically widest apart would, according to the colour, be the nearest "related." The valve and harpe, however, are in *parmatus* exactly as in the externally different *aristeus* from the Moluccas, which again could be interpreted as being an expression of close relationship. In *hermocrates* from Burma, Borneo, the Philippine Islands, and the lesser Sunda Islands, the genital armature stands somewhat intermediate between that of *anticrates* and *parmatus,* but agrees in the form of the ventral dentate ridge obviously better with that of the first. As *hermocrates* and *aristeus* are connected by intergradations (occurring in Burma) in the wing-pattern as well as in the genital armature, an arrangement of the four allied forms according to the similarity in pattern would be thus: *parmatus* (Queensland and New Guinea)—*anticrates* (India)—*hermocrates* (Borneo, Philippines, lesser Sunda Islands)—*aristeus* (Moluccas); while the arrangement according to the similarity in the organs of copulation would be this: *anticrates*—*hermocrates*—*aristeus*—*parmatus.*

The representative occurring on the island of Celebes (and some islands south of it) is treated as a distinct species (by Eimer even as belonging to a widely different group). In the wing-markings it agrees best with *hermocrates,* but has mostly one band less; in the forewings being falcate it comes again closer to *hermocrates,* which is also geographically the nearest form, than to *aristeus, anticrates,* or *parmatus.* As *rhesus* is an exaggerated development of *P. aristeus,* differing from it in a similar way as *androcles* (Celebes) does from *antiphates* (India and Malayan Islands, Philippines, Moluccas), it sounds very reasonable to consider it to have developed from that form of *P. aristeus* to which it comes nearest in the before-mentioned points, namely from *P. aristeus hermocrates*. An examination of the valve of the *male* shows, however, that there is also very good reason to deduce *rhesus* from *P. aristeus aristeus* or *parmatus*. The ventral dentate ridge is in the eastern forms (*rhesus, aristeus, parmatus*) half-ring-shaped (f. 83, 84), and at a glance distinguishable from the smaller, only slightly curved, ridge of the western forms (*anticrates* and *hermocrates*, f. 73, 74); as intergradations in the two kinds of ridge are unknown (although we have examined a great many individuals), the constant difference in the ridge would be regarded by many systematists as justifying a specific separation of the western forms from the eastern ones, in which case *P. rhesus* agreeing in the ridge very well with the eastern forms, apart from minor differences, would appear to be closer allied to *aristeus* and *parmatus* than to *hermocrates* and *anticrates*, and hence could be considered a descendant of *P. aristeus aristeus* from the Moluccas.

If we take A, B, C as three geographical representatives, a_1 and a_2 as the different degrees of development of the wings of these forms, and b_1 and b_2 as the different degrees of development of the organs of copulation, the several cases above adduced, which we think sufficiently illustrate the questions we are to deal with, can be put diagrammatically as follows:—

A	B	C
a_1	a_1	a_2
b_1	b_2	b_2

The dilemma arising from the contradiction of the characters of geographical representatives allows a satisfactory solution, if we take into account, firstly, that according to the theory of evolution the peculiar modifications of the organs of a certain animal are partly inherited and partly acquired, and that therefore a similarity between two forms, and a dissimilarity, *a priori* neither prove nor disprove a close phylogenetic connection of the forms; and secondly, that when the similarity between two forms is due to inheritance the character common to the two forms is inherited either by both independently from the common ancestor, from which also other forms which have lost that character have descended, or by one of the two forms from the other. The conclusions generally deduced in systematic and other works from the similarity and dissimilarity of geographical representatives in respect to the Geographical Distribution of Animals as a science (not as a mere statement of facts) are to our mind mostly based on the erroneous assumptions, firstly, that every similarity is due to inheritance, and secondly, that the presence of a peculiarity in two forms must necessarily be due to the common character being inherited by one of the two forms from the other. Let us, then, briefly inquire into the question of the probable origin of the similarities and dissimilarities of allied forms, and try to arrive at a correct estimate of their actual value in our judgment of the relationship of the forms in which they are observed.

If we recall to mind that the development of a species into more species is possible only by means of isolated transforming factors associated with a more or less complete prevention of the affected portion of the species from interbreeding with the original stock, the distinguishing characters of locally separated allied forms, which we must regard as the outcome of the transformation of a common ancestral form, must be due to the effect of isolated evolutionistic factors of any kind present in the district inhabited by each single representative form. It is quite possible that when a species separates thus, first into subspecies and then into representative species, one of the descendant forms remains identical with the original form; but considering that in all widely distributed groups of representative forms a considerable time must have elapsed before the insects could spread over the whole area—for instance, from North-West India to the Solomon Islands—it is at least in such cases more probable that the form now inhabiting the country where the ancestral form lived has also more or less changed in characters. And this is the more likely to be true as we know from experiments that the time required to bring about a change in colour in *Lepidoptera* is very short, and as we further know that a change, within a short time, has taken place in the famous Porto Santo rabbits, some birds, and a great number of plants. From the comparison of seasonal forms, which are the more different the greater the contrast of the seasons is, with artificially produced varieties, we can with safety conclude that the degree of divergency of representative forms largely (not entirely) depends on the intensity of the transmuting factors, and that therefore a wider gap between two otherwise allied forms does not necessarily imply that the two animals or plants have been separated for a longer time than less different allies. As it is also well known that the intensity of the biological factors does not gradually increase or decrease as we proceed from one (geographical) extremity of the range of a group of representative forms to the other extremity, it is evident that the degree of diversity of two forms can be independent of the geographical position of the areas inhabited, and that the more similar forms can live widely apart, while the more dissimilar forms may live closer together.

Though the differences between representative forms are the effect of the action of some kind of local biological factors, it does not follow with necessity that the respective factors are found in every place where a certain form now occurs; on the contrary, we have instances that species when living under different conditions do not show any change in their characters. For example, the island of Kalao, south of Celebes, is inhabited by the Celebensian *Papilio rhesus*, and the Kalao individuals are to our knowledge not distinguishable from the Celebes *rhesus*. The other species of *Papilio* found on Kalao (and Djampea, which is close by) are all conspicuously different from the Celebensian representatives, and have mostly developed to peculiar forms not known (as such) from anywhere else; hence there must be biological factors peculiar to those islands which have not had any effect on *P. rhesus*. That *rhesus* is a true Celebensian insect (*i.e.* originated from the ancestor of *P. aristeus* on Celebes) is virtually proved by its characters being quite analogous to those of many other Celebensian forms. Curiously enough, on the island of Djampea, which lies north of Kalao and therefore is somewhat closer to Celebes, Mr. A. Everett obtained no specimens of *rhesus*, but a long series of the Malayan representative *P. aristeus hermocrates*, during the same month when he found a good series of individuals of *P. rhesus* and no *hermocrates* on Kalao. The fact that *rhesus* is the same on Celebes and Kalao admits three explanations: (1) The biological (transmuting) factors are in respect to *rhesus* the same on both islands, which would mean that on Djampea, as inhabited by a representative, the corresponding factors must be different. (2) The biological factors are different on Celebes and Kalao, but the characters acquired by *rhesus* under the action of the Celebensian factors have become inheritable. (3) The characters of *rhesus* and its representatives are not acquired under the influence of biological factors.

If, however, a change in the biological factors does not necessarily imply that a change in the characters of every living being will take place, it is intelligible that in geographical representatives, which live under the influence of different biological factors, one or the other character can remain unaffected. When this takes place independently in several forms which represent each other, the preserved character of the common ancestor renders those forms more similar to each other than to the other representatives which live under conditions that have modified the respective character, while the actual blood-relationship is the same between all the forms. On the other hand, it could also be thought possible that, when a species gradually spreads over a larger area and develops into a number of representative forms, now and again a character acquired by a new form A occurring in locality M will remain unaltered in form B living in locality N which is a descendant of A; in this case the presence of the same character in forms A and B would imply that the forms are closer connected with one another than B is with any other ally except its own descendants. However, if we admit this case to occur, we should maintain that an acquired character is inheritable when the conditions which have brought it into appearance are absent, and we thus should decide offhand the much-contested question whether acquired characters are or are not inheritable. Notwithstanding we believe that ultimately the inheritance of acquired characters will be proved, the contest clearly shows that it is in every case a mere assumption, if we conclude from the presence of the same character or similar characters in two representative forms that one of the two is the parent, the other the daughter form.

If divergency is the effect of transmuting factors, could not similarity also be the outcome of biological factors in different districts? Many facts, for instance the similarity of desert forms, point to an affirmative answer; but we have an almost certain proof furnished us by the experiments * of Standfuss, who succeeded in breeding under artificial conditions from ordinary European *Lepidoptera* such forms as come close to their geographical representatives. As Standfuss was able to produce from our common *Vanessa antiopa* specimens similar to the Central American form (*thomsoni*), there can no longer be any doubt that a similarity in a character of the representative forms of a species can come about in widely separated countries, and will come about when the conditions are favourable.

Taking into account all the points here adduced, it seems to us obvious that all the evidence points to one end, namely, that the similarities and dissimilarities exhibited by representative forms are not an expression of closer or less close phylogenetic connection, but an expression of the similar or dissimilar effect of the action of the different and similar biological factors on the organs in every locality. If we apply this conclusion, which is a necessary consequence of the assumption that the theory of evolution is correct, to the facts observed in the Papilios, there is no longer any contradiction in the characters of the wing and the characters of the organs of copulation. That *Papilio aristeus anticrates* from India agrees with the Papuan representative in the colour, but disagrees with it entirely in the harpe, that the Chinese *P. cloanthus* is in the harpe more similar to the Sumatran than to the North Indian form, that *P. sarpedon teredon* from Ceylon and South India agrees best in the wings and genital armature with the forms occurring on the lesser Sunda Islands and disagrees widely with the form inhabiting all the interjacent countries, that the Nicobar form of *P. agamemnon* is in the presence of a series of red spots on the underside of the hindwing similar to the representative from the

---

* Standfuss, *Handbuch für Schmetterlingssammler* 1896.

Solomon Islands (!), and so on, is no more miraculous than the fact that there *is* seasonal and local variation.

Now, when we observe, for instance, that the Chinese *P. eurypylus* agrees best with the Andaman form of that species, and that the Chinese *P. antiphates* comes close to the Ceylon form, while in both cases the representative inhabiting North India, Burma, and Tenasserim is different, it would certainly sound preposterous to explain the similarity in characters by assuming that there was at a former period a land-connection between Ceylon, the Andaman Islands, and China independent of Continental India. If we concede that this explanation cannot possibly be accepted, and that consequently the similarity finds a correct explanation in the assumption that it is merely the consequence of the similar effect of the biological factors—the effect is positive in respect to the characters which are modified, and negative in respect to the characters which are not modified—of Ceylon, the Andamans, and China, it is not difficult to perceive what great bearings the questions here discussed have on the Geographical Distribution of Animals. Before proceeding to draw the consequences, let us for a moment consider what is the aim of the study of the Distribution of Animals.

"If we keep in view . . . that the present distribution of animals upon the several parts of the earth's surface is the final product of . . . the wonderful revolutions in organic and inorganic nature," says Wallace,* . . . "it will be evident that the study of the distribution of animals and plants may add greatly to our knowledge of the past history of our globe. It may reveal to us, in a manner which no other evidence can, which are the oldest and most permanent features of the earth's surface, and which the newest. It may indicate the existence of islands or continents now sunk beneath the ocean, and which have left no record of their existence save the animal and vegetable productions which have migrated to adjacent lands. It thus becomes an important adjunct to geology. . . . Our present study may often enable us, not only to say where lands must have recently disappeared, but also to form some judgment as to their extent, and the time that has elapsed since their submersion." Indeed, the distribution of animals and plants, especially if the extinct forms are also taken into account, illustrates wonderfully the past history of the earth's surface, and the main object of the study of the distribution of animals and plants is to arrive at conclusions as to the probable past changes in the geological features of the earth. This has so plainly been recognised by nearly all recent systematists, as a glance in a volume of the *Proceedings of the Zoological Society* or of the *Ibis* will show, that, when dealing with a certain group of animals from a certain locality, they endeavour to draw from the affinities in the fauna of the respective district with that of other districts conclusions relating to the geological history of the locality, and it is very generally assumed that similarity in the components of the fauna of two areas means close geological connection, while great dissimilarity means long geological separation. So true this is in a great many cases, so erroneous is the generalisation. Nevertheless it has very often been entirely lost sight of that the present distribution of animals and plants is, to use Wallace's words, "the final product of revolutions in inorganic *and organic* nature," and that accordingly the differences in the fauna of neighbouring districts and the similarities in the fauna of geographically widely separate areas can be quite independent of the geological history of the region in question. If we now recall to mind what we have said above about the similarity

* *Distribution of Animals* 1876. p. 7.

and dissimilarity in the characters of representative forms, it will be evident that the difference in the fauna of neighbouring districts is due to two kinds of entirely different factors, *geological* and *biological* factors, and that the study of the geographical distribution of animals and plants consists of two branches—the *geological branch*, which has to do with such discrepancies in the fauna and flora as allow conclusions to be drawn as to the geological transformation of the *districts* in question, and the *biological branch*, which has to do with those other differences from which we can draw conclusions as to the transformation of the *animals* and *plants* in question. And now we can extend our former conclusion that differences in the characters of representative forms which stand in close phylogenetic connection are an expression of the differences in the biological factors of the districts concerned, to those forms which are representatives in a biological sense and exclude each other in consequence of a similarity in habits, and again to those which cannot exist beside one another in the same district, because they are so different in habits that they require a respectively different environment. It is certainly conceivable that the absence of woodpeckers, which abound in the Indo-Malayan Region, from the lesser Sunda Islands (and all the islands farther east) could be due to the presence of cockatoos, which are, like those, hole-breeders, though we do not maintain that this similarity in habits is the actual reason of the two widely different groups of birds excluding each other in that region ; while, again, the discrepancy in the composition of the West African from the East African fauna is readily explained by the former being a wet forest country, the latter an open and dry country, and the affinities which the West African forest region has in the fauna with the Indo-Malayan Region, and the analogous affinities of East Africa with Western India, are consequences of similar physiographical conditions of the countries, just as the difference in the characters of the West African, East African, and Malagassic morphologically allied representative forms—which I shall call *morphological* representatives, as distinctive of *biological* representatives—are due to differences in the transmuting factors of the respective districts. The contrast in which these explanations of faunistic discrepancies stand to the explanation of those discrepancies which are caused by geological factors is strikingly illustrated by a comparison of the fauna of Madagascar with that of Africa, in so far as we have rightly to conclude from the absence of the large *Carnivora* and *Ungulata*, so abundant in Africa, from Madagascar that this island must already have been isolated from Africa by a wide sea-arm at that time when those animals immigrated from the North into the Aethiopian Region.

The difference in the fauna of various districts is further dependent on the fact that different animals are, as said before, affected by the transmuting factors of a certain locality in a different degree, so that often some forms are, while some are not, modified in characters. We have consequently to take into account also the physiological constitution of the animals when treating upon zoogeographical questions. Hence it is evident that the result of the division of the earth's surface in zoogeographical areas must be different according as we take the faunistic discrepancies depending on the geological factors as the base of division, or those differences which are caused by the biological factors; and again, if we take the biological division, it is obvious that the extent of the areas must be quite different according as to which animals are taken into consideration.

Although *Wallace* has emphasised, throughout his *Geographical Distribution of Animals*, the high importance of the geological branch of zoogeography relating to the past history of the earth and its inhabitants, the minor zoogeographical districts have by most authors been delimited according to differences which are the effect of biological factors. Considering, however, that different environment, such as forest, open land and desert, fresh and salt water, naturally gives subsistence to different animals, and that therefore physiographically different districts a priori are known to have a different fauna, the division of the earth's surface in such zoogeographical districts as are identical with physiographical districts (Sharpe's West African Subregion = district covered with forest; Merriam's Sonoran Subregion of North America = arid country) is as such of very little value, concerning more the geographer than the zoologist. And taking further into account that we now know perfectly well that in every district where the physiographical (and meteorological) conditions are different from those of other districts these differences are accompanied by modifications of the characters of some of the animals, the simple statement that a certain number of the inhabitants are different from the representative forms of the adjacent districts is, like the mere record of the discrepancy in the composition of the fauna, of little consequence for science, as from those statements as such no new general conclusions can be drawn. The biological branch of the study of the geographical distribution of animals, therefore, must have another aim, and that is to be to Zoology what the geological branch is to Geology, namely an adjunctive science which, by a comparative study of the differences in the environment and those in the animals, may help greatly to find the causal connection between the modifications in the organs of the animals and in the environment (in the widest sense), and thus could reveal to us the history of the descent of the forms of animals.

For the sake of illustration let us apply these conclusions to the *Papilio* fauna of a small area, for instance the lesser Sunda Islands. We select this group of islands for two reasons: firstly, because the discrepancy existing between their fauna and that of the larger Sunda Islands has been accounted for by Wallace by geological factors (*Wallace's line*); and secondly, because our conclusions derived from the Papilios stand in opposition to those which Wallace derived from the avifauna. As a complete exposition of the zoogeographical relations of the islands between Lombok and Timor* would be much too extensive for our present purpose, we will restrict the discussion chiefly to the two questions: (1) what conclusions can we draw from the composition of the *Papilio* fauna of the lesser Sunda Islands with respect to the geological history of the group of islands? and (2) is there anything in the distinguishing characters of the Papilios which can serve to solve a question relating to the phylogeny of animals?

The islands of Lombok, Sambawa, Sumba eastward as far as Letti and Moa, are inhabited by fifty odd different forms of *Papilio*, which we can arrange in eighteen series of closely allied representatives. Of these eighteen series one (*canopus*-series) is distributed farther east over the Tenimber Islands and North Australia to the New Hebrides, but has also some allies in the Indo-Malayan Region; the *peranthus*-series occurs from Java to New Guinea, but is absent from Australia; another, *haliphron*-series, is found in Celebes and not in the Indo-Malayan Region; a fourth, *antiphates*-series, is found all over the Indo-Malayan Region, Celebes, and the Northern Moluccas, but is absent from the Southern Moluccas, Australia, New Guinea, and the islands farther east. Seven series reappear in the Indo-Malayan

---

* Compare also Doherty, *The Butterflies of Sumba and Sambawa*, in *Journ. As. Soc. Beng.* 1891: Pagenstecher, *Ueber die Lepidopteren von Sumba und Sambawa*, in *Jahrb. Nass. Ver. Nat.* 1896.

and Papuan Regions; while again seven series are found in the Indo-Malayan Region, three of them also in Celebes, all seven being absent from the Papuan Region.

The lesser Sunda Islands have therefore not a single series of representatives which is purely Australian. From this fact we have to conclude that the immigration of the *Papilio* fauna cannot have taken place from Australia ; and as the physiographical and meteorological conditions of the islands resemble in many respects more those of Northern Australia than those of the larger Sunda Islands, the reason why no Australian types immigrated cannot lie in the biological conditions of the islands, and hence must be accounted for by the assumption of another barrier against immigration. This barrier most probably was, at the time when the lesser Sunda Islands became first populated by Papilios, the same which we perceive at the present epoch, namely the wide Timor Sea separating Timor from Australia.

If we leave out of consideration as being indifferent the types which occur in the Indo-Malayan and Papuan Regions, there remain eight series of representatives which the lesser Sunda Islands have in common with the Indo-Malayan Region, and which are absent from Australia and New Guinea, four of them being represented also on Celebes, and one also on the Northern Moluccas ; opposite these eight types stands one single Celebensian type that is absent from the Indo-Malayan Region. The Indo-Malayan element, therefore, is so strongly predominating that we must assume that there never existed any mechanical barrier of consequence preventing migration of the Papilios from the greater to the lesser Sunda Islands (or in the opposite direction).

Nevertheless there is a marked discrepancy in the fauna of the nearest of the larger Sunda Islands, Java, and the lesser Sunda Islands, as fourteen Javan types are not represented on any of the islands east of the Strait of Lombok, while six types occurring on the lesser Sunda Islands (except Lombok) are not known to have a representative on Java. From the fact that out of the eighteen series of representatives of the lesser Sunda Islands twelve occur on Java it is evident that the absence of fourteen Javan types from the lesser Sunda Islands, and the absence of six lesser Sunda Islands types (five of which are represented in other parts of the Indo-Malayan subregion) from Java, cannot acceptably be explained by the suggestion that there existed at a former period a very much wider strait between the lesser Sunda Islands and Java than there is now. And taking into account the *Papilio* fauna of Lombok alone,* which island has not a single *Papilio* type that is absent from Java, it is obvious that there is no evidence in favour of the assumption that the Strait of Lombok has changed in width since the appearance of the Papilios on the Malayan islands. If we exclude Lombok from the lesser Sunda Islands, and compare the *Papilio* fauna of the rest of the islands with that of Java, there is an important numerical discrepancy which amounts to more than 60 per cent. of the total number of the series of representative forms (as opposed to the total number of representative forms) inhabiting Java and the lesser Sunda Islands. As this amount † is very high compared with the numerical discrepancy

---

* The island has recently been visited by three explorers (W. Doherty, A. Everett, and H. Fruhstorfer).

† As Flores is lepidopterologically almost a blank the numbers may become quite different in consequence of the exploration of this large island. The absence of many types from small islands which are found on neighbouring large islands is quite natural, considering that the small islands do not afford so great a variety in the conditions of life as would be required by a multitude of diverse inhabitants.

between Sumatra and Java (about 38 per cent. of the total number), and that between Sambawa and Timor (33 per cent. of the total number), and is not much inferior to the amount of discrepancy between the lesser Sunda Islands and Australia (65 per cent.), the lesser Sunda Islands can be considered as a faunistic district in opposition to the greater Sunda Islands and Australia; and it is very interesting to note that almost the same amount of numerical discrepancy must in the case of the lesser Sunda Islands and Java be accounted for by biological factors, while in the case of the lesser Sunda Islands and Australia the only acceptable explanation is the assumption of the existence of a geological barrier between Australia and Timor at the time when the lesser Sunda Islands became populated by Papilios.

Turning now our attention to the second question proposed above to be discussed, we note that the characters of the forty-seven specifically or subspecifically different representative forms of *Papilio* occurring on the lesser Sunda Islands are such that forty-one of the forms are not found (as such) outside the group of islands, and hence are "peculiar" to the group. As these forty-one forms are so distributed that every island has some of its own and others in common with other islands of the group, while not a single form is identical on all the islands and at the same time distinguished by some peculiarity from the representative occurring on Java, Borneo, or in Australia, each island is in fact a small faunistic district, and the question arises whether the faunistic peculiarity exhibited by each island is not due, instead of to differences in the biological conditions existing on each island, to the mechanical separation of the islands by arms of sea, and hence finally would have to be referred to the action of geological factors. In the introduction we have endeavoured to show on *a priori* grounds that mechanical geographical separation as such cannot give rise to a new form; let us now consider the *a posteriori* reasons which speak against the origin of forms by mechanical isolation.

The widely distributed *Papilio sarpedon* has on the lesser Sunda Islands developed into three subspecies, one inhabiting Adonara, Sambawa, and Lombok—specimens from the latter island come very close to those from Java—a second found on Sumba, and a third occurring on Timor and Wetter. The external characters of the Timor form are such that it agrees in the shape of the hindwing best with the representative from Ceylon, in the breadth of the band of the forewing with the Australian form, and in the special shape of the anterior portion of the band with *monticolus* from Celebes, while it bears also great affinities to the forms from Sumba and Adonara. Setting apart all the reasons brought forward against the theory of isolation in the introduction, and conceding, for the sake of argument, that geographical isolation of aberrational specimens could lead to the origin of a new form of animal without the help of any biological factor, the combination of different affinities in the wing-characters of the Timorese, Sumbanese, or Adonaranese *sarpedon* forms might be thought due to the first specimens immigrated into each island having accidentally possessed the respective combination of characters. As a combination of characters such as exhibited by *P. sarpedon timorensis* is not found even in a less obvious degree in any specimen of *sarpedon* from the larger Sunda Islands and Australia we have seen, and therefore must be very rare, it is certainly scarcely probable that just such specimens which had that combination of characters should have happened to be the first to come to Timor; and this applies to all those localised forms, not only of the lesser Sunda Islands, but also of all other districts, the characters of which point in different directions. The improbability, however, changes into impossibility, if we take further into account the characters of the organs of copulation. We know from the researches laid down in the body of this paper that the genital armature varies quite independently from the wings and their markings, and that therefore the association of distinguishing characters relating to the wings with such found in the organs of copulation in morphological representatives has nothing to do with correlative development; in not a single case have we observed that a certain aberrational wing-character constantly reappearing among the individuals of a certain geographical form of *Papilio* is accompanied by an aberrational character in the organs of copulation. As, however, most geographical races are distinguished by characters of the wings and characters of the genital armature, it is evident that the explanation of the peculiarities in the wing-markings of a representative form by mechanical isolation of parent specimens which accidentally possessed those peculiarities (in a lower degree) is not an explanation of the presence of the peculiarities in the organs of copulation of the respective form. In order to account for the fact that *P. sarpedon timorensis* is (in the sexual armature of the *male*) different from all other forms of *sarpedon*, and comes in these organs nearest to the Ceylonese, Sumbanese, and Australian forms, and disagrees considerably with the representatives from Adonara, Sambawa, Lombok, and the Indo-Malayan Region, the theory of isolation would have to assume that the ancestral specimens of *timorensis*, no matter whether they immigrated into Timor from Australia or from the Malayan islands, must have been distinguished by a rare mixture of external peculiarities combined with an equally rare character in the organs of copulation. The rarity of such a combination in aberrations—in fact we never have found an aberrational individual in which the characters of both the wings and genital armature pointed in the direction of one representative form, instead of in different directions—is directly opposed to the fact that most geographical races exhibit that combination; this contradiction can only be solved by abandoning the assumption that the association of distinguishing characters of the wings with such of the sexual organs in localised forms is due to the geographical isolation of ancestral individuals which accidentally possessed such a combination of characters, and by accepting as the final cause of the presence of those characters the modifying action of isolated biological factors peculiar to the district where the peculiar form has originated.

If we thus attribute the existence of differences in the characters of the specimens of the same species inhabiting different districts to the presence of a difference in the biological factors of the districts, the degree of divergency of the forms is dependent on the degree of sensitiveness of the species to the action of the external biological factors, as maintained by Darwin, and after him by Weismann and others, in several places; further, on the intensity of the factors and the time they have been active; and thirdly, on the degree and duration of mechanical isolation as a prevention of the annihilating effect of intercrossing. As a difference in the physiological constitution of various species can be proved only by rearing them under exactly the same external conditions, and as we do not know in the case of species found in the same small island whether they actually do exist under the same external influences, it is not possible for us to decide with any degree of certainty whether the difference in the amount of divergency exhibited by the forms occurring in the same district is due to the different physiological constitution of the species differently affected, or to differences in the special external conditions under which each species perhaps exists. Nor are we able to say whether the greater diversity exhibited by a certain form is due to the form having come under the influence of the biological factors of the respective district at an earlier period than another form which shows a less high degree of divergency, unless the sensitiveness of the forms has been tested by experiment. And again, the higher or lesser degree of geographical isolation, though often corresponding to the greater or lesser diversity of the isolated forms, is a factor the actual influence of which it is in a given case scarcely possible to estimate rightly. The great diversity of the representative of *Papilio peranthus* on the island of Sumba compared with the lesser diversity between the representatives inhabiting Java, Sambawa, Timor, Kalao and Djampea, Celebes, etc.; the high degree of specialisation of the representative of *Papilio memnon* on Timor compared with the lesser degree of divergency of the forms occurring on Sumba, Sambawa, Lombok, Java, etc.; the absence of a difference between some Lombok and Java forms of *Papilio* which on the other lesser Sunda Islands have developed divergently, compared with the presence of a difference in other Lombok forms; the great similarity between the specimens of *Papilio sarpedon* found in the area extending from N.W. India to Lombok, the gradual transition to the more divergent Adonara form, and the obvious divergency in the pattern and shape of the wings and the sexual armature (of the $\male$) of the forms inhabiting Sumba, Timor, and Wetter respectively, compared with the great diversity of so many forms in N. India, Sumatra, Java, Lombok, Sambawa, and Adonara; the identity of the Malayan form of *Papilio aristeus* on the Philippines, Borneo, the lesser Sunda Islands, and the island of Djampea, compared with the great diversity exhibited in the same area by *Papilio memnon* and other species, and so on, are phenomena which seem to us most readily explainable by the surmise that the diverse species are sensitive in a different degree to the external biological factors; while the surprisingly great divergency of so many Celebensian Papilios must be attributed to a high intensity of the biological factors on that island and to a high degree of geographical isolation. However, it does not appear to us very probable that the narrow sea between Celebes and Borneo is a sufficiently effective barrier to prevent passive migration * of the widespread and common low-land Papilios of Borneo, such as *P. agamemnon, sarpedon, polytes*, etc., from Borneo to Celebes; and the same could be said of New Guinea, New Britain, and other islands. In fact the absence of intergraduate specimens between two representative forms in localities which are geographically close together—the intergradations may occur in other localities—makes it manifest to us that, as the effect of the biological factors would in these cases be annihilated, or at least much lessened, by specimens of the old stock coming into the country in consequence of the insufficiency of the geographical barrier, there must be another factor active to assist the external biological factors to such a degree that the migration of fresh individuals, if it does not take place very often, is not able to prevent divergent development. Now, recalling to mind the kind of variation in the scent-organs found to exist in a certain moth, as mentioned on p. 435, it is evident that where such a variation takes place in the geographical forms of a species the difference in the organs of smell and scent must be of great influence as preventing (to a certain extent) intercrossing between racial

---

* Passive migration by means of hurricanes seems to occur often among birds, as the records of foreign birds found in England prove; in the case of *Lepidoptera* the power of resistance is certainly slighter, but the effect of migration in respect to intercrossing with specimens of the same species in the new locality will often be annihilated in consequence of the new-comers having too much suffered to be fit for propagation.

forms, so that the distinguishing characters can accumulate in spite of the occasional immigration of specimens of the old stock. As we shall have to deal with this kind of physiological selection, which in fact is a kind of sexual selection, in another place when treating upon the variation of glandular organs in *Lepidoptera*, we do not now enter upon a discussion of that factor. The variation of *Lepidoptera* explained in the present paper points, however, to a further factor being active, the importance of which as a means of prevention of intercrossing has been recognised as far back as 1849, and we shall try now, on the ground of our researches, to estimate the influence of geographical polymorphism in the sexual organs upon the divergent development of the *Lepidoptera*.

C. Mechanical Selection.

The genital armature of *Lepidoptera* has the function of a special prehensile apparatus during copulation. As such apparatus is found in all *Lepidoptera*, we can conclude that its function is of high importance for the act of copulation. When, however, a physiologically important organ is found to be of a different structure in every two species, it is likewise correct to conclude that the difference in the structure has some physiological significance. Hence we think that the inference to be drawn from the fact that the copulatory organs are different in different species has to be that the specialisation in structure means a specialisation in the function as prehensile organ—*i.e.* that, as with the help of the apparatus the *male* has to take a hold on the *female* and the *female* on the *male*, the organs of the sexes of the same species are best adapted to each other. This inference implies that specimens of diverse species can less easily copulate than specimens of the same species, and that there might be cases in which copulation between different species would be impossible in consequence of the highly divergent specialisation of the organs of copulation. A similar view has been held by the earlier writers on the subject, Siebold* and Dufour,† the latter giving expression to his opinion in the well-known sentence that "*l'armure copulatrice . . . est la garantie de la conservation des types, la sauvegarde de la légitimité de l'espèce.*"

The different development of the organs of copulation is here alleged to be a means of prevention of intercrossing between specifically different specimens, and the same has more recently been said by Escherich,‡ Hoffer,§ and others. Though we know from experience that Escherich's assertion that an effective copulation between diverse species is never possible is erroneous, and that therefore Verhoeff ‖ is right in rejecting this part of Escherich's theory, it remains nevertheless evident that under the premiss that the *male* and *female* of one species really are adapted to each other in respect to the genital armature, specimens which are not adapted to each other cannot so easily unite in copulation, and that even when a union is effected the penis will sometimes be prevented from entering the vagina in consequence of the vaginal armature, to which the organs of that *male* are not adapted, being a mechanical barrier. Before, however, this conclusion can be accepted, it has to be ascertained whether the actual structure of the *female* and *male* genital armature is in accordance with the above premiss.

Gosse¶ was quite right in expecting to find that "every peculiarity in the

---

* *Vergl. Anat.* Berlin, 1848. § 354. note 2.
† *Ann. Sc. Nat.* (3). I. 1844. p. 253.
‡ *Verh. z. b. Ges. Wien.* 1892. p. 234.
§ *Mitth. Nat. Ver. Steiermark.* 1888.
‖ *Ent. Nachr.* 1893. p. 44.
¶ *Trans. Linn. Soc. Lond.* (2). II. 1883. p. 279.

prehensile organs of the *male* would have a corresponding peculiarity in that part of the *female* body which they were formed to grasp," but he failed to discover " the corresponding peculiarity in the *female*," as he laboured under the erroneous assumption that "it is the *exterior* of the final segments of the *female* abdomen that are seized *in coitu*." We have been able to show in the body of this paper that the *male* claspers seize in copulation the protruding vaginal bulb, taking hold on chitinised folds and ridges, and not the terminal segments; and we have further been able to demonstrate the following two important points:—

(1) We have a pair of *Papilio memnon* united *in coitu*, from which we see clearly that the harpe of the *male* fits exactly the lateral ridge of the vaginal bulb of the *female*, while the high and broad processes at the mouth of the vagina, besides being a guide to the penis, take hold on the internal parts of the ninth segment of the *male*. When uniting artificially *males* and *females* of other species, we find that the organs of one sex fit those of the other of the same species, while the copulatory organs of different species, for instance of a *male* of *P. machaon* and a *female* of *P. podalirius*, or of *memnon* and *helenus*, do not fit each other. We are therefore justified in concluding that the organs of the sexes of the same species are better adapted to each other than to the organs of other species.

(2) The vaginal armature is in every species of *Papilio* examined different from that of every other species examined. As our researches, only part of which have been laid down in this paper, relate to a great many species of various groups of *Papilio* and other *Lepidoptera*, we can safely say that the *females* of every species are, like the *males*, in the sexual armature distinguished from the *females* of every other species. A comparison of the figures on Pl. XIX. will show that the divergency between the apparatus of representatives of diverse groups is in every respect so great, that it is readily conceivable that a *male* the apparatus of which is adapted to such organs as represented by f. 181 cannot effectively copulate with a *female* the organs of which are so different as those represented by f. 163, 179, or 182.

The position of the vaginal orifice and the lateral fold of the vaginal bulb, the number, length, and position of the ridges and processes at the orifice of the vagina, and the development of the eighth segment of the *female*, as well as the special structure of the clasping organs of the *male*, are in diverse groups of *Papilio* so widely different that there can be no doubt that the divergency in the organs of copulation has rightly been interpreted by Siebold as a means of prevention of intercrossing. It is obvious that the more the organs of different species resemble each other the less they will be able to prevent copulation. In closely allied species, therefore, the effect of this mechanical barrier against intercrossing will generally be less great than in the case of more diverse forms; but we must recall to mind that even in the most closely related species, such as *Papilio euchenor* and *depilis* (f. 51 and 65), the organs of copulation, though built up after the same plan, can in the position and direction of the spines, hooks, and ridges of the harpe be so different that the prehensile organs of the *male* come in contact during copulation with quite different points of the vaginal region of the *female*, and that it is accordingly also in such cases obvious that an intercrossing is not possible without violence. In groups of *Lepidoptera* in which the genital armature is very simple the effect as a mechanical barrier must necessarily be very slight; this seems to us to be one reason why specimens of diverse species of *Saturnidae*, for instance, copulate much easier than is the case with specifically different individuals in other groups.

Notwithstanding it is manifest that the difficulty to obtain hybrids is in many cases to be attributed to the diverse development of the copulatory organs, the question is of very little importance as long as we take into account only an intercrossing between species and species. According to our definition, specifically distinct animals or plants cannot fuse together; the appearance of hybrids has no influence on the evolution of the species which have produced them, though they may obscure the fact of the specific distinctness of the parents.* The barrier between species is the degree of diversity already attained, and as this barrier is absolute, the presence of any other barrier must be indifferent. The question as to the evolutionistic influence of the peculiar development of the copulatory organs, therefore, does not relate to those forms which have already attained that degree of diversity which renders a fusion with other forms impossible, but to those lower degrees of diversity which are still able to fuse together under favourable circumstances. Hence we have to inquire whether the variation of the copulatory organs within the limits of a species is such that the difference in the organs in question between varieties of the same species can be thought to be a barrier against the sexual intercourse between the varieties.

The effect of the divergency of the copulatory organs in respect to prevention of intercrossing can very well be compared with that of geographical isolation, as the effect of both the morphological and the geographical factor depends on the extent of the mechanical barrier. The Atlantic Ocean is certainly an effective barrier between the Nearctic and Palaearctic Regions, although even such a wide sea cannot absolutely prevent a migration from one region to the other; the Ganges Valley which separates the North Indian from the South Indian fauna, and the straits between the various islands of the Indo-Australian Archipelago, are barriers which render the occurrence of an intercrossing between specimens born on the opposite sides of the barrier, though by no means impossible, doubtless very rare; in fact every barrier which isolates to any degree, however small the degree may be, a number of specimens *A* from other specimens *B* renders the intercourse of the two sets of individuals with one another less probable than the copulation of the specimens of each set *inter se*, and hence is a barrier against intercrossing. If, therefore, the varieties of the same species exhibit in both sexes any diversity in the the organs of copulation, we can rightly infer that this diversity will act like a geographical barrier, isolating the varieties from one another to a certain extent according as the anatomical diversity is great or slight. As our researches have proved that the premiss of this conclusion is correct as regards the Eastern Papilios, namely that there are varieties which differ in the genital armature from other varieties of the same species in both sexes, it is manifest that the diversity in the genital armature of the Papilios has a great bearing upon the divergent development of the varieties.

We have found that there is individual and geographical variation in the genital armature of the Eastern Papilios. As regards individual variation, it is obvious that specimens which have any marked peculiarity in the prehensile organs, additional hooks or ridges, in which they deviate from the normal, will be at a disadvantage in

---

* As besides the mechanical barrier against intercrossing there is, in most species with highly developed organs of sense, a psychological barrier active, hybrids must in the state of nature be very rare in comparison with the number of specimens which are not hybrids. Specimens connecting two supposed distinct species have not rarely been treated as hybrids, though the natural conclusion from the regular occurrence of such intergradations would be that the supposed distinct species form one, di- or polymorphic, species, provided that the reverse has not been *proved*.

respect to propagation compared with the normal specimens in which copulation is more facilitated ; monstrosities and sports in respect to the genital armature, in short every specimen with an anomalous development of the organs of copulation is less favoured than the rest of the individuals. Such abnormal individuals have, therefore, the same position in the struggle for propagation, as in the struggle for existence those specimens have which are less adapted to the circumstances of life than their cospecific rivals. If it is right to conclude in the latter case that the struggle for existence leads in the long run to a survival of the fit (*Romanes*), we can apply the same inference to our special question and say that the specimens abnormal in the genital organs will in the struggle for propagation succumb, and that consequently the anomalies in the organs of copulation will become rarer and rarer, the individuals acquiring gradually by breeding in and in the same form of prehensile organs, comparatively few specimens deviating widely from the normal. It is apparent that the mechanical selection thus effected will end in a comparatively great constancy of the prehensile apparatus.* A number of the forms of *Papilio* dealt with in this paper have attained a remarkably high degree of constancy in the genital armature, for instance several forms of *P. sarpedon* and *P. aristeus*, *P. alcinous confusus*. As the conclusion applies to every segregate form of *Papilio*, it is evident that, when a species develops divergently into geographical races, each race (or some of them) can, and will, acquire a special normal form of the genital armature, if the evolutionistic factors of the locality affect the organs of copulation. Many species have not yet attained that degree of divergent ramification. The geographical forms of *Papilio aegeus, machaon*, some of the races of *P. sarpedon*, namely *sarpedon sarpedon* and *sarpedon semifasciatus*, the two eastern races of *P. aristeus*, *P. aristeus aristeus* and *P. aristeus parmatus*, and others, we have found to be identical or almost so with one another in the genital armature. In the geographical forms of other species the specialisation in the prehensile organs is already visible, but not very conspicuous, as in *P. cloanthus clymenus* and *P. cloanthus cloanthus* ; while in others again the degree of divergency in the organs of copulation of geographical races is so high that the normal of one form differs from the normal of the other so much that the one stands far outside the usual limits of variation of the other. Now, when it happens that forms so diverse in the genital armature inhabit neighbouring districts, as is the case with *P. sarpedon teredon* (S. India), *P. sarpedon sarpedon* (N. India to the Philippines and Java), and *P. sarpedon milon* (Celebes); *P. alcinous confusus* (China and Formosa) and *P. alcinous loochooanus* (Loo Choo Islands) ; *P. bathycles chiron* (N. India to Burma) and *P. bathycles bathycloides* (Malacca), it is manifest that the occasional (passive or active) migration of specimens from one district to the other, which we must concede to occur, as otherwise the wide range of many species would not be explainable, can have no great retarding effect on the divergent development of the forms, since an effective copulation of the immigrants with the occupants of the district is improbable, and as further, if copulation should occasionally be effective, the characters of the descendants of the new-comers will soon get swamped away by mechanical selection. Thus it is conceivable that, when a certain degree of divergency is attained by geographical races, mechanical selection will greatly help to accelerate evolution, and that it is able to prevent retrograde development when the geographical barrier which formerly separated two forms has, in consequence of

---

* Marked dimorphism in the organs of copulation to which the variation of the organs could also lead does not seem to occur ; see p. 499.

geological revolutions, become so changed that the isolation is much more incomplete than it formerly was. And when the comparatively great diversity in the organs of copulation has become so constant that intergradations in these organs between two respective subspecies seldom occur in the area occupied by these forms, though the subspecies may be connected with one another by intergraduate forms inhabiting other districts, it is possible that, when the districts of the two subspecies, if geographically close together, like the Bornean and Philippine Islands, become so extended that they overlap, the two diverse subspecies may live together without immediate and complete fusion.

It will have been noted that the effect attributed to mechanical selection is twofold. The variation of the genital armature caused by transmuting factors of any kind is so guided by mechanical selection that every incipient species will acquire a special armature of its own. As the variation of the organs of copulation is independent of that of the wing-pattern, mechanical selection does not directly affect the latter; but its indirect influence must be of importance, when the special genital armature acquired by mechanical selection has become so specialised that it is, like local isolation, a means of prevention of intercrossing.

In consequence of the demonstration of individual and geographical variation of the organs of copulation it becomes very probable that Eimer's * interesting suggestion is correct, that the variation of a species concerns also egg and sperm. If geographical and individual variation in these cells would be demonstrable as in the case of the ectodermal sexual organs, we should most probably find that the variation in the size and form of the sperm and in the structure of the micropyle apparatus † of the egg is independent of the characters of the wing and sexual armature, and that accordingly similar conclusions could be drawn as in the case of the organs of copulation. As diversity in sperm and egg would partly or totally prevent fecundation, infertility between certain individuals of the same race and between the specimens of diverse races would be the result of the variation, and hence the demonstration of the occurrence of such variation would give an actual base to Romanes's theory of physiological selection, and thus supply one explanation of that hitherto unsolved question how the general infertility between species has come about.

"The real difficulty," says Darwin,‡ "in our present subject is not, as it appears to me, why domestic varieties have not become mutually infertile when crossed, but why this has so generally occurred with natural varieties, as soon as they have been permanently modified in a sufficient degree to take rank as a species. We are far from precisely knowing the cause; nor is this surprising, seeing how profoundly ignorant we are in regard to the normal and abnormal action of the reproductive system."

We observe that the cause of infertility adduced by us would not be physiological diversity, but mechanical impossibility of fertilisation. Mechanical selection acting in a similar way as in the case of the organs of copulation in the direction of adapting within each race sperm to egg by excluding extreme varietal specimens from propagation, would raise a barrier between the diverse subspecies which would admit an independent development ending in specific distinctness of mutually infertile forms. Mechanical selection acts upon variation caused by other factors, and is, therefore, like mechanical geographical isolation, a preservative, not a productive factor.

---

* *Artb. bei Schmett.* II. 1895. p. 15.
† Standfuss, *Handb. f. Schmetterlingssammler* 1896.
‡ *Origin of Species* 6th ed. London, 1888. p. 36.

LIST OF INSECTS MENTIONED IN THIS PAPER.

Acherontia atropos, in Hübner, *Samml. Eur. Schm.* II. *Sphing.* f. 63 (1798-1805).
Callimorpha dominula dominula, in Esper, *Eur. Schm.* IV. t. 83. f. 1. 2 (1786).
,, *dominula persona*, in Hübner, *Samml. Eur. Schm.* III. f. 319-322 (1798—1805).
Euploea rhadamanthus, in Distant, *Rhop. Mal.* 1885. t. 4. f. 4. 5 (as *diocletianus*).
Gonopteryx cleopatra, in Hübner, *Samml. Eur. Schm.* f. 445. 446 (1798—1805).
,, *rhamni*, *ibid.* f. 442-444.
Hepialus humuli hethlandicus, in *Entomol.* 1880. t. 3.
,, *humuli humuli*, in Esper, *Eur. Schmett.* IV. t. 80. f. 1 (♂). 2 (♀). (1786).
Ornithoptera, see *Troides*.
Papilio aegeus, in Novit. Zool. 1895. p. 304.
,, *aegeus adrastus*, in *Reise Novara, Lep.* I. t. 16. f. a (1865).
,, *aegeus aegeus*, in Staud. & Schatz, *Exot. Schmett.* I. t. 4 (1884).
,, *aegeus bismarckianus*, in Novit. Zool. 1895. p. 308.
,, *aegeus keianus*, in Novit. Zool. 1896. p. 422.
,, *aegeus ormenus*, in Novit. Zool. 1895. p. 306.
,, *aegeus ormenus* ♂-ab. *ormenus*, in *Voyage Coquille, Ins.* t. 14. f. 3 (1829).
,, *aegeus ormenus* ♂-ab. *othello*, in Novit. Zool. 1894. p. 332. n. 3.
,, *aegeus ormenus* ♂-ab. *pandion*, in *Trans. Linn. Soc. Lond.* XXV. p. 56. n. 72 (1865).
,, *agamemnon*, in Novit. Zool. 1895. p. 447.
,, *agamemnon argynnus*, in *Ann. Mag. N. H.* (6). II. p. 235 (1888).
,, *agamemnon neopommeranius*, in *Berl. Ent. Zeitschr.* XXXI. t. 6. f. 4 (1887).
,, *alcinous*, in Novit. Zool. 1895. p. 267. t. VI. f. 1-20, 32-38.
,, *alcinous alcinous*, in Klug, *Neue Schmett. Ins. Samml. Berlin* 1836. t. 1 (♂, ♀).
,, *alcinous confusus*, in Novit. Zool. 1895. p. 269.
,, *alcinous loochooanus*, in Novit. Zool. 1896. p. 421.
,, *ambrax*, in Novit. Zool. 1895. p. 354.
,, *ambrax ambrax*, in Dehaan, *Verh. Nat. Gesch. Ned. ovez. bez.* 1840. t. 7. f. 1 (♂). 2 (♀).
,, *antiphates*, in Novit. Zool. 1895. p. 410.
,, *aristeus*, in Novit. Zool. 1895. p. 418.
,, *aristeus anticrates*, in Gray, *Cat. Lep. Ins. B. M.* I. t. 3. f. 3 (1842).
,, *aristeus aristeus*, in Cramer, *Pap. Exot.* IV. t. 318. f. E. F (1782).
,, *aristeus hermocrates*, in *Reise Novara, Lep.* I. t. 12. f. E (1865).
,, *aristeus parmatus*, in Gray, *l.c.* t. 3. f. 2 (1842).
,, *aristolochiae*, in Novit. Zool. 1895. p. 245.
,, *aristolochiae antiphus*, in Donovan, *Ins. of India* t. 15. f. 2 (1800).
,, *aristolochiae aristolochiae*, in Distant, *Rhop. Mal.* 1885. t. 31. f. 6. 7 (as *diphilus*).
,, *aristolochiae austrosundanus*, in Novit. Zool. 1895. p. 249.
,, *bathycles*, in Novit. Zool. 1895. p. 437.
,, *bathycles bathycles*, in *Nov. Act. Ac. Nat. Cur.* 1831. t. 14. f. 6. 7.
,, *bathycles bathycloides*, in *Berl. Ent. Zeitschr.* XXVIII. t. 10. f. 3 (1884).
,, *bathycles chiron*, in *Trans. Linn. Soc. Lond.* XXV. p. 66. note (1865).
,, *bathycles chiron* ab. *chironides*, in *Berl. Ent. Zeitschr.* XXVIII. t. 10. f. 4 (1884).
,, *canopus*, in Novit. Zool. 1895. p. 341.
,, *caunus*, in Novit. Zool. 1895. p. 376.
,, *caunus aegialus*, in Distant, *Rhop. Mal.* 1885. t. 27B. f. 5.
,, *caunus caunus*, in Westwood, *Cab. Or. Ent.* 1848. t. 9. f. 2. 2*.
,, *caunus mendax*, in Haase, *Unters. üb. Mimicry* 1893. t. 8. f. 53.
,, *cloanthus*, in Novit. Zool. 1895. p. 445.
,, *cloanthus cloanthus*, in Westwood, *Arc. Ent.* 1841. t. 11. f. 2.

Papilio cloanthus clymenus, in Leech, *Butt. from China and Japan* 1893. t. 32. f. 2.
,, *cloanthus sumatranus*, in *Iris* VI. 1894. p. 27.
,, *clytia*, in Novit. Zool. 1895. p. 364.
,, *deiphobus*, in Clerck, *Icones* II. t. 25. f. 1 (1764).
,, *deiphontes*, in Staud. & Schatz, *Exot. Schmett.* I. t. 5 (♂,♀) (1884).
,, *deipylus*, in *Verh. z. b. Ges. Wien.* 1864. p. 323. n. 455.
,, *depilis*, in Novit. Zool. 1895. p. 339.
,, *depilis depilis*, in Novit. Zool. 1895. p. 340 (N. Britain, *nec* N. Ireland).
,, *depilis novohibernicus*, in Novit. Zool. 1896. p. 422.
,, *echidna*, in Dehaan, *l.c.* t. 8. f. 6.
,, *euchenor*, in Novit. Zool. 1895. p. 339.
,, *euchenor euchenor*, in *Voyage Coquille, Ins.* t. 13. f. 3 (1829).
,, *euchenor godarti*, in *Ann. Sc. Phys. Nat. Lyon.* 1856. p. 398 (♂, *nec* ♀).
,, *euchenor obsolescens*, in Novit. Zool. 1895. p. 339.
,, *eurypylus*, in Novit. Zool. 1895. p. 429.
,, *gambrisius*, in Cramer, *l.c.* II. t. 157. f. A. B (1779).
,, *inopinatus*, in Smith & Kirby, *Rhop. Exot., Pap.* 1893. t. 12. f. 1 (♂). 2 (♀).
,, *lowi*, in *P. Z. S.* 1873. t. 33. f. 6 (♂).
,, *macfarlanei*, in Novit. Zool. 1895. p. 446.
,, *machaon*, in Novit. Zool. 1895. p. 272.
,, *machaon flavus*, in Tutt, *Brit. Butt.* 1896. p. 218; as *machaon* in Barrett, *Lep. Brit. Isl.* 1893. t. 1. f. 1. 1a.
,, *machaon hippocrates*, in Pryer, *Rhop. Nih.* 1886. t. 1. f. 1A. 1B.
,, *machaon machaon*, in Roesel, *Ins. Belust.* I. 2. 1746. t. 1.
,, *machaon sphyrus*, in Novit. Zool. 1895. p. 275.
,, *mandarinus*, in Novit. Zool. 1895. p. 408.
,, *mayo*, in *P. Z. S.* 1873. t. 63. f. 1 (♂).
,, *memnon*, in Novit. Zool. 1895. p. 312.
,, *memnon agenor*, in Distant, *Rhop. Mal.* 1885. t. 27B. f. 7; t. 28. f. 1-7; t. 29. f. 1. 4. 5; under various names.
,, *memnon memnon*, in *Trans. Linn. Soc. Lond.* XXV. t. 1. f. 1. 2. 3. 4 (1865).
,, *nomius*, in Novit. Zool. 1895. p. 421.
,, *paradoxus*, in Novit. Zool. 1895. p. 371.
,, *paron*, in Smith & Kirby, *Rhop. Exot., Pap.* 1893. t. 31. f. 3. 4.
,, *peranthus*, in Novit. Zool. 1895. p. 391.
,, *phestus*, in *Voyage Coquille, Ins* 1829. t. 14. f. 2. A. B.
,, *podalirius*, in Novit. Zool. 1895. p. 402.
,, *polytes*, in Novit. Zool. 1895. p. 343.
,, *polytes alcindor*, in Oberthür, *Et. d'Ent.* IV. 1879. t. 6. f. 4 (♀); Novit. Zool. 1895. p. 350.
,, *polytes alphenor*, in Cramer, *l.c.* I. t. 90. f. B (♀) (1779); ♂ as *ledebouria* in Kotzebue's *Reise* III. t. 3. f. 7 (1821).
,, *polytes borealis*, in Novit. Zool. 1895. p. 348.
,, *polytes borealis* ab. *thibetanus*, in Oberthür, *l.c.* XIII. 1886. p. 14.
,, *polytes nicanor*, in Staud. & Schatz, *Exot. Schmett.* I. 1884. t. 3 (♂, ♀).
,, *polytes nikobarus*, in *Verh. z. b. Ges. Wien.* 1862. p. 483. n. 112.
,, *polytes perversus*, in Novit. Zool. 1895. p. 353.
,, *polytes polytes*, in Distant, *l.c.* t. 33. f. 7-10.
,, *polytes theseus*, in *Trans. Linn. Soc. Lond.* XXV. t. 2. f. 2. 4. 7 (1865).
,, *rhesus*, in Eimer, *Artb. u. Verwandtsch. Schmett.* 1889. t. 4. f. 6.
,, *rumanzovius*, in Kotzebue's *Reise* III. 1821. t. 2. f. 4a. 4b (♀); ♂ as *krusensternia*, *ibid.* t. 3. f. 5a. 5b.

Papilio *sarpedon*, in Novit. Zool. 1895. p. 441.
,,   *sarpedon adonarensis*, Novit. Zool. 1896. p. 324.
,,   *sarpedon anthedon*, in Staud. & Schatz, *Exot. Schmett.* I. 1884. t. 6.
,,   *sarpedon choredon*, in Scott, *Austr. Lep.* II. 1890. t. 17.
,,   *sarpedon dodingensis*, in Novit. Zool. 1896. p. 323.
,,   *sarpedon impar*, in Novit. Zool. 1895. p. 443.
,,   *sarpedon imparilis*, in Novit. Zool. 1895. p. 443.
,,   *sarpedon isander*, in Smith & Kirby, *Rhop. Exot.* I. 1888. *Pap.* t. 6. f. 3.
,,   *sarpedon jugans*, in Novit. Zool. 1896. p. 324.
,,   *sarpedon milon*, in *Trans. Linn. Soc. Lond.* XXV. 1865. t. 7. f. 2 (as *miletus*).
,,   *sarpedon monticolus*, in *Societas Entomol.* 1896. p. 20.
,,   *sarpedon sarpedon*, in Distant, *Rhop. Mal.* 1885. t. 22. f. 6.
,,   *sarpedon semifasciatus*, in *Trans. Ent. Soc. Lond.* 1889. t. 7. f. 2 (as *sarpedon* var.)
,,   *sarpedon teredon*, in Moore, *Lep. Ceyl.* I. 1882. t. 62. f. 1.
,,   *sarpedon teredon* ab. *thermodusa*, in *P. Z. S.* 1885. p. 146. n. 145.
,,   *sarpedon timorensis*, in Novit. Zool. 1896. p. 323.
,,   *semperi*, in Semper, *Tagf. Philipp.* 18.
,,   *tydeus*, in *Reise Novara, Lep.* I. 1865. t. 16. f. c; t. 17. f. a.
Pieris rapae ab. loc. *novangliae*, in Scudder, *Butt. New Engl.* II. 1889. p. 1207.
Plusiotis aurora ab. *aurora*, in *P. Z. S.* 1875. t. 23. f. 7.
,,   *aurora* ab. *chrysopedila*, in *Biol. Centr. Amer., Col.* II. t. 16. f. 12 (1888).
Polyommatus phlaeas, in Hübner, *Samml. Eur. Schm.* I. f. 362. 363 (1798—1805).
Troides croesus, in *P. Z. S.* 1859. t. 68. 69.
,,   *lydius*, in *Reise Novara, Lep.* I. 1865. t. 3. f. a. b.
,,   *priamus*, in Novit. Zool. 1895. p. 183.
,,   *priamus euphorion*, in Rippon, *Icon. Ornith.* 1890. t. 2.
,,   *priamus poseidon*, in Westwood, *Cab. Or. Ent.* 1848. t. 11 ($\male$).
,,   *priamus priamus*, in Cramer, *Pap. Ex.* I. 1775. t. 23. f. A. B.
,,   *priamus urvillianus*, in *Voyage Coquille* 1829. *Ins.* t. 13. f. 1. 2.
,,   *tithonus*, in Dehaan, *l.c.* t. 1. f. 1.

CYCLE IN THE LIFE OF THE INDIVIDUAL (ONTOGENY) AND IN THE EVOLUTION OF ITS OWN GROUP (PHYLOGENY).*

ALPHEUS HYATT.

THE organic cycle, as generally understood both by laymen and scientists, and as usually described in literature, is, as a rule, considered from a physiological rather than structural point of view. The development of the young, and the attainment of the adult or comparatively permanent, stage completes the progressive stages. Old age, accompanied by losses of characteristics and functions and consequent weakening of the body, is retrogressive and brings on second childhood, thus completing the cycle in the ontogeny.

My purpose to-night is to show that the cycle is also represented in the life history of the individual by definite structural changes, and that these have direct correlations with the history of the changes in the forms of the group while evolving in time.†

The fundamental discoveries that are

* This paper was in large part read as a general summary of the phenomena of cycles, before the American Academy in Boston, but does not assume to be an exhaustive or even complete account of the literature or theoretical views treated of.

† These correlations have been more fully stated in a number of publications by the author, especially 'Genesis of the Arietidæ,' Smithsonian Contribution, 673, and Mem. of Mus. of Comp. Zoology, Vol. XVI.; 'Bioplastology and the Related Branches of Scientific Research,' Proc. Bost. Soc. Nat. Hist., XXVI.; and 'Phylogeny of an Acquired Characteristic,' Proc. Am. Phil. Soc., XXXII., No. 143.

more than any other directly useful in the study of the phenomena of the cycle, both in ontogeny and phylogeny, may be briefly noticed as follows:

The opinion that the higher animals are complex, colonial aggregates of cells, which in structure are equivalents to the lowest and minutest adult forms of the animal kingdom, the unicellular bodies of Protozoa, has been steadily gaining in probability since it was first announced by Oken in 1805, in 'Die Zeugung,' Frankfurt bei Wesche, 8vo. This work we have not yet seen, but in the first edition of the Naturphilosophie, Jena, 1809, II., XII. Buch, Zoogenie, he describes protoplasm as 'Punktsubstanz' and as giving rise to the 'Blasenform or Zellform' in both animals and plants. Oken considered the lower animals 'Polypen, Medusen, Beroen, kurz alle Gallertthiere' to be composed of Punktsubstanz.' The nerves, the cartilage, bones of higher animals, were considered as modifications of this form of 'protoplasm,' but the skin and fleshy parts, including the viscera, were described as cellular, ' dem Fleisch liegt die Bläschenform zu Grunde;' again on p. 30, ' die Eingeweide welche am meisten aus Zellengewebe bestehen.' Oken in XII., VIII. Buch, treats of the subject we are more immediately interested in and writes as follows: " Pflanzen and Thiere können nur Metamorphosen von Infusorien sein," " im kleinsten sind sie nur infusoriale Bläschen die durch verschiedene Combinationen sich verschieden gestalten und zu höheren Organismen aufwachsen," and also adds on p. 29, in anticipation of one of the points advanced by the author in his 'Larval Theory of the Origin of Cellular Tissues,'* ' auch besteht der Samen aller Thiere aus Infusorien.'

This author directly compares his cystic or intestinal animals, Infusoria, with ova,

* Proc. of the B. S. N. H., Vol. XXIII., March 5, 1884.

and speaks of them as oozoa, and in the preface to the English edition of his Physiophilosophy, Lond. 1847, Roy. Society, he writes that all organic beings originate from and consist of vesicles or cells. "Their production is nothing else than a regular agglomeration of Infusoria; *not, of course, of species previously elaborated or perfect, but of mucous vesicles or points in general which first form themselves by their union or combination* into particular species." Oken's view was based on the resemblances existing between the Protozoa and the cells in the tissues of the Metazoa, and it is evident he is entitled to be considered the first teacher of the unicellular doctrine, an honor now universally given to von Siebold.

However imperfect and imaginative the results as compared with the more objective statements of later observers, the author who wrote such sentences as these had as clear ideas as the knowledge of his time permitted and was the Haeckel of the early part of this century, and like him a great and successful leader, making many errors but also many discoveries and 'blazing out' some of the paths that we are still following.

Meckel* seems to have been the first author who brought together and stated in a clear way the scattered observations and ideas with regard to the correlations existing between the transient stages of development of the individual and the so-called permanent modifications represented by the similar characters in the adult stages of similar forms.

Meckel says: " Es giebt keinen guten Physiologen, den nicht die Bemerkung frappirt hätte, dass die ursprüngliche Form aller Organismen eine und dieselbe ist, und dass

* Meckel. 'Entw. e. Darstellung der Embryonalzustände d. höheren Thiere u. d. Perman. d. zu d. niedern stattfindenden Parallele.' Beitr. z. vergeich. Anat., II., Leipzig, 1811, pp. 1–148; Meckel speaks of his publications as only preparatory to more extended researches.

aus dieser einen Form sich alle, die niedrigsten wie die höchsten so entwickeln, dass diese die permanenten Formen der ersten nur als vorübergehende Perioden durchlaufen. Aristoteles, Haller, Harvey, Kielmeyer, Autenrieth und mehrere andere haben diese Bemerkung entweder im Vorübergehen gemacht oder, besonders die letzen, hervorgehoben und für die Physiologie ewig denkwürdige Resultate daraus abgeleitet.

" Von diesen niedrigsten Wirbelthieren an bis zu den höchsten Geschlechtern lässt sich die Vergleichung zwischen dem Embryo der höhern Thiere und den niedern im vollkommenen Zustande vollständiger und treffender durchführen.

" In der That giebt es ja eine Periode wo der Embryo des höchsten Thieres, wie schon Aristoteles sagt, nur die Gestalt einer Made hat, wo er ohne äusere und innere Organisation, bloss ein kaum geformtes Klümpschen von Polypensubstanz ist. Ungeachtet des Hervortretens von Organen bleibt es doch noch wegen des gänzlichen Mangels eines innern Knochengerüstes eine Zeitlang Wurm und Mollusk und tritt erst später in die Reihe der Wirbelthiere, wenngleich Spuren der Wirbelsäule schon in den frühesten Perioden seinen Anspruch auf diese Stelle in der Reihe der Thiere beglaubigten."

It is very obvious, from these statements of Meckel's, that the correlations of embryology and the epembryonic stages of the individual with the permanent modifications of animals of simpler construction was understood, as far as was possible with existing knowledge, from the time of Aristotle and that it was, to a greater or less extent, a working hypothesis at that time and, as declared by him, had been helpful in giving a clearer understanding of the development of the individual and of the relations of the individual to the whole animal kingdom.

The next step was taken by von Baer, in dividing the animal kingdom into four types and in limiting this general statement to animals occurring within each of these types. He also considered it highly probable (not barely possible, as it is quoted by some writers) that the earliest stages of the embryo resemble in aspect the adult stages of the lowest grade of forms in the animal kingdom. He had in mind in this statement the modern view of the affinities of the earliest stages of the embryo or its repetitions of the characteristics of Protozoa,* so far as the knowledge of his time permitted.

Von Baer endeavored to prove that each of the four types had similar embryos and that the type characters were determinable at early stages in the ontogeny. Both von Baer and Louis Agassiz were pupils of Ignatius Dollinger, an embryologist who published nothing. Both of these eminent men have recognized him as their master in embryology, but have given no definite statement of what they were taught by him. Louis Agassiz accepted von Baer's opinions and subsequently enlarged them, when he published on his fossil fishes by the introduction of the element of succession in time and thus laid the basis for all more recent work.

Agassiz gave the fullest expression of his views in ' Twelve Lectures on Comparative Embryology,' Lowell Institute, Boston, 1848 –49, subsequently published in pamphlet form. One wonders as he reads how any man holding such views could have held his mind closed to the conclusion that animals were evolved from simpler or more primitive forms. The effect of theoretical preconceptions in closing the mind to the reception of new ideas never had a stronger illustration. Louis Agassiz, in 1849, had all the facts essential for building up a hypothesis of evolution that would have

* Entwickelungsgesch. d. Thiere, Scholion V., p. 199, p. 120, etc.

placed him in the history of science on the same line with Lamarck and Darwin.

He states four lectures, p. 26, as follows: "The results thus far obtained in the lectures which I have delivered can be expressed as follows: There is a gradation of type in the class of Echinoderms, and indeed in every class of the animal kingdom, which, in its general outlines, can be satisfactorily ascertained by anatomical investigation; but it is possible to arrive at a more precise illustration of this gradation by embryological data. The gradation of structure in the animal kingdom does not only agree with the general outlines of the embryonic changes. The most special comparison of these metamorphoses with full grown animals of the same type leads to the fullest agreement between both, and hence to the establishment of a more definite progressive series than can be obtained by the investigation of the internal structure. These phases of the individual development are the new foundations upon which I intend to rebuild the system of zoology. These metamorphoses correspond, indeed, in a double sense, to the natural series established in the animal kingdom: first, by the correspondence of the external forms, and secondly, by the successive changes of structure, so that we are here guided by the double evidence upon which the progress in zoology has, up to this time, generally been based.

"Their natural series again correspond with the order of succession of animals in former geological ages, so that it is equally as true to say that the oldest animals of any class correspond to their lower types in the present day as to institute a comparison with the embryonic changes, and to say that the most ancient animals correspond with the earlier stages of growth of the types which live in the present period. In whatever point of view we consider the animal kingdom, we find its natural series agree with each other; its embryonic phases of growth correspond to its order of succession in time, and its structural gradation, both to the embryonic development and the geological succession, corresponds to its structure; and if the investigations had been sufficiently matured upon this point, I might add that all these series agree also in a general way with the geographical distribution of animals upon the surface of our globe, but this is a point upon which I am not yet prepared to give full and satisfactory evidence. So much for the views referring to embryology in its bearing upon zoological classification."

And again on p. 27:

"However, another step had to be made to show a real agreement between the earlier types of animals and the gradual development of the animal kingdom, which has been the last progress in our science of fossils: namely, to show that these earlier types are embryonic in their character—that is to say, that they are not only lower in their structure when compared with the animals now living upon the surface of our globe, but that they actually correspond to the changes which embryos of the same classes undergo during their growth. This was first discovered among fishes, which I have shown to present, in their earlier types, characters which agree in many respects with the changes which young fishes undergo within the egg. Without entering into all the details of these researches, I will conclude by saying it can now be generally maintained that earlier animals correspond not only to lower types of their respective classes, but that their chief peculiarities have reference to the modifications which are successively introduced during the embryonic life of their corresponding representatives in the present creation. To carry out these results in detail must now be, for years to come, the task of paleontological investigations."

Perhaps, in consequence of pressure of other work or of his theoretical views, Louis Agassiz seemed to have lost sight of the great importance of continuing his researches upon the meaning and correlations of the epembryonic stages. These were referred to in his publications, but were not made as prominent as they deserved after the lectures at the Lowell Institute in 1849, and in his personal talks with his students or in his lectures I cannot remember that they were ever treated directly by anything more than incidental references, although embryology was very often the principal theme.

Nevertheless, I must have got directly from him, subsequently to 1858, the principles of this branch of research, and through this and the abundant materials furnished by the collections he had purchased and placed so freely at my disposal, I soon began to find that the correlations of the epembryonic stages and their use in studying the natural affinities of animals was practically an infinite field for work and discovery.

Although within a year after the beginning of my life as a student under Louis Agassiz I had become an evolutionist, this theoretical change of position altered in no essential way the conceptions I had at first received from him, nor the use we both made of them in classifying and arranging forms. This experience demonstrated to my mind the absurdity of disputing the claims of any author to the discovery of a series of facts and their correlations because of his misinterpretation of their more remote relations or general meaning. It is of some importance to notice this because it is the rule now to attribute von Baer's and his predecessors' and Louis Agassiz's discoveries in this line to Haeckel. This eminent author has, indeed, given one of the most modern definitions of this law and has named it the ' law of biogenesis.' Haeckel's discoveries in embryology are sufficiently great without swelling the list with false entries, but it will probably be a long time before naturalists realize and acknowledge this error. Some of the most eminent embryologists in this country have adopted the Haeckelian nomenclature without sufficient critical examination of the term under discussion. The so-called Haeckelian (' law of biogenesis ') is really Agassiz's law of embryological recapitulation restated in the terms of evolution.

It has surprised me that serious objections to the use of the word ' biogenesis ' in this connection have not been made. This word has been long employed in another sense as antithetical to ' abiogenesis.' The latter has been for many years applied to the theory of the generation of living from inorganic matter, and the former to the theory asserting that living matter can originate only from living matter; the use of the phrase ' the law of biogenesis ' is consequently inappropriate, since neither did Agassiz's nor Haeckel's discoveries cover so much ground. The former gave us a law for the correlations of the earlier stages of ontogeny with phylogeny. This cannot be called ' the law of biogenesis,' since that has been long ago stated as the law of the origin and continuity of organism, or in other words, the genesis and continuity of life from and through living matter only. There are two different manifestations of Agassiz's law, which Haeckel defined and named ' palingenesis ' and ' conoegenesis,' the former referring to the ordinary as regular mode in which the characteristics of ancestors are repeated in the development of the individual and the other to what is frequently called the abbreviated mode, etc.

These two modes are by no means all, but at present only the first or simplest manifestations of the phenomena need be treated of. This, or what Haeckel very appropriately calls 'palingenesis,' was what Louis Agassiz had studied and, so far as all the essential facts were concerned, thoroughly understood, and it was this that he taught his students, so that it became, at any rate in my own case, the foundation of all my subsequent work in determining the mutual relations of forms. If then, as I have proposed in former publications, the term 'law of palingenesis' be adopted this expressly states just what Louis Agassiz discovered.

Observations upon this ground made especially upon Cephalopoda have led to the discovery of correlations between the latter or epembryonic stages and the adult stages of extinct ancestors which have greatly enlarged the field of application of Agassiz's law of palingenesis and given it an exactitude that has made it of surpassing importance in the study of evolution. Beecher has been able to point out the single species of Brachiopod from which the whole of the vast number of distinct forms of this great group have originated. He has established this fact not only by showing that the young of the existing and fossil forms all repeat more or less at one stage the form of the adult of the initial species, but has also found a very near affinity of this single ancestral species as a fossil appearing in one of the earliest of the fossil-bearing formations.

Dr. R. T. Jackson has done the same work for the Pelecypoda, tracing all to one genus, *Nucula*, and has treated the Echinodermata in the same way, tracing them by the use of Agassiz's law to the genus *Bothriocidaris*.

Although the evidence is perhaps less conclusive with reference to the ancestor of Cephalopoda as a whole, this class has furnished the means of showing the action of this law in smaller groups with great accuracy. It has been possible to trace the origin of a number of smaller groups to single ancestors within the class by carefully studying the correlations of the epembryonic stages with the adults of the same group that have preceded them in time, and this study has also led to further discoveries. It has been found that the new characters were first introduced in the later stages of ontogeny, usually in the full-grown stage; then, as old age approached, certain losses of the characters of the adult took the place, or, if additional growths were acquired, these were of a peculiar kind. These senile stages had been noticed by D'Orbigny and Quenstedt, but these authors did not attempt to show that any correlations existed between any stages of the ontogeny and the gradations occurring in the full-grown forms during their evolution in time, or what is called phylogeny. The oldest stage of the shell in Cephalopoda, Brachiopoda and Pelecypoda is commonly marked by a series of retrogressive changes, which have been fully described elsewhere. These changes have a similar nature to those found in the old age of man, but they are more noticeable because they are recorded in the permanent characters of the hardened shell. The old man returns to second childhood in mind and body, and the shell of the cephalopod has in old age, however distinct and highly ornamented the adult, very close resemblance to its own young. This resemblance is a matter of form and aspect only, since there can be no close comparison in minute structure, nor functions between organs and parts at these two different ends of life. Such analogies, however, have their own meaning and are of great importance when properly translated.

In the first place they show that the cycle of life as manifested in man is found also in the ontogeny of other animals and more pefectly in proportion to the perfection of the record. They are consequently among shell-bearing animals, especially those that carry their embryonic shells and all their subsequent stages of development throughout their lives, more perfect, more decisive, as well as more obvious, than in animals, like the vertebrata, which carry no such burden of hard, dead parts upon, and in which their stages of development are recorded. The cycle of the ontogeny is, therefore, not only physiological, but it is also a definite series of structural changes and is often accompanied by transformations of remarkable and sometimes startling character.

These retrogressive transformations in old age of the shells of Cephalopoda, Brachiopoda and Pelecypoda have been found to have decided correlations with the adult characters of species that appear simultaneously or later in time. If one traces any group through its evolution in time it has, as stated by many authors, a period of rise called the epacme, a second period of greatest expansion in numbers of forms and species called the acme, and then usually a movement towards contraction called the paracme. All three of these terms were first proposed by Haeckel, who used them largely in a physiological or dynamical

The paracme is the decline, and this takes place through the reduction and actual loss of structures and characteristics that have been built up by evolution during the epacme. This is no ideal picture, but a simple statement of the experiences of those paleontologists who have patiently traced the history of groups through geologic time. Agassiz's law enables one to follow the epacme of the evolution of a species, or genus, or order, or larger group, but further correlations between the cycle of individual life and those in the evolution of its own genetic group must be sought in the correlations existing between the older retrogressive stages of the ontogeny and the paracme of each group.

The importance and peculiar nature of these corelations led me, in one of my papers, to introduce, for this branch of research, the term Bioplastology, which will be found convenient by those interested in this class of work.

The following table of terms is useful here to explain the relations of the cycle of development in the individual to that of the group to which it belongs.

TERMS OF BIOPLASTOLOGY EXPLAINING THE CORRELATIONS BETWEEN STAGES OF THE ONTOGENY AND THOSE OF PHYLOGENY.

Ontogeny or Development			Phylogeny or Evolution of the Phylum		
Structural Conditions	Stages		Structural Conditions	Stages	Dynamical
Anaplasis	Embryonic............Embryo or Foetal Nepionic.............Baby Neanic..............Adolescent		Phylanaplasis	Phylembryonic Phylonepionic Phyloneanic	Epacme
Metaplasis	Ephebic..............Adult		Phylometaplasis	Phylephebic	Acme
Paraplasis	Gerontic............Senile		Phyloparaplasis	Phylogerontic	Paracme

sense. The epacme of any group, large or small, is usually a process of evolution by addition of new structures or characteristics based on older structures and thus leading to greater and greater complication of the primitive organization. The acme represents the time of greatest complication in structure and greatest expansion in numbers of forms for any group, large or small.

The dynamical terms are quoted from Haeckel and were used by him to designate the phenomena of the rise and decline of types, and also the terms anaplasis and metaplesis. He, however, used 'cataplasis' in place of paraplasis, which is here preferred on account of the faulty derivation of cataplasis.

He realized the importance of these phenomena and also the significance of the structural characteristics of decline, but did not trace out the distinct correlations which are claimed as fundamentals in bioplastology.

The terms anaplasis, etc., and their correspondence, phylanaplasis, are the structural correlatives of dynamical terms, epacme, etc., and will be found useful when the statical phenomena or structures are mentioned or contrasted with the dynamical phenomena, or with periods of time in which they occur, since the terms epacme, acme and paracme also refer to time. Terms of the ontogeny are placed opposite to their correlatives in the column of phylogenetic terms, but in reading the table it should be clearly understood that the individual whose life history is represented by the first three columns is supposed to have been taken from the midst of those that lived during the acme of the phylum and belonged to a phylephebic species. In studying the development of such an individual it has been repeatedly observed that the embryo repeated the adult characters of the most ancient representatives of the phylum, which are here called in accordance with this evidence, phylembryonic.

It has also been ascertained that there are full-grown types in the epacme and acme of groups which correspond to the transient nepionic or baby stage of those that occur later in time; these are the phylonepionic; others have similar correspondences with the neanic stages and are properly designated as phyloneanic types or forms. The structures of the ephebic (adult) stage are essentially the differentials of the time and fauna in which they occur, and necessarily have no correlations with the past. Their relations are obviously and wholly with the present, except in so far as they represent the consummations of evolution in structures. The structural changes in the gerontic stage of the individual are repeated with sufficient accuracy in the adult, and often even in the neanic stages of types that occur in the paracme of the evolution of a phylum, so that one is forced to consider seriously whether they may not have been inherited from types that occur at the acme of the same group. The fact that these changes occur first in the ontogeny during the gerontic stage does not necessarily imply that they first make their appearance after the reproductive period. No gerontic limit is known to the reproductive time in the lower animals, and it may well be that the continual recurrence of gerontic stages in individuals during the epacme of groups may lead to their finally becoming fixed tendencies of the stock or hereditary in the phylum, and thus established as one of the factors that occasion the retrogression or paracme of groups. The paracme may also be considered as occasioned by changes in the surroundings from favorable, as they must have been up to acmatic time to unfavorable during the succeeding paracmatic period in evolution. Still a third supposition is also possible, viz., that the type, like the individual, has only a limited store of vitality, and both must progress and retrogress, complete a cycle and finally die out, in obedience to the same law.

All of these views can be well supported, but, whatever may be the true explanation, it is obvious that there are plenty of paracmatic types, which, in their full-grown and even in their neanic stages, correlate in characters and structures with the characters and structures that one first finds in the transient gerontic stages of acmatic forms of the same type. These can, therefore, be truthfully and accurately described as phylogerontic in the phylum.

In other words, one is able to apply gerontic changes in the ontogeny to the deciphering of the true relations, the arrangement and classification forms occurring in the paracme, just as Agassiz's law of palingenesis can be used to explain the relations of the links in the chain of being forming the epacme of groups.

The cycle of the ontogeny is, therefore, the individual expression and abbreviated recapitulation of the cycle that occurs in the phylogeny of the same stock, and, while the embryonic, nepionic and neanic stages give us, in abbreviated shape, the record of the epacme, the gerontic stages give, in a similar manner, the history of the paracme.

The difference between the nature of the two records is, however, necessarily as great as between the beginnings and the endings of existence. The successive stages of the individual are derived from the past, and simply point backwards along the track traversed by the phylum; the successive changes of the gerontic stage on the other point to the future, and are prophetic of what is to come in the decline of the type. The retrogressive decline of the individual and of its type are along parallel lines and the two are in direct correlation, so that the former becomes an abbreviated index of the latter.

One of the most useful results of these studies has been the method of work developed, the mode of study by series. To follow it out successfully one must trace the terms of series from the first, or most primitive, grade to the last, through perhaps long periods of time and, if upon the same level, through many gradations of structure.

The histologist and embryologist picks out a convenient form here and there for thorough investigation, but does not seem as yet to see the importance of the point of view here insisted upon, viz., that the only method of getting at the correlations of ontogeny and phylogeny is by following out the history of representative series of genetically connected embryos, and the same is true of the experimentalist. While, consequently, their results have been in the highest degree instructive and progressive along other lines of research, they throw no very strong light on the laws of evolution, and the best modern works on embryology, zoology and experimentation neglect the only proper and efficient mode of studying one very important side of their subject.

One of the results of this mode of study has been the discovery of the law of acceleration in the inheritance of characters, or tachygenesis. Thus it has been found that characteristics are inherited in successive species or forms in a given stock at earlier and earlier stages in the ontogeny of each member of the series. These characteristics, as a rule, disappear from the ontogeny altogether in the terminal, or last-occurring, members of a series, and terminal forms thus become very distinct in their development. This law I habitually illustrate as the crawling, walking, hopping, skipping and jumping law.

Another result of this mode of study is the discovery that, in most genetic series, primitive forms exhibit much greater indifference to geologic changes, persist with comparatively unchanged structures through longer periods of time than those that occur at the acme of groups, and paracmatic forms, if widely distributed, are apt to be particularly short lived, and are very often narrowly localized in origin and duration. Primitive forms are also less changeable in their ontogeny; the adult differs less from either the young or the old than in acmatic forms. The same is true of phylogerontic forms; their old age and youth are less distinct as stages from each other than in acmatic forms. Primitive forms are less affected by gerontic changes in their ontogeny, i. e., they have shorter old-age stages than acmatic forms. Paracmatic forms have much longer old-age or gerontic stages than acmatic forms.

Lastly, it has been found that at the beginning of the evolution of any stock the progress was not only very rapid, but the departures in structures much more marked between the diverging lines of different species, genera or families, and so on, than those that subsequently occurred in any one of these. This rapidity of expansion is also marvellously sudden in every series near its point of origin, and it is equally so in the whole animal kingdom, which appears with the larger proportion of all its principal divisions in the earliest known fossil-bearing rocks. Each series or type appears to have had a more or less free field, and its first steps in evolution were obviously not affected by natural selection. Subsequently, in the acme of the same series or type, the departures became less marked, and the divergences took place in less important structures; in other words, as stated above, the evolution is slower.

On the other hand, after the acme is passed and the paracme sets in, there is a sensible quickening of evolution during decline.

Phylogerontic forms become more and more numerous, and there are wider departures in the structures from the acmatic forms than any of the divergences that occur within the acmatic forms themselves.

The hopping, skipping and, at last, the jumping begins in the extremes of the series, so that it becomes difficult, as has been shown by the author in a number of series and by Cope when giving illustrations of the action of the law of tachygenesis, to connect one of these extreme forms with its nearest congeners.

The characters of the cycle in the ontogeny are here again similar to those of the phylogeny; thus the final substages of the gerontic stage are wider departures from the ephebic substages than these are among themselves and when compared with each other. The analogy of the old with the young shows this most conclusively and with the similarity of phylogerontic forms in the same stock occurs a parallelism in the phylogeny.

In fact, there is no end to the homological and analogical similarities and parallelisms of ontogeny and phylogeny wherever both are found complete.

There are types in which the ontogeny is incomplete, as among insects and other purely seasonal animals, and in these it becomes difficult, if not impracticable, to study the gerontic stages, and thus translate the phylogerontic types if they occur. These same types, and others also, present difficulties in their larval stages, owing to their indirect modes of development, which have been discussed by the author in Insecta and other publications, and need only be referred to here.

One of the bearings of these researches is of interest on account of the discussions between biologists, geologists and mathematicians with regard to the length of time that life has existed on this planet and the bearing of this upon calculations with regard to the age of the earth. It cannot be assumed that the time ratio was the same during the eozoic or pre-Paleozoic as during the Paleozoic or the Mesozoic, so far as the evolution of forms is concerned. The evidence is very strong that great structural differences were evolved much more quickly in these early times, and the probabilities are that the progressive steps of the evolution of the primitive types of organisms took place with a rapidity unexampled in later ages. If the laws of bioplastology are true the evolution of these forms must have occurred more quickly than those of their descendants, except perhaps some isolated phylogerontic types and phylopathic forms.*

* The phrase 'evolution by saltation' has been used for the sudden appearance of divergent types by several authors, first by Dr. W. H. Dall; but this seems

The author in other publications has claimed that this must have been the law, and explained the phenomena as parallel with that which takes place at the beginning of every series arising in the Paleozoic and Mesozoic, and also according to Minot's law of growth and other phenomena of the earlier stages in the ontogeny of every animal

All inferences with reference to the length of time that life has existed upon the earth are consequently defective, since, as far as known to the author, they do not take into consideration the differing rates of evolution at different times in the history of organisms.

EVOLUTION IN MENDELIAN POPULATIONS
SEWALL WRIGHT

EVOLUTION IN MENDELIAN POPULATIONS

SEWALL WRIGHT

University of Chicago, Chicago, Illinois

Received January 20, 1930

TABLE OF CONTENTS

	PAGE
Theories of evolution	97
Variation of gene frequency	100
Simple Mendelian equilibrium	100
Mutation pressure	100
Migration pressure	100
Selection pressure	101
Equilibrium under selection	102
Multiple allelomorphs	104
Random variation of gene frequency	106
Rate of decrease in heterozygosis	107
The population number	110
The distribution of gene frequencies and its immediate consequences	111
No mutation, migration or selection	111
Nonrecurrent mutation	116
Reversible recurrent mutation	121
Migration	126
Irreversible recurrent mutation	128
Selection	129
General formula	133
The distribution curves	134
Dominance ratio	137
The mean and variability of characters	139
The evolution of Mendelian systems	142
Classification of the factors of evolution	142
Lability as the condition for evolution	147
Control of evolution	151
Agreement with data of evolution	153
"Creative" and "emergent" evolution	154
SUMMARY	155
LITERATURE CITED	158

THEORIES OF EVOLUTION

One of the major incentives in the pioneer studies of heredity and variation which led to modern genetics was the hope of obtaining a deeper insight into the evolutionary process. Following the rediscovery of the Mendelian mechanism, there came a feeling that the solution of problems of evolution and of the control of the process, in animal and plant breeding and in the human species, was at last well within reach. There has been no halt in the expansion of knowledge of heredity but the advances in the field of evolution have, perhaps, seemed disappointingly small. One finds the subject still frequently presented in essentially the same form as before 1900, with merely what seems a rather irrelevant addendum on Mendelian heredity.

The difficulty seems to be the tendency to overlook the fact that the evolutionary process is concerned, not with individuals, but with the species, an intricate network of living matter, physically continuous in space-time, and with modes of response to external conditions which it appears can be related to the genetics of individuals only as statistical consequences of the latter. From a still broader viewpoint (compare LOTKA 1925) the species itself is merely an element in a much more extensive evolving pattern but this is a phase of the matter which need not concern us here.

The earlier evolutionists, especially LAMARCK, assumed that the somatic effects of physiological responses of individuals to their environments were transmissible to later generations, and thus brought about a directed evolution of the species as a whole. The theory remains an attractive one to certain schools of biologists but the experimental evidence from genetics is so overwhelmingly against it as a general phenomenon as to render it unavailable in present thought on the subject.

DARWIN was the first to present effectively the view of evolution as primarily a statistical process in which random hereditary variation merely furnishes the raw material. He emphasized differential survival and fecundity as the major statistical factors of evolution. A few years later, the importance of another aspect of group biology, the effect of isolation, was brought to the fore by WAGNER. Systematic biologists have continued to insist that isolation is the major species forming factor. As with natural selection, a connection with the genetics of individuals can be based on statistical considerations.

There were many attempts in the latter part of the nineteenth century to develop theories of direct evolution in opposition to the statistical viewpoint. Most of the theories of orthogenesis (for example, those of EIMER and of COPE) implied the inheritance of "acquired characters." NÄGELI postulated a slow but self contained developmental process within protoplasm; practically a denial of the possibility of a scientific treatment of the problem. Differing from these in its appeal to experimental evidence and from the statistical theories in its directness, was DE VRIES' theory of the abrupt origin of species by "mutations." A statistical process, selection or isolation, was indeed necessary to bring the new species into predominance, but the center of interest, as with Lamarckism, was in the physiology of the mutation process.

The rediscovery of Mendelian heredity in 1900 came as a direct consequence of DE VRIES' investigations. Major Mendelian differences were naturally the first to attract attention. It is not therefore surprising that the phenomena of Mendelian heredity were looked upon as confirming DE VRIES' theory. They supplemented the latter by revealing the possibilities of hybridization as a factor bringing about an extensive recombination of mutant changes and thus a multiplication of incipient species, a phase emphasized especially by LOTSY. JOHANNSEN's study of pure lines was interpreted as meaning that DARWIN's selection of small random variations was not a true evolutionary factor.

A reaction from this viewpoint was led by CASTLE, who demonstrated the effectiveness of selection of small variations in carrying the average of a stock beyond the original limits of variation. This effectiveness turned out to depend not so much on variability of the principal genes concerned as on residual heredity. As genetic studies continued, ever smaller differences were found to mendelize, and any character, sufficiently investigated, turned out to be affected by many factors. The work of NILSSON-EHLE, EAST, SHULL, and others established on a firm basis the multiple factor hypothesis in cases of apparent blending inheritance of quantitative variation.

The work of MORGAN and his school securely identified Mendelian heredity with chromosomal behavior and made possible researches which further strengthened the view that the Mendelian mechanism is the general mechanism of heredity in sexually reproducing organisms. The only exceptions so far discovered have been a few plastid characters of plants. That differences between species, as well as within them, are Mendelian, in the broad sense of chromosomal, has been indicated by the close parallelism between the frequently irregular chromosome behavior and the genetic phenomena of species crosses (FEDERLEY, GOODSPEED and CLAUSEN, etc.). Most of DE VRIES' mutations have turned out to be chromosome aberrations, of occasional evolutionary significance, no doubt, in increasing the number of genes and in leading to sterility of hybrids and thus isolation, but of secondary importance to gene mutation as regards character changes. As to gene mutation, observation of those which have occurred naturally as well as of those which MULLER, STADLER, and others have recently been able to produce wholesale by X-rays, reveals characteristics which seem as far as possible from those required for a directly adaptive evolutionary process. The conclusion nevertheless seems warranted by the present status of genetics that any theory of evolution must be based on the properties of Mendelian factors, and beyond this, must be concerned largely with the statistical situation in the species.

VARIATION OF GENE FREQUENCY

Simple Mendelian equilibrium

The starting point for any discussion of the statistical situation in Mendelian populations is the rather obvious consideration that in an indefinitely large population the relative frequencies of allelomorphic genes remain constant if unaffected by disturbing factors such as mutation, migration, or selection. If $[(1-q)a+qA]$ represents the frequencies of two allelomorphs, (a, A) the frequencies of the zygotes reach equilibrium according to the expansion of $[(1-q)a+qA]^2$ within at least two generations,[1] whatever the initial composition of the population (HARDY 1908). Combinations of different series are in equilibrium when these are combined at random, but as WEINBERG (1909) and later, in more detail ROBBINS (1918) have shown, equilibrium is not reached at once but is approached asymptotically through an infinite number of generations. Linkage slows down the approach to equilibrium but has no effect on the ultimate frequencies.

Mutation pressure

The effects of different simple types of evolutionary pressure on gene frequencies are easily determined. Irreversible mutation of a gene at the rate u per generation changes gene frequency (q) at the rate $\Delta q = -uq$. With reverse mutation at rate v the change in gene frequency is $\Delta q = -uq+v(1-q)$. In the absence of other pressures, an equilibrium is reached between the two mutation rates when $\Delta q=0$, giving $q=v/(u+v)$.

Migration pressure

The frequency of a gene in a given population may be modified by migration as well as by mutation. As an ideal case, suppose that a large population with average frequency q_m for a particular gene, is composed of subgroups each exchanging the proportion m of its population with a random sample of the whole population. For such a subgroup, $\Delta q = -m(q-q_m)$.

The conditions postulated above are rather artificial since, in an actual species, subgroups exchange individuals with neighboring subgroups rather

---

[1] This statement assumes that there is no overlapping of generations which may bring about some delay in the attainment of equilibrium.

than with a random sample of the whole species and the change in q will be only a fraction of that given above. The fraction is the average degree of departure of the neighboring subgroups toward the population average. The formula may be retained by letting q_m stand for the gene frequency of immigrants rather than of the whole species.

Selection pressure

Selection, whether in mortality, mating or fecundity, applies to the organism as a whole and thus to the effects of the entire gene system rather than to single genes. A gene which is more favorable than its allelomorph in one combination may be less favorable in another. Even in the case of cumulative effects, there is generally an optimum grade of development of the character and a given plus gene will be favorably selected in combinations below the optimum but selected against in combinations above the optimum. Again the greater the number of unfixed genes in a population, the smaller must be the average effectiveness of selection for each one of them. The more intense the selection in one respect, the less effective it can be in others. The selection coefficient for a gene is thus in general a function of the entire system of gene frequencies. As a first approximation, relating to a given population at a given moment, one may, however, assume a constant net selection coefficient for each gene. Assume that the genes a and A tend to be reproduced in the ratio $(1-s):1$ per generation. The gene array $[(1-q)a+qA]$ becomes $[(1-s)(1-q)a+qA]/[1-s(1-q)]$. The change in the frequency of A is $\Delta q = [sq(1-q)]/[1-s(1-q)]$ or with sufficiently close approximation $\Delta q = sq(1-q)$ if the selection coefficient is small.

A second approximation may be obtained by considering the zygotic frequencies. Assume that the types aa, Aa, and AA reproduce in the ratio $(1-s'):(1-hs'):1$ per generation. The change in the frequency of A to a sufficiently close approximation is $\Delta q = s'q(1-q)[1-q+h(2q-1)]$ (WRIGHT 1929). In the case of selection for or against a complete recessive ($h=0$, s' negative or positive respectively), $\Delta q = s'q(1-q)^2$.

The case of no dominance ($h=\frac{1}{2}$) is the same as the case of genic selection except that the selection against the gene is $s'/2$ instead of s.

The two factor case in which the phenotypes aabb, aaB−, A−bb and A−B− reproduce at the rates $(1-s_{ab}):(1-s_a):(1-s_b):1$ respectively yields (for low values of the selection coefficients):

$$\Delta q_A = q_A(1-q_A)^2[s_a + (s_{ab} - s_a - s_b)(1-q_B)^2].$$

The frequency of A depends on the frequency and selection of B, becoming independent only if $s_{ab} = s_a + s_b$, that is, if the two series of genes are cumulative with respect to selection. It does not seem profitable to pursue this subject further for the purpose of the present paper, since in the general case, each selection coefficient is a complicated function of the entire system of gene frequencies and can only be dealt with qualitatively. Attention may, however, be called to HALDANE's (1924–1927) studies of selection rates and of the consequent number of generations required for unopposed selection to bring about any required change in gene frequency under various assumptions with respect to mode of inheritance and system of mating.

Equilibrium under selection

There may be equilibrium between allelomorphs as a result wholly of selection, namely, selection against both homozygotes in favor of the heterozygous type. Putting $\Delta q = s'q(1-q)[1-q+h(2q-1)] = 0$ gives $q = [1-h]/[1-2h]$ as the condition.[2] This includes the case of selection against both homozygotes and also that in favor of them, but examination of the signs of Δq above and below the equilibrium point shows that only the former is in stable equilibrium in agreement with FISHER (1922). The linkage of a favorable dominant with an unfavorable recessive of another series is a case in which selection would be against both homozygotes as JONES (1917) has pointed out, and stressed as a factor in the vigor of heterozygosis. In a population produced by the intermingling of types in which different deleterious recessives have become fixed, there will be a temporary selection in favor of the heterozygotes even without any linkage at all. Unless the linkage is very strong, however, this effect does not persist long enough to have much effect on gene equilibrium. The extreme case of equilibrium of the sort discussed here is, of course, that of balanced lethals, found in nature in Oenothera.

In the two factor case, discussed in the preceding section,

$$\Delta q_A = 0 \text{ if } q_B = 1 - \sqrt{\frac{s_a}{s_a + s_b - s_{ab}}}$$

$$\text{and } \Delta q_B = 0 \text{ if } q_A = 1 - \sqrt{\frac{s_b}{s_a + s_b - s_{ab}}}.$$

---

[2] The condition can be expressed in a more symmetrical form by using a different form of statement of the selection coefficients. Assume that the rates of reproduction of the three types aa, Aa and AA are as $(1-s_a):1:(1-s_A)$. The value of q at equilibrium comes out $q = \dfrac{s_a}{s_A + s_a}$ with stable equilibrium only for positive values of the two selection coefficients.

There may be equilibrium here, if s_a and s_b are alike in sign, and s_{ab} is either opposite in sign or of the same sign and smaller, but it is an unstable equilibrium. Of more general importance, perhaps, is the equilibrium reached by a deleterious mutant gene. For mutation opposed by genic selection $\Delta q = -uq + sq(1-q) = 0$, $q = 1 - u/s$. For mutation opposed by zygotic selection (aa, Aa and AA reproducing at rates $(1-s'):(1-hs'):1$ it is easily shown that $q = 1 - u/hs'$ (WRIGHT 1929), unless h approaches 0. Thus with no dominance, $q = 1 - 2u/s'$, and for selection against a dominant mutation, $q = 1 - u/s'$. The important case of selection against a recessive is that in which $h = 0$. The formula becomes $q = 1 - \sqrt{u/s'}$. All of these cases are illustrated in figure 1 in which the ordinates show the selection pressure as related to factor frequency, under the different conditions of selection. The intersections with the straight line representing mutation pressure give the points of equilibrium. If deleterious dominant and recessive mutations occur with equal frequency and are subject to the same degree of selection, the frequency of the recessive mutant genes will be greater than that of the dominant ones in the ratio $\sqrt{u/s'}$ to u/s'. The corresponding figure for factors lacking dominance is $2u/s'$, where s' is the selection against the homozygote. These considerations alone should lead to a marked correlation in nature between recessiveness and deleterious effect. This correlation is further increased by the greater frequency of recessive mutation which seems to be a general phenomenon. It is this correlation which gives the theoretical basis for the immediate degeneration which usually accompanies inbreeding, a process which increases the proportion of recessive phenotypes.

The amount to which gene frequency in a subgroup may depart from the species average as a result of local selection held in check by population interchange with other regions may be calculated by solving the quadratic $\Delta q = sq(1-q) - m(q-q_m) = 0$. If the local selection coefficient is much greater than the proportion of migration $(s > m)$, $q = 1 - \frac{m}{s}(1 - q_m)$ or $-mq_m/s$ depending on the direction of selection, formulae analogous to those for the equilibrium between mutation and selection. If, on the other hand, selection is weak compared with migration $(s < m)$, the departure from q_m is small and $q = q_m[1 + \frac{s}{m}(1-q_m)]$. This case is doubtless the more important in nature. Large subgroups living under different selection pressures should show gene frequencies clustering about the average according to this expression. The effect of small size of the subgroups in bringing about random deviation in this and other cases is not here considered. The case in which s and m are of the same order of magnitude may be illustrated by the case of exact equality. Here $q = \sqrt{q_m}$ or $1 - \sqrt{1-q_m}$ depending on the direction of selection.

Multiple allelomorphs

The foregoing discussion has dealt formally only with pairs of allelomorphs, a wholly inadequate basis for consideration of the evolutionary pro-

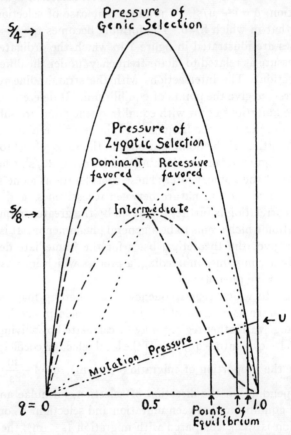

FIGURE 1.—Rate of change of gene frequency under selection or mutation. Genic selection (A, a reproducing at rates $1:1-s$); Zygotic selection: dominant (B-, bb at rates $1:1-s$), recessive (cc, C— at rates $1:1-s$), intermediate (DD, Dd, dd at rates $1:1-\frac{1}{2}s:1-s$), Mutation such that $u = -0.05\ s$. Intersections of line of mutation pressure with those for selection pressure determine the equilibrium frequencies.

cess unless extension can be made to multiple allelomorphs. Among the laboratory rodents some 40 percent of the known series of factors affecting coat color are already known to be multiple. The number of multiple series is large in other organisms, for example, Drosophila (MORGAN, STURTEVANT and BRIDGES 1925). It is not unlikely that further study will indicate that all series are potentially multiple. In this case, each gene has a history which is not a mere oscillation between approximate fixation of two conditions but a real evolutionary process in its internal structure. Presumably any particular gene of such an indefinitely extended series can arise at a step from only a few of the others[3] and in turn mutate to only a few. Since genes as a rule have multiple effects and change in one effect need not involve others, it is probable that in time a gene may come to produce its major effects on wholly different characters than at first. Continuing this line of thought, it indeed seems possible that all genes of all organisms may ultimately be traced to a common source, mitotic irregularities furnishing the basis for multiplication of genes.

The relative frequencies of all allelomorphs in a series tend, of course, to remain constant in the absence of disturbing forces. The zygotes reach the equilibrium of random combination of the genes in pairs by the second generation from any initial constitution of the population. The effects of the various kinds of evolutionary pressure on the frequency of each gene may be treated as before by contrasting each gene with the totality of its allelomorphs. In the binomial expression $[(1-q)a+qA]$, A may be understood as representing any gene, and a as including all others of its series. Such treatment, however, requires further qualification with regard to the constancy of the various coefficients. It may still be assumed that the rate (u) of mutational breakdown of the gene in question (A) is reasonably constant, but its rate (v) of mutational origin from allelomorphs must be expected to change. This may be expected to rise to a maximum, as genes closely allied to A in structure become frequent, and to fall off to zero as changes accumulate in the locus. Even at its maximum, however, the rate of formation should in general be of the second order compared with the rate of change to something else, simply because it is one and its alternatives many. Moreover, there is an indication that the genes which become more or less established in a population are not a random sample of the types of mutations which occur. It has been the common experience that mutations are usually recessive. Recessiveness is most simply interpreted physiologically as due to inactivation which may well be the commonest type of mutational change. But the evolutionary process presumably involves in-

[3] Those most closely related genetically, however, need not always be closest in effect. The complete inactivation of a gene in a particular respect may for example occur more freely than a partial inactivation.

crease in activity of genes at least as frequently as inactivation with the consequence that the rate of formation (v) of genes of evolutionary significance becomes negligibly small in comparison with rate of breakdown (u) of such genes. It should be said that FISHER has advanced an alternative hypothesis according to which genes originally without dominance become dominant through a process of selection of modifiers (FISHER 1928, 1929, WRIGHT 1929, 1929a).

The selection coefficient, s, relating to a gene A cannot be expected to be constant if the alternative term a includes more than one gene. The coefficient should rise to a maximum positive value as A replaces less useful genes but should fall off and ultimately become negative as the group of allelomorphs comes to include still more useful genes. But as already discussed, even if A has only one allelomorph, the dependence of the selection coefficient on the frequencies and selection coefficients of non-allelomorphs keeps it from being constant. The existence of multiple allelomorphs merely adds another cause of variation.

Random variation of gene frequency

There remains one factor of the greatest importance in understanding the evolution of a Mendelian system. This is the size of the population. The constancy of gene frequencies in the absence of selection, mutation or migration cannot for example be expected to be absolute in populations of limited size. Merely by chance one or the other of the allelomorphs may be expected to increase its frequency in a given generation and in time the proportions may drift a long way from the original values. The decrease in heterozygosis following inbreeding is a well known statistical consequence of such chance variation. The extreme case is that of a line propagating by self fertilization which may be looked upon as a self contained population of one. In this case, 50 percent of the factors with equal representation of two allelomorphs (that is, in which the individual is heterozygous) shift to exclusive representation of one of the allelomorphs in the following generation merely as a result of random sampling among the gametes. From the series of fractions given by JENNINGS (1916) for the change in heterozygosis under brother-sister mating (population of two) it may be deduced that the rate of loss in this case is a little less than 20 percent per generation. A general method for determining the decrease in heterozygosis under inbreeding has been presented in a previous paper (WRIGHT 1921). It can be shown from this that there is a rate of loss of about $1/2N$ in the case of a breeding population of N individuals whether equally divided between males and females or composed of monoecious individuals, assuming pairs of allelomorphs. HAGEDOORN (1921) has urged the importance of such random fixation as a factor in evolution.

Another phase of this question was opened by FISHER (1922) who attempted to discover the distribution of gene frequencies ultimately reached in a population as a result of the above process. He studied a number of conditions relative to mutation and selection. He does not state the rate of decrease in heterozygosis (where any) which would follow from the solutions which he reached but this can be deduced very directly from them. It comes out 1/4N for a population of N breeding individuals in the absence of selection or mutation. This is just half the rate indicated by the method referred to above.

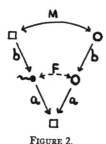

FIGURE 2.

Rate of decrease in heterozygosis

The following symbols and formulae were used in the previous paper in determining the consequences of systems of inbreeding. Primes were used to indicate the number of generations preceding the one in question. Only pairs of allelomorphs are considered here.

M correlation between genotypes of mates

b $(=\sqrt{\frac{1}{2}(1+F')})$ path coefficient, zygote to gamete

a $\left(=\sqrt{\frac{1}{2(1+F)}}\right)$ path coefficient, gamete to fertilized egg

F $(=b^2M)$ correlation between uniting egg and sperm, also, total proportional change in heterozygosis.

P $(=2q(1-q)(1-F))$ proportion of heterozygosis.

The general formula for the correlation between uniting gametes is easily deduced and has been used as a coefficient of inbreeding in dealing with complex livestock pedigrees (WRIGHT 1922, 1923, 1925, MCPHEE and WRIGHT 1925, 1926),

$$F = \Sigma[(\tfrac{1}{2})^{n_s+n_d+1}(1 + F_A)].$$

Here F_A is the coefficient of inbreeding of any common ancestor that makes the connecting link between a line of ancestry tracing back from the sire and one tracing back from the dam. The numbers of generations from sire and dam to such a common ancestor are designated n_s and n_d respectively. The contribution of a particular tie between the pedigrees of sire and dam is $(\frac{1}{2})^{n_s+n_d+1}(1+F_A)$ and the total coefficient is simply the sum of all such contributions. This formula makes it possible to compare quantitatively the statistical situation in actual populations with that in ideal populations.

In dealing with regular systems of mating the method of analysis consists in expressing the correlation between mated individuals in terms of path coefficients and correlations pertaining to the preceding generation $(M = \phi(a, 'b, 'M'))$ and from this obtaining expression for F in terms of the F's of the preceding generations.

Consider a population in which there are N_m breeding males and N_f breeding females, and random mating. The proportion of matings between full brother and sister will be $1/(N_m N_f)$, that between half brother and sister $(N_m+N_f-2)/(N_m N_f)$, and that between less closely related individuals $(N_m-1)(N_f-1)/(N_m N_f)$. The correlation between mated individuals may be written as follows, giving due weight to these three possibilities:

$$M = a'^2 b'^2 \left[\frac{1}{N_m N_f}(2 + 2M') + \frac{N_m + N_f - 2}{N_m N_f}(1 + 3M') \right.$$
$$\left. + \frac{(N_m - 1)(N_f - 1)}{N_m N_f} 4M' \right].$$

This leads to the following formula for proportional change in heterozygosis since the foundation period:

$$F = F' + \left(\frac{N_m + N_f}{8 N_m N_f} \right)(1 - 2F' + F'').$$

The proportion of heterozygosis may be written, relative to that of preceding generations:

$$P = P' - \left(\frac{N_m + N_f}{8 N_m N_f} \right)(2P' - P'').$$

It is to be expected that the proportional change per generation will reach approximate constancy. This rate may be found by equating P/P' to P/P''

$$-\frac{\Delta P}{P'} = \tfrac{1}{2}\left(1 + \frac{N_m + N_f}{4 N_m N_f}\right) - \tfrac{1}{2}\sqrt{1 + \left(\frac{N_m + N_f}{4 N_m N_f}\right)^2}$$

This gives $(1/8N_m+1/8N_f)(1-1/8N_m-1/8N_f)$ as a close approximation even for the smallest populations while for reasonably large ones the form $1/8N_m+1/8N_f$ is sufficiently accurate.

The simplest case is that of continued mating of brother with sister ($N_m=N_f=1$). The rate of loss of heterozygosis comes out $\frac{1}{4}(3-\sqrt{5})$ or 19.1 percent per generation. The formula for proportion of heterozygosis takes the form $P=\frac{1}{2}P'+\frac{1}{4}P''$ as given in the previous paper, with results in exact agreement with those derived by JENNINGS (1916) by working out in detail the consequences of every possible mating from generation to generation.

Another simple case is that in which one male is mated with an indefinitely large number of half-sisters. This is approximately the system of breeding continuously within one herd, headed always by just one male. In this case $N_m=1$, $N_f \doteq \infty$, with rate of loss of heterozygosis of 11.0 percent per generation in agreement with previous results (WRIGHT 1921).

With a relatively limited number of males but unlimited number of females, the rate becomes approximately $1/8N_m$.

An especially important case is that in which the population is equally divided between males and females. Here $N_m=N_f=\frac{1}{2}N$ and the rate of loss is approximately $1/2N$ (or somewhat more closely $1/(2N+1)$) where N is the total size of the breeding population.

It is not, perhaps, clear at first sight that a population of N monoecious organisms, in which self fertilization is prevented, should show a decrease in heterozygosis exactly equal to that in a population of the same size equally divided between males and females. The chance that uniting gametes come from full sisters is $2/[N(N-1)]$, the chance that they come from half sisters is $4(N-2)/[N(N-1)]$ while the chance that they come from less closely related individuals is $(N-2)(N-3)/[N(N-1)]$ giving

$$M = \frac{a'^2 b'^2}{N(N-1)}[2(2+2M')+4(N-2)(1+3M')+(N-2)(N-3)4M']$$

$$P = P' - \frac{1}{2N}(2P' - P'')$$

exactly as in the preceding case.

The somewhat arbitrary case in which the gametes produced by N monoecious individuals unite wholly at random is that which can be compared directly with FISHER's results. The gametes have a chance $1/N$ of coming from the same individual and of $(N-1)/N$ of coming from different individuals. The correlation between uniting gametes may thus be written

$$F = \frac{1}{N}b^2 + \left(\frac{N-1}{N}\right)4b^2a'^2F'$$

$$P = \frac{(2N-1)}{2N}P'.$$

As might be expected, the result does not differ appreciably from that of the preceding case. The rate of loss of heterozygosis is exactly $1/2N$ instead of merely approximately this figure. The simplest special case is, of course, continued self fertilization in which $N=1$ and the formula gives the obviously correct result of 50 percent loss of heterozygosis per generation.

From the mode of analysis it might be thought that the loss in heterozygosis is wholly the consequence of the occasional matings between very close relatives. This, however, is not the case. If instead of random sampling of the gametes produced by the population it is assumed that all individuals reproduce equally and that inbreeding is consistently avoided as much as possible, the percentage of heterozygosis still falls off. The rate of loss is, however, only about half as rapid (approximately $1/4N$) in a reasonably large population equally divided between males and females. The cases of $N=2, 4, 8$ and 16 have been given previously (WRIGHT 1921).

In dealing with heterozygosis in the foregoing, it has been assumed for simplicity that each locus was represented by only two allelomorphs in the population in question and that either complete fixation or complete loss of a particular gene means homozygosis of all individuals with respect to the locus. But in any case beyond that of self fertilization, more than two allelomorphs may be present and complete loss of the gene no longer implies homozygosis of the locus. The initial rate of loss of heterozygosis in a large population may thus be only $1/4N$ with gradual approach to $1/2N$ as the number of loci with only two remaining allelomorphs increases. The rate of decay of the distribution of gene frequencies is $1/2N$ regardless of number of allelomorphs.

The population number

It will be well to discuss more fully, before going on, what is to be understood by the symbol N used here for population number. The conception is that of two random samples of gametes, N sperms and N eggs, drawn from the total gametes produced by the generation in question ($N/2$ males and $N/2$ females each with a double representation from each series of allelomorphs). Obviously N applies only to the breeding population and not to the total number of individuals of all ages. If the population fluctuates greatly, the effective N is much closer to the minimum number than to the maximum number. If there is a great difference between the number of mature males and females, it is closer to the smaller number than to the larger. In fact, as just shown, a population of N_m males and an indefinitely large number of females is approximately equivalent to a population of $4N_m$ individuals equally divided between males and females.

The conditions of random sampling of gametes will seldom be closely approached. The number of surviving offspring left by different parents may vary tremendously either through selection or merely accidental causes, a condition which tends to reduce the effective N far below the actual number of parents or even of grandparents. How small the effective N of a population may be is indicated by recent studies of SMITH and CALDER (1927) on the Clydesdale breed of horses in Scotland, in which they find a steady increase in the degree of inbreeding (Coefficient F) equivalent to that in population headed by only about a dozen stallions. Even more striking is the rapid increase in the coefficient of inbreeding in the early history of the Shorthorn breed of cattle (MCPHEE and WRIGHT 1925).

THE DISTRIBUTION OF GENE FREQUENCIES AND ITS IMMEDIATE CONSEQUENCES

No mutation, migration or selection

On making a cross between two homozygous strains a population is produced in which the members of each pair of allelomorphs in which the strains differ are necessarily equally numerous. The proportional frequency of each allelomorph in unfixed series is $q = 0.50$. In an indefinitely large population, there should be no change in this frequency in later generations (except by recurrent mutation or selection). In any finite population, however, some genes will come to be more frequent than their allelomorphs merely by chance. This means a decrease in heterozygosis, since the proportion of heterozygosis under random mating is $2q(1-q)$, and this quantity is maximum when $q = 0.50$. As time goes on, divergences in the frequencies of factors may be expected to increase more and more until at last some are either completely fixed or completely lost from the population. The distribution curve of gene frequencies should, however, approach a definite form if the genes which have been wholly fixed or lost are left out of consideration. This can easily be seen by considering a case opposite in a sense to that considered above. Suppose that a large number of different mutations occur in a previously pure line. The frequency ratio of mutant to type allelomorphs is initially $(1/2N):(2N-1)/2N$ where N is the number of individuals. The great majority of such mutations will be lost, by the chances of sampling, as FISHER (1922) points out. Those which persist are largely those for which there has been a chance increase in frequency. The distribution curve of frequencies of persisting mutations will thus continually spread toward higher frequencies. There must be a position of equilibrium as far as form is concerned between this situation and that first considered, although a uniform decline in absolute numbers.

As noted above, decrease in heterozygosis takes place in the early generations following a cross without any appreciable fixation or loss of genes. But after equilibrium has been reached in the form of the distribution curve, further loss in heterozygosis must be identical in rate with fixation plus loss.

In simple cases, the equilibrium distribution of gene frequencies can easily be worked out directly. Under brother-sister mating, for example, the following relative frequencies of the 4 possible types of mating involving unfixed factors are in equilibrium although the absolute frequencies of all are falling off 19.1 percent $(=\frac{1}{4}(3-\sqrt{5}))$ each generation as new genes enter the fixed states. AA × AA or aa × aa.

Mating	Relative Frequency
	Percent
AA×Aa	$7-3\sqrt{5}=$ 29.2
Aa ×Aa	$-22+10\sqrt{5}=$ 36.1
AA×aa	$9-4\sqrt{5}=$ 5.6
Aa ×aa	$7-3\sqrt{5}=$ 29.2
	100.1

Similarly in populations of 2 and 3 monoecious individuals with random union of gametes, the following relative frequencies are in equilibrium although the absolute frequencies are decreasing in each generation by exactly 25 percent and 16⅔ percent respectively verifying the 1/2N of theory.

Gene Frequency		Class Frequency
	Percent	
3A:1a	.32	Case of 2 monoecious
2A:2a	.36	individuals per generation
1A:3a	.32	

Gene Frequency		Class Frequency
	Percent	
5A:1a	18.3	
4A:2a	21.0	Case of 3 monoecious
3A:3a	21.4	individuals per generation
2A:4a	21.0	
1A:5a	18.3	
	100.0	

In order to determine generally the distribution of gene frequencies, consider the way in which genes (A) with frequency q are distributed after one generation of random mating. In a population of N breeding individuals, each of the specified genes will have 2Nq representatives among the zygotes and their allelomorphs $2N(1-q)$. A random sample of the same size will be distributed according to the expression $[(1-q)a+qA]^{2N}$. The contribution of this sample to the frequency class with an allelomorphic ratio of $q_1:(1-q_1)$ will be in proportion to the $2Nq_1$'th term of the above expression and to the number of genes included in the contributing class (f). The sum of contributions from all such classes should give the $2Nq_1$'th term an absolute frequency smaller than its value in the preceding generation (f_1) by the amount $1/(2N+1)$ deduced above. Following is the equation to be solved for f as a function of q.

$$f_1\left(1 - \frac{1}{2N+1}\right) = \frac{\lfloor 2N}{\lfloor 2Nq_1 \ \lfloor 2N(1-q_1)}\Sigma q^{2Nq_1}(1-q)^{2N(1-q_1)}f$$

Replacing summation by integration and letting $f = \phi(q)/2N = \phi(q)dq$ we have[4]

$$\frac{\phi(q_1)}{2N+1} = \frac{\lfloor 2N}{\lfloor 2Nq_1 \ \lfloor 2N(1-q_1)} \int_0^1 q^{2Nq_1}(1-q)^{2N(1-q_1)}\phi(q)dq.$$

The cases of 2 and 3 monoecious individuals as worked out by simple algebra suggest an approach to a uniform distribution. As a trial let $\phi(q) = C$. It will be found that this makes the right and left members of the equation identical and is thus a solution.

$$\frac{C}{2N+1} = \frac{C\lfloor 2N}{\lfloor 2Nq_1 \ \lfloor 2N(1-q_1)} \frac{\Gamma(2Nq_1+1)\Gamma(2N-2Nq_1+1)}{\Gamma(2N+2)} = \frac{C}{2N+1}.$$

The case of loss at rate $1/2N$ should not differ appreciably from that at rate $1/(2N+1)$. It would appear that after a cross, the gene frequencies will spread out from 50 percent toward fixation and loss until a practically uniform distribution is reached. The frequencies of all classes will then slump at a rate of about $1/2N$ as $1/4N$ of the genes become fixed and the same number lost per generation. Figure 3 is intended to illustrate this situation.

[4] f must be equated to $\phi(q)/2N$ here, rather than $\phi(q)/(2N-1)$, if the convenient limits 0 and 1 are to be used for integration in place of the limits $1/2N$ and $(2N-1)/2N$ of the summation with its $2N-1$ terms.

Before finally accepting this solution, however, it will be well to examine the terminal conditions. The amount of fixation at the extremes if N is large can be found directly from the Poisson series according to which the chance of drawing 0 where m is the mean number in a sample is e^{-m}. The contribution to the 0 class will thus be $(e^{-1}+e^{-2}+e^{-3}\cdots)f = \dfrac{e^{-1}}{1-e^{-1}} f, = 0.582f.$

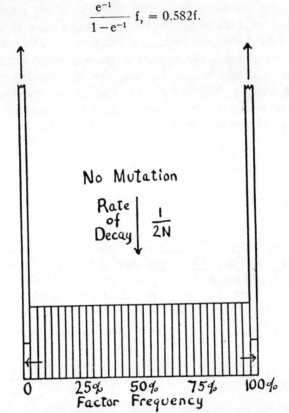

FIGURE 3.—Distribution of gene frequencies in an isolated population in which fixation and loss of genes each is proceeding at the rate $1/4N$ in the absence of appreciable selection or mutation. $y = L_0 e^{-T/2N}$.

This is a little larger than the $\frac{1}{2}f$ deduced above and indicates a small amount of distortion near the ends due to the element of approximation involved in substituting integration for summation. The nature and amount of this distortion are indicated by the exact distributions obtained in the extreme cases of only 2 and 3 monoecious individuals.

Letting L_0 be the initial number of unfixed loci (pairs of allelomorphs) and T the number of generations we have approximately

Unfixed loci in the T'th generation $L_T = L_0 e^{-T/2N}$

An analogous formula holds for genes in multiple series, but in this case, as previously noted, the rate of fixation of loci is only half that given above.

The amount of genetic variation with respect to cumulative characters is easily calculated assuming for simplicity pairs of allelomorphs. The contribution of each factor pair to variance, in the case of no dominance, is $2a^2q(1-q)$ where a is the average difference in effect between plus and minus allelomorphs. The general formula for variance in this or any other distribution is thus $\sigma^2 = 2a^2 \int_0^1 q(1-q)\phi(q)dq$. In the present case in which $\phi(q) = L$ this reduces to $\sigma^2 = \frac{1}{3}La^2$. PEARSON's β_2 comes out with a value 2.8 a slightly platykurtic distribution. Since the percentage of heterozygosis for a given factor frequency, q, is $2q(1-q)$, the formula for heterozygosis is the same as that for variance except that a^2 is to be omitted.

Similarly in the case of dominance, the contribution of a single factor pair to variance is $4a^2(1-q)^2(2q-q^2)$ where a is half the average difference in effect between dominant and recessive zygotes. The total variance with perfect dominance is thus in general

$$\sigma^2 = 4a^2 \int_0^1 (1-q)^2(2q-q^2)\phi(q)dq.$$

In the case of a uniform distribution this gives $\sigma^2 = 8/15\, La^2$.

FISHER (1918) has emphasized the importance of a characteristic of the population which he calls the dominance ratio. He analyzes the variance of characters into three portions, that due to genetic segregation (τ^2) that due to dominance, as something which causes deviations of the phenotype from the closest possible linear relation with the genotype (ϵ^2), and that due to environment. Assuming environment constant, $\sigma^2 = \tau^2 + \epsilon^2$. The simple formulae for the correlations between relatives, to be found if there is random mating and no dominance, must be modified, if dominance is present depending on the value of the dominance ratio defined as ϵ^2/σ^2. Following are examples which he gives:[5]

---

[5] The author wishes here to correct an error in his 1921 paper which was written without knowledge of FISHER's results cited above. In this paper it was assumed that the correlation with no dominance needed merely to be multiplied by the squared correlation between genotype and phenotype, the same as FISHER's $\tau^2/\sigma^2 = (1-\epsilon^2/\sigma^2)$, to obtain that with dominance. This gives correct results (if there is no assortative mating) in the case of offspring with parents, all other ancestors and also in the case of collaterals where one of the individuals is related to the other through only one parent but it is more or less in error in other cases, the fraternal correlation being that most affected. The reasoning followed was not exact because a correlation in the deviations due to dominance in the cases indicated was overlooked.

| | Correlation | |
	No dominance	Dominance
Parent and offspring	$\frac{1}{2}$	$\frac{1}{2}(1-(\epsilon^2/\sigma^2))$
Brothers	$\frac{1}{2}$	$\frac{1}{2}(1-(\epsilon^2/2\sigma^2))$
Uncle and nephew	$\frac{1}{4}$	$\frac{1}{4}(1-(\epsilon^2/\sigma^2))$
Double first cousins	$\frac{1}{4}$	$\frac{1}{4}(1-(3\epsilon^2/4\sigma^2))$

FISHER has shown that the contribution of a single factor to ϵ^2, if there is complete dominance, may be written $\delta^2 = 4q^2(1-q)^2 a^2$ where q is the frequency for dominant allelomorphs. Whether a particular dominant gene has a plus or minus effect on the character under consideration is immaterial. The contribution due to genetic segregation he gives as $\beta^2 = 8q(1-q)^3 a^2$ thus

$$\frac{\epsilon^2}{\sigma^2} = \frac{\Sigma \delta^2}{\Sigma(\delta^2 + \beta^2)} = \frac{\int_0^1 q^2(1-q)^2 \phi(q) dq}{2\int_0^1 q(1-q)^3 \phi(q) dq + \int_0^1 q^2(1-q)^2 \phi(q) dq}.$$

In the present case this reduces to $\frac{1}{4}$ as given by FISHER who also obtains a uniform distribution of factor frequencies for the case of no mutation or selection, although a different rate of decay.

Nonrecurrent mutation

If mutation is occurring, however low the rate, the decline in heterozygosis, following isolation of a relatively small group from a large population, cannot go on indefinitely. There will come a time when the chance elimination of genes will be exactly balanced by new genes arising by mutation. The equation to be solved is obviously as follows:

$$\frac{\phi(q_1)}{2N} = \frac{\underline{|2N}}{\underline{|2Nq_1} \; \underline{|2N(1-q_1)}} \int_0^1 q^{2Nq_1}(1-q)^{2N(1-q_1)} \phi(q) dq.$$

It may be found by trial that the expression $\phi(q) = C_1 q^{-1} + C_2(1-q)^{-1}$ is a solution. The terminal condition, reduction of the class of fixed genes ($q=1$) by an occasional mutation (contributing to the class $q=(2N-1)/2N$ necessarily involves the appearance of new genes (contributing to the class $q=1/2N$) and therefore means that only the symmetrical solution $\phi(q) = Cq^{-1}(1-q)^{-1}$ can be accepted as descriptive of the distribution of the entire array of genes at equilibrium (under the rather arbitrary postulated condition, no selection, no migration, no recurrence of the same mutations). Letting $f = (C/2N) q^{-1}(1-q)^{-1}$ and making $\Sigma f = 1$,

$$C = \frac{1}{2[(0.577 + \log(2N-1))]}$$

or approximately $C = 1/(2 \log 3.6N)$ (compare figure 5).

Before attainment of equilibrium with respect to heterozygosis the distribution will pass through phases of approximately the form $\phi(q) = C_1 q^{-1}(1-q)^{-1} + C_3$ in which the term C_1 gradually displaces C_3 as the number of temporarily fixed genes approaches equilibrium with mutation.

Each particular gene has a probability distribution for the future which spreads in time from the initial frequency in curves which are at first approximately normal in form but later (if the initial q was not too close to 1) become flat, the chances of complete fixation or complete loss each increasing by $1/4N$ each generation. As the chances of complete fixation increase, the chance of mutation must be taken into account. The distribution passes through phases of the type $C_2(1-q)^{-1} + C_3$, C_2 gradually displacing C_3, relatively, but itself beginning to decline as the chance of complete loss increases. With initial q equal (or close) to 1, equilibrium with mutation, and hence the hyperbolic distribution, is reached directly. The ultimate result in any case is complete loss of the gene in question (still assuming no recurrence of the same mutation and hence mutation *of* the gene but not *to* it). If there is reverse mutation, but at very low rate, a term $C_1 q^{-1}$ must be added to the formula, and an equilibrium will be reached in the form $Cq^{-1}(1-q)^{-1}$. This last formula means that in the long run (assuming no disturbances from selection, migration, etc.) the gene will usually be found either completely fixed or completely absent from the population (with frequencies proportional to the mutation rates to and from the gene respectively) but that occasionally fixation or absence will not be quite complete and that at extremely rare intervals the gene will drift from one state to the other.

The turnover among genes in equilibrium in the distribution $Cq^{-1}(1-q)^{-1}$ can be determined from consideration of the variance of q, and independently by application of the Poisson law.

Let $\sigma_q^2 = \Sigma(q-\tfrac{1}{2})^2 f / \Sigma f$ be the variance of q, excluding the terminal classes, the summation including $2N-1$ terms. This variance is increased in the following generation by the spreading out of each frequency class as a result of random sampling. The variance from the spreading of a single class is $q(1-q)/2N$ and the average is thus

$$\Delta \sigma_q^2 = \frac{\Sigma q(1-q)f}{2N \Sigma f} = \frac{1}{2N}\left(\frac{1}{4} - \sigma_q^2\right) = \frac{2N-1}{(2N)^2} C.$$

The sum $\sigma_q^2 + \Delta\sigma_q^2$ includes the newly fixed factors whose contribution is $\frac{1}{4}K$ where K is the rate of fixation, plus loss, but excludes mutation.

Digressing for a moment to the case of no mutation but equilibrium of form, we have at once

$$\sigma_q^2 + \Delta\sigma_q^2 = K\tfrac{1}{4} + (1-K)\sigma_q^2$$

$$\left(K - \frac{1}{2N}\right)\left(\sigma_q^2 - \frac{1}{4}\right) = 0 \quad \text{giving an independent demonstration}$$

that the rate of decay is $1/2N$ in this case.

Returning to the case of equilibrium under mutation, the contribution of new mutations to variance is $K(N-1)^2/(2N)^2$.

$$\sigma_q^2 + \Delta\sigma_q^2 - \tfrac{1}{4}K + K\left(\frac{N-1}{2N}\right)^2 = \sigma_q^2$$

$$K = C = \frac{1}{2 \log 3.6N}.$$

The proportion exchanged at each extreme is thus about half the corresponding subterminal class where N is large ($f_1 = f_{2N-1} = 2NC/(2N-1)$ by this method. This compares fairly well with the proportion as determined by the Poisson law, which is 0.46 times the subterminal class instead of 0.50.

The equilibrium frequencies can be worked out algebraically in simple cases. The figures below give the results in the case of a population of 3 monoecious individuals for comparison with the theoretical values deduced above. The rate of exchange at each extreme is actually 10.8 percent in comparison with 11.0 percent as $\frac{5}{12}\left(=\frac{2N-1}{4N}\right)$ the subterminal class, or 11.4 percent from the formula $\frac{1}{4(.577+\log 5)}$. The case of irreversible mutation is also given.

	Reversible Mutation		Irreversible Mutation	
Gene frequency	Exact equilibrium	$Cq^{-1}(1-q)^{-1}$	Exact	$C(1-q)^{-1}$
5A:1a	27.5	26.3	47.7	43.8
4A:2a	15.4	16.4	20.6	21.9
3A:3a	14.1	14.6	14.1	14.6
2A:4a	15.4	16.4	10.2	10.9
1A:5a	27.5	26.3	7.3	8.8
Totals	99.9	100.0	99.9	100.0
Terminal exchange	10.8	11.0	18.0	18.25
Loss			3.6	3.65

The number of unfixed loci (L) which a given mutation rate per individual (μ) will support in a population is easily found, assuming only pairs of allelomorphs. The number of mutations is KL as well as $N\mu$. Therefore $L = N\mu/K = 2N\mu \log 3.6N$. The variance of cumulatively determined characters worked out as in the preceding case comes out $2N\mu a^2$ in the case of no dominance and $10/3\ N\mu a^2$ in the case of dominance, in both cases, directly proportional to the size of population[6] and to the mutation rate. In view of the piling up of new mutations, one might perhaps, expect to find a leptokurtic distribution for characters. This, however, turns out not to be the case: PEARSON's β_2 comes out exactly 3 in the case of no dominance on substitution in the general formula

$$\beta_2 = 3 + \frac{\int_0^1 q(1-q)[1 - 6q(1-q)]\phi(q)dq}{\left[\int_0^1 q(1-q)\phi(q)dq\right]^2}$$

FISHER's dominance ratio comes out 1/5 in this case.

The preceding results differ somewhat from those presented by FISHER (1922). The latter's analysis was based on a transformation of the scale of factor frequencies designed to make the variance due to random sampling uniform at all points. The variance at a given value of q is $q(1-q)/2N$. FISHER assumes that if the ratio of small differences on the q scale to the corresponding differences on a new θ scale be made proportional to the varying standard deviation of q, the standard deviation on the θ scale will be uniform. Letting $dq/d\theta = \sqrt{q(1-q)}$ leads to the transformation $\theta = \cos^{-1}(1-2q)$ with uniform variance of factor frequencies of $1/2N$. Letting $y = F(\theta)$ be the distribution of factor frequencies in one generation, he wrote that in the next as

$$y + \Delta y = \int_0^\pi \frac{1}{\sigma\sqrt{2\pi}} e^{-\delta\theta^2/2\sigma^2}\left(y + y'\delta\theta + \frac{\delta\theta^2}{\lfloor 2} y'' \cdots\right)$$

and measuring time in generations (T) he reached the equation

$$\frac{\partial y}{\partial T} = \frac{1}{4N} \frac{\partial^2 y}{\partial \theta^2}.$$

After noting that the solution for the symmetrical stationary case is

---

[6] These estimates of number of unfixed loci and of variance depend, of course, on the validity of the conditions on which the formula of the distribution curve is based. How far the mutation rate per locus can be considered negligibly small as size of population increases is discussed later.

$y = L/\pi$, he proceeded to derive the formulae for increasing and decreasing y. Considering the latter, $dy/dT = -Ky$ where K is the rate of decay, giving $1/4N \, d^2y/d\theta^2 = -Ky$ as the equation to be solved. In the symmetrical case this yields $y = C \cos[\sqrt{4NK}(\theta - \pi/2)]$ where $C = \sqrt{4NK}/[2 \sin(\frac{1}{2}\pi\sqrt{4NK})]$ in order to give a total frequency of unity and is to be multiplied further by $L_0 e^{-KT}$ to show change from the initial frequency of L_0.

The maximum value which K can take without giving negative frequencies within the range is obviously $1/4N$ and FISHER found reason for accepting this as the value in the case of no mutation. The formula for the distribution in this case reduces to $y = \frac{1}{2} \sin \theta$. FISHER transformed these equations to the scale $Z = \log[q/(1-q)]$ in which the case of no mutation becomes $y = \frac{1}{4}|\operatorname{sech}^2 \frac{1}{2}Z$ and the case of loss balanced by mutation becomes $y = 1/2\pi \operatorname{sech} \frac{1}{2}Z$. This transformation brings the curves into an approach to the form of the normal probability curve. For our present purpose it is preferable to transform to the scale of actual factor frequencies. The case of steady decay becomes $y = 1$ with which my results are in agreement, although in disagreement as to rate of decay. In the case of loss balanced by mutation, FISHER's formula transforms into $y = 1/[\pi\sqrt{q(1-q)}]$ instead of $1/[2(\log 3.6N)q(1-q)]$ as developed in the present paper. FISHER obtained $\sqrt{\pi}/2N^{3/2}\mu$ for the number of unfixed factors, in contrast with $2N\mu \log 3.6N$; and $\sqrt{\frac{2}{\pi N}}$ for the factor turnover in contrast with $1/[2 \log 3.6N]$.

It will be seen that FISHER's solution leads to a smaller number of unfixed factors with more rapid turnover in very small populations (less than 81) but to a larger number of such factors with slower turnover in larger populations. In a breeding population of one million with one mutation per 1000 individuals, FISHER's formula gives 1,250,000 unfixed factors with a turnover of 0.08 percent while I get 30,000 unfixed factors with a turnover of 3.3 percent.

The exact harmonizing of the results of the two methods of attack has been a somewhat puzzling matter, but Doctor FISHER, on examination of the manuscript of the present paper, has written to me the following which I quote at his suggestion. ".... I have now fully convinced myself that your solution is the right one. It may be of some interest that my original error lay in the differential equation

$$\frac{\partial y}{\partial T} = \frac{1}{4N} \frac{\partial^2 v}{\partial \theta^2}$$

which ought to have been

$$\frac{\partial v}{\partial T} = \frac{1}{4N} \frac{\partial}{\partial \theta}(y \cot \theta) + \frac{1}{4N} \frac{\partial^2 y}{\partial \theta^2}$$

the new term coming in from the fact that the mean value of δp in any generation from a group of factors with gene frequency p is exactly zero,[7] and consequently the mean value of δθ is not exactly zero but involves a minute term $-1/4N \cot \theta$. With this correction, I find myself in entire agreement, with your value 2N, for the time of relaxation and with your corrected distribution for factors in the absence of selection."

Reversible recurrent mutation

It only requires a very moderate mutation rate in a large population for the number of unfixed loci to become enormous. This raises the question as to the effect of a limitation in the number of mutable loci, and recurrence of mutations.

Consider now the case of genes with uniform rates of recurrence of mutation and reverse mutation. Let u be the rate per generation for break down of the gene A and v that for origin from allelomorphs. A class of genes with frequency q (that of all allelomorphs, $1-q$) will be distributed in the following generation under random sampling according to the expansion of the expression

$$\{[(1-q) - v(1-q) + uq]a + [q + v(1-q) - uq]A\}^{2Nf}.$$

Equating the total contribution to a given class, to the frequency of this class in the parent generation, reduced by the proportion K, if there is a uniform rate of decay, gives as the equation to be solved:

$$\phi(q_1)\frac{1-K}{2N} = \frac{\underline{|2N}}{\underline{|2Nq_1}\ \underline{|2N(1-q_1)}} \int_0^1 [q(1-u-v) + v]^{2Nq_1}$$

$$[1 - q(1-u-v) - v]^{2N(1-q_1)}\phi(q)dq.$$

It will be found by trial that the right and left members became identical in certain cases in which $\phi(q)$ is of the form $q^s(1-q)^t$.

Let $x = q(1-u-v) + v$

$$q = \frac{x-v}{1-u-v} \qquad dq = \frac{dx}{1-u-v}$$

$$q^s = \frac{x^s - svx^{s-1}\cdots}{(1-u-v)^s} \qquad (1-q)^t = \frac{(1-x)^t - tu(1-x)^{t-1}}{(1-u-v)^t}.$$

---

[7] p is the q of the present paper. Since the above was written, FISHER has published this revision of his results in *The genetical theory of natural selection*, 1930.

The small amount of spread from a given class will justify retention of the untransformed limits of integration.

Noting that $\Gamma(c+s+1) = \underline{|c}\ c^s \left(1 + \dfrac{s(1+s)}{2c}\right)$ approximately when c is an integer and s is small compared with c, and making use of the following derived relation

$$\dfrac{2N\underline{|2N}}{\underline{|2Nq_1}\ \underline{|2N(1-q_1)}} \int_0^1 x^{2Nq_1+s}(1-x)^{2N(1-q_1)+t} dx$$

$$= \dfrac{4N + s(s+1)q_1^{-1} + t(t+1)(1-q_1)^{-1}}{4N + (s+t+1)(s+t+2)} q_1^s(1-q_1)^t$$

the equation may be written as follows for small values of s and t (compared with N) and values of u and v of a still lower order of size.

$$(1-K)q_1^s(1-q_1)^t$$
$$= \dfrac{1}{1-(u+v)(s+t+1)}\left[\dfrac{4N}{4N+(s+t+1)(s+t+2)} q_1^s(1-q_1)^t\right.$$
$$+ \left(\dfrac{s(s+1)}{4N+(s+t+1)(s+t+2)} - \dfrac{4Nsv}{4N+(s+t)(s+t+1)}\right)q_1^{s-1}(1-q_1)^t$$
$$\left.+ \left(\dfrac{t(t+1)}{4N+(s+t+1)(s+t+2)} - \dfrac{4Ntu}{4N+(s+t)(s+t+1)}\right)q_1^s(1-q_1)^{t-1}\right]$$

The coefficients of $q_1^{s-1}(1-q_1)^t$ and of $q_1^s(1-q_1)^{t-1}$ must equal 0 either under complete equilibrium or equilibrium merely in form of distribution. Neglecting small quantities:

$$s = 0 \quad \text{or} \quad s = 4Nv - 1$$
$$t = 0 \quad \text{or} \quad t = 4Nu - 1.$$

In the case of complete equilibrium $(K=O)$, it turns out that the coefficients of $q_1^s(1-q_1)^t$ in the left and right members are also satisfied to a first approximation by $s = 4Nv-1$, $t = 4Nu-1$. They are also satisfied by letting $s=0$, $t=0$ provided that $u+v = 1/2N$. The relation between the fixed terminal and the unfixed subterminal classes, however, requires that $u = v = 1/4N$ in this case, which thus becomes merely a special case of the first solution. Similarly, the solutions $s=0$, $t=4Nu-1$ and $s=4Nv-1$, $t=0$ require that $v=1/4N$ and $u=1/4N$ respectively and thus also reduce to special cases of the first solution. It appears then that the distribution of gene frequencies in equilibrium under mutation and reverse mutation may be represented approximately by curves of PEARSON's Type I,

$$\phi(q) = \frac{\Gamma(4Nu + 4Nv)}{\Gamma(4Nu)\Gamma(4Nv)} q^{4Nv-1}(1-q)^{4Nu-1}.$$

The terminal conditions are of interest in this and other cases to be considered. The factor turnover at each extreme may be written

$$K_0 = \tfrac{1}{2}f_1 = 2Nvf_0$$
$$K_{2N} = \tfrac{1}{2}f_{2N-1} = 2Nuf_{2N}$$

where the subterminal classes have the frequencies

$$f_1 = \frac{1}{2N}\phi\left(\frac{1}{2N}\right)$$
$$f_{2N-1} = \frac{1}{2N}\phi\left(1 - \frac{1}{2N}\right).$$

In the present case, the terminal classes have the frequencies $f_0 = C/[4Nv(2N)^{4Nv}]$ and $f_{2N} = C/[4Nn(2N)^{4Nu}]$ where C is the coefficient in the expression for $\phi(q)$.

It will be seen that the form of the curve depends not only on the rates of mutation of the genes but also on the size of the breeding population. With small populations or rare recurrence of mutations, the distribution approaches the symmetrical form $y = 1/(2 \log 3.6N) q^{-1}(1-q)^{-1}$ already discussed (figure 5). The ratio of the class of temporarily fixed genes (f_{2N}) to the class of complete absence (f_0) must be approximately $v:u$ in this case in order that the number of mutations at each extreme of the symmetrical distribution of unfixed factors may be equal.

With increase in size of the population, the gene frequencies tend in general to be distributed in asymmetrical U– or even I– or J–shaped curves. For example, if the size of population reaches $1/4u$ and v is much smaller than u, the distribution will be the hyperbola $\phi(q) = Cq^{-1}$ with a piling up of factors with few or no plus representatives.

With sufficient increase in the size of population, the distribution at length takes a form approaching that of the normal probability curve; centered about the point $\bar{q} = v/(u+v)$ which, indeed, is always the mean

$$\left(\bar{q} = \int_0^1 q\phi(q)dq = \frac{v}{u+v}\right).$$

The variance of gene frequencies, $\sigma_q^2 = \int_0^1 (q-\bar{q})^2 \phi(q)dq$ is

$$\frac{\bar{q}(1-\bar{q})}{4N(u+v)+1}.$$

The amount of genetic variation of cumulative characters may be calculated as before. In the case of no dominance and paired allelomorphs

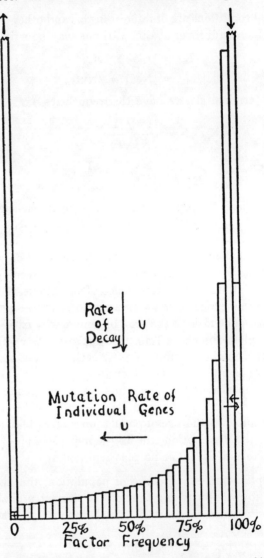

FIGURE 4.—Distribution of type genes in an isolated population in which equilibrium has been reached with destructive mutation but has not been approached with respect to formative mutation. $y = 4NuL_0 e^{-uT}(1-q)^{4Nu-1}$ with $4Nu$ much smaller than 1 and the formula approximately $\dfrac{L_0(1-q)^{-1}}{\log 3.6N}$.

it is $\sigma^2 = 2La^2\, 4Nuq/(4Nu+4Nv+1)$ or $2La^2[\bar{q}(1-q)-\sigma_q^2]$. Where u is much greater than v, this can be written approximately $\sigma^2 = 2La^2 4Nv/(4Nu+1)$ approaching $2La^2\bar{q}$ as a limit, as N increases and L comes to include all loci.

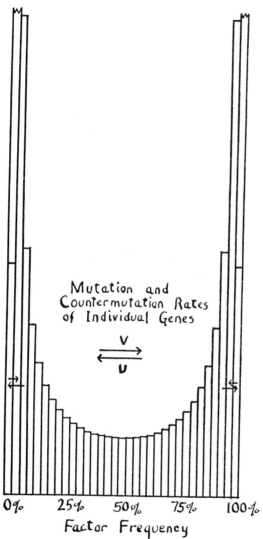

FIGURE 5.—Distribution of gene frequencies (or probability array of gene) where equilibrium with mutation has been attained. Population so small that the terms $4Nu$ and $4Nv$ are both much smaller than 1. $y = Cq^{4Nv-1}(1-q)^{4Nu-1}$, approximately $\dfrac{q^{-1}(1-q)^{-1}}{2\log 3.6N}$.

As the formula for this case was derived on the assumption of small values of u and v, it is desirable to obtain an independent test of its applicability to large values. This can be done as follows: the increase in variance of q due to random sampling is

$$\frac{1}{2N}\int_0^1 q(1-q)\phi(q)\,dq = \frac{\bar{q}(1-\bar{q})-\sigma^2_q}{2N}.$$

Letting $\Delta q = -uq + v(1-q)$ be, as before, the change in q due to mutation, $q + \Delta q - \bar{q} = (q-\bar{q})(1-u-v)$. Thus the effect of mutation is as if all deviations from the mean were reduced in the proportion $(1-u-v)$. The decrease in the variance of q, due to mutation is therefore $\sigma_q^2[1-(1-u-v)^2]$. At equilibrium the increase in σ_q^2 due to random sampling is exactly balanced by the decrease due to mutation yielding:

$$\sigma^2_q = \frac{\bar{q}(1-q)}{4N(u+v)-2N(u+v)^2+1}.$$

The term $-2N(u+v)^2$ in the denominator is important only when $(u+v)$ has a large absolute value. Omitting this, the formula is identical with that deduced by the first method and thus gives an independent demonstration of its validity. As mutation approaches its maximum value $(u+v=1)$, the variance of q approaches $\bar{q}(1-\bar{q})/2N$, that due to random sampling alone.

Migration

The distribution of gene frequencies in an incompletely isolated subgroup of a large population can be obtained immediately from the preceding results. The change in gene frequency per generation under migration $\Delta q = -m(q-q_m)$ can be written $-m(1-q_m)q + mq_m(1-q)$ which is in the same form as the change of q under mutation, $\Delta q = -uq + v(1-q)$. We may write at once for the distribution under negligible mutation rates:

$$\phi(q) = \frac{\Gamma(4Nm)}{\Gamma[4Nmq_m]\Gamma[4Nm(1-q_m)]} q^{4Nmq_m-1}(1-q)^{4Nm(1-q_m)-1}.$$

The mutation terms $4Nu$ and $4Nv$ can be inserted, if mutation rates are not negligible.

Figure 6 shows how the form of the distribution changes with change in m or N. Where m is less than $1/2N$ there is a tendency toward chance fixation of one or the other allelomorph. Greater migration prevents such fixation. How little interchange would appear necessary to hold a large population together may be seen from the consideration that $m = 1/2N$

means an interchange of only one individual every other generation, regardless of the size of the subgroup. However, this estimate must be much qualified by the consideration that the effective N of the formula is in general much smaller than the actual size of the population or even than the breeding stock, and by the further consideration that q_m of the formula refers to the gene frequency of actual migrants and that a further factor must be included if q_m is to refer to the species as a whole. Taking both of these into account, it would appear that an interchange of the order of thousands of individuals per generation between neighboring subgroups of a widely distributed species might well be insufficient to prevent a considerable random drifting apart in their genetic compositions. Of course,

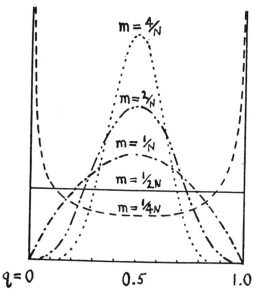

FIGURE 6.—Distribution of frequencies of a gene among subdivisions of a population in which $q_m = 1/2$ (or probability array of gene within a subdivision) under various amounts of intermigration. $y = Cq^{4Nmq_m-1}(1-q)^{4Nm(1-q_m)-1}$.

differences in the condition of selection among the subgroups may greatly increase this divergence. It appears, however, that the actual differences among natural geographical races and subspecies are to a large extent of the nonadaptive sort expected from random drifting apart. An interesting example, apparently nonadaptive, is the racial distribution of the 3 allelomorphs which determine human blood groups (BERNSTEIN 1925).

The variance of distribution of values of q among subgroups (in the ideal case) is $\sigma_q^2 = q_m(1-q_m)/(4Nm+1)$ by substitution in the formula for the preceding case.

The zygotic distribution $[(1-q)a+qA]^2$ cannot be expected to hold in a population made up of isolated groups among which gene frequency varies. WAHLUND (1928) has shown that the proportions in each homozygous class are increased at the expense of the heterozygotes by the amount of the variance of the gene frequencies among the subgroups,[8] the proportions becoming $[(1-q)^2+\sigma_q^2]aa + [2q(1-q)-2\sigma_q^2]$ $Aa + [q^2+\sigma_q^2]$ AA. By substituting the expression for σ_q^2, given above, in WAHLUND's formula one might determine empirically the effective value of $4Nm$ for the population, except that it would be difficult to rule out the possibility that some of the variance of gene frequencies might be due to differences in the selection coefficients among the subgroups instead of merely to random drifting apart.

Irreversible recurrent mutation

The solution $s=0$, $t=4Nu-1$ for the equation reached in the case of recurrent mutation satisfies the conditions for equilibrium of form under irreversible mutation ($v=0$), with decay at rate $K=u$.

$$\phi(q) = 4NuL_0e^{-uT}(1-q)^{4Nu-1}.$$

The proportional frequency of the unfixed subterminal class which is not replenished by mutation is $f_1/(L_0e^{-uT}) = 2u$, twice the rate of decay and thus approximately satisfying the necessary terminal condition.

For values of u as small as $1/(2N \log 3.6N)$ the coefficient in the expression for $\phi(q)$ must be calculated to a closer approximation $\dfrac{4NuL_0e^{-uT}}{1-\left(\dfrac{1}{2N}\right)^{4Nu}}$ which approaches $\dfrac{L}{\log 3.6N}$ as u approaches zero.

The distribution of gene frequencies under irreversible mutation is illustrated in figure 4.

This case is of most interest as representing for a long time the distribution of type genes in a small group isolated from a large one in which all type genes are close to fixation. The release of deleterious mutation pressure from equilibrium with selection will result in approximate equi-

---

[8] The percentage of heterozygotes is $2\int_0^1 q(1-q)\phi(q)dq$ where $\phi(q)$ is the distribution of values of q among the subgroups. As shown above this reduces to $2\bar{q}(1-\bar{q})-2\sigma_q^2$, thus demonstrating WAHLUND's principle.

librium of the form described above. With decay at the rate u, it may be a very long time before effects of reverse mutation become appreciable and the final equilibrium $y = Cq^{-1}(1-q)^{-1}$ approached. Assuming that type genes are dominant, the dominance ratio in this case is 1/3.

Selection[9]

Using $\Delta q = sq(1-q)$ as the measure of the effect of genic selection, the class of genes with frequency $(1-q)a:qA$ is distributed after one generation according to the expression:

$$\{(1-q)(1-sq)a + q[1+s(1-q)]A\}^{2N}.$$

The distribution of gene frequencies which is in equilibrium may be obtained from the following equation which represents the total contribution to class q_1 after one generation, as equal to its previous frequency.

$$\frac{\phi(q_1)}{2N} = \frac{\lfloor 2N}{\lfloor 2Nq_1 \ \lfloor 2N(1-q_1)} $$

$$\int_0^1 q^{2Nq_1}(1-q)^{2N(1-q_1)}(1+s(1-q))^{2Nq_1}(1-sq)^{2N(1-q_1)}\phi(q)dq$$

To a first approximation, the selection terms approach the value $e^{2Ns(q_1-q)}$. The introduction of a factor e^{2Nsq} into the previously reached formula for $\phi(q)$ gives a solution of the equation (for very small values of s) since it cancels the new term e^{-2Nsq} in the integral, and leaves e^{2Nsq_1} as a factor in $\phi(q_1)$. This was the basis for the formula published (WRIGHT 1929a) as $\phi(q) = Ce^{2Nsq}q^{4Nv-1}(1-q)^{4Nu-1}$ intended to exhibit in combination the effects of selection, mutation in both directions and size of population. Further consideration reveals that this solution is the correct one only for the case of irreversible mutation and then only when the selection coefficient is exceedingly small, less than 1/2N in fact. FISHER (1930) in his recently published revision of the results of his method of attack on this problem has given a formula for a special case of selection, equilibrium of flux from an inexhaustible supply of mutating genes. This is given as accurate as long as Ns^2 is small. Assuming one mutation per generation, he writes:

$$y = \frac{2dp(1-e^{-4anq})}{pq(1-e^{-4an})}.$$

[9] This and the following section have been rewritten since submission of the manuscript in order to take account of the correction of my formula, suggested by FISHER's results in *The genetical theory of natural selection*, 1930 as noted herein.

In this formula, $a(=-s)$ is the selection coefficient, $p(=1-q)$ is frequency of mutant genes and dp may be taken as $1/2N$ numerically. This agrees with my previous formula for irreversible mutation, $y = Ce^{2Nsq}(1-q)^{-1}$ only when s is less than $1/2N$, above which value my formula rapidly leads to impossible results. On reexamination of my method, however, I find that the same degree of approximation can be reached by it. The expansion of $[1+s(1-q)]^{2Nq_1}[1-sq]^{2N(1-q_1)}$ yields series of terms which condense into the expression $e^{2Ns(q_1-q)}\{1-Ns^2[q_1(1-q_1)+(q_1-q)^2]\}$ taking into account terms in Ns^2, N^2s^3, N^3s^4, N^4s^5 as well as those in which N and s have the same exponent. Since the random deviations of q have a variance of $q_1(1-q_1)/2N$ the term (q_1-q) is of the order $\sqrt{1/2N}$. A second order approximation should be obtainable by retaining the term $Ns^2q_1(1-q_1)$ while that in $Ns^2(q_1-q)^2$ may be dropped. The equation to be solved can now be written.

$$\phi(q_1) = \frac{2N\,\lfloor 2N}{\lfloor 2Nq_1\ \lfloor 2N(1-q_1)} e^{2Nsq_1}[1 - Ns^2q_1(1 - q_1)]$$

$$\int_0^1 q^{2Nq_1}(1 - q)^{2N(1-q_1)} e^{-2Nsq}\phi(q)dq.$$

Let $\phi(q) = e^{2Nsq}q^{-1}(1-q)^{-1}(a+bq+cq^2+dq^3\cdots)$.

The exponential term in the integral being cancelled, it becomes possible to carry out the integration by means of the approximate formula already used in the case of mutation (page 122).

$$\frac{2N\,\lfloor 2N}{\lfloor 2Nq_1\ \lfloor 2N(1-q_1)} \int_0^1 q^{2Nq_1+z-1}(1 - q)^{2N(1-q_1)-1}dq$$

$$= \frac{4N + z(z - 1)q_1^{-1}}{4N + z(z - 1)} q_1^{z-1}(1 - q_1)^{-1}.$$

The resulting coefficients of the powers of q_1 on the right side of the equation may now be equated separately to those of $\phi(q_1)$. To a sufficient approximation it turns out that $c = \frac{(2Ns)^2}{\lfloor 2}a$, $d = \frac{(2Ns)^2}{\lfloor 3}b$, $e = \frac{(2Ns)^4}{\lfloor 4}a$, $f = \frac{(2Ns)^4}{\lfloor 5}b$, $g = \frac{(2Ns)^6}{\lfloor 6}a$, etc.

Letting $C_1 = a/2$ and $C_2 = \frac{b}{4Ns}$

$$\phi(q) = 2e^{2Nsq}q^{-1}(1 - q)^{-1}[C_1 \cosh 2Nsq + C_2 \sinh 2Nsq].$$

From considerations of symmetry, it is obvious that another solution may be obtained by replacing q by $(1-q)$ and s by $-s$. The full solution may be written in the form

$$\phi(q) = q^{-1}(1-q)^{-1}[C_1(e^{4Nsq}+1) + C_2(e^{4Nsq}-1) \\ + C_3(1+e^{-4Ns(1-q)}) + C_4(1-e^{-4Ns(1-q)})].$$

The relative values of the coefficients in the case of equilibrium can be obtained by setting up the equation for the absence of flux. Each group of genes, $f = \phi(q)dq$ tends to be shifted by the amount $\Delta q = sq(1-q)$ in a generation. There is thus a total flux measured by $\int_0^1 \phi(q)\Delta q dq$ unless there is counterbalancing mutation. The amount of mutation in each direction (assuming the rates of recurrence to be very small compared with $1/4N$) is approximately half the respective subterminal classes, as demonstrated in the preceding cases.

$$f_1 = 2C_1 + 2sC_2 + (1+e^{-4Ns})C_3 + (1-e^{-4Ns})C_4$$
$$f_{2N-1} = (e^{4Ns}+1)C_1 + (e^{4Ns}-1)C_2 + 2C_3 + 2sC_4.$$

Since mutation moves genes from the fixed classes to the subterminal classes with gene frequencies of $1/2N$ and $\left(1 - \dfrac{1}{2N}\right)$ respectively, it creates a net flux of $\dfrac{f_{2N-1}}{4N} - \dfrac{f_1}{4N}$ which at equilibrium should balance that due to selection

$$\int_0^1 \phi(q)\Delta q dq - \frac{f_{2N-1}}{4N} + \frac{f_1}{4N} = 0.$$

Substitution of the values given above leads to the condition $C_1 - C_2 + C_3 + C_4 = 0$. Under this condition the formula simplifies greatly, becoming for all values of s $\left(\text{of lower order than } \dfrac{1}{\sqrt{N}}\right)$

$$\phi(q) = Ce^{4Nsq}q^{-1}(1-q)^{-1}.$$

The effect of selection in this case is perhaps best exhibited in the ratio of the classes of alternative fixed genes in the highly artificial case of equality in the rates of mutation in opposite directions. This ratio is e^{4Ns}. More generally, $f_0 = \dfrac{C}{4Nv}$ and $f_{2N} = \dfrac{Ce^{4Ns}}{4Nu}$ where u and v, both assumed to be very small compared with 1/4N, are the opposing mutation rates. The number of unfixed loci (pairs of allelomorphs) takes the form

$$L = \frac{2N\mu}{f_1 + f_{2N-1}} = \frac{2N\mu}{C(e^{4Ns} + 1)}$$

where μ is the mutation rate per individual and C is chosen so that $\int_0^1 \phi(q)dq = 1$. The effect of selection on the variance of cumulative characters (pairs of allelomorphs) may be seen by comparing the formula

$$\sigma^2 = 2\mu a^2 \left(\frac{e^{4Ns} - 1}{2s(e^{4Ns} + 1)} \right)$$

with the previously given form $2N\mu a^2$ which it approaches as s approaches 0.

In the case treated by FISHER, there is assumed to be irreversible mutation at the rate of one per generation from an inexhaustible supply. As each new gene becomes fixed, it may be considered as transferred to the type class, ready to mutate to new allelomorphs in its series. Thus in place of a return flux of $\frac{f_1}{4N}$, due to reversible mutation, we must write $\frac{f_1}{2}$ (if $v = 0$)

$$\int_0^1 \phi(q)\Delta q\, dq - \frac{f_{2N-1}}{4N} + \frac{f_1}{2} = 0.$$

This is solved if $C_1 = C_3 = C_4 = 0$ and

$$\phi(q) = C_2(e^{4Nsq} - 1)q^{-1}(1 - q)^{-1}.$$

In case the direction of mutation coincides with that of selection ($u = 0$), the mutational terms must be written $\frac{f_{2N-1}}{2} - \frac{f_1}{4N}$ giving the solution

$$\phi(q) = C_4(1 - e^{-4Ns(1-q)})q^{-1}(1 - q)^{-1}.$$

These are identical with FISHER's results on proper choice of the coefficient.

An interesting question which FISHER has discussed, is the chance of fixation of a single mutation. This is given by the ratio of the subterminal classes in the formulae just considered. Where selection opposes mutation, $\dfrac{f_1}{f_{2N-1}} = \dfrac{2s}{e^{4Ns} - 1}$, always less than 1/2N. In the case of favorable mutations, $\dfrac{f_{2N-1}}{f_1} = \dfrac{2s}{1 - e^{-4Ns}}$, or approximately 2s. FISHER also gives an independent derivation of the last figure.

General formula

It is of especial importance to assemble the effects of all evolutionary factors into a single formula. Unfortunately, the equation of equilibrium of class frequencies becomes rather complicated and has not yet been worked through. Presumably the form is given at least approximately by a formula of the type $Ce^{aq}q^{4Nv-1}(1-q)^{4Nu-1}$ in the case of reversible mutation. In order that there may be no flux, $\int_0^1 \phi(q)\Delta q dq = 0$. It is not necessary to consider the terminal classes in this case. Thus

$$C \int_0^1 e^{aq} q^{4Nv-1}(1-q)^{4Nu-1}[-uq + v(1-q) + sq(1-q)]dq = 0.$$

Integration of the first term (that in $-uq$) by parts gives an expression which is immediately solved by letting $a = 4N$. Thus the selection term appears to be e^{4Nsq} regardless of the rates of mutation provided there is reversibility. It is approximately of this value in the case of irreversible mutation, discussed above, provided that s is considerably larger than 1/4N. The conclusions based on the previously presented value e^{2Nsq} still hold,[10] except that they should be applied to selection intensities just half as great.

The position of the mode of the I-shaped distribution curve given when u and s are greater than 1/4N can be found by equating the differential coefficient of the logarithm of the formula to zero.

$$4Ns + \dfrac{4Nv - 1}{q} - \dfrac{4Nu - 1}{1 - q} = 0.$$

When v is small but u and s are both large, q approaches the value $1 - \dfrac{u}{s}$ already given as the equilibrium point. The mean would be somewhat below this point, as expected from the curvilinear relation of selection

---

[10] These conclusions were presented at the meeting of the AMERICAN ASSOCIATION FOR THE ADVANCEMENT OF SCIENCE for 1929 and were summarized in the abstracts (WRIGHT 1929b).

pressure to gene frequency and in contrast with the case of equilibrium between opposing mutation pressures (but no selection) in which the mean is always the equilibrium point $\frac{v}{u+v}$ and the mode is more extreme than this figure, $\frac{4Nv-1}{4N(u+v)-2}$.

Migration pressure introduces no other complications. Combining all factors:

$$\phi(q) = Ce^{4Nsq}q^{4N(mq_m+v)-1}(1-q)^{4N[m(1-q_m)+u]-1}.$$

The selection coefficient refers here to the difference between the selection in the group under consideration and that in the species as a whole, the effect of the latter being taken account of in the mean gene frequency of the species q_m.

The distribution curves

Some of the forms taken by the probability array of gene frequencies, in cases involving selection, are illustrated in figures 7 to 14. Figures 7 to 10 deal with the case in which the rates of mutation are negligibly small compared with $1/4N$. The curves are thus all variants of the form $y = Ce^{4Nsq}q^{-1}(1-q)^{-1}$. Figure 7 illustrates the relatively slight effect of selection below a certain relation to size of population. All conditions are the same in figure 8 except that the populations are four times as large as in figure 7. Thus while the absolute intensities of the selection coefficients are the same, the relations to size of population are altered.[11] The curves bring out the great effect of selection beyond the critical point, $s = \frac{1}{2N}$ (where mutation rates are low). Figures 9 and 10 are intended to show the effects of change in size of population where the intensity of selection remains constant (low in figure 9, four times as severe in figure 10). Up to a certain point $\left(N = \frac{1}{2s}, \text{figure 9}\right)$ increase in population raises the middle portion of the curve. Above this point (figure 10) increase in population depresses the middle portion. In the former case, the increase in unfixed factors brings increased variability of cumulative characters, in the latter there is little change of variability in relation to population size, the depression among middle frequencies being balanced by the accumulation of nearly but not quite fixed factors. All of these fig-

[11] The probability that increase in the number of unfixed genes would react on the individual gene selection coefficients, reducing them, is here ignored.

ures (7 to 10) may be taken as representative of conditions in small inbred populations, which have been isolated sufficiently long to reach equilibrium in relation to mutation. It will be recalled that figures 3 and 4 represent successive stages preceding the attainment of such equilibrium.

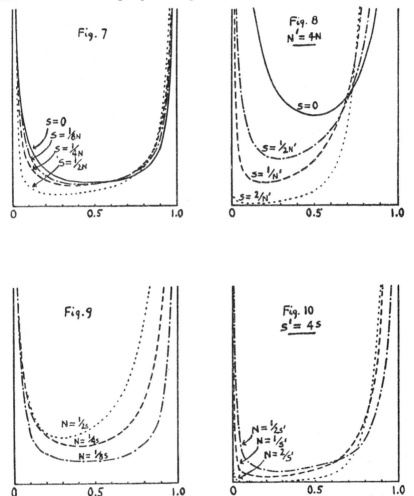

FIGURES 7, 8, 9, and 10.—Distribution of gene frequencies in relation to size of population and intensity of selection where rates of mutation and migration are small compared with $1/4N$. Formulae all of type $y = Cy^{4Nsq-1}(1-q)^{-1}$.

Figure 7. Small population, four degrees of selection. Figure 8. Population four times as large as in figure 7 under the same four (absolute) degrees of selection. Figure 9. Three sizes of population under given weak selection. Figure 10. Same three sizes of population as in figure 9, under selection four times as severe.

Figures 11 to 14 present exactly the same series of comparisons as figures 7 to 10, for small populations that are not completely isolated from the main body of a species.[12] In all cases the gene frequency (q_m) of the

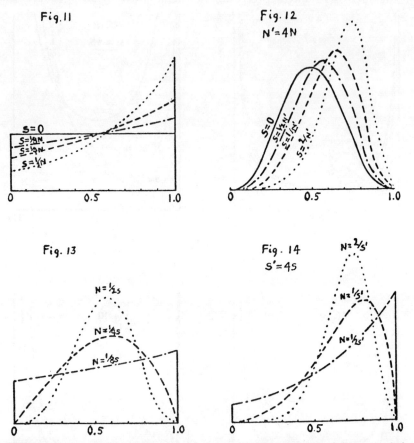

FIGURES 11 to 14.—Distribution of gene frequencies in subgroups of large population (mean frequency $q_m = 1/2$) in relation to size of population and intensity of selection. Formulae all of the type $y = Ce^{4Nsq}q^{4Nmq_m-1}(1-q)^{4Nm(1-q_m)-1}$ Same comparisons as in figures 7 to 10.

Figure 11. Small subgroups ($2Nm=1$), four degrees of selection. Figure 12. Subgroups four times as large as in figure 11, under same four (absolute) degrees of selection. Figure 13. Subgroups of three sizes under given weak selection. Figure 14. Same three sizes as in figure 13 under selection four times as severe.

The figures may also be used to illustrate cases of equal mutation to and from a gene ($u=v$).
$y = Ce^{4Nsq}q^{4Nv-1}(1-q)^{4Nu-1}$

[12] These figures also represent the distribution of gene frequencies in population in which mutation and reverse mutation are equally frequent, but this seems to be so exceptional a case especially under multiple allelomorphism, as to be of little importance.

whole species is assumed to be 1/2. The relations of migration to size of population are such that there is very little complete fixation of genes. In figure 11, m = 1/2N and the purely exponential curves show how increasing intensity of genic selection shifts a uniform distribution in the direction favored by the selection. The fourfold greater population of figure 12 brings about concentration, in curves approaching the normal in form. Figure 13 brings out the concentrating effect of increase in population in the case of weak selection while figure 14 does the same for the case of selection four times as severe.

The important case in which mutation is balanced by selection in a moderately large population (both s and u large compared with 1/4N) is illustrated in figure 19. The four curves represent four degrees of selection, rising by doubling of severity at each step from a case in which mutation pressure practically overwhelms the effect of selection to the reverse situation. The limiting condition in populations so large that 1/4N is very small compared with both s and u is that of concentration of factor frequency almost at a single value (figure 20, page 148).

Dominance ratio

The form of the distribution of the frequencies of the dominant genes affecting a character is of interest in connection with the dominance ratio. Since different genes have different mutation rates and selection coefficients, this distribution is a composite of curves of the types discussed. In small populations which have reached equilibrium, all of these arrays and hence their composite are of the type $Cq^{-1}(1-q)^{-1}$. The dominance ratio is 1/5 in so far as dominance is complete. FISHER gives the value as $3/13 = 0.23$ for the case "when in the absence of selection, sufficient mutation takes place to counteract the effect of random survival." The difference from the value 0.20 given above is due solely to the difference in the formula for the curve, discussed earlier.

In the case of the isolation of a small part of a large population, the dominance ratio takes the value 1/3 in so far as dependent on dominant type genes in equilibrium with recessive mutation but not with reverse mutation $(y = C/(1-q))$. Where following isolation both fixation and loss are substantially irreversible $(y = 1)$ the dominance ratio is 1/4 in agreement with FISHER's result. In both of these cases, of course, the dominance ratio falls to 1/5 when equilibrium is finally attained.

The foregoing discussion applies practically only to very small completely isolated populations. In large populations where the distribution of gene frequencies, even in partially isolated subgroups, tends to approach the normal type, the dominance ratio comes to depend mainly on the mean gene frequency which depends on the relation of selection to mutation, or on selection against both homozygotes in favor of heterozygotes. In the extreme case in which the gene frequency is reduced to a single value, the dominance ratio is $q/(2-q)$. Values close to unity should not be uncommon, especially where gene frequency is controlled by the balance of selection and mutation. Such a dominance ratio has rather surprising effects on the correlation between relatives. The correlation between parent and offspring approaches 0 although that between brothers may remain as high as 0.25. However, the occurrence of an appreciable number of genes at lower frequencies, for example, held in equilibrium by selection favoring the heterozygote against both homozygotes would greatly lower the dominance ratio.

All of these figures are on the assumption that dominance is complete. Dominance, however, is frequently not complete. Among 22 heterozygotes in the guinea pig which have been studied with some care, at least 9 or about 40 percent are to some extent intermediate. Most of these have to do with color characters. It is not unlikely that incomplete dominance will be found to be even more frequent on careful study of size characters.

FISHER (1922) comes rather definitely to the conclusion that the dominance ratio is typically in the neighborhood of 1/3. This was based primarily on a distribution of factor frequencies which he reached for the case of selection against a recessive,[13] with which the results of the present study are not at all in agreement. He also finds, however, that the differences between fraternal and parent-offspring correlations in data which he analyzes indicate the same figure. The analysis of a large number of correlations of these sorts would undoubtedly furnish valuable information with regard to the statistical situation in populations. It is to be noted, however, that similarity in the environment of brothers as compared with parent and offspring may also contribute to a higher fraternal correlation and that in any case one cannot reason from the dominance ratio deduced from correlations to the distribution of factor frequencies without making some assumption as to the prevalence of dominance.

About all that seems justified by the present analysis, is the statement that for permanently small populations under low selection the value should

[13] This was given as $df \propto \dfrac{d\theta}{\sin \frac{1}{2}\theta \cos^3 \frac{1}{2}\theta}$ or $\phi(q) = Cq^{-1}(1-q)^{-2}$ on transformation of scale.

FISHER does not discuss dominance ratio in connection with his recent revision of his results in *The genetical theory of natural selection*, 1930.

be less and probably considerably less than 0.20 but that this figure may be raised by severe selection (favoring dominants) and especially by increase in size of population. It may even approach unity in very large populations under severe selection, if complete dominance is the rule.

Mean and variability of characters

In the case of genes which are indifferent to selection (s less than u), the mean frequency $\bar{q} = v/(u+v)$ remains unchanged through all transformations from a U-shaped distribution in small populations to an I-shaped one in large populations. The variance, due to such genes, is small in small populations, rises in nearly direct proportion to size of population up to a certain critical point (about $N = 1/4u$) and then approaches a limiting value. For the case in which mutations in one direction (u) occur at a much greater rate than in the other (v), the general formula reduces to $\sigma^2 = \sigma_\infty^2 \left(\frac{4Nu}{1+4Nu} \right)$, in which $\sigma_\infty^2 (= 2La^2v/u)$ is the limiting value. This case is illustrated in figure 15. The dotted lines represent mean gene frequencies and the line of dashes the variance.

Actual changes in the size of a given population are not of course accompanied by instant adjustment of the distribution of gene frequencies. A decrease in size to a point well below the critical value is followed by decrease in heterozygosis and variance at a rate between $1/4N$ and $1/2N$ per generation depending on the number of allelomorphs. This may be a fairly rapid process in terms of geologic time but the recovery of heterozygosis through growth of the population to its original size occurs more slowly, since this depends on mutation pressure. On the other hand, the intercrossing of a number of isolated strains, in each of which the reduction of variance has occurred, is followed by immediate recovery of the original statistical situation (except with respect to factor combinations in which there is some delay).

In the opposite case of genes under vigorous selection (s much greater than u) mean frequency as well as variance is affected by size of population and by severity of selection. As in the preceding case, variance is small in small populations, rises in nearly direct proportion to growth of population[14] until a critical point is approached (here about $N = 1/4s$) and then rapidly approaches a limiting value $\sigma^2 = \sigma_\infty^2 \left(\frac{e^{4Ns} - 1}{e^{4Ns} + 1} \right)$ where

---

[14] As before, the probability that the increase in variance, due to growth of population, would react on the selection coefficients of the individual genes, reducing them, requires some qualification of this statement in application to actual populations.

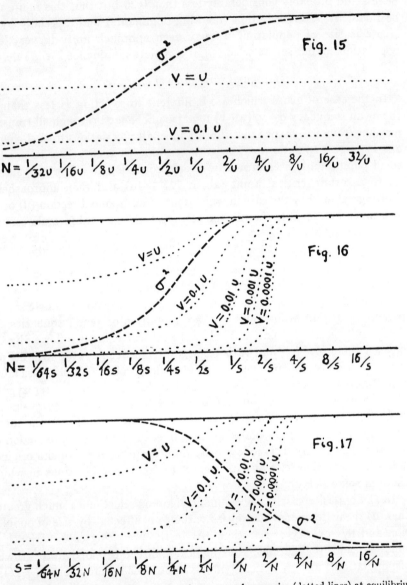

FIGURES 15 to 17.—The variance (σ^2) and mean gene frequencies (dotted lines) at equilibrium under various conditions of mutation, selection and size of population. Figure 15. Effects of increasing population where selection is negligible relative to mutation. Figure 16. Effects of increasing population where mutation rates are small compared with 1/4N. Figure 17. Effects of increasing selection where mutation rates are small compared with 1/4N.

$\sigma_\infty^2 = \mu a^2/s$. The mean factor frequency in large populations ($\bar{q} = 1 - u/s$) is close to 1, nearly complete fixation of the favorable gene. In small populations, on the other hand, the equilibrium point approaches that of the opposing mutation pressures ($\bar{q} = v/(u+v)$) and hence practically 0, with complete loss ($v = 0$) as the inevitable ultimate fate in an extended multiple allelomorphic series. Up to the point at which mutation pressure seriously disturbs the form of the distribution curve, the mean gene frequencies are simply the ratios of the chances of fixation at each extreme, namely, $ve^{4Ns}/(u + ve^{4Ns})$.

The relations of mean frequency and variance to size of population in this case are both shown in figure 16, the former for various relative values of u and v. Inspection of figures 9 and 10 may also be of assistance in understanding this situation.

As in the other case, actual change in size of population is not accompanied by immediate attainment of the new equilibrium. Decrease in population to a number well below the critical point is followed by decrease in heterozygosis at the rate described, bringing with it at the same rate the well known inbreeding effects, loss of variance and, in general, decline in vigor toward a new level. This immediate decline in vigor is not due to change in mean gene frequency, but merely to the greater proportion of recessive phenotypes as homozygosis increases, and thus comes to an end when the degree of homozygosis has reached equilibrium. The change in mean gene frequency proceeds more slowly since it depends on mutation pressure. Long continued isolation should thus involve two distinct degeneration processes, a rapid but soon completed process of fixation and a very slow process of accumulation of injurious genes. The recovery on increase in size of population is slow in both cases, depending on mutation pressure. The intercrossing of isolated lines, on the other hand, is followed by immediate return to the original status of the population if only the immediate inbreeding effect has occurred, but must wait on favorable mutations if there has been time for the slower process.

The effects of different intensities of selection on mean gene frequency and variance (population size constant) are illustrated in figure 17, still assuming that the selection coefficient is of higher order than mutation rate. Figures 7 and 8 showing the distribution of gene frequencies in this case may also be of assistance here. Selection has little effect on variability until it reaches about the value $1/8N$, about half the variance is eliminated when selection reaches $1/N$ and most of it at $s = 4/N$. The formula is $\sigma^2 = \sigma_0^2 \left(\dfrac{e^{4Ns} - 1}{2Ns(e^{4Ns}+1)} \right)$. Selection, of course, affects the mean gene frequency, the formula being the same as that given above under the effect of size of population. On actual increase in the intensity of selection, the rate of change toward the new equilibrium both in mean and variance is controlled by selection pressure and may thus be fairly rapid in terms of geologic time in a large population. This is the case in which HALDANE'S formulae for progress under selection are most applicable. The increase in variance and in the proportion of unfavorable genes following relaxation of selection, on the other hand, are controlled by mutation pressure and thus approach equilibrium relatively slowly. A shift in gene frequency at rate uq may well mean no more than 0.000,001 per generation.

The type of result where the selection coefficient is of the same order of magnitude as mutation rate can be inferred, qualitatively, at least from the preceding extreme cases. Inspection of figure 19 may also be of assistance here.

THE EVOLUTION OF MENDELIAN SYSTEMS
Classification of the factors of evolution

In attempting to draw conclusions with respect to evolution one is apt, perhaps, to assume that factors which make for great variation are necessarily favorable while those which reduce variation are unfavorable. Evolution, however, is not merely change, it is a process of *cumulative* change: fixation in some respects is as important as variation in others. Live stock breeders like to compare their work to that of a modeller in clay. They speak of moulding the type toward the ideal which they have in mind. The analogy is a good one in suggesting that in both cases it is a certain intermediate degree of plasticity that is required.

The basic cumulative factor in evolution is the extraordinary persistence of gene specificity. This doubtless rests on a tendency to precise duplication of gene structure in the proper environment. The basic change factor is gene mutation, the occasional failure of precise duplication. Since the time of LAMARCK, a school of biologists have held that the primary changes in hereditary constitution must be adaptive in direction in order to account for evolutionary advance. Unfortunately, the results of experimental study have given no support to this view. Instead, the characteristics of actually observed gene mutations seem about as unfavorable as could be imagined for adaptive evolution. In the first place, is their fortuitous occurrence. No correlation has been found between external conditions and direction of mutation, and those few agents which have been found to affect the rate (X-ray, radium, and to a relatively unimportant extent, temperature) merely speed up the rate of random mutation. The great majority of mutations are either definitely injurious to the organism or produce such small effects as to be seemingly negligible. MULLER has graphically compared the range of mutations to a spectrum in which the nonlethal conspicuous mutations form a narrow field between broad regions of individually inconspicuous mutations on the one hand and of sublethal and lethal mutations on the other. In addition, the great majority of mutations are more or less completely recessive to the type genes from which they arise. These effects are easily understood if mutation is an accidental process. Random changes in a complex organization are more likely to injure than to improve it, and with respect to the immediate products of the gene, random changes are more likely to be of the nature inactivation (and hence probably recessive) than of increased activation. Finally is to be mentioned the extreme rarity of gene mutation. Even in Drosophila, mutation rates as high as $u = 10^{-5}$ per locus seem to be exceptional and 10^{-6} or less more characteristic. This infrequency seems unfavorable to rapid evolution, yet it is a necessary corollary of the usually injurious effect, if life is to persist at all. Moreover, the more advanced the evolution, the slower must be the time rate of mutation. In one-celled organisms, dividing several times a day, a rapid time rate of mutation will not prevent the production of sufficient normal offspring to maintain the species. The same time rate in Drosophila with an interval of some two weeks between generations would mean such an accumulation of lethals in every gamete that the species would come to an abrupt end. The time rate of lethal mutation in Drosophila (7 per 1000 chromosomes per month under ordinary conditions according to MULLER (1928)) would be quite impossible in the human species. The problem is to determine how an adaptive evolutionary process may be derived from such unfavorable raw material as the infrequent, fortuitous and usually injurious gene mutations.

It will be convenient here to classify factors of evolution according as they tend toward genetic homogeneity or heterogeneity of the species. They are grouped below in more or less definitely opposing pairs.

Factors of Genetic Homogeneity	*Factors of Genetic Heterogeneity*
Gene duplication	Gene mutation (u, v)
Gene aggregation	Random division of aggregate
Mitosis	Chromosome aberration
Conjugation	Reduction (meiosis)
Linkage	Crossing over
Restriction of population size $(1/2N)$	Hybridization (m)
Environmental pressure (s)	Individual adaptability
Crossbreeding among subgroups (m_1)	Subdivision of group $(1/2N_1)$
Individual adaptability	Local environments of subgroups (s_1)

The first pair have been discussed above. They enter into the formulae through the mutation rates u and v. MULLER has pointed out the necessary similarity, in order of size, of genes and of filterable viruses and has suggested the possibility that the latter may consist of single genes. If so, their evolution rests wholly on a not too high rate of mutation, and selection, which seems possible enough in organisms as simple as these presumably are, especially as in this case where the gene is the organism, the mutation of the gene need not be expected to be as fortuitously related to the activities of the organism as in more complex cases.

Presumably the first step toward higher organisms is the aggregation of such genes with multiplication of the aggregate by random division. Given occasional gene mutation, this leads to a new kind of variation, that in proportional abundance of the different kinds of genic material. The larger the aggregate, the less violent the variation. Large aggregates present a labile system capable of quantitative variation in response (perhaps physiologically as well as through selection) to changing conditions. As far as observation goes, the bacteria, and blue green algae have no mechanism of division beyond a random division of the protoplasmic constituents. Such apportionment of more or less autonomous materials may also be important in the differentiating cell lineages of multicellular organisms, but, except for a few plastid characters, seems to play no important role in heredity from generation to generation, as far as has been determined by experiment. There seems here an adequate basis for an evolutionary process in organisms so simple that the handing on of a few different protoplasmic constituents can determine all of the characteristics of the species but the conditions are not favorable for an extensive cumulative process.

Mitosis provides the mechanism by which an indefinitely large number of qualitatively different elements may be maintained in the same proportions. But it provides so perfectly for the persistence of complex organization that further change is difficult. Irregularities in mitosis provide a source of variation but of so violent a nature for the most part as to be of infrequent evolutionary importance, although the differences in chromosome numbers of related species demonstrate that they play a genuine rôle. Complete duplication (tetraploidy) is important in doubling the possible number of different kinds of genes. Other aberrations, especially translocations, are probably more important in isolating types, than for the character changes which they bring. Gene mutation remains the principal factor of variation, but seems inadequate as the basis of an evolutionary process under exclusively mitotic (asexual) reproduction.

The most important factor in transcending the evolutionary difficulties inherent in the characteristics of gene mutation is undoubtedly the attainment of biparental reproduction (EAST 1918). This involves two phases, conjugation, a factor which makes the entire interbreeding group a physiological unit in evolution, and meiosis, with its consequence, Mendelian recombination which enormously increases the amount of variability within the limits of the species. Each additional viable mutation in an asexually reproducing form merely adds one to the number of types subject to natural selection. The chance that two or more indifferent or injurious mutations may combine in one line to produce a possibly favorable change is of the second or higher order. Under biparental reproduction, each new mutation doubles the number of potential variations which may be tried out. The contrast is between $n+1$ and 2^n types from n viable mutations.

Biparental reproduction solves the evolutionary requirement of a rich field of variation. But by itself it provides rather too much plasticity. It makes a highly adaptable species, capable of producing types fitted to each of a variety of conditions, but a successful combination of characteristics is attained in individuals only to be broken up in the next generation by the mechanism of meiosis itself.

An excellent illustration of the principle that a balance between factors of homogeneity and of heterogeneity may provide a more favorable condition for evolution than iether factor by itself may be found in the effects of an alternation of a series of asexual generations with an occasional sexual generation. Evolution is restrained under exclusive asexual reproduction by the absence of sufficient variation, and under exclusive sexual reproduction by the noncumulative character of the variation, but, on alternating with each other, any variety in the wide range of combinations provided by a cross may be multiplied indefinitely by asexual reproduction. The selection of individuals is replaced by the much more effective selection of clones and leads to rapid statistical advance which, however, comes to an end with reduction to a single successful clone. On the other hand a new cross (before reduction to a single clone) may provide a new field of variation making possible a repetition of the process at a higher level. This method has been a favorite of the plant breeder and is perhaps the most successful yet devised for human control of evolution in those cases to which it can be applied at all. Under natural conditions, alternation of asexual and sexual reproduction is characteristic of many organisms and doubtless has played an important rôle in their evolution.

The demonstration of the evolutionary advantages of an alternation of the two modes of reproduction seems to prove too much. Asexual reproduction is practically absent in the most complex group of animals, the vertebrates, and is rather sporadic in its occurrence elsewhere. The purpose of the present paper has been to investigate the statistical situation in a population under exclusive sexual reproduction in order to obtain a clear idea of the conditions for a degree of plasticity in a species which may make the evolutionary process an intelligible one.

First may be mentioned briefly a modification of the meiotic mechanism which has been introduced only qualitatively into the investigation where at all. This is the aggregation of genes into more or less persistent systems, the chromosomes. Complete linkage cuts down variability by preventing recombination. Wholly random assortment gives maximum recombination but does not allow any important degree of persistence of combinations once reached. An intermediate condition permits every combination to be formed sooner or later and gives sufficient persistence of such combinations to give a little more scope to selection than in the case of random assortment. Close linkage, moreover, brings about a condition in which selection tends to favor the heterozygote against both homozygotes and so helps in maintaining a store of unfixed factors in the population.

Resrtiction of size of population, measured by $1/2N$, is a factor of homogeneity and conversely with increase of size. The effects of restricted size may also be balanced by those of occasional external hybridization, measured by m.

Environmental pressure on the species as a whole is a factor of homogeneity. It has been urged by some that because natural selection is a factor which reduces variability, and most conspicuously by eliminating extreme types, it cannot be the guiding principle in adaptive evolution. From the viewpoint of evolution as a moving equilibrium, however, the guiding principle may be found on the conservative as well as on the radical side. The selection coefficient, s, depends on the balance between environmental pressure and individual adaptability. High development of the latter permits the survival of genetically diverse types in the face of severe pressure.

Subdivision of a population into almost completely isolated groups, whether by prevailing self fertilization, close inbreeding, assortative mating, by habitat or by geographic barriers is a factor of heterogeneity with effects measured by $1/2N_1$, N_1 being here the size of the subgroup. This factor may be balanced by crossbreeding between such groups, measured by m_1.

It is interesting to note that restriction of population size is a factor of homogeneity or of heterogeneity for the species, depending on whether it relates to the species as a whole or to subgroups and conversely with the crossbreeding coefficient. Similarly the selection pressures of varied environments within the range of the species (s_1) constitute factors of heterogeneity, restrained from excessive genetic effect by the same individual adaptability which appears in the opposite column in relation to the general environment of the species. Individual adaptability is, in fact, distinctly a factor of evolutionary poise. It is not only of the greatest significance as a factor of evolution in damping the effects of selection and keeping these down to an order not too great in comparison with $1/4N$ and u, but is itself perhaps the chief object of selection. The evolution of complex organisms rests on the attainment of gene combinations which determine a varied repertoire of adaptive cell responses in relation to external conditions. The older writers on evolution were often staggered by the seeming necessity of accounting for the evolution of fine details of an adaptive nature, for example, the fine structure of all of the bones. From the view that structure is never inherited as such, but merely types of adaptive cell behavior which lead to particular structures under particular conditions, the difficulty to a considerable extent disappears. The present difficulty is rather in tracing the inheritance of highly localized structural details to a more immediate inheritance of certain types of cell behavior.

Lability as the condition for evolution

The statistical effects of the more important of these factors in a freely interbreeding population are brought together in the formula

$$y = Ce^{4Nsq}q^{-1}(1-q)^{4Nu-1}.$$

The term $4Nv$ in the exponent of q is here assumed to be negligible and the terms applicable in case of external hybridization are also omitted.

Consider first the situation in a small population in which $1/4N$ is much greater than u and than s (figure 18). Nearly all genes are fixed in one phase or another. Even rather severe selection is without effect. There is no equilibrium for individual genes. They drift from one state of fixation to another in time regardless of selection, but the rate of transfer is extremely slow. Such evolution as there is, is random in direction and tends toward extinction of the group.

Consider next the opposite extreme, a very large undivided population under severe selection. Assume that s is in general much greater than u and that the latter is much greater than $1/4N$. There is almost complete

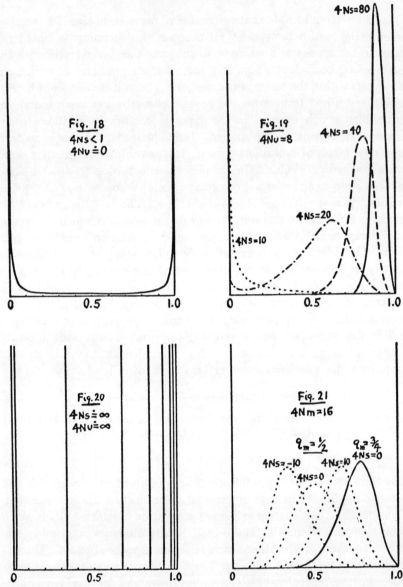

FIGURES 18 to 21.—Distributions of gene frequencies in relation to size of population, selection, mutation and state of subdivision. Figure 18. Small population, random fixation or loss of genes $(y = Cq^{-1}(1-q)^{-1}$. Figure 19. Intermediate size of population, random variaton of gene frequencies about modal values due to opposing mutation and selection $(y = Ce^{4Ns}q^{q-1}(1-q)^{4Nu-1}$. Figure 20. Large population, gene frequencies in equilibrium between mutation and selection $(q = 1 - u/s$, etc.). Figure 21. Subdivisions of large population, random variation of gene frequencies about modal values due to immigration and selection. $(y = Ce^{4Ns}q q^{4Nmq_m-1}(1-q)^{4Nm(1-q_m)-1}$.

fixation of the favored gene for each locus. Here also there is little possibility of evolution. There would be complete equilibrium under uniform conditions if the number of allelomorphs at each locus were limited. With an unlimited chain of possible gene transformations, new favorable mutations should arise from time to time and gradually displace the hitherto more favored genes but with the most extreme slowness even in terms of geologic time.[15]

Even if selection is relaxed to such a point that the selection coefficients of many of the genes are not much greater than mutation rates, the conditions are not favorable for a rapid evolution (figure 20). The amount of variability in the population may be great, maximum in fact, but if the distributions of gene frequencies are closely concentrated about single values, the situation approaches one of complete equilibrium and hence of complete cessation of evolution. At best an extremely slow, adaptive, and hence probably orthogenetic advance is to be expected from new mutations and from the effects of shifting conditions.

It should be added that a relatively rapid shift of gene frequencies can be brought about in this case by vigorous increase in the intensity of selection. The effects of unopposed selection of various sorts and in various relations of the genes has been studied exhaustively by HALDANE, with regard to the time required to bring about a shift of gene frequency of any required amount. The end result, however, is the situation previously discussed. The rapid advance has been at the expense of the store of variability of the species and ultimately puts the latter in a condition in which any further change must be exceedingly slow. Moreover, the advance is of an essentially reversible type. There has been a parallel movement of all of the equilibria affected and on cessation of the drastic selection, mutation pressure should (with extreme slowness) carry all equilibria back to their original positions. Practically, complete reversibility is not to be expected, and especially under changes in selection which are more complicated than can be described as alternately severe and relaxed. Nevertheless, the situation is distinctly unfavorable for a continuing evolutionary process.

Thus conditions are unfavorable for evolution both in very small and in very large, freely interbreeding, populations, and largely irrespective of severity of selection. We have next to consider the intermediate situa-

[15] This, nevertheless, seems to be the case which FISHER (1930) considers most favorable to evolution. The greatest difference between our conclusions seems to lie here. His theory is one of complete and direct control by natural selection while I attribute greatest immediate importance to the effects of incomplete isolation.

tion in which s is not much greater than u for many genes and the latter is not much greater than $1/4N$. Such a case is illustrated in figure 19. The size of population is sufficient to prevent random fixation of genes, but insufficient to prevent random drifting of gene frequencies about their mean values, as determined by selection and mutation. It is to be supposed that the relations of the selection and mutation coefficients vary from factor to factor. The more indifferent ones drift about through a wide range of frequencies in the course of geologic time while those under more severe selection oscillate about positions close to complete fixation. In any case, all gene frequencies are continually changing even under uniform environmental conditions. But the selection coefficients themselves are in general to be considered functions of the entire array of gene frequencies and will therefore also be continually changing. The probability arrays of some genes will travel to the right and close up as their selection coefficients stiffen, while some of the genes which have been nearly fixed will come to be less severely selected and their probability arrays will shift to the left and open out or even move to the extreme left under displacement by another allelomorph. A continuous and essentially irreversible evolutionary process thus seems inevitable even under completely uniform conditions. The direction is largely random over short periods but adaptive in the long run. The less the variation of gene frequency about its mean value, the closer the approach to an adaptive orthogenesis. Complete separation of the species into large subspecies should be followed by rather slow more or less closely parallel evolutions, if the conditions are similar, or by adaptive radiation, under diverse conditions, while isolation of smaller groups would be followed by a relatively rapid but more largely nonadaptive radiation.

As to rate, since the process depends mainly on the value of $1/4N$, assumed to be somewhat less than u (and s) the process cannot be as rapid as one due temporarily to either unopposed selection or unopposed mutation pressure. Hundreds of thousands of generations seem to be required at best for important nonadaptive evolutionary changes of the species as a whole; while adaptive advance, depending on the chance attainment of favorable combinations would be much slower. Even so the process is much the most rapid non-self-terminating one yet considered.

In reaching the tentative conclusion that the situation is most favorable for evolution in a population of a certain intermediate size, one important consideration has been omitted. This is the tendency toward subdivision into more or less completely isolated subgroups in widely distributed populations. Within each subgroup there is a distribution of gene frequencies dependent largely on its own size (N_1), amount of crossing with the rest of the species (m_1), selection due to local conditions (s_1) and the mean gene frequency of migrants from the rest of the species (q_m). It may be assumed that $1/4N_1$ is so much larger than u that mutation pressure (and also the average selection coefficient, s, for the whole species) can be ignored.

$$y = Ce^{4N_1 s_1 q} q^{4N_1 m_1 q_m - 1}(1 - q)^{4N_1 m_1 (1 - q_m) - 1}.$$

Gene frequency in each subgroup oscillates about a mean value, which is that of the whole species only if conditions of selection are uniform. Figure 21 represents various cases. The random variations of gene frequency have effects similar to those described above within each group. The result is a partly nonadaptive, partly adaptive radiation among the subgroups. Those in which the most successful types are reached presumably flourish and tend to overflow their boundaries while others decline, leading to changes in the mean gene frequency of the species as a whole. In this case, the rate of evolution should be much greater than in the previous cases. The coefficients $1/4N_1$ and s_1 may be relatively large and bring about rapid differentiation of subgroups, while the competition between subgroups will bring about rapid changes in the gene frequencies of the species as a whole. The direction of evolution of the species as a whole will be closely responsive to the prevailing conditions, orthogenetic as long as these are constant, but changing with sufficiently long continued environmental change.

A question which requires consideration is the effect of alternation of conditions, large and small size of population, severe and low selection. The effects of changes in the conditions of selection have already been touched upon. Persistence of small numbers or of severe selection for such periods of time as to bring about extensive fixation of factors compromises evolution for a long time following, there being no escape from fixation except by mutation pressure. Many thousands of generations may be required after restoration to large size and not too severe selection, before evolutionary plasticity is restored. Short time oscillations in population number or severity of selection, on the other hand, probably tend to speed up evolutionary change by causing minor changes in gene frequency.

Control of evolution

With regard to control of the process, it is evident that little is possible either within a small stock or a freely interbreeding large one. Even drastic selection is of little effect in the former, and in the latter, while it may bring about a rapid immediate change in the particular respect selected, this must be at the expense of other characters, and in any case, soon leads to a condition in which further advance must wait on the occurrence of mutations more favorable than those fixed by the selection. The limitations in this case have been well brought out in a recent discussion by KEMP (1929). Maximum continuous progress in a homogeneous population requires an intensity of selection for each of the more indifferent genes not much greater than its mutation rate and also a certain size of population. Even so, the direction of advance is somewhat uncertain and the rate to be measured in geologic time.

If infrequency of mutation is the limiting factor here, it would seem that a considerable increase in the rate of evolution should be made possible by a speeding up of mutation, as by X-rays. There is a limit, however, imposed by the prevailingly injurious character of mutations. Even the most rigorous culling of individuals means in general, only a low selection coefficient (in absolute terms) for each of the presumably numerous unfixed genes, which are not in themselves lethal or sublethal in effect. Such culling would become insufficient to hold mutation pressure in check when the latter had increased beyond a certain point ($u > s$). Moreover, as the number of unfixed genes becomes greater under an increased mutation rate, the smaller becomes the separate gene selection coefficients, making it certain that mutation rate could not increase very much before the possibility of effective selection (in all respects at once) rather than infrequency of mutation would become the limiting factor. With respect to lethal mutations, it has already been noted that the observed natural time rate in Drosophila is such as would mean immediate extinction, if transferred to the human species. It is clear that an evolution in the direction of increased gene stability, rather than mutability, has been a necessary phase, in the evolution of the longer lived higher animals. This makes it unlikely that a general increase in mutation rate would increase the rate of evolutionary advance along adaptive lines.

The only practicable method of bringing about a rapid and non-self-terminating advance seems to be through subdivision of the population into isolated and hence differentiating small groups, among which selection may be practiced, but not to the extent of reduction to only one or two types (WRIGHT 1922a). The crossing of the superior types followed by another period of isolation, then by further crossing and so on *ad infinitum* presents a system by means of which an evolutionary advance through the field of possible combinations of the genes present in the original stock, and arising by occasional mutation, should be relatively rapid and practically unlimited. The occasional use of means for increasing mutation rate within limited portions of the population should add further to the possibilities of this system.

Agreement with data of evolution

We come finally to the question as to how far the characteristics of evolution in nature can be accounted for on a Mendelian basis. A review of the data of evolution would go far beyond the scope of the present paper. It may be suggested, however, that the type of moving equilibrium to be expected, according to the present analysis, in a population comparable to natural species in numbers, state of subdivision, conditions of selection, individual adaptability, etc. agrees well with the apparent course of evolution in the majority of cases, even though heredity depend wholly on genes with properties like those observed in the laboratory. Adaptive orthogenetic advances for moderate periods of geologic time, a winding course in the long run, nonadaptive branching following isolation as the usual mode of origin of subspecies, species and perhaps even genera, adaptive branching giving rise occasionally to species which may originate new families, orders, etc.; apparent continuity as the rule, discontinuity the rare exception, are all in harmony with this interpretation.

The most serious difficulties are perhaps in apparent cases of nonadaptive orthogenesis on the one hand and extreme perfection of complicated adaptations on the other. In so far as extreme degeneration of organs is concerned, there is little difficulty—this is to be expected as a by-product of other evolutionary changes. Because of their multiple effects, there can be no really indifferent genes, whatever may be true of organs which have been reduced beyond a certain size. Zero as the value of a selection coefficient is merely a mathematical point between positive and negative values. It is common observation that mutations are more likely to reduce the development of an organ than to stimulate it. It follows that evolutionary change in general will have as a by product the gradual elimination of indifferent organs. Nonadaptive orthogenesis of a positive sort, increase of size of organs to a point which threatens the species, constitutes a more difficult problem, if a real phenomenon. Probably many of the cases cited are cases in which the line of evolution represents the most favorable immediately open to a species doomed by competition with a form of of radically different type or else cases in which selection based on individual advantage leads the species into a cul-de-sac. The nonadaptive differentiation of small subgroups and the great effectiveness of subsequent selection between such groups as compared with that between individuals seem important factors in the origin of peculiar adaptations and the attainment of extreme perfection. It is recognized that there are specific cases which seem to offer great difficulty. This should not obscure the fact that the bulk of the data indicate a process of just the sort which must be occurring in any case to some extent as a statistical consequence of the known mechanism of heredity. The conclusion seems warranted that the enormous recent additions to knowledge of heredity have merely strengthened the general conception of the evolutionary process reached by DARWIN in his exhaustive analysis of the data available 70 years ago.

"Creative" and "emergent" evolution

The present discussion has dealt with the problem of evolution as one depending wholly on mechanism and chance. In recent years, there has been some tendency to revert to more or less mystical conceptions revolving about such phrases as "emergent evolution" and "creative evolution." The writer must confess to a certain sympathy with such viewpoints philosophically but feels that they can have no place in an attempt at scientific analysis of the problem. One may recognize that the only reality directly experienced is that of mind, including choice, that mechanism is merely a term for regular behavior, and that there can be no ultimate explanation in terms of mechanism—merely an analytic description. Such a description, however, is the essential task of science and because of these very considerations, objective and subjective terms cannot be used in the same description without danger of something like 100 percent duplication. Whatever incompleteness is involved in scientific analysis applies to the simplest problems of mechanics as well as to evolution. It is present in most aggravated form, perhaps, in the development and behavior of individual organisms, but even here there seems to be no necessary limit (short of quantum phenomena) to the extent to which mechanistic analysis may be carried. An organism appears to be a system, linked up in such a way, through chains of trigger mechanisms, that a high degree of freedom of behavior as a whole merely requires departures from regularity of behavior among the ultimate parts, of the order of infinitesimals raised to powers as high as the lengths of the above chains. This view implies considerable limitations in the synthetic phases of science, but in any case it seems to have reached the point of demonstration in the field of quantum physics that prediction can be expressed only in terms of probabilities, decreasing with the period of time. As to evolution, its entities, species and ecologic systems, are much less closely knit than individual organisms.

EVOLUTION IN MENDELIAN POPULATIONS

One may conceive of the process as involving freedom, most readily traceable in the factor called here individual adaptability. This, however, is a subjective interpretation and can have no place in the objective scientific analysis of the problem.

SUMMARY

The frequency of a given gene in a population may be modified by a number of conditions including recurrent mutation to and from it, migration, selection of various sorts and, far from least in importance, mere chance variation. Using q for gene frequency, v and u for mutation rates to and from the gene respectively, m for the exchange of population with neighboring groups with gene frequency q_m, s for the selective advantage of the gene over its combined allelomorphs and N for the effective number in the breeding stock (much smaller as a rule than the actual number of adult individuals) the most probable change in gene frequency per generation may be written:

$$\Delta q = v(1-q) - uq - m(q-q_m) + sq(1-q)$$

and the array of probabilities for the next generation as $[(1-q-\Delta q)a + (q+\Delta q)A]^{2N}$. The contribution of zygotic selection (reproductive rates of aa, Aa and AA as $1-s^1:1-hs^1:1$) is $\Delta q = s^1 q(1-q)[1-q+h(2q-1)]$. In interpreting results it is necessary to recognize that the above coefficients are continually changing in value and especially that the selection coefficient of a particular gene is really a function not only of the relative frequencies and momentary selection coefficients of its different allelomorphs but also of the entire system of frequencies and selection coefficients of non-allelomorphs. Selection relates to the organism as a whole and its environment and not to genes as such. The mutation rate to a gene (v) can usually be treated as of negligible magnitude assuming the prevalence of multiple allelomorphs.

In a population so large that chance variation is negligible, gene frequency reaches equilibrium when $\Delta q = 0$. Among special cases is that of opposing mutation rates $\left(q = \dfrac{v}{u+v}\right)$, of selection against both homozygotes $\left(q = \dfrac{1-h}{1-2h}\right)$, of mutation against genic selection $\left(q = 1 - \dfrac{u}{s}\right)$, of mutation against zygotic selection $\left(q = 1 - \dfrac{u}{hs^1}\right.$ unless h approaches 0, when $q = 1 - \sqrt{\dfrac{u}{s}}\right)$, of selection and migration $\left(q = 1 - \dfrac{m}{s}(1-q_m)\right)$ or

$-\dfrac{mq_m}{s}$ if s is much greater than m, $q = q_m\left(1 + \dfrac{s}{m}(1-q_m)\right)$ if s is much smaller than m, while the values $q = \sqrt{q_m}$ or $1 - \sqrt{1-q_m}$ when $s = \pm m$ illustrate the intermediate case).

Gene frequency fluctuates about the equilibrium point in a distribution curve, the form of which depends on the relations between population number and the various pressures. The general formula in the case of a freely interbreeding group, assuming genic selection, is

$$y = Ce^{4Nsq}q^{4N[mq_m+v]-1}(1-q)^{4N[m(1-q_m)+u]-1}$$

The correlation between relatives is affected by the form of the distribution of gene frequencies through FISHER's "dominance ratio." It appears that this is less than 0.20 in small populations under low selection but may even approach 1 in large populations under severe selection against recessives.

In a large population in which gene frequencies are always close to their equilibrium points, any change in conditions other than population number is followed by an approach toward the new equilibria at rates given by the Δq's. Great reduction in population number is followed by fixation and loss of genes, each at the rate $1/4N$ per generation, where N refers to the new population number. This applies either in a group of monoecious individuals with random fertilization or, approximately, in one equally divided between males and females (9.6 percent instead of 12.5 percent, however, under brother-sister mating, $N=2$). More generally with an effective breeding stock of N_m males and N_f females, the rates of fixation and of loss are each approximately $(1/16N_m + 1/16N_f)$ until mutation pressure at length brings equilibrium in a distribution approaching first the form $y = C(1-q)^{-1}$ with decay at rate u and ultimately $Cq^{-1}(1-q)^{-1}$. The converse process, great increase in the size of a long inbred population, is followed by a slow approach toward the new equilibrium at a rate dependent in the early stages on mutation pressure.

With respect to genes which are indifferent to selection, the mean frequency is always $q = v/(u+v)$. The variance of characters, dependent on such genes, is proportional (at equilibrium) to population number up to about $N = 1/4u$. Beyond this, there is approach of variance to a limiting value.

In the presence of selection (s considerably greater than 2u) the mean frequency at equilibrium varies between approximate fixation of the favored genes ($q = 1 - u/s$) in large populations and approximate, if not complete, fixation of mutant allelomorphs ($q = v/(u+v)$) in small populations, the rate of change from one state to the other being the mutation rate (u). A consequence is a slow but increasing tendency to decline in vigor in inbred stocks, to be distinguished from the relatively rapid but soon completed fixation process, described above as occurring at rate $1/2N$. The variance of characters in this as in the preceding case, is approximately proportional to population number up to a certain point (N less than $1/4s$) and above this rapidly approaches a limiting value. Variance is inversely proportional to the severity of selection in large populations unless the selection is very slight but in small populations is little affected by selection unless the latter is very severe (s greater than $1/4N$).

Evolution as a process of cumulative change depends on a proper balance of the conditions, which, at each level of organization—gene, chromosome, cell, individual, local race—make for genetic homogeneity or genetic heterogeneity of the species. While the basic factor of change—the infrequent, fortuitous, usually more or less injurious gene mutations, in themselves, appear to furnish an inadequate basis for evolution, the mechanism of cell division, with its occasional aberrations, and of nuclear fusion (at fertilization) followed at some time by reduction make it possible for a relatively small number of not too injurious mutations to provide an extensive field of actual variations. The type and rate of evolution in such a system depend on the balance among the evolutionary pressures considered here. In too small a population ($1/4N$ much greater than u and s) there is nearly complete fixation, little variation, little effect of selection and thus a static condition modified occasionally by chance fixation of new mutations leading inevitably to degeneration and extinction. In too large a freely interbreeding population ($1/4N$ much less than u and s) there is great variability but such a close approach to complete equilibrium of all gene frequencies that there is no evolution under static conditions. Change in conditions such as more severe selection, merely shifts all gene frequencies and for the most part reversibly, to new equilibrium points in which the population remains static as long as the new conditions persist. Such evolutionary change as occurs is an extremely slow adaptive process. In a population of intermediate size ($1/4N$ of the order of u) there is continual random shifting of gene frequencies and a consequent shifting of selection coefficients which leads to a relatively rapid, continuing, irreversible, and largely fortuitous, but not degenerative series of changes, even under static conditions. The rate is rapid only in comparison with the preceding cases, however, being limited by mutation pressure and thus requiring periods of the order of 100,000 generations for important changes.

Finally in a large population, divided and subdivided into partially isolated local races of small size, there is a continually shifting differentiation among the latter (intensified by local differences in selection but occurring under uniform and static conditions) which inevitably brings about an indefinitely continuing, irreversible, adaptive, and much more rapid evolution of the species. Complete isolation in this case, and more slowly in the preceding, originates new species differing for the most part in nonadaptive respects but is capable of initiating an adaptive radiation as well as of parallel orthogenetic lines, in accordance with the conditions. It is suggested, in conclusion, that the differing statistical situations to be expected among natural species are adequate to account for the different sorts of evolutionary processes which have been described, and that, in particular, conditions in nature are often such as to bring about the state of poise among opposing tendencies on which an indefinitely continuing evolutionary process depends.

LITERATURE CITED

BERNSTEIN, F., 1925 Zusammenfassende Betrachtungen über die erblichen Blutstrukturen des Menschen. Z. indukt. Abstamm.-u. VererbLehre 37: 237–269.

CALDER, A., 1927 The role of inbreeding in the development of the Clydesdale breed of horse. Proc. Roy. Soc. Edinb. 47: 118–140.

EAST, E. M., 1918 The role of reproduction in evolution. Amer. Nat. 52: 273–289.

FISHER, R. A., 1918 The correlation between relatives on the supposition of Mendelian inheritance. Trans. Roy. Soc., Edinb. 52 part 2: 399–433.

1922 On the dominance ratio. Proc. Roy. Soc., Edinb. 42: 321–341.

1928 The possible modification of the response of the wild type to recurrent mutations. Amer. Nat., 62: 115–126.

1929 The evolution of dominance; reply to Professor Sewall Wright: Amer. Nat. 63: 553–556.

1930 The genetical theory of natural selection. 272 pp. Oxford: Clarendon Press.

HAGEDOORN, A. L., and HAGEDOORN A. C., 1921 The relative value of the processes causing evolution. 294 pp. The Hague: Martinus Nijhoff.

HALDANE, J. B. S., 1924–1927 A mathematical theory of natural and artificial selection. Part I. Trans. Camb. Phil. Soc. 23: 19–41. Part II. Proc. Camb. Phil. Soc. (Biol. Sci) 1: 158–163. Part III. Proc. Camb. Phil. Soc. 23: 363–372. Part IV. Proc. Camb. Phil. Soc. 23: 607–615. Part V. Proc. Camb. Phil. Soc. 23: 838–844.

HARDY, G. H., 1908 Mendelian proportions in a mixed population. Science 28: 49–50.

JENNINGS, H. S., 1916 The numerical results of diverse systems of breeding. Genetics 1: 53–89.

JONES, D. F., 1917 Dominance of linked factors as a means of accounting for heterosis. Genetics 2: 466–479.

KEMP, W. B., 1929 Genetic equilibrium and selection. Genetics 14: 85–127.

LOTKA, A. J., 1925 Elements of Physical Biology. Baltimore: Williams and Wilkins.

MCPHEE, H. C., and WRIGHT S., 1925, 1926 Mendelian analysis of the pure breeds of livestock. III. The Shorthorns. J. Hered., 16: 205–215. IV. The British dairy Shorthorns. J. Hered. 17: 397–401.

MORGAN, T. H., BRIDGES, C. B., STURTEVANT, A. H., 1925 The genetics of Drosophila. 262 pp. The Hague: Martinus Nijhoff.

MULLER, H. J., 1922 Variation due to change in the individual gene. Amer. Nat. 56: 32–50.

1928 The measurement of gene mutation rate in Drosophila, its high variability, and its dependence upon temperature. Genetics 13: 279–357.

1929 The gene as the basis of life. Proc. Int. Cong. Plant Sciences 1: 897–921.

ROBBINS, R. B., 1918 Some applications of mathematics to breeding problems. III. Genetics 3: 375–389.

SMITH, A. D. B., 1926 Inbreeding in cattle and horses. Eugen. Rev. 14: 189–204.

WAHLUND, STEN., 1928 Zusammensetzung von Populationen und Korrelationserscheinungen vom Standpunkt der Vererbungslehre aus betrachtet. Hereditas 11: 65–106.

WEINBERG, W., 1909 Über Vererbungsgesetze beim Menschen. Z. indukt. Abstamm.-u. Vererb-Lehre 1: 277–330.

1910 Weiteres Beiträge zur Theorie der Vererbung. Arch. Rass.-u. Ges. Biol. 7: 35–49, 169–173.

WRIGHT, S., 1921 Systems of mating. Genetics 6: 111–178.

1922 Coefficients of inbreeding and relationship. Amer. Nat. 61: 330–338.

1922a The effects of inbreeding and crossbreeding on guinea-pigs. III. Crosses between highly inbred families. U. S. Dept. Agr. Bull. No. 1121.

1923 Mendelian analysis of the pure breeds of live stock. Part I. J. Hered. 14: 339–348. Part II, J. Hered. 14: 405–422.

1929 FISHER's theory of dominance. Amer. Nat. 63: 274–279.

1929a The evolution of dominance. Comment on Doctor FISHER's reply. Amer. Nat. 63: 556–561.

1929b Evolution in a Mendelian population. Anat. Rec. 44: 287.

WRIGHT, S., and McPHEE, H. C., 1925 An approximate method of calculating coefficients inbreeding and relationship from livestock pedigrees. J. Agric. Res. 31: 377–383.

THE LAW OF GEMINATE SPECIES

PRESIDENT DAVID STARR JORDAN

STANFORD UNIVERSITY

In "Evolution and Animal Life," by Jordan and Kellogg (page 120), the following words are used:

"Given any species, in any region, the nearest related species is not to be found in the same region nor in a remote region, but in a neighboring district separated from the first by a barrier of some sort or at least by a belt of country, the breadth of which gives the effect of a barrier."

Substituting the word "kind" for species in the above sentence, thus including geographical subspecies, or nascent species, as well as species clearly definable as such, Dr. J. A. Allen accepts this proposition as representing a general fact in the relations of the higher animals. To this generalization Dr. Allen, in a late number of *Science*, gives the name of "Jordan's Law." The present writer makes no claim to the discovery of this law. The language above quoted is his, but the idea is familiar to all students of geographical distribution and goes back to the master in that field, Moritz Wagner.

This law rests on the fact that the minor differences which separate species and subspecies among animals are due to some form of segregation or isolation. By some barrier or other the members of one group are prevented from interbreeding with those of another minor group or with the mass of the species. As a result, local peculiarities arise. "Migration holds species true, localization lets them slip," or rather leaves them behind in the process of modification. The peculiarities of the parents in an isolated group become intensified by in and in breeding. They become modified in a continuous direction by the selection induced by the local environment. They are possibly changed in one way or another by germinal reactions from impact of environment. At last a new form is recognizable. And this new form is never coincident in its range with the parent species, or with any other closely cognate form, neither is it likely to be in some remote part of the earth. Whenever the range of two such forms overlaps in any degree, the fact seems to find an explanation in reinvasion on the part of one or both of the forms. The obvious immediate element in the formation of species is, therefore, isolation, and behind these are the factors of heredity, of variation, of selection, and others as yet more or less hypothetical involved in the effect of impact of environment on the germ cells themselves. The formation of breeds of sheep as noted by Jordan and Kellogg (p. 82), seems exactly parallel with the formation of species in nature. In like manner, the occasional development of breeds arising from the peculiarities of individuals is possibly parallel with the "mutations" of the evening primrose. Such breeds are the Ancon sheep in Connecticut and the blue-cap Wensleydale[1] sheep in Australia. The ontogenetic species—groups in which many individuals are simultaneously modified in the same way by like conditions of food or climate—show no permanence in heredity. Such forms, however strongly marked, should, therefore, have no permanent place in taxonomy. The recent studies of Mr. Beebe on the effects of moist air in giving dusky colors to birds serve to illustrate the impermanence of the groups or subspecies characterized by dark shades of color developed in regions of heavy rainfall.

It may also be noted in passing that one cause of the

[1] Blue Cap, a ram of Leicester-Teeswater parentage, having a blue shade on his head, was the progenitor of a breed having this peculiarity, known as the Wensleydale, in Australia.

potency of artificial selection among domesticated animals or cultivated plants is that such selection is always accompanied by segregation. The latter is taken for granted in discussions of this topic and hence its existence as a factor is usually overlooked. While poultry or pigeons can be rapidly and radically changed by artificial selection, in isolation, no process of selection without isolation is likely to have any permanent result. For example, we know no way of improving the breed of salmon, because the salmon we have selected for reproduction must be turned loose in the sea, where they are at once lost in the mass.

New forms of gold-fish and carp can be made easily in domestication, because these fishes can be kept in aquaria or in little ponds, but new forms of mackerel or herring are beyond the control of man and the species actually existing have been of the slowest creation, their origin lost in geologic times.

One of the most interesting features of "Jordan's law" is the existence of what I may term *geminate* species—twin species—each one representing the other on opposite sides of some form of barrier. In a general way, these geminate species agree with each other in all the respects which usually distinguish species within the same genus. They differ in minor regards, characters which we may safely suppose to be of later origin than the ordinary specific characters in their group. Illustrations of geminate species of birds, of mammals, of fishes, of reptiles, of snails, or of insects, are well known to all students of these groups, and illustrations may be found at every hand.

Each island of the West Indies, which is well separated from its neighbors, has its own form of golden warbler. Each island in the East Indies has its geminate forms of reptiles or fishes. Each island of the Hawaiian group has its own representative of each one of the types or genera of Drepanidæ. Each group of rookeries in Bering Sea has its own species of fur seal.

One of the most remarkable cases of geminate species

is that of the fishes on the two sides of the isthmus of Panama. Living under essentially the same conditions, but separated since the end of the Miocene Period by the rise of the isthmus, we find species after species which has been thus split into two. These geminate species, a hundred or more pairs in number, were at first regarded as identical on the two shores of the isthmus. Later one pair after another was split into recognizable species. The latest authority on the subject, Mr. C. T. Regan, seems to doubt if any species of shore fishes are actually identical on the two sides of the isthmus.

To make this clear, though at the risk of being tedious, I give below a partial list of these geminate species about the isthmus of Panama:

Atlantic Coast	Pacific Coast
Harengula humeralis	*Harengula thrissina*
Clupanodon oglinus	*Clupanodon libertatis*
Centropomus undecimalis	*Centropomus viridis*
Centropomus pedimacula	*Centropomus medius*
Centropomus affinis	*Centropomus ensiferus*
Epinephelus adscensionis	*Epinephelus analogus*
Alphestes afer	*Alphestes multiguttatus*
Dermatolepis inermis	*Dermatolepis punctatus*
Hypoplectrus unicolor	*Hypoplectrus lamprurus*
Lutianus cyanopterus	*Lutianus novemfasciatus*
Lutianus apodus	*Lutianus argentiventris*
Lutianus analis	*Lutianus colorado*
Lutianus synagris	*Lutianus guttatus*
Hæmulon album	*Hæmulon sexfasciatum*
Hæmulon parra	*Hæmulon scudderi*
Hæmulon schrancki	*Hæmulon steindachneri*
Anisotremus suriamensis	*Anisotremus interruptus*
Anisotremus virginicus	*Anisotremus tæniatus*
Conodon nobilis	*Conodon serrifer*
Pomadasis crocro	*Pomadasis branicki*
Calamus macrops	*Calamus taurinus*
Xystæma cinereum	*Xystæma simillimum*
Encinostomus pseudogula	*Encinostomus dowi*
Kyphosus incisor	*Kyphosus analogus*
Isopisthus parvipinnis	*Isopisthus remifer*
Nebris microps	*Nebris zestus*
Larimus fasciatus	*Larimus pacificus*

Atlantic Coast	Pacific Coast
Odontoscion dentex	*Odontoscion xanthops*
Corvula sialis	*Corvula macrops*
Bairdiella verœ-crucis	*Bairdiella armata*
Micropogon furnieri	*Micropogon ectenes*
Umbrina broussoneti	*Umbrina xanti*
Menticirrhus littoralis	*Menticirrhus elongatus*
Eques acuminatus	*Eques viola*

This list may be greatly extended, but the series noted will illustrate the point in question. Whenever a distinct and sharply defined barrier exists, geminate or twin species may be found on the two sides of it, unless, as sometimes happens, the species has failed to maintain itself on one side of the barrier. So far as Panama is concerned, we have evidence that the barrier was raised near the end of Miocene time with no trace of subsequent depression. We can thus form some estimate of the age of separation in at least a small number of closely related species. In this and similar cases it is not possible to conceive of the formation of these species by sudden mutation, or that they would retain their separate existence were the element of segregation removed. While segregation or isolation is not a force, and perhaps not strictly a cause in species formation, it is a factor which apparently can never be absent, if the species retains its independent existence.

There is no doubt that the distribution of higher animals in general is in accord with "Jordan's Law." Examples by the thousand come up from every hand. If we had a hundredth part of the amount of available evidence in support of mutation theories, these theories would pass from the realm of hypothesis into that of fact. But the application of this law or rule to plants and to one-celled animals has been questioned. So far as rhizopods are concerned, Dr. Kofoid finds that the species are in general sharply defined and of the widest distribution in the sea, so that we can hardly state laws as defining their geographical distribution. To these minute floating animals, the sea scarcely offers barriers at all, and the recognized species do not seem to be products of geographical isolation. Doubtless these species in duration and in nature correspond more nearly to genera or families of higher animals than to actual species. Perhaps minor specific differences such as we note among arthropods or vertebrates are intangible or non-existent. The effects of isolation may be tangible only among forms which possess more varied relations with their environment.

The application of this law to plants has also been denied. But geminate species are just as common in botany as in zoology, and the effects of isolation in species-forming are just as distinct. The law is just as patent in the one case as in the other. It is merely obscured by other laws or conditions which obtain among plants.

In the nature of things, most physical barriers are more easily crossed by plants than by animals. The possibilities of reinvasion are thus doubtless much increased. The plant is limited by climate, rainfall, nature of soil, and the same species is likely to occupy all suitable locations within a large area. Animals are more mobile than plants within their range, a fact which tends to keep the interbreeding masses more uniform. In the struggle for existence, the plant is pitted against its environment. Whether a plant survives or not depends not much on the nature of the seed, but mainly on its relation to the spot on which it falls. There is little selection within the species due to the choice of one individual as against another. This can only happen where plants are overcrowded, and there the survival is mainly that of the seed whose roots run deepest. There is little room for struggle between closely related species. Each individual grows—if it can—on the spot where it falls. The variations among plants are great, but these variations are mostly lost unless reinforced by segregation. There is no likelihood of the survival of DeVries' mutants of the evening primrose if these forms are left free to mix in the same field.

Among plants we often notice the fact—rare though not unknown among animals—of numerous species of the same genus occupying the same area. In some cases these species are closely related, suggesting mutants, and in other cases the relation indicates the existence of hybrids. In California, for example, there are in the same general region many species of Lupinus, of Calochortus, of Ceanothus, of Arctostaphylos, of Eschscholtzia, of Godetia, of Œnothera, and Opuntia. Eucalyptus, Acacia and Epacris in Australia are examples even more striking. But I have never seen very closely related or geminate forms in any of these genera actually growing together. I suspect that they do so sometimes and that the explanation is found in reinvasion. But "growing together" is an indefinite statement as applied to plants. The elder, the alder and the madroño (arbutus) abound in the Santa Clara Valley. But no one ever saw any two of these trees standing side by side. Each has its limitations, as to soil and moisture.

Setting aside these genera which are represented by many species in a limited area, and among which mutation and hybridism may be conceivably factors in species-forming, we find the law of geminate species applying to plants as well as to animals. Crossing the temperate zone anywhere on east and west lines, we find species after species replaced across the barriers by closely related forms. Illustrations may be taken anywhere among the higher plants—equally well, no doubt, among lower ones. Many genera are local in their distribution, monotypic— with a single species, the origin of which can not be traced. But many other genera belt the earth or come very near doing so, each form or species being geminate as related to its next neighbor. This fact is illustrated in Rubus, Alnus, Sambucus, Platanus, Fagus, Veratrum, Symplocarpus, Symphoricarpus, Castanea, Quercus, Pinus, Tsuga, Acer, Rhus, Pyrus, Prunus, Lonicera, Ranunculus, Trientalis, Lilium, Trillium, Veronica, Aquilegia, Gentiana, Viola, Epilobium, Pteris, Mimulus, Trifolium, Solidago, Aster. All these genera and many others furnish an abundance of examples.

We may, therefore, say that with plants as well as animals geminate species as above defined owe their distinctness to some form of isolation or segregation, and that, broadly speaking, with occasional exceptions, given any form of animal or plant in any region, the nearest related form is not to be found in the same region nor in a remote region, but in a neighboring region, separated from the first by a barrier of some sort, not freely traversable.

A law, that is, an observed relation of cause and effect is not invalidated by the presence of other effects due to other causes, in the same environment. The actual conditions in nature are everywhere not products of single and simple forces, but resultants of many causative influences, often operative through the long course of the ages.

It may be urged that these geminate groups or forms are not true species, because they often intergrade one into another, and they would probably be lost by intermingling if the barriers were removed. It is sometimes claimed that only physiological tests of species can be trusted, as true species will not blend and their hybrids, if formed, will be sterile. All this is purely hypothetical and impracticable to the systematic zoologist, and not of much value to the botanist. Closely related species can usually be readily crossed. As the relation becomes less close, partial sterility of all grades and then total sterility appear.

Species as we find them in nature are real *species* if that term has any definition. And real species have, as a rule, indefinite boundaries, shading off into subspecies, geminate species, ontogenetic forms and the like. And if we are to understand the significance of nature, we have to describe these facts and relations as they actually are. Then we have to find out what changes we can work in individuals and in species by such alterations of conditions as experiment can give.

We do not know actually any species of animal or plant until we know all changes that would take place in its individuals under all conditions of environment.

THE TEST OF THE SUBSPECIES
By P. A. Taverner

THE TEST OF THE SUBSPECIES

By P. A. Taverner

In the Journal of Mammalogy, Vol. 1, No. 1, pp. 6–9, appears an article by Dr. C. Hart Merriam advocating that the amount of differentiation be used as the test of specific or subspecific status, rather than the generally accepted one of intergradation. It is with some trepidation that I dare take direct issue with so eminent an authority but the case seems so clear against the proposal and its acceptance is so fraught with possibilities of confusion that I feel justified in lodging a protest.

Doctor Merriam makes much of the uncertainty of human judgment in estimating the probability of intergradation when direct evidence of it is lacking. For the sake of these minority cases where the human element may give varying results, he advocates the recognition of the amount of difference exhibited rather than the presence or absence of intergradation as the test for specific status; thus throwing open each and every case, instead of an occasional one, to the uncertainty of personal standards of judgment. It looks like out of the frying pan into the fire and the choosing of the greater instead of the lesser of two evils. Under the one standard we have numerous cases where intergradation can be demonstrated and subspecific status fixed. Under the other all are equally uncertain. It would seem more logical to go to the other extreme and ascribe every difference to the specific that cannot be demonstrated to intergrade with others. This would however be carrying logic to an extreme and I see no real reason why we should not continue to rely upon the good judgment of experience to assume the probability of intergradation where data is incomplete, readjusting mistakes according to new evidence. Finality can thus be gradually approached even though it may never be perfectly attained. The occasional transference of species to subspecific status and the converse are not serious disturbances so long as we keep the fundamental differentiation in mind and remember that in many cases intergradation is hypothetical and still awaits demonstration.

It is also more than probable that a strict adherence to the proposed criterion would land us in greater confusion than we experience now. Under it, on the evidence of specimens on which most of our conclusions are based, we would class the gray-cheeked and olive-backed thrushes as mere subspecies whilst the extremes of such forms as song sparrows, fox sparrows and horned larks we would raise to full specific status.

It is beside the question that in some particular cases such a proceeding might conceivably be an advantage for it is recognized that extremes of obvious subspecies sometimes differ more in apparent characters than do other distinct species. It is thus apparent I think, from the standpoint of mere expediency, that the amount of divergence as a test of the lower systematic units is open to serious question.

The only logical ground for applying a quantitative rather than an intergradational test to the subspecies is that of the instability of species. If species are liquid quantities flowing imperceptibly into each other the amount of difference by which they are characterized is the only practical means for their recognition. If, however, the species is a definite entity it must be cut off sharply from all other similar entities and degree of divergence becomes unimportant and isolation (discontinuity) its final test. Herein lies a conflict of ideas.

Those who concentrate their attention on the paleontological evidence are prone to regard the species as a mere concept, an ever varying quantity in constant state of development, adopted for convenience in referring to arbitrary points along a continuous line of progress. The modern zoologist however finds species the termination of lines of descent, and each sharply marked off from the other. As both of these reasonings are demonstrable it is apparent that in the word "species," as generally accepted, we have lumped two separate concepts. Certainly contemporaneous and consecutive species bear fundamentally different relations to each other and between themselves, and eventually will probably have to be differentiated by systematists. It is only the fragmentary nature of our geological evidence that has heretofore concealed the essential difference between species merging into each other along a line of descent and species the outcome of independent lines of descent. Intergradation is a concomitant of the first but incompatible with the latter.

It can be urged that evolution is an always present activity, that the processes of the past are continuing in the present and any system founded upon the stability of the species is doomed to eventual confusion. This may be correct philosophically, but in practice need hardly be considered in dealing with modern material. Within historical times we have absolutely no evidence of serious evolutionary change. A system that would have sufficed for three thousand years in the past will probably do for an equal time in the future. By the time evolutionary change introduces serious disturbance in the present scheme of things it is probable that our whole classification system will have been scrapped for something better or else altered beyond recognition. In the meantime I think we are safe in basing our working system on the convention that existing relationships are practically stable.

Though all standards of taxonomic measurement are not mutually transferable between paleontology and modern zoology it does not follow that paleontological evidence should be neglected by modern systematists. As its evidence increases and its lessons become plainer, paleontology must be, even more in the future than it has been in the past, the rule and guide of our classification. We should however bear in mind that concepts that apply to the one may require modification before they can be transferred to the other.

However otherwise it may have been in the past or may be in the future, at the present moment or on any one given geological horizon, the species is a definite entity and its essential character is its genetic isolation. Absence of intergradation with other forms is the only test of the species as it exists at present. There is a barrier that isolates modern specific groups one from another, individualizes contemporary species and prevents wholesale mongrelization. Just what this barrier may be we cannot say with confidence, nor is it altogether necessary to the present argument to do so, but the agent that seems most capable of producing present results is the degree of fertility between such groups. When fertility between divergent forms breaks down, when differentiation progresses to the reproductive processes sufficiently to form a handicap to crossbreeding, genetic isolation ensues that forever separates the varying branches of a common stock and a new species is born. Whatever the mechanics may be that tend to hold a species true and prevents promiscuity, subspecies are incipient species, and I do not see what they can be but variations tending towards, but not reaching, specific status until connection with other forms (intergradation) is broken down and isolation established. Thus intergradation is not only an indication of a condition but it is the condition itself and the refusal to recognize it as the essential quality of the subspecies seems to be a denial of fact. It should be admitted, and can be without discrediting the fundamental argument, that intergrades will not always be discovered. Material from critical localities may not be available or connecting distributions may be obliterated through geographical or ecological changes. Subspecific variations may appear in disconnected communities and give rise to discontinuous distribution where even intergradation through individual variation may conceivably be obscured. without in any way denying the relative status of the forms concerned. The accidental absence of intergrades in these cases complicates the demonstration but cannot alter the fundamental facts. No system of classification has ever entirely done away with the necessity of exercising some judgment and probably none ever will, and the best we can arrive at is to reduce the human equation to its lowest possible terms consistent with the facts of nature. The *possibility* of intergradation where contact between races is physically impossible must necessarily be estimated under the guidance of what evidence we have. The test of intergradation or its possibility where physically prevented gives a far more definite basis of judgment than unmeasurable generalities expressed as vague comparatives of difference. That such proceeding does in practice and in some cases, approximate the criterion laid down by Doctor Merriam is beside the point as the resemblance is superficial and not fundamental. In one case it is frankly an expedient, a suggestion or means to an end, in the other it is the end itself and final.

In this argument I do not forget such cases as the hybrid flicker nor Lawrence's and Brewster's warblers. These if anything substantiate the view that degrees of sterility form the specific boundary lines. That the parent forms of these anomalies are not mongrelized is evidence that such cross breeding is under a handicap as against purer lines of descent; for it is a mere matter of mathematics to prove that otherwise species that hybridize regularly, even if only occasionally, would eventually merge. I have little doubt that the hybrid flicker which shows no appreciable evidence of sterility is only continued through fresh crossings of original stock and that should either parent form be exterminated, it would in a few generations die out through inherent weakness and inability to compete with either of its more virile parent forms.

Therefore, for reasons of both expediency and philosophy, stability of nomenclature and the teachings of evolution, I respectfully submit that the fact of intergradation is the only proper and workable test of subspecific status and should be firmly held to by all students of speciation.

THE ORIGIN OF SPECIES, V: SPECIATION AND MUTATION

HENRY FAIRFIELD OSBORN

RESEARCH PROFESSOR OF ZOOLOGY, COLUMBIA UNIVERSITY; HONORARY CURATOR OF VERTEBRATE PALEONTOLOGY, AMERICAN MUSEUM OF NATURAL HISTORY

AROUSED by the late William H. Bateson's statements in his South African and Toronto addresses to the effect that we know neither the *modes* nor the *causes* of the origin of species, the writer opened in 1925 a series of papers, enumerated below (I–V), of which the present is the final one, ranking as No. V in the series:

THE ORIGIN OF SPECIES AS REVEALED BY VERTEBRATE PALEONTOLOGY (I). Address delivered before National Academy of Sciences, April 28, 1925, and published in *Nature*, June 13 and 20.

THE ORIGIN OF SPECIES, II: DISTINCTIONS BETWEEN RECTIGRADATIONS AND ALLOMETRONS. Read before National Academy of Sciences, November 11, 1925, and published in its *Proceedings*, Vol. 11, No. 12.

THE ORIGIN OF SPECIES, 1859–1925 (III). Read at dedication of new Yale Museum, December 29, 1925, and published in its Bulletin I, Number 1.

THE PROBLEM OF THE ORIGIN OF SPECIES AS IT APPEARED TO DARWIN IN 1859 AND AS IT APPEARS TO US TO-DAY (IV). Read before Section D (Zoology) of the British Association at Oxford, August 5, 1926, and published in *Nature*, August 21, 1926. Republished with corrections in *Science*, October 8, 1926.

THE ORIGIN OF SPECIES, V: SPECIATION AND MUTATION.

Renewed support of the exclusive origin of species by *mutation* with lack of proof of environmental influence has recently been given by Crampton (1925, p. 35), by Calkins (1926, p. 350) and by Morgan (1926, p. 106), who says:

Darwin rested his case for evolution on the observed small differences that all animals and plants show. Now, while we realize to-day that many of these slight differences are not inherited, we also know that amongst them there are some that are inherited and that these, so far as we know, have arisen as mutations. At present there is no evidence that will stand the test of criticism in favor of any other origin than that all known variations owe their appearance to the same process of mutation that also produces the larger differences; and, I repeat, that there is much explicit evidence to show that very small differences, that add to or subtract from characters already present, do appear by mutation.

The present paper (V) presents in more detail the zoological data on which the fourth paper of the series was based, including a fairly complete bibliography of the papers referred to. The writer is again indebted to his colleagues in the American Museum for their more careful scrutiny of the facts and figures in their previous statements; to R. I. Pocock for his note on the zebras; to W. H. Osgood and F. B. Sumner for revision of their previous statements.

The present paper is divided as follows:

(1) 1859–1926. Discovery and naming of Subspecies and Geographic Varieties.

(2) Authority for the Word "Speciation."

(3) Principles of Continuous Geographic Isolation, Intergradation and Speciation in the genus Io.

(4) 1859–1926. Discoveries relating to the Origin of Species in Fishes.

(5) Discoveries relating to the Origin of Species in Amphibia and in Reptilia; (a) Intergrades in Amphibia; (b) Intergrades in Reptilia.

(6) 1859–1926. Discoveries relating to the Origin of Species, Subspecies and Mutations in Birds; (a) Species-origin environmental and mutational in birds; (b) Stresemann on Mutations; (c) Chapman on Zonal Speciation; (d) Speciation and Mutation of the South American House Wren.

(7) 1859–1926. Discoveries relating to the Origin of Species in Mammals; (a) Speciation of the Deer Family around the World on the Northern Parallels; (b) Probable Intergradation in Zebras and Foxes; (c) Speciation of the Deer-mice in Contiguous North American Areas; (d) Isolation in the Origin of Localized Races of *Peromyscus* and of *Thomomys*; (e) Speciation in the Gopher (*Thomomys*).

(8) Summary of Speciation and Mutation as observed in the Living Vertebrates.

1859–1926. DISCOVERY AND NAMING OF SUBSPECIES AND GEOGRAPHIC VARIETIES

In Darwin's own library in the Botany School at Cambridge we find some volumes which cover the mammalogy of his day; for example, Lesson's "Manuel de Mammalogie" (1827), in which 1,124 fossil and living species are listed by number, and Agassiz' "Nomenclatoris Zoologici Index Universalis" (1848), in which 889 genera of mammals are listed. Also accessible to Darwin were the complete works of Linnaeus, Buffon, Bonaparte, Audubon and other great zoologists. Darwin was little concerned about the *number* of species known in his day; he remarked that *transitions between species and varieties* are rare and he would have been the first to welcome the numerous transitions now observable among fishes, reptiles, birds and mammals. Remember we are here speaking only of the period 1837–1859, when "The Origin" was being written, not of the subsequent period of ardent exploration.

Vertebrates. Günther and Shipley list the number of vertebrate species known in the years 1830 to 1881 and 1924 as below. The lists of Sharpe, Chapman and Anthony include species and subspecies.

	1837–1859	1881–1886	1924	1926	1909	1926
	Lesson-Agassiz Darwin	Günther-Leunis Shipley	Poole and others	American Museum Anthony	Sharpe	Chapman Estimated
Mammals	1,200	2,300	12,540	13,000		
Birds	3,600	11,000	16,000		18,939	24,000
Reptiles and Amphibians	543	3,400	9,000			
Fishes	3,500	11,000	20,000			

Mammals. Several authorities, including Sherborn, the conservative historiographer of species, agree that not more than 1,200 species of mammals were known when Darwin was writing "The Origin of Species"; the increase to the present day is over tenfold, and the present number is being rapidly expanded through explorations in Asia, Africa and South America. The present estimated total of 13,000 living species and subspecies of mammals is geographically subdivided as follows:

Asia Estimated after Allen and Anthony	Europe Estimated after Miller, 1912, and subsequent discovery	Africa Estimated after Herbert Lang	Australia Estimated after H. C. Raven	N. America Estimated after Miller, 1924, and subsequent discovery	S. America Estimated after Anthony
3,550	1,000	1,000–2,000	350	2,600	3,500

To summarize and contrast successive lists, the estimates of mammals are approximately as follows:

```
1827  Lesson: Living and fossil species ............................ 1,124
1924  Poole: Living species and subspecies ..................... 12,540
```

Accordingly, tenfold is a conservative estimate of the increase of our knowledge of mammals alone since Darwin's volume of 1859, but the enumeration is not on equal terms, because we include not only species but numberless subspecies listed in trinomials and still largely regarded as geographical variations. It has, however, recently been shown by experiment (Sumner, 1924, III) that many, if not all, subspecies *are stable under changed conditions of environment.* Consequently, while under suspicion as to reality, often vexatious and inconvenient, and always annoying to the systematist, *well-authenticated subspecies are of priceless value to the biologist who seeks to ascertain the conditions under which new species arise.* In fact, many true "subspecies" and "geographic variations" are now known to be germinal transitions, intermediates and intergradations from "Species" to "Species" of the higher kind known to Darwin.

AUTHORITY FOR THE WORD "SPECIATION"

In the classic shades of Oxford the writer hesitated to introduce the etymologically doubtful word "speciation" now being quite widely used in less classic America. On this linguistic point (1) D'Arcy Thompson, the eminent zoological classicist of St. Andrews, writes, July 16, 1926:

> *Speciation* will really not do for "the making of species." Obviously the word is, or ought to have been, *specification*, but that has long ago taken unto itself a different meaning. *Speciation* is not Latin; it is, however, *understandable*. You are therefore on the horns of a dilemma, as it seems to me: you may throw conservative scholarship to the winds and, heading straight for practical utility, talk of speciation as much as you please; or, alternately, you must be content with plain English and talk of the *making*, or the *origin*, of species, or any similar descriptive phrase. I feel pretty sure that, speaking in Oxford, the atmosphere of the place will exercise its conservative influence on you.

(2) L. Davies Sherborn, July 20, 1926, advises:

> The word "speciation" has no English usage. I suppose it might be derived from *speciatus-a-um*. John Evelyn, 1664, in Sylva, p. 1679, uses "specify" for breaking into species—hence "specification" (now used in another sense) but this has priority over your "speciation."

In a second communication, however, Dr. Sherborn writes as follows:

> The word is justified, but ugly. We can only suggest "differentiation into species," more clumsy but more precise and in accordance with the English language, which is inelastic as compared with many others. "Speciation" has already received two meanings (as below) but the first example seems to be the use you want: "Speciation and Host Relationships of Parasites," (Chandler, 1923. *Journ. Parasitol.*, XV, 3, p. 326). "Speciation" here seems to be used in the sense of the formation or evolution of species. "Speciation and Specificity in the Nematode genus Strongyloides," (Sandground, 1925. *Journ. Parasitol.*, XII, 2, p. 59). Here "speciation" appears to be used in the sense of "systematic differentiation between species" (*i.e.*, in the mind of the systematist). "Specificity" appears to mean adaptation to particular hosts.

(3) P. Chalmers Mitchell, secretary of the Zoological Society of London, referring to previous American usage, writes, July 25, 1926:

> "Speciation" says what you wish to say; the word is required; the word is already used in America in the sense you propose. There are rough analogies: variation and varieties, glaciation and glaciers. Language is dynamic, etymology is static. Therefore use "speciation."

PRINCIPLES OF CONTINUOUS GEOGRAPHIC ISOLATION, INTERGRADATION AND SPECIATION IN THE GENUS Io[2]

A striking parallel to the subspeciation of *Troglodytes* and *Peromyscus* is seen in the fresh-water gastropod

[2] Willard G. Van Name: Notes supplied for this paper, 1926.

genus *Io* of the family Pleuroceridae observed by Adams[3] in the Tennessee River and its branches, where it is con-

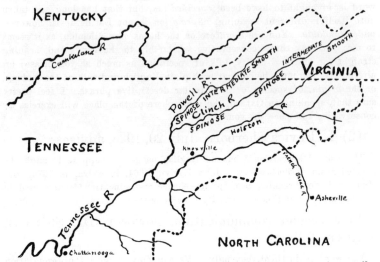

FIG. 1. Headwaters of the Tennessee River, showing the northern tributaries, the Powell and the Clinch, from which Adams collected subspecies of the fresh-water gastropod genus *Io*.

fined, chiefly above Chattanooga (Fig. 1). In two of the main branches, the Powell River and the Clinch River, as we proceed upstream from outlet to source, we pass from the extremely spinose, C, through the sub-spinose, B, into the smooth, A, condition (Fig. 2) without interruption (*i.e.*, by intergradation). In headwaters of both these rivers the shells are smooth or nearly so; as we proceed downstream from source to outlet the shells gradually intergrade from smooth to spinose, first through the development of conical tubercles, which in shells from the lower reaches of the river become strong conical spines. Collections of smooth specimens from the headwaters are quite uniform and are totally unlike collections of spinose specimens from the lower parts of the streams, which are also quite uniform. At intermediate points

[3] C. C. Adams: "Variation in *Io*." *Proc. Amer. Assoc. Adv. Sci.* 1900, pp. 208–225, Pls. I–XXVII.

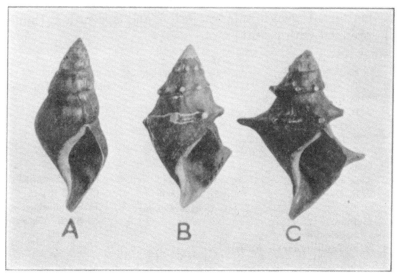

FIG. 2. Intergradation of shells in the genus *Io* of the family Pleuroceridae; collected in the Powell and Clinch rivers (C. C. Adams, *Proc. Amer. Assoc. Adv. Sci.*, 1900, pp. 208–225). A, smooth intergrades at the river source; B, sub-spinose intergrades in midstream; C, extremely spinose intergrades downstream, near outlet.

along the streams specimens intermediate in character prevail with some admixture of smoother and more spiny individuals; so complete is the intergradation that Adams considers them all one species, *Io fluvialis* (Say), 1825, though several species had been recognized by most authors.

Conclusion, Adams, 1900. The want of complete geographic isolation between A, smooth, and C, spinose, obviously prevents a division of the genus *Io* into a smooth species A and an extremely spinose species C.

1859–1926. DISCOVERIES RELATING TO THE ORIGIN OF SPECIES IN FISHES

The outstanding ichthyologists whose writings bear on our problem are arranged by the continents in which their chief observations have been made. The list includes recent systematists and classifiers of fishes, as well as those who have especially directed our attention to the matter of geographic variation and speciation (names prefixed with a star):

North America:
E. D. Cope
T. N. Gill
*C. L. Hubbs
*D. S. Jordan
B. Dean ⎰ fossil
C. R. Eastman ⎱ fishes

Asia:
*L. S. Berg
F. Day

Europe:
J. Schmidt
R. Traquair (fossil fishes)

Africa:
G. A. Boulenger
J. Pellegrin

South America:
*C. H. Eigenmann

Australia:
E. R. Waite
W. MacLeay
D. G. Stead
A. R. McCullough

Indo-Australian Archipelago:
P. Bleeker
Max Weber

World at large:
D. S. Jordan
A. C. L. Gunther
C. Tate Regan
F. Steindachner
A. Smith Woodward (fossils)

In marine fishes geographic isolation exhibits largely different phases from the isolation of mammals and birds on the continents. The complete isolation in streams and rivers of fresh-water fishes corresponds with island isolation or insulation among birds and mammals. The most recent treatment of this great subject is that of C. Tate Regan in his notable address, "Organic Evolution," as president of Section D, Zoology, before the British Association at Southampton last year. He treats the very phenomena we are considering in their bearing on the older and more recent theories of the causes of evolution, which are not under consideration in this paper. G. K. Noble (1926) remarks that, from the genetic standpoint, isolation brings about inbreeding and would tend to change a heterozygous community into a homozygous one; he believes that there is no experimental proof for the action of isolation in the manner Regan assumes, but that, on the other hand, there is good proof that *isolation produces a change in fixing incipient races which have arisen genetically and which would otherwise be swamped.*

Amplifying Regan's summary of the *effects of physical environment* on the origin of species in fishes is the record of observations especially prepared for this paper by Dr. E. W. Gudger, bibliographer and associate in the department of fishes of the American Museum, as follows:

From an early period the number of vertebrae in fishes has been used in the description and classification of fishes and in more recent time an increasing number of observations on temperature, salinity and skeletal speciation has been made:

(1) Günther in 1862 showed that the *Labridae* of temperate regions had more vertebrae than those living in warm water.

(2) In the following year Gill reviewed Günther and confirmed this observation for *Acanthopterygii* and *Malacopterygii*, and in 1864 he illustrated the doctrine by citations of species of the genus *Sebastes*.

(3) The field was widened by Jordan in 1885, by Jordan and Goss in 1889 in classifying the flounders and soles (*Pleuronectidae*), and by Jordan and Evermann in 1891 in classifying the *Serranidae*.

(4) In 1891, 1893 and 1894 Jordan made the following generalizations of forms with the most vertebrae:

Primitive forms:
  Amphioxus 50–80 segments
  Lampreys 100–150
  Sharks   120–150
  Ganoids   60–110

Cold water and tropical Teleosts:
  Herring   56–41 vertebrae
  Blennies 38–40 to 23–25, etc.
  Flounders   60–30

Shallow water and deep sea Teleosts:
  Gadidae   ca 50
  Grenadiers   65–80

Pelagic forms more than shallow water:
  Mackerels   31–50
  Littoral forms 24–26

Freshwater more than salt:
  Centrarchidae   30
  Serranidae   24

To sum up: Northern, cold, deep waters (1, low temperature, 2, darkness, 3, monotonous life)—many small primitive vertebrae; southern, warm, shore waters—few, large, specialized vertebrae.

(5) In 1905 Jordan published an article, "The Origin of Species through Isolation," in which he vigorously argued for this principle, backing up his contention with a wealth of illustrations not merely from fishes but from other animal forms as well.

(6) Among the data supporting Jordan's contentions are:
  (a) The Pacific coast *Leptocottus* shows constant increase in vertebrae from California to Alaska.
  (b) Atlantic salmons and trouts form a marked group with few species, due to few barriers in a country geologically old. We note that of the Char (salmonoid fishes of the genus *Salvelinus*) and of *Coregonus*, different species occur in each isolated lake in Scotland, Germany and Scandinavia. But there are exceptions [Regan, 1923].
  (c) Landlocked salmon is a new species due to isolation in the eastern United States.
  (d) Among Pacific *Salmonidae* a new species arises in each creek and lake, especially in a geologically new country with many new barriers.
  (e) The supposed more numerous American species are due to the energy of American ichthyologists in giving names [Regan, 1926].

(7) Hubbs experimentally found seasonal variations in large broods of minnows, with more caudal vertebrae and more fin rays in the cold season.

(8) Regan finds evidence that slight differences in average number of vertebrae in closely related species may be in part directly due to environmental influences.

(9) A definite illustration of the selection principle is that adduced by Thienemann in 1911 and 1912 of *Coregonus fera*[4] from Bodensee in Laacher See, Germany, which after forty years disappeared, leaving a new species, *C. sancti-benedecti*, with gill rakers twice as long and thickset. In the Bodensee this species feeds on crustacea and mollusca living on the shallow lake floor; Laacher See is in the crater of an extinct volcano, with steeply falling-away shores and no shore-dwelling fauna to serve as food, so that *Coregonus* had to feed on plankton; the modified gill rakers which developed to catch this food are sufficient, with certain outward bodily modifications, to constitute a new species.

Supplementing the above notes by Dr. Gudger is a summary by John T. Nichols, associate curator of the American Museum of Natural History:

(1) *Marine.* As a rule, isolation pertains among fishes to a less extent than among other groups, hence we have very wide ranges for faunal groups and, in many cases, for individual species. Taking the world as a whole, tropical marine fishes are less isolated than are those of the northern oceans. Essentially the same tropical fauna extends from the east coast of Africa across the Indian Ocean and into the Pacific. A warm current which passes the Cape of Good Hope against the wind has also given it access to the warmer parts of the north and south Atlantic, so that very similar fishes occur in the tropics around the world. The wide expanse of landless ocean between the most easterly islands of Polynesia and the American coast has been comparable as a barrier with the more or less temporary isthmus between North and South America, so that certain Atlantic species are found either identical or little changed on the Pacific side of the American continent which are not found further westward in the Pacific. In most cases isolation has developed in such species slight or average differences which may perhaps best be looked upon as racial.

(2) *Colder Marine.* In contrast, the faunae of colder north Atlantic and north Pacific are quite unlike. Several families are abundant in the Pacific which are not represented in the Atlantic (such as Greenlings) and the numerous genera of Pacific Sculpins and Flounders are for the most part peculiar to that ocean. That this is the result of isolation is emphasized by the fact that as one goes northward in the Pacific to sub-Arctic and Arctic waters, far northern forms are again identical with, or very close to, those of the Atlantic, by reason of free water connection around the Pole.

(3) *Deep Sea.* Though highly specialized for their peculiar environment, fishes of the depths show great similarity the world over, due to a lack of barriers between the ocean depths of one part of the world and of another. As we approach high latitudes in the southern hemisphere, marine shore fishes are peculiar—entirely different from those found anywhere else. There are many species which show a certain similarity to the northern Sculpins, but this is now understood to be parallelism due to parallel environment. They belong to the entirely unrelated family Nototheniidae.

[4] The LaacherSee *Coregonus*. In Regan's lecture, "Museums and Research," published in "Animal Life and Human Progress," 1917, he gave this as an example of the need for correct determination, the supposed evolution being due to the selection of the wrong species as ancestor.

(4) *General.* Specialized fishes of peculiar specialized environments, such as the deep-sea or the torrents or mountain regions, show two things: First, a high degree of specialization correlated with the environment; second, clear relationship to less specialized forms in less specialized environments. The deduction is obvious: They are descended from these less specialized forms; their migration to these environments has been the cause of evolutionary change.

(5) *New Environment.* We have not far to seek for cases where fishes have invaded a new environment, and, where they have become thoroughly established, there they are different. Marine Gobies and marine Herrings have invaded fresh-water, the Gobies notably in the West Indies, the Herrings in Africa (genus *Pelonula*, etc.), and we have fresh-water genera of these groups. In such cases, however, it is often not possible to say whether the new environment has brought about the change or has preserved the fresh-water forms from a certain amount of change which is doubtless continually in progress, due to competition in the old salt-water environment. All we see is that the fresh-water forms which have unquestionably come from the salt water are now different from the salt-water forms.

(6) *Salmonidae.* Of fresh-water fishes the Trout are peculiarly plastic, so that those in any one river are recognizably different from those in an adjoining piece of water. Isolation probably plays a part, but the slightly different environment does also, and how firmly their characters are implanted on the germ-plasm it would be difficult to say.[5] The retreat of the ice left descendants of the Arctic Char (such as the Golden Trout) insulated here and there in New England lakes; this insulation enables us to recognize them as distinct species or at least sharply marked races which, however, for all we know, may be duplicated somewhere in the wide northern range of the parent fish if the environment is exactly duplicated. As trout run into the swift cold water of high altitudes, their scales decrease in size and they become more highly colored, a condition which is found in its extreme in the one or more species of Golden Trout of the California Sierras. The condition has doubtless been emphasized by the fact that these upper waters are here isolated by falls. What we do not know is how much, if any, tendency these fish would show to revert to the parent form were they removed for any considerable period of time from their specialized environment.

Discoveries relating to the Origin of Species in Amphibia and Reptilia

Owing to restricted habitat and persistent intermediate function between aquatic water-breathing, and terrestrial air-breathing, life, *external specific* and *generic characters in amphibia are far more sensitive to local* or *geographic isolation than the higher vertebrates.*

[5] The transfer of eastern brook trout to western waters, as of Colorado, produces no immediate environmental change of color or form. (H. F. O.)

The following are the herpetologists who have made outstanding contributions on the continents in which they are interested:

North America :
E. D. Cope
John Van Denburgh

Africa:
G. A. Boulenger
F. Nieden

Asia:
L. Stejneger
O. Boettger

South America:
E. D. Cope
G. A. Ruthven

Indo-Australia:
P. N. Van Kampen

Europe:
G. A. Boulenger
E. G. Boulenger

Australia:
Dene B. Fry
F. Werner

As summarized by Noble in notes supplied for this paper, in Amphibia from the standpoint both of the field naturalist and the geneticist:

(1) Isolation produces species origin or speciation. This isolation is especially subtle under certain conditions. If species are found in the same locality they may differ (a) physiologically, as, for example, have different breeding habits: (b) in habit preference, one selecting drier ground, etc., than another; or (c) in size, as exemplified in *Plethodon*. (2) Whenever a species crosses a barrier it meets new conditions; *i.e.*, the sieve of natural selection changes. It is remarkable that in spite of the different selection sieves on the various East Indian and West Indian Islands the same species ranges through many of them. In the West Indies the same species of lizards and snakes (*e.g., Typhlops, Mabuya*, etc.) range through many islands without subspecific change, whereas other groups (especially terrestrial and arboreal ones) such as the frogs (*Hylodes*) tend to split up on each island; thus some insular species are stable and others plastic. (3) Species may arise within the range of an ancestral stock by physiological isolation. For example, female toads are attracted towards the male by his voice; if a change in the voice of the male should occur (as a mutational, a genetic difference) he probably would not attract a female, but if he should happen to seize a female, their offspring could only interbreed, for the voice of the offspring of the mutant would not be attractive to females of the original stock. (4) Most species are genetically complex and comprise hundreds of small mutations.

Intergrades in Amphibia

Noble (September, 1926) adds the following observations on intergradation and speciation in amphibians: Many wide-ranging species of Amphibia break up into different races or subspecies in different parts of the species range. In almost all cases the various subspecies intergrade with the other races of the species occupying contiguous ranges. A good example of this is to be seen in the commonest salamander in eastern North America, *Desmognathus fuscus*. This species breaks up into the five well-recognized forms: *fuscus, auriculatus, brimleyorum, carolinensis* and *ochrophaeus*. In parts of the Adirondacks exact *intermediates* [intergrades] between *fuscus* and *ochrophaeus* occur. At Flatrock, N. C., perfect intermediates between *fuscus* and *carolinensis* are found, as shown by the collections in the American Museum. Similarly, the intergrades between the other subspecies occur in regions where the ranges of two forms meet.

Intergrades between the races of other species of urodeles have often been described. This is especially true of the subspecies of the Red Salamander, *Pseudotriton ruber,* of the Worm Salamander, *Batrachoseps attennatus,* and of the races of both the European and American newts. In fact many systematists distinguish a subspecies from a species solely on the basis that it intergrades with its closest relatives.

In the Salientia subspecies, or as we may now define them, *intergrading races,* are numerous but not always distinguished by a name. One of the most careful studies on the subspecies of a frog is that made by Boulenger[6] on the edible frog, *Rana esculenta*. Boulenger recognizes five intergrading races extending across Europe, Asia and North Africa. All the races except *R. lessonae* have a more or less definite range, but intergrading and apparently interbreeding occur between the different subspecies where their ranges meet or overlap. This is true even of the most clearly defined races. Boulenger[7] states: "The typical form is completely connected with the *var. ridibunda,* and where the two coexist in a locality, annectant individuals may be regarded as the result of crossing, such as undoubtedly must take place."

MUTATIONS IN *Amphibia*

Not all the species of Amphibia have arisen by a gradual evolution from subspecies occupying different ranges.

[6] 1918, *Ann. Mag. Nat. Hist.* (9) II: 241–257.
[7] *Loc. cit.* p. 254.

Very often two species differing only in a few pronounced structural differences will be found to occupy the same range. In these cases an examination of the group as a whole reveals that the derived form has arisen *in situ* by one or more sudden changes (*i.e.*, mutations). Thus the species of *Dendrobates* differ from those of *Phyllobates* merely by lacking the maxillary teeth. The species of *Nyctimantis* are merely hylas with a vertical instead of horizontal pupil. The salamander *Leurognathus marmoratum* agrees closely in external form and color with its immediate ancestor *Desmognathus quadramaculatus*, but it has suffered a great flattening of the skull, a loss of the vomerine teeth and the reduction of various skull foramina. *Manculus* is merely a dwarf form of *Eurycea bislineata cirrigera* which has lost the fifth or outer toe. There is abundant evidence in both the Plethodontidae and the Hynobiidae that the outer toe is usually lost at a single step (*i.e.*, mutation) and not by a gradual dwindling away. There are also many cases where maxillary or vomerine teeth may be lost at one step, for both ancestral and derived forms may live side by side and agree otherwise in almost the minutest detail (as, for example, in the species of Cacosternum).

Conclusion, Noble. Evolution in the Amphibia has not always been gradual and progressive in time and space. Pronounced and sudden changes, whatever be their origin, have also occurred. This is in full agreement with the geneticists' data. Large mutations of the magnitude of generic differences occur in the laboratory but are far less numerous than the small mutations—the material out of which subspecies are made.

Intergrades in Reptilia

Among the outstanding cases of intergradation in reptilia is that among the lizards of the genus *Chalcides*, as reviewed by E. G. Boulenger. Referring to Major Stevenson-Hamilton's paper on the geographic distribution of African mammals, it is pointed out that *one would be justified in treating some of the varieties as distinct spe-*

cies were it not for the existence of intermediate forms. Boulenger continues with notes on the classification and distribution of the Skink *Chalcides ocellatus,* "a species inhabiting southern Europe, northern and northeastern Africa and southwestern Asia, which present an extraordinary amount of variation: in fact, the structural difference between the two extreme forms is so great, that, *were it not for the wonderfully complete manner in which they are connected, they could not possibly be denied specific rank"* (italics my own). In other words, because of the existence of intergrades, specific rank is denied.

Continuing, Boulenger writes:

> In papers written nearly 30 years ago my father, dealing with the matter, came to the conclusion that this species could be divided into five distinct varieties or subspecies, characterized mainly by the coloration and by the number of scales round the body, which was found to vary between 24 and 40—a range of variation far greater than is to be found in any other lizard. The five forms then described were the *forma typica,* and the varieties *ragazzii, tiligugu, vittatus,* and *polylepis.* To these must be added the var. *occidentalis* (*Ch. simonyi* Stdr.).

The intergradations are observed in the following specific characters: (1) position of nostril, (2) slenderness of the body, (3) breadth of the scales, (4) proportions of limbs and body. *Isolation* is observed in the insular form on the island of Linosa distinguished from those of Tunisia and Malta by the following characters: (1) spotting, (2) paler or more intense colors, (3) number of scales; and Boulenger concludes that "these lizards are undoubtedly distinct from all the other forms of the species *C. ocellatus"* and proposes for them the varietal name of *C. o. linosae.* As to the effect of isolation and insulation in *Chalcides* he observes:

> In all, therefore, we have, apart from the typical form, seven varieties of the lizard *C. ocellatus,* and it is interesting from the evolutionary point of view that they are geographically connected, it being possible to trace every link in the chain from the short and stout variety with as many as 40 scales from Morocco, which must be regarded as the most generalized form, to the long and slender type with only 22 scales round the body from Abyssinia and Somaliland.

He proceeds to define the eight varieties or subspecies into which *C. ocellatus* may be divided, *e.g., C. o. poly-*

lepis, C. o. occidentalis, C. o. vittatus, C. o. tiligugu, C. o. typica, C. o. linosae, C. o. ragazzii and *C. o. bottegi*, and concludes with a map showing isolation and insulation distribution and with comparative figures of the two extreme forms and supposed relationship and distribution.

1859–1926. DISCOVERIES RELATING TO THE ORIGIN OF SPECIES, SUBSPECIES AND MUTATIONS IN BIRDS

Due to the early interest of naturalists, we note that birds were comparatively well known in Darwin's time. A brief estimate of the increase in our knowledge of birds of the world since Darwin's time is as follows:

1871—Gray's list:	1,500 genera, 11,162 species
1909—Sharpe's list:	2,810 genera, 18,939 species and subspecies
1926—Chapman's list:	24,000 " " "

Ornithologists who have contributed data on the problem of speciation as affected by isolation and insulation are the following:

North America:
  J. A. Allen
  R. Ridgway
  C. H. Merriam
  J. Grinnell

West Indies:
  C. B. Cory

Central America:
  Salvin and Godman
  R. Ridgway
  Bangs
  Griscom

South America:
  P. L. Sclater
  O. Salvin
  von Berlepsch
  Chapman
  C. E. Hellmayr
  Todd

Africa:
  R. B. Sharpe
  G. E. Shelley
  A. Reichenow
  W. L. Sclater
  Chapin

Madagascar:
  E. Oustalet

Indo-Malaysia:
  W. T. Blanford
  E. W. Oates
  Rothschild and Hartert
  Stuart Baker
  T. H. D. La Touche
  Oberholser
  Bangs
  McGregor
  Stresemann
  Robinson and Kloss
  Delacour

Australia:
  J. Gould
  G. M. Mathews

New Zealand:
  W. L. Buller
  Rothschild and Hartert

Melanesia:
  Rothschild and Hartert

Polynesia:
  Rothschild and Hartert
  Murphy

Palaearctic Region:
  H. E. Dresser
  E. Hartert
  E. Stresemann
  Witherby
  Kleinschmidt
  P. P. Sushkin
  Kuroda

World at large:
  J. Gould
  R. B. Sharpe
  A. R. Wallace

Species-Origin Environmental and Mutational in Birds

Chapman and Stresemann have recently discussed independently the subject of the factors of environment and mutation in relation to the origin of species in birds. Referring to the principle of direct environmental action Chapman, 1923[8] (pp. 243–4, 272–3), observes in a passage freely quoted, italicized and emended by the present author:

With enormous series of wide-ranging species at his command, the ornithologist is in possession of material to determine the character and extent of the variations in color and in size exhibited by what is obviously the same species. When to this intensive laboratory investigation he adds a knowledge of the environmental conditions under which the species exists, he can often definitely correlate effect and cause.

Thus, the ornithologist finds large forms occupying colder areas, dark ones humid areas, and pale ones arid areas; and *as the [environmental] conditions which obviously produce these variations in size and color merge one with the other, so do the [subspecific] forms themselves intergrade.* That these variations are inherent [*i.e.*, germinal, constitutional, hereditary] and not merely the temporary impress of [physical] environment on the individual is apparently shown by the fact that they are often as well marked in the nestling as in the adult.

While my experience as a collector and "museum man" has convinced me of the profound influence exerted by observable environmental factors (chiefly climatic) on the species, it has also brought to my attention certain instances in which I believe species have arisen by what is termed "mutation," that is, the appearance of characters, great or small, which apparently are the expression of an inherent tendency to vary rather than of environment. Such characters are commonly termed "individual variations." They may be manifested by a greater or lesser number of individuals in widely separated parts of the range of the species, and their perpetuation evidently depends upon their frequency and, especially, on an environment which affords sufficient isolation to ensure their preservation. . . .

Briefly, we have two groups of birds in the genus *Buarremon*, the members of which are distinguished from one another, primarily, by having a chestnut or a black and gray or black crown and, secondarily, by the presence or absence of a black pectoral band. It is this black collar which is the principal mutant character and which . . . appears or disappears purely as an individual variation and without relation to external influences. Its perpetuation or establishment as a specific mark does, however, depend upon environment expressed in what is doubtless the *most important external agent in promoting evolution—that is, isolation.* . . .

[8] Frank M. Chapman: "Mutation Among Birds in the Genus *Buarremon*." *Bull. Amer. Mus. Nat. Hist.*, Vol. XLVIII, Art. IX, pp. 243–278, October 15, 1923.

Summary. 1. The range of species of the genus *Buarremon* having a black and gray head and a breast-band are, with one possible exception discontinuous. Between the ranges of the northern and southern species, forms without a breast-band occur. 2. Although representative and closely related, none of the forms of this section is known to intergrade and, taxonomically, they therefore may rank as species. . . .

Conclusions. 1. That, although differing from the members of Section A of Group II in possessing a black breast-band, the members of this section are nevertheless representatives of the members of that section. 2. That the differences between the members of Section A and Section B are due to mutation. 3. That the differences between the members of Section B are in part due to the direct action of environment, in part to mutation, both being made effective by isolation.

Summary. With the extensive collections of South American birds at his command, carefully recorded as to exact locality and life-zone, Chapman observes a distinct case of mutation within the genus *Buarremon* in the presence of a black pectoral feather-band, which is established as a specific character by isolation or geographic discontinuity. Like the single case of black dorsal barring observed within the genus *Troglodytes*, the mutation does not intergrade and is thus recognized as of discontinuous or mutational origin.

Stresemann on Mutations

Stresemann, in his "Mutationsstudien I–XXV," also finds partial melanism as well as albinism arising as not infrequent mutational characters in *Anser, Accipiter, Egretta, Chlorophoneus* (*e.g., C. multicolor,* mut. = *C. nigrithorax* of Togo), *Lanius* of south China, *Sylvia* of Madeira, *Rhipidura* of New Zealand and Australia, *Accipiter* of New Guinea, an albinistic mutant. Stresemann concludes:

Die nunmehr erfolgende Zusammenfassung der wichtigsten Ergebnisse meiner vor drei Jahren begonnenen Mutationsstudien soll keineswegs ankündigen, dass der Stoff erschöpft ist und die Untersuchungen des Gegenstandes eingestellt werden müssen. . . .

Je länger wir uns mit diesem Gegenstande beschäftigten, desto mehr hat sich in uns die Ueberzeugung befestigt, dass den grossen Mutationen für die Veränderung des Artbildes eine hohe Bedeutung zukommt. Von Rotbraun zu Schwarz (= Grau = Blau), von Schwarz (Grau, Blau) zu Weifs kann nur ein Sprung führen; viele auffälige Zeichnungen sind höchstwahrscheinlich ebenfalls plötzlich in aller Vollkommenheit (und nicht allmählich durch gehäufte Wirkung vieler kleiner Schrittchen) aufgetreten. . . .

Eines lehren den Systematiker und den Zoogeographen die gegenwärtig dichromatischen Vogelarten mit aller Eindringlichkeit: dass die Grösse des Färbungsabstandes geographischer Rassen durchaus keinen Massstab für den Grad der Verwandtschaft und die Dauer der räumlichen Trennung bildet; und diese Erkenntnis muss dazu führen, den Artbegriff weiter zu fassen, als dies in der Regel bisher geschehen ist.

Chapman on Zonal Speciation

While still a firm believer in the importance of isolation as a factor in the development of new forms (*i.e.*, subspecies), Chapman has now come to feel that isolation as a rule also implies change of environment and that the two combined are the most potent factors in speciation. His most recent views are presented in his "Distribution of Bird-Life in Ecuador," and may be summed up as follows:

(1) Where the environment remains the same, the fauna shows comparatively little change; where the environment is altered or a new environment is offered, evolution proceeds at a correspondingly rapid pace.

(2) Even when the life of one zone has been derived from that of an adjoining zone their faunas differ more widely than that of separated areas belonging in the same zone. . . . Change of environmental conditions without isolation is more productive of evolution than isolation without change of conditions.

(3) What more adequate tribute to the power of environment can one ask than to discover within this comparatively small but marvelously diversified country [Ecuador] one-fourth the birds of all South America and one-twelfth those of the entire world!

(4) Whether environment originates or whether, in the language of the day, it merely "starts something" by arousing latent potencies within the germ, the fact remains that organisms do respond to change and furthermore, that certain conditions, for example, humidity and aridity, almost invariably provoke the same type of response.

Speciation and Mutation of the South American House Wren (Troglodytes)

Corresponding with the field work of Osgood in North America on subspecies of the mammal *Peromyscus* is the field work of Chapman and Griscom on the House Wren (*Troglodytes musculus*) of Central and South America. The chief centers of speciation and subspeciation of this ubiquitous little bird are indicated in a zoogeographic map of South America (Fig. 3) which represents years

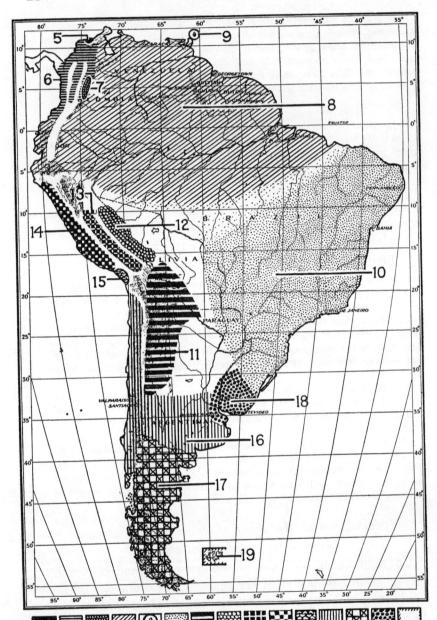

Fig. 3. Distribution of South American House Wrens. After Chapman and Griscom. 5, *Troglodytes m. atopus*; 6, *T. m. striatulus*; 7, *T. m. columbæ*; 8, *T. m. albicans*; 9, *T. m. tobagensis*; 10, *T. m. musculus*; 11, *T. m. rex*; 12, *T. m. carabayæ*; 13, *T. m. puna*; 14, *T. m. audax*; 15, *T. tecellatus*; 16, *T. m. chilensis* and, from the valley of Copiapo northward, *T. m. atacamensis*; 17, *T. m. magellanicus*; 18, *T. m. bonariæ*; 19, *T. cobbi*.

of close analysis (Chapman-Griscom, 1924) of very extensive collections. The total number of *Troglodytes* specimens collected and examined is 1,500, as compared with 30,000 of *Peromyscus*. Altogether, nineteen subspecies have thus far been recognized.

Those familiar with the marked physiography of South America, with (a) its cold and arid Pacific coastal belt, (b) its Andean Mountain chain subdivided by Chapman into four horizontal life-zones each clearly demarcated, (c) its vast Amazonian forest plain, (d) its pampas and plains of the Southeast and (e) its humid forests bordering the Straits, may anticipate wide speciation in this continent.

It is noteworthy that observers have discovered in *Troglodytes* only one instance of mutation in the sense of DeVries, which I shall speak of as D. Mutations, because there are no intergrades leading to it; this case is No. 15 on the South American chart and the supposed mutation consists of *broad bars crossing the dorsal feathers*. In other subspecies this character is only faintly indicated, absent entirely or a matter of individual variation.

At the writer's request in the month of May, 1926, Ludlow Griscom, assistant curator of birds, American Museum of Natural History, summed up his own opinions and theories as to isolation and speciation as follows:

(1) Granting the mutational possibilities existing in the germ-plasm, it can not be questioned that *isolation* is a most important factor in the speciation of birds, especially in its extreme form of *insulation* or island life. Isolation in birds brings about some degree of variation, even if not sufficient to give taxonomic or descriptive value.

(2) Insulation almost invariably produces speciation in direct proportion to the distance of an island from the mainland.

(3) Climatic isolation is a very important factor, but does not cause such extreme speciation as the insulation factor.

(4) In both factors of isolation and insulation, the time element from the geological point of view is of controlling importance.

(5) A second factor of similar importance is what might be called the relative plasticity of the particular bird. As a general principle, the more recent orders and families exhibit far more relative plasticity than older and more widely distributed groups.

(6) Isolation factors are more pronounced in the tropics than in temperate regions, due to the sedentary habits of tropical birds..

(7) Care must be taken to distinguish speciation within a faunal area due to isolation factors, from the speciation observed by passing from one faunal area to another.

(8) The evidence as far as birds are concerned is to the effect that a change of habits is a secondary and never a primary cause in speciation.

Griscom illustrates the above principles as follows:

(1) (a) In the classic Darwinian case of the finches (*Geospiza*) of the Galapagos Islands, the degree of speciation probably depends on the geologic time that has elapsed since the islands were separated from each other as well as from the mainland. (b) The very slight differences recorded in some of the smaller land-birds of Newfoundland indicate a relatively brief period of geologic isolation, combined with the migratory habits of the birds themselves, which further reduces the possible influence of their insular breeding-grounds. (c) In the case of Cozumel Island off the coast of Yucatan, we are certain that the insulation and isolation began in Pliocene time at the latest; the consequent speciation is very pronounced, and we are here dealing with birds of absolutely sedentary habits.

(2) Geographic isolation is the main element even where climatic conditions remain similar; *e.g.*, the climate of the recently studied Island of Cozumel is exactly similar to that of the adjoining mainland of Yucatan, yet insular speciation has advanced to a remarkable degree. Similar observation is that of Miller on the *Traguline* deer of the East Indies.

(3) The factor of plasticity is well illustrated in the House Wrens of South America, already cited, and in the Song Sparrow (*Melospiza melodia*), which in North America divides into over twenty subspecies. This may be contrasted with the more stable Killdeer (*Oxyechus*), which in the same region presents no subspecific variation. Plasticity, however, would seem to be of less importance in the Tropics than in temperate regions.

In conversation with Griscom and with the eminent French ornithologist, Jean Delacour, some of the exceptions were noted where geographic isolation is not attended by geographic variation or speciation:

(1) Land birds and perching birds of widely migrating habits do not tend to speciation; *e.g.*, the bank swallow or sand martin, the only small Passerine bird common to both the New and the Old Worlds which exhibits no appreciable variation.

(2) Certain species of ducks of world-wide and discontinuous distribution are excellent examples of birds which do not show any specific variation. In general, older and more primitive birds do not tend to speciation; *e.g.*, tree ducks (*Dendrocygna*), a subfamily now dying out.

1859–1926. Discoveries relating to the Origin of Species in Mammals

Before and after Darwin there is an honor roll of mammalogists, who have been observing and collecting mammals all over the world and whose publications have furnished and will furnish data for purely biological treatment, although perhaps they themselves have not made biological use of their material. For convenience of reference they may be listed by countries in alphabetical order, without any question of seniority or importance:

	J. A. Allen	J. Grinnell	E. W. Nelson
	G. M. Allen	E. Heller	W. H. Osgood
	H. E. Anthony	N. Hollister	T. S. Palmer
	J. J. Audubon	A. H. Howell	E. A. Preble
NORTH AMERICA	J. Bachman	H. H. T. Jackson	S. N. Rhoads
	V. Bailey	M. W. Lyon, Jr.	F. B. Sumner
	E. Coues	E. A. Mearns	W. P. Taylor
	D. G. Elliot	C. H. Merriam	F. W. True
	E. A. Goldman	G. S. Miller, Jr.	

Of these, the most important philosophical contributions have been made by Joel A. Allen on geographic distribution, isolation, climatic evolution; by C. Hart Merriam on principles of humidity and temperature control and subspecific variation; by Gerrit S. Miller on geographic isolation, insulation, variation, speciation; by Wilfred H. Osgood on subspeciation, especially in *Peromyscus;* by Francis B. Sumner on biological analysis of subspeciation.

South America. Following the work in South America of such pioneers as Azara and Humboldt are the important observations on geographic isolation, variation, and speciation, of the following authorities, supplemented by the field work of Edmund Heller and others:

	Einar Lönnberg
J. A. Allen	W. H. Osgood
H. E. Anthony	R. A. Phillippi
Outram Bangs	O. Thomas

Europe. All the earlier contributors to European mammalogy are listed in Miller's "Mammals of Western Europe." The fullest analytical treatment of geographic isolation and variation among European mammals is found in the same work. Europe, with its relatively restricted area and fauna, devastated by civilization, does not lend itself to biological discussion in the manner of Africa, Asia, North America or Australia, where the natural wild faunae may still be studied. Among leading contributors are the following mammalogists:

K. Andersen	R. Lydekker	R. I. Pocock
A. Cabrera	P. Matschie	O. Thomas
W. H. Flower	Gerrit S. Miller, Jr.	E. L. Trouessart
E. Lönnberg		H. Winge

Asia. The field naturalists and museum workers who have made and will make important contributions in the way of collections and of taxonomic analyses are chiefly the following:

W. L. Abbott	C. M. Hoy	N. Nassonov
G. M. Allen	C. B. Kloss	H. C. Raven
R. C. Andrews	G. S. Miller, Jr.	P. P. Sushkin
N. Hollister	A. Milne-Edward	O. Thomas
		R. C. Wroughton

Philosophic discussion of speciation among Asiatic mammals after the manner of P. P. Sushkin's work on the birds is still to come.

Africa. The earlier explorers, collectors and naturalists in Africa enumerated in the volumes of Sclater include many priceless observations of now extinct species like the Quagga, which must be considered in the marked *north to south* speciation characteristic of Africa, in contrast to the *east to west* speciation characteristic of Asia, North America and Australia. Among earlier and more recent contributors to mammals are the following:

J. A. Allen	E. Heller	A. Roberts
J. Anderson	N. Hollister	T. R. Roosevelt
J. V. B. du Bocage	H. Lang	K. H. Satunin
J. L. Bonhote	E. Lönnberg	P. L. Sclater
J. F. Brandt	R. Lydekker	Stephenson-Hamilton
A. Cabrera	P. Matschie	O. Thomas
W. E. De Winton	A. Milne-Edwards	E. L. Trouessart
G. Dollman	R. I. Pocock	R. C. Wroughton

Of these, Lydekker, Matschie, Pocock, Hamilton and Thomas have made some of the most notable contributions to geographic and climatic variation and speciation.

Speciation of the Deer Family around the World on the Northern Parallels

On a new equal-area map of the world arranged for zoogeographic and paleogeographic plotting, it may be shown as a first principle that *geographic separation, migration and isolation frequently but not invariably lead to subspeciation and thus finally to speciation.*

To Darwin in 1837–1859 were known only two species of stag (*Cervus elaphus* and *Cervus canadensis*), one species of moose (*Alces machlis*) and one species of reindeer (*Rangifer tarandus*). Then, as now, these species occupied three great transverse geographic belts, *Cervus* in the southern, *Alces* in the central, *Rangifer* in the northern parallels, extending from North Temperate to Arctic regions. As collected and observed from Scotland across Eurasia and North America to New Brunswick, *Cervus* yields twenty-three species and subspecies, *Alces* yields eight species and subspecies, *Rangifer* yields twenty-one species and subspecies. The more uniform forest and swamp environment of *Alces* yields fewer species than the highly varied circumpolar and boreal environment of *Rangifer* in its woodland and barren-ground groups, or than *Cervus* in its plains-bordering uplands, forests humid and arid, and high plateaus.

Zoologists disagree as to the taxonomic rank of these twenty-three subspecies of *Cervus,* but close museum comparison or field observation would prove that they are germinally separate, as in the case of *Peromyscus*. A notable subspecific case in *Alces* is *Alces americanus shirasi* Nelson, occupying the mountain sides bordering the western slopes of the Yellowstone Park in Idaho, Wyoming and Montana, and with conspicuous adaptations in hoof, horn, limb and pelt.

In discussing these not very obvious subspecific differences, Anthony observes that many of the alleged subspecific differences seem to be man-made, dependent on knowledge of actual geographic locality, not clearly separable, without geographic record even breaking down entirely. He points also to instances of uniformity in very wide range of species like *Bison americanus*, which does not split up into many subspecies, or like *Felis concolor*, in which subspecific differences have been determined with great difficulty. Quite different is the case of the American *Ursus*, in which certain of Merriam's subspecies live side by side in the same region and may represent hybrids. We are, however, speaking only of subspecies which are widely separated geographically.

By contrast, the really germinal as well as somatic differences between the far-flung species of *Cervus*, *Alces* and *Rangifer* can not be challenged. As listed and plotted geographically by Anthony for the present paper, these three great antlered types of the North Temperate and Arctic plains, forests, mountains, barren lands and tundras, described as fifty-two species and subspecies in all, are distributed as to geography and as to climate as shown in the following table and map (Fig. 4).

CERVUS: STAG, DEER, WAPITI. EAST TO WEST RANGE

1. *Cervus barbarus*	Barbary deer	Morocco; Algiers; North Africa (Palearctica)		Chiefly arid
2. *Cervus corsicanus*	Corsican stag	Corsica and Sardinia		Semi-arid
3. *Cervus scoticus*[9]	Red deer	Great Britain and Ireland		Humid
4. *Cervus elaphus hispanicus*	Southern Spanish Red deer	Southern Spain		Humid an semi-arid
5. *Cervus elaphus bolivari*	Central Spanish Red deer	Central Spain		
6. *Cervus elaphus germanicus*	Red deer	Middle Europe		Humid
7. *Cervus elaphus atlanticus*	Norwegian Red deer	West coast Norway		Humid
8. *Cervus elaphus elaphus*	Red deer	Sweden		Humid
9. *Cervus maral*	Maral	Persia; Crimea; Caucasus	Chiefly forest	
10. *Cervus bactrianus*		Russian Turkestan		Chiefly arid
11. *Cervus cashmiriensis*	Kashmir deer; hangul	Vale of Kashmir and adjacent mountains	Chiefly forest, with open parks	Humid
12. *Cervus wardi*	Ward's stag	Tibet		
13. *Cervus wallichi*		Tibet		
14. *Cervus albirostris*		Tibet		
15. *Cervus macneilli*	Kansu stag	Kansu and Szechuan border		
16. *Cervus yarkandensis*	Yarkand stag	Eastern Turkestan		Mostly arid
17. *Cervus songaricus*	Tian-shan stag	Tian-shan Mts.	Forest and alps	Mostly arid
18. *Cervus sibiricus*	Altai maral	Baikal, Saiansk, and Altai Mts., Southern Siberia and northern Mongolia	Formerly forests and open timberless country—even in open alps on desert mts. Now restricted to forests and meadows	Extreme h to ext arid

[9] Scotch red deer taken to New Zealand has increased in size and in spread of antler—almost a species in less than a century.—P. C. M.

Cervus xanthopygus	Bedford's deer; Manchurian stag	Manchuria, etc.— adjoining parts of Siberia	Forests	Humid
Rangifer greenlandicus	Olympic elk	Wash., Oreg., Calif., formerly s. to San Francisco Bay. Vancouver Is.?	Chiefly forested regions, with open meadows	Humid
Cervus nannodes	Dwarf elk	San Joaquin Valley, Calif., and adjoining foothills	Plains and tule swamps	Generally arid
Cervus cervus merriami	Merriam's elk	New Mexico and Arizona	Mts. and plateaus, forests and meadows	Generally arid, but wet forests
Cervus canadensis	American elk; wapiti	Southern Canada and U. S., including N. Y., N. J., Central States (Missouri, Nebraska, Dakotas, etc.), west beyond the Rockies, south to Virginia and Carolinas.	Open plains, badlands, sand hills; forests and meadows	Humid to extreme arid

ALCES: MOOSE. EAST TO WEST RANGE

1. *Alces alces germanica*   East Prussia and marshes of Pinsk, Poland.
2. *Alces alces niger*   European Russia and forest regions of the Yenesi.
3. *Alces alces alces*   Europe.
4. *Alces alces bedfordiae*   East Siberia.
5. *Alces gigas*   Alaska.
6. *Alces columbae*   British Columbia.
7. *Alces americana shirasi*   Wyoming.
8. *Alces americana americana*   Northeastern North America.

RANGIFER: REINDEER, CARIBOU. EAST TO WEST RANGE

1. *Rangifer tarandus tarandus*   Swedish Lapland.
2. *Rangifer platyrhynchus*   Spitzbergen.
3. *Rangifer fennicus*   Finnland.
4. *Rangifer tarandus pearsoni*   Novaya Zemlya.
5. *Rangifer tarandus sibiricus*   Siberia.
6. *Rangifer granti*   Alaskan Peninsula.
7. *Rangifer excelsifrons*   Northern Alaska.
8. *Rangifer arcticus ogilvyensis*   Ogilvie Mts., Yukon, Canada.
9. *Rangifer stonei*   Kenai Peninsula, Alaska.
10. *Rangifer arcticus arcticus*   Dist. of Mackenzie, Canada.
11. *Rangifer mcguirei*   Alaska—Yukon Boundary.
12. *Rangifer dawsoni*   Graham Island, Queen Charlotte Islands, Canada.
13. *Rangifer montanus*   Selkirk Range, British Columbia.
14. *Rangifer osborni*   Cassiar Mts., British Columbia.
15. *Rangifer fortidens*   Alberta, Canada.
16. *Rangifer pearyi*   Ellesmere Island.
17. *Rangifer caribou sylvestris*   Southwestern shores of Hudson Bay.
18. *Rangifer caribou caribou*   Eastern Canada.
19. *Rangifer arcticus caboti*   Greenland.
20. *Rangifer greenlandicus*   Newfoundland.
21. *Rangifer terranovae*   Labrador.

Probable Intergradation in Zebras and Foxes

R. I. Pocock writes, July 30, 1926:

I do not suppose that any one museum possesses a sufficient number of skins actually to prove ocularly the existence of intermediates [intergrades] between all the described forms of Burchell's Zebra. Personally I am, and have long been, of the opinion that such intermediates exist at the present time—the evidence is strongly in its favour. The intergradation between Grant's Zebra of British East Africa and Burchell's of the Orange Free State (I am here writing from memory, having no books to check

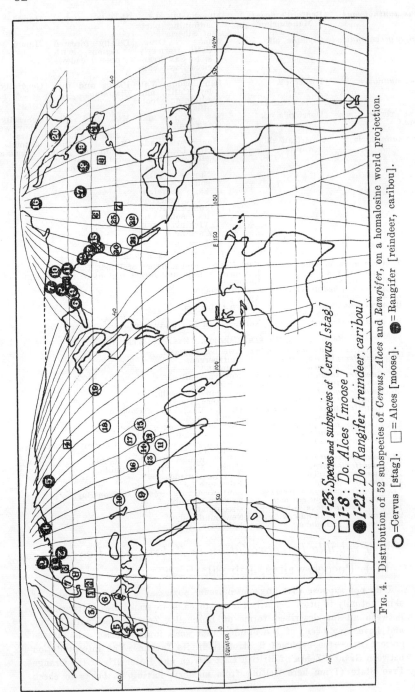

FIG. 4. Distribution of 52 subspecies of *Cervus*, *Alces* and *Rangifer*, on a homalosine world projection.

exact localities, and it is thirty years since I worked at the group) was almost complete, judging from the comparatively small amount of material I had then to judge from, and the gradation ran approximately as follows: Grant's, Selous's, Chapman's, Wahlberg's, Burchell's.

Since then one or two additional subspecies have been described, all serving to confirm the hypothesis. So far as living forms are concerned, we are here brought to a standstill. Nevertheless the differences between Grant's and Burchell's are very marked, more marked than the differences between Burchell's and the typical extinct Quaggas shown in Edward's coloured plate taken from life. From the evidence I was able to gather from photographs and other illustrations, I came to the conclusion that there were probably several subspecies of Quagga ranging in coloration from forms hardly differing from Burchell's to forms in which the obliteration of the stripes was almost completed by pigmentation of the intermediate areas and epigmentation, or depigmentation, of the stripes. So I did not hesitate to refer all these plateau zebras to *Equus quagga*—Grant's quagga, with its conspicuous banding of black and white, being the most primitive existing type.

If I had known you were after existing subspecific intergradations, I think I could have shown you some interesting cases. For instance, from skull measurements and skins I came to the conclusion a year or two ago that the small Punjab fox, *Vulpes leucopus*, is linked up through the Himalayan fox *V. montana* with typical *V. vulpes* of Europe.

Speciation of the Deer-Mice in Contiguous North American Areas

In contrast with the three far-flung genera of deer, the geographic and germinal separation of which is partly due to natural extinction or to elimination by man of intermediate forms, let us examine the Deer-mice or Prairie Mice of the genus *Peromyscus* of the central North Temperate region of North America, *in which certain of the intermediates or intergrades have not been eliminated.* This little animal is the best-known zoologically and biologically of any species of mammal in the world to-day, because of the monumental field work of Osgood, who has collected no fewer than 30,000 specimens in a decade, and the experimental biological work of Sumner, who approaches the problem of speciation and mutation from the highly trained genetic and experimental point of view of the school of Wilson and Morgan, with no preconceptions, with no affiliations, with no particular sympathy for American systematists.

Osgood's color map (*N. A. Fauna,* No. 28) exhibited the entire range of this polyform and polychrome rodent. The area which presents the best example of contiguous subspecies with intergrades is that of the central United States bordering on Canada and British Columbia, as

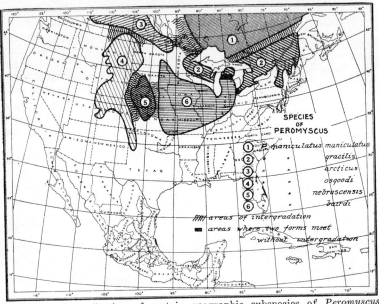

FIG. 5. Distribution of certain geographic subspecies of *Peromyscus maniculatus.* After Osgood, May 19, 1926. ∫∫∫∫∫ = areas of intergradation. ■ = areas where two subspecies meet without intergradation.

shown in Fig. 5, wherein the intergrades are explained by Osgood as follows:

"Typical *Peromyscus maniculatus* (No. 1) is of good size, stockily built and dark colored; in the east it intergrades only with *P. m. gracilis* (No. 2), which has a longer tail; in the west *P. m. m.* intergrades with *P. m. arcticus* (No. 3), which is about the same color but with slightly different proportions. *P. m. arcticus* then intergrades with *P. m. osgoodi* (No. 4), which is of paler color. *P. m. osgoodi* turns eastward and intergrades with *P. m. nebrascensis,* which is decidedly smaller, paler and shorter tailed. *P. m. nebrascensis* then grades into No. 6 (*P. m. bairdi*), which retains the characters of small size and short tail but takes on a decidedly darker color. Thus, by various stages *P. m. m.* (No. 1), a forest-dwelling form, evolves via Nos. 3, 4, and 5 into a prairie form (No. 6) with somewhat different habits.

"It is probable that under aboriginal conditions the Prairie Mouse was confined to the natural prairie and did not extend eastward beyond Illinois.

With the clearing and cultivation of the land, however, it has worked across Indiana and into Ohio, Michigan, and southern Ontario. Here it has met the natural range of *P. m. gracilis* (No. 2), and these two occur in different ecological niches throughout a considerable area. Within this area they appear as distinct species although, by different routes, they may be traced back to the same *P. m. maniculatus* form (No. 1)."

The area studied biologically by Sumner is that of the contiguous forests and arid regions of northern, southern and central California; recently he has also studied Florida and Alabama. He has also collected, but not reared, much material from Arizona. The numerous subspecies observed by Merriam and by Osgood have been accepted with great reluctance, especially by naturalists not familiar with the excessively sharp geographic and climatic barriers of the western United States. My own reluctance to accept these subspecies as of real germinal or genetic value has been entirely removed by the persistent, penetrating observations and experiments of Sumner, largely in the identical collecting grounds of Merriam and Osgood. We now welcome Osgood's observations as the most *convincing demonstration of the principle of continuity thus far afforded in zoological series, namely, continuity in subspeciation, which gives convincing proof of continuity in speciation.*

We may append the most recent expression of Osgood's views, June 8, 1926, taking the liberty of changing certain of his terms to conform with the language used in the rest of this paper:

(1) If there are any mammals which do not show differentiation [speciation] connected or coincident with geographic boundaries, they have not come to my notice. (2) With certain limitations on account of their constitution, I have little doubt that other vertebrates and at least the majority of plants show the same relation between differentiation and distribution as mammals. (3) I have tried to express this briefly by saying, "The existence of diverse inosculating units [intergrades] correlated with geography is characteristic of terrestrial vertebrates." This statement, of course, requires proof, but if made to the comparatively few zoologists who have wide acquaintance with animals in nature, does not encounter much opposition. (4) It is obvious, of course, that some animals are more wide-ranging without differentiation than others but most of those, like the puma (*Felis concolor*), which have been regarded as inflexible types, are now known to be subject to the same kind, if not the same number, of geographic variations [subspeciation] as the others, largely dependent upon the amount of

material available for study. Even the Polar Bear (*Ursus maritimus*) with its relatively uniform environment, is credited with four or five recognizable differentiations [subspeciations]. (5) The failure of some of the larger carnivores to develop numerous geographic races is perhaps not because they are less susceptible to change, but because their greater powers of locomotion offer greater opportunities for diverse matings and thus for the "swamping effect" of breeding. It is possible that Merriam's numerous subspecies of bears illustrate this point. Perhaps the large number of slightly characterized but still recognizable forms [subspecies] he is able to distinguish are only the result of an enforced isolation brought about by the extermination of the animals formerly inhabiting intervening areas and providing opportunity for cross-breeding [hybridizing] on a much larger scale than at present. It seems to me that the kinds of bears Merriam describes are different from the other species and subspecies we know, and if they require a special explanation, this may be offered as a possible one.

Isolation in the Origin of Localized Races of Peromyscus

Sumner's observations, beginning in 1915 and extending over eleven years, are of the utmost importance in adding the principles of genetics and experimentation to those of field zoology. While the so-called Merriam-Osgood subspecies are based on real and obvious differences, classification becomes like dividing the spectrum and depends upon the standard set (Sumner, 1917–Osgood). The Sumner-Osgood observations supplement each other. Sumner removes the suspicion which had prevailed as to the germinal reality of Osgood's subspecies and proves that we have real species-in-the-making, by continuous intergradations which link one climatic or environmental region with another, as follows:

(1) Both by repeated experiment and by the genetic analysis of hybridizing, the reality of subspecies is tested.

(2) Color mutations occur (Sumner, *Genetics*, Vol. 2, May, 1917; Sumner-Collins, *Journ. Exp. Zool.*, Vol. 36, Oct., 1922) but not in the main-line of evolution, which is continuous.

(3) Anatomical, proportional, and pigmental or color differences are found to be differences of degree (Sumner, 1918).

(4) Sumner, 1924.3. (1) The summary of eight years' experiment proves the comparative stability of subspecies of the Deer-mice (*Peromyscus*) under very marked new environmental physical conditions. (2) The Merriam-Osgood subspecies are proved to be stable under changed conditions of environment, by transplantation experiments, *i.e.*, a desert subspecies, *P. m. sonoriensis*, reared for eight years in the relatively humid environment of the southern California coast (La Jolla), is entirely unmodified in the direction of the local subspecies *P. m. gambeli*. This doubly proves that (a) characters of the desert *P. m. sonoriensis* are germinal, not

environmental, and (b) that humid environment makes no modification whatever toward increased depth of color, in eight years and in seven to twelve generations. (3) Similar results are obtained by the transplantation of *P. m. rubidus* from the highly humid northwestern coast of California to the relatively less humid coast of southern California. (4) These two introduced subspecies, therefore, fail to converge when reared alongside one another in the latter (intermediate) type of environment.

(5) Sumner, 1924.1. (1) Protective coloration is less urgent in nocturnal than in diurnal animals. (2) Protective coloration (*i.e.*, Selection) does not appear to explain color differences between nocturnal desert and humid subspecies. (3) Optical properties of background (*i.e.*, black lava rocks) are only doubtfully less responsible for color peculiarities which have been attributed to them by some naturalists (*e.g.*, Merriam, Osgood, Goldman). (4) Atmospheric humidity is prevailingly correlated with color depth in birds and mammals.

(6) Sumner, 1925.1, Huestis. Crossing of the desert *P. eremicus eremicus* and the humid *P. e. fraterculus* subspecies shows that all the subspecific characters follow the ''blending'' type of inheritance.

(7) Sumner, 1925.2. (1) Color differences used in defining subspecies are shown by experiment to be hereditary, consequently part of the speciation process. (2) There are strong reasons for doubting whether paleness of desert animals is chiefly due to Selection for concealment purposes. (3) There is some reason for attributing paleness or intensity of color to the direct or indirect effect of climatic environment on pigment formation acting through many generations.

(8) *Independent Characters.* Sumner in a recent paper (*Journ. Mam.*, 1926) and in a letter dated September 29, 1926, appears to have been forced to modify his early views regarding the inheritance of subspecific characters. He no longer takes the position ''that the characters which distinguish subspecies from one another are fundamentally different from those which distinguish the various 'mutant' stocks which chiefly interest Mendelian workers. There is, to be sure, an obvious and striking *apparent* difference, which has been stressed by some taxonomists, and which greatly influenced me during the earlier stages of these studies. The characters of 'natural' races *seem* to be inherited in a strictly continuous fashion, while the characters of mutants are plainly inherited in discontinuous fashion. But I have been (I may say reluctantly) forced to the belief that there is probably no fundamental distinction here. It now seems to me likely that the 'multiple factor' theory satisfactorily accounts for the mode of inheritance of subspecific characters. Individuals indistinguishable in color characters from one or another of the parent races have already been recovered, even in a rather limited series of F_2 hybrids and back-crosses.

(9) ''It is plain that these 'subspecies' have diverged from one another in respect to characters which have varied quite independently [of one another]. There is no single graded series for all the characters, which would lead us to suppose that they are in some way correlated or 'linked' together (AMER. NAT., April–May, 1918, pp. 205, 206). Certain differential characters of the component subspecies may, it is true, undergo simultaneous change throughout a part of the range of the species, but the asso-

ciation between these same characters is found to be entirely dissolved when we pass to other parts of its range (AMER. NAT., May–June, 1923, p. 239). Most of the elements comprised in the total subspecific complex have been found to be independent of one another in inheritance (*Journ. Exp. Zool.*, Oct., 1923, p. 272). As in my previous studies, I find no general tendency for the elements of the complex which distinguish one race from another to vary together within the single race. They may or may not do this. Thus, tail length and foot length, which have undergone parallel modifications in the evolution of some geographic races, are found to vary together in the same direction, even when the influence of body size has been eliminated from consideration. And certain of the pigmental characters, though not all, are likewise rather strongly correlated with one another. On the other hand, neither tail length nor foot length appears to be correlated with any pigmental character (*Journ. Mam.*, Aug., 1926). It seems to me obvious that these facts bear directly on the nature of the process by which one species is transformed into another."

Of our real ignorance of the causes of species formation Sumner[10] says: "We are not yet prepared to frame any adequate general hypothesis as to species formation. The Mendelian Mutation system of facts and theories appears to me to be no more successful in this respect than its predecessors."

Speciation in the Gopher, Thomomys

The most recent study of speciation under the influence of geographic distribution is that of Grinnell, who has made an intensive study of the pocket gophers of the genus *Thomomys* of California; he divides the state geographically into thirty-three specific and subspecific areas as shown in figure 6. He remarks:

> The most universally distributed type of rodent in California is the pocket gopher. It is found thriving at and below sea level, around the southern end of Salton Sea in Imperial County, and above timber line, at 11,500 feet altitude in the vicinity of Mount Whitney; it is found from the arid desert mountain ranges of the Inyo region, such as the Panamint Mountains, to the rainy and foggy coast strip at Humboldt Bay and Crescent City; it is found in the yielding sands of the Colorado River delta at the Mexican line and on the Modoc lava beds at the Oregon line.

As to the seeming correlation of the distribution area of each race with a relative uniformity of environmental conditions, he concludes that (1) the inferior powers of locomotion of *Thomomys* has produced a condition of ex-

[10] "Discussion of the Origin and Inheritance of Specific Characters," 1923, p. 254.

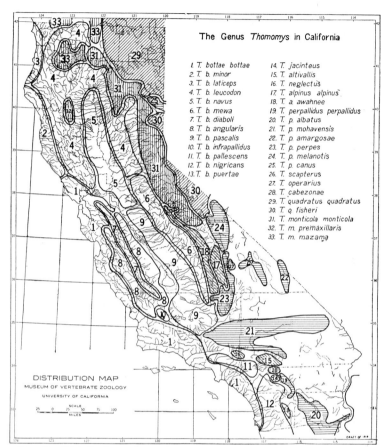

FIG. 6. Distribution of species and subspecies of the Pocket Gopher, genus *Thomomys*, in California. After Joseph Grinnell, July, 1926.

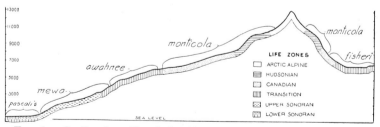

FIG. 6a. Sectional profile of the Sierra Nevada through the region of Yosemite, showing the location of the species and subspecies of the Pocket Gopher according to altitude and life zone.

treme provincialism, as it were; (2) influenced by the more or less impassable barriers here and there, each of the many and diverse specific and subspecific areas has impressed its *stamp,* namely, a peculiar combination of adaptive characters best fitting that subspecies to its restricted area; (3) the evolution of the geographic habitats, or "differentiation areas," must have preceded the differentiation [speciation] of the *Thomomys* stocks which subsequently came under their impress.

Regarding the color gradations shown in the Plate used as the frontispiece to this article, Grinnell observes:

> The matter of coloration of gophers presents a rather baffling problem. Referring to the accompanying plate, which depicts eight of the thirty-three gophers of California, it becomes apparent that the paler colored forms are generally associated with arid habitats; the darker colored with humid habitats. *Bottae* of the coast belt as compared with the almost white *albatus* of the Colorado delta presents an extreme amount of difference. Is the factor which has to do with these diverse conditions of coloration, light, or temperature, or is it humidity of the air?

SUMMARY OF SPECIATION AND MUTATION AS OBSERVED IN LIVING VERTEBRATES

A summary of the conclusions drawn from the independent observations of many kinds of living vertebrates is practically contained in the writer's previous article, "The Problem of the Origin of Species as it appeared to Darwin in 1859 and as it appears to Us To-day." It appears that the problem of the age-long process of the origin of species can best be studied under the natural conditions of past and present time; it appears that all vertebrates which range into a new set of natural conditions, whether by geographic isolation or otherwise, in the course of time give rise to new specific forms; it appears that, where vertebrates can be traced from one geographic range as it merges into another, "transitional" or "intermediate" forms are observed, so that if placed side by side one subspecies passes gradually into another by *intergrades,* and thus the supposed barriers existing between species and subspecies disappear.

Speciation is a normal and continuous process; it governs the greater part of the origin of species; it is apparently always adaptive. Mutation is an abnormal and irregular mode of origin, which while not infrequently occurring in nature is not essentially an adaptive process; it is, rather, a disturbance of the regular course of speciation.

BIBLIOGRAPHY

Arthur E. Shipley.
Address to the Zoological Section, British Association for the Advancement of Science, Winnipeg, 1909.

C. Tate Regan.
Address to Zoological Section, British Association for the Advancement of Science, Southampton, 1925.

Henry Fairfield Osborn.
"The Origin of Species as revealed by Vertebrate Palaeontology" (I). Address before National Academy of Sciences, Washington, D. C., April 28, 1925.
"The Origin of Species" (II.) Address before National Academy of Sciences, November 11, 1925.
"The Origin of Species, 1859–1925" (III). Address at Peabody Museum of Natural History, Yale University, December 29, 1925.
"The Problem of the Origin of Species as it appeared to Darwin in 1859 and as it appears to Us To-day" (IV). Address before Section D, Zoology, British Association, Oxford, August 5, 1926.

Frank M. Chapman.
"The House Wrens of the Genus Troglodytes." With Ludlow Griscom. *Bull. Amer. Mus. Nat. Hist.*, New York, July 8, 1924.
"Mutation among Birds in the Genus Buarremon." *Bull. Amer. Mus. Nat. Hist.*, Vol. XLVIII, Art. IX, pp. 243–278, October 15, 1923.
"Distribution of Bird-Life in Ecuador." *Bull. Amer. Mus. Nat. Hist.*, Vol. LV.

F. B. Sumner.
"Studies on the races of California deer-mice (*Peromyscus*)," 1916–1923, Scripps Institution, La Jolla, California, and a series of papers in THE AMERICAN NATURALIST, *Journal of Heredity, Ecology, Biological Bulletin* and *Journal of Mammalogy*.

    1924   I. "The Supposed Effects of the Color Tone of the Background upon the Coat Color of Mammals."
              II. The Partial Genetic Independence in Size of the Various Parts of the Body."
              III. "The Stability of Subspecific Characters under Changed Conditions of Environment."
    1925   I. "Studies of Coat-Color and Foot Pigmentation in Subspecific Hybrids of *Peromyscus Eremicus*."
              II. "Some Biological Problems of Our Southwestern Deserts."
    1926   I. "An Analysis of Geographic Variation in Mice of the Peromyscus Polionotus Group from Florida and Alabama." *Journal of Mammalogy*, Vol. 7, No. 3, August, 1926.

Erwin Stresemann.
Mutationsstudien I–XXV. *Journal für Ornithologie*, LXXXIV. Jahrgang, pp. 377–385. 4, 1. Juli 1923.

Joseph Grinnell.
"Geography and Evolution in the Pocket Gopher." *Univ. of Cal. Chronicle*, July, 1926.

G. A. Boulenger.
"The Tailless Batrachians of Europe" (2 parts). The Ray Society, London, 1897.
"A Contribution to the Knowledge of the Races of *Rana Esculenta* and their Geographical Distribution." *Proc. Zool. Soc. of London*, 1891.

E. G. Boulenger.
"On Some Lizards of the Genus *Chalcides*." *Proc. Zool. Soc. of London*, 1920, pp. 77–87.

Wilfred H. Osgood.
"Revision of the Mice of the American Genus *Peromyscus*." *North American Fauna* No. 28, Washington, 1909.

Gerrit S. Miller, Jr.
"Two New Malayan Mouse Deer." *Proc. Biol. Soc. of Washington*, Vol. XV, pp. 173–175, August 6, 1902.
"Tragulus napu Miller." *Proc. Wash. Acad. Sci.*, II, p. 227, August 20, 1900.
"Mammals collected by Dr. W. L. Abbott in the Region of the Indragiri River, Sumatra." *Proc. Acad. Nat. Sci. of Phila.*, Vol. LIV, 1902, pp. 143, 144.
Catalogue of the Mammals of Western Europe. British Museum, 1912.
"Mammals collected by Dr. W. L. Abbott on the Natuna Islands." *Proc. Wash. Acad. Sci.*, Vol. III, 1901, pp. 111–138.

C. Hart Merriam.
"Results of a Biological Survey of the San Francisco Mountain Region and Desert of the Little Colorado, Arizona." With Stejneger. *North American Fauna* No. 3, 1890.
"Revision of North American Pocket Mice." *North American Fauna* No. 1, 1889.
"Results of a Biological Survey of Mt. Shasta, California." *North American Fauna* No. 16, 1899.

T. H. Morgan.
"Evolution and Genetics." *Amer. Journ. Phys. Anthr.*, Vol. IX, No. 1, Jan.–March, 1926, pp. 105, 106.

Gary N. Calkins.
"Organization and Variation in Protozoa." *Scientific Monthly*, April, 1926.

Henry Edward Crampton.
"Contemporaneous Organic Differentiation in the Species of Partula living in Moorea, Society Islands." AMER. NAT., Vol. LIX, Jan.–Feb., 1925.

Ludlow Griscom.
"The Ornithological Results of the Mason-Spinden Expedition to Yucatan." In 2 parts. *Amer. Mus. Novitates*, Nos. 235, 236, Nov. 18, 19, 1926.

L. S. Berg.
"Nomogenesis, or Evolution determined by Law." London, 1926.

SUPRA–SPECIFIC VARIATION IN NATURE AND IN CLASSIFICATION

ALFRED C. KINSEY

SUPRA-SPECIFIC VARIATION IN NATURE AND IN CLASSIFICATION[1]

FROM THE VIEW-POINT OF ZOOLOGY[2]

ALFRED C. KINSEY
INDIANA UNIVERSITY

During the last three years biologists of every field have shown renewed interest in problems of organic evolution. This is clearly the result of recent developments in genetics, taxonomy, paleontology and some other sciences which are basic to an understanding of species. Not many decades ago a symposium on evolution seemed to call for declarations of scientific creeds, on the basis of which biologists were split into rival schools and independent sciences. To-day we are more interested in collating and correlating sound data from all the fields which have anything to contribute.

In the wide range of living forms, there is more than one kind of species. Entirely aside from nomenclatorial difficulties, species differ in different groups; and even within single groups they vary with the age and the size of the population, the mutation rate, the recency of mutation, the chance for hybridization with related species and in still other ways. These factors are so diverse in different groups as to account for the different or even contradictory emphases placed by different investigators upon mutation, hybridization, polyploidy, sterility mechanisms, other isolating factors and selection in the evolution proc-

[1] Read at a symposium of the American Society of Naturalists in joint session with the American Society of Zoologists, the Botanical Society of America, the Genetics Society of America, the American Phytopathological Society and the Ecological Society of America. The American Association for the Advancement of Science, Atlantic City, N. J., December 30, 1936.

[2] Contribution from the Department of Zoology, Indiana University, No. 264 (Entomological Series No. 18). The detailed data, of which the present paper is a summary, are to be found in the author's 1930 and 1936 publications on *Cynips* (Ind. Univ. Studies 84–86, and Ind. Univ. Sci. Ser. 4). The last-named publication contains a more extended treatment of higher categories.

ess. But where observations are accurate and sufficient, they can be accepted as applicable to the material studied; and it can not be over-emphasized that there is no reason for expecting simple and uniform explanations of evolutionary processes among all kinds of organisms.

The picture which I shall present is based upon an intensive study of the gall wasp family Cynipidae and on only a more casual acquaintance with species in other groups. Whether this material is translatable to other portions of the living world must be determined through comparable studies of other organisms.

Definition of Higher Categories

Confusion will be avoided if we call the basic taxonomic unit the species. It is the unit beneath which there are in nature no subdivisions which maintain themselves for any length of time or over any large area. The unit is variously known among taxonomists as the species, subspecies, variety, Rasse or geographic race. It is the unit directly involved in the question of the origin of species, and the entity most often indicated by non-taxonomists when they refer to species. Systematists often introduce confusion into evolutionary discussions by applying the term to some category above the basic unit. Anything above this unit, even though it may be called a species, is, in reality, a higher category whose evolutionary history is essentially the same as the history of such other classificatory composites as are commonly called complexes, subgenera, genera, etc. All categories above the basic taxonomic unit present a single problem, which is to determine the nature and the manner of origin of a group of units, in contrast to the nature and origin of the units themselves.

Nature of Generic Characters

A higher category is traditionally defined as a group of species with at least some characteristics which extend throughout the group. The opinion is current that the morphologic or physiologic characters which define these categories are usually different from those which distinguish species. In those instances where the same characters are involved in specific and generic differentiation, it is currently understood that the differences between species are not as great as the differences between higher categories. Just as some of the geneticists have insisted that the laboratory genetics may explain the nature and origin of Mendelian races, but not of natural species, so others indicate that the qualities of higher categories must be explained on bases other than those involved in species.

On the contrary, recent taxonomic studies and such experimental analyses as have been made of natural species indicate that Mendelian genetics provides all the hereditary mechanism necessary for the evolution of species as well as for laboratory races; and I shall undertake to show that the same genetics is all that is involved in the origin and development of the characteristics of higher categories.

First, as to the constancy of generic characters: This should mean, genetically, that the genes which differentiate species are more liable to change than those which define genera. With many plant and animal groups, Miocene, Oligocene or older in origin, we are asked to think of generic characters as those which have undergone little or no mutation for twenty, thirty or even hundreds of millions of years. While the geneticists have shown that some genes are more mutable than others, laboratory data can hardly be counted on to prove that there are genes that remain unchanged for such long periods of time. The impression that there are such long-stable characters is, therefore, based primarily on the evidence of current classifications. But this is arguing in a circle. It proves nothing as to the stability of generic characters to exhibit genera that have been delimited by stable characters. It would be more important to know how many characters are stable throughout groups that are evolutionary units.

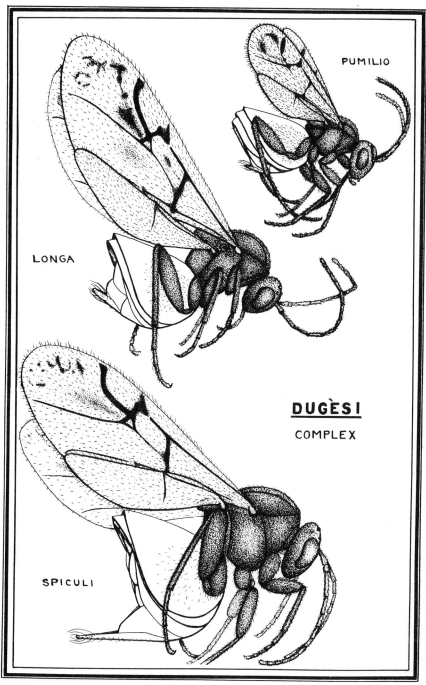

FIG. 1. THREE CLOSELY RELATED SPECIES
Mexican gall wasps. Note differences in wing length, abdominal spines, etc. Major mutations obscure the really close relations of such species.

Too often classifications are built on a limited number of characters and relationships recognized by the sole criterion of similarity. Where only minor mutations have been involved, this leads to no difficulties, but where major mutations have occurred, closely related species may differ so radically as to obscure their relationships. It has been one of the major contributions of genetics to show that similarity is not wholly safe as a criterion for the recognition of relationships. In modern taxonomic studies we have many cases of closely related but dissimilar species which have hitherto been placed in apparently divergent groups. It is only when phylogenetic interpretations are based upon a variety of morphologic, physiologic and distributional data that true relationships in such instances become evident.

To illustrate from the gall wasp family Cynipidae, it is to be noted that the genera and higher groups in the accepted classifications have been defined for more than half a century by remarkable differences in wing length, relative proportions in thoracic and abdominal structures, antennal segmentation, body sculpture and clothing and other clear-cut characteristics. The book-classification furnishes a typical or even extreme example of higher categories discretely defined by stable characters. And yet if the genera of this family are rearranged on modern phylogenetic bases, the stability of the generic characters proves to be quite illusory and the instability of practically all characters the rule. So-called genera that have been delimited by peculiar and constant wing lengths, distinctive thoracic or abdominal structures, etc., prove to be artificial aggregations of species that are phylogenetically remote. A phylogenetic classification gives genera which are difficult to define because none of the characters is fixed throughout the species that must be recognized as related. In *Cynips,* there are nearly a hundred characters which are available for classificatory purposes, but not a single one of these extends uniformly throughout the genus.

Whether the Cynipidae are a fair sample of other organisms must be determined by more studies than we yet have. That there is evidence of the same variability of the characters in other groups is strikingly attested by the wide-spread tendency in present-day systematics to multiply the number of genera at the expense of the size of the groups. Such small or monophyletic genera as are commonly recognized by vertebrate taxonomists, and notably by the ichthyologists, merely evidence the unwarranted faith which systematists have in the stability of characters as generic criteria and their hesitancy to believe that major mutations may occur among vertebrates as well as among fruit flies. If the stability of generic characters can be demonstrated only by limiting genera to a few or to single species, we are in reality establishing the fact that there are few if any characters which are stable through any long time or through any large series of units.

From the foregoing it may be concluded that exactly the same characters differentiate Mendelian races, species and the best-defined genera. The accompanying table shows that this is what we do find when we make modern phylogenetic interpretations, instead of depending on such traditional arrangements as are current in elementary keys and catalogues. In genetic terms, this means that mutations developing within the heart of a population and not prevented by some sort of isolation factor from interbreeding with the parental stock bring about an extension of the individual variation within the species. If, on the other hand, the very same mutation, large or small, is isolated or selected from out of the parental stock, the new species which ensues is differentiated by the same characters which were the basis of the Mendelian race in the first instance. But, finally, if the specific differentiation involves major mutations which are continued through any series of species, we ordinarily consider that two genera have evolved. Since exactly the same characters are concerned in all these cases, there is no need for believing

CATEGORICAL SIGNIFICANCE OF CHARACTERS IN CYNIPIDAE

Characters	Genera e.g.	Diagnostic for: Species within complexes e.g.	Individuals within species e.g.
EYES: protrusion	Disholcaspis	X. volutel.	B. eburnea
ANTENNAE: no. of segm.	Cynips	X. crystal.	
THORAX			
Size: proportions	Disholcaspis	C. villosa	C. anceps
Pronotum: dors. width	Aulacidea	Philonix	Philonix
Paraps. groove: length	Besbicus	C. weldi	B. eburnea
Mesopleuron: sculpt.	Xystoteras		
Fov. groove: subdivis.	Diplobius	X. crystal.	C. fulvic.
Sculpture, clothing	Disholcaspis	X. crystal.	
ABDOMEN			
Size: proportions	Disholcaspis	C. villosa	B. eburnea
Shape	Heteroecus	X. crystal.	B. eburnea
Segments: rel. sizes	Heteroecus	Philonix	B. eburnea
Pubescence: extent	Disholcaspis	C. villosa	Acraspis
HYPOPYGIAL SPINE	Disholcaspis	C. villosa	C. bulla
WINGS			
Length: 0.10–2.00	Besbicus	Acraspis	B. eburnea
Radial cell: closure	Cynips	Lytorhod.	C. pictor
Venation	Feron	C. mellea	C. pictor
TARSAL CLAWS: toothing	Disholcaspis	C. mellea	C. mellea
GALL			
Location: on plant	Cynips	C. aggreg.	Neur. pacif.
Larval cells: number	Cynips	Acraspis	C. cruenta
Tissues involved	Atrusca	C. arida	
Separability	Cynips	Diastroph.	Liposthenes
Special covering	Atrusca	Acraspis	C. erinacei
HOST: restriction			
To single species	Heteroecus	C. dugesi	C. unica
To related species	Cynips	C. bulboid.	C. incompta
LIFE HISTORIES			
Larval life: duration	Amphibolips	D. pernic.	C. fulvic.
Season of emergence	Disholcaspis	C. dugesi	C. sierrae
Sex ratios	Aulacidea	Diplolepis	D. rosae
Heterogeny	Aulacidea		

that different mechanisms of heredity are at the base of the evolution of different categories.

THE ORIGIN OF HIGHER CATEGORIES

In the course of evolution, old species are related to the new as links in a chain. It is unfortunate that the classic analogy should have been with a tree, for the picture is that of an infrequently dividing chain. Minor mutations in a few characters are the usual bases by which one link differs from the next; major mutations occasionally occur; but nowhere in the genetic, taxonomic or paleontologic data is there sound justification for believing that great changes in whole groups of characters are the sources of the higher categories.

The phylogenetic trees which substantiate the major catastrophe idea are most tree-like when the fewest fossil or existent species are at hand to limit the imagination of the artist. The known species are too often united by dotted extensions of the lines representing the existent data. An experienced paleontologist or taxonomist should be able to take the sticks from any woodpile, no matter how diverse their origin, and unite them so skilfully that they will not only look like a tree, but pass among biolo-

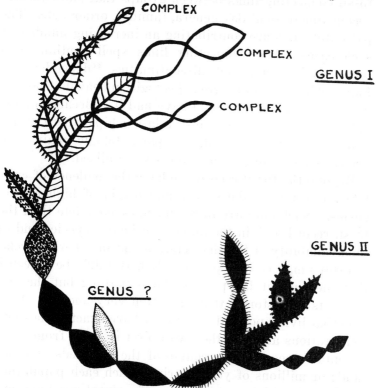

FIG. 2. THE EVOLUTIONARY PATTERN IS THAT OF A CHAIN
There is no discontinuity in such a chain, although the diverse ends may represent different genera.

gists as derivatives of the root system of a single dicotyledonous plant. The only restriction is that no species, living or fossil, shall be derived from any other known species, although both, according to the traditional rules, may be derived from a hypothetical "common ancestor." Such a procedure has the merit of putting embarrassing questions far enough into the background to allow us to ignore them when we wish. But if we rub out the dotted lines and hypothetic connections from these phylogenetic trees, what do we know about the origin of higher categories?

We know, first of all, that there is an increasing list of fossil connecting links between groups that subsequently became the present-day genera, families, orders, etc. The paleontologists are contributing an increasing number of such cases. And connecting links spell continuity in the organic chain, not discontinuities "dependent on Lamarckian effects of great geologic upheavals," as the older paleontologists would have had it. Because of the difficulty of recognizing close relationships where major mutations have occurred, the paleontologic chains are probably more continuous than is yet realized.

Beyond the fossil record, we have the evidence of existent species as to the nature and origin of higher categories. Such data are in reality more complete than the fossil record both in the number of known species and in the availability of the characters by means of which relationships may be interpreted. It may fairly be objected that the present-day picture shows nothing but the end products of histories that are often very ancient. It is true that living populations do not have exactly the same constitutions as when they were first isolated from their parental types. In the course of the hundreds of thousands or millions of years during which each population has been in existence, intra-specific mutations have extended the range of variation, hybridization with related species may have introduced new allelomorphic genes, and chance elimination or selection may have dropped some factors out of the inheritance of the group. The complete picture in any case should show a third dimension representing intra-specific evolution; but it is only when a detailed fossil record is available that the vertical can be correlated with the horizontal evolution of a group.

Meanwhile the present-day species tell so much of the story that it is surprising that they have not more often been called upon in studying the history of evolution. The neglect of these data is very largely due to the preoccupation of taxonomists with local faunas and floras. Among plant and animal groups of any size, there are hardly twenty which have had the monographic treatment necessary for an understanding of the relationships of higher categories. European studies rarely get out of Central Europe, though the evolutionary histories of those

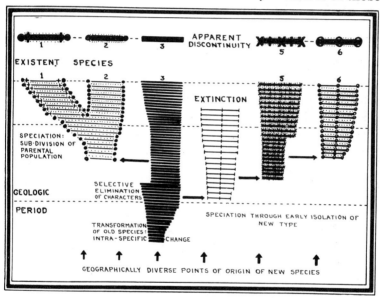

FIG. 3. THE ORIGIN OF EXISTENT CHAINS
Present-day species represent cross-sections of fluctuating populations which originated at different times and in different places.

forms must be followed through Mediterranean, North African and Asiatic groups if one is to find the links between complexes and genera. American studies too often stop at international boundaries. In consequence, it is concluded that American genera are sharply limited, discrete groups, although the pursuit of those same evolutionary lines into the north or into Mexico and Central America might lead to very different conclusions.

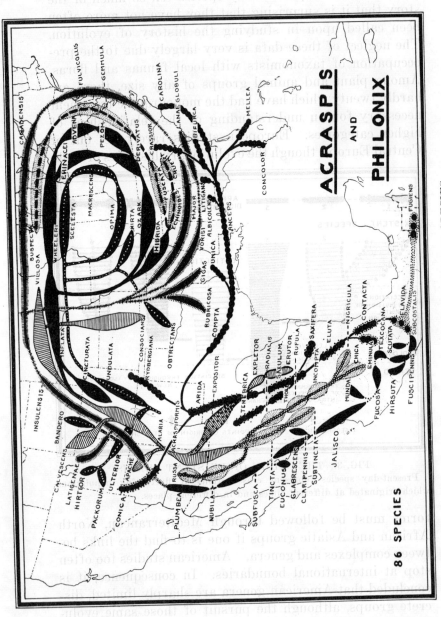

FIG. 4. A CHAIN OF EXISTENT SPECIES
Gall wasps of the *Cynips* group of genera. See Fig. 5.

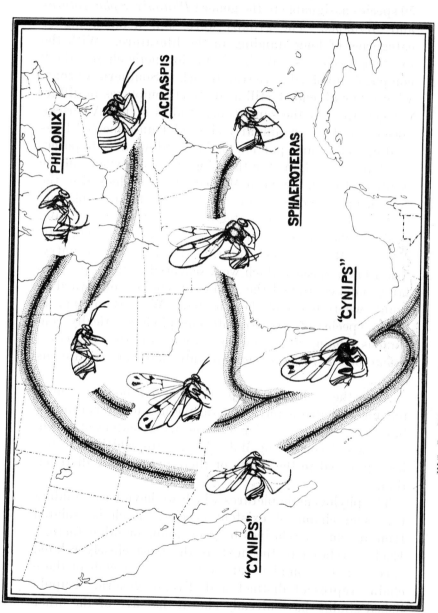

FIG. 5. EXTREME TYPES IN A CONTINUOUS CHAIN

See Fig. 4 for the 86 species involved. The diverse ends of this chain represent the distinct "genera," *Philonix, Acraspis, Sphaeroteras* and *Cynips*; but the limits of the genera are indefinable because of the continuity of the chain.

Illustrating again from the Cynipidae: In 1930 we knew 50 species assignable to the genera *Philonix, Sphaeroteras* and *Acraspis*. The groups were well defined, higher categories of long standing in the literature. With the existent species as links in the phylogenetic chains, the 9 complexes involved were traced back to Southern Arizona, where they all ended, still as distinct complexes in different genera. At the same time, studies of the American species of *Atrusca* similarly showed 4 unconnected lines leading to the same portion of the Southwest. I hypothesized ancient origins for the 13 groups somewhere in the Southwest or in northern Mexico, and even looked to the spectacular geologic history of that area for possible factors in the origin of the categories! But in 1931–32, and again in 1935–36, we carried on fieldwork over some 18,000 miles of Mexican and Central American back-country. The 70 new species which we discovered in those particular groups continued the American chains southward to points where many of them united. We can now put 44 of the species of *Atrusca* into one phylogenetically continuous chain, and the 86 species of the *Acraspis-Sphaeroteras-Philonix* group into another continuous chain.

As in the building of a jig-saw puzzle, too many missing pieces leave a phylogenetic picture uninterpretable; but there is no excuse for hypothesizing centers of origin and ancient connecting links for groups which have not been pursued to the limits of their present-day distribution.

The phylogenetic pictures which we have been building represent chains of existent species. Each is isolated from its closest relatives by geographic or other factors. Each is related to the next as the most closely related species in any complex—and yet the diverse ends of these chains represent distinct and diverse genera. Unpublished studies have already shown that five of the other genera in the family will be united in these same chains, via Mexican and Rocky Mountain material that we have collected. Ultimately some four or five hundred species may fall into a continuous chain. Similar collapses of generic definitions are not unknown among other groups, but how many of the living forms of plants and animals may ultimately be placed on continuous chains, there is no way of predicting. That many of the connecting links are no longer in existence is evidenced by every fossil species which is different from a living form and by the present-day absence of whole groups from large areas (deserts, tropic lowlands or mountain divides) across which the original paths of migration must have extended. The higher the taxonomic category, the less, in general, is the chance of finding living connecting links; but for many complexes and genera the living record may be more detailed than that supplied by fossils.

In these chains there are no real discontinuities that may be taken as the limits of higher groups. Apparent discontinuities due to our lack of knowledge or to the actual extermination of once-existent forms may allow for some agreement among taxonomists as to the limitations of the higher groups; but such breaks in the data do not indicate real discontinuities in the history of the chain. While there are points at which particular characters show major mutations, these do not coincide with the points of major mutation in other characters; and phylogenetically sound classifications must be based on all the heritable characters that are taxonomically available. Without coincidence in the extent of the several characters, the limitation of a higher category appears to be an arbitrary procedure, no matter how diverse the ends of the chains may be. As matters of classificatory convenience, we shall undoubtedly continue to utilize higher categories; but in so doing we must recognize the artificial nature of their limitations.

Higher categories are, then, definable as arbitrarily delimited sections of phylogenetically continuous chains of species. They are not necessarily groups of similar species or groups of species with constant characters.

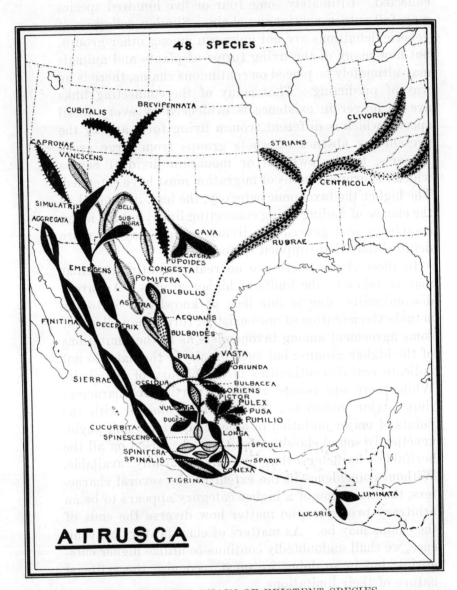

FIG. 6. ANOTHER CHAIN OF EXISTENT SPECIES
Previous to the discovery of the Mexican species, the ends of this chain appeared as well defined and not closely related complexes.

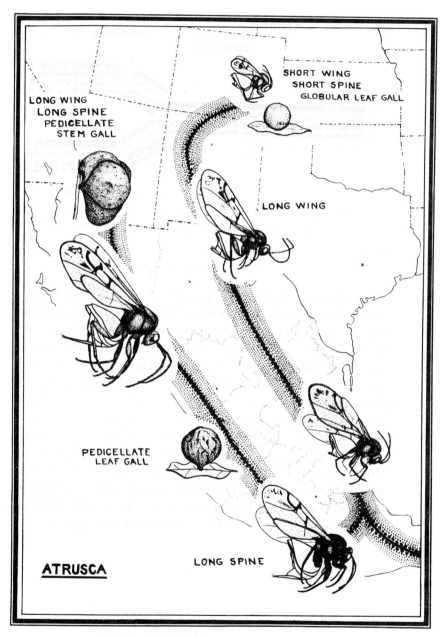

FIG. 7. EXTREME TYPES IN THE CHAIN

See Fig. 6 for all the species involved. Compare wing, abdominal spine, gall shape and location of galls in the extreme types. The major mutations in these several characters occurred at different points in the chain.

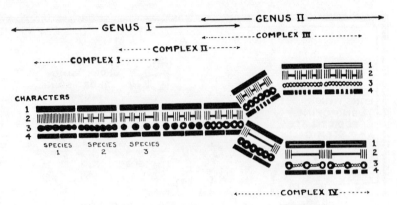

FIG. 8. THE ORIGIN OF "GENERIC CHARACTERS"
In such a chain, each character originates (by mutation) and disappears independently of the other characters of the species. There is no concomitant mutation in the several characters, although the extreme ends of such chains are the higher categories of taxonomic classifications.

They are not necessarily groups of species which have originated by radiate evolution from a common ancestor. But higher categories are series of phylogenetically related species. The same system of genetics which explains Mendelian races and the origin of species will explain the nature and origin of any higher category, for there are no characters in such categories apart from the characters of the species of which the group is composed. Evolution is never more than a process of change in single genes, rearrangements of gene complexes and aggregations of genes, and the modification of gene frequencies in the development of specific populations. There is, after all, no evolution apart from the modification of existent species or the differentiation of new species.

WHERE SUBSPECIES MEET.

By KARL JORDAN.

(With text-figures 48–53.)

IN a paper read at the 6th International Congress of Entomology which met at Madrid, September 6th to 12th, 1935, I dealt at some length with this subject. In March 1937 the General Secretary of the Congress informed me that 400 pages of the *Transactions* had been printed before war broke out and that the other manuscripts were set up in type. No progress has been made since. The General Secretary being involved in the clash of factions, it is most unlikely that the *Proceedings* and *Transactions* of the 6th Congress will be published before the end of the struggle in Spain. As my investigations of 1936 into the subject were a continuation of those recorded in the Congress paper, I repeat here some of the earlier observations before referring to my later results.

Innumerable species live side by side. However close they may agree in their morphology, there is a barrier between them which prevents interspecific promiscuity, keeping the specific populations pure. In a large percentage of species the area of distribution is subdivided into minor areas, each subarea being inhabited by a subspecific population. In those groups of animals where the systematist has advanced to the study of subspecies of the dead material at his disposal in museums, the diagnoses of subspecies are based on morphological distinctions, which are of the same kind as in the case of species. The two systematic categories being in this respect essentially alike, differing only in degree, the question arises as to whether the aversion to intercrossing which keeps the species apart also exists among the subspecies of one and the same species wherever they come into contact with one another. Where this aversion does not exist, the subspecies, so closely related to each other morphologically, must be expected to hybridize and to form hybrid populations in districts of contact. And conversely, the absence of hybrid populations may be taken as evidence that the subspecies in question are kept apart by a physiological barrier as are the species. Such conclusions may, in a given instance, be correct or not, being based on the evidence of circumstances which are not known in their entirety. Field researches and experimental breeding will have to supplement the investigations of the museum's systematist in order to make the results conclusive.

A most suitable subject for this kind of research in Europe is the widely distributed mouse-flea, *Ctenophthalmus agyrtes* Heller 1896, which is common on *Apodemus sylvaticus* and *Evotomys glareolus* and has developed into a number of subspecies. The species offers the student the great advantage that the differences between the subspecies are structural and definite, at least in the male. Moreover, a survey with the object of ascertaining the exact distribution of each subspecies of this flea would also be a survey of the fauna of mice, extensive collecting thus obtaining results in two classes of animals. I have collected most of the subspecies of *Ctenophthalmus agyrtes*; but it was only in 1931 that I for the first time devoted my holiday to obtaining an answer to the question: where do subspecies of *Ct. agyrtes* come into contact with one another? I concentrated on trapping mice; collecting other hosts and other fleas was incidental. The fleas were studied at the hotel, so as to be certain of the progress of the investigation. The distribution of the three subspecies concerned was known at that time to be as follows :

(1) The central European *Ct. agyrtes agyrtes* Heller 1896.——Norway, Denmark, Germany, southward to the Lake of Geneva and westward to Seine Inférieure.

(2) The north-western *Ct. agyrtes nobilis* Roths. 1898 (= *Ct. a. celticus* J. & R. 1922).——Great Britain and Ireland, Brittany and southward to the Puy-de-Dôme district and Charente.

(3) The French alpine *Ct. agyrtes provincialis* Roths. 1910.——Valescure, Hautes Alpes, Zermatt and Bex-les-Bains in the Swiss Rhône valley.

My first investigation aimed at the discovery of the districts where *Ct. a. agyrtes* and *Ct. a. provincialis* meet. As I had previously collected *agyrtes* above Montreux and *provincialis* at Bex-les-Bains, I expected to find both subspecies somewhere in between the two places. I went in the spring, which proved to be a mistake ; mice were very rare from Villeneuve, at the upper end of the Lake of Geneva, to Aigle, while slugs and snails abounded and sprung the traps. At Villeneuve I found only *Ct. a. agyrtes* ; but at Aigle, a little east of the town, on the left side of the river coming down from Les Ormondes, I obtained *Ct. a. provincialis*; and much higher up the right bank of the Rhône valley towards Leysin, the only specimen caught was a ♂ intermediate between the two and suggesting that the plateau north of the main chain of the Bernese Oberland is inhabited by a special intermediate subspecies or by a hybrid race. We have no fleas from these higher districts of Vaud and Valais ; on the north side of the mountains *Ct. a. agyrtes* is found. Poor as the results of several weeks' trapping were, they proved that the two subspecies meet round about Aigle. A renewed comparison of the fairly extensive material from Bex with the specimens from Zermatt and the French Alps has shown us that some of the Bex males are less typical, inclining a little towards *Ct. a. agyrtes*, which agrees with the suggestion that the two subspecies hybridize.

On the south side of the lake I collected at La Roche-sur-Foron, where I found only *Ct. a. agyrtes*, while in the Chamounix district only *Ct. a provincialis* was obtained, in evidently typical specimens. The two fleas probably come into contact on the escarpment from the lower to the higher level, and good centres for this investigation would be Servoz in the Arne valley and St. Germain in the Montjoie valley. On the south side of the Col de Bonhomme *Ct. a. provincialis* may be expected to occur at higher altitudes (as it does on the north side), and *Ct. a. verbanus* J. & R. 1920 farther down. Courmayeur would appear to be a convenient place for trapping up and down the Dora Baltea in search of the two subspecies.

As stated above (p. 104), two subspecies were known from Northern France, and it was evident from the occurrence of the one in Seine Inférieure and the other in Brittany that their areas of distribution must meet somewhere in Normandy, and Bagnoles-de-l'Orne, nearly in the centre of Normandy, seemed indicated as a good starting-point for the investigation. Trapping behind the bathing establishment on a wooded knoll covered with boulders produced *Ct. a. agyrtes* ; but outside the wood, in the hedges and meadows, I found *Ct. a. nobilis*. The results remained constant during the fortnight's stay in the place ; only *agyrtes* in the wood and only *nobilis* in the open country : a few hundred yards from the wood to the nearest trap in a hedge. The intermediate ground was occupied by houses and gardens where I could not put up any traps. The restriction of the two subspecies each to a definite biotope at Bagnoles is interesting and instructive. If the specimens had not been studied on the spot, but been labelled in the usual way with locality, date and host, and then been examined at home, the fact that the two insects are separate in space at Bagnoles would have been concealed. The areas of distribution do not overlap, but dovetail. Whether that is so in other districts of Normandy remains to be ascertained. There is no indication in the morphology of the Bagnoles specimens that intercrossing occurs.

Ct. a. nobilis was found by Mr. J. F. Cox at Ruffec, Charente, and in the Puy-de-Dôme district in 1931, and the Hon. Miriam Rothschild obtained it at Royat-les-Bains. In these districts it occurs together with another flea, Ctenophthalmus arvernus Jord. 1931, which is nearly related to Ct. agyrtes, but evidently specifically distinct. Miss Rothschild found the two insects actually on the same host at the same spot. There is no trace of hybridization in the fairly long series obtained.

As we had no collections from farther south and west in France, we did not know what happens to Ct. agyrtes south of the Auvergne. Does Ct. a. nobilis occur there or is it replaced by another subspecies ? To provide an answer to this question I visited South-west France in 1936, selecting four widely separate places at which to spend some time : Figeac (Lot), Brassac (Tarn), St. Jean-Pied-de-Port (Basses Pyrénées) and Gavarnie (Hautes Pyrénées).

As I had found in North Spain, in 1935, a subspecies of Ctenophthalmus agyrtes which is in some details nearer to Ct. a. agyrtes than is Ct. a. nobilis, I expected to discover in one or the other of these French localities a race like the Spanish one or intermediate between it and Ct. a. nobilis, and was much surprised that I obtained no subspecies of Ct. agyrtes at all, but only Ct. arvernus in all four places. I have certainly not covered the whole country, and there exist possibly pockets between the Auvergne and the Pyrenees where Ct. a. nobilis may occur ; but so much is certain that the ordinary flea of Apodemus and Evotomys in South-west France is Ct. arvernus, which takes here the same place in the fauna as Ct. a. nobilis in Brittany and Great Britain and Ct. a. agyrtes in Central Europe. A connection between the area of Ct. a. nobilis and that of the Spanish subspecies may be discovered to exist along the coast of the Bay of Biscay ; or the pine woods of Les Landes may harbour a race of their own. We do not know. Our knowledge of the distribution in Europe, not only of fleas but of large numbers of other animals, is so meagre and it would be in many instances so easy to fill in the lacunae that I here appeal to local naturalists to take an interest in this kind of research. I suppose it is too much to hope for a well-organized survey of the fauna of Western and Central Europe.

I append here the list of Siphonaptera collected in South-western France, to which is added the description of the Spanish race of Ctenophthalmus agyrtes above referred to, and that of another flea from Sardinia.

Collecting was concentrated on Apodemus sylvaticus, Evotomys glareolus and Talpa europaea. As the Spanish race occurs on the Mole, there was a possibility that the same or a similar race of Ct. agyrtes was parasitic on Talpa in the Pyrenees ; the result was negative. Shrews were taken incidentally, but there were hardly any fleas on them, which was contrary to the usual experience.

No attempt was made to obtain a large and varied collection; nevertheless, several subspecies and one species are new to science. Some of the mice collected are still rare in museums. The specimens from Figeac, Brassac and St. Jean-Pied-de-Port were collected in June and those from Gavarnie and neighbourhood in July.

1. Citellophilus occidentis sp. nov. (text-fig. 48).

Gave d'Ossoue, Gavarnie, 1,460 m., 13.vii.1936, on *Microtus nivalis aquitanius*, one ♀.——The mouse was caught in a place where *Eliomys quercinus*

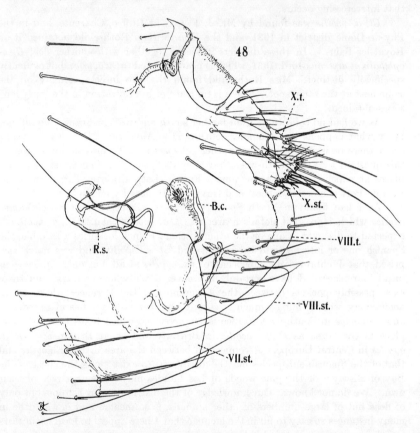

occurred; this is possibly the true host of the new flea. The only example of *Eliomys* caught was dead in the trap, and there were no fleas on it.

♀. Somewhat aberrant in this genus. *C. danubianus* Roths. 1909 is geographically the nearest species we have in the collection. The new species differs from it particularly in having 3 antepygidial bristles, no row of bristles on the outer surface of mid- and hindfemora, a small stigma on tergite VIII, and a long body and short tail to the spermatheca. In Wagner's Catalogue of the Palaearctic fleas *C. danubianus* is placed as a synonym of *C. simplex* Wagn. 1902; but this can hardly be right, if Wagner's figures are absolutely correct as he affirms. The digitoid of *C. danubianus* bears a spiniform bristle not shown in Wagner's figures and the other bristles of the digitoid are much longer in his figures than in our specimens of *C. danubianus* from Rumania and Bulgaria.

Rostrum extending to middle of trochanter, shorter than in *C. danubianus*. In front of eye a row of 3 long bristles, no other bristles on frons; on occiput a long bristle above middle of antennal groove, subapical row widely interrupted above the long bristle placed at the lower angle, above the wide interspace only 2 or 3 bristles in the row each side. Eye with hardly any pigment except at periphery (accidental ?). Bristles of segment II of antenna much longer than club.

Pronotal comb with 19 spines inclusive of a small ventral one. On mesonotum a posterior row of 10 long bristles (the two sides together) and a row of 11 small ones, no further lateral bristles except at anterior margin, and a row of 3 or 4 dorsal ones a short distance behind the basal ones, on inner surface 11 bristle-like spines; on mesopleura 7 long and longish bristles and anteriorly 10 or 12 small ones. Bristles on metanotum 15, 10; on metepimerum 6.

Apical spines on abdominal terga (the two sides together): I 3, II 4, III 3; bristles in two rows, no additional dorsal bristles: I 13, 10, II 10, 15, III 9, 14, IV 7, 14, V 7, 14, VI 8, 13, VII 4, 12; on sterna: II 0, 2, III 0, 7, IV, 0, 8, VI 0, 8, VII 12, 10.

Apart from the subapical subventral bristle, on outer surface of forefemur 8 lateral bristles (and 2 near apex), on inner surface 2 lateral ones; on midfemur on outside no lateral bristles, on inside a row of 5; on outside of hindfemur no lateral bristles, on inside a row of 7 on one femur and 9 on the other. On hindtibia 7 dorsal notches (inclusive of apical notch), third notch with a single heavy bristle, between fifth and sixth a single subdorsal heavy bristle; on outer surface of hindtibia 9 dorsolateral bristles, including an apical one; the longest apical bristle of segment II of hindtarsus reaching to apex of III; in all tarsi of *C. danubianus* segment V with 2 apical ventral bristles and proximally of them 1 or 2 similar but smaller bristles, these latter absent in the new species. Three antepygidial bristles, upper one $\frac{3}{8}$ of middle bristle, lower one $\frac{3}{4}$, approximately.

Modified Segments.—Sternum VII narrowing apicad, the upper margin slanting-incurved (text-fig. 48, VII st.), apex broad, rounded-truncate, upper apical angle a little over 90°. Above stigma of VIII t. 3 bristles each side; stigma much smaller than in *C. danubianus*, below stigma 2 short and 3 long bristles, on lower portion of outer surface 17 bristles, inclusive of those at apical margin, on inner surface 2. Sensilium measured dorsally shorter than its distance from stylet. Some of the bristles of anal sternum (X. st.) stout and curved. Spermatheca (R.s.): body twice as long as broad, almost of even width, ventrally concave, dorsally convex, orifice terminal-subdorsal, tail measured in a straight line a very little shorter than body; bursa copulatrix with conspicuous glandular tissue, as in *C. danubianus*, blind duct long.

Length 3·0 mm.; hindfemur 0·48 mm.

2. Nosopsyllus fasciatus Bose 1801.

Brassac, on *Apodemus sylvaticus*, 1 ♂, and on *Mus musculus*, 1 ♂.

3. Ctenophthalmus arvernus Jord. 1931 (text-fig. 49).

Figeac, a series on *A. sylvaticus* and *Evotomys glareolus*.——The best collecting ground at Figeac was a steep damp lane above the railway, with rank vegetation and a hedge each side. Here the two mice and *Crocidura russula* were rather abundant. On the dry higher ground of each side of the valley hardly any mice were found; but I obtained on the right side well above the town a specimen of *Pitymys pyrenaicus*, the second record of the species from Central France.

Brassac, a series on *A. sylvaticus* and *E. glareolus*, and 1 ♂ on *Mus musculus*.

St. Jean-Pied-de-Port, on *A. sylvaticus*, a small series.——Mice were very scarce; the hills too dry and the valley too populated. No *Evotomys* were found.

Gavarnie, on *A. sylvaticus*, 2 ♂♂, 1 ♀.——Only one specimen of this mouse was trapped and no *Evotomys*.

Gave d'Ossoue, above Gavarnie, on *Pitymys pyrenaicus*, 4 ♂♂, 2 ♀♀.——One specimen of this mouse taken in a mole run.

The marginal projection of VII. st. of ♀ is always short (text-fig. 42, the segment spread out), in *Ct. a. nobilis* twice as long or over.

4. Ctenophthalmus nivalis nivalis
Roths. 1909.

Gave d'Ossoue, above Gavarnie, on *Microtus nivalis aquitanius*, 1 ♀.——As sternum VII of the specimen is injured, I cannot say with certainty whether the specimen agrees in this segment with *Ct. nivalis nivalis* from the Hautes Alpes and Haute Savoie. I obtained only two Snow Mice; one, without fleas, was trapped among the large boulders a short distance beyond the first bridge over the brook, the other at the other side of the water in the wood near a meadow beyond the second bridge.

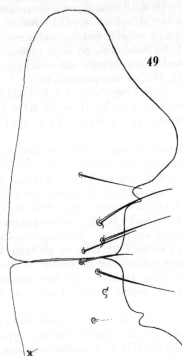

5. Ctenophthalmus agyrtes hispanicus subsp. nov. (text-fig. 50).

A single ♂ from the nest of *Talpa*, near Lake Enol, above Covadonga, Asturia, 3 Sept. 1935.——Mole-hills were plentiful in a meadow, and here and there the feet of cattle had sunk into the ground, a sure sign that there was a nest underneath. I took several nests, but contrary to my expectation (and usual experience) there were no flea-larvae in them and only one imago; the ground had been perhaps too wet.

♂. Nearest to *Ct. a. nobilis* Roths. 1898, but differing in the genitalia, particularly in the ventral lateral lobe of the phallosome. Manubrium (M) of clasper strongly curved upwards, nearly as in *Ct. baeticus* Roths. 1910 from Portugal. The ventral rounded corner of clasper more projecting downwards than in *Ct. a. nobilis* and its lowest point farther distant from digitoid F. Process P² of clasper obliquely truncate-sinuate as in some specimens of *Ct. a. nobilis*, the posterior angle but little projecting ; in the majority of French specimens of *Ct. a. nobilis* the apex of P^2 is more deeply incurved, the lower angle therefore projecting considerably ; this is also the case to the same extent in a minority of British specimens, an indication of the commencement of subspecific separation. Ventral, horizontal, arm of sternum IX as narrow as in *Ct. a. nobilis*, but its ventral margin feebly angulate at some distance from tip at the point

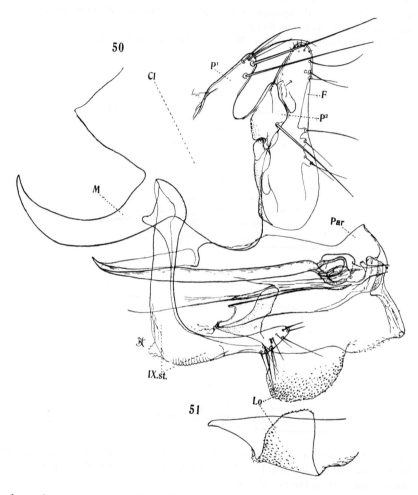

where there is a cluster of bristles. Phallosome with large, rounded, ventral flap each side (Lo), densely covered anteriorly and ventrally with semiovate, somewhat scale-like, swellings directed frontad ; in *Ct. a. nobilis*, on the other hand, the flap is turned upwards (occasionally turned down in mounted specimens) and its upper margin very distinctly denticulate (text-fig. 51) ; *nobilis* differs therein from all other subspecies of *Ct. agyrtes*, and from the purely morphological point of view therefore has some claim to specific distinctness.

6. Rhadinopsylla mesa Jord. 1920.

Gave d'Ossoue, Gavarnie, 1,460 m., 13.vii.36, 1 ♂ on *Talpa europaea*.—— The specimen differs slightly from the only Swiss ♂ (the type) we have; the difference is either individual or geographical; at least one more ♂ is necessary for a tentative decision. The genal comb of the Pyrenean ♂ consists of 5 spines on each side; in the Swiss ♂ there are 5 spines on one side and 6 on the other; in our two Swiss ♀♀ 6 on each side. The uppermost spine of this comb is, in the Pyrenean ♂, very little broader at the base than the lowest spine and a trifle longer, not shorter as in the Swiss ♂.

In the Key to the species of *Rhadinopsylla* published in Nov. Zool., xxxv,

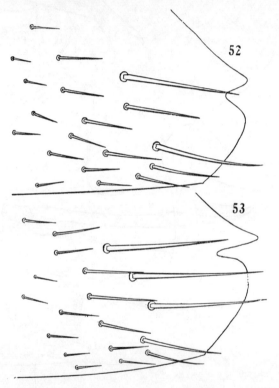

pp. 183, 184 (1929) it is stated under a^4 that the bristles of VIII. st. are apical in *Rh. isacanthus*; it should read subapical, the row of bristles being placed some distance from the apical margin.

7. Typhloceras favosus rolandi subsp. nov. (text-fig. 52).

Basses Pyrénées: St. Jean-Pied-de-Port, 150 m., on *Apodemus sylvaticus*, 28.vi.36, one ♀.

♀. Agrees with our two Algerian ♀♀ in the eye being smaller than in *T. poppei* Wagn. 1903, thorax and abdomen more distinctly reticulated, bristles of metanotum less numerous, many of the bristles of the body less pointed, the interspace between eye and upper genal spine broader than third genal spine, hindtibia with about 25 lateral bristles (in *T. poppei* over 40). The Pyrenean specimen differs from the Algerian ones in the pronotal comb containing 22 spines instead of 20, and in the sinus of sternum VII not being deeper than in *T. poppei* : triangular lobe above this sinus narrower than in *T. poppei* (narrower on right side than on left figured) and not longer than lower lobe, which is broad and rounded (text-fig. 52).

8. Typhloceras favosus asunicus subsp. nov. (text-fig. 53).

The above-described subspecies is connected with *T. favosus favosus* J. & R. 1914 by an intermediate one from Sardinia, of which we have also but one ♀, collected by Dr. A. H. Krausse at Asuni off a mouse (and mentioned in Nov. Zool., xxi, p. 236 (1914)). Pronotal comb with 22 spines as before. Sinus of sternum VII narrow (text-fig. 53), the lobe above it decidedly longer than the broad lower lobe and twice as long as in the preceding subspecies, but one-third shorter than in *T. f. favosus*, in which the lower lobe, moreover, is nearly as long as the upper. As VII. st. is injured on the left side, we figure the right side ; the long bristles are the same on both sides.

CRITERIA FOR GENERA, SPECIES, AND SUBSPECIES IN ZOOLOGY AND PALEOZOOLOGY

By

GEORGE GAYLORD SIMPSON
The American Museum of Natural History, New York, N. Y.

INTRODUCTION

This paper does not present the consensus of vertebrate paleontologists, nor is it confined to vertebrate paleontology. The criteria used by various students, or by each of them at various times, for recognizing genera, species, and subspecies among fossil vertebrates are so different that it would be a gigantic task to summarize them all and an impossible task to find among them procedures in universal use. It is certainly more practicable and it may be more useful to approach the problem more broadly and at the same time more individually. This study, then, presents an individual opinion as to the general principles that do or that should underlie the selection and use of criteria for these taxonomic categories.

The viewpoint is paleozoological throughout and the differences between paleozoological and neozoological problems and criteria are discussed, but these two branches of zoology have much in common and a search for general principles cannot be confined to one branch. Although there is thus some invasion of the fields of other contributors to this symposium, it may be hoped that even here the difference of viewpoint will have some value and will prevent mere repetition.

DEFINITIONS

The Species

The title of this symposium was deliberately chosen to avoid the direct question "What is a species?"—a subject of such long, frequently futile, and sometimes acrimonious debate. The theme of the present discussion is not what genera, species, and subspecies are in a theoretical sense, but how specialists in the different branches of vertebrate zoology should set about recognizing and delimiting such units in the practice of classification. The distinction is not merely verbal. One of the points that I want to emphasize is that the species in nature is something different from the species in classification. It is, however, impossible to consider criteria for recognition of a taxonomic group without having a reasonably clear idea of the nature of such a group. Assuming, for the present, agreement with the belief that the species is the fundamental unit of classification, its nature may first be discussed and those of subspecies and genera may be related to the concept of species. Categorical statement is necessary for brevity and does not indicate lack of awareness of radically different opinions or unwillingness to grant validity to other points of view.

A species in nature is a group of organisms. It is not a process, as some geneticists say; or an infinite mathematical abstraction, as some statisticians maintain; or a collection of individual specimens, as no one is likely to say but as many, perhaps most, working taxonomists seem unconsciously to assume. The group arises by dynamic genetic processes, it can be described and interpreted by statistical methods, and it is composed of individuals, but it is the group itself that is a species.

There are not literally an infinite, but certainly an enormous number of possible and real groups of organisms differing both in content and in kind. In modern taxonomy it is a basic concept that the species in nature is a genetic group. The kind of genetic group that should be called a species grades into kinds that are given other names and this gradation, sometimes denied, is the most fruitful source of misunderstanding and disagreement. Nevertheless one idea underlies most of the definitions given by evolutionary taxonomists and is clearly involved in recent discussions by such an able zoologist as Mayr (1940, p. 256) and by such an able geneticist as Dobzhansky (1941, p. 373). The idea is expressed in various ways and with different qualifications and exceptions made necessary by special situations, but it is fundamentally this: *a genetic species is a group of organisms so constituted and so situated in nature that a hereditary character of any one of these organisms may be (possibly, but not necessarily) transmitted to a descendant of any other.*[*]
To the paleozoologist there is a large, important field in which definitions involving this idea are inapplicable: vertical species, dynamic temporal sequences. Discussion of this point is deferred to a later page, and for the present consideration is limited to the field in which this genetic concept is appropriate.

The neozoologist could conceivably apply this concept directly to the observation of nature and use it as his criterion for species in classification. In reality he does no such thing. What he does instead, I leave

---
[*]Obviously a primary qualification is that definitions implying this can apply only to sexually reproducing organisms. Since all vertebrates normally reproduce sexually, discussion of asexual or parthenogenetic species is not pertinent in this symposium.

to the other contributors to this symposium, except as his practice is related to that of paleozoology. For paleozoologists the direct use of such a criterion is not conceivable, even aside from the difficulty pointed out in the last paragraph. The neozoologist, by custom and for practical reasons, and the paleozoologist, from necessity, both define their species by morphology and not by the transmission of heredity or breeding habits and potentialities. Again subject to exception and modification for special cases, and again inapplicable to successive stages in vertical sequences, a species may be defined morphologically more or less as follows: *a morphological species is a group of individuals that resemble each other in most of their visible characters, sex for sex and variety for variety, and such that adjacent local populations within the group differ only in variable characters that intergrade marginally.*

This morphological definition is merely a description of the usual result of the situation involved in the genetic definition. Therefore, morphological species tend to correspond closely to genetic species, although it cannot be expected that the correspondence will be exact and universal. This treatment of present theory really reverses the historical sequence. The objective effect of the genetic situation was observed long before there was any clear idea of that situation or of phylogenetic processes in general. Species were defined morphologically before the concept of a phylogenetic species was achieved, but just because the species are real groups and because the phylogenetic situation does have definite morphological effects, it turns out that essentially the same groups are called species under both definitions.

Now that it is admitted that the phylogenetic unit, the species of evolutionary theory, is best defined by breeding or genetic structure, the morphologically defined species is not to be considered the species proper, in the strictest sense, but it is what most classifiers agree to call a species. This does not at all mean that the morphological species is a subjective or artificial concept. It is a real group that nearly corresponds with and is taken as a sufficient approximation of the genetic species.

Even the morphological definition does not bring us down to what the taxonomist really observes or to the species that he actually defines. What is observed and described is a series of individual specimens. These specimens do not constitute the species, and their description and differentiation from other series certainly do not constitute the description or diagnosis of a species. That this is what nine authors out of ten call a definition of a species does not alter the fact that they are mistaken or that they deal with species only by implication and not in their literal expressions. The series of specimens in hand is a sample drawn from a natural population (here assumed to be a species) but never completely representative of that population. If a species is defined, the process is to infer from the sample the characters and limits of the morphological species from which the sample was drawn. This inference is the species that really is used in taxonomy. The taxonomic species is a subjective concept and it cannot be exactly equivalent to the morphological species (which is an objective group), but the taxonomic species approximates the morphological species more or less closely according to the adequacy of the sample and the skill of the taxonomist. Thus we need a third definition of a species: *a taxonomic species is an inference as to the most probable characters and limits of the morphological species from which a given series of specimens has been drawn.*

What is really done in classifying organisms is to base, on a series of specimens, a taxonomic species which is a subjective estimate of a morphological species, which in turn is a group of organisms so defined as to approximate a genetic species. Practical classifiers will probably object to the complexity of this statement, but the complexity is only revealed, not created, by the analysis. This is the process and it is complex.

The Subspecies

It is true of all other categories of classification, as of species, that three distinct entities are involved, one taxonomic, one morphological, and one phylogenetic. There are three main sorts of categories: the species, the subspecies, and all higher categories, from "species groups" or subgenera up to kingdoms.

Genetically a subspecies may be approximately defined as a group of organisms throughout which active interbreeding occurs regularly, or so that the average hereditary repertory is approximately the same in the various local subgroups. This is by no means as clear-cut a thing as a species, even granting that the latter also shows considerable intergradation. Even the smallest local groups, below any level reasonably called subspecies, almost always show some differences in average genetic composition and adjacent "good" subspecies do usually interbreed normally where they are in contact. There are species in which fairly homogeneous* local groups have well-marked genetic differences from their neighbors and intergrade with the latter only along narrow zones

---

*Homogeneity does not mean that all individuals are alike; on the contrary they may be markedly different within a homogeneous population. It means that any two samples drawn from the population at random will not differ significantly in composition, or in representation of diverse variants. An analogy: a mixture of salt and pepper is homogeneous if the proportion of salt to pepper is about the same throughout; there is no implication in the word homogeneous (as here used) that it must be all salt or all pepper.

of contact. Here there is little doubt as to the reality and extent of the subspecies (provided the genetic facts are known). On the other hand, there are species in which only the most restricted groups—point rather than area groups—are at all homogeneous and the intergradation is essentially continuous from one end of the species to the other, although the ends may be quite different. In such groups, the clines of Julian Huxley (1938), there is no natural number of subspecies and no way to delimit them that is not arbitrary. The situation has an interesting analogue in continuous vertical sequences, to be discussed later. The practice in such cases is to make arbitrary divisions such that the differences between their median characters are approximately of the order of those between subspecies with limits observable in nature. An alternative, perhaps more logical but as yet little used in practical classification, would be to define the terminal points and to designate intermediate conditions by their distance from the ends. Finally there are species in which the whole population is fairly homogeneous. In such circumstances the species is commonly said to lack subspecies. Logically, it might be preferable to say that the species then includes only one subspecies, an example of monotypy, which also has special importance for the paleontologist and will be discussed later.

The relationship of taxonomic and morphological to genetic subspecies is closely analogous to that of taxonomic and morphological to genetic species and is sufficiently obvious on that basis. Given the difference in genetic definition, the important distinction between species and subspecies in practical work lies chiefly in the differences in methods of inference from specimen-series to taxonomic species and subspecies.

The Genus

The basic element in phylogenetic subspecies and species is a form of genetic continuity, actual or potential; that is, hereditary characters are being or can be transferred from one part of the population to another. By definition, such transfer cannot normally occur between different species,* and above the rank of species there may be genetic discontinuity within the group defined as a taxonomic unit. These higher categories are all essentially alike in this respect. They differ from each other in scope and, for the paleontologist, to some extent in their balance between horizontal (geographic) and vertical (time) dimensions. Thus in theory there is an absolute distinction between species and genus, and the dis-

---

*It is not necessary here to take up the old question of fertile interspecific and intergeneric hybrids. The isolating mechanism does not have to be reproductive but can be geographic, mechanical, psychological, etc. *Bison* and *Bos* can produce fertile hybrids, but they did not, in fact, interbreed in the undisturbed natural conditions under which the genera arose.

tinction can usually be made in practice, but there is no theoretical qualitative difference between genus and family. The two intergrade and it is a matter of custom and taste where the line is drawn.

According to universal agreement, a genus is one of the lower supraspecific categories. It includes either a cluster of species of not long antecedent common origin, or a single species that differs as much from other known species as if it were more or less central in a cluster of which the other members are missing. This is a vague definition and provides no rule of thumb, but no better can be expected. Unlike a species, the genus is a member of a continuously intergrading hierarchy of categories and therefore cannot be defined exactly. This does not mean that a genus is not "real" or "natural," but that the distinction of one member of the hierarchy as a genus requires art as well as science.

Some students insist that the genus be the lowest category employed above a species. The *reductio ad absurdum* of this principle is to place every species in a separate genus, a tendency that is operative in the work of a number of specialists. To name only one of many examples, the successive revisions of Severtzow (1858), Pocock (1917), J. A. Allen (1919) and others, resulted in recognizing at least 23 genera of living felines and placing every well-established species in a different genus.* Obviously, such classification makes the genus useless as a category in taxonomy. On the other hand, some students tend to use the genus for the largest group of species that can be demonstrated, with reasonable certainty, to have a common and exclusive ancestry. Placing all living felines in the single genus *Felis* is an example of this. The arguments for lumping and splitting will probably be covered adequately by the neozoologists on this program and it suffices for me to suggest that both extremes should be avoided, but that the over-inclusive use of the genus is less harmful and more acceptable than the tendency to equate it with the species.

Well-balanced use of the genus does leave a practical need within some polytypic genera for a lower collective (supra-specific) category. The need is well filled by the use of subgenera. There has been an unfortunate recent tendency to neglect the subgeneric category, but there are signs that it is being revived either as such or in the form of "species-groups," "supra-species" and the like.

---

*These authorities did, it is true, list more than one species in some of their "genera," but most of the additional species were either poorly known to them or must be judged by a dispassionate reviser to be subspecies by the best recent standards.

PROBLEMS OF INFERENCE

The erection of a taxonomic subspecies, species, or genus by inferring the nature and limits of corresponding morphological groups from a given series of specimens is essentially a statistical problem. Many zoologists —until recently it would have been fair to say "most zoologists"—are strongly resistant to this statement and maintain that they neither need nor use statistics in their work. However, the need is not open to any question and most zoologists whose taxonomic work is sound are using statistical methods whether they realize it or not. The misunderstanding is not primarily the fault of the zoologists, because it has been fostered by the statisticians. In a narrow sense, to which many statisticians adhere, statistics comprises a certain set of mathematical operations carried out with numerical data. Many of these operations are useful in zoology, but good taxonomic work can be and is being done without them, or with only the most elementary of them. In a broader sense statistics is the science of: (a) estimating the characteristics of populations from samples; and (b) describing groups, as such, rather than individuals taken singly. Since these two things are precisely what a systematist does when he sets up a species (or other taxonomic group) on the basis of a series of specimens, it follows that he is using statistics in a broad sense. Of course his statistics may be good or bad, rational or intuitive.

Although the species is the basic and, despite disagreement, the most easily defined genetic unit, the subspecies, at its best, is the simplest taxonomic unit because it is the one in which there is or may be a direct, simple statistical relationship between sample and population. Given a sample such as to make the assumption permissible, it is assumed that it is a random representation of a population homogeneous in the sense previously defined. It is then possible to estimate the probable characteristics and limits of variation in that hypothetical homogeneous population. In the simplest case, that estimate is a taxonomic subspecies. The usual methods of estimation are more or less familiar to all working zoologists and space need not be taken to discuss them here. They vary from the loose and often mistaken use of intuition, through the usual and pragmatically valid empiricism of experience derived from handling many such samples, to the relatively exact and reliable use of statistical methods in the narrower mathematical sense.

Since the results of such inference are in any case probabilities and not hard and fast limits, properly conservative interpretations of this kind automatically allow for a reasonable amount of heterogeneity in the morphological subspecies of which the taxonomic subspecies is a mental image. Present accumulations of data suggest that so-called subspecies so heterogeneous as to make this image importantly false are in most cases not morphological subspecies in the strict sense by clines, or segments of clines. In this case the taxonomic units derived by inference on the hypothesis of essential homogeneity do not correspond with subspecies as areas on the plane of distribution but correspond with points on a line or on a plane. The taxonomic treatment of this rather common situation is still in its infancy, but it is quite clear that these taxonomic points can be used to approximate and define morphological lines and planes just as well as the taxonomic areal unit, called a subspecies, approximates morphological subspecies. Such approximation of clines must be a secondary inference based on two or more primary or point inferences.

If a species includes only one subspecies, or if only one of its subspecies is known, the description of that subspecies describes the whole species. It might then be proper to say that the species does not need to be and indeed cannot be defined in terms of itself and on any objective basis. The only way in which such definition can be made different from that of the subspecies is by attempting to allow for the subsequent discovery of other subspecies, an attempt that may make us look very wise or very foolish if the discovery is made, but that is unlikely to have much value.

Such monotypic species can indeed be diagnosed. I wish here to establish a distinction between *definition*, a description of the characters and limits of variation of a single group (of any sort or scope), and *diagnosis*, a statement of the differences between adjacent groups. Definition may be said to tell what a thing is and diagnosis to establish what it is not. Etymologically "definition" means to set limits and "diagnosis" means to know apart or know [the differences] between. Obviously diagnosis plays a large role in zoological taxonomy, generally the major role in routine procedure of sorting specimens. Yet definition is more basic and it seems best to concentrate primarily on this aspect. No two organisms are precisely alike and the important point is to determine how much alike they must be to belong to the same group. This is definition, and it is impossible to make a proper diagnosis between groups without first knowing by definition what belongs within each group. It is beside the point that convenience often dictates the publication and subsequent use of a diagnosis rather than of the definitions on which it is based.

The species is the fundamental *genetic* unit, but polytypic species are already secondary or multiple *taxonomic* units. From the point of view of erection of taxonomic species by inference as to morphological species, if more than one subspecies is present the methods of inference do not differ from those involved in taxonomic genera. This underlies the rather common feeling among zoologists that the difference between species and genera, like that between genera and subfamilies, is simply a matter of scale and convenience, whereas to the geneticist there is a profound qualitative difference. Granting that phylogenetic classification should attempt to approximate the genetic ideal, the problem of the zoologist is to determine whether the different results of like methods of inference are more nearly harmonious with genetic species or genera.

The polytypic species and genera of taxonomy cannot, like the subspecies, be set up by simple statistical inference from a sample to a hypothetical homogeneous population. Such first-order inferences must be made for each of a number of different populations and then a second-order inference must be made from these data to define a larger and admittedly heterogeneous group. It is customary to set up a combined definition and diagnosis of the heterogeneous group by listing characters believed to be common to all its members and not to be common, either alone or in combination, to members of morphologically contiguous groups. This usually suffices for the rapid labeling of specimens, a legitimate goal but a very limited one. It certainly does not suffice for any real understanding of the nature and significance of the group. On higher taxonomic levels, especially as these are revealed to the paleontologist, such definition-diagnosis by common and exclusive characters is more clearly revealed as inadequate and sometimes as quite impossible. For instance it is instructive to try to list characters common to all Equidae and excluding all members of any other group. This may be possible, but I have been unable to make or to find such a list, although the Equidae certainly form a very real and valid taxonomic and genetic group. It is important to list characters in common, but it is equally important to define polytypic groups by the nature and sequence or arrangement (in space or in time) of the differences between the included lesser groups.*

The most practical criterion for judging whether a polytypic morphological group, in which there is no qualitative distinction between genera and species, corresponds more nearly with a generic or specific genetic group in which there is such a distinction, is by intergradation. If adjacent contemporaneous subgroups within a larger morphological group

---

*A point that may here be mentioned parenthetically, because it seems to be misunderstood in some otherwise excellent recent work, is that a polytypic group does not usually have a mean or average condition in the same sense as a monotypic group. The mean in a single population is a point of central tendency around which variations tend to cluster. The same figure derived from several different populations has no such significance and may, indeed, be a value that does not occur, or tend to occur, in any of the included groups. In a polytypic group a more or less valid analogue of the mean is the "condition in common" and an approximate analogue of variations from the mean is the range of differing means in the different subgroups.

differ in their means but intergrade in all known characters, it is a reasonable inference that the subgroups are subspecies and the group a species. If the subgroups do not intergrade in one well-marked character, or preferably in several characters, it is proper to infer that the subgroups are species and the group a genus. Allowance must be made for adequacy of sampling: a gap may represent a missing subgroup that would overlap both of two that do not themselves intergrade. It is also to be remembered that such intergradation of the populations may not be visible in small samples. It is not to be judged on the basis of the sample in hand but on the basis of the taxonomic concept derived from that sample. On the other hand, if only a few characters are available for study, the differentiating, non-intergrading characters of separate species may be missed.

This criterion does not assure full agreement between taxonomic and genetic groups. There are known certainly valid genetically distinct species—that is, populations that do not or cannot interbreed—that intergrade completely in morphology. These are, nevertheless, uncommon and it is not a serious error to reduce two such species, certainly very closely related, to subspecific status or even to fuse them completely in the taxonomic system. Moreover, it seems probable that such a situation can only be temporary. I do not know any example of the converse situation—that is, of two populations definitely belonging to one genetic species but not intergrading in a morphological character, either directly or through intervening groups.* Probably cases can occur, but they must be rare.

In general the use of this criterion will rarely, if ever, result in confusing subspecies or mere variations within a subspecies for species or species for genera, if valid methods of inference are used, but it may result in failure to distinguish valid genera and species or in giving them less than their genetic rank. When it fails, the criterion is on the side of caution, which seems desirable. Of course the number of such failures, which are not failures to recognize good morphological groups but to make these equivalent to genetic groups, will be in inverse ratio to the number of variable characters observed and to the closeness of inference permitted by the samples.

CATEGORICAL RANK OF CHARACTERS

It would be incalculably valuable to taxonomists if characters or, more strictly, the differences between them could be assigned fixed categorical

---

*There are of course striking unit *characters* that cannot have any intermediate condition and so cannot intergrade in this sense, but the *populations* exhibiting them can still intergrade through a population showing both or all conditions. Such a character is the dextral and sinistral coiling of shells and Crampton's classic study (1932) shows that intermediate populations do occur.

value; for instance, if a difference between individuals of 20 per cent in size could be taken as prima-facie evidence that they belong to different species, or a difference in tooth formula that they belong to different genera. Many zoologists have believed that they could assign such values and have proceeded on this belief. The search is reminiscent of that of the alchemists for the philosopher's stone. It has been an attempt to find an easy way to do something that can, indeed, be done, but not by easy methods.

In the first place, the belief that a morphological distinction of given degree or kind constitutes in itself a fixed rank of diagnostic distinction in taxonomy is inconsistent with modern taxonomic principles because it approaches the problem on inadmissible premises. Morphological differences are used to describe and distinguish taxonomic groups, but it is the groups of organisms that are being classified, not the morphological characters themselves. That the distinction is important and not particularly subtle is readily evident from a simple example. Suppose that one were introduced to all the inhabitants of a village and asked to determine their relationships to each other. Obviously, possession of the same shade of hair would not indicate siblings; possession of slightly different shades of one color, cousins; and of different colors, members of unrelated families. But, just as obviously, hair color would be a datum essential for solution of the problem.

Every working taxonomist knows that some morphological differences do tend to be diagnostic of certain levels of classification, but the problem of determining this correspondence is essentially empirical and the values are properly assigned only *a posteriori*. Once so determined, there may be a degree of probability, not certainty, that similar values can be assigned *a priori* in studying allied forms. The basic data for the problem are extensive observations of the variation that can and does appear in defined groups. Many such observations are available, but many more are needed and most of those that are available need better analysis and presentation.

A major distinction is often made between quantitative and qualitative characters and examination of many diagnoses shows that subspecies and species are usually defined in quantitative, but genera and higher groups often in qualitative terms. This has been advanced as a rule, and there has even been argument as to whether "numerical" or "morphological" characters are better or more reliable in taxonomy (e. g., Ehrenberg, 1928). The distinction between quantitative and qualitative characters is real but by no means absolute. The difference is often a mere matter of convenience. For instance the absence of a premolar in a mammal is customarily treated as qualitative, but the closely analogous absence of a scale in a reptile is often designated quantitatively. Many characters are treated qualitatively—e. g., strength or prominence of a crest on a tooth—simply because an easy means of measurement has not been devised, although the character may be more quantitative than qualitative in its real nature. With some exceptions it might be said that structures differ only quantitatively and that the only true qualitative differences are in the appearance of wholly new structures or the total loss of old. The practical distinction in study may be rather a matter of technique than of any real biological difference.

The characters of high taxonomic categories may show more obvious and constant distinctions than those of genera or smaller groups. The higher categories originate in ways that assure that such distinctions will exist genetically and the sharpness of the distinction arises mainly by slow accretion from species to species and by the disappearance of the transitional stages. On the level of genera close enough to each other not to represent different monotypic higher categories, this process is just beginning and the morphological distinctions may differ little in degree and not at all in kind from those that may appear between species, subspecies, or even individuals in one subspecies.

The fundamental point is not so much the distinctiveness of unit characters as their distribution in the groups; these groups, and not the characters, being the objects of classification. For instance, such a character as union or disunion of ectoloph and metaloph in a mammalian molar usually becomes widespread and relatively invariable within species, genera, or even families; hence, it is a character of those categories and is often felt to guarantee high categorical rank, but it may also vary within one interbreeding population and hence not be a so-called taxonomic character at all (Simpson, 1937a). Or, again, exactly the same morphological character may have different systematic value at different times in the history of one phylum. For instance, the most striking diagnostic character between the early Miocene horses, primitive species of *Parahippus*, and those of the preceding stages, *Miohippus* and advanced species of *Mesohippus*, is the uniform presence of a crochet on the upper molars of *Parahippus*. This is a generic character of that genus. But in transitional species of *Mesohippus-Miohippus* and in *Miohippus* the same character may appear as a variation within highly localized samples, purely individual variants of one subspecies, or may be a variable but useful average character of a subspecies or a species (Schlaikjer, 1935; Stirton, 1940).*

If subspecies become species by isolation and species become genera by divergence and diversification, it is inevitable that diagnostic characters should thus appear as individual variations and tend gradually to become subspecific, specific, then generic characters, and that no particular kind of character should be characteristic of a particular taxonomic level. Paleontological examples of this process demonstrate that this sort of sequence does certainly occur at times and may be typical of evolutionary history. They thus demonstrate that geneticists of the cataclysmic school, like Goldschmidt (1940), are wrong for some cases, and therefore are wrong in general, because they claim that this process never occurs. There are, nevertheless, at least three frequent distinctions between intra-specific and extra-specific diagnostic characters. First, certain differences rise above the specific level less often than others; second, the accumulation of intra-specific differences normally gives rise to extra-specific differences of greater degree although not of different kind; and third, the whole number of significant differences is likely to be larger for higher categories than for lower.

As an example of the first sort, the most obvious of all animal characters, that of gross individual size, may be cited. Genera do usually differ in the size range covered by their species, just as species and subspecies usually differ in the size range of the included individuals. Thus size is frequently a generic as well as a specific character, but as a rule it is not a good diagnostic character for genera and so is not often used in defining them. An important technical reason for this is that genera (unless monotypic) do not have a true average size. From a given sample of a subspecies it is possible to observe and to measure a central tendency as regards size. Such a tendency does not necessarily exist for a genus and if it does exist it cannot be measured or estimated from one sample or in any other simple way. A more obvious reason is that related genera often tend to overlap very widely in size range, each with species more or less covering the optimum size ranges over a variety of local conditions, so that the sizes confined to one genus tend to be typical only of one or a few of its species and hence to be used for specific, not generic, diagnosis. This real but limited phenomenon is perhaps the origin of the too generalized dictum that quantitative characters are specific and qualitative, generic. Yet practical tribute to the fact that

---

*Of course it is possible to argue in the other direction and to say that an individual *Miohippus* with a crochet belongs *ipso facto* to a different genus because a crochet is a generic character. This is a common tendency among practicing zoologists and is, indeed, what Schlaikjer did in this particular instance. For reasons expressed in this paper, I am convinced that this reasoning is fallacious.

genera do differ in this respect is paid by the zoologist who, when he finds a species far beyond the size range of known genera, immediately looks for, and usually finds, justification on other grounds for placing the aberrant-sized species in a new genus.

The quantitative increase of differences by increments, the number and hence the total of which is roughly proportional to taxonomic rank, is a factor known to and used by every zoologist. Its cause is related to such factors as sizes of mutations and rates of evolution. By experience, it is learned that differences of a certain degree cannot arise, or at any rate have not arisen in any known case, in a single step or within an interbreeding population. It is therefore justifiable to conclude that, when such differences are observed, they represent an accumulation of smaller differences such as accompanies divergence of specific or greater rank. Brachyodonty and hypsodonty in mammals provide a clear example. Relative height of cheek-tooth crowns varies in all groups, but only within narrow limits in one population. As far as has been determined empirically (and up to now such determinations are necessarily empirical), a fully hypsodont tooth cannot arise from a brachyodont tooth in one or a few steps. Hypsodont and brachyodont mammals never belong to the same genus. This implies not that height of crown is inherently a "generic character," but that the terms designate a high degree of difference that has generic or greater diagnostic value but is only the extreme on a continuous scale, on which lesser differences may be specific, subspecific, or individual. The generic degree is a simple sum of lesser degrees of difference.

The same element of rate of change is involved in the belief that adaptive characters are diagnostic of lower taxonomic groups than inadaptive, or habitus characters than heritage characters (e. g., Gregory 1936). The higher taxonomic value of heritage characters is, in fact, a matter of definition rather than of any independent or esoteric biological relationship. Characters that have evolved slowly or not at all and that have been passed on from a more or less remote common ancestry to diverse descendents are by definition heritage characters and, also by definition, are characters of high taxonomic rank—indeed, "heritage" in this sense is merely another name for characters of high rank.

Adaptive characters are by no means confined to the diagnostic level of low taxonomic groups. For instance, the streamlined contour of cetaceans is obviously adaptive and it characterizes a whole order (or whole cohort). The point involved is not directly the adaptive nature of the character so much as its rate of evolution, in which the adaptive relationship to the environment is one of several important determinants

If a potentially variable character is rather narrowly adapted to an environmental condition and if the environment changes or the group invades a new environment, then the character may also change relatively rapidly and differences in it may be diagnostic of minor units such as subspecies or still more localized groups. Pelage in protectively colored rodents (Dice, 1940) and other mammals is a typical example and such characters are largely responsible for the impression of zoologists that subspecies tend to differ in "superficial" ways of no higher taxonomic significance. If, however, the response of an adaptive character is more sluggish—for instance, from rarity of mutation, slight variation, or breeding structure crossing a very large population—a difference that does appear is likely to be of higher taxonomic value. The same will be true, as in cetacean body form, if the adaptation corresponds with a larger range of environmental conditions or if the pertinent environmental condition does not change.

The fact that higher taxonomic categories tend to differ in more characters than do lower categories is another aspect of much the same sort of evolutionary phenomena. Groups like subspecies that have some genetic transfer are not likely to differ markedly unless in the average conditions of characters of more immediate survival value. Once genetic transfer ceases and species become distinct, even pure chance tends to multiply the number of differences and this will become more extensive the longer the separation. This is a restatement of one of the reasons why morphological categories do tend to approximate genetic categories and why the foundation of a so-called phylogenetic classification on purely morphological data is justified.

NEOZOOLOGICAL AND PALEOZOOLOGICAL MATERIALS

The rest of this paper is devoted to some special problems and procedures more particularly related to the study of fossil vertebrates. Paleontology is as much a part of zoology as is the study of recent animals—a point here emphasized by using the names paleozoology and neozoology for the two major divisions of the subject. This subdivision arises from the different nature of available observational data, not from any fundamental difference in the aims of study or from any logical dissection of the science into "dead" and "living" parts. The principal differences in materials and data are as follows:

1.—Paleozoological specimens are all dead. So, in most cases, are the neozoological specimens used in taxonomy. Studies on the physiology and genetics of living animals have been made for such a very small number of species that they do not constitute the data of neozoological taxonomy but are only examples useful in interpreting data derived from dead specimens. They serve the same purpose for both paleozoological and neozoological taxonomists.

2.—With unimportant exceptions, paleozoological specimens comprise only bones and teeth, and generally only a fraction of these for each individual. Neozoological data could include the whole anatomy of each animal, but they seldom do in practice. At least through the generic level, neozoologists base classification almost exclusively on external characters. Ichthyologists and herpetologists do, as a rule, collect whole animals, but rarely use internal characters on these levels of classification. Ornithologists collect only skins for ordinary taxonomic purposes. Mammalogists collect skins and skulls. The use of mainly external characters by neozoologists and of exclusively internal characters by paleozoologists is the most striking difference in their materials.

3.—Neozoologists sometimes have larger samples than paleozoologists. The difference is not as great as might be supposed. There is no absolute criterion as to what constitutes an adequate sample for taxonomic purposes, but I would judge that samples of more than ten specimens will usually suffice to establish the reality and basic distinctions of a subspecies, if efficiently used and analyzed. More than fifty specimens, analyzed with equal efficiency, will usually permit a virtually complete definition. Such definition is exceptional both in neozoology and in paleozoology. As a fairly typical example, in a recent neomammalogical revision competently based on all the pertinent materials of several great museums (G. M. Allen, 1938, 1940), the samples studied by the author were inadequate (ten or fewer specimens) for 58 per cent of the unit groups (subspecies and monotypic species), adequate but incomplete (eleven to fifty specimens) for 31 per cent, and fully sufficient (over fifty specimens) for only 10 per cent. There does not happen to be a closely comparable recent paleomammalogical revision giving such data as to sizes of samples, but a census of materials for a characteristic Tertiary fauna (Lebo) shows: inadequate, 65 per cent; adequate, 31 per cent; fully sufficient, 4 per cent. Collections for some of the well-known later Tertiary faunas have fewer inadequate unit samples and those for some recently discovered or poorly collected faunas have more. In general it is evident that paleomammalogists have about the same percentage of adequate samples but have fewer large, relatively complete samples.*
In both fields the sizes of available samples are steadily increasing. For

---

*It is only fair to add that the "classical" methods of mammalogy are so inefficient that they do not, as a rule, obtain any more information from samples of two or three hundred specimens than could be obtained by efficient use of twenty or thirty. Thus the neomammalogists have derived little real advantage from their large samples. Of course there will be improvement in this respect and the large samples are available for it.

some of the groups other than mammals, students of fossils are at a greater disadvantage. For instance there are only one or two species of dinosaurs for which adequate samples have been collected. For this reason, if for no other, the specific taxonomy of such groups is highly unreliable or practically non-existent.

4.—Neozoological samples can and frequently do cover the whole areal range of the included groups and the whole number of distinct groups present in a given area. Paleozoological samples practically never do either. This is much the most important permanent disability of paleozoology. It means that paleozoology can contribute relatively little to some taxonomic problems, especially those having to do with geographic variation and horizontal subspecies.

5.—Paleozoological samples can cover long periods of time. This is the great advantage of paleozoological data. The neozoologist's universe has no time dimension. For all practical purposes his subjects are all contemporaneous. If the time observable suffices for any significant change in his populations, this change does not proceed beyond the lowest levels of subspecies or still smaller groups. He may thus take the discontinuities between species, genera, and higher units as absolute, a fact that greatly simplifies the taxonomic task but that removes from his direct observation a fundamental goal of research: the mode of origin of these discontinuities. The paleozoologist, on the other hand, is as much concerned with temporal as with geographic sequences. He is so constantly dealing with time that his whole manner of thought is affected by it. The added dimension greatly complicates his practical task of classification, but at the same time gives him a direct approach to major evolutionary patterns and processes.

These differences in materials for study not only necessitate differences in procedures of classification but also induce differences in attitudes toward classification and toward the broader problems of taxonomy. They emphasize the desirability of an understanding of both neozoological and paleozoological contributions to these problems and the absolute necessity of a synthesis of the two fields for further progress toward their solution.

With differences caused by personal taste and knowledge, all zoologists recognize the same classes, orders, and families. These can all be defined either in paleozoological or in neozoological terms and a family of fossils is the same sort of thing as a family of recent animals. When it comes to the levels being discussed in this symposium, genera, species, and subspecies, this equivalence can no longer be taken for granted. As regards these low taxonomic ranks there is real room for question whether a single, unified zoological system is an attainable ideal or whether paleozoological and neozoological classifications must forever be different. The problem of vertical units enters into this, but it is distinct and will be discussed separately. The present question is whether the horizontal genera, species, and subspecies of paleozoology and neozoology are or can be units of about the same real taxonomic rank. The question is an old one. More than a century ago Hitchock (1836) wrote: "When I speak of species here, I mean species in oryctology [paleontology], not in ornithology. And I doubt not, that in perhaps every instance, what I call a species in the former science, would be a genus in the latter." Zoologists are still wondering whether this may not be true.

Genera

For genera, it is possible to give a definite answer: paleogenera and neogenera (if I may be permitted a self-explanatory barbarism) can be exactly equivalent and usually are approximately so. I have yet to see a genus of recent mammals, or, at least, one that had any good chance of being valid and having this rank in a reasonable classification, that could not be recognized from a single specimen of the skull. Most of them can be recognized from a single jaw, and many from a single tooth. Thus neither incompleteness of individual specimens nor small size of samples prevents a paleozoologist from recognizing neozoological mammalian genera. Studies of Pleistocene faunas, in which neozoological genera occur as fossils, provide an experimental check on this and confirm the equivalence and recognizability. Among some of the lower vertebrates identification of recent genera sometimes requires more complete knowledge of the skeleton, but even in these groups the usual paleozoological data generally seem to suffice for this purpose. Paleozoologists frequently discover that they can recognize genera from skeletal characters that are not used by (and are often unknown to) the neozoologists who founded the genera. For instance, neornithologists never define genera on osteological characters, but paleornithologists have no serious difficulty in referring Pleistocene bird bones to the proper recent genus (e. g., Howard, 1930).

It is, nevertheless, probable that paleozoologists tend to draw generic lines somewhat more broadly than do neozoologists, although in both fields the tendency is obscured by great differences between individual workers. If real, the tendency is not a matter of necessity, and certainly not a matter of erroneous interpretation, but one of taste. The paleozoologist has, on an average, fewer species and subspecies within a genus of given scope but has a greater variety of generic and, especially, higher groups and is much concerned with clear expressions of relationship on upper taxonomic levels. The use of rather inclusive genera is therefore more convenient and natural for him. The neozoologist has greater numbers of subspecies in a given morphological range and is mainly concerned with the arrangement of these and other low taxonomic groups, so he tends to use smaller and smaller genera. For instance, the great number and small scope of genera of recent murid and cricetid rodents commonly recognized by neozoologists seem to the paleozoologist unjustified by the morphological (or probable genetic) facts.

My own sympathies in this battle of the lumpers and the splitters naturally incline toward the broader, paleozoological sort of inclusive genera, but a compromise is possible and may eventually prove acceptable to both. I am informed (by Dr. Mayr) that the ornithologists, having once broken all their genera into small fragments, are now engaged in putting them together again. Perhaps the neomammalogists will soon enter this new lumping phase. As for the paleomammalogists, they may have the splitting phase still to suffer, but may happily be spared its most extreme form.

Species

For many years paleozoologists have gone on naming hundreds and thousands of species, some of them never doubting that these were true species in the neozoological sense and most of the rest not caring whether they were or not. A note of doubt has been heard from time to time, more frequently in recent years. Some of the best paleozoological taxonomists (e. g., Jepsen, 1933; Scott and Jepsen, 1936) have even suggested that it may not be possible to recognize true species (in the neozoological sense) among fossil vertebrates, at least for the present.

There are really two questions. Are paleontological species real groups? Are they approximately equivalent to taxonomic species as defined on a previous page from a more neozoological point of view? Like neozoologists, and with more excuse, paleozoologists have created an enormous number of synonyms and have also given many names to what may have been but cannot be shown to be real and newly discovered groups. Fossil species are usually first described from fragments, sometimes not homologous with the known parts of related forms, and many supposed species have been based originally on single specimens. These are inevitable results of the nature and history of paleontological discovery. Earlier paleontologists had no real idea of the extent of morphological variation that can occur in a single species and workable criteria have only slowly been achieved, hand in hand with similar work by neozoologists and with experimental work. It is conservative to guess that among previously proposed species of fossil vertebrates, aside from types of currently recognized genera, not more than a quarter represent natural and distinct groups. The fraction of valid species is probably much lower.

This situation is certainly deplorable, but it exists to greater or less degree in every field of zoology. It can be and is being cleared up by revision with enlarged samples and by the use of more rational and objective criteria for intraspecific variation, especially those based on the statistical relationships of samples and populations. No matter how many mistakes may have been made in the attempt to do so, there is little real doubt that the groups called species in competent recent paleozoological studies tend to correspond with real units of population in nature.

Strict application of competent modern methods to adequate samples should seldom or never result in the recognition of false groups (aside from the element of human error by the student), but it may fail to distinguish two groups properly of specific rank and it may result in calling groups species when their true rank may be higher, e. g., genera, or lower, e. g., subspecies. The extent to which this may result from peculiarities special to paleontology is a measure of the degree in which the species of paleozoological taxonomy are likely to differ from those of neozoological taxonomy.

The fact that a neozoologist could and that a paleozoologist could not have recourse to actual breeding experiments in a critical case has little practical significance at present. How many vertebrate species have, in fact, been defined experimentally in a way importantly different from previous morphological definitions? All will admit that the number is insignificantly small. Larger size of unit samples for recent animals is also a minor consideration. Although it does impede the paleozoologist in certain particular cases, it is neither a general nor a necessary disability. The better spatial distribution of neozoological samples is more pertinent to the recognition of subspecies, but as regards species it may tend to have an effect opposite to that usually supposed: it is likely to make the paleozoological species a smaller, not larger, unit than the neozoological species. Terminal groups that are shown by intervening samples to be subspecies of a single species may frequently be mistaken for species if the intervening samples are lacking, which is more likely to be true in paleozoology than in neozoology.

The important difference between paleozoological and neozoological taxonomic species—if such difference does necessarily exist—must arise from the availability and use of different morphological characters. It is a common belief, on both sides, that paleozoologists cannot recognize neozoological species because the latter are defined by external characters. This statement contains a glaring (but sometimes unnoticed) fallacy: it rests on unstated and unproved postulates. These postulates are that external and internal characters are fundamentally different in taxonomic value, that their variations do not tend to be closely correlated, and that their population distributions are significantly dissimilar. Making these assumptions by implication and without conscious analysis is unscientific. They demand more careful statement and study than has yet been given them.

There is, as far as I know, no evidence that external and internal characters have any different genetic basis or involve any different hereditary processes. On the contrary, all the pertinent data suggest that they are exactly analogous in this respect. Can two organisms differ only in external, or only in internal, characters? Since it is theoretically possible for organisms to differ only in a single character, the answer to this question is evidently "yes," but the question does not mean very much for taxonomy. Taxonomy is more pragmatic and is not concerned so much with what can as with what does happen. Distinct populations that differ in only one or two characters are so unusual as to have almost no bearing on practical taxonomy. Even a single difference in alleles commonly results in more than one phenotypic "unit character" difference. In nature most distinct populations differ in average values of dozens or of hundreds of morphological peculiarities.

The chances that all such multiple differences will be either external or internal can be worked out for various sets of permissible postulates. For instance, if only ten differences are involved and these are as likely to be internal as external, there is less than one chance in one thousand that all will fall into a particular one of these categories. From such considerations, it is easily shown that the chances of all such differences being external are quite negligible unless the differences are very few in number, unless there are many more distinguishable external than internal characters, or unless external characters are inherently much more likely to vary than internal characters. It is an established matter of experience that such conditions, if they ever obtain for the characters available for distinction of species, are certainly highly unusual.

Regardless of these factors related to the genetic incidence of differential morphological characters, it could still happen that the taxonomic significance of external and internal characters was different. First, the segregation of the two could conceivably be different, so that populations defined in terms of one would have distinctly different boundaries from those defined in terms of the other even though comparable in scope. This possibility need not be considered in detail because it follows from the similar hereditary determination of the two and from the basic definition of species in terms of breeding structure that no consistent tendency toward such a difference can exist, even though some difference of the sort might occasionally arise by chance and at random. Second, it is conceivable that some non-genetic factor, notably natural selection, might operate so much more strongly on one than on the other that populations defined in terms of the first would be more sharply delimited and smaller in scope than those defined in terms of the second.

This is the most essential point involved in the present problem. It deserves a great deal more attention than it has ever received and more detailed discussion than can be included here. Data for checking it empirically and objectively are scanty, but some are available. TABLE 1

TABLE 1

VARIATIONS OF CHARACTERS IN *Peromyscus*

A. THREE TYPICAL "INTERNAL" CHARACTERS

Stock	Linear Bone Dimensions					
	Mandible		Condyle-Premaxilla		Bullar Width	
	M	SR	M	SR	M	SR
Alexander, Iowa....	15.36	14.04–16.68	22.81	20.50–25.12	10.26	9.17–11.35
Moville, Iowa......	17.39	16.05–18.73	25.62	23.79–27.45	11.25	10.47–12.03
Greenland, N. H....	16.72	15.42–18.02	25.05	23.24–26.86	11.02	8.45–13.59
Vineyard Haven, Mass............	17.72	16.43–19.01	26.26	23.95–28.57	11.40	10.71–12.09

B. TYPICAL "EXTERNAL" CHARACTER

Stock	Tint Photometer Readings on Dorsal Stripe					
	Red		Yellow		Green	
	M	SR	M	SR	M	SR
Alexander, Iowa....	5.10	1.21–8.99	4.25	1.01–7.46	3.67	1.08–6.29
Moville, Iowa......	8.59	3.43–13.75	7.07	2.57–11.57	5.75	1.44–10.06
Greenland, N. H....	8.65	5.63–11.07	7.44	5.02–9.86	6.36	3.94–8.78
Vineyard Haven, Mass............	8.03	3.13–12.93	6.67	2.55–10.79	5.27	2.44–8.10

M is the mean, as given by Dice and SR is standard range from standard deviation (see Simpson, 1941), calculated from Dice's data, and here expressed by limits rather than by span.

is a good example, based on Dice's observations (1937a, b, 1939) on uniformly laboratory bred stocks of *Peromyscus* from different localities.

Although it is unlikely that there is any genetic or any selective correlation between skull dimensions and color of pelage, it is evident that the inferences as to resemblances and differences of these four populations would be the same whether based on the internal or on the external characters. On the basis of either sort of character, the two geographically close stocks from Iowa differ more than do the three geographically scattered Moville, Greenland, and Vineyard Haven stocks. The Moville and Greenland stock are particularly similar, almost indistinguishable, and the Vineyard Haven stock is a little more distinctive. The mammalogists' taxonomic expression of these facts is to place the Alexander population in *Peromyscus maniculatus* and the other three in a different species, *Peromyscus leucopus*. The Moville and Greenland populations are local variants of one subspecies, *Peromyscus leucopus noveboracensis*, and the Vineyard Haven group has been placed in a different subspecies, *P. leucopus fusus* (although it may be noted that this arrangement rests more on the relatively sudden transition to adjacent *P. l. noveboracensis* than on the degree of difference—a point in subspecific definition not particularly pertinent at this point in the discussion).

Data on other variates of these samples, on many other samples of allied and intervening populations, and on a wide variety of other mammals support the same generalization: natural populations of mammals tend to differ about as much and about as sharply in skeletal characters, such as dimensions of skull and teeth, as in external characters, such as pelage color, and the groups based on the two sorts of characters by analogous methods of inference tend closely to coincide. There are a few real exceptions to this generalization and there are supposed exceptions that are more apparent than real. Among the latter may be counted such classic cases as the lion and tiger or the horse and zebra, so obviously different when seen in the flesh and so apparently similar when seen as skeletons. But this is only a matter of ease of observation. The species are really distinct in osteology also, when the skeletal characters are closely and correctly analyzed, and the osteological species are the same in scope as the pelage species.

It may be concluded that paleozoologists can, as a rule, recognize the same sort of species as do neozoologists. It is undoubtedly true that they frequently do not do so. This is due in part to personal factors— certainly not all neozoologists recognize the same species; in part to the fact that specific taxonomy is a retarded study in paleontology, only now slowly emerging from the days of rule-of-thumb and excessive subjectivity; and in part to the paucity of well-distributed paleozoological samples. I believe that the tendency in the two sciences is for their specific concepts to become more and more similar, although this question of sample distribution does involve an important difference more explicitly stated at the end of the following discussion of subspecies.

Subspecies

Few vertebrate paleontologists ever propose subspecies and those who use them at all do so sparingly. Checking the last 25 papers on fossil mammals to reach my desk, I find that 24 of them do not so much as mention subspecies, although they include some large faunal revisions based on excellent collections, and that 49 new species are proposed and only 2 subspecies. In contrast, 25 recent taxonomic papers on living mammals, taken at random, propose 6 species and 6 subspecies. The disproportion would have been greater except for the accident that one of these papers is by one of the few students who habitually proposes subspecific names for fossils. It is also significant that both the supposed new subspecies that are described are recorded as occurring in direct association with much more abundant remains of the typical subspecies of the same species. They receive separate designation only because they seemed to the author too far from the average condition to be called typical, the criteria used being entirely subjective as far as shown. The chances are great that these are not subspecies at all but are either artificial groups based on individual variants within the typical groups or are distinct species.

The number of true subspecies in the paleontological literature, that is, of subspecies that are really analogous in structure and scope to those of sound neozoology, is certainly extremely small. Practically speaking, it is not too much to say that paleozoology does not treat with subspecies in this sense. We may briefly discuss why this is true now and whether it is necessarily and permanently true.

From the considerations summarized in the preceding discussion of species, it is fairly well established that paleozoologists could recognize, and would tend to arrive at, the same sort of subspecies as neozoologists if they had samples analogous in size and distribution and if they treated them in a similar way. Neither of these conditions has been commonly satisfied in the past.

The very fact that their materials usually consisted of a few isolated specimens during the early and classical periods of their science meant that paleozoologists perforce developed the attitude and methods of comparison of individual specimens. Their development of group concepts and of the methods of population inference from samples has been retarded; indeed even now many active paleozoologists hardly understand what these words imply. In some respects the higher the taxonomic group, the less subtle and difficult are the methods of group inference. Moreover work on the higher groups seems more important and it is a field that neozoologists have recently tended to underemphasize. Thus paleozoologists have not made any serious effort to study such small groups as subspecies, partly because they did not know how and partly because they did not want to spend time on what they felt to be a relatively unimportant subject.

These differences of method and aim are mainly historical and psychological and are not likely to persist except as they are forced by necessity. The fact is, however, that paleozoologists do not have the materials for recognition of the horizontal subspecies of most of their species and that they are not likely ever to have them except in a minority of special cases. Such subspecies can only be truly defined on the basis of a considerable sequence of adequate, contemporaneous samples scattered over most of the range of the species. Paleozoologists have few sample sequences of this sort (with an increasing number of exceptions mostly in the Pleistocene), and it follows from the conditions of deposition and preservation of fossils that they will never have adequate sequences for a majority of their species.

They do have now a few and certainly will have many more series of scattered samples that cover not the whole areal and ecological range of a species but at least the range of more than one subspecies. These permit and will eventually necessitate the use of the subspecific category in paleozoological taxonomy, work that certainly can be but has not yet been properly done. Even in such cases, however, it is quite impossible to guarantee the contemporaneity of dispersed samples within a few millenniums. This introduces temporal as well as spatial variation and hence there is an element involved in the relationship of taxonomic to phylogenetic paleozoological subspecies that is necessarily somewhat different in kind from this relationship in neozoological subspecies.

It results from the sampling conditions that a majority of fossil species, as they are known and definable, are essentially monotypic for any one time, that only one subspecies is usually represented by the available demonstrably contemporaneous samples of a species. It follows, with apparent paradox, that the paleozoologist, who rarely mentions subspecies, deals almost exclusively with subspecies rather than truly dealing with species. The group that he can and does really envision from his samples is usually a subspecies, but since no other subspecies of the same species is usually known to him, this inferred subspecific group is taken to represent a species. It is, in fact, the species of most paleozoological taxonomy. Since it is probable that most extinct species did have two or more subspecies in nature, this involves a systematic difference between paleozoological and neozoological taxonomy. Having only one subspecies in each species (as a rule), the paleozoologist has no occasion to distinguish subspecies of one species, but only to distinguish subspecies each of a different species. Thus he may be and usually is defining subspecies, although this has so little evident importance or practical significance that it is not even noticed, and he is diagnosing (not really defining) species.

VERTICAL UNITS

The great, special problem of paleozoological taxonomy is the definition and subdivision of units that have an extension in time as well as in space. If species arose discontinuously, as claimed by Goldschmidt and a few others, the problem would not exist, or at least it would be no different from the recognition of the *de facto* discontinuous species of horizontal classification. In such a case, there would be a definite point in time when each new species arose and the specific boundaries would be real and visible, as much so as if species were invariable units resulting from divine creation. Then subspecies would also be without gradations into any units except other subspecies and, with sufficient samples in hand, should be absolutely definable as single phylogenetic branches even though they changed slightly in time. Although somewhat less clear-cut, genera, too, would be rather easily definable; they might not arise full-blown at one step, but could at least be separated at a sharp interspecific boundary.

From a practical point of view this pleasingly simple situation does actually exist for a considerable part of paleozoological taxonomy, whether or not this theoretical reason for it be accepted. A large number of species and genera do appear suddenly, without closely similar predecessors, in the paleontological record and do disappear just as sharply, without known immediate descendents. As long as this condition of the evidence exists, classification is simple and requires no special consideration of the time dimension, whether the condition is supposed to arise from the non-existence or from the non-discovery of intermediate stages. A century ago, this was so universally the case with known species that paleontology was regularly cited as evidence against the theory of organic evolution. The apparent sudden origin of species is no longer universally true but it is still true often enough to be cited—not, as a rule, by paleontologists—as evidence against the continuity of morphological evolution. Whatever may be held as to the causes of the breaks or as to the universality of continuity, there are now many known examples of longer or shorter sequences in which there is no definite and real discontinuity, each population more or less overlapping in variation those preceding and those following it, but which change so much that every taxonomist places the earliest and latest members of the line in different species or genera. Clearly a species as a subdivision of such a temporal, or vertical, succession is quite a different thing from a species as a spatial, or horizontal, unit and cannot be defined in the same way. The difference is so great and, to a thoughtful paleozoologist, so obvious that it is proper to doubt whether such subdivisions should be called species and whether vertical classification should not proceed on an entirely different plan from the basically and historically horizontal Linnaean system.

In line with these theoretical or philosophical misgivings, various distinct terms have been proposed for successive stages within a single vertical line. Of these "mutation," as proposed by Waagen (1869), is most important and it alone has been very widely used. Unfortunately the geneticists, most of whom seem to be unaware that this is the prior and historically correct use of the word "mutation," have used the same term with a sharply different and yet obliquely related meaning. A great deal of misunderstanding and of mutual irritation between geneticists and paleozoologists has resulted. Now the paleozoologists may just as well abandon their priority and admit that the geneticists have carried the day and have succeeded in purloining the word. We have no more chance of retrieving it than the Indians have of retrieving Manhattan and the sooner we abandon it completely, the sooner we will be able to express ourselves intelligibly to the geneticists, and in turn to understand them without unnecessary difficulty.

So far none of the varied proposals for non-Linnaean arrangement and nomenclature of vertical units and their successive subdivisions has been widely accepted and none seems promising at present. This is not the place to go into detail as regards the reasons for this failure or the serious and intricate difficulties of the problem, and it is emphatically not the occasion for making any new proposal of this kind.

We do, in fact, use the same system of nomenclature for subdivisions of a vertical line as for subdivisions of a horizontal distribution. The tendency of paleozoological practice is this: *successive taxonomic units are inferences as to morphological units such that the net difference in morphology between corresponding parts of those units is of the same order as that between horizontal units of the same rank in the same or allied groups.* For instance, in the main line of horse evolution, the average difference in structure between successive genera tends to be of about the same order as the average difference between closely allied but distinct contemporaneous genera of perissodactyls. No claim can be made that this practice is perfect or even that it is theoretically desirable, but it is what paleozoologists really tend to do, it works fairly well for them, and no one has yet proposed a system generally believed to be more promising.

The line between such units within essentially continuous sequences must, of course, be arbitrary. Although the average difference between successive genera or the difference between their medial or central (not necessarily typical in a technical sense) species is comparable to that between contemporaneous genera, the difference between the last population placed in one genus and the first placed in the next may be, and in the postulated circumstances must be, comparable to that between contemporaneous subspecies or still more nearly related groups. Although specialists usually seem to reach some sort of working agreement as to the most convenient approximate position for these artificial boundaries, their exact position is determined by no general rule or criterion and naturally remains subject to doubt and dispute.

In practice the boundary is usually placed in one of two ways. Quite commonly, the sequence in question was not really continuous when its successive units were first defined—that is, knowledge of it was not continuous. The gaps in knowledge were then used to separate the units. When the series is filled in, the boundaries are inevitably placed somewhere within what were the gaps. If the accidents of discovery had first revealed other points along the line, the eventual subdivisions of the completed continuous sequence would have been in quite different positions on it. If the gap was large, there is a strong tendency for the worker whose material fills the gap to define a new unit rather than to refer his novelties to one or the other of the adjacent units, even when such a step is not warranted by the pragmatic rule of comparable morphological scope in units of the same rank. This human tendency to seek the solution of problems by evading them makes more difficulty in the end because eventually it only means that two boundaries have to be settled where only one existed before, or was necessary. Hence arises much of the needless splitting and complication in paleozoological nomenclature.

The second and somewhat more rational but not always more practical way of arriving at such boundaries is to select some feature or features of essential importance characterizing a genus or species and to draw the line where this character becomes dominant or universal in the evolving population. The character may be selected because it is easily observed and helpful to a hurried taxonomist in his capacity as sorter and cataloguer of specimens, e. g., the crochet in *Parahippus*. Or it may be a character of primary selective value in the economy of the animals concerned, e. g., cement in *Merychippus*. Usually both factors operate in varying degree. Even such characters do not arise all at once and they still define a region rather than a point in the sequence, but if they are well-chosen they certainly provide the best possible basis for such subdivision. This sort of criterion is not, however, available in all cases, or even in most. It generally involves the appearance of a new structure or character and this is a rare event in evolution compared with the gradual modification of an existing structure. Thus the criterion can seldom be used to distinguish successive species, which are much more numerous than are the new structures involved in their evolution— another reason why so-called qualitative characters are commonly believed to have inherent superspecific rank.

Seen in narrower perspective, the taxonomically troublesome continuous vertical sequences of paleozoology are closely analogous to the horizontal sequences of neozoology called clines by Huxley (1938). Although no new principle or discovery is involved in this restatement of known facts, the terms of this restatement do involve a new way of looking at the facts and this will, I think, prove to have far-reaching and even fundamental effects on zoological theory. Sufficient study of this aspect of the matter requires considerable detail and I have in hand a special paper on it that I plan to present elsewhere. It is, however, sufficiently pertinent to the present subject to demand mention here and to warrant the inclusion of one illuminating example.

As originally proposed, the idea of a continuous cline is that of a sequence of contemporaneous populations arrayed geographically, and intergrading in progressively changing morphological characters. Although many such sequences probably also show some gradation that is solely phenotypic, the gradation is normally maintained by genetic exchange, actual interbreeding of adjacent segments of the population. Now if a sequence of successive populations is arrayed temporally, it normally shows a similar sort of continuous intergradation in progressively changing morphological characters. The cause is different (although not unrelated), but the objective sequences are so precisely analogous that I see no reason why these may not usefully be given the same name: clines. Clines may, then, be distinguished according to the variate that is used to define the array. In one case the arrangement is geographical and these may be called choroclines, i. e. "space clines." In the other, the arrangement is temporal, and these may be called chronoclines, i. e. "time clines."

The following example illustrates a typical chronocline and serves to point the contrast between defining a vertical chronocline, with various successive horizontal subdivisions, and defining "vertical species" in the usual sense. Data are given for only a single variate, but other variates in the same samples confirm the conclusion, some more and some less clearly. The named stratigraphic subdivisions from which the samples come are arranged in temporal sequence from left, Clark Fork, oldest, to right, Lost Cabin, youngest. *Ectocion* is a genus of primitive ungulates, condylarths, typical of this part of the North American faunal sequence. (See also Simpson, 1937b.)

The "vertical species" of the table represent routine identification by

TABLE 2
Ectocion IN THE UPPER PALEOCENE AND LOWER EOCENE OF WYOMING
AMERICAN MUSEUM SAMPLES

Length of M_1 in mm.	Frequency in Each Horizon				So-Called "Vertical Species"
	Clark Fork	Sand Coulee	Gray Bull	Lost Cabin	
5.3–5.6	1*	0	0	0	*E. parvus*
5.7–6.0	1	0	0	0	
6.1–6.4	6*†	3	1	0	*E. ralstonensis*
6.5–6.8	4	4†	7	0	
6.9–7.2	1	3	2	0	*E. osbornianus*
7.3–7.6	0	0	2*†	0	
7.7–8.0	0	0	0	0	*E. superstes*
8.1–8.4	0	0	0	1*†	
Ascending stages in unit phylum or chronocline.	*E. osbornianus ralstonensis*	*E. osbornianus complens*‡	*E. osbornianus osbornianus*	*E. osbornianus superstes*	True Shifting Vertical Species or Chronocline *Ectocion osbornianus*

*Type of "vertical species."
†Type of subspecific stage.
‡Completion of the example unfortunately requires the proposal of a new name at this time: **Ectocion osbornianus complens.** Type: Amer. Mus. No. 22498. Hypodigm: American Museum specimens of *Ectocion* from the Sand Coulee of Granger. Diagnosis: Ranges of all characters overlapping those of *E. o. ralstonensis* and *E. o. osbornianus* but means intermediate between those two groups; length of M_1 (ten specimens) 6.72 ± 0.11.

my predecessors at the American Museum and by me in earlier years. They follow what was, and still is to a large extent, standard practice in paleontology. Such practice is supposed to produce a classification in which the vertical, time element is taken into consideration, because the species are commonly given such a dimension and shown as running vertically through various horizons. In this case, and innumerable others similar in character, this supposition now seems to me naive. Study of the distributions as a whole shows beyond much doubt that the "species" of this system are purely subjective size groups. They do not tend to correspond with any real, defined populations that existed in nature and therefore they are not really species in any sense of the word. Instead of taking the effect of the time dimension into consideration, they ignore and conceal it.

Although it is always possible that some extraneous specimen has crept in, what seems really to have happened in this case is that the sample from each horizon was derived from an essentially homogeneous population. Each of these (more or less) contemporaneous populations appears to have been derived from that preceding it in the same general area and to have given rise to that following it. At any one time there was then only one species, apparently only one subspecies, and the genetically continuous, ancestral-descendant, series of populations gradually changed in morphology, most noticeably in size as shown by the exemplifying data of the table. The resulting picture is typically that of a cline, and in this case a chronocline.

Such a cline as a whole might be given any taxonomic rank in the Linnaean system, depending on how great is the morphological difference between known early and late members. In the example, my judgment is that the chronocline is of specific scope. This is supported by the fact that all the ranges of the unit populations overlapped. (The samples do not overlap in every character, but some of the samples, especially that from the Lost Cabin, are very small and the populations almost certainly did overlap.) Even in Lost Cabin times there were undoubtedly some Ectocions of this lineage that were as small as some of the larger individual variants in the long precedent Clark Fork and that could indeed hardly be distinguished from the latter if they were compared as individuals.*

---

*Here is a point that is a subject for criticism, not to say ridicule, among some paleontologists of the old school. If two specimens cannot be distinguished, except by trivial variations admittedly less than can occur within a subspecies, how can one maintain that they are taxonomically distinct? If a fossil can only be identified when its horizon is known, what becomes of the whole basis of paleontological correlation of horizons by the identification of their fossils? Both objections are based on the fallacious tendency to compare *individuals* when the correct comparison is of *groups*. The groups as such are here readily distinguishable even though some individuals are not. Valid inference of group characters requires some homogeneity as to time, and hence some specification as to horizon.

The chronocline is thus essentially continuous and is definable as a species, for which in this example the valid name is *Ectocion osbornianus*. This natural chronocline species is of course quite different in character and limits from the wholly artificial "vertical species" hitherto given that name. Since the species consisted of morphologically different populations at different times, it is both theoretically and practically valuable to have some means of subdividing it and of designating different stages in its development. It appears that each of the samples, sorted by collector's specifications, is distinguishable from the others in average characters, so that the number and character of the chronocline subdivisions are automatically determined by the stratigraphic subdivisions recognized and recorded by the collector. If he had used different stratigraphic units, the taxonomic units would also be different. The lines drawn between the different taxonomic subdivisions are in this sense arbitrary. It does not follow that the taxonomic groups are artificial or unreal: they are natural groups approximating populations that once existed in nature. In this they differ profoundly from the completely artificial "species" of the old classification.

Since the chronocline has been designated by a Linnaean binomial, its subdivisions may conveniently be designated by trinomials. This is also justified by the fact that their average scope, resemblances, and differences are fairly analogous to those of subspecies along a chorocline in neozoology. Although paleontologists only very exceptionally have materials permitting the recognition of subspecies such as enter into choroclines and are recognized by neozoologists, they frequently have materials fully adequate, when carefully analyzed by group methods, for the recognition of analogous units in chronoclines.*

In studying choroclines, some are found to have almost even slope (in graphic terms) from one end to the other, while some have distinct plateaus bounded by shorter steep slopes or narrow transition zones. Ideally, it is the latter phenomenon that permits the definition of well-defined and homogeneous subspecies. There is considerable evidence that a similar phenomenon occurs in chronoclines on a much larger scale. The chronocline analogue of the steep transition zone in a chorocline is a relatively brief period of relatively rapid evolution. Even within an essentially continuous sequence, an acceleration of this sort provides a definite and natural boundary zone (although not a line or point). For many reasons too complex to list here, it is clear that these zones of acceleration are least likely to be represented by fossils, and are almost sure

---

*And also, still more commonly, they have low-rank units, subspecific in morphological scope and distinctiveness, in which both time and space are a factor. This is a complication that need not be considered in the present summary.

to be more poorly represented than are the periods of more even and slower evolution. Thus it happens that the inevitable and at first sight merely accidental division of vertical units by gaps in the record does probably tend to approximate a real and important sort of division in phylogeny. This phenomenon apparently has little bearing on the lower levels of taxonomy but it has probably been active in relation to some genera and may usually be involved in the delimitation of higher taxonomic categories.

LITERATURE CITED

Allen, Glover M.
1938, 1940. The mammals of China and Mongolia. Natural History of Central Asia **11**: Pts. 1-2. Amer. Mus. Nat. Hist.

Allen, J. A.
1919. Severtzow's classification of the Felidae. Bull. Amer. Mus. Nat. Hist. **41**: 335-340.

Crampton, H. E.
1932. Studies on the variation, distribution, and evolution of the genus *Partula*. The species inhabiting Moorea. Pub. Carnegie Inst. Washington, No. **410**.

Dice, Lee R.
1937a. Additional data on variation in the prairie deer-mouse, *Peromyscus maniculatus bairdii*. Occas. Pap. Mus. Zool. Univ. Michigan, No. **351**.
1937b. Variation in the wood-mouse *Peromyscus leucopus noveboracensis*, in the northeastern United States. Occas. Pap. Mus. Zool. Univ. Michigan, No. **352**.
1939. Variation in the wood-mouse, *Peromyscus leucopus*, from several localities in New England and Nova Scotia. Cont. Lab. Vert. Genet. Univ. Michigan, No. **9**.
1940. Intergradation between two subspecies of deer-mouse (*Peromyscus maniculatus*) across North Dakota. Cont. Lab. Vert. Genet. Univ. Michigan, No. **13**.

Dobzhansky, Theodosius
1941. Genetics and the origin of species. 2nd ed. New York, Columbia Univ. Press.

Ehrenberg, K.
1928. Betrachtungen über den wert variations-statistischer Untersuchungen in der Paläozoologie nebst einigen Bemerkungen über eiszeitliche Bären. Pal. Zeits. **10**: 235-257.

Goldschmidt, R.
1940. The material basis of evolution. Yale University Press.

Gregory, William K.
1936. Habitus factors in the skeleton of fossil and recent mammals. Proc. Amer. Phil. Soc. **76**: 429-444.

Hitchock, Edward
1836. Ornithichnology. Description of the footmarks of birds (*Ornithichnites*) in New Red sandstone in Massachusetts. Amer. Jour. Sci. (1) **29**: 307-340.

Howard, Hildegarde
    1930. A census of the Pleistocene birds of Rancho La Brea from the collections of the Los Angeles Museum. Condor **32**: 81–88.

Huxley, Julian S.
    1938. Species formation and geographical isolation. Proc. Linnaean Soc. London **1937-38**: 253–264.

Jepsen, G. L.
    1933. American eusmiloid sabre-tooth cats of the Oligocene epoch. Proc. Amer. Phil. Soc. **72**: 355–369.

Mayr, Ernst
    1940. Speciation phenomena in birds. Amer. Nat. **74**: 249–278.

Pocock, R. I.
    1917. The classification of existing Felidae. Ann. Mag. Nat. Hist. (8) **20**: 328–350.

Schlaikjer, Erich M.
    1935. Contributions to the stratigraphy and paleontology of the Goshen Hole area, Wyoming. IV. New vertebrates and the stratigraphy of the Oligocene and early Miocene. Bull. Mus. Comp. Zoöl. Harvard Coll. **76**: 97–189.

Scott, W. B., & Jepsen, G. L.
    1936. The mammalian fauna of the White River Oligocene—Part I. Insectivora and Carnivora. Trans. Amer. Phil. Soc., n. s. **28**: 1–153.

Severtzow, M. N.
    1857–1858. Notice sur la classification multisériale des carnivores, specialement des felidés, et les études de zoologie générale que s'y rattachent. Rev. Mag. Zool. (2) **9**: 387–391, 433–439; (2) **10**: 3–8, 145–150, 192–199, 241–246, 385–393.

Simpson, George Gaylord
    1937a. Super-specific variation in nature and in classification from the view-point of paleontology. Amer. Nat. **71**: 236–267.
    1937b. Notes on the Clark Fork, Upper Paleocene, fauna. Amer. Mus. Novitates, No. **954**.
    1941. Range as a zoological character. Amer. Jour. Sci. **239**: 785–804.

Stirton, R. A.
    1940. Phylogeny of North American Equidae. Univ. California Pub., Bull. Dept. Geol. Sci. **25**: 165–198.

Waagen, W.
    1869. Die Formenreihe des Ammonites subradiatus. Benecke Geog.-pal. Beitr. **2**: 179–257.

KARL JORDAN'S CONTRIBUTION TO CURRENT CONCEPTS IN SYSTEMATICS AND EVOLUTION.

By Ernst Mayr.

It is a particular pleasure to be able to contribute an essay to this volume which honours one of the great biological thinkers of our time. I have admired Karl Jordan and his work since my student days and I am most grateful to the editors of this volume for this opportunity to express my admiration. It would seem particularly fitting that a non-entomologist should discuss Jordan's contributions to systematics and to the study of evolution, since Jordan has been at all times, first and foremost a biologist, even though working with insects as study material.

There is no need for me to stress Jordan's contributions to the field of entomology. They are universally appreciated and his election as Honorary Life President of the International Congresses of Entomology is one of many visible tokens of his prestige in the entomological world. What is not known among entomologists, however, and even less among most general biologists, is the fact that Jordan has made a number of highly important conceptual contributions in the fields of systematics and evolution. Yet his name is rarely mentioned in histories of the development of these fields. Possible reasons for this neglect are manifold, but it is evident that the basic one is that Jordan was far ahead of his time and thus was destined to suffer the fate of so many pioneers. A second reason is that he worked with a material (natural populations) and with methods (non-experimental) that were unpopular among the laboratory biologists of his time. He has always been averse to the spectacular which, as the history of biology shows, very often monopolises attention, no matter how soon afterwards it is shown to be wrong. It is a curious paradox that the name of the investigator who finds the right answers is likely to be forgotten, because his findings are incorporated anonymously into the treasury of common knowledge and accepted theory. That this has been Jordan's fate to a considerable extent, will be revealed in the subsequent analysis.

It has been pointed out recently that most " histories " of evolution are actually " pre-histories ". Their detailed treatment stops around 1860, precisely at the time when the development of the modern concepts began subsequent to the publication of Darwin's *Origin of Species* in 1859. The reluctance of historians to deal with the modern developments is understandable, since only a specialist can truly evaluate the significance of various contributions. With respect to systematic and evolutionary concepts there is the additional difficulty that much of the documentation is hidden away in taxonomic monographs or in rather specialised writings of naturalists. Furthermore, each author stands on the shoulders of his predecessors and with certain concepts it is hard to decide who deserves the greater credit : he who first vaguely mentioned a new idea, he who first clearly formulated it, or he who supplied final proof. Also there has been much parallelism, so that many conceptual advances have been made by several authors independently. In view of this it would be unwise to assert too dogmatically which particular concept was pioneered by Jordan. To determine this unequivocally would require an analysis of the writings of all of Jordan's contemporaries and predecessors similar to this one. The time has not yet come for such an analysis and this evaluation of Jordan's contribution is limited to the available evidence. It may have to be modified after the detailed history of various branches of natural history has been written.

Jordan is still working actively and has published only recently highly competent papers in the special fields in which he is an acknowledged master. Those publications, however, in which he made his most important conceptual contributions were published principally during the period from 1895 to 1911. In addition to a number of shorter papers (which are quoted in the bibliography), they include several major treatises which have since become classics. Among these are two general papers (on mechanical selection, 1896, and on geographical versus nongeographical variation, 1905) and a number of entomological monographs, prepared in joint authorship with Lord Rothschild, in which most of the general discussions were written and are usually signed by K. Jordan (eastern Papilios 1895, Sphingidae 1903, and American Papilios 1906).

The turn of the century was intellectually an exciting period in the history of systematics and evolution. The theory of evolution had by then been almost universally accepted by biologists, yet they were still groping for the correct interpretation of the causes of evolution. As far as the "material of evolution" is concerned the year 1900, as the birthdate of the science of genetics, must be considered the starting point of our knowledge. The battles about the mode of speciation, the existence of natural selection, and the meaning of mimicry and of polymorphism were at their height. Jordan entered all these battles with enthusiasm and conviction, and one is filled with admiration bordering on awe when one compares Jordan's discussions with those of most of his contemporaries. In almost every case it is Jordan who is right, and his formulations and solutions of biological problems are very often superior to those offered by other biologists during the succeeding fifty years.

Jordan's Biological Philosophy

Jordan has had the happy faculty of synthesising new concepts of biology by fusing the best elements of various schools of thought. Through his education at the University of Göttingen, he received a training in the best tradition of German zoology. Among his teachers was Ehlers, a competent comparative anatomist of invertebrates, who brought him in contact with the battles then raging between Weismann, Haacke, Eimer, Haeckel, and others, concerning the causes of evolutionary change. Through his entomological interests Jordan came in contact with an experienced group of insect collectors and breeders. Equally, if not more important, after his transfer to Tring, was his steady contact with Rothschild and with Hartert. Rothschild's great enthusiasm, particularly for the study of island faunas, was infectious, and Hartert at that period had become the leader of that group of ornithologists who were fighting for the consistent application of trinomials, for the downgrading of isolated allopatric populations from the rank of species to that of subspecies (wherever possible), and for a broadening of the species concept (Stresemann 1951.) Hartert (1859–1934), a great naturalist and practical systematist, never wrote a major paper on principles of evolution or on the new sytematics, but there is no denying his influence on Jordan, particularly during the formative years 1893-1895. Yet, a careful study of the publications of the Tring triumvirate (Rothschild, Hartert, Jordan) shows quite clearly that Jordan, perhaps as a consequence of his more complete university training and his greater interest in general biology, soon surpassed his teachers, when it came to developing new concepts in systematics and evolution.

Jordan's entire work in the fields of entomology, systematics, and evolution is founded on a few basic beliefs. They relate to the position of systematics in biology, the application of the concept of evolution to taxonomy, the meaning of variability, and the working methods of the scientist :

Systematics and biology.—There have always been two kinds of taxonomists : those for whom species of animals and plants are like postage stamps to be collected or described merely for the sake of rarity or novelty, and those others for whom systematics is an important branch of biology. Jordan has always included himself enthusiastically in the latter camp. He expressed repeatedly (e.g. 1905 : 150) his regret over the short-sightedness of so many laboratory biologists who thought that nothing could be learned from systematics because they themselves knew nothing about this field and had contact only with the stamp collecting type of taxonomists. " Classification, as we know, has the reputation of being as dry as our cabinet specimens—if not mouldy—and of having an interest only for those who work at the special group of animals classified. There are even biologists of fame who, in their misguided wisdom, scoff at systematics and look down upon this kind of work as more or less fruitless . . . I take the opportunity . . . of stating emphatically that sound systematics are the only safe basis upon which can be built up sound theories as to the evolution of the diversified world of live beings " (1911 : 385.).

It cannot be denied, admitted Jordan, that systematics is unthinkable without painstaking attention to a great deal of detail. Without this there is the danger of committing glaring errors such as Darwin's statement that the birds of Madeira and the Canary Islands have not been modified in comparison to European birds. The subsequent careful studies of Hartert and other bird taxonomists have proved the error of this statement. " A precise knowledge of the lowest systematic units which only the systematist can produce is not only the foundation for a valid theory of evolution but also a necessity for the correct placing of the many units into a system " (1905 : 153). It was the lesson which Jordan had learned as a zoology student at the university, that the taxonomist has a great deal of information of vital significance to the general biologist, which Jordan stressed throughout his career. For long it seemed like a hopeless endeavour to bring this point home, but the increasingly closer collaboration between taxonomists and experimental biologists during the past two decades proves that the point has finally been made. Jordan has contributed his share toward the victory of the idea that every taxonomist must have a broad training in general biology, and that every biologist must know and understand the important generalisations that come from the field of systematics.

Taxonomy and evolution.—Jordan has been an enthusiastic evolutionist from the beginning of his career. One might go so far as to say that all his research is ultimately directed toward adding further light on the processes of evolution.

The ambiguous rôle of the theory of evolution in the development of the theories of systematics is often commented on. Prior to Darwin the system of animals was merely a system of similarities. But ever since 1859 there has been an argument, which has not yet been unequivocally decided, whether the system should only serve the purely practical aims of the proper cataloguing and pigeon-holing of specimens or whether, as a consequence of the theory of evolution, animals should be classified on the basis of presumed relationships. Jordan is quite emphatically of the second opinion. He cites (1905) one case after the other where a modification of the classification on the basis of presumed relationships has lead to a considerable improvement of the distributional picture and to a better understanding of biological phenomena. For instance, if the West African butterfly *Papilio zalmoxis* is removed from the large Oriental *Aristolochia* feeders where this species was placed by all specialists on account of a superficial resemblance, and associated with various small African species with which it is really related, the remarkable fact is at once apparent that in spite of the presence of food plants there are no *Aristolochia* feeders in Africa. Jordan cites numerous cases from the writings of Wallace, Scharff, Eimer, and others where complex zoogeographic situations are at once clarified as soon as the respective species are classified not on the basis of superficial resemblance, but on that of relationship as revealed by totality of characters including internal ones. " The aim of the systematist as such is the establishment, on reasoned evidence, of the degree of relationship between the forms with which he is concerned, the evidence being furnished by the specimens and the bionomics of the species and varieties to which the specimens belong " (1911 : 386).

The meaning of variability.—One of the most revolutionary changes of concept in biology has been the replacement of typological thinking by thinking in terms of populations. According to this concept no two individuals or biological events are exactly the same and processes in biology can be understood only by a study of variation. The greatest of Jordan's papers (1905) was devoted to a discussion of the meaning of variation and the light it sheds on an understanding of evolutionary phenomena. The study of variation is an absolute necessity in the work of the taxonomist. Jordan, partly influenced by Rothschild and Hartert, came to understand very early how important it was to study adequate samples of as many natural populations as possible " . . . . Nowadays systematics comprehend more and more that a few specimens of each species are insufficient for a serious study, and hence try to bring together long series from every locality " (1896 : 447) (*see also* 1905 : 182).

Methods of the scientist.—Jordan, who himself has had such a broad background, has always emphasised a broad approach and a balanced analysis. He himself, superb master though he has become of certain groups of insects, has always tried to remain broad in his interests. In the field of entomology he was equally at home among beetles, butterflies, moths and siphonaptera. When analysing a given taxonomic situation he has stressed the importance of utilising the greatest number of characters, as we shall discuss below. He believes that by analogy the same working principle should be applied to the study of evolution. He warned the evolutionist not to be deceived by an artificial system of factors just as the taxonomist must beware of an artificial system of characters. " Scientific systematics utilises so far as possible all the attributes of the forms that are to be classified in order to determine their real relationship.

If there is a contradiction between one organ and another, there is either an error in observation or in interpretation. The research on one organ serves as a control for the research on the other, and this permits the recognition and elimination of errors. The same method must be used in the investigation of the causes of evolution if one does not want to be satisfied with an artificial explanation of the origin of the gaps between species which corresponds to an artificial system of classification " (1905 : 210). The common sense and the mature balance of mind which is expressed in this quotation has always been characteristic of Jordan. It has been characteristic of his taxonomic decisions and of his philosophy of nomenclature. It is not my task here to discuss Jordan's contribution toward the theory of nomenclature, and more specifically toward the stabilisation of nomenclature which he so ardently desires, but mention of his attitude toward nomenclature must be made in this context. " The philosophic aspect of systematics is unfortunately much obscured . . . due in a large measure to an unduly great importance being attributed to the mere giving of names. Nomenclature is the servant of science, but has in many houses the position of master " (1911 : 386).

CONTRIBUTIONS TO CONCEPTS OF SYSTEMATICS.

It has become fashionable to apply the label " new systematics " to the set of concepts that is now prevailing in the field of taxonomy. The silent assumption is made that these concepts are new. Actually their roots go back to the middle of the nineteenth century or earlier and it would be difficult to name a single concept that was not formulated at least thirty or forty years ago. What may come as a revelation to some of the younger students in this field is the discovery of the extent to which this " modern " philosophy is present in all of its completeness in the classical publications of Jordan. While much of it had been expressed casually or timidly by earlier authors, it is in the writings of Jordan that we find many of these concepts for the first-time stated fully and in detail. His broad knowledge of the contemporary zoological literature and his contacts with students of other kinds of animals as well as with general biologists, permitted him to develop many concepts in advance of his contemporaries. If any one deserves to be called the father of the new systematics it is Jordan. His influence on entomologists and taxonomists in general cannot be exaggerated. More specifically this influence is apparent in the development of thorough taxonomic techniques, a new evaluation of the infra-specific categories and his contribution toward the development of the modern species concept.

Taxonomic techniques.—We have mentioned above that Jordan has repeatedly stressed the need for the study of large series of specimens and for the utilisation of as many additional taxonomic characters, both external and internal ones, as can possibly be found. In his studies of butterflies he emphasised the taxonomic importance of such structural characters as scales and wing venation, and his extensive studies of the individual and geographic variation of the sexual armatures of both male and female butterflies (1896, 1905) have been hardly surpassed since. He is undoubtedly the pioneer in studying the variability of these structures in a large number of specimens from a single population and in following up their geographical variation. But he was not satisfied to use this evidence for purely taxonomic purposes, but utilised it, as we shall presently see, as an elegant proof of geographic speciation.

His concepts of the validity of taxonomic characters were far ahead of his time and many contemporary taxonomists might learn from what Jordan said in 1905 (p. 171). " One speaks in phylogeny often of generic characters as opposed to specific ones. There is, however, no valid definition of generic morphological characters as there is none of specific ones. To call the above described variability a change in generic characters because the diagnosis of genera in our somewhat superficial taxonomy of butterflies is frequently based only on differences in wing venation, would be circular reasoning. Differences in wing venation may be either individual, specific, generic, or even a taxonomic criterion for subfamilies and higher systematic categories ". He is convinced that one cannot hope to understand the variation of taxonomic characters by studying individuals. In order to understand the variability of species, it is necessary, he says, to study the offspring of a single female or " the individuals flying [he speaks of butterflies] in the same locality," in other words local populations (1896 : 430).

The subspecies.—In the field of systematics the name Tring or Tring School is associated with definite concepts such as a liberal use of subspecies and an attempt to arrange natural populations into polytypic species. There is little doubt that Hartert was the original leader of this team since he had fought for these concepts in print as early as 1891 (before coming to Tring). Little of this philosophy was noticeable in Jordan's early papers on beetles; for instance, among several hundred descriptions of new beetles published in 1894, there are only a few subspecies. But as soon as he started working on the species of *Papilio* with their immense geographic variation, he applied wholeheartedly the new subspecific concept as explained in his introduction (pp. 168–182) of Rothschild's revision of the Papilios of the eastern hemisphere (1895). Up to that time there had been great confusion as to the correct usage of the terms subspecies, variety and aberration. When the Tring triumvirate started publishing the *Novitates Zoologicae* they found it necessary to clear up this confusion in the following editorial preamble (1894, *Novit. Zool.* 1 : 1) :—

" The term ' variety,' especially among entomologists, has been indiscriminately used to denote an individual variation within a species as well as climatic or geographical races. We therefore, to avoid all possible errors, have determined to discard the term ' variety ' altogether. To denote individual variations we shall in this periodical, employ the word aberration, and for geographical forms, which cannot rank as full species, the term subspecies.

Editors."

At this period the introduction of subspecies into taxonomy was vigorously opposed by the older generation. The subspecies was still a relatively new concept and no unanimity was yet achieved as to its definition or application. Jordan took a stand on both these points. " A subspecies is a localised group of individuals of a species, the mean of the characters of which is different from the mean of the characters of all the other localised groups, and which will, under favourable circumstances, fuse together with other groups " (1896 : 447). He realises that this definition fails to make a clear distinction between populations that are sufficiently distinct to deserve to be named as subspecies and other populations that are not. " The above definition has not had regard to the degree of divergence attained by the localised form. Now, we ask, which then is the lower limit of application of the term 'subspecies'?" On the basis of cases where one sex is clearly different in two places while the other one is identical, " we shall have to use the term ' subspecies ' when a localised variation is such that about half of the individuals belong to the varietal form. All lower degrees of localised variation may be termed ' localised aberration ' " (1896 : 447–448). It is evident that at this date there was still some difficulty in distinguishing between individual and geographic variation as well as in specifying the degree of the distinctness of that half of the form which is different. In subsequent years Jordan has been particularly interested in the biological problem of the meeting of two subspecies with special emphasis on a most intriguing and puzzling case among fleas which he had discovered (1938, 1940). As far as isolated subspecies are concerned, he adopted Hartert's viewpoint . . . " that geographical separation of different forms cannot be an *a priori* criterion of specific distinctness, though this has often enough been alleged" (1896 : 431).

The species concept.—Again and again in his writings Jordan comes to grips with the species concept. He realises clearly that this concept is not only fundamental for the practical work of systematics, but equally crucial for studies of speciation and evolution, as is so painfully evident from the writings of those who have ignored the species problem in their discussions. In order to understand and evaluate Jordan's own viewpoint and his contributions to the subject, it will be necessary to outline the current status of this problem. It is rather evident from the fact that the controversy is still continuing (Gregg, 1950 ; Burma, 1954 ; Burma and Mayr, 1949) that a full understanding of the species problem has not yet been reached. After a thorough study of most of the literature on this subject, I have come to the startling conclusion that the disagreement is due to the fact that there is in existence not merely one but actually three entirely different species concepts. All past arguments and discussions have suffered from the fact that an author has either championed one of the three concepts against the others or has wavered between two of these without realising it.

The first of these concepts is the typological one which goes back to the Eidos (εἶδος) of Plato. Such a species is a " different thing ". Implicit in this concept is that variation as such is unimportant since it represents only the " shadows " of the Eidos. Translated into biology the typological concept becomes the morphological species concept. Certain individuals are a different species " because they are different ". When a minerologist speaks of " species of minerals " or a physicist of " nuclear species " he has this typological species concept in mind. This concept is still widespread in certain branches of invertebrate zoology and in paleontology. The objections against this concept are manifold and have been stated by Jordan with great clarity and vigour. First of all it is a strictly subjective concept. " It would almost appear, in fact, as if a ' species ' is that which a respective author chooses to consider a ' species ' " (1896 : 426). Inevitably this concept cannot be applied without all sorts of inconsistencies. " If it is the presence of morphological difference which leads us to split up into the one case, and the absence of such difference in the other case to unite, why then are *Distomum, Redia,* and *Cercaria ; Rhabdomena* and *Rhabditis ; Vanessa levana* and its offspring *prorsa*, the same species?

Morphological difference alone is not a criterion of specific distinctness " (1896 : 434). After citing numerous other cases which show the worthlessness of a typological-morphological species concept, he comes " to the conclusion that morphological differences of any kind and degree are not decisive criteria as to specific distinctness; the systematist actually sinks his species in spite of distinguishing characters as soon as it is proved that the morphologically different forms appear among the offspring of the same female. The most general case of bodily difference which is not regarded as being specific is the difference between males and females; notwithstanding the great dissimilarity which the sexes so often exhibit, not only in the reproductive organs, but also in other morphological characters, the systematist puts male and female together in one species, and hence makes at once the concession that his term ' species ' is not a purely morphological one, but that the higher criterion of the term is of a physiological kind " (1896 : 436). One has only to read the papers of De Vries and other contemporary plant breeders and early Mendelians to appreciate how far ahead of them Jordan was in his thinking. Having laid to rest the morphological concept so decisively, Jordan was ready to investigate what the physiological criteria are that characterise species. This led him to a consideration of the second species concept.

This is altogether different from the typological species. It does not deal with things, describing their degree of difference, but specifies a relationship. It is a concept like the word " brother " which has a meaning only with reference to some other object. This " biological " or " non-dimensional " species concept describes the relationship of two natural populations that co-exist at the same locality and specifies this relationship as " non-interbreeding ". The concept has been designated as " objective " because it can be defined as objectively as the word "brother" or as many similar designations of relationship. It has nothing to do with the objectivity of the concept itself that its *application* to taxonomic situations may occasionally run into difficulties just as it is occasionally difficult to establish the father of a baby even though there is nothing subjective about the biological concept "father" itself.

The revolt against the typological species concept and the attempt to replace it by a more biological concept had started long before Jordan. Indeed it is this concept of the non-dimensional species, expressing the relationship of non-interbreeding among sympatric populations of a single locality, which was the foundation of the original biological species concept of John Ray (1692), of that of Linnaeus, and of that of all local naturalists since. Although they had arrived at this concept empirically they had not thought it through sufficiently to describe it and define it unequivocally. That a species is a reproductively isolated population had also been previously expressed. Eimer, for instance, had spoken of species as " groups of individuals modified in such a manner that interbreeding between them and other groups no longer takes place or is not possible with success indefinitely " (1889 : 16). Poulton (1903 : 94) had said that " the idea of a species as an interbreeding community, as syngamic, is I believe the more or less acknowledged foundation of the importance given to transition " in geographic variation. Yet in all my search through the literature I have not found a single statement by any other author which indicates an understanding equal to that of Jordan. He was the first to point out the completely objective nature of the species concept as a measure of relationship in a non-dimensional situation (1905 : 157) : " . . . Individuals connected by blood relationship form a single faunistic unit in an area, to which unit we must add all the other individuals of the area which resemble them. This is the cornerstone of the building of the systematist and the starting point for an exact analysis of the correctness of the theory of evolution." And speaking of the single locality of the neighbourhood of Goettingen, his alma mater, he continues : " The three common *Pieris* of the gardens of Goettingen, the carab beetles of the Hainberg, the physopodes in the flowers of the Botanical Garden, the mice on the fields of the Weend, the bembids on the sand of the shore of the Leine, they all prove that the living inhabitants of a region are not a chaotic mass of intergrading groups of individuals, but that they are composed of a finite number of distinct units which are sharply delimited against each other and each of which forms a closed unit . . . The units, of which the fauna of an area is composed, are separated from each other by gaps which at this point are not bridged by anything. This is a fact which can be tested by any observer. Indeed all faunistic activity begins with the searching out of these units. A list of the species that occur in a region is an enumeration of such independent units which with Linneaus we call species " (1905 : 157). This masterful statement, published in 1905, is as true today as it was then. If it had not been ignored by so many taxonomists and non-taxonomists since that time, we could have saved ourselves much useless argument.

The third species concept is that of the multi-dimensional species. While the non-dimensional species defines a relationship of two populations, the multi-dimensional species is a grouping of populations. It is thus essentially the same as any higher category, a grouping of units of the next lower category, as the genus is for instance, a grouping of species. This third species concept then is a collective concept and, as all collective concepts, it lacks in principle the precision and objectivity which is generally inherent in relational concepts. If we group a number of geographically representative populations under the heading of a single polytypic species it is inevitable that we have to make a number of subjective judgements. As long as these populations are interbreeding and intergrading with each other there is usually little conceptual difficulty. In such cases it is generally granted that the combined populations conform to the criterion of the non-dimensional species concept and form a single reproductive community. As soon as populations are added to this polytypic species that are separated in space or time the judgement becomes truly subjective. The classifying taxonomist is faced here with the same situation as he is on the generic level when he must make a decision on the status of an aberrant species which is considered congeneric by some authors and generically different by others.

It was an inevitable consequence of the great period of geographical exploration that the non-dimensional species of the local naturalist had to be expanded into a multi-dimensional species, a geographically variable species. Early authors were confused as to what to do with geographical varieties nomenclaturally. Some, like Pallas, described the differing populations of new localities as " varieties," and so did Linnaeus himself in some cases. Other authors, limiting themselves rigidly to the classical binominal system of Linnaeus, called each of these different populations a separate species. Eventually a system of triple names won out, with the geographical varieties listed as subspecies.

That this new method produced a basic conflict of two very different species concepts was overlooked either consciously or unconsciously. The first author, I believe, who clearly realised that the non-dimensional species of Linnaeus and the multi-dimensional species that was so rapidly gaining favour in the second half of the nineteenth century, are not one and the same thing was O. Kleinschmidt. He restricted the term species to the non-dimensional relation and coined for the collective species the term " Formenkreis," later emended by Rensch into " Rassenkreis." This action, though inconvenient for practical purposes, was entirely logical. It also presented the historical developments correctly. The recent attempt to restrict the term species to the multi-dimensional concept and to apply the group symbol of symbolic logic to the concept species, as was done by Gregg (1950), is based on a misunderstanding of the historical developments.

It is only natural that Jordan, as a member of the Tring team, endorsed wholeheartedly the application of the concept of polytypic species to natural populations. Hartert already earlier had come to the conclusion that isolated populations should not be excluded from polytypic species if they seemed to belong to them on the basis of corroborating evidence. Jordan repeatedly explained why such a decision is inevitable. Since it is known from experiments and observation that strikingly different individuals may be members of a single interbreeding population " it is *a priori* evident that also geographically separated forms, in spite of their being morphologically distinct and in spite of their not being connected with one another by intergradations, can very well be subspecies of one species, i.e. that they can under favourable circumstances fuse into one form. The actual proof of specific distinctness the systematist as such cannot bring . . . we work with the mental reservation that the specific distinctness of our species novae deduced from morphological differences will be corroborated by biology . . . " (1896 ; 450–451). The viewpoint of the Tring school was at first not well received among taxonomists, and Jordan found it necessary to return to this argument in 1905. He uses a discussion of the geographic variation in the genitalic structures and other taxonomic characters of the butterfly *Papilio dardanus* to prove in a brilliant, closely reasoned argumentation that complete continuity of populations is not necessary in order to permit combining geographically isolated subspecies into polytypic species (1905 : 196–197). The tremendous simplification of classification which these arguments have made possible is now a matter of history. Yet there are some contemporary biologists who still fail to make a distinction between the *reproductive* isolation of sympatric populations, which proves that they are different species, and the *geographic* isolation of allopatric populations, which does not necessarily prove it. I recommend to them reading Jordan's arguments which have as much weight to-day as they had when he first stated them.

In retrospect we begin to understand why taxonomists have had so much difficulty during the past 100 years to apply the species concept to the purely practical task of classifying natural populations. They did not realise that there is no such thing as " the species concept " but rather three different concepts, each of which may permit a different conclusion in a particular situation. Jordan himself was no exception in this respect. In his various attempts at a species definition he wavered between the three stated concepts and finally attempted to include in his species definition criteria taken from all three.

It is still of interest to study his discussion of the species definition as given in 1896 (p. 438). His definition then reads " a species is a group of individuals which is differentiated from all other contemporary groups by one or more characters, and of which the descendants which are fully qualified for propagation form again under all conditions of life one or more groups of individuals differentiated from the descendants of all other groups by one or more characters." It is evident that as a taxonomist he feels that in his practical work he cannot do with a species definition that omits reference to morphological criteria. "The question of specific distinctness or non-distinctness is therefore two-fold : first, one of morphological, and second, one of physiological difference " (1896 : 442). He continues by pointing out that in most cases the systematist is not able to test by experiments the presence of physiological distinction and that therefore he can never prove with certainty from specimens alone whether the distinguishing morphological characters are of specific value or not. However, experience shows that physiological isolation and degree of morphological difference are so closely correlated that one is permitted, in the case of isolated populations, to determine their taxonomic status by comparing their morphological distinctness with that between related species and subspecies. Therefore, " If in a given case we have to decide whether A and B, which live together, are two different species, or two forms of one species, the morphological characters of A compared with those of B and the geographical representatives of B will have to guide us in our judgement " (1896 : 453). " The same kind of evidence we may employ when we have to come to a decision as to the specific distinctness of geographically separated forms which are not connected by intergradation. We must accept as a general law that forms which are connected by all intergradations, or forms which overlap in characters, are specifically identical ; . . ." This statement, made in 1896, (p. 454) is still somewhat vague. By 1905 Jordan had considerably clarified his ideas. He states (1905 : 157) that the working taxonomist is usually faced with two potential difficulties. The first is to decide whether a group of variable specimens from one locality belongs to one species or to several, and the other one whether somewhat different samples from two different localities belong to one or to two species. How can we tell, he asks, whether a certain number of specimens collected at one locality belong to one species or to several. Unfortunately, he answers this question, it is quite impossible to give a general answer. Species criteria must be determined anew for each group of organisms. In some cases only the breeding in the laboratory can supply the final proof, yet, much as the specific criteria change from group to group, there is no evidence that the species concept as such varies from one taxonomic group to another. The study of living populations and the breeding in captivity indicate that the morphological species criteria are correlated with a physiological condition. " We find that the morphological gap between individuals of two different species is accompanied by a physiological difference which is missing in the case of the morphological gap between individuals of one and the same species. This physiological difference has two consequences, namely :

" First, that individuals of a species, no matter how different from each other morphologically, will produce only individuals of their own species and,

" Second, that different species can co-exist in the same area without fusing into a single species " (1905 : 159).

He then continues to summarise his viewpoint in one of the most important passages in the history of biology. " The criterion of the concept species is thus a triple one and each of its three aspects can be investigated : A species has certain morphological characters, does not produce individuals belonging to a different species, and does not fuse with other species. As in 1896 I place great emphasis on this latter point. The non-fusion is the explanation for the immense number of existing species. Nothing keeps the species of an area separated but their own organisation. Individuals of one species live side by side with those of other species so genetically independent as if there had never been any genetic connection between them, as if each species has been created separately. The discovery of this fact led, during Linnaeus' time, to the dogma of the constancy of species. The experience which the observer of the individuals of his district and of his time made was erroneously extended to the individuals of all times and all districts." Here again Jordan emphasises the validity of the species concept for the non-dimensional system. He continues " If non-fusion is the principal criterion of specific difference it follows that one can prove specific distinctness only for those related forms which co-exist in the same area. The work of the systematist must begin with these and when dealing with all presumptive species from different regions he must ask himself whether the differences between them justify the conclusion that these forms could co-exist in the same area " (1905 : 159–160).

There is one among the conventional species criteria in insects to which Jordan paid special attention, difference in the structure of the genitalic armatures. Dufour had asserted that the male genital armatures were different in every species and had proposed the hypothesis that this difference serves as a mechanism to prevent hybridisation between different species and thus preserves the purity of each species. During the fifty years preceding Jordan's paper (1896) on mechanical selection, the idea had become quite universal among entomologists that each species could be diagnosed on the basis of its genitalia and that there was no individual variability with respect to this character. Only few authors such as Kolbe (1887), Perez (1894), and Edwards (1894) had disagreed with this hypothesis. Jordan, however, had *a priori* doubt against this concept " because if the concept of evolution is correct then species differences in the genital organs must have evolved, and the beginning of such differences must be visible already within the species " (1905 : 164). It is the main object of his important 1896 paper to demonstrate individual and geographical variation in these structures. An analysis of twenty-seven species of *Papilio* proved conclusively the existence of geographic variation. Jordan illustrates this on four plates with 189 figures. Each of the examined species of *Papilio* had a diagnostic difference in the structure of the genitalia from every other species of the genus. In 1896 Jordan was therefore convinced that Dufour was at least in part right. However, further studies convinced him that species specific difference in genital morphology is not a necessity. For instance, among 698 studied species of hawk moths (Sphingidae) there were forty-eight which did not differ in their genitalic structures; at least Jordan was not able to establish differences on the basis of the available material (1905). Jordan thus was the first to prove conclusively that reproductive isolation between species of insects was not necessarily supported by mechanical isolating mechanisms and that the genitalic structures are subject to the same laws of variability as all other taxonomic characters. The aspect of this investigation most pleasing to a modern worker is the method of approach. Jordan did not single out one convenient example to demonstrate his case; no, he made a complete statistical analysis of all the available species of a whole family. Jordan's work on the variation of genitalic structures in insects is outstanding in its formulation of the problem, procedure and conclusions.

We shall now leave our discussion of Jordan's contributions to the development of the modern species concept and to other phases of the new systematics. Although he left his successors the opportunity to straighten out a few details, such as the relation between non-dimensional and multi-dimensional species concepts, it is evident that he had already developed in broad outlines all the modern views. After Jordan's discussion of the species concept in 1905 substantially no progress was made in the literature of the succeeding thirty years. In fact, in the entire genetic literature I have not found a single discussion of the non-dimensional species which is as penetrating as that of Jordan.

CONCEPTS OF EVOLUTION.

Jordan at all times has been of the firm conviction that the taxonomist can make an important contribution to the understanding of the processes of evolution, indeed that no full understanding of evolution is possible without taking into account the conclusions of the taxonomist. For instance, " my object is to demonstrate that there is more in classification than lies at the surface by proving that there exists a very intimate connection between, on the one hand the vexed question of specific distinctness, and on the other hand the elucidation of the factors of evolution to which is due the great diversity of the animal world " (1911 : 386). And among all the evolutionary problems he realised that the systematist was best equipped to solve that of the origin of gaps between species. Such an investigation, he says (1905 : 160), has two aims. First to explore the steps which lead to these gaps, and second to find out what causes are responsible. As far as the method of investigation is concerned he believes that the systematist must find out whether the differences which exist among species are something quite unique or whether they are merely a more advanced stage of the kind of differences which one may find among individuals within a species. His aim then was to solve the problem of speciation by means of the inductive method.

Modes of speciation.—The particular problem as to how one species breaks into several has been the source of bitter argument ever since the theory of evolution was proclaimed. At the time of Jordan's major papers a bitter battle was waging over the right theory of speciation. The general biologists, the geneticists, animal and plant breeders and a minority of the taxonomists were in one camp and the other taxonomists in the other. The argument, in short, was whether or not speciation was possible without geographical segregation. Both Karl Jordan (1905 : 162) and D. S. Jordan at about the same time stated that the theory of sympatric speciation was virtually universally accepted among the biologists of that period. According to this theory individual variants can develop into full species without geographical isolation. It is a historical fact that Bateson, De Vries, and their powerful schools of contemporary geneticists endorsed and postulated this mode of speciation. In the discussion of this problem Jordan notes that at first sight the concept of sympatric speciation appears to be abundantly supported by various facts. One of them is that there are many places on the earth where a great number of closely related species co-exist, as the lemurs on Madagascar, certain groups of insects in the Hawaiian Islands, the marsupials in Australia, and the humming birds in America. However, says Jordan, some authors dissent and have come to the conclusion that one species can split into several only by way of a geographic variety. This concept was first clearly expressed by Moritz Wagner and had since been adopted by a considerable number of taxonomists. " My own investigations completely confirm this theory as I want to demonstrate in the subsequent discussion of the great difference between geographic and non-geographic variation " (1905 : 163).

The theory of evolution postulates that species evolve from incipient species. Darwin and his successors, following up this thought, came to the conclusion that varieties within species are these incipient species. No agreement, however, was reached on the question whether or not all of the many kinds of varieties within species are equally important as incipient species. Jordan distinguishes three kinds of such varieties, or, as he expresses it, three types of polymorphism within species: individual, seasonal and geographic polymorphism. It is interesting to note, incidentally, that there has been some recent improvement in the precision of our terminology. The term polymorphism is now applied only to variation within a population, while differences between populations are no longer included under this term. A species that breaks up into subspecies is referred to as a polytypic species and not as a polymorphic species. The clear distinction between variation within a population and between populations which we now make was not fully appreciated by Jordan. Fortunately it did not affect the validity of his argument.

Full species of insects, Jordan concludes, differ from each other in coloration and other external morphological characters and usually also in the structure of the genitalia. Incipient species then should differ from each other either in external morphological characters or in the structure of the genitalia or in both. A variety, in order to qualify as incipient species, must vary consistent with this postulate. Accordingly Jordan studied the mode of variation of the three kinds of varieties (individual, seasonal, and geographical).

Speciation by macromutation.—De Vries had recently emphasised the evolutionary importance of those individuals that lie completely outside the normal variation of a population, his so-called mutations. Such forms, Jordan says, are not infrequent among butterflies and are eagerly sought by the amateur collector. They are generally referred to as " sports ". Jordan examined the genitalia of many such sports and this is what he found (1905 : 167) : " In not a single aberration, either caught in nature or artificially produced by temperature shocks, have I found that the aberrant colouration was correlated with any differences in the genital organs. The sports remain with respect to these organs within the normal limits of variation. They are not new ' species ' which have separated themselves from the mother species by a saltation and which now could co-exist independently." It is curious how long Jordan's clear-cut refutation of De Vries' theory of speciation by mutation was ignored among the Mendelians. It is perhaps not out of place to cite in this historical survey what De Vries actually said on speciation. " The theory of mutation assumes that new species and varieties are produced from existing forms by certain leaps. The parent type itself remains unchanged throughout this process and may repeatedly give birth to new forms " (1906 : vii). " The origin of species may be seen as easily as any other phenomenon. It is only necessary to have a plant in a mutable condition " (1906 : 26). De Vries came to this viewpoint by not believing in the species of the naturalist " as Linnaeus has proposed them. These units are not really existing entities; they have as little claim to be regarded as such as genera and family " (1906 : 12). Jordan came back repeatedly in his writings to this theory of speciation by mutation. In 1911 (p. 401) he pointed out that even polymorph types are not nearly as clear-cut as is usually maintained, provided one has sufficiently large collections. The modern geneticist might say that additional modifying factors may help to bridge the phenotypic gap between two polymorph alleles. And Jordan continues " and it is perhaps also not superfluous to reiterate in this connection that the geographical races, too, are nearly always found to be connected by intergradations, if a sufficiently large number of specimens is examined ".

Sympatric speciation.—The fact that many similar but ecologically slightly different species may occur in the same geographical area was, at an early time, interpreted as evidence for speciation without geographic isolation. One author after the other tried to discover the mechanism which would make such sympatric speciation possible. The more we learn about the co-adaptation of the gene complexes of local populations the more such schemes of sympatric speciation appear impossible. Jordan, however, without this knowledge of modern genetics had come precisely to the same conclusion merely through the study of incipient speciation and by an astonishingly lucid and logical analysis of the speciation process. Although he had disproved sympatric speciation quite conclusively in 1896, he came back to it whenever a new paper was published which tried to revive the dead theory. Those who postulated such a process with the help of the hypothesis of homogamy (the mating of the most similar individuals within a population) he countered by stating that the observations on which these claims were based were ambiguous, if not mistaken, and that even if homogamy occurred it could not lead to a splitting of a single population into two (1905 : 173). When H. M. Vernon (1897) proposed a new version of this theory entitled evolution by " Reproductive Divergence," Jordan (1897, 1898) demonstrated its fallacy so conclusively that this hypothesis never again entered the biological literature. It is historically interesting that Jordan's premises are essentially the same as those that led, subsequently, to the establishment of the Hardy-Weinberg Formula. Another contemporary who had proposed a theory of sympatric speciation was Romanes (" physiological selection ") (1897), but he had already been refuted so decisively by Wallace, in his book *Darwinism* (1889), that Jordan could afford to dismiss his thesis rather summarily (1896 : 443).

The striking morphological difference sometimes found between seasonal forms of insects had lead to the establishment of still another hypothesis of sympatric speciation by some entomologists. They argued that new species might arise if these seasonal forms became more and more different. Jordan admits that seasonal variants differ from ordinary individual variants, because they frequently " ... separate breeding communities. ... In the species in which the first generation of mature individuals has disappeared before the next one hatches, these generations are reproductively isolated from each other as much as is one species from another" (1905 : 174). Jordan then asked himself whether or not this really could lead to species formation. He concludes finally that the odds are all against it. First of all the summer generation is normally the direct descendant of the spring generation and secondly, there is no other evidence except phenotypic difference that these are incipient species. Jordan finds the best proof for this in the variation of the genitalia. Geographical races, which are incipient species, often deviate quite markedly in the structure of the genitalia from each other while, in the case of a large number of seasonal forms investigated by him, he found only a single species (*Papilio xuthus*) in which there is even the slightest genitalic difference. There is thus no evidence that seasonal forms are incipient species (1905 : 174).

There are two authors in particular whose hypothesis of sympatric speciation Jordan rejected emphatically. One is Dahl (1889) who had outlined an imaginary case of sympatric speciation in the genus *Gonopteryx* by postulating differences of incipient food plant races. "At the time when selection sets in in Dahl's illustration there were already two 'species,' and hence the specific distinctness of the two is not the outcome of psychological selection" (1896 : 427). Dahl had failed to explain the first crucial step which can be explained only by geographical speciation. The other of his opponents was Petersen (1903) who had proposed quite a fantastic hypothesis based on blending inheritance, correlated host and mating preference, and various other improbable or actually erroneous assumptions. It is a real pleasure to read Jordan's answer (1903) with its superb knowledge of the evolutionary literature of the period and his critical emphasis on the essential points most of which he had already listed in his 1896 paper (p. 444) : "The physiological selection will, therefore, in no case result in divarication of a species into two, but the outcome of the physiological variation will be either dimorphism of the species, when both the normal and the varietal form are equally favoured in respect to the circumstances of life, or extinction of that form which is the least favoured. If however the most favourable kind of variation does not lead to the origin of a new species beside the parent one, no other variation will lead to this end. Hence we must conclude that a divarication of a species into two or more species cannot come about so long as the divergent varieties live so together that a direct or indirect intercrossing is not prevented."

His rejection of sympatric speciation is decisive. He says that there is no possibility of the origin of a new species without the help of some kind of local separation. Nor is the hypothesis of sympatric speciation needed to explain the geographical co-existence of closely related forms. Those who postulate it overlook two factors : The ability of animals for active or passive changes of their range and the time factor. In fact, says Jordan, it can be shown quite clearly in many cases that the present co-existence of two closely related species is obviously the result of a comparatively recent invasion of each others ranges (1905 : 203).

Geographic speciation.—By showing that the various schemes of sympatric speciation do not work, and that neither mutations nor individual variants nor seasonal races qualify as incipient species, Jordan prepared the way for a discussion of geographic speciation. The questions he poses are whether or not geographic races are incipient species, and whether or not the great diversity in nature can be explained as having arisen through geographic speciation.

These questions can be answered only through a study of geographic variation.
The old concept that species are fixed, invariable types was undermined particularly by the work of Wallace in the East Indian Archipelago and that of Bates in the basin of the Amazon. Bates had pointed out that a comparison of individuals of a species from different localities revealed a graded series, ranging from such slight differences that they are not worth mentioning, to such great differences that it was doubtful whether they were still conspecific. Jordan comments that (1905 : 181-182) " these grades of variability, which Bates found on the Amazon River can be found by the systematist also in Europe. It is a universally valid fact that there are all degrees between virtual identity of individuals from two areas and morphologically constant and well defined differences. A complete identity of all the individuals of a species in two localities occurs rarely, if ever. Most often that which is normal ($=$ most frequent) at one locality, is not normal at the other. The curve of variability has shifted at the upper or lower end or at both. It also may occur that one organ remains the same and another is different or that a species is monomorph at one locality and dimorph at another."

Jordan takes this observation of geographic variability as his starting point and asks next whether it affects only unimportant, superficial characteristics or also those which distinguish good species. It was this question in particular which induced him to undertake his historical researches in the geographic variation of genitalic structures in Lepidoptera (1896, 1905). As a conclusion of his researches he states that since many geographic races of butterflies differ from each other in the structure of their genitalia in the same manner as good species and since such geographic races vary also in all the other characters of colour and structure which distinguish species, it must be concluded that geographic races are to be considered incipient species. He continues to point out that the various characters of a species may be independent in their geographic variation. Two subspecies, which differ in colouration or wing venation, do not necessarily differ in their genitalia, the more so since even a certain number of good species is not distinguishable on this basis. The conclusion that species originate by geographic variation can be extended by analogy to animals other than insects. Jordan says that one must apply the same method to other species characters where species do not differ in the structure of the genitalia. This had been done already by Wagner, Seebohm, Hartert, Rothschild and others.

Jordan was not satisfied merely to point out the fact of geographic variation, but attempted to establish general laws governing this phenomenon. We mentioned already his discovery of the independence of different organs with respect to their variation. In the African *Papilio phorcas*, for instance, specimens from the Congo agree in colouration with such from Sierra Leone, but in the structure of the genitalia with the populations from East Africa. Other rules established by Jordan are the following : The degree of geographic variation is different from species to species and even among closely related species one may be highly constant over large areas while the other is subject to a great deal of geographic differentiation. " The amount of geographic variation depends on the nature of the environment (including the living one) and on the nature of the species. But not only on these two factors. It is also obvious that two areas between which there is an active inter-change of individuals are actually only a single area for the species because the individuals of the two areas form a single inter-breeding unit as if they were co-existing in the same area. One must therefore conclude that geographic variation depends furthermore on the degree of separation of the various geographically localised interbreeding units. The degree of separation by one and the same geographic barrier in turn will be different according to the nature of the animal. What is a barrier for a terrestrial animal like a snail or flightless insect, and therefore separates various breeding communities, may not be any barrier at all for a winged animal. And finally it is easy to comprehend that the amount of time which has passed since the separation is a factor in the origin of a geographic variety " (1905 : 187). There was no other biologist in 1905 who saw and expressed the situation as clearly as did Jordan.

" That species may differ according to locality has been known to systematists as far back as the 18th Century " (1905 : 184). Jordan's major contribution to the subject has been to treat the subject quantitatively by attempting to determine the amount of geographic variation in every species of an entire group, such as the Papilios or the Sphingidae, and by being the first to extend the investigation from a few superficial characters to such internal structures as the genital armatures. His treatment of the geographic variation of genitalic structures in the species *Papilio dardanus* is still outstanding as an example of such a quantitative treatment. Even the percentage occurrence of certain variants is recorded for each population : a spine on the valve occurs in 100 per cent. of the East African specimens, in 13 per cent. of Uganda individuals and in less than 3 per cent. of those from the west coast.

As a true biologist Jordan is not satisfied merely with describing the fact of geographic speciation, he also wants to determine its causes. Wagner, the founder of the theory of geographic speciation, had claimed that isolation in itself is an active factor of trans-mutation. This is rejected by Jordan (1896 : 445) : " Isolation as such is not an active factor which produces a character but is a factor which merely preserves a character produced by some other factor ; isolation has, therefore, no direct effects." In support of this conclusion he cites numerous facts. How can isolation itself account for the phenomenon, he asks, that the butterflies and moths of Sumatra and Borneo are quite universally dark and those from Queensland pale? Since isolation itself cannot account for this ... " there is only one way possible by which the divarication of a species into two or more can come about—that is, the combination of isolation and transmuting factors " (1896 : 446). I believe that with this statement Jordan was the first to state clearly that speciation is the joint product of mutations and isolation. Wagner had neglected one of these factors while the contemporary Mendelians and the Darwinians ignored the other. And how, asks Jordan, do these transmuting factors act during this period of isolation : " The geographical races thus produced we must assume to be first inconstant, to become more and more constant and divergent by the incessant influence of the transmuting factors, and to develop finally into a form which is so modified that it never will fuse either with the parent form or the sister forms, and that it therefore agrees with the definition of the term ' species ' " (1896 : 446). It is quite evident from the context that Jordan understood under " transmuting factors " both genetic changes and natural selection. This was written in 1896 before the birth of genetics and a certain vagueness in terminology, like " constant " or " inconstant " is excusable. The modern geneticist would speak of the development of reproductive isolating mechanisms and of the evolution of co-adapted and integrated gene complexes. What is important is that Jordan at this early date described the process of the change of geographically isolated populations just as we do to-day. As far as the nature of the genetic changes is concerned he was still on the fence, allowing even the possibility of Lamarckian changes. However, he pointed out repeatedly, and as we must admit in retrospect correctly, that this point did not really make any difference so far as the theory of speciation is concerned. In this he was far more penetrating than any of his contempories or any other author during the fifty years that followed him.

Again and again Jordan emphasised the importance of the study of geographic variation. In 1896 (p. 446) he said with respect to geographic speciation " as this kind of divarication of species is the only possible one . . . the study of localised varieties is of the greatest importance in respect to the theory of evolution ; *the study of geographical races, or subspecies, or incipient species, is a study of the origin of species.*" And in 1903 (p. 664) he summarised his views as follows : " Geographic variation is the basis of species formation and it alone gives the explanation for the mutual sterility of species for which one has searched in vain since Darwin's day. Spatial separation alone permits the gradual divergence in morphology and physiology. It alone can prevent fusion with parental and sister populations of the varieties which develop in a new environment and become gradually constant. This permits that the originally very small, unimportant and inconstant differences in genitalia and other characters increase by accumulation to such an extent that a fusion is no longer possible."

Although the taxonomist normally works with morphological characters, it is evident from this quotation and from Jordan's other writings how highly he ranked physiological characters in the speciation process. " For us systematists that kind of psychological variation is of more [= greatest] interest which is the immediate outcome of morphological differences in the organs of sense and in the organs which are destined to affect the senses " (1896 : 434). In view of the fact that males discriminate between the scent of females of their own species and that of other species it should be important to look for geographic variation in this character. One such case, he continues, has already been found, namely that of Standfuss on geographic variation in smell preference between Swiss and Italian individuals of *Callimorpha* [= *Panaxia*] *dominula*. This, he says, proves how potent a factor in speciation a geographic variation in scent preference would be (1896 : 434). The relatively greater constancy of colouration in a given population of birds as compared to the much greater variability of most butterflies is due to the fact that the visual sense is the primary sense in birds while in most insects it is the olfactory sense.

Factors of Evolution.

A study of this problem shows where the principal advance in concepts has occurred since the turn of the century. The rôle of the environment was not yet clearly analysed at that time. Nothing was known as to the nature of genetic factors. Jordan was a firm believer in natural selection, but was in doubt as to its precise rôle. Again and again he comes back to the great importance of natural selection in the evolutionary process. This is particularly evident in his repeated discussions of mimicry (1897a, 1897b and 1911). " ... There is here no difficulty at all for the theory of natural selection. The theory offers a very simple explanation by assuming that the variable ancestor of the mimicking species has been gradually modified into a di- or polymorphic species by the weeding out of those intermediate forms which did not resemble protected species already existing in the country. The frequent rarity of intermediates is direct evidence for this theory " (1911 : 403). He continues to say that this theory is not in conflict with other theories dealing with the origin of variation itself and those which interpret adaptation to local environment. Jordan then presents in full detail the genetic theory of mimetic polymorphism which has since been adopted by the majority of geneticists.

How far ahead of his contemporaries Jordan was in his thinking on natural selection is evident from two notes on mimicry he published in 1897 in *Nature*. In order to account for the usual rarity of mimics as compared to the models some students of mimicry had advanced the exceedingly anthropomorphic explanation that mimics are rare, because it is to their advantage to be rare in order that the predators will not become conditioned to them. This explanation which, as a matter of fact, is still maintained in much of the current literature, does not provide for a genetic mechanism as was clearly realised by Jordan. Instead he advanced a theory which—anticipating the recent selection experiments of Mather, Lerner, and others—has a remarkably modern ring. If the mimic " for instance, *Papilio alcidinus*, has acquired that wonderful similarity in colour and form to its model, an Uraniid moth, in consequence of a continued selection in the one direction, it is obvious that the result of such a one-sided selection will not only be similarity to the immune model, but also physiological one-sidedness. The more rigorous the selection is, the better will the mimetic species become adapted to its model, and the more it will lose its adaptability to new biological factors. ... Consequently, the most striking ' mimics,' in spite of, or rather in consequence of, the resemblance to immune species, are, in the long run, the less favoured in the struggle for existence, which means that they will become relatively scarce. From this consideration it is apparent to me that the selection of those specimens which are the very fittest in any *special* direction is in itself a danger to the species, and can lead to destruction " (1897a : 153). This was written at a time when natural selection was being attacked by environmentalists and mutationists alike. Almost fifty years have passed by before the analysis of gene complexes by Mather and others have led to a genetic confirmation of Jordan's views of the detrimental effects of one-sided selection.

In 1905 (pp. 205–209) he once more tried to analyse the causes of speciation. His analysis largely leads him to reject various erroneous theories without permitting a constructive answer on the basis of the genetic facts known at that time. " The morphological facts on which we have based our conclusions give us unfortunately only the above negative answers to our question which factor or which aggregate of factors produces the difference between subspecies and species : that which better adapts an individual for the West African environment is not a selection of specimens according to a special wing pattern or according to a shift or shortening of spurs or crests on the genital valves ;

it is not the direct effect of humidity or temperature on wings or genital organs nor the release of latent characters by such factors " (1905 : 209). He was willing to wait for the eventual solution with an open mind.

CONCLUSIONS.

The preceding discussion makes use of only a small portion of Jordan's life work as presented in his early major publications. Yet I believe it is enough to show that by and large Jordan was far ahead of his times, sometimes by as much as fifty years, and that in his writings and through his contacts he exerted an influence on the development of the new systematics and on evolutionary thought that cannot be over-estimated. Even though Jordan's name may not be cited as widely in textbooks of evolution and zoological systematics as it deserves, at least he has the satisfaction that most of the concepts which he pioneered and for which he fought have now been accepted by virtually every worker in the field. As one of those who have benefited from Jordan's efforts to clarify evolutionary thought and to enhance the prestige of systematics I want to record on this occasion my feeling of immense gratitude.

REFERENCES.

BURMA, BENJAMIN H., 1954, Reality, existence, and classification : a discussion of the species problem. *Madroño, S. Francisco* **12** :193-209.

——, and Ernst MAYR, 1949, The species concept : a discussion. *Evolution* **3** : 369-373.

DAHL, FR., 1889, Die Bedeutung der geschlechtlichen Zuchtwahl bei der Trennung der Arten. *Zool. Anz.* **12** : 262-266.

DE VRIES, HUGO, 1906, *Species and varieties. Their origin by mutation.* Chicago and London.

EDWARDS, WM. H., 1894, Notes on " A revision of the genus Oeneis " (Chionobas). *Canad. Ent.* **26** : 55-64.

EIMER, G. H. T., 1889, *Artbildung und Verwandtschaft bei Schmetterlingen.* Jena. **1** : 16.

GREGG, JOHN R., 1950, Taxonomy, language and reality. *Amer. Nat.* **84** : 419-435.

JORDAN, KARL, 1896, On mechanical selection and other problems. *Novit. zool.* **3** : 426-525.

——, 1897a, On mimicry. *Nature, Lond.* **56** : 153.

——, 1897b, On mimicry. *Ibid.* **56** : 419.

——, 1897c, Reproductive divergence : a factor in evolution ? *Natural Science*, **11** : 317-320.

——, 1898, Reproductive divergence not a factor in the evolution of new species. *Ibid.* **12** : 45-47.

——, 1903, Bemerkungen zu Herrn Dr. Petersen's Aufsatz : Entstehung der Arten durch physiologische Isolierung. *Biol. Zbl.* **23** : 660-664.

——, 1905, Der Gegensatz zwischen geographischer und nichtgeographischer Variation. *Z. wiss. Zool.* **83** : 151-210.

——, 1911, The systematics of some Lepidoptera which resemble each other, and their bearing on general questions of evolution. *Int. Congr. Ent.* I (Brussels, 1910) **2** : 385-404.

——, 1938, Where subspecies meet. *Novit. zool.* **41** : 103-111.

——, 1940, Where subspecies meet. *Int. Congr. Ent.* VI (Madrid, 1938) **1** : 145-151.

KOLBE, H. J., 1887, Carabologische Auseinandersetzung mit Herrn Dr. G. Kraatz. *Ent. Nachr.* **13** : 132–144.

PEREZ, J., 1894, De l'organe copulateur mâle des Hymenoptères et de sa valeur taxonomique. *Ann. Soc. ent. Fr.* **63** : 74–81.

PETERSEN, WILHELM, 1903, Entstehung der Arten durch physiologische Isolierung. *Biol. Zbl.* **23** : 468–477.

POULTON, EDWARD, B., 1903, What is a species ? *Proc. ent. Soc. Lond.* **1903** : xciv.

ROMANES, GEORGE J., 1897, *Darwin, and after Darwin.* **3** : 41–100. Chicago. [Summary of earlier publications.]

ROTHSCHILD, W., 1895, A revision of the Papilios of the eastern hemisphere, exclusive of Africa. *Novit. zool.* **2** : 167–463. [A considerable part of the introductory matter and synonymy is by K. Jordan.]

——, and K. JORDAN, 1903, A revision of the lepidopterous family Sphingidae. *Ibid.* **9** : Suppl.

——, 1906, A revision of the American Papilios. *Ibid.* **13** : 411–752.

STRESEMANN, ERWIN, 1951, *Die Entwicklung der Ornithologie.* Berlin. Pp. 249–270.

VERNON, H. M., 1897, Reproductive divergence : An additional factor in evolution. *Natural Science,* **11** : 181–189.

WALLACE, ALFRED RUSSELL, 1889, *Darwinism.* London and New York.

POSTSCRIPT

1916, Jordan, K. Notes on Arctiidae. *Novit. zool.* **23** : 124-150. Besides being an excellent revisionary study, this paper concludes with probably the first clear statement of the subject of sympatry and allopatry in their relation to systematics, apparently overlooked hitherto, due to the deceptive title. *Editor.*

The Popular Names of Birds

Ernest E. T. Seton

CORRESPONDENCE.

[*Correspondents are requested to write briefly and to the point. No attention will be paid to anonymous communications.*]

The Popular Names of Birds.

To the Editors of The Auk:—

Sirs: The 'powers that be,' I understand, are preparing a 'Check List,' and revising the scientific and *popular* names of our birds.

There is no doubt that scientific names are entirely in the hands of scientists, but it seems to be overlooked that popular names are just as completely in the hands of the people. Scientists may advise, but not dictate on this point. A short analysis of the principle of common names may place the matter in a new light.

A bird's name, to be popular, must be distinctive, and in accordance with the genius of our language. Examples of such are Thrush, Rail, Heron, Hawk, Crane, Night-Jar, and many others. These are truly popular names, evolved originally out of a description, handed down and condensed and changed until they have assumed their present terse, abrupt, and, to a foreign ear, uncouth forms, but, nevertheless, forms in accordance with the pervading spirit of the Saxon tongue; or, in other words, they are *really* popular.

On the other hand, look at the so-called popular, but really translated, scientific or spurious English names given to our birds, taking as examples the following: Baird's Bunting, Leconte's Sparrow, Wilson's Green Black-capped Flycatching Warbler, Bartram's Sandpiper, Sprague's Lark, Wilson's Thrush, Black Ptilogonys, Semipalmated Tattler, Fasciated Tit, Florida Gallinule, etc.

Surely, the gentlemen whose names are applied to these birds have not so slight a hold on fame as to require such aids as these to attain it, if indeed aids they be, which I question; for such nomenclature *cannot* stand the test of time.

If you show to an 'out-wester' the two birds mentioned above as Baird's Bunting and Leconte's Sparrow, and tell him that these are their names, he will probably correct you, and say one is a 'Scrub Sparrow,' the other a 'Yellow Sparrow.' Convince him that he is wrong, and in a month he will have forgotten all but the names he formerly gave them; they are so thoroughly appropriate and natural that they cannot be forgotten The next name in the list above given is clumsy enough to strangle itself with its own tail. A lad on the Plains once brought me a *Neocorys spraguei*, and asked its name. I replied that it was Sprague's Lark. Soon afterward he came again; he could not remember that name; so I told him it was a 'Skylark,' and he never forgot that. On the Big Plain that seed was sown, and not all the scientists in America can make, or ever could have made, the settlers there call that bird anything but 'Skylark.' And I consider that lad precisely represented the English-speaking race; he rejected the false name, and readily remembered the true one, and was aided by that which was apt and natural. No better illustration could be given of the fact, that phraseology may be the life or death of a cause, according as it is happy or unfortunate.

A similar instance is the case of 'Bartram's Sandpiper.' Ever since Wilson's time this name has been continually thrust into the face of the public, only to be as continually rejected; 'Upland Plover' it continues to be in the east, and 'Quaily' on the Assiniboine, in spite of Bartram and Wilson, and will continue so until some name, answering all conditions, is brought forward; for here, as elsewhere, the law of the survival of the fittest rigidly prevails. As an example of the fit ousting the false, note how, in spite of scientists, 'Veery' is supplanting 'Wilson's Thrush' throughout the length and breadth of the land.

The spurious English names scarcely need comment, they so evidently contain in themselves the elements of their own destruction. Imagine a western farmer being told that a certain songster was a 'Ptilogonys.' In spite of the books, the other three examples cannot hold ground against 'Willet,' 'Ground Wren,' and 'Waterhen,' respectively.

The purpose of a Check List that includes English names is, I take it, not to attempt the impossible feat of dictating to our woodmen what names they shall give their feathered friends, but rather to preserve and publish such names as are evolved in the natural way,—names which are the outcome of circumstances. Only in case of egregious error is a common name to be superseded; and in doing this it must be remembered that no name can be popular unless true to the principles of the English tongue. It must be short, distinctive, and, if possible, descriptive. Of this class are Veery, Junco, and Vireo. These are the only successful artificial names that I can at present recollect. Among natural English names for American birds are Bobolink, Chewink, Kingbird, and many others. Such as these not only more than hold their own, but are as great aids to the spread of knowledge as the Ptilogonys kind are hindrances; while such as Wilson's Thrush can only be accepted as provisional, until the better knowledge of the bird and its surroundings shall result in the evolution of an English name founded on true principles.

ERNEST E. T. SETON,
of Manitoba.

Glen Cottage, Howard Street,
  Toronto, March 21, 1815.

THE VARIATIONS AND MUTATIONS OF THE INTRODUCED SPARROW. *PASSER DOMESTICUS.*

(A SECOND CONTRIBUTION TO THE STUDY OF VARIATION.)

HERMON C. BUMPUS.

IN the preface to the second volume of these Lectures it is stated that one of the leading objects of the course is "to bring forward the *unsettled* problems of the day, and to discuss them freely." The question of the adequacy of natural selection is one that at the present time still divides two schools of speculative biology, and is a question that can be solved only by those inductive methods which it is the function of a Biological Laboratory to suggest, adopt, and execute.

The principle of "Panmixia," or the "suspension of the preserving influence of natural selection," has formed an integral part of the speculative writings of Weismann, and, as part of his theory of "the continuity of the germ-plasm," is presumed to explain adequately the reduction of useless organs, and the occurrence, especially among domesticated animals, of "the greater number of those variations which are usually attributed to the direct influence of the external conditions of life."

This view of the regressive power of natural selection was, at the time of the original presentation of Weismann's essay ('83), not entirely new to science. Lankester ('90) calls attention to the fact that, eleven years earlier, in 1872, Darwin, in the sixth edition of the *Origin of Species*, had the identical principle in mind when he wrote: "If under changed conditions of life a structure before useful becomes less useful, its diminution will be favored, for it will profit the individual not to have its nutriment wasted in building up a useless structure." Shortly after this Romanes advanced a not totally dissimilar idea in his theory of the "Cessation of Selection" ('74).

In 1890 Romanes revised his earlier views, calling especial attention to the points in which they differed from those of Darwin and Weismann, and in 1895, in his posthumous work, the salient features of his theory are again indicated. Cope carried the application from structures to species when he wrote ('96): "In other cases it is to be supposed that extremely favorable conditions of food, with absence of enemies, would have occurred, in which the struggle would have been *nil*. Degeneracy would follow this condition also."

But, without entering into the conflicting claims of originality and of priority, all the disputants are agreed that the withdrawal of the supporting influence of natural selection from an adapted organ or organism must or may, directly or indirectly, lead to a condition of degeneration. That the arguments, however, are too speculative in character is generally admitted, and there is consequently demand for inductive evidence to prove:

(1) That in a specific case, and in respect to certain characters, the operation of natural selection has been suspended.

(2) That, when the operation of natural selection has been suspended, increased variation occurs.

(3) That, on the occurrence of (1) and (2), there is a departure from a previously maintained and presumably high standard, and

(4) That, unless a new equilibrium is established by adaptation to the new environment, degeneration and perhaps final elimination ensues.

It would also be of incidental interest to learn from observed facts whether the suspension of the action of natural selection is felt immediately by an organ or organism; whether there is any indication of "self-adaptation" tending to the establishment of a new equilibrium; and whether this self-adaptation, if detected, follows one or several definite lines. Of course, if the evidence can be gathered from animals in a state of nature, and if it can be checked by a large number of examples, so much the better.

In 1850 the first house sparrows of Europe were introduced into this country, and from that time to 1870 upwards of 1500 birds are said to have been brought from the Old World (Merriam-Barrows, '89). To these introduced birds the environment has been novel. They have found abundant food, convenient and safe nesting places, practically no natural enemies, and unrivaled means of dispersal. Aside from an early and brief period of fostering care, they have been left to shift for themselves; natural agencies have since been at work, and in the relatively short space of forty years a continent has been, not merely invaded, but inundated by an animal which, in its native habitat, has been fairly subservient to the regulations imposed by competing life.

It seems to the speaker that here is an excellent example of the suspension of natural selection, for here, at least as far as certain external factors of selection are concerned, Nature does not select. Nearly all the young birds reach maturity; variations in color and structure, unless most extreme, are apparently not disadvantageous to their possessor; and if these variations are heritable, they do not seriously handicap the individuals of the next generation. A considerable departure in nesting and breeding habits does not jeopardize the domestic interests, and the simple mode of life permits even the weak individuals to endure. We conclude, then, that there is evidence to prove the first proposition, *viz.*, in a specific case and in respect to certain characters, the operation of natural selection has been suspended.

For a proper discussion of propositions 2, 3, and 4, it was my first purpose to collect a large number of the American birds and compare them directly with an equal number collected in England; but the labor and expense involved made this procedure inexpedient. The egg of the bird, however, is easy to secure, readily preserved, and can be purchased from European dealers for a relatively small price. It presents a remarkable range of variation, both in shape, size, and color, and offers certain fixed and readily measurable features which are not presented by the bird itself. Moreover, my observations lead me to think that it is a structure which indicates departures from "normality" in a remarkable way. At all events, the variations, though they may present greater amplitude, are of the same inductive value, qualitatively, as variations of the skeleton, feathers, or other adult structures. The egg may be taken, then, as a convenient and inexpensive means for the solution of at least some of the questions bearing on the subject of Panmixia.

At first, one hundred eggs, imported from an English dealer, were compared with an equal number collected in Providence, R. I. The dissimilarity in the two lots of eggs was so striking that I felt there must be some mistake, and at once imported another hundred from a different locality, collecting in the meantime a second hundred of American specimens. On comparing the two enlarged collections, such interesting variations were found that I ordered all the English eggs that could be procured, and collected extensively from certain localities at home. At the close of the summer, 1896, I had 1736 eggs, one half of which were European, the other half American. These eggs, 868 foreign and 868 native, were compared (*a*) with respect to length, (*b*) ratio of length to breadth, (*c*) general shape, and (*d*) color. These comparisons ought to reveal any tendency towards increase of variation on the withdrawal of natural selection, that is, they ought to yield evidence in support of the second proposition. The data may be conveniently arranged in "curves of frequency."

If we erect on a base line (Diagram I), extending from 18 mm., which represents the shortest egg, to 26 mm., which represents the longest egg, a series of ordinates representing in sequence the added increment of ½ mm., and arrange on these ordinates the eggs that measure respectively 18 mm., 18.5 mm., 19 mm., 19.5 mm., etc., it is evident that the mean ordinates will be occupied by a considerably larger number of specimens than the extreme, and that the ascending and descending curve will indicate the general plan of the distribution of variation around the mean. Now if a species or structure is stable and shows only a slight tendency to vary, the base of the curve obviously will be short. If, on the con-

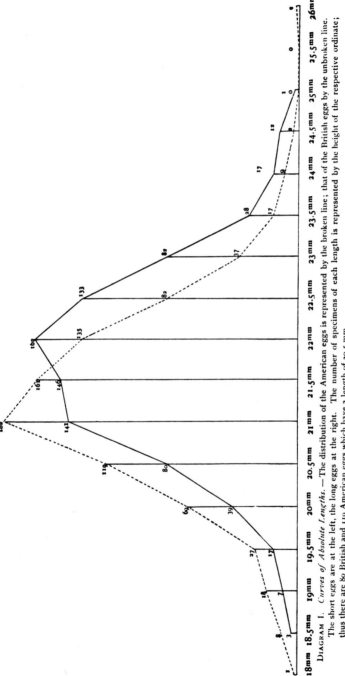

DIAGRAM 1. *Curves of Absolute Lengths.*—The distribution of the American eggs is represented by the broken line; that of the British eggs by the unbroken line. The short eggs are at the left, the long eggs at the right. The number of specimens of each length is represented by the height of the respective ordinate; thus there are 80 British and 119 American eggs which have a length of 20.5 mm.

trary, a species is unstable and has a general tendency to vary, the base will be long.

The 868 American eggs arrange themselves in respect to lengths as represented by the broken line on Diagram I. The base of this curve is long. Its summit coincides with the ordinate of 21 mm. Its interest, of course, lies chiefly in the relationship it bears to the curve of British eggs.

The latter curve is represented by an unbroken line. Its base extends from the ordinate of 18.5 mm. to the ordinate of 25 mm., and its point of greatest altitude is upon the ordinate of 22 mm.

A moment's examination of these curves reveals not only the fact that the American eggs are more variable, *i.e.*, the base of the dotted curve is broader, but it also yields data appropriate to the third and fourth propositions; for it will be observed that the American eggs have undergone a striking reduction in their average length, that is, they show a departure from a previously maintained higher standard, *viz.*, 22 mm. in length, and they are also tending to gather about a new point of equilibrium, *viz.*, 21 mm. in length.

Without commenting upon these observations, which are based upon *absolute* measurements, let us see if the *ratio* of the breadth of the egg to the length, that is, the *shape* of the egg, has also been affected by the withdrawal of natural selection.

The curves on Diagram II are designed to represent the distribution of eggs according to the ratio of their major and minor diameters. When an egg approaches sphericity, the ratio is higher; when it is elongated, the ratio is lower. The more elongated eggs are arranged at the right of the diagram; the short, stumpy ones are arranged at the left. Oval and ellipsoidal eggs naturally occupy positions along the middle ordinates. The broken line, as before, represents the distribution of American eggs, the unbroken line, of British.

On this diagram it will be noted that the American eggs again show a greater amplitude of variation, the base of the dotted curve being nearly one-fifth broader than that of the entire curve. It will also be noted that, appropriate to the third proposition, the American eggs have undergone a striking

THE INTRODUCED SPARROW.   7

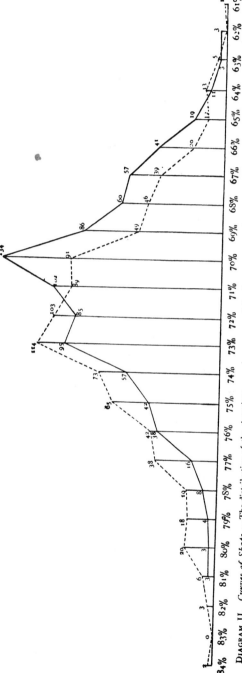

DIAGRAM II. *Curves of Shape.*—The distribution of the American eggs is represented by the broken line; that of the British eggs by the unbroken line. The more nearly spherical eggs are at the left, the more elongated at the right. For example: the figure 84% indicates that the shorter axis of the egg is to the longer as 84 : 100. The number of specimens of each particular shape is represented by the height of the respective ordinate; thus, there are 95 British and 114 American eggs in which the breadth is 73% of the length.

change in shape, as indicated by the ratio of breadth to length; and, appropriate to the fourth proposition, that the American eggs are not indiscriminately distributed, but tend to gather about a mean type. This type is located on or near the ordinate of 73%, and is removed some little distance toward the side of sphericity, and away from the correlative ordinate (70%) of the British specimens.

The second curves, then, bring out in a more emphatic way the same general facts that were shown on the first diagram.

But it is quite evident that the mere ratio of breadth to length is not an adequate index of variation in shape. On this ratio alone, an egg that is conical, or pear-shaped, may not appear in any way different from one that is ellipsoidal or lemon-shaped. I have made several attempts to bring out these extreme variations in some practical arithmetical manner, but have felt each time that the eggs varied far more than the numerical results indicated.

For want of a better method, I finally adopted the following:

Having placed upon each American egg a secret mark, the eggs of both countries were thoroughly mixed together in a single tray. A disinterested person was then requested to select, from the mixture of 1736 eggs, one hundred eggs which appeared to him to present extremes of shape-variation. If eggs from the two countries are equally variable, it is clear that approximately the same number from each would be selected; and, of course, if the American eggs are more variable, more American eggs would be selected. The result of this experiment was most striking, and in harmony with the evidence derived from the comparison of lengths and the ratios of breadth to length. Eighty-one of the selected eggs were American, while only nineteen were English; *over four times as many of the former as of the latter*.

As before mentioned, the colors of both European and American eggs are subject to variation, arising from modifications of the ground color and from the color and distribution of the spots or blotches. Some are of a somber color, much like the eggs of our common song sparrow; others resemble the eggs of the kingbird; and still others have the delicate ivory white of certain vireos. An attempt was made to arrange the colors in sequence, but after many fruitless efforts the plan of disinterested selection, above mentioned, was adopted.

The British and American eggs were thoroughly mixed together and the request was made that twenty-five eggs which presented the greatest variation toward the kingbird type should be selected first; then twenty-five of the somber type; third, twenty-five of extremely light color; and, fourth, twenty-five anomalous varieties. Some hours were spent in making the selection of one hundred eggs, and with the results indicated on Diagram III, where *b* represents the British eggs and *A* represents the American.

Kingbird Type.	Somber Type.	Light Type.	Anomalous.
b	b	b	b
b	b	b	A
b	b	A	A
b	b	A	A
b	b	A	A
b	b	A	A
b	b	A	A
b	A	A	A
A	A	A	A
A	A	A	A
A	A	A	A
A	A	A	A
A	A	A	A
A	A	A	A
A	A	A	A
A	A	A	A
A	A	A	A
A	A	A	A
A	A	A	A
A	A	A	A
A	A	A	A
A	A	A	A
A	A	A	A
A	A	A	A
A	A	A	A

Diagram III.—This diagram is designed to illustrate the preponderance of extreme color variation on the part of American eggs. A indicates American, b indicates British eggs.

Of the kingbird type and of the somber type there were over twice as many American as British eggs. There were among the light eggs nearly twelve times as many departures from the mean of color on the part of the American as on the part of the British eggs, and among the anomalous eggs there were twenty-four times as many American extremes as British. (It may also be of interest to add that the single British egg was the last egg to be selected, that is, it presented the least departure from the mean of the twenty-five anomalous variations.)

Eighty-two of the examples of extreme color-variation were thus found to be American and eighteen British. That so large a proportion of extreme variation in *color* was found among the American eggs is interesting in itself, but a comparison with the relative amount of extreme variation in *shape*, enhances the significance of both results, for not only is the preponderance of variation among American eggs very obvious, but in both cases, in length and in shape, it is almost precisely the same (81:19 in the first, 82:18 in the second).

Our data, then, whether it be gathered from comparisons of length, ratio of breadth to length, shape, or color, all point in one direction; and, granting that the sparrow since its introduction has been comparatively free from the action of natural selection, we may conclude that the predicted results of Panmixia have been realized.

The collection of a series of facts, for the mere support of some favorite theory, ought not to be the purpose of biological investigation. The relation that the facts may have to other facts and the bearing that they may have upon collateral theories should, at least, be indicated.

The following questions naturally arise:

Apart from the tendency to vary, is the new form, adopted by the American egg, the result of the selection of adaptive adventitious or fortuitous variations, or is it "determinate," the result of the direct action of a new environment? If due to the direct molding influence of a new environment, is the variation *ontogenic*, that is, does it occur anew and repeatedly in each successive generation, in obedience to reiterated environmental demands; or have the directive influences of the mechanism of heredity been so affected that the variation becomes established as *phylogenic?* Is the mechanism of heredity affected immediately, through the action of the new environment on the germ itself, or mediately, through the influence of ontogenic somatic change?

I think it improbable that the new form adopted by the American egg can be the result of the selection of adaptive fortuitous variations.

Fortuitous variation means *chance*-variation, and, although it is mathematically *possible* for the same particular variation to appear fortuitously in all or nearly all of the American eggs, it is absurd for us to suppose that this has actually happened. We cannot believe that the new form and shape, which are so universally presented by the American species, are variations which have arisen by mere chance. Again, even admitting for the sake of argument that a *chance*-variation has simultaneously appeared in nearly all the American individuals, what have we to show that this variation is adaptive, that it has selective value? Who will say that the shorter egg is a superior egg, or that the more spherical egg is, in the new environment, an improvement on the European type?

In the third place, even admitting the all-sufficiency of natural selection, there has not been sufficient time for the establishment of a new type of egg, that is, for the conclusion of the struggle between "Nature and Nurture." Neo-Darwinians deal with centuries and ages. Forty years can accomplish nothing.

If we again refer to the curves, we shall find other reasons for the belief that the American type of egg is not to be explained by the principle of adaptive fortuity.

Although the American eggs are unquestionably more variable, as is shown by the more elongated base lines, the curves rising to the culminating points of American variation are no less regular than those rising to the culminating points of British variation. This means that the new type is definitely established and that nearly all the eggs tend towards this type. Now, is it likely that mere chance-variation would yield an American curve so nearly parallel to the British curve? If the selection favors those eggs which are located on ordinates 21 and 73 (Diagrams I and II), that is, favors a certain type, why do other eggs on distinct ordinates and of an entirely different type arrange themselves in an orderly manner?

This brings us to another point. The curves show that the British influence is still felt in America. There are distinct elevations in the American curves as they cross the ordinates of 22 and 70. These elevations, which may represent the conservatism of certain individuals which still retain British instincts, are perhaps of less interest than the elevations on the British curves which lie immediately under the American culminating points. One wonders why ruthless natural selection should have spared these particular individuals.

There has been a *general* reduction in the shape of practically all the eggs since the introduction of the birds into this country, and this reduction has taken place not only in the neighborhood of the new mean, but also at the extremes. Not only has the old culminating point been shifted, but the entire curve has been shifted. The larger eggs have become smaller, the medium eggs have become smaller, the smaller eggs have become smaller; and all the eggs, whether of the ellipsoidal or spheroidal type, have become more nearly spherical.

Concluding, then, that the evidence does not favor the view that the American egg is the result of the action of natural selection upon fortuitous variations, let us examine the alternative, that is, the variations are due to the molding influence of a new environment.

A new environment, offering new food, peculiar climatic conditions, etc., might affect a large number of individuals in certain peculiar and definite ways, and it is evident that the respective curves of variation given in Diagrams I and II are in harmony with such a conception of the march of transformation. It is, indeed, a phenomenon that is seemingly of the nature of a "mutation" (Scott, '94). This view, moreover, is not contrary to the later ideas of Darwin, who distinctly stated that the greatest error which he had committed was in not allowing sufficient weight to the direct action of environment independent of natural selection.

Moreover, if the new environment is directly responsible for the new variations, the question of time is no longer a disturbing factor, and it is perfectly natural that certain less plastic individuals should, through the influence of heredity, continue loyal to the British standard; for the tendencies toward the establishment of a new type are not the result of the selection of the fit nor the elimination of the unfit, but, rather, the result of a direct influence upon all.

The questions remain to be answered: Are the new variations the result of the influence of the environment reiterated in the case of each particular individual, or has the mechanism of heredity been affected so that the American birds are producing new eggs through its directive influence? Has "Buffon's factor" (Osborn, '94), the direct action of environment, produced definite and adaptive variations which are merely "contemporary individual differences" (Cunningham, '93), or are these variations approved and adopted as a part of the constitution of a phyletic series? In brief, is the new variety merely ontogenic, or is it phylogenic?

The maturating as well as the developing ovum must be looked upon as an organism, and "as such must dominate its own development" (Whitman, '94). The ovarian ovum gathers to and about itself certain constituent parts and incorporates them according to its individual peculiarities. As it leaves the ovary, laden with yolk, it gathers about itself the envelopes of albumen, shell-membrane, and shell which it is the function of the oviducal walls to secrete. To assume that the organized ovum has no control, exercises no influence over the development and arrangement of these secondary envelopes, is like assuming that the presence of an ovum in the mammalian uterus exercises no influence upon the uterine walls. But the material submitted to the ovum by the somatic cells is not necessarily always qualitatively and quantitatively the same, and, on the other hand, there is no reason to suppose that any two ova, even of the same parent, have precisely the same peculiarities. The entire bird's egg is the result of the centrifugal influence of the ovum exerted upon the surrounding tissue no less than the centripetal influence of the surrounding tissues exerted upon the ovum; of the keimplasm exerted upon the soma no less than of the soma exerted upon the keimplasm, and, in dealing with a portion of the resulting structure, *viz.*, the shell, we are dealing perhaps somewhat more directly with the influence of heredity and its vehicle than we would be, if the subject of our discussion were a more distant somatic product, such as a bone or a feather.

The relation of the ovum to the complete egg is practically the same as that of a "caddis-worm," to its "case." The preferred material may be bits of straw, but, in the absence of straw, small pieces of wood may be made to answer. The "worms" in the "cases" of wood are themselves not different from their, perhaps more fortunate, neighbors in straw "cases." It is only when they adopt the wood in preference to the straw that an ontogenic makeshift becomes a phylogenic variation. New building material does not make a new architect.

In America the materials supplied for the developing ovum are different from those supplied in England, and the resulting structure is consequently different. To what extent the new materials have won the favor of the keimplasm cannot be determined by merely allowing American birds to breed again in England, for in England there would be a prejudice in favor of local material, and under the revival of an ancient environment palingenic variation might also deceive. Both English and American birds should be placed in some third locality which combines equally or eliminates the prejudicial environmental conditions of the two countries. Then, and not until then, shall we know to what extent the ontogenic variations in either country have really become phylogenic.

REFERENCES.

'72. DARWIN, CHARLES. The Origin of Species by Means of Natural Selection. (Sixth edition.)
'74. ROMANES, GEO. J. *Nature.* Vols. ix and x.
'83. WEISMANN, AUGUST. Inaugural Lecture as Pro-Rector of the University of Freiburg. (Reprinted in '89 as the second of the " Essays.")
'89. WEISMANN, AUGUST. Essays upon Heredity and Kindred Biological Problems. Oxford.
'89. MERRIAM, C. HART, and BARROWS, WALTER B. The English Sparrow in North America. United States Department of Agriculture. (*Division of Economic Ornithology and Mammalogy, Bulletin I.*)
'90. ROMANES, GEO. J. Panmixia. *Nature.* Vol. xli.
'90. LANKESTER, E. RAY. The Transmission of Acquired Characters, and Panmixia. *Nature.* Vol. xli.
'93. CUNNINGHAM, J. T. The Problem of Variation. *Natural Science.* Vol. iii.
'94. SCOTT, W. B. On Variations and Mutations. *Am. Jour. Sci.* Vol. xlviii.
'94. WHITMAN, C. O. Evolution and Epigenesis. *Biological Lectures.* (Wood's Holl.) 1894.
'94. OSBORN, H. F. The Hereditary Mechanism and the Search for Unknown Factors of Evolution. *Biological Lectures.* (Wood's Holl.) 1894.
'95. ROMANES, GEO. J. Post-Darwinian Questions, Heredity and Utility. Chicago, 1895.
96. COPE, E. D. The Primary Factors of Organic Evolution. Chicago.
'97. BUMPUS, H. C. A Contribution to the Study of Variation. *Journal of Morphology.* Vol. xii.

CRITERIA FOR THE DETERMINATION OF SUBSPECIES IN SYSTEMATIC ORNITHOLOGY.[1]

BY FRANK M. CHAPMAN.

I AM fully aware that an adequate presentation of this subject would require far more time than could well be accorded me on this occasion. Nor in any event could I expect to treat it in a manner which would meet with the approval of systematists generally. I feel, however, that it is a subject which demands discussion. The principles governing the procedure of systematists are now so diverse that until they have been harmonized it is useless to expect uniformity in method. This paper, therefore, is offered with a hope that it may lead to a general exchange of views and thereby prepare the way to the establishment of common standards in determining the status of representative forms.

The cytologist reaches definite conclusions concerning the status of a species through a study of its germ-cells; the experimental biologist determines the relationships of allied forms by the results of his attempts to breed them, but the systematic ornithologist defines species in terms of their external characters of size (including relative proportion of parts), color, and pattern of coloration plus their distributional relationships and, in some instances, their habits and voice.

The fundamental test of specific standing is non-intergradation with other forms. The indisputable proof of the specific distinctness of two or more forms is their occurrence together when breeding without intergradation.

The Greater and Lesser Scaup Ducks, the Downy and Hairy Woodpeckers, the Alder and the Acadian Flycatchers, the Common Crow and Fish Crow, the Gray-cheeked and Olive-backed Thrushes are more like each other than are many subspecies; but wholly aside from the absence of intermediates, the fact that they occupy in whole or part the same area when breeding and still maintain

---

[1] Read before the Annual Meeting of the American Ornithologists' Union, Cambridge, Mass., October 10, 1923.

their distinguishing characteristics is, to the ornithologist, final proof of their specific distinctness. The cytologist may discover the same number of chromosomes in the germ cells of the Crow and the Fish Crow, the experimental biologist may produce fertile offspring from a union of the Greater and Lesser Scaups, but they will still remain species in the eyes of the systematist simply because they are associated in nature when breeding but do not intergrade.

In cases of this kind, therefore, the ornithologist is not at loss for a satisfactory test of specific distinctness without regard to degree of difference. It is when forms which resemble each other more or less closely are not associated while breeding but replace one another in areas which may or may not be connected that, lacking specimens to demonstrate their exact relations, he must decide whether to treat them as species or subspecies.

The easiest way out of this difficulty is to follow one or the other of two general rules in more or less current use.

1. Within certain limits of differentiation to consider all presumable representative forms as subspecies whether or not they are known to intergrade and whether or not their breeding ranges are contiguous or are widely separated.

2. To treat all obviously representative forms as species until their intergradation is proven.

While the adoption of either of these rules has the merit of consistency, their use often leads to such biologically incorrect results that no mere question of expediency can, in my opinion, pardon their acceptance.

Better be inconsistent than deliberately to handicap one's experience and discrimination by blind adhesion to a law which, however convenient it may be in practice, inevitably leads to false and misleading representations. Furthermore, it seems perfectly logical to insist that if a systematist refuses to rank certain forms as subspecies until their intergradation is proven, he should also refuse to treat them as species until the fact of their non-intergradation is established. Certainly in many cases we have no more right to assume that intergradation does not occur, than we have to assume that it does occur. Errors we are bound to make in any event, but there is assuredly no excuse for making them by unreasoning adherence to a purely arbitrary man-made law.

Is it not more scientific to treat each case on its merits, basing our conclusions on due consideration of all the available pertinent evidence? Systematic zoology has a higher end to serve than mere classification, and its nomenclature, should, so far as its limitations permit, express our knowledge of relationships. In the study of both physical and geographic origin of species it is of fundamental importance for us to know whether given forms are species or subspecies. It is evident, therefore, that their value as factors in problems of evolution and zoogeography is measured by the correctness of our classification. To the best of his ability and in the light of his material and experience the systematist should attempt to supply this information. If his labors have brought him into disrepute among biologists is it not because he has so often treated his specimens as dried skins rather than as biologic facts?

Our problem seems deceptively simple. Briefly, it requires us to decide when to treat representative forms as species, and when as subspecies. "Form," it should be explained, is the indefinite term which the systematist employs when he is in doubt whether to write "species" or "subspecies." Representative forms, then, are species or subspecies which so closely resemble one another that one evidently replaces the other. Whether they have descended from a common ancestor by different lines of descent, or whether one has arisen from the other, whether their differentiation is due to environmental or external factors or to mutational and internal causes are all subjects for inquiry, and our findings are, in a measure, expressed by our nomenclature.

The systematist knows that many proved subspecies differ far more from one another than do many species. It is not necessary therefore for us to distinguish between them when we list the characters on which species and subspecies are usually based. These are:

1. Size, including relative proportion of parts.

2. Color, including variations in intensity of color and actual differences in color.

3. Pattern of coloration, including variations in area, due primarily to increased or decreased pigmentation; variations in marking of individual feathers and the presence or absence of such "unit

markings" as wing-bars, pectoral crescents, superciliary lines, etc.

4. Shape, as it may be affected by increase or decrease in size or be expressed in the form of certain feathers.

In spite of the fact that these characters are present in both species and subspecies, we inquire which of them are most commonly considered of specific, which of subspecific value?

VARIATIONS IN SIZE.

Wide-ranging species so frequently differ in size geographically that laws have been formulated to express the normal trend of this type of variation. Thus J. A. Allen in 1876, writing on 'Geographical Variation among North American Mammals, Especially in Respect to Size,' advanced the following generalization[1].

"The maximum development of the individual is attained where conditions of environment are most favorable to the life of the species." Generally speaking it is the individuals of a species from the highest latitudes and altitudes which are the largest. Wide-ranging species, therefore, are sure to encounter conditions which usually produce variation in size and if their range be continuous the change in these conditions—and hence the change in size—is as gradual as the change in latitude or altitude itself. Variation in size, therefore, is the most common type of geographic differentiation. It is exhibited by great numbers of species which in other respects are essentially alike throughout their range. Moreover, as an expression of environment it keeps pace with range, and when a species is more or less continuously distributed we expect a complete intergradation in size between its largest and smallest members.

The assumption therefore seems warranted that when representative forms differ from each other only in size they will intergrade if their ranges are connected. The question, however, arises if the ranges of such forms are not connected and when their differences in size are not bridged by individual variation, shall we treat them as different species or shall we assume that they are organic units which would fuse were their ranges to be joined?

---

[1] Bull. Geol. and Geogr. Surv., II, No. 2, p. 310.

Our action here may be governed by either one or the other of the procedures above mentioned or it may be determined by the light of our experience. On the one hand we know that many representative forms differing only in size intergrade and are therefore properly classed as subspecies. On the other, we know that size alone is in some cases a specific character and that species differing only in size nest in the same area without intergrading. From Bolivia to eastern Colombia two Toucans (*Ramphastos cuvieri* and *R. culminatus*), which are minutely alike in color and differ only in size,[1] occur together without intergrading. The difference between them is not extremely pronounced, but it is obviously sufficient to keep the two apart as distinct species. Now it happens that two other Toucans (*Ramphastos ambiguus* and *R. abbreviatus*) are found which also differ from each other only in size. In this case, however, their ranges do not coincide, the larger bird occurring in the Subtropical Zone of eastern Colombia, the smaller in the Tropical Zone of western Colombia. Whether or not their ranges are connected I do not know. Meanwhile shall we treat them as species or assume that if their ranges were connected they would intergrade and hence rank them as subspecies?

If the two Toucans first mentioned had not shown us that to them mere size may make a species, I confess I should treat these two representative Colombian Toucans as races on the assumption that they did or would intergrade, but reasoning by analogy I accord them specific rank.

In Central America two species of Flycatchers of the genus *Myiarchus* are found associated which differ only in size, and that so little that Ridgway,[2] after naming one of them as distinct from the other expressed doubt of its validity. Bangs,[3] however, has shown that they are distinct, and in this belief he is supported by Miller and Griscom who have met both birds in life. Possibly both Toucans and Flycatchers may possess notes and habits which distinguish them.

A more familiar illustration of representative species which differ only in size, plus the presence in one of a slight whitish

[1] *R. culminatus* averages, wing, 199; tail, 146; culmen, 123 mm.
[2] *R. cuvieri* averages, wing, 242; tail, 155; culmen, 172 mm.
[3] *Myiarchus nuttingi brachyurus* Ridgw., Bull. 50, IV, U. S. N. M., 1907, p. 630.
[4] Proc. Biol. Soc. Wash., 1909, p. 34.

margin on the wing-coverts which is absent in the other, is afforded by our Black-capped and Carolina Chickadees. In spite of the fact that these two birds more closely resemble each other than do hundreds of subspecies, and that their ranges are not separated, they do not intergrade. Here we know that there are differences in voice markedly apparent to our ears and doubtless more so to Chickadees'.

Other well-known cases of species which are distinguished wholly or almost wholly by size are supplied by the Greater and Lesser Yellow-legs,[1] Hairy and Downy Woodpeckers, Common Crow and Fish Crow, and here also we know that there are recognizable differences in voice.

Notwithstanding these and similar instances it is certainly customary for representative forms differing only in size to intergrade if their ranges are connected, and when intergrades are lacking, we are, I think, more warranted in ranking them as subspecies than as species. A distinction, however, should be made between difference in size and difference in proportion. In the first, all the parts commonly used in comparison, bill, wing, tail, tarsus, etc., may show the same relative variation in size. In the latter, their variations may be disproportionate, or one member may be larger, another smaller, than those of the allied form, and such instances are more complex, less explicable than those in which the variations are all in the same ratio.

VARIATIONS IN COLOR.

The color characters which distinguish representative forms from one another may be roughly classified as differences of degree and differences of kind. The first includes those variations in intensity of color which we have learned to associate with climatic environment and which are so strikingly illustrated by the classic cases of the Song Sparrows and the Horned Larks.

A very large proportion of our subspecies are based on these differences in degree of color. Increase in the color of plastic, responsive species so commonly accompanies increase in rainfall,

---

[1] In this connection, however, reference should be made to J. T. Nichols' description of the skeletal differences between these two species and particularly to his suggestive comment that they are not so closely related as their "similarity in plumage would lead one to suppose" (Auk. Oct. 1923, p. 594).

and the reverse so frequently accompanies decrease in rainfall, that we have come to consider the relation between humidity and dark colors, and aridity and pale colors, as one of cause and effect. Thus the systematist can tell with some approximation to truth the annual precipitation in a given region from an examination of its breeding Song Sparrow. The characters separating representative forms of this kind are apparently the expression of an existing environment. When, therefore, the environments, so to speak, intergrade, we may be reasonably sure that the forms to which they have given rise will also merge; and under such conditions it is customary to rank them as subspecies without the confirmation of intergrading specimens. But again, I ask, when the ranges of such forms are separated, prohibiting fusion by contact, and when the birds are too unlike each other to intergrade by individual variation, how shall we classify them?

The systematist to whom the fact of non-intergradation is a sufficient test replies "as species;" but I am convinced that it is often biologically incorrect and misleading to follow this course.

Is the Towhee of Guadalcupe Island any less a race of *Pipilo maculatus* because its range is insular and hence isolated? Is the Horned Lark of the Bogotá, Colombia, Savanna any less a race of *Otocoris alpestris* because its range is separated from its nearest relative by all Central America?

To rank these birds as species is, to my mind, not only biologically false but it results in the adoption of a nomenclature which to an extent conceals their origin and relationships.

In assuming that representative forms of this kind would intergrade if their ranges adjoined, we are on safer ground than with forms differing only in size. Birds which resemble each other in everything but size we have seen may live together as distinct species, but I do not recall an instance of two birds differing only in degree of color being associated as species. It appears, therefore, that this type of differentiation is racial or subspecific rather than specific in character and representative forms distinguished by it may be expected to intergrade.

It is sometimes difficult to distinguish between the cumulative effect of differences of degree and differences of kind in color, as well as in pattern. Increased pigmentation may result in such excessive deepening of tone as to give an apparently new color, and in such change of area as to produce an essentially different pattern. Examples of these types of differentiation are shown by the Song Sparrow (*Melospiza*), Horned Lark (*Otocoris*), Seaside Sparrow (*Passerherbulus*), and many other species.

Such characters, though far more pronounced than many which separate unquestionably distinct species, are still subspecific in their origin and nature, and their possessors may be expected to intergrade if directly or indirectly their ranges are connected.

When the characters distinguishing representative forms become so pronounced as to be actually different colors they are usually accompanied by change in pattern as well. It is therefore difficult to draw a line between these two types of differentiation. However, among North American birds, the Magpie and Yellow-billed Magpie, the Myrtle and Audubon's Warblers, the Maryland and Belding's Yellow-throats, are examples of representative species which differ in color, but agree essentially in pattern of coloration.

Variations in Pattern and Form.

Familiar illustrations of representative forms among North American birds, which differ from each other chiefly in pattern of marking, are the eastern and the western Towhees (*Pipilo erythrop-thalmus* and *P. maculatus*), the Mourning and Macgillivray's Warblers (*Oporornis philadelphia* and *O. tolmiei*), the Canada Jay and Oregon Jay (*Perisoreus canadensis* and *P. obscurus*). Although each of these birds more nearly resembles its representative than do many undoubted subspecies, the character of their distinguishing differences has won for them recognition as species.

It seems evident that the evolutionary influences which produce these qualitative differences of color and form are not the same as those to which we may attribute quantitative variations in the same color. The former are mutational in character and appear to be the external expression of internal or germinal processes stimulated by unknown factors, past or present; the latter seem to be the obvious product of an existing climatic environment.

The differences in color or pattern separating two species may be less in quantity than those which distinguish two subspecies, but they are obviously unlike in quality and the birds exhibiting them give proof of the truth of this belief by living together, or in adjoining ranges without intergrading.

This fact, therefore, must be taken into consideration when in the absence of specimens and detailed information in regard to range, we attempt to determine the probable relationships of two representative forms.

OTHER FACTORS IN THE PROBLEM.

Aside from the extent and nature of the characters separating representative forms there are other factors to be considered in forming an opinion of their relationships.

Differentiation in but one sex.—The differentiating characters of both species and subspecies may be shown by only one sex, either the male or the female, without affecting the status of the form. Thus the males of representative forms may be exactly alike but if the females show those differences in color and pattern which we commonly consider of specific value, then, notwithstanding the similarity of the males, the two birds should rank as species. The same rule applies when the males are unlike and the females alike.

Intergradation through a common ancestor.—It sometimes happens that two forms which have departed from a still existing common ancestor by different geographic routes, and have developed different characters subsequently meet without intergrading. Let us assume, for the sake of illustration, that a still existing Meadowlark of the Rio Grande region is the common ancestor of both the eastern *Sturnella magna* and the western *Sturnella neglecta*. These two forms we know meet in the Mississippi Valley without intergrading.[1] Intermediates are occasionally found, perhaps the result of crossing, but there is no fusion, both forms retaining their distinctive markings and uttering their quite unlike calls and songs even when their breeding ranges overlap.

In such cases birds may be subspecifically related through a third form in one part of their range, and specifically distinct in another. Our system of nomenclature is not sufficiently comprehensive and adaptable adequately to express this kind of relationship.

---

[1] The relationships of these birds is treated at length in Bull. Am. Mus. Nat. Hist., XXII, 1900, pp. 297-320.

In the case I have mentioned the circumstances are pronounced because the two forms actually meet in nature. There are, however, many others in which the biological problem involved is essentially the same, but in which the extremes of a connected series of intergrading races are geographically remote from each other. Under this circumstance we are less apt to ask whether they would meet as subspecies or species.

Assuming, again for the sake of illustration, that our twenty-odd races of Song Sparrows form a connected series of subspecies, it seems improbable that the comparatively small, pale *Melospiza melodia fallax* of southeastern California would intergrade, for example, with the large, dark *Melospiza melodia insignis* of Alaska.

It is not, of course, to be expected that every subspecies in a large, widely differentiated group is subspecifically related to every other race in the group. Nevertheless, even when through the exigencies of distribution they meet without intergrading and thus conform to our chief requirement for specific standing, it seems proper that we should recognize the continued existence of a common ancestor by treating them trinomially.

Intergradation by hybridization.—Intergradation by hybridization unlike intergradation by geographic variation is accomplished regardless of climatic conditions along the line of contact of the ranges of the hybridizing forms. In racial intergradation occasioned by the action of environment, it is customary to find only intergrades in the area of intergradation.

When intergradation is due to hybridization it is customary to find typical specimens of each of the parent forms as well as hybrids between them in the intergrading area. The Bronzed Grackle, for example, breeds from Texas north to Great Slave Lake and Newfoundland without exhibiting any appreciable variation but in a comparatively narrow strip from southern New England southwest along the Alleghanies, where its range meets that of the Purple Grackle, it evidently intergrades with that species by hybridization.[1]

Again, the Yellow-shafted and Red-shafted Flickers hybridize where their ranges meet throughout an area extending from Texas

---

[1] For the data on which this statement rests see Bull. Amer. Mus. Nat. Hist. IV, 1892, pp. 1–20.

along the western border of the Plains to Canada and northwestward to British Columbia.[1] In at least a portion of this area (Wyoming and Montana) hybrids are the prevailing type. Here, although complete intergradation occurs, the fusion of the two forms has not yet produced a uniform intergrade.

On a lesser scale hybridization also occurs with more or less regularity between the Blue-winged and Golden-winged Warblers chiefly at the junction of their ranges from northern New Jersey to the Connecticut Valley, and casually to eastern Massachusetts. The Black-capped and Carolina Chickadees also hybridize, Mr. W. De W. Miller tells me, where their breeding ranges meet in central New Jersey.

In these and similar instances both parents and their hybrid offspring are found in the area of intergradation, and they thus conform to the conditions which distinguish intergradation by hybridization from intergradation by geographic or environmental variation where only the connecting intermediates are found in the area of intergradation.

Of the cases mentioned no ornithologist would question the specific distinctness of the two Flickers and two Warblers. But it is obvious that their marked differences in color and in pattern are accorded greater significance by the systematist than by the birds themselves. Certainly they are not sufficient to prevent these species from freely mating and producing fertile offspring. It is clear therefore that they are organically more closely related than their superficial unlikeness indicates. In notes and habits the Flickers are alike, the Warblers much alike, facts which no doubt have an important bearing on their mating.

The whole subject is far too wide to be adequately treated in this connection and it is introduced chiefly to illustrate the difficulties which the systematist encounters in attempting to employ nomenclature consistently. If we are to be governed by our own definition of a subspecies as an intergrading form, the Red-shafted Flicker would be known as *Colaptes auratus cafer*, the Golden-winged Warbler as *Vermivora pinus chrysoptera*. But whatever the birds may think, it would be difficult to convince the systematist that both Flickers and Warblers are not distinct species.

---

[1] See Allen, Bull. A. M. N. H. IV, 1892, pp. 21–44.

Accepting therefore our artificial standards we may treat these comparatively rare cases as exceptions to the rule and continue to rank the hybridizing parents as specifically distinct. But there will always be cases of intergradation which some ornithologists will attribute to hybridization, others to the action of environment, and their nomenclatural treatment will vary accordingly.

SUMMARY.

Following this superficial review of the more important evidence to be considered in the case of species *vs.* subspecies, I append a summarized statement of the criteria which, in default of specimens, may aid us in reaching a conclusion regarding the status of representative forms.

First, the nature of their differentiations; whether they are positive or comparative in character.

Second, the degree of difference attained.

Third, the relations in space and time of their ranges; whether they are connected or separated, and if separated the extent and nature of such separation.

Fourth, the relations in their respective ranges of the environmental factors which appear to be responsible for the differentiations exhibited whether or not they merge.

Fifth, the relative plasticity of the species and of the group to which it belongs.

Sixth, relative adaptability in habit permitting continuity of range under a widely varying environment, as with the Song Sparrows, for example.

Seventh, information to be derived from the study of other birds and other organisms in the areas concerned.

Eighth, probable lines of descent with relation to the existence or non-existence of a common ancestor.

Ninth, similarity or divergence in habit and in voice.

Finally, while admitting that the relationships of representative forms can be learned conclusively only by the study of adequate collections from throughout their ranges, I nevertheless maintain that the experienced systematist employing the criteria here mentioned can, in most cases, more correctly predicate their status than if he were to act in conformance with certain rules.

And I am convinced that the systematist who will thus treat his specimens as exponents of their environment will make a far more valuable contribution to biology than he who regards them merely as objects to be classified and named.

American Museum of Natural History, New York.

SPECIATION IN THE AVIAN GENUS JUNCO

BY

ALDEN H. MILLER

(Contribution from the University of California Museum of Vertebrate Zoölogy)

INTRODUCTION

THE PRIME OBJECT of this treatise on the genus *Junco* is not nomenclatural or taxonomic. The purpose has been to make a thorough analysis of races and species as they occur in nature in order to determine the degree of unity of each, and to trace differentiation from individual variants through successive stages of group differentiation to the species. Particular attention has been directed to hybridization and intergradation, to the effects of different kinds of isolation, to recombination of racial stocks, and to the spreading of comparatively large mutations in populations. From such investigations certain assumptions concerning the evolution of natural units may be made, and some knowledge can be gained of the genetics of characters.

Juncos do not lend themselves to experimental breeding on a scale that will permit the procedure followed in the study of geographic variation in mice (*Peromyscus*) by Sumner (1932). The method must be largely observational, with resort to various analytical devices. Among birds, the genus *Junco* is especially rich in instances of natural interbreeding of well-marked forms, and it shows great variety in degree and kind of differentiation. Although experiments may not be set up and controlled, there is reasonable assurance that some of the conditions of natural "experiments" can be discerned at the same time that the results are emerging.

The author feels confident that, in spite of limitations in method, a significant body of information about the finely graded stages in the process of evolution of species has been gained. Full knowledge of such stages reveals part of the natural mechanism whereby progressive change takes place. Furthermore, it shows what actually is taking place and not what might take place through human experimentation. There is something presumptuous about the current expression "experimental evolution." The extremely important and essential experiments that pertain to certain aspects of evolutionary process have significance according to the degree to which they relate directly to known conditions of natural evolution. Perfection in our knowledge of natural situations is far from realization. The convergence of various approaches upon the problems of evolution will lead to success only as the weight of various types of evidence is felt by investigators whose backgrounds are of different kinds. The approach in this work on *Junco*, although bearing the earmarks of taxonomy, has at least attempted freely to utilize genetic concepts and the ecologic background.

The taxonomic and nomenclatural contributions of this paper do not embody radical departures from previous systems of classification. Ridgway (1901) and Dwight (1918) have contributed the best information on the taxonomy of juncos to date. No new forms are named herein, but some are found to be rather differently constituted than is generally conceived. Emphasis throughout has been placed on breeding series, the previous inadequacy of which Dwight admitted curtailed his solution of many problems. The attention given to intergradational areas in the search for principles of speciation has clarified nomenclatural treatment in some measure. The customary criterion of intergradation, either geographic or morphologic, has been emphasized in arbitrarily defining species and subspecies. Thus the point of view coincides with that of current workers in vertebrate taxonomy and with the rassenkreis principle. But the application of such a seemingly simple criterion as intergradation encounters pitfalls in insular groups, as Rensch (1934, pp. 47–51) has especially well pointed out. Also, even when the full story of natural interrelations of forms is known, I have found that intergradation is not a matter of simple presence or absence. Some kinds of intergradation are in reality partly hybridization and merge over into all degrees and kinds of limited and sporadic hybridization which defy simplification. The more there is known about the variety of interrelations of forms, the less easy is the use of any one criterion.

Nomenclatural matters and lists of breeding localities have been segregated in the appendixes (pp. 381–425) in order to relieve the discussions of races and species of material not essential to the biologic problems. These items are vital to the proper understanding of classification in the genus, and the breeding localities provide information essential to workers concerned with local distributional problems. It is regretted that detailed localities for migrant and winter birds cannot be presented. The cost of printing lists so extended as these is prohibitive. Winter ranges and distributions are sufficiently well outlined to reveal all important features of winter dispersal.

SURVEY OF THE GENUS

The genus *Junco* ranges from the arctic tree line in North America south to western Panama and comprises twenty-one forms. The forms are all geographically complementary and constitute what Rensch terms an "artenkreis." Some distinct subgroups occur within the genus that are fully isolated and strongly differentiated from one another. Thus artenkreise of lesser scope within the entire group may be conceived. The following outline serves to introduce the forms of the genus and suggests the relationships which are detailed later:

Streaked-backed, dark-tailed junco; *Junco vulcani* (1), Costa Rica, Panama
Plain-backed, partly white-tailed juncos
    Yellow-eyed juncos
        alticola artenkreis
            Junco alticola (2), Guatemala
            Junco fulvescens (3), Chiapas, Mexico
        phaeonotus artenkreis
            Junco bairdi (4), Cape district, Lower California
        phaeonotus rassenkreis
            Junco phaeonotus phaeonotus (5), southern Mexican highlands
            Junco phaeonotus palliatus (6), northern Mexico, southern Arizona

Dark-eyed juncos
Red-backed juncos
  caniceps rassenkreis
    Junco caniceps dorsalis (7), central Arizona, New Mexico
    Junco caniceps caniceps (8), Nevada, Utah, Colorado
  oreganus artenkreis
    oreganus rassenkreis
      Junco oreganus mearnsi (9), Wyoming, parts of Montana and Idaho
      Junco oreganus montanus (10), interior of British Columbia, Washington, and Oregon
      Junco oreganus shufeldti (11), coastal Washington and Oregon
      Junco oreganus oreganus (12), southern Alaska, central coastal British Columbia
      Junco oreganus thurberi (13), Sierra Nevada and mountains of southern California
      Junco oreganus pinosus (14), central coastal California
      Junco oreganus pontilis (15), Sierra Juárez, Lower California
      Junco oreganus townsendi (16), Sierra San Pedro Mártir, Lower California
    Junco insularis (17), Guadalupe Island, Lower California
  hyemalis artenkreis
    Junco aikeni (18), Black Hills, western South Dakota
    hyemalis rassenkreis
      Junco hyemalis cismontanus (19), interior of northern British Columbia
      Junco hyemalis hyemalis (20), Boreal forests of continent, Alaska to New England
      Junco hyemalis carolinensis (21), Appalachian region

    It is the belief of the author that a thorough understanding of the natural history of a group is prerequisite to sound consideration of its evolutionary problems. A large body of information relating to the life requirements of juncos has been gathered from a variety of sources, but it is not yet ready for presentation. Several field seasons have been spent in search of factors in distribution, reproductive rate and cycle, foraging method, migration, molt, territoriality, flock behavior, concealing coloration, and recognition marks. Although little special mention of such items will be found, they naturally have had much weight in attacking the problems of speciation. Especially significant have been observations on tolerance in breeding between members of strongly differentiated forms. In some juncos, fundamental differences in behavior coincide with structural differentiation. Particularly is this true of the yellow-eyed juncos (at least, of *palliatus*), which contrast with dark-eyed forms in manner of terrestrial locomotion, song and notes, general temperament, and method of concealment. Such considerations strengthen one's impression of fundamental distinction.

    Juncos vary from strongly migratory types to permanently resident forms. This variation involves differences in physiological adjustment which may have important effects in isolating stocks and thus furthering structural differentiation. Nevertheless, the tendency of migration to break down the geographic segregation of populations is apparent in this genus as in other genera of birds (see Miller, 1931, pp. 121-124) where similar conditions prevail. Even in resident species the habits of moving about and flocking in winter operate to prevent extreme local isolation.

    Speaking broadly, juncos are a widely adaptable and tolerant type of sparrow that can survive and breed under a variety of local conditions in Boreal and sub-Boreal areas. Usually they are a dominant constituent in point of numbers of Boreal avifaunas (Linsdale, 1932, pp. 223–224). Although they migrate from regions of extreme winter climate, they are among the hardiest of ground-foraging northern fringillids. Differentiations are entirely of a horizontal type, there being essentially no instances of ecologic or zonal segregation in the same geographic region. They do not display striking examples of protective coloration, and the present adaptive value of many of their racial and specific characters is at least not self-evident.

MATERIALS

Field expeditions.—In 1931 active field work was begun. The plan was to visit breeding areas of the various forms of juncos occurring in the Rocky Mountain area. From May 22 to August 1, eighteen successive collecting stations were established, beginning in southern Arizona (Catalina Mountains) and ranging north through the San Francisco Mountain region, along the Wasatch Mountains of Utah and into eastern Idaho to the Yellowstone Plateau. The stations were from 20 to 100 miles apart, in accordance with the problems of local distribution and intergradation that were encountered. The following forms were studied and the boundaries and interrelations of each in this sector were determined: *palliatus, dorsalis, caniceps,* and *mearnsi.*

In 1932, from June 15 to August 1, a similar plan was pursued in Oregon and Idaho. The fifteen collecting stations were distributed in an east–west line from the range of *shufeldti* in the Cascades of central Oregon through the *montanus* complex to *mearnsi* in central Idaho, and in a north–south line from the British Columbian border to the Snake River, within the range of *montanus*. Also, the relation of *thurberi* and *shufeldti* in the Cascades of southern Oregon was studied.

In 1934, from May 13 to June 13, a trip was made to the Fraser River Valley of interior British Columbia, especially the Bowron Lake region near Barkerville. The northern division of the race *montanus* and occasional hybrids with *J. hyemalis* were met with.

In 1937, during a month in the field in Lake and Deschutes counties, Oregon, much attention was directed to the interrelations of *shufeldti, montanus,* and *thurberi.*

In May and June, 1935, an expedition to the mountains of northwestern Nevada served to clarify the status of juncos in the marginal breeding areas there. In 1936, further time was spent in late May with *Junco phaeonotus palliatus* in the Chiricahua Mountains of Arizona. In California, field work in the coastal area from Humboldt County to San Luis Obispo County has been carried on irregularly, as opportunity offered, in the course of the last six years. All materials obtained on field expeditions are now in the California Museum of Vertebrate Zoölogy.

Total number of specimens examined.—The grand total of specimens used in connection with the work was 11,774. The working collection assembled at the Museum of Vertebrate Zoölogy numbered 4486. However, totals in themselves mean little, and it must be stressed that the examination of many hundreds of winter-taken *hyemalis*, for example, was cursory in the extreme and in keeping with their slight value. The critical material representing birds on their breeding grounds totaled 4552. This number, compared with the 500 breeding birds available to Dwight in 1918, indicates the opportunities which have been presented to me.

Collections studied.—In the summer of 1933 the juncos in the collections listed below were examined:

Colorado College, Colorado Springs (kindness of Mr. E. R. Warren)
Colorado Museum of Natural History, Denver (kindness of Mr. J. D. Figgins)
University of Colorado, Boulder (kindness of Mr. H. G. Rodeck)
University of Kansas, Museum of Birds and Mammals, including the Alexander Wetmore Collection, Lawrence (kindness of Mr. C. D. Bunker and Mr. W. S. Long)
Carnegie Museum, Pittsburgh (kindness of Mr. W. E. C. Todd)
United States National Museum, Washington, D.C. (kindness of Dr. Alexander Wetmore and Dr. Herbert Friedmann)
Biological Survey Collection, U.S. Nat. Mus. (kindness of Dr. H. C. Oberholser)
Philadelphia Academy of Sciences (kindness of Dr. Witmer Stone)
American Museum of Natural History, New York, including the collections of J. Dwight, Jr., L. C. Sanford, and J. T. Zimmer but not the Rothschild Collection (kindness of Dr. Frank M. Chapman and his staff)
Museum of Comparative Zoölogy, Harvard, including the collections of John H. Thayer, C. F. Batchelder, A. C. Bent, and F. H. Kennard (kindness of Mr. J. L. Peters and Mr. Ludlow Griscom)
Field Museum of Natural History, Chicago (kindness of Dr. Wilfred Osgood and Mr. Rudyerd Boulton)

In the summer of 1934 the juncos in the following collections were examined:

National Museum of Canada, Ottawa (kindness of Mr. P. A. Taverner)
Royal Ontario Museum of Zoölogy, Toronto (kindness of Mr. J. H. Fleming)
J. H. Fleming Collection, Toronto
British Museum, London (kindness of Mr. N. B. Kinnear)
Berlin Museum, Berlin (kindness of Professor Dr. Erwin Stresemann)

Certain juncos have been examined in the Leland Stanford Junior University Collection (kindness of Professor Willis H. Rich) and in the Louis B. Bishop Collection at Los Angeles.

Entire collections or important parts of collections have been lent to me for extended examination in connection with the museum material at Berkeley:

Biological Survey, through Dr. Harry C. Oberholser
D. E. Brown
California Academy of Sciences, through Mr. Harry S. Swarth and Mr. James Moffitt
California Institute of Technology (Donald R. Dickey Collection), through Mr. A. J. van Rossem
John Cushing
William B. Davis
Ralph Ellis
Reed Ferris
Laurence M. Huey
Randolph Jenks
Stanley G. Jewett
J. Eugene Law
Los Angeles Museum of History, Science and Art, through Mr. George Willett

J. and J. W. Mailliard
T. T. and E. B. McCabe
D. D. McLean
University of Michigan (including Max Minor Peet Collection), through Dr. Josselyn Van Tyne
Museum of Northern Arizona, through Mr. Lyndon L. Hargrave
Provincial Museum of British Columbia, through Dr. Ian McTaggart Cowan
San Diego Society of Natural History, through Mr. Laurence M. Huey
United States National Museum, through Dr. Herbert Friedmann

Acknowledgments.—To the persons represented in the foregoing lists who have sent specimens or who have courteously made collections available for study I wish to express my very great appreciation. Many have contributed information in letters or in conversation that has been of distinct importance. Without the interest and coöperation of this large group of ornithologists the scope of the work would have been seriously curtailed.

It is a pleasure to acknowledge special assistance received from several sources. Financial support of field work in 1931 and 1932 was provided by grants from the Board of Research of the University of California. Assistance from the Work Projects Administration, Project A.P. 165–03–6079, has been had in the preparation of maps and manuscript. Mr. Lyndon L. Hargrave has, among other courtesies, sent me live juncos from Arizona for purposes of experimental breeding. Mr. Harry S. Swarth gave me the benefit of his wide experience with juncos by contributing criticism and counsel. Dr. Lewis W. Taylor, of the Division of Poultry Husbandry, University of California, coöperated in arranging facilities for captive birds. Mr. T. T. McCabe for a number of seasons directed special attention to the assembling of critical material from British Columbia; his intelligent and energetic pursuit of my problems in the field has been highly valued. Mr. A. J. van Rossem carefully examined and furnished reports on types of juncos in Old World museums. Dr. William B. Davis has obtained some especially significant material for me in Idaho. The continued support of the Museum of Vertebrate Zoölogy through its former director, the late Dr. Joseph Grinnell, was most encouraging, and a variety of facilities was placed at my disposal. The assembling of data from specimens in museums visited in 1933 was possible on the scale attempted because of the assistance of Virginia D. Miller; also in matters of field work and measurements her efforts have been a large factor in the bringing of this work to completion.

METHODS OF ANALYZING CHARACTERS

Pigmentary characters were analyzed in two ways. The total color effect in a given area of plumage was matched with a graded series of samples, selected to represent approximately equal degrees of difference in color. For example, twelve grades of black and gray were recognized, and typical samples were selected for use in designating variants in head color in a number of the northern forms of the genus. It was not feasible to employ a color wheel for standardizing the analyses, because of the variety of places where examinations were conducted. Type samples of variants were always carried with me when determinations were attempted. Much effort was directed toward establishing equivalents of variants in worn and fresh plumage. Careful comparison of stages in wear, consideration of feather structure and pigment distribution, and study of molting birds has made possible a fairly safe identification of colors of the breeding plumage with their initial appearance in fresh plumage. The majority of critical analyses of coloration have been made on breeding birds in spring plumage before the feathers were badly worn. Ideally, all the basic comparisons should be made on fresh-plumaged birds taken while still on their breeding grounds in late summer; but practically this is impossible, since so far as many northern races are concerned the period over which such birds might be collected would be only two weeks.

A second phase of color analysis has been simple microscopic examination of feathers to determine types of pigment present in various parts of the feather. In juncos there appear to be no lipochromes in the plumage, but only melanins. The melanins apparently fall into the two major classes defined by Görnitz (1923, pp. 468–469): eumelanins, dark brown, gray, or black pigments, frequently peripherally distributed in barbules; and phaeomelanins, reddish brown, pink, and yellow. Although Görnitz points out that there is no absolute distinction between these two kinds, in *Junco* the pigments almost always fall obviously into one class or the other.

The types and distribution of pigments differ in important ways. For example, eumelanin may be entirely absent from an area of the plumage in one species, but present in that region in another. Nevertheless, microscopic study has not revealed a sound basis for much of Dwight's (1918) system of quantitative and qualitative color differences. In general, fewer basic qualitative differences are found than he indicated. If eumelanin grades into phaeomelanin by successive degrees of oxidation, as Görnitz's work indicates, is not this a finely graded quantitative change? Even the fundamental distinction in these two classes of melanins is not, in one sense, qualitative.

The annual molt of juncos is the only one of any importance. The prenuptial molt is nearly obsolete and effects no material change in appearance. The postjuvenal, or first fall molt, is incomplete; primaries, secondaries, and greater primary coverts are not replaced, and the tail usually is not molted at this time. Juvenal wing and tail feathers carried through the first year are indistinguishable from those of birds a year or more old. The body plumage of the fall immatures in a number of races is different in average coloration from that of adults. But no absolute differentiation has been revealed—a situation also true of sexual dimorphism.

Because of this situation and the fact that the age of but few measurable birds taken on their breeding grounds has been dependably determined from the skull, no segregation of age groups could be made in the analysis of wing and tail lengths. In one group of fall *thurberi* of known age, the maximum for first-year birds was nearly the same as that for adults. The factor of age composition of samples does not appear to be as important, therefore, as in some other genera, and of necessity is here neglected.

The tail pattern of juncos has been variously regarded in previous work on the genus. No one can reasonably deny that there is a large amount of significant variation. The narrow taxonomic approach minimizes the variations in tail white because great individual variation makes it unusable in practical diagnoses. It appears, however, that tail pattern is relatively stable in an individual. The variation in successive feathers grown from a given feather germ usually is not great, though it is appreciable. This does not prove that the details of pattern are genetically fixed, but only that the feather germ is fairly stable once developed. The variations between right and left sides of the tail of the same bird would not likely be of genetic origin. In analyzing tail variants, only differences larger than those ordinarily encountered between the two sides of the same individual have been considered. Probably they represent in fair measure hereditary differences. Accordingly, each feather was classified separately for the two halves of the vane in respect to whether it was entirely black, black with a white spot, ¾, ½, or ¼ black, white with a black spot, or entirely white. Finer distinctions usually would be meaningless.

Linear measurements from skins, in millimeters, were taken as follows: wing, the chord; tail, from posterior base of oil gland; bill length, from anterior rim of nostril to tip; bill depth at base of culmen; tarsal length; middle toe without claw, from junction with inner toe; hind toe without claw, to base of inner toe (digit II). The greatest instrumental errors are made in the toe measurements. Bill depth was not taken if the mandible was out of place. Juvenal birds were measured for wing, tail, and feet if the flight feathers showed no sheaths. Feather growth has at that time ceased and these same feathers would be carried throughout the first year. The feet attain full growth before the wing feathers. It was found that bill dimensions, especially length, increased through part of the fall in immatures. Bill length is especially subject to alteration from abrasion (see p. 183).

Weights of some series of breeding birds have been available. When seasonably comparable, they may in males be the best practical means of registering the general size of the bird. None of the linear measurements satisfactorily indicate body size, although they reflect differences in size.

VARIATIONS AND INTERRELATIONSHIPS OF RACES AND SPECIES

Treatment of the different forms of juncos may for convenience begin with *Junco caniceps*. It is a central type geographically, and clarification of its relationships lays the ground work for consideration of a number of other species and races. Although related to the brown-eyed northern juncos, in a sense it forms a link with the yellow-eyed juncos. There is no reason to suppose that *J. caniceps* is phylogenetically more important than other rassenkreise.

Rassenkreis Junco caniceps
Gray-headed Junco

The group is composed of two sharply marked races, a southern *dorsalis* and a northern *J. c. caniceps*. They replace each other geographically and interbreed freely in certain isolated areas that are intermediate geographically. The rassenkreis occupies the southern Rocky Mountains and the Great Basin ranges. Migration is irregular, but is definite and fairly extensive in certain members of *J. c. caniceps*.

Principal characters of the group.—Iris dark brown; lower mandible flesh-colored, upper mandible black or flesh-colored. Back with sharply defined mahogany red area, confined normally to interscapular region; red color results from phaeomelanins deposited in barbules, basally in distinct granules, distally uniformly; eumelanins lacking in barbules; tips of feathers grayish because of reduced amount of phaeomelanin pigment. Lores and ocular region black, contrasting with neutral grays of head. Sides gray without line of demarcation separating them from upper breast. Tail always with two, and usually three, outer feathers partly or completely white. Size fairly large, wing of males averaging about 82 mm.

Sexual differentiation in plumage slight. Females average slightly lighter in colors of head and side in race *caniceps* only. Fourth rectrix more often pure black, fifth and sixth rectrices less often pure white in females than in males. Degree of sexual differentiation in rectrices variable, depending on population involved. Size differentiation about 5 per cent in wing and tail, 2 per cent in bill and feet. Females often have traces of buff on tips of feathers of sides, and reddish- or buff-tipped wing coverts. A distinctive immature or retarded plumage not recognizable; occasionally young males have buff-tipped sides, and this feature predominates in young females.

The red of the back alters with the season, becoming brighter and yellower to approach burnt sienna and Sanford's brown. The gray tips are worn off early, but this does not seem to affect the tone of color importantly. Brightening is not pronounced until late April, but in the succeeding six weeks the greatest alteration takes place. Since there are no dark eumelanistic deposits in the ends of the barbules, wearing away of these parts cannot account for the change. Seemingly there is an alteration in the phaeomelanin itself under the action of light. Abrasion of the surfaces of the barbs and barbules does not alone explain adequately the brightening of the color. The grays of the head pale slightly, so that the mass effect in worn plumage is that of lighter hood, except in the extreme state of wear of midsummer, when exposure of the basal downy barbules gives a sooty appearance.

Junco caniceps caniceps

Racial characters.—Features present in all individuals that distinguish them from *J. c. dorsalis* are: (1) darker neutral gray (light neutral gray and neutral gray) of hood, especially that of throat, and (2) flesh-colored upper mandible. Averages for the following measurements are lower: tail length 3 per cent, bill length 3 to 6 per cent, bill depth 3 to 6 per cent, tarsal length 2 to 3 per cent, and middle toe length 2 to 6 per cent. The amount of white in the tail averages less (see p. 207).

Nongeographical variation.—The race *caniceps* is relatively uniform. Especially is this true with regard to dorsal coloration, which in seasonally comparable specimens displays none of the variation found in the backs of *J. hyemalis* and *J. oreganus*. Hood color and white in the fourth and fifth rectrices varies individually, but also regionally, and so will be considered separately. White of the outer rectrix is not geographically variable within the race; 7.2 per cent lack an immaculate white outer web; 1.7 per cent lack a completely white inner web.

Reddish color of variable extent is not uncommon on the pileum. It appears in all populations. In the total of 772 *J. c. caniceps* examined, 48, or 6.21 per cent, have some feathers distinctly red, not merely buff tipped. Definitely delimited white spots on wing coverts occur in 1.29 per cent.

Wing, tail, and tarsal lengths are so nearly the same (average differences no greater than 1 per cent and unreliable statistically) in the Nevada, Utah,

TABLE 1
WING, TAIL, AND TARSAL LENGTHS IN JUNCO CANICEPS CANICEPS

Measurement	Sex	No. of specimens	Mean with prob. error	Standard deviation	Coeff. of var.
Wing............	♂	159	81.96 ± .11	1.97	2.40
	♀	81	76.83 ± .17	1.71	2.22
Tail.............	♂	145	74.37 ± .13	2.44	3.28
	♀	72	70.02 ± .18	1.93	2.75
Tarsus..........	♂	157	20.31 ± .03	0.58	2.85
	♀	81	20.01 ± .05	0.67	3.30

Correlation for wing and tail is .71 ± .027.

Colorado, and New Mexico subregions of the breeding area that all breeding birds may be combined for the statistical treatment of these features. (See table 1.)

Geographical variations within the race.—Color of hood, varying between neutral gray and light neutral gray, averages the same in Utah, Colorado, and New Mexico. But in Nevada there is a greater proportion of dark-headed types and a very few individuals that are darker than the extremes in the other three regions. Most of the Nevada specimens are from the Toyabe Mountains, western frontier of the *caniceps* rassenkreis. The Nevada group includes birds from the rest of the eastern part of the state except the northern half of Elko County, and all of Clark County, where hybridization with other juncos occurs. The samples from the Ruby Mountains and Snake Mountains, comprising most of the individuals aside from the Toyabe group, differ in no respect from the latter series.

To determine the degree of differentiation in hood character in Nevada examples, 62 breeding males, 30 from southern and central Utah, and 32 from the Toyabe Mountains, were mixed and then seriated according to color of the ventral aspect of the hood. In the darker third, 3 out of 21 were Utah birds; in the lighter group, 4 out of 21 were Toyabe birds. Forty-four per cent could not be segregated into geographic groups on the basis of hood color. In a group of 30 females from the same localities 43 per cent could not be segregated. One bird from the Toyabe Mountains has a peculiar black nuchal crescent, representing the posterior part of the hood dorsally. The rest of the hood is dark gray of the type encountered in other birds of this region.

At least three individuals from central and eastern Nevada have some color in the sides of *J. oreganus* type. One of these is in all respects *shufeldti*. All were taken in the breeding season. There are some other individuals of similar character, not taken on dates that would completely preclude migration, that might also have remained in central Nevada throughout the summer. The significance of these variants will be discussed later, particularly in connection with the relationships of *caniceps* to adjoining species.

The white of the tail tends to be more extensive in rectrices 4 and 5 in Nevada

TABLE 2
PERCENTAGE OF POPULATION OF J. C. CANICEPS WITH INNER WEBS OF RECTRICES SHOWING BLACK OR WHITE

Sex	Colorado	New Mexico	Utah	Nevada
Percentage with inner web of rectrix 4 all black				
♂	7	5	3	0
♀	10	17	..	6
Percentage with inner web of rectrix 5 all white				
♂	53	52	39	90
♀	24	33	20	53

birds. This is best shown in the inner webs of these feathers. Segregation of more than half of the Nevada population from the other *caniceps* groups would be impossible on this character alone.

The difference between the Colorado–Nevada and the Utah–New Mexico groups in bill length is five times the probable error of the difference and hence is significant statistically. However, it was noticed in comparing the longer bills of the Utah group with the shorter ones of the Nevada group that the tip of the bill was more attenuated in the former. This extreme terminal part is subject to much variation in captive juncos, owing to wear. Most of the Nevada series was taken in the Toyabe Mountains in May, 1930. In this season the birds were delayed in breeding by late snows. Unlike the Utah group, which was nesting and feeding upon insects, they, of necessity, fed upon seeds on the barren ground. Such a factor might account for stubbier and shorter bills. No known seasonal factors can be adduced to explain the short bills of the Colorado specimens. The Colorado, New Mexico, and Utah series are completely comparable seasonally. Nevertheless, the uncertainty regarding inherent differences in bill length in Nevada throws doubt upon differences among these other groups.

No reservations need be made concerning bill depth. This measure is reasonably free from influences attendant upon wear. Depth is greater in Nevada and New Mexico than in Colorado and Utah. The differences are three to five times the probable errors and are equally manifest in both sexes.

Length of the middle toe is greater in Nevada and New Mexico than in Colorado and Utah. The differences are only two to three times the probable error, but because they occur in both sexes and parallel differences in bill depth they are thought to be significant.

TABLE 3
MEASUREMENTS OF DIFFERENT POPULATIONS OF J. C. CANICEPS

A. MALES

Measurement	Pop.	No. of specimens	Mean with prob. error	Standard deviation	Coeff. of var.
Bill..............	Colo.	49	8.28 ± .03	0.36	4.34
	N. M.	18	8.56 ...	....	....
	Utah	37	8.57 ± .03	0.29	3.38
	Nev.	50	8.22 ± .04	0.35	4.25
Bill depth.........	Colo.	38	6.27 ± .03	0.29	4.62
	N. M.	16	6.50 ...	....	....
	Utah	37	6.26 ± .03	0.23	3.67
	Nev.	46	6.43 ± .03	0.30	4.66
Middle toe.........	Colo.	41	11.52 ± .05	0.45	3.90
	N. M.	21	11.79 ...	....	....
	Utah	37	11.59 ± .05	0.46	3.96
	Nev.	51	11.81 ± .04	0.41	3.47

B. FEMALES (*averages*)

Pop.	No. of specimens	Bill	Bill depth	Middle toe
Colo...............	23	8.16	6.12	11.23
N. M...............	12	8.33	6.40	11.41
Utah...............	12	8.24	6.20	.....
Nev................	35	8.22	6.34	11.41

In summary, partial differentiation geographically within the race involves head color, tail white, and bill depth, less certainly bill length and toe length. The color characters as well as the dimensional characters blend from one extreme to the other. Probably each is inherited through some mechanism of multiple factors—a supposition which is further supported by evidence from naturally occurring hybrids with other races. One population contains a greater number of individuals that approach one extreme (homozygous for many allelomorphic pairs?) than do other populations, but all populations contain occasional extreme individuals. The Nevada population has a higher frequency of variants with dark head and extremely white tail. Both the Nevada group and the New Mexico group show greater average bill depth and toe length. Nowhere is the geographical segregation of characters high enough to make practical any nomenclatural innovations. These incipient differentiations are, however, significant stages in racial evolution.

Distribution in breeding season.—The areas here mentioned are those in which *J. c. caniceps* breeds in pure form, that is, in populations showing no decisive evidence of hybridization or geographic intergradation.

The spotted distribution of this junco must be emphasized. It inhabits for the most part a series of mountaintop islands above 7000 feet in the arid Rocky Mountain and Great Basin ranges. Associations in which it breeds include coniferous forest types dominated either by spruce (*Picea*), *Pseudo-*

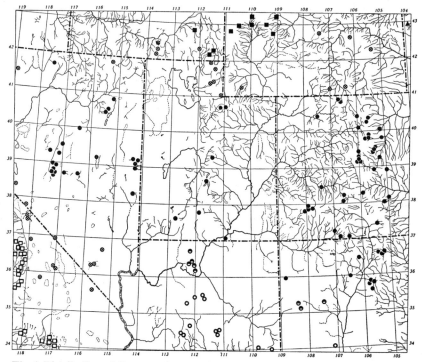

Fig. 1. Distribution of *Junco caniceps caniceps* and neighboring forms in the breeding season. Great Basin and central Rocky Mountain region of United States (Utah, and parts of eastern California and Nevada, southern Idaho and Wyoming, northern Arizona and New Mexico). Each symbol represents a locality from which specimens have been examined. Dots, *J. c. caniceps*; circles, *J. c. dorsalis*; half dots, intermediates between *caniceps* and *dorsalis*; circles enclosing cross, hybrids of *J. c. caniceps* and *J. o. thurberi* (in west), or *J. o. mearnsi* (in north); solid squares, *J. o. mearnsi*; open squares, *J. o. thurberi*; triangle, *J. aikeni*.

tsuga, Pinus contorta, Pinus ponderosa, Pinus flexilis, or fir (*Abies*). It also breeds in pure stands of aspen (*Populus tremuloides*) and of mountain mahogany (*Cercocarpus ledifolius*). Compared with *Junco oreganus*, it shows high tolerance for arid forest and ground cover. It may occupy aspen groves where there is no surface water within two to five miles of the nest site. Unshaded forest floor with the ground poor in humus and nearly lacking in green plant cover is unsuitable for breeding.

Colorado.—Chiefly in Boreal Zone (Cary, 1911) in mountains west of longitude 105°, including the ranges forming the continental divide and such outlying mountains as the Medicine Bow Range, Williams Fork Mountains (Routt County), San Miguel Mountains,

Pike's Peak, Sangre de Cristo Range, Culebra Range, and the Wet Mountains; probably also on the Uncompahgre Plateau.

New Mexico.—Chiefly in Boreal Zone (Bailey, 1928, pp. 7, 8, 742, pl. 1) in the Sangre de Cristo, Jemez, San Juan, and Chuska mountains in the northern part of the state.

Arizona.—Mountains of northeastern Apache County, adjoining the Chuska Mountains of New Mexico.

Utah.—Uinta and Wasatch mountains from southern Summit County southward through the mountains of the central part of the state to the Pine Valley, Parowan, and Escalante mountains; in the southeastern section in the Navajo and La Sal mountains (Tanner and Hayward, 1934, p. 232), and presumably also in the Henry and Abajo mountains; probably also in the Deep Creek Mountains on the Nevada border.

Nevada.—Restricted areas in mountains of east-central section of the state: Ruby, Snake (Wheeler Peak), Shell Creek, White Pine, Roberts, Monitor, Toquima, Toyabe, Shoshone and Cedar ranges. Also in Santa Rosa Mountains of Humboldt County, although additional samples from this range might reveal a hybrid population.

Wyoming.—Mountain ranges along the Colorado and Utah borders such as the northern Medicine Bow and Sierra Madre ranges and the Uinta Mountains in the vicinity of Fort Bridger have breeding populations that show traces of intermixture of character with *mearnsi*. Although birds in this area are usually typical *caniceps*, no pure population is known from the state.

Distribution in winter.—Winters abundantly in the mountainous sections of Arizona and New Mexico. Also occurs in Colorado along the east base of the mountains and south through western Texas (Chisos Mountains) and Sonora and Chihuahua to northern Durango. The bird rarely moves eastward any great distance onto the plains. There are records for south-central Nebraska (Red Cloud), Kansas (no. 24040, Acad. Nat. Sci. Phila., W. S. Woods, 1859, no locality), and Oklahoma (sight record, Nice, 1931, Norman, Cleveland County). Probably *caniceps* winters in parts of Nevada, but winter-taken specimens have not come to hand. They occur at least in southern Utah (Presnall, 1935, p. 209). Occasionally winters in coastal southern California (6 specimens), as far south as San Diego County. A report from Michigan (Allen, 1879, p. 123) is not well authenticated. In Arizona the birds usually frequent the foothill districts, especially the oak association.

Relationships with adjacent forms: J. c. dorsalis.—Populations of intermediate and mixed character exist on the Kaibab Plateau in northern Arizona and in the Zuni Mountains in New Mexico. Twenty-one breeding males are available from the Kaibab region. They show that *dorsalis* and *caniceps* interbreed freely. In collecting this series in the summer of 1931, I saw no evidence of segregation of the two types in the field. Each bird reacted to other juncos in the same fashion irrespective of bill and head color.

The Kaibab population affords opportunity for analysis of a kind of intergradation frequently found between geographic races of birds. Seven characters are worth considering in these racial hybrids: color of upper mandible; color of head; length of tail, bill, tarsus, middle toe; and depth of bill. *Dorsalis* is significantly greater than *caniceps* in these measurements and possesses a black upper mandible rather than a flesh-colored one (see fig. 2). The head of *dorsalis*, especially the ventral surface, is lighter than in *caniceps*. Samples of *caniceps* from points in southern Utah nearest (35 miles) the Kaibab Plateau and of *dorsalis* from the plateau south of the Grand Canyon of the

Colorado (50 miles distant) give no suggestion of a gradient in character toward the region of intermediacy. In each adjacent region the characters average the same as for the rest of the race of which the birds are a part; in *caniceps*, the same as for the Utah region. The question arises whether the intermediates are an intermediate stage in differentiation, having evolved *in situ*, or are the result of interbreeding of the two fully differentiated parental types. We may assume that the latter is the true state of affairs, since both parental types are present, and because the characters of these races show no certain correlation with environmental conditions that would suggest origin of intermediates at this spot.

Head color in the Kaibab series was classifiable into five groups. The birds first were seriated and certain faint breaks in the seriation determined. The range of color variation was slightly greater than that in the Utah–Nevada comparison (see p. 182) and of course involved different absolute values. The five classes are: (1) darkest group, equivalent to the average of Utah birds and equal to the darker members of the lighter third of the Utah–Nevada seriation; (2) equivalent to the four lightest birds of the Utah group; (3) intermediate birds, lighter than any *caniceps* and slightly darker than any *dorsalis;* (4) equivalent to *dorsalis*, but darker than average; (5) equivalent to average *dorsalis*, or lighter than average.

Frequency distribution of these types in order was: 3:4:6:4:2. This is remarkably suggestive of phenotypic ratios derived in hybrid generations subsequent to the F_1 where two or three pairs of multiple factors are in operation. The ratio for three pairs reduced to comparable figures is $\frac{1}{3}:2:5:7:5:2:\frac{1}{3}$. The extremes appear only once in 64 times and probably would not occur in a sample of 20. The ratio with two pairs of factors would be 1:4:6:4:1. It might be assumed that the terminal members of the ratio in the intergrades are augmented by immigration of a few pure *dorsalis* and *caniceps* types. There is no evidence of simple dominance of either head type. Apparently the two parental stocks have contributed about equally to the population.

Bill color in these same individuals was classified on the basis of field notebook sketches of the color areas made when the birds were in the flesh. These sketches have been checked with the appearance of the bill in the dried skins. Certain alterations may occur in drying, and the original drawings then prove vital to an accurate recording of the bill color. Figure 2 shows successive stages in the pigmentation of the upper mandible. The first three are types that occur normally in *caniceps*. The first, without any black except for the pit of the nostril, is rare. Types 2 and 3 are equally abundant. Types 8 to 10 are characteristic of *dorsalis;* most individuals are solid black. Types 4 to 7 are found, with rare exceptions, only on the Kaibab Plateau and in the Zuni Mountains. Progressive pigmentation takes place by increase in the size of the basal and terminal black areas until they meet on each side, lateral to the culmen. The basal part of the culmen may or may not be pigmented. The Kaibab birds could be classified into the stages as figured, although there was some variation in the manifestation of each type.

The genetic mechanism is not readily apparent from inspection of the frequency of occurrence of these bill types. Types 2 to 9 occurred on the Kaibab Plateau as follows: 2:6:3:2:2:3:3:1. Pigmentation of the basal part of the culmen occurs independently of the general extension of pigment elsewhere. Thus type 6 is like 4, with the culmen dark, and similarly, 9 is comparable to 7. A single independent pair of factors may control pigmentation of this area. Only 3 out of 22 birds possess black culmens; hence black seems to be recessive to flesh color.

Ruling out the culmen, extension of black on the rest of the bill can be best expressed by the merging of type 6 with 4, and of 9 with 7. The frequency distribution from light to dark is then: 2:6:5:2:4:3. Probably several factors are involved to produce such results. The distribution is bimodal, how-

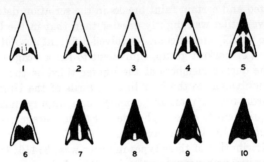

Fig. 2. Dorsal surfaces of bills, showing variation in distribution of pigment. 1–3, *J. c. caniceps;* 8–9, *J. c. dorsalis;* 4–7, intermediates from Kaibab Plateau, Arizona.

ever, and the several classes might be the variable manifestation of two genetic types corresponding to the two modes. The proportion of these two is 13:9. This approaches the 9:7 phenotypic ratio derived in the F_2 with two pairs of complementary factors. Irrespective of the exact mechanism, there is partial dominance of the light-billed *caniceps* type in the same group of individuals that shows no dominance of *caniceps* head color.

Bearing out the evidence for dominance derived from this wild population is the result of a cross of *dorsalis* with *J. o. pinosus* (light-billed), the offspring of which possesses a flesh-colored bill like that of *J. c. caniceps*. The F_1 individual is of type 3 and thus shows the dominance of flesh-colored culmen over black, as also that of the light-billed condition generally.

The average values for the mensurable characters fall about halfway between the values for *caniceps* and *dorsalis*, with the exception of tail length, which is equal to that for *caniceps*. The distribution of size values (fig. 3) shows a small group of three long-tailed birds, equivalent to *dorsalis*, that are separate from the rest of the individuals. This suggests that the long-tailed condition is recessive. There are enough individuals of the dominant, short-tailed *caniceps* type so that in this small sample of intermediates the average happens to be similar to *caniceps*, in fact slightly lower. It should be noted that although wing and tail are correlated in intraracial variations (.71), the differentiation of *dorsalis* from *caniceps* involves a lengthening of tail without correlated lengthening of the wing. It would appear that there are factors governing the ratio of wing and tail and, in addition, factors independent of them which determine a correlated variation in the length of the two.

Bill length shows a distinctly bimodal curve in the Kaibab hybrids, with the greater mode representing the *caniceps* type. This approximates a 9:7 ratio of F_2 hybrids with two pairs of complementary factors, just as in the extension of the dark areas on the bill. Length and color are not correlated in these hybrids and therefore cannot depend on the same factors, nor upon linked factors.

Variations in the other three measurements display normal curves, with

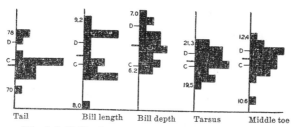

Fig. 3. Individual variations in dimensions (mm.) in intermediates between *J. c. caniceps* and *dorsalis* from the Kaibab Plateau, Arizona. D, average value for *dorsalis*; C, average for *caniceps*; heavy horizontal marker, average for intermediate population.

averages intermediate between the two races. They conform to the pattern of variation encountered in the inheritance of quantitative characters by multiple factors. No appreciable correlation was found between these characters, nor between any of them and the color of the head.

On the Zuni Mountains of New Mexico three significant specimens have been taken by E. A. Goldman. One of these is in all particulars *caniceps* and another is just as distinctly *dorsalis*. The third is intermediate in head color and of type 6 bill. These three specimens, perhaps supplemented by observations of other birds, seemingly formed the basis for Cooke's statement (Bailey, 1928, p. 743) that *dorsalis* and *caniceps* bred at the same locality and hence were "full species." Doubtless the intermediate bird was referred to *caniceps* because it lacked a completely dark upper mandible. In the light of our knowledge of the Kaibab situation, these three birds give every indication of a similar mixed population on the Zuni Mountains, and they fail to argue for specific distinctness of the two forms.

There are specimens of *dorsalis* at hand from localities near the Zuni Mountains and from the San Mateo Mountains of Valencia County in the same latitude. Larger samples might reveal mixtures at these localities also.

Junco oreganus thurberi.—The interbreeding of *caniceps* with the strikingly different *thurberi* was first suggested by van Rossem (1931, pp. 330–331) in his account of the peculiarly heterogeneous populations of the Charleston and Sheep mountains of southern Nevada. Many of the essential features of these groups were remarked by him. He indicated that head color and side color varied from a *caniceps*-like condition to that of *thurberi* in his series, but that there was no correlation of these characters in individuals. "The most constant character in the series is the red back," which is close to *caniceps* in tone. Van Rossem made two proposals which he thought merited serious consideration in accounting for the variations within these colonies which are out of contact with other populations of breeding juncos. These were that "there may be here a blending of races because of the simultaneous occupation of formerly unoccupied territory by two or more distinct types, or else because of an invasion of the range of an established form by radiations from another area," or that "the unit may be a formerly stable race which is now changing by a series of mutations into another type." In connection with the first proposal, he states that "this population was probably derived either from *Junco caniceps*, or the ancestral stock of which *caniceps* is the present day expression. The dilution of this original stock may have come from *thurberi*, which is resident in the mountains some 150 miles to the west . . . or from *Junco oreganus shufeldti* Coale [= *montanus*] . . . which is the common wintering one in this region. While on geographical grounds *mutabilis* [= the Charleston Mountain population] could conceivably be the result of a fusion of *caniceps* with *thurberi* and while there is evidence of such fusion in the variability of head and side color, the uniformly red backs in both sexes, in which no suggestion of the back color of either *thurberi* or *shufeldti* is apparent, rather militates against this supposition." This latter argument concerning back coloration I do not consider valid. Evidence which I now have of interbreeding of these races in other regions, such as the White Mountains of California, points to the dominance of the *caniceps* back in hybridization. Further, a close study of the backs of van Rossem's Charleston Mountain series reveals that certain of the birds, chiefly those with *thurberi* head, have some mixture of back character. The *caniceps* back does predominate, but seemingly because of genetic dominance.

Van Rossem appears to favor the mutational hypothesis partly because of this supposed lack of blending in the back. He uses mutation in the older sense of the term—as applying exclusively to large or striking alterations. Such a veritable explosion of mutations, locally concentrated as would be necessary to produce the Charleston Mountain group, is so much at variance with conditions now existent elsewhere in the genus and with natural conditions among avian species generally as to be highly improbable. Even in the genetics laboratory such a situation arises only under excessively abnormal treatment such as that imposed by artificial radiation. Mutations, though undoubtedly the basic units out of which species of juncos have been built, are, as usual, infrequent in occurrence in this group.

It is also difficult to explain as a coincidence the fact that all these supposed mutations of the Charleston Mountain group are characters of a kind to be found normally in the adjacent species. None of them is really new or incapable of being produced by various assortments of factors in interbreeding. Other populations similarly diverse in character are found in the genus only in areas between races or species. The mutation argument recalls the classic example of *Oenothera* expounded by De Vries, in which frequent mutations of a striking sort later proved to be in large part the result of an earlier hybridization of species.

Van Rossem's first proposal holds much merit and, in the light of substantiating information on similar populations not available to him, will be adopted as the most tenable view. Contributory evidence will be apparent in the analyses of *caniceps–thurberi* mixtures in the Grapevine, Panamint, Inyo, White, Walker River, and Pine Forest mountains along the western side of the Great Basin.

The characters which differentiate *caniceps* and *thurberi* and which are to be considered in the analysis of interbreeding are color of head, distinctness of hood mark, color of sides, color of back, length of wing, tail, bill, and tarsus, and depth of bill.

The Charleston and Sheep mountains, Nevada.—The nearest known breeding colony of typical *caniceps* is in the Pine Valley Mountains of Utah about 150 miles from the Charleston Range, 100 miles from the Sheep Mountains. The nearest breeding colony of *thurberi*, and this of a somewhat mixed nature, is in the Panamint Mountains, 85 miles distant from the Charleston Mountains. Pure populations of *thurberi* are 150 miles distant, west of Owens Valley. Scattered pairs of birds breed in isolated Transition areas to the south of Charleston Peak on Potosi Mountain, in the same range, and on Clark Mountain, San Bernardino County, California (Miller, 1940, p. 163). Because these birds are similar to those of the Charleston area, they may be considered as part of the Charleston–Sheep Mountains group. With such distances involved, these annectent colonies present a situation somewhat different from that between *caniceps* and *dorsalis* on the Kaibab Plateau. Chance for equal increments from adjacent pure forms would be less, although *caniceps, thurberi, shufeldti,* and *montanus* may all winter in the area or pass through in migration, affording opportunity for hybridization should they by chance remain to breed. The characters of the Charleston group furthermore suggest not a stock to which *thurberi* and *caniceps* contributed equally, but, even discounting dominance, one in which the *caniceps* type is the more prevalent one. More of the *caniceps* type entered into the original implantation, or else, as van Rossem suggests, there was an original *caniceps* stock which has had *thurberi* or *shufeldti* [*montanus*] stock added to it. However, there is the possibility that an originally equal mixture, when isolated, may have had the averages of characters shifted toward one extreme or the other, either through selection or through periodic reduction in numbers with repopulation from small residua.

The Charleston series of 23 summer resident males, plus 2 from the Sheep Mountains and 1 from Clark Mountain have been sorted with respect to head color. The color categories are: (1) average male *caniceps* of Utah; (2) dark male *caniceps* of central Nevada; (3) intermediate tone of gray, equivalent to *mearnsi;* (4) light-headed female *thurberi;* and (5) average female *thurberi*. All male *thurberi* are somewhat darker than this fifth type, so that classes

3–5 are in reality intermediates; no types equivalent to male *thurberi* occurred. Frequency of occurrence of the head types in the southern Nevada series was 6:7:6:2:4. Several genetic factors would seem to be involved without dominance of one type over the other. The predominance of *caniceps*-like heads is in keeping with the general prevalence of the characters of this race in the sample. The second category is a type of *caniceps* head encountered almost exclusively, but not uniformly, in Nevada.

Females of the Charleston–Sheep Mountains series do not attain head color quite equal to pale female *thurberi*. Individuals fell into three classes: (1) average female *caniceps* of Utah; (2) average female *caniceps* of Nevada; and (3) intermediates near *thurberi*, slightly darker than male *caniceps*. Frequency occurrence was 6:5:4. Thus 73 per cent are indistinguishable from *caniceps*, whereas in males only 54 per cent are indistinguishable. The greater sexual dimorphism of the head in *thurberi* may have some bearing on this difference.

The pigments of the sides of *caniceps* and *thurberi* do not blend in the hybrids. The mass effect may appear intermediate or blended, but the pigment granules are either of one type or the other. In *caniceps* the barbules are transparent under the microscope, except distally, where there is blackish eumelanin. In *thurberi*, terminal eumelanin is lacking and faint yellowish phaeomelanin is scattered, sometimes quite regularly in granules throughout the barbule, being quite as prominent basally as distally. In hybrids the areas of a feather that are yellow-pigmented (mass effect pinkish) are usually lacking in eumelanin in the distal parts of the barbules. One half of the vane may be predominantly pinkish, the other gray. The basal part of the vane is pink more often than are the tips. Certain individuals have a wood brown color such as is found in *montanus* and *shufeldti*. In these all the barbules have yellow pigment but also black tips. This condition might be the result of *montanus* intermixture, but it also could be a product of *caniceps* × *thurberi* with retention of both types of pigment throughout each feather. Even then the pigments do not appear blended under the microscope. The brown color is a summation effect of differently colored parts.

The summer resident males were classified as follows with respect to side color: (1) *caniceps* type, eumelanin terminally, including those with faintly buffish tips that are frequently found among *caniceps*; (2) those with a few feathers with areas of phaeomelanin; (3) those with considerable areas of phaeomelanin; (4) those with full extent of phaeomelanin, evenly distributed, as in *thurberi*, including those of *montanus* type as described above. Frequency occurrence of these types is 12:4:3:4. Preponderance of *caniceps* types is evident here as in the hood, and again it appears to be not a case of dominance of one character but of greater proportion of *caniceps* individuals entering into the parentage. It should be noted that some birds attain full *thurberi* character, whereas this was not true of hood color.

The females, classified in the same way, are distributed as follows: 4:2:7:3. Here the distribution favors *thurberi*. This is in agreement with the general trend of sexual dimorphism in the genus, in which phaeomelanins are augmented or unmasked and eumelanins diminished in females. Side-color determiners in these crosses are in some way interactive with sex factors or with sex hormones, just as in the hood. With respect to the sides, this interaction accentuates the *thurberi* character (phaeomelanin), making it really dominant; with respect to the head, it reduces the *thurberi* character (eumelanin).

The foregoing analysis of side color takes little account of the disappearance of eumelanin and the intensity of both pigments. In individuals with hood darker than in *caniceps* and with sides gray, the sides take on the increase in phaeomelanin, in agreement with the hood. This results in a combination of characters resembling to a marked degree certain hybrids between *Junco hyemalis* and *Junco oreganus*. The only distinctly different feature in such birds is the brilliant red back of *caniceps*. This darkening of the sides correlatively with the hood occurs only when the *thurberi* side factors are absent, but then invariably. It suggests an independent factorial control of intensity of eumelanin which may apply to any part of the body plumage carrying such pigment, but which is inoperative for a region when other factors suppress or replace that type of pigment. One set of controls determines whether the side pigmentation shall be eumelanin or phaeomelanin. If eumelanin is determined exclusively, a general factor for intensity of eumelanin may exert an influence.

In the brownish, *montanus*-like sides, the eumelanin is never abundant and does not increase with head intensity. The fact that this type of color occurs in *caniceps–thurberi* hybrids close to the range of *thurberi* substantiates my belief that it is in large measure a result of segregation and interaction of factors in the *caniceps–thurberi* cross. Of the 11 males which have phaeomelanin in the sides 4 have this *montanus* color, and of 12 females 4 are of this type.

The phaeomelanin, when present without eumelanin, is often of a redder type than in *thurberi,* so that it resembles *mearnsi*. This initially suggested to me that *mearnsi* migrants in the spring had remained to breed and had contributed these sides. However, if this were the fact, why should so many individuals bear this character when there were many with heads approaching *thurberi* that could not possibly have been derived from *mearnsi* parentage? Also, why was there none showing the greatly extended area of flank color typical of *mearnsi* which is so frequent in *caniceps–mearnsi* crosses in northern Utah? A better explanation is that the phaeomelanins of the side correspond in some measure with the reddish phaeomelanin of the brilliant *caniceps* back. There might be a factor for increased redness of phaeomelanins, much as there was for eumelanin, that affects all areas of that pigment to some degree. All birds with ruddier sides have pure or partly red backs. Not all red-backed birds that have phaeomelanin in the sides have the *mearnsi*-like hue, so that the case is not completely comparable with intensification in the eumelanins. *Mearnsi* as a race does not fall in with the hypothesis of correlated intensity of phaeomelanin throughout the body, for the back is poor in this pigment, whereas the sides are especially well supplied. But it must be remembered that any such intensification factor is subject to modifications by factors controlling each of the special areas in which the pigment appears. I conceive it as not impossible that the *mearnsi* resemblance of certain hybrids is also a result of segregation and interaction of factors in the *caniceps–thurberi* mixture.

It should be stressed that the margin of difference in hue of normal *thurberi* and *mearnsi* is often very slight and that the importance of this slight difference should not be emphasized out of proportion to the truly great differences involving the presence or absence of phaeomelanin in the sides. Of 10 males which have phaeomelanin in the sides 5 have *mearnsi* hue, and of 12 females 6 are so colored.

The demarcation of the hood ventrally is in large part a product of the pigmentation of head and of sides. Either pronounced *thurberi* pigmentation of the head with gray sides, or *caniceps* head with sides well pigmented with phaeomelanin, will yield a fair line of demarcation. *Thurberi* pigmentation of both regions naturally produces the strongest line. If the phaeomelanin of the sides is slight and mixed, it contributes nothing to the demarcation. In eumelanin intensification on the side the hood mark is nevertheless still visible; hence there may be some factor which affects demarcation alone.

Back color, like side color, involves two pigments. *Caniceps* pigmentation consists of an abundant, ruddy phaeomelanin with no eumelanin. *Thurberi* has a slightly paler, less abundant phaeomelanin, with variable amounts of eumelanin distally on the barbules. The pigments of the two types do not blend, the situation being comparable to that in the flank feathers. Intermediate birds often have the lateral webs of the feathers red, the inner webs tannish brown, much as in *thurberi*, owing to the addition of blackish tips on the barbules of this web. Another form of intermediacy is one with the paler phaeomelanin of *thurberi* present without eumelanin, so giving a yellowish color which may or may not be accompanied on the same feathers with areas of normal *thurberi* color (eumelanin tips). Back pigmentation is not correlated with head or side color.

Birds were classified as follows: (1) *caniceps* back; (2) backs with both red (phaeomelanin) and tan-brown (phaeomelanin plus eumelanin) areas; (3) backs with both pale yellowish and tan-brown areas (an uncommon variant in *thurberi*); (4) backs entirely yellow; and (5) *thurberi* back (somewhat variable in hue). Distribution of types in males and females combined was 26:11:1:1:0. When this ratio is compared with those for head and side color, the predominance of the *caniceps* character is seen to be distinctly greater. One may assume from this, in spite of unequal contributions from parental races, that there is dominance of the red-backed type. The same inequality of contributions of course precludes exact comparison with standard genetic ratios.

Nevertheless, it has been possible to devise a Mendelian mechanism that might produce these results and those noted in the hybrid population of the Panamint–Walker River series. Independently segregating factors, one pair for presence or absence of eumelanin (AA'), and another pair for red and yellow (Rr), interact to some degree. The AA zygomatic formula produces no eumelanin; A′A′ produces the *thurberi* complement of eumelanin; AA′ produces the partial deposit of eumelanin in mixed pattern. RR and Rr produce red, and rr yellow. RR interacts with AA′ to reduce the eumelanin and produce an unmixed red. Such an assumed interaction is called for in this theoretical mechanism by the high proportion of pure reds relative to mixed reds. The proportion expected in a normal dihybrid is 3:6, but with this interaction it is 5:4. The actual ratio in the Charleston Mountains is 26:11. Since in other characters the pure *caniceps* types are twice the expectation in a case of equal *caniceps* and *thurberi* parentage, the 26:11 ratio is roughly equivalent to the 5:4 with the first member doubled (= 10:4 or 27½:11). *Caniceps* back would accordingly be determined by the genetic formulae AARR, AARr, or AA′RR; mixed eumelanin and red, AA′Rr; mixed eumelanin and yellow, AA′rr; pure yellow, AArr; and *thurberi* A′A′RR, A′A′Rr, and A′A′rr. *Thurberi* usually would be A′A′rr (rarely AA′rr), but a red phaeomelanin masked by a full complement of eumelanin would in all probability pass as a *thurberi* back even though it should be of a slightly ruddier hue. Absence of *thurberi* types from the Charleston Mountains may be attributed to the small proportion of pure *thurberi* entering into the cross. A mating of AARR with AA′Rr would, in successive interbred generations, produce chiefly *caniceps* back and mixed red with the occasional appearance of yellows, mixed yellows, and *thurberi* about in the proportion (except for *thurberi*) in which they are actually found on the Charleston Mountains. A true *thurberi* back should appear in larger samples.

A single experimental hybrid of *dorsalis* × *pinosus* (see p. 372) involves back characters comparable, if not identical, with *caniceps* × *thurberi*. This F_1 individual is red rather than yellow and thus appears to confirm the dominance of red, for its makeup should be Rr. The eumelanin is not mixed, but is largely suppressed. This may be the result of AA′ in the *pinosus* parent or of a different gene for eumelanin of the back in *pinosus* where this pigment normally is less in amount than in *thurberi*.

Size characters, with the exception of bill depth, have average values lying between *caniceps* and *thurberi*. In all these but the wing the average is somewhat closer to *caniceps*, corresponding to the situation in those color characters in which there is no dominance. Distribution of individuals with respect to size categories presents no departures from a simple curve, except for bill depth. In the *caniceps–thurberi* hybrids the distribution suggests a segregation of the parental values of *caniceps* and *thurberi* (note positions of averages in fig. 4). The coefficient of variability here is accordingly high—5.83 compared with 3.67 to 4.62 in *caniceps* populations. This is remindful of the mode of inheritance of tail length in the *caniceps–dorsalis* crosses. In bill depth in *caniceps–thurberi* hybrids a single pair of factors with *caniceps* dominant is suggested; the ratio is 15:4. In *caniceps–dorsalis* crosses, the bill depth presented a simple curve, not a bimodal curve as here, but the difference in the parent stocks was much less. It is not unreasonable to suppose that multiple factors may regulate small size differences within the two major types and between rather similar races, but that there may be a different factorial arrangement, in this instance a single pair, which determines the rather distinct bill types of *caniceps* and *thurberi*.

Other size characters have coefficients of variability no greater than the parent populations, so that, unlike the situation with respect to color characters, there is no evidence of segregation. This situation was met by Sumner (1932, p. 48) in his analysis of inheritance of linear measurements in interracial crosses in *Peromyscus*. The increase in variability in F_2 may not be apparent if many factors are involved (Babcock and Clausen, 1918, p. 185), unless the sample is very large.

None of the dimensions in the Charleston group correlates with color characters—a circumstance which points to complete intermixture of the birds and to completely independent segregation of characters. Wing and tail correlation is $.750 \pm .060$, which is about the same as within *caniceps* ($.71 \pm .027$).

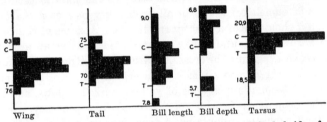

Fig. 4. Individual variation in dimensions (mm.) in hybrids of *J. c. caniceps* and *J. o. thurberi* from the Charleston and Sheep mountains, Nevada. C, average value for *J. c. caniceps*; T, average for *thurberi* of the Sierra Nevada; heavy horizontal marker, average for hybrid population.

Wing and bill, bill and tarsus, and wing and tarsus are in no way correlated. However, bill length and depth are correlated at $.61 \pm .098$, whereas they are not within *caniceps*. Sumner (1929, p. 488) has reported similar phenomena in which characters are correlated in interracial crosses but not intraracially. Sumner concludes that the intraracial variation is to a less degree genetic than that in interracial hybrids. This has already been suggested in regard to bill length of juncos by the variable state of wear of the tip, yet there must be some inherited intraracial variation in the bill apart from wear. It might be that such a difference in the matter of correlation, intraracially and interracially, is due to different sets of factors determining dimensions in each case, one group of factors being linked or common factors, the other not.

Panamint, Inyo, White, and Walker River mountains.—These four mountain ranges lie in a northwest-southeast line, 20 to 50 miles distant from the nearest junco-inhabited territories of the Sierra Nevada. Each is to some degree isolated from the others in the series, so far as junco habitat is concerned. The similar degree of separation from territory occupied by *thurberi* probably accounts for similarity in the amount of departure from the characters of that race. In these mountains, much more so than in the Charleston Mountain population, the characters of the birds are mostly those of one race, *thurberi*, with no typical *caniceps* types occurring. But many characters are in evidence, which are the same as those of the Charleston hybrids. The birds appear not as though *caniceps* had entered into their immediate parentage, but as though some birds of mixed blood from a mountain range such as the Charlestons had crossed with pure *thurberi*. Another possibility is that continual influx from the Sierra Nevadan area has in large part swamped the *caniceps* characteristics.

A classification of characters is employed here which is similar to that used with the Charleston group. Head color for males is never any paler than type 5 (average female *thurberi*). To this series of head categories still darker classifications, type 6 (dark female, extraordinarily light male *thurberi*) and type 7 (average male *thurberi*) are added. Occurrence of head color of males with respect to these three categories was (5) 3: (6) 7: (7) 3. Although the 6th type is not unknown in males of *thurberi*, it would usually occur in about the ratio of 1:10 of the darker type, instead of 7:3.

The females are of head categories 4 to 6 in the following proportions: (4) 5: (5) 4: (6) 2. Class 4 is exceedingly rare in *thurberi*. There is a high proportion of intermediate head types in these samples. It might be suggested that some but not all of the several factors governing head color are heterozygous in these populations.

Of the four side-color types only the first (pure gray eumelanin with intensity increased in correlation with the hood) and the fourth (normal *thurberi*) occur. There are but two of the former. It is curious that there is none with mixed areas of eumelanin and phaeomelanin, but possibly the sample (19 males and females) is too small to include them. The tone of the phaeomelanin pigment is usually normal for *thurberi*, but two approach *montanus* (see p. 192) and two resemble *mearnsi*.

Back color shows decisively the *caniceps* mixture. No pure reds occur, but mixed yellows, yellows, and one mixed red appear. Original parentage of AArr or AA'rr with A'A'rr (*thurberi*) would yield these phenotypes, except for the single mixed red, according to the mode of inheritance previously hypothesized. *Thurberi*-type, mixed yellows (rare in *thurberi*), and pure yellows occur as follows: 10:10:3.

The potentialities in this situation for a new race (or species) are extremely suggestive. If by chance the AArr (yellow) had populated any one of these mountains exclusively and had been sufficiently isolated from influx of other kinds, a uniform race of yellow-backed, blackish-headed juncos of hybrid origin would have appeared. Such a race would be quite comparable to *Junco dorsalis, Junco hyemalis cismontanus*, or *Junco o. pontilis*. This mode of origin is, by itself, no argument against considering such groups as races. The hybrid must be stable and must occupy a definite area to the exclusion of other types if a race is to be established. Such has not been the case on the Charleston Mountains or on the ranges under consideration.

The size characters of the Panamint–Walker River group cannot be fully analyzed, because of the small sample of each sex. All these characters, except bill length, average greater than in *thurberi*, that is, toward *caniceps;* no character attains the average value found in the Charleston series. The averages for these quantitative characters substantiate the view that *thurberi* parentage predominates in this group.

	Wing	Tail	Bill length	Bill depth	Tarsus
Average *thurberi*, ♂	76.54	68.65	7.99	5.58	19.62
Average Panamint–Walker River series, ♂	78.47	70.12	7.76	5.75	19.73

Grapevine Mountains, Nye County, Nevada.—Subsequent to the study of the Panamint–Walker River group a hybrid population was discovered in the Grapevine Mountains on the California-Nevada line east of Death Valley. The series of 21 adults obtained there resembles most the Panamint–Walker River material but contains at least one member that is typical *caniceps*. The group is thus decidedly heterogeneous, possessing extreme parental types of both *caniceps* and *thurberi* and serving to link and further explain the hybrid complexes in the Charleston and Panamint–Walker River series.

Most critical in the series are three birds that show preponderance of *caniceps* characters. None of these has any pink in the sides. One male is typical *caniceps* in all characters and is type 1 head. Another male shows mixed red back and type 4 head, approaching 3. Neither male was taken with a mate. The female *caniceps*-like bird is typical except that the back is mixed red. She was mated to a male with black, type 6 head (light male *thurberi*), typical *thurberi* sides, and mixed yellow back. The general aspect of the pair was that of opposite extremes and if juncos take account of such matters they must have appeared to each other as foreign species.

The pair had two short-tailed juveniles which give some suggestion of segregation of character. The back color of one is definitely red as in young *caniceps*, that of the other more yellow. The genotype of the adults, judging from their external appearance, should have been ♂ rr and ♀ Rr. The segregation in the juveniles is exactly the result expected from such parents. One of the juveniles is more suffused with yellow on the underparts than is the other. This may presage the yellowish side color of adult *thurberi*.

The series of adults, all members included, shows the following phenotypic ratios: Heads of males, (1) 1: (4) 1: (5) 5: (6) 5: (7) 3; females (1) 1: (4) 1: (5) 4. Sides, (1) 3: (3) 1: (4) 17, of which 4 are of *mearnsi* hue, 4 *montanus* hue. Back, red 2: mixed red 2: yellow 1: mixed yellow 4: *thurberi*, 12. Each of the ratios shows more diversity of type than in the Panamint–Walker River series. Fundamentally the population may be said to consist of a group with the same array of genes as in the Panamint–Walker River series but with the addition of a few distinctly *caniceps*-like individuals. The presence of these probably represent recent invasions from the east, either on the part of these birds themselves or their immediate parents. A new injection of *caniceps* blood is currently in evidence, which increment has not yet been thoroughly mixed through the population. Irrespective of the frequency of such an event, there is demonstrated by this infiltration the way in which mixtures of species types undoubtedly have been set up in mountains along the western edge of the Great Basin. The main barrier to this infiltration, which has probably been bilateral, is the series of low desert basins that run more or less continuously from the Amargosa desert northwestward toward the Carson sink.

The Pine Forest Mountains, Humboldt County, Nevada.—This range of mountains provides a limited area suitable for juncos. Three meadows and the adjoining pine-covered slopes, comprising an area of not more than a square mile, appear to be suitable junco habitat. The range is isolated from the Warner Mountains of California, where *thurberi* breeds, by 85 miles of desert. The Santa Rosa Mountains, 50 miles eastward, constitute the nearest station where breeding *caniceps* has been taken; only a single male, active sexually, was procured in the course of three days' special search there. A true population of *caniceps* is not known to exist closer than 150 miles southeast in the Toyabe Range.

Evidence for interbreeding of *caniceps* and *thurberi* on the Pine Forest Mountains is found in a pair that was taken June 5, 1934. These birds were foraging together, the male singing. The female had yellow ova 2 to 3 mm. in diameter. The male was a hybrid *caniceps* × *mearnsi*, with the side color of *mearnsi* and the back of *caniceps*. The female had a dull type of *thurberi* back (which is not, I think, evidence of *montanus* influence), typical *thurberi* sides, and the palest type of female *thurberi* head. On the same date a female *shufeldti* equivalent to the breeding birds in western Washington or northwestern Oregon was taken; it was in breeding condition. Whether this individual would have bred on these mountains is problematical, but it was not a cripple and was behaving like the other female taken on this day. It was foraging in a meadow partly covered with snow, a suitable breeding locality.

The juvenal *thurberi* reported by Taylor (1912, p. 395) from this same locality in 1909 I believe to have been raised on the mountain (Miller, 1935, p. 468). Identification of juvenal individuals is dubious, but certainly the bird is more like young *thurberi* than *caniceps*, although not in all respects typical of the former.

The Pine Forest Mountains occupy a position between the breeding ranges of *thurberi*, *caniceps*, *mearnsi*, and *montanus*. *Shufeldti* may pass through in migration. Very probably the breeding population of the mountains is subject to much seasonal fluctuation and occasional extermination. Stragglers from adjoining races or from zones of hybridization may repopulate it or mix with the resident population. The situation is a duplication in miniature of what took place in the history of the Charleston and Grapevine Mountain groups. Under present biotic conditions there cannot be built up so large a self-perpetuating group of hybrids as in these southern areas.

The facts ascertained relative to the *caniceps–thurberi* contact, and the presence of *shufeldti* in the breeding season on the Pine Forest Mountains and on the Toyabe Mountains (in the latter, not breeding) as stragglers, afford some support for the view that the slight difference in the Nevada population of *caniceps*, consisting chiefly of a larger proportion of individuals with darker type of head, is the result of occasional mixtures with the black-headed juncos to the north and west. Endemic races of other birds in the Great Basin become even lighter colored than Rocky Mountain forms. Why should the juncos be slightly darker headed? Where a multiple factor mechanism is involved, a very small admixture of black-headed birds would be diluted in the fairly large population to produce a number of slightly darkheaded individuals. Dying out of some of the initial hybrids from such a mixture might eliminate certain of the factors for blackness so that heterozygosity in only one pair of the set of multiple factors would prevail. Only one or two phenotypes, closest to the gray extreme, would then appear.

Junco oreganus mearnsi. Northern Wasatch Mountains.—The section of the Wasatch Mountains north of the Uinta Mountains in Utah has a breeding population of hybrids, the proportion of *mearnsi* characters increasing northward until nearly pure populations exist near the Idaho line. Only typical *mearnsi* are known from north of the line in these mountains. Parts of the Wasatch Range have restricted junco habitat and there are occasional gaps in distribution of from 2 to 5 miles.

The first of several localities visited in this section of Utah in 1931 was Kamas (10 miles east), Summit County, at the western end of the Uinta Range. Junco habitat in lodgepole pine and aspen forest was extensive here. Juncos were numerous and all were of pure *caniceps* type. This group may be taken as a starting point in considering the hybrids immediately to the north.

The next locality was Porcupine Ridge, Summit County, 20 miles in an air line north of Kamas and within 3 miles of the southwestern corner of Wyoming. The mountains are low here, consisting of outlying ranges of the Uintas. They are the highest mountains of the uplift extending north to the northern division of the Wasatch. Junco habitat is scarce, with aspens constituting the only cover. Occasional pairs breed in the denser part of this timber. *Mearnsi* characters begin to appear at this point.

A locality 14 miles southwest of Woodruff, Rich County, which was visited, is 30 miles north of Porcupine Ridge. Here there was a moderate amount of aspen timber and a few patches of Douglas fir near the summit of the Wasatch divide. Juncos were somewhat more numerous than at Porcupine Ridge and their habitat more extensive. Between Porcupine Ridge and this region are the relatively low Echo and Weber River canyons, forming an east–west pass. This pass constitutes the most distinct break in breeding habitat along these mountains. From the highway it was observed that in Echo Canyon suitable aspen groves were almost lacking on the mountainsides up to distances of 3 to 5 miles north and south. The junco population in this sector must be very sparse, even more so than at Porcupine Ridge. The Woodruff juncos were exceedingly mixed in character, the proportions of *mearnsi* and *caniceps* types being about equal.

Twenty miles farther north at a point 8 miles west of Randolph, Rich County, another population was sampled in the eastern foothills of the Wasatch Range. Junco habitat is continuous to the south toward the Woodruff locality. Juncos were abundant in the aspen and lodgepole pine forests. The general character of the population was more like *mearnsi* than at Woodruff.

The divide 12 miles west of Garden City, Cache County, was the last locality visited. It is 20 miles northwest of the Randolph region. Abundant aspen, lodgepole pine, and Douglas fir habitat was found here along the crest of the mountains. Habitat appears to be continuous from here southward. The proportion of characters here was the same as at Randolph, with no fewer *caniceps* characters appearing.

Collections from these four stations—Porcupine Ridge, Woodruff, Randolph, and Garden City—show well the shift in frequency of occurrence of characters from *caniceps* to *mearnsi* along this narrow belt of connectant breeding range. But, because there is clearly a varying proportion of the two parental types entering into the mixture at these stations, the genetic situation is difficult to interpret. No one sample is large enough to give reliable proportions of phenotypes. The Randolph and Garden City groups are so nearly alike in proportion that they may be combined, and the Woodruff group might justifiably be added; but the Porcupine Ridge population is evidently so different in parentage that it confuses ratios of occurrence of the separate characters to combine it with the others.

Characters which distinguish *mearnsi* are: blacker head, complete replacement of the gray sides by broad areas of pinkish cinnamon, and dull brown instead of mahogany red back. Size characters and white markings in the tail differ but slightly. The three color characters are not correlated in the hybrids, and any combination of parental and hybrid characters may appear in individuals.

Head color was classified as (1) average Utah *caniceps*, (2) darker Nevada *caniceps*, not entirely absent in Utah but rarely found in *mearnsi*, and (3) typical *mearnsi*. Females were classed in three equivalent categories. Separation of the birds into these categories was difficult because some of them showed extreme wear. The ratio in the Woodruff–Garden City group was 7:10:15. There is evidently a blend of character as in the *dorsalis–caniceps* and *caniceps–thurberi* crosses, the *mearnsi* stage of pigmentation equaling one of the *thurberi–caniceps* intermediate types. Undoubtedly, there are fewer multiple factors than in the *caniceps–thurberi* cross. The proportion of types probably should be about 4:6:4 in an equal mixture of *mearnsi* and *caniceps*. The excess of *mearnsi* in the 7:10:15 ratio may be taken as an index of predominance of *mearnsi* parentage; or about 2 *mearnsi* to 1 *caniceps*. This sample is as predominantly *mearnsi* as the Charleston sample was *caniceps*.

Back color behaves in a manner analogous to that in the *caniceps–thurberi* cross, yet with some significant modifications. Just as in *thurberi*, the *mearnsi* color results partly from eumelanin in the tips of the barbules, which is absent in *caniceps*. The extent of the eumelanin is about twice that in *thurberi*, the mass effect being a much darker, more neutral brown. The phaeomelanin of *mearnsi* is of a pale yellowish character and is not abundant. A factor for *mearnsi* eumelanin may be assumed to be paired with the *caniceps* factor A for absence of eumelanin, much as was the *thurberi* factor. We may call it A^2. As before, AA gives absence of eumelanin; AA^2, mixed areas on the feather; and A^2A^2, uniform eumelanin tips of *mearnsi* type.

The factor governing the yellow pigment of *mearnsi* gives evidence of being dominant to the red, instead of recessive like the factor governing the yellow of *thurberi*. It may be designated R'. The evidence for dominance is found in the prevalence of mixed yellows and yellows over mixed reds and reds in the phenotypic ratios of the Woodruff–Garden City series, and also, significantly, in the appearance of 2 pure yellow to 4 pure red in the Porcupine Ridge group, which in other respects is almost exclusively *caniceps*. A single pure *mearnsi* back also occurs in this latter group. Dominant yellow may alter some AA^2 individuals that otherwise would have had mixed eumelanin to produce uniform eumelanin, just as it was thought the dominant red caused AA to show no melanin in *caniceps* × *thurberi*. The ratio of 20 *mearnsi* to 10 of the mixed type is twice as many *mearnsi* as would be expected, even with the known predominance of *mearnsi* parentage. But such a ratio is expected if an interaction of R'R' with the AA^2 takes place to complete the eumelanin deposit.

Although the pigments of the back of *mearnsi* resemble those of *thurberi*, their different relative quantities bespeak different factors controlling them. It is not surprising, then, that the dominance with respect to red and yellow and the interaction of factors for eumelanin and phaeomelanin is different where *mearnsi* pigments are involved.

In accordance with the foregoing hypothetical mechanism, *mearnsi* back (1) results from genotypes $A^2A^2R'R'$, $A^2A^2R'R$, $AA^2R'R'$ or A^2A^2RR (ruddy brown, of which few occur); (2) yellow with mixed areas of eumelanin, $AA^2R'R$; (3) red with mixed areas of eumelanin AA^2RR (the difference between this latter and the pure red $AA'RR$ of *caniceps* × *thurberi* lies in the new factor A^2, which produces more eumelanin than A'); (4) pure yellow $AAR'R$ and $AAR'R'$; (5) pure red $AARR$. The ratio in the F_2 generation would be 6:4:2:3:1. The ratio in this population with probable 2:1 predominance of *mearnsi* parentage was 20:7:3:2:0.

Mearnsi side color, like back color, appears to be dominant over the gray of *caniceps*. Eumelanin is entirely lacking in the pinkish cinnamon of *mearnsi*, and the latter extends medially onto the breast farther than does the gray of *caniceps*; it may even meet medially along the margin of the hood. The sides of hybrids tend to be either of one type or the other, and in the entire Wasatch group there is only one individual with partly gray and partly pinkish feathers. This is quite different from the *caniceps–thurberi* cross in which mixed feather types were common. Among winter-taken hybrids of *caniceps* and *mearnsi* and in a group of hybrids from northern Nevada several mixed types occur. Perhaps there is some variability in dominance, so that heterozygotes at times are mixed, at other times possessed of the dominant character. If there was no dominance, it hardly seems possible that only one mixed bird would appear in the Woodruff–Garden City series in which other characters show evidence of *mearnsi* blood and in which 4 pure gray *caniceps* occur. The ratio of pinks, mixtures, and grays is 27:1:4. A 27:4 ratio of pinks and grays is to be expected in a cross with a single pair of characters with dominance where *mearnsi* parentage outweighs *caniceps* 2:1. Phaeomelanin and eumelanin are never combined in the same barbule to produce a wood brown color such as is found in certain *thurberi–caniceps* hybrids.

The northern section of the Wasatch Mountains may be conceived of as a narrow connecting bridge, 90 miles long, between the breeding ranges of *caniceps* and *mearnsi*. On the north slopes of the Uinta Mountains *mearnsi* influence appears. North of the most sparsely populated part of the bridge at Echo Canyon a population of mixed origin is found, with *mearnsi* predominating 2:1. The prevalence of *mearnsi* appears even greater because of genetic dominance of certain *mearnsi* characters. In the Salt River Range in Wyoming, pure *mearnsi* occurs. Every year *mearnsi* mingles with *caniceps* in winter and migrates north along this bridge. Which, if either, was the original occupant of the area cannot be determined, nor can it be said whether one or the other is gaining ground now. Hybrids comparable to those found at Porcupine Ridge were taken seventy-seven years ago at Fort Bridger, which is situated similarly at the north base of the Uinta Mountains but is 40 miles east of Porcupine Ridge and not adjoining a breeding area extending northward like that of the Wasatch Range. In terms of years this *mearnsi–caniceps* contact has been of considerable duration without any obvious alterations in distribution of the pure parental races. If migration often carried *caniceps* too far north, or *mearnsi* not far enough north, much more intermixture should result. That such abnormal migration does occur rarely is evidenced by one breeding bird with *mearnsi* sides taken in the range of *caniceps* in eastern Nevada, and by a fully characterized *caniceps* taken in May at Glendive, Montana.

The complete mutual tolerance of the two kinds of birds is shown by the following mated pairs taken in the zone of hybridization: Porcupine Ridge, ♂ pure red back × ♀ pure brown back with young not determinable with respect to back color, and ♂ pure yellow back × ♀ pure red; Garden City, ♂ mixed red × ♀ brown. A pair of birds about my camp west of Garden City was feeding young on July 22. The male was collected at 5 A.M.; it was pure *mearnsi* on back and sides, but with head intermediate. The female, which could be seen to have normal *mearnsi* color on the sides and back, was left with the young. At 10 A.M. the small young were in the same group of bushes with the female, and a new male was on hand, singing, following her with tail fanned, and twittering in characteristic mating behavior; she did not drive him away. This bird was taken and found to have a mixed yellow back, *mearnsi* sides and intermediate head. Female X, as she now became known, had another male attached to her party shortly after 11 A.M. This male was *mearnsi* in all characteristics except for intermediate tone of head. A fourth male came to female X at noon and proved to have pure *mearnsi* back, pure *caniceps* sides, and intermediate head. I am doubtful that these males were all unattached previous to their interest in female X. Males of various sparrows are known to be polygamous on occasion. There was no doubt of the attraction of the female for all of them, however. Not knowing the history of the case, an observer would have considered each to have been her normal mate. No intolerance was evidenced by the female. Some of the males gathered food for the young. This indicates disregard on the part of the junco for differences in colors of sides and backs.

The following list shows the intermixture of back and side types in hybrids, and the predominance of one type or another along the Wasatch Mountains.

Porcupine Ridge	Randolph
4 red back, gray sides	2 yellow back, pink sides
2 yellow back, gray sides	1 mixed red back, pink sides
1 *mearnsi* back, gray sides	2 mixed yellow, pink sides
Woodruff	7 *mearnsi* back, pink sides
1 mixed red back, mixed sides	1 *mearnsi* back, gray sides
1 mixed red back, gray sides	Garden City
1 mixed yellow back, pink sides	4 mixed yellow, pink sides
3 *mearnsi* back, pink sides	7 *mearnsi* back, pink sides
	1 *mearnsi* back, gray sides

Southern Wyoming.—Evidence of intermixture is found in a few birds with some mixed areas in the back or with pure *mearnsi* back from the vicinity of Fort Bridger, and from the northern Medicine Bow Range and the Sierra Madre. Other pure *caniceps* from these localities show that the mixture is probably comparable in degree to that in the Porcupine Ridge group.

The Laramie Range and other ranges in south-central Wyoming provide some disconnected junco habitat that forms an incomplete bridge north toward *mearnsi* breeding territory in the Wind River and Big Horn mountains. Evidence of intermixture is at hand from the following mountains in that area (for detailed localities see Appendix and map, fig. 9) : Laramie Peak, Rattlesnake Mountains, Casper Mountains.

Northern Nevada and southern Idaho.—There are several mountain ranges north of the Humboldt River drainage of Nevada and the Great Salt Lake desert and south of the Snake River of Idaho in which intermixture occurs. A breeding hybrid bird has been taken just west of the Bear River at Swan Lake, Idaho.

In the Goose Creek and Raft River mountains of Idaho, both pure *caniceps* and hybrids occur. Of 9 breeding birds, 6 are typically *caniceps*, 1 has pink sides, light *caniceps* head, and red back; 1 mixed sides, dark head and red back; and 1 gray sides, light head, and mixed red back.

There is a single breeding hybrid from the Jarbidge Mountains of northern Nevada. Eighteen early September birds taken in these mountains were certainly not far from their breeding grounds. Of these, five have some evidence of *mearnsi* back, six have partly, or entirely, *mearnsi* sides, and only three approach the minimum of light-headed *mearnsi* variants. There are many pure *caniceps*. This suggests that the breeding population of these mountains is principally *caniceps*. A single hybrid is known from the Pine Forest Mountains. Nowhere south of the Snake River, in this sector, are there pure populations of *mearnsi*. *Caniceps* probably was the original occupant of this area, which *mearnsi* has to some degree invaded. Since *mearnsi* characters are dominant, it is less likely that the reverse has been true.

Summary.—The race *caniceps* is a relatively uniform aggregation of individuals. As in other races, certain peculiar characters unrelated to hybridism or intergradation occur occasionally. These are not correlated geographically. The only geographic variation is found in the Nevada area. This slight difference in Nevada may be ascribed to occasional hybridization in the past with *thurberi*, perhaps also currently. Evidence is at hand of occasional *caniceps* individuals outside of their normal range in the breeding season and of members of adjacent races occurring within the range of *caniceps*. *Caniceps* will cross with any of the adjacent forms if opportunity is presented. Its geographic isolation from these forms is incomplete in the fact that restricted intervening areas are occupied by mixed populations.

The differences between *caniceps* and *dorsalis* are all either differences in average dimensions or in intensity and extent of the same type of pigment. Regardless of the mode of origin of *dorsalis* or possible complete separation of the race at one time, this race and *caniceps* now behave as do most races at points of interbreeding. Accordingly, they have been grouped in one rassenkreis.

The meeting of *caniceps* with *mearnsi* and *thurberi* results in no less interbreeding than with *dorsalis*. Because of partial isolation, interbreeding cannot take place over large areas, and mixed populations usually are predominantly of one type or the other, owing to proximity to the breeding range of one or the other kind. The characters differentiating these forms involve totally different pigments in certain areas of the body. Although I disclaim degree of difference or qualitative characters as wholly adequate criteria for species, I think they have some weight in these instances. The differentiation of *caniceps* from *mearnsi* and *thurberi* I am compelled to feel is a matter of long standing, and I believe that at times they probably have been completely isolated from one another. It is not easy to visualize their present degree of differentiation as evolving while they have been in continual or even occasional contact. The actual relations of these forms have been fully detailed. The subspecies-species distinction is arbitrary. I believe it will be more useful to consider *caniceps* a separate species from *Junco oreganus* (*mearnsi* and *thurberi*). They are in many ways comparable to the "species" of flicker, *Colaptes cafer* and *C. auratus*, which, however, hybridize over much more extensive areas. At the same time it is most useful to designate the *caniceps–dorsalis* relation as a racial one, despite fairly complete isolation.

Junco caniceps dorsalis

Racial characters.—For characters distinguishing *dorsalis* from *J. c. caniceps*, see page 181.

Dorsalis and *J. phaeonotus palliatus* of southern Arizona are not so closely related as was formerly supposed. The distinctness of the two is much more apparent in birds in the flesh than in skins, since coloration of plumage is similar. Interbreeding of the two, which had been assumed from the variable amount of red on the wings of *dorsalis* (a *palliatus* character), has not been found.

Characters of *dorsalis* which differentiate it from *J. phaeonotus palliatus* are dark brown instead of yellow iris, flesh-colored instead of yellow lower mandible, and flesh-colored instead of faintly yellowish feet. The yellow lipochrome which is lacking in these soft parts of *dorsalis* is a constant feature of all forms from Arizona to Central America. In some kinds of birds such pigment is highly variable, but its constancy in the southern juncos seems to mark a major division of the genus.

Other less constant characters that distinguish *dorsalis* from *palliatus* are absence, usually, of red on wings; clearer neutral grays of head, rump, and underparts; slightly darker red of back; longer wing (3 per cent); longer bill (4 to 5 per cent); deeper bill (3 to 7 per cent); and less white in tail (see p. 212). The tail, tarsus, and toes are not significantly longer.

Nongeographical variation.—Dwight was of the opinion (1918, p. 300) that *dorsalis* "... showed a good deal of variations in plumage...." I cannot concur in this, for the coloration of all parts but the wings is as constant as in *J. c. caniceps*. The bearing of the problem of uniformity of *dorsalis* upon its possible hybrid origin calls for a careful analysis of intraracial variation.

Mahogany red areas on the outer webs of the inner secondaries and greater secondary coverts appear in 9 out of 230 *dorsalis* examined, or in 3.9 per cent. This occurrence of a *phaeonotus* character is not so great as that of red pileum in *J. c. caniceps* and is only slightly greater than that of white wing-covert spots in *J. h. hyemalis*, neither of which characters has been thought to point to hybridization in these races. The reddish secondaries, though suggestive of *phaeonotus*, may not be the result of any recent hybridization. The condition is not limited to areas adjacent to the range of *J. phaeonotus palliatus*. Of the 9 individuals displaying red on the wings, 6 were on their breeding ground. One of these was from the Sierra Ancha, nearest point to the range of *palliatus*, three from the White Mountains, a region somewhat more remote from *palliatus* and inhabited by a large population of *dorsalis*, whereas one was from the San Francisco Mountains, northern limit of *dorsalis*, and one from east central New Mexico at the northeastern limit of *dorsalis*.

The same parts of the wing feathers that occasionally are reddish, much more often are bright buff; 21 per cent are so marked. This character occurs in about the same percentages in *caniceps* and is most prevalent in first-year birds and in females.

Red on the pileum occurs in 1.7 per cent, compared with 6.2 per cent in *caniceps* and 1.7 per cent in *palliatus*. White in the wing coverts appears in 0.8 per cent, hence somewhat less frequently than in *caniceps*.

A white basal segment of the culmen is a *caniceps* character, inherited independently of the pigmentation of the remainder of the upper mandible (see p. 188), which occurs in 4.7 per cent of *dorsalis*. Five such birds have been taken in the White and Mogollon mountains and five in the San Francisco Mountains. Thus it is not limited to the regions closest to *caniceps*. Two birds, one a winter specimen, have incomplete black pigment on the upper mandible apart from the culmen. One is type 7 of the Kaibab hybrid series, the other type 6. This latter was breeding in the Mogollon Mountains (Pivot Rock Spring) at the southern border of the range of the race and far removed from *caniceps*. Two breeding birds with comparable hybrid bill patterns were taken in Utah, one 10 miles east of Kamas, and the other at Cedar Breaks. These two *caniceps* and the one *dorsalis* are the only individuals of either race with this intermediate pattern that have been taken outside of the areas where there is interbreeding.

The range of variation in hood color is less than in *caniceps* and the difference in the sexes is lacking. The lower surface of the head is either pale or

TABLE 4

OCCURRENCE OF RECTRIX PATTERN IN J. C. DORSALIS AND J. C. CANICEPS (IN PERCENTAGES)

	Male		Female	
	dorsalis	*caniceps* (four separate populations)	*dorsalis*	*caniceps* (four separate populations)
Rectrix 4				
Inner web:				
pure black..................	5	0–7	6	6–17
mixed......................	95	93–100	94	83–100
Outer web:				
pure black..................	72	40–57	66	61–70
mixed......................	28	41–58	34	25–39
pure white..................	0	2–5	0	0–8
Rectrix 5				
Inner web:				
mixed......................	79	10–61	80	47–76
pure white..................	21	39–90	20	24–53
Outer web:				
pure black..................	1	0	4	0
mixed......................	80	38–70	81	66–83
pure white..................	19	30–62	15	17–34
Rectrix 6				
Inner web:				
mixed......................	6	0–4	14	0–10
pure white..................	94	96–100	86	90–100
Outer web:				
mixed......................	7	4–15	34	0–10
pure white..................	93	85–96	66	90–100

pallid neutral gray. No geographic correlations could be detected. The red color of the back is highly uniform.

No geographic variation in the white areas of the tail could be found within *dorsalis*, but the individual variation is comparable in range to that in *caniceps*. Pure white outer webs on rectrix 4 are lacking in *dorsalis*, but present in a few *caniceps*. Pure black outer webs on rectrix 5 are present in a few *dorsalis*, but lacking in *caniceps*. Apart from these, the same variants are present in the two forms. The color variants of the inner web of rectrix 4 average the same in occurrence in *dorsalis* and *caniceps*, but the outer web is more often black in *dorsalis* (at least in males) and never is pure white. Rectrix 5 is oftener pure black or mixed on both webs in *dorsalis* than in *caniceps*. Rectrix 6 is not significantly different in the two races, except that females of *dorsalis* show more black. Average sexual difference is found in rectrices 5 and 6 in *dorsalis*, and in rectrices 4 and 5 in *caniceps*. In *dorsalis* all webs of rectrices 4 to 6, except the inner web of 4, show in one sex or the other a greater average amount of black in *dorsalis* than in *caniceps*.

The differences between the average measurements of *dorsalis* (table 5) and those of *caniceps* are given on page 180. All those mentioned are significant statistically, that is, they are five times the probable error of the difference. Coefficients of variability are similar to those for *caniceps*, except for the

TABLE 5
MEASUREMENTS OF J. C. DORSALIS

Measurement	Sex	No. of specimens	Mean with prob. error	Standard deviation	Coeff. of var.
Wing	♂	95	82.38 ± .15	2.23	2.70
	♀	47	77.81 ± .15	1.53	1.96
Tail	♂	87	76.70 ± .19	2.59	3.37
	♀	43	72.74 ± .20	2.02	2.77
Bill length	♂	101	8.87 ± .02	0.38	4.34
	♀	48	8.79 ± .04	0.39	4.45
Bill depth	♂	98	6.83 ± .02	0.27	3.98
	♀	46	6.70 ± .03	0.31	4.61
Tarsus	♂	103	20.85 ± .04	0.60	2.88
	♀	50	20.70 ± .06	0.60	2.89
Middle toe	♂	102	11.98 ± .04	0.63	5.25
	♀	49	12.01 ± .05	0.53	4.44
Hind toe	♂	87	8.47 ± .03	0.48	5.63
	♀	51	8.27 ± .05	0.50	6.02

middle toe, and hence show no evidence of greater heterogeneity in this race that would suggest hybridization. The ranges in value in all size characters overlap to some degree the measurements for *J. c. caniceps*.

Coefficients of correlation are: wing and tail, .721 ± .035; tarsus and middle toe, .287 ± .058.

Geographical variations within the race.—*Dorsalis* seems devoid of any internal variations correlated with geography. Characters of forms to the north and to the south, when present in *dorsalis*, are not confined to marginal areas (see p. 206).

Distribution in breeding season.—There is a large area in the Mogollon Mountains of Arizona and New Mexico over which distribution is nearly continuous. In addition to this, numerous isolated mountain ranges are occupied, much as in the breeding range of *J. c. caniceps*. Plant associations in which *dorsalis* breeds consist of coniferous forests wherein the following types predominate: *Pinus ponderosa*, *Abies*, and, less commonly, *Pseudotsuga*. Birds also breed in pure groves of aspens and in coniferous forests with some oaks intermixed. Compared with the habitats of *J. c. caniceps*, those in which I have found *dorsalis* are, if anything, less arid, more luxuriant forests, although occasionally birds will breed in the dry lower portions of the yellow pine belt. Ground cover is rarely as poor in grass, low bushes, and humus as in many parts of the range of *J. c. caniceps* in Utah and Nevada. However, the presence of good stands of yellow pine in the Transition Zone provide junco habitat at lower zonal levels than in Utah and Colorado.

New Mexico.—Occurs chiefly in high Transition and Canadian zones (Bailey, 1928, p. 740), above 7000 feet, between latitudes 32° and 35°, in the central and southern parts of the state. The following mountains are included: Capitan, Sierra Blanca, Sacramento, Guadalupe, Datil (probably), Magdalena, San Mateo of Socorro County (probably), Black Range, Mimbres, Pinos Altos, and the Mogollon group of Catron County. A single

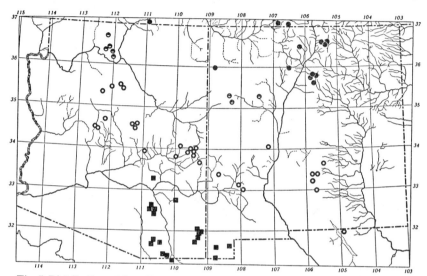

Fig. 5. Distribution of breeding or permanently resident juncos in Arizona and New Mexico. Each symbol represents a locality from which specimens have been examined. Dots, *J. c. caniceps*; circles, *J. c. dorsalis*; half dots, intermediates between *J. c. caniceps* and *J. c. dorsalis*; squares, *J. p. palliatus*.

dorsalis is at hand from the San Mateo Mountains of Valencia County. Since the Zuni Mountains at the same latitude have a hybrid population, such may also be the condition in the San Mateo Mountains. Sight records from Colfax County (Bailey, *loc. cit.*) are not substantiated by specimens and doubtless pertain to *J. c. caniceps*, which is known to breed there.

Texas.—Known only from the Guadalupe Mountains, Culberson County, just south of the New Mexico line. A juvenile taken here by V. Bailey, August 11, 1901, indicates breeding. Such migration as does occur in *dorsalis* does not begin until September, after the postjuvenal molt is finished.

Arizona.—Occurs in high Transition and Boreal zones (Swarth, 1914, p. 93, map) in mountains south of the Colorado and Little Colorado rivers and north of the Gila River (except Pinal Mountain in Gila County). This includes the mountains of the Mogollon group from Baker's Butte, southeast through the White Mountains into New Mexico, and the Sierra Ancha. It also includes much of the Coconino Plateau from Bill Williams Mountain east to the San Francisco Peaks and north, irregularly, to the south rim of the Grand Canyon. Also breeds on Mingus Mountain and on the mountains near Prescott.

Distribution in winter.—Winters on the breeding grounds and at lower elevations in the mountains within the breeding range. Few individuals stray south of the southern limits of the breeding range of the race. Several winter records to the southward have been based upon *caniceps–dorsalis* intermediates or upon birds incorrectly identified. From the Chircahua Mountains, Arizona, a single January-taken bird is at hand. This is in contrast to 58 *J. c. caniceps* taken from the same locality in winter. Other unequivocal winter or migration records based on specimens taken south of the breeding range are: Big Hatchet Mountains, Hidalgo County, New Mexico (1), May 18, 1892; Playa, Hidalgo County, New Mexico (1), February 17, 1885; "southwestern New Mexico" (1), November 5, 1873; Chisos Mountains, Texas (1), January 28, 1914; Huachuca Mountains, Cochise County, Arizona (1), January 7, 1905; and Pinal County [=Santa Catalina Mountains], Arizona (2), October 26, 1884, and February 11, 1886. Reports of *dorsalis* in Chihuahua and northeastern Sonora by Ridgway (1901, p. 298) apparently were based on skins in the American Museum. In Allen's report (1893, p. 39) covering this matter, only one *dorsalis* from San Diego, Chihuahua, is mentioned. The only junco now at the American Museum from San Diego is no. 56861, which is *J. c. caniceps;* no *caniceps* from this locality was recorded by Allen. Records south of the United States border therefore are lacking. That the race occasionally ranges south of the line is not improbable in view of occurrences in the Huachuca Mountains and other ranges along the Mexican boundary.

Relationships with adjacent forms: J. c. caniceps (see p. 186), *J. phaeonotus palliatus.*—The only important character of *palliatus* which occurs in *dorsalis* is the red on the wing. This has been shown to be variable, and when occurring in *dorsalis* it is not limited to regions adjacent to *palliatus. Palliatus* occasionally may lack this red. I have never seen evidence of intermediacy in the striking character of yellow iris and yellow lower mandible of adult *palliatus.* No intermediate or mixed populations are known, nor is it apparent where they might occur. The closest points of approach of the ranges of the two forms in southeastern New Mexico are the Animas Mountains and the mountains near Silver City. Sixty miles of Lower Sonoran desert intervenes and the total distance is probably about 90 miles.

In Arizona from the top of the Sierra Ancha where I was collecting breeding *dorsalis,* Pinal Mountain, 37 miles to the south, could be seen, where *palliatus* in pure form had been taken the week before. Between lay the Salt River Valley. This is the closest point of approach known, although the distance is not much greater along the Gila River Valley to the southeast. Pinal Mountain is the northernmost outpost of the yellow-eyed juncos, which in various forms extend south to Panama. This 37-mile break in distribution is one of great significance; it divides the genus into two major groups, geographically complementary, between which there is at present no interbreeding.

Rassenkreis Junco phaeonotus
Mexican Junco

This group consists of two geographic differentiates, *J. p. palliatus* in the north and *J. p. phaeonotus* in the south. Between them is an extensive gradient which occupies a great part of central Mexico along the Sierra Madre

Occidental. It follows that sharp interracial boundaries do not exist, either geographically or morphologically. The rassenkreis occupies the mountains of southeastern Arizona, extreme southwestern New Mexico, and Mexico south to the Gulf of Tehuantepec. There is no migration.

Principal characters of the group.—Iris yellow; lower mandible pale yellow; upper mandible black. Back area reddish brown (Kaiser brown to hazel), sharply defined anteriorly and posteriorly with lateral margins more or less confluent with reddish areas of wing coverts and inner secondaries; reddish color produced by phaeomelanins, as in *J. caniceps* and *J. bairdi;* eumelanin practically or entirely absent from the area; tips of feathers grayish because of lack of phaeomelanin. Lores and ocular region black, contrasting with grays of head. Sides smoke gray to buffy brown and snuff brown, but not sharply set off from pale smoke gray of breast and throat. Side color produced by combination of terminal eumelanin and proximal deposits of phaeomelanin in barbules. Tail with two, rarely three, outer feathers partly white. Size moderately large, weight 19.1 gm. average; wing of males averaging 77 to 79 mm.; feet fairly large.

Sexual differentiation in coloration is nearly lacking. Females display slightly more phaeomelanin in the sides and on the rump. The heads of females are not lighter than those of males. The amount of black on the fifth rectrix averages greater in females. Size differentiation in wing and tail is about 5 per cent, as in *Junco caniceps;* but unlike *caniceps* and *J. alticola, phaeonotus* shows no sexual dimorphism in bill and tarsus and apparently none in the toes. An immature plumage is not distinguishable.

The effect of wear upon the reddish areas is the same as in *Junco caniceps.* The smoke gray and even the buffy browns of the sides are greatly reduced by wear and fading in the course of the year. The color of the head and breast is not altered greatly, except that it becomes slightly darker through staining and exposure of the basal parts of the feathers late in the breeding season.

Junco phaeonotus palliatus

Racial characters.—*Palliatus* and *J. p. phaeonotus,* at the geographic extremes, are constantly differentiated by the darker grays and buffs in the plumage of the latter. In *J. p. palliatus* the anterior upper parts are pale mouse gray rather than mouse gray; rump light grayish olive rather than deep olive; sides smoke gray rather than buffy brown. Less constant characters are slightly lighter gray of breast, greater amount of white on fifth rectrix, more frequent white spotting of fourth rectrix and less prevalent black spotting of sixth, and paler, less abundant phaeomelanin in the barbules of the back, producing a lighter red-brown than in *J. p. phaeonotus.* Extreme populations of *palliatus* are on the average longer winged (2 per cent), longer tailed (2 per cent), longer billed (4–5 per cent), and have longer middle toes (2 per cent) and shorter tarsi (2 per cent) than *J. p. phaeonotus.*

For differences between *palliatus* and *J. c. dorsalis* see page 205.

Nongeographical variation.—Red on the crown occurs in 1.7 per cent of 400 individuals, the same as in *dorsalis.* Occurrence of white wing coverts in but 0.5 per cent is likewise similar to *dorsalis*. There is no red on the wing coverts in 1.5 per cent; all birds lacking red were from the Pinal, Chiricahua, and Santa Catalina mountains. Even so, I hesitate to consider it geographically limited to these areas, for with a character of so infrequent occurrence extremely large samples would be necessary to prove its absence. In addition

TABLE 6

OCCURRENCE OF RECTRIX PATTERN IN J. P. PHAEONOTUS, J. P. PALLIATUS AND J. C. DORSALIS

	Male			Female		
	J. p. phaeonotus	palliatus	dorsalis	J. p. phaeonotus	palliatus	dorsalis
Rectrix 4						
Inner web:						
pure black..................	83	33	5	80	51	6
mixed.....................	17	67	95	20	39	94
Outer web:						
pure black..................	100	95	72	100	97	66
mixed.....................	0	5	28	0	3	34
Rectrix 5 (always mixed in *J. phaeonotus*)						
Inner web:						
⅛ or less black.............	10	59	(21 pure white)	8	27	(20 pure white)
⅛ or less white.............	59	7		80	23	
Outer web:						
pure black..................	98	47	1	100	69	4
trace of white..............	2	27	(19 pure white)	0	21	(15 pure white)
mixed.....................	0	26	80	0	10	81
Rectrix 6						
Inner web:						
mixed, or trace of black......	93	84	6	96	86	14
pure white.................	7	16	94	4	14	86
Outer web:						
mixed, or trace of black......	93	76	7	100	77	34
pure white.................	7	24	93	0	23	66

to individuals that are entirely lacking in red, certain ones have the color reduced so that it involves only one or two secondary coverts. Such individuals occur in Chihuahua as well as in Arizona. The red never occurs on more than the greater secondary coverts and on the three innermost secondaries. No birds were found with white base of the culmen, such as occurs in 4.7 per cent of *dorsalis*. This would point to the origin of this feature in *dorsalis* from *J. c. caniceps*.

The tone of gray on head, rump, and sides varies within narrow limits in any one population; the important variations in these features are geographically correlated.

No geographic variation in white of tail within the Arizona area was detected. In *J. p. phaeonotus, palliatus,* and *J. c. dorsalis* successively greater amounts of white occur in each web of all three feathers. The difference is much greater in each instance than it is between *caniceps* and *dorsalis.* Pure white vanes on rectrix 5 are lacking in *J. phaeonotus;* only *J. p. phaeonotus* lacks mixed as well as white variants of the outer webs of rectrices 4 and 5. Sexual differences are pronounced in both vanes of rectrix 5 in all three forms. It also is distinct in the inner vane of 4 in *palliatus* and in rectrix 6 in *dorsalis.*

The significant differences between the average measurements and those for *dorsalis* are given on page 206. Variability of wing is 7 to 18 per cent

TABLE 7
MEASUREMENTS OF J. P. PALLIATUS OF ARIZONA

Measurement	Sex	No. of specimens	Mean with prob. error	Standard deviation	Coeff. of var.
Wing.............	♂	113	79.34 ± .14	2.28	2.87
	♀	78	75.47 ± .15	1.92	2.54
Tail..............	♂	111	75.16 ± .15	2.39	3.17
	♀	71	72.17 ± .18	2.22	3.07
Bill length.......	♂	105	8.43 ± .02	.36	4.27
	♀	59	8.50 ± .03	.40	4.70
Bill depth.........	♂	89	6.33 ± .02	.32	5.05
	♀	52	6.47 ± .03	.32	4.94
Tarsus............	♂	117	20.94 ± .03	.57	2.72
	♀	77	20.95 ± .04	.57	2.72
Middle toe........	♂	111	12.06 ± .03	.52	4.31
	♀	76	11.86 ± .04	.52	4.38
Hind toe..........	♂	114	8.55 ± .03	.41	4.78
	♀	78	8.44 ± .04	.48	5.68

higher than in the rassenkreis *caniceps,* approaching the degree of variability found in the tail. This is not the result of assembling birds from different mountain ranges for statistical treatment, for a similar variability was found in each locality. No increase in tail variation occurs, at least in males. Bill depth is 6 to 8 per cent more variable; the middle toe is 10 per cent more variable than in *caniceps,* but is similar to *dorsalis.* Other characters have similar coefficients of variability. The occurrence of increased variability in mensurable characters is interesting in a race in which plumage variation is relatively slight. Intraracial correlation of characters is essentially the same as in *Junco caniceps.*

Geographical variation in Arizona.—Since the gradient between *palliatus* and *J. p. phaeonotus* will be considered at length, there remain to be mentioned here only such variations as have been detected within the range in Arizona and New Mexico. Large series from the Chiricahua and Huachuca ranges in Arizona were compared. The Huachuca group had bills 3 per cent longer and deeper. This difference is reliable statistically and occurs in both males and females. These two isolated colonies are otherwise similar except for a small difference in wing length. Comparable differences may exist between groups from other of the mountain ranges, but the samples are not large enough to permit establishment of small average differences.

Distribution.—In the absence of any evidence for migration, records of occurrence for all months of the year are used to indicate range. Areas in New Mexico, Arizona, Sonora, and Chihuahua are included, for despite the beginning of a gradient in Chihuahua toward *phaeonotus,* the resemblance of these northern Mexican populations is close to the Arizona extreme.

Palliatus is restricted chiefly to isolated mountain peaks and ranges in Arizona and to the Sierra Madre in Chihuahua and Sonora. It occurs in the pine association, Transition Zone (Nelson and Goldman, 1926, p. 585), usually above 7000 feet, but occasionally as low as 6000 feet. There is a winter record at 4000 feet, below the pine belt, for the Santa Catalina Mountains of Arizona. The pine association may have a considerable mixture of deciduous oak. Prominent forest associates are *Pinus ponderosa* var., *Pinus chihuahuana, Pseudotsuga taxifolia,* and *Abies religiosa.* The forest floor which the junco occupies, though perhaps more arid on the average than that in the range of *dorsalis,* usually consists of pine needles and oak leaves, with some green bushes such as *Symphoricarpos, Garrya, Rhus, Ceanothus,* and brake ferns.

New Mexico.—Found in the Animas, San Luis, and Big Hatchet mountains of the extreme southwestern section of the state.

Arizona.—Occurs on the Graham, Chiricahua, Santa Catalina, Santa Rita, Whetstone, and Huachuca mountains in the region south of the Gila River. Occurs on Pinal Mountain north of the river.

Sonora.—Known from the San José Mountains just southeast of the Huachuca Mountains of Arizona. Almost certainly to be found along the west slopes of the Sierra Madre.

Chihuahua.—Occurs in the pine belt of the Sierra Madre from at least latitude 30° 15' to the southern end of the state. Also occurs in the San Luis Mountains just south of the United States boundary.

Intergradation between palliatus and J. p. phaeonotus.—The changes in characters along the north–south mountain ranges of Mexico are gradual, beginning in Chihuahua and terminating in the opposite extreme south of the Rio Grande de Santiago in Jalisco. The Mexican material has not been assembled at one place for examination. This circumstance and the fact that the specimens often are not seasonally comparable have made detailed analysis difficult. Accordingly, the changes in the coloration of the three important body areas, rump, crown, and flanks, cannot be analyzed separately. It was evident in studying the intermediates that variation in these three areas did not correlate, various shades of browns and grays in these regions being combined on the same bird.

Each bird was classed according to the total color effect: (1) typical *palliatus;* (2) slightly darker and browner; (3) intermediate; (4) slightly paler than *J. p. phaeonotus;* (5) typical *J. p. phaeonotus.* The Arizona populations are composed chiefly of type 1, with less than 10 per cent of type 2 and none any darker. Mexican birds south of the Rio Grande de Santiago and the Rio Pánuco are of type 5, except for 11 per cent of type 4.

Mensurable characters which show statistically significant differences between extremes are wing, tail, bill length, and middle toe. In the intervening area a progressive alteration is to be noted. Tarsal length changes significantly, but it is not importantly different at the two geographic extremes. The white areas of the tail show pronounced change, almost entirely in the northern section of the intergradational area.

Chihuahua region.—Out of ten birds from localities close to latitude 30° in the Sierra Madre, two are of type 2, the remainder of type 1. A group of breeding birds in worn plumage from Pinos Altos near latitude 28° were not noticeably different. However, a group of 18 fresh-plumaged birds from the same vicinity (30 miles west of Miñaca) showed plainly an increase in the darker plumage types, with types 3 and 4 appearing for the first time.

TABLE 8
OCCURRENCE OF PLUMAGE TYPES IN INTERGRADES

Locality	Type				
	1	2	3	4	5
Chihuahua (lat. 30°)................	8	2	0	0	0
Chihuahua (lat. 28°), fresh plumages..	10	5	2	1	0
Coahuila-Durango..................	3	5	3	1	0
Tamaulipas–northern Jalisco.........	1	5	11	27	10
Mexico south of Rio Grande de Santiago and Rio Pánuco.............	0	0	0	9	80

Wing and tail diminish in average values in the Chihuahua group in both males and females. All birds from localities between 28° and 30° have been combined in these averages. The diminution is no greater than the slight colonial differentiation between the Huachuca and Chiricahua groups, but it is significant as a step in differentiation toward *phaeonotus*. Middle toe length diminishes to a point halfway between the extremes; bill length remains the same as in the Chiricahua Mountains. The tarsus is remarkably short, reaching a figure distinctly lower than in either extreme type. This is a case of local differentiation unrelated to other geographic trends. I do not think it is the result of interbreeding of the two extremes, as no similar tendency is seen in intergrades in Durango, Coahuila, and Zacatecas. It would seem to be a case of colonial differentiation, extremely significant as a possible step toward local racial evolution.

The white in the tail is diminished in Chihuahua to essentially the same value as in southern Mexico. The graph (fig. 6) indicates the increase in black areas on the outer tail feathers of males, the same individuals as those for which size characters were determined.

Coahuila-Durango region.—Twelve birds from scattered localities in the Sierra Madre of Durango and the mountains of southern Coahuila are as predominantly of intermediate type as any sample at hand. Few extreme pale types appear, although the mean of the group is still closest to *palliatus*. The sample is not adequate for analysis of size characters, as averages would be unreliable where such small differences are involved.

Tamaulipas-Tepic region.—Included in this group are birds from many localities in the Sierra Madre Occidental and Sierra Madre Oriental, as also some from mountains in the intervening plateau of this section. All localities in Tamaulipas, Nuevo León, San Luis Potosí, Aguascalientes, Zacatecas, and Tepic are included, and those in Jalisco north of the Rio Santiago.

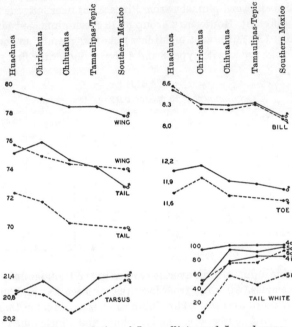

Fig. 6. Intergradation of *J. p. palliatus* and *J. p. phaeonotus*. Average measurements (mm.) of populations (see text for numbers of specimens used). Tail white represented as percentage of occurrence of variants in males of each population: number represents rectrix; *i*, inner web; *o*, outer web; 4*i*, 4*o*, 5*o*, pure black; 6*o*, mixed rather than black; 5*i*, two-thirds or more black. First vertical series of dots for tail white represents all *palliatus* from Arizona.

Although all color types are found in the assemblage of 54 birds, for the first time types 4 and 5 predominate. An excellent group of 12 fresh-plumaged birds from western San Luis Potosí (Jesús María) are nearly all of type 4 and only one is a typical *phaeonotus*. Other groups, those from the Sierra Nayarit and Zacatecas and from the Sierra Madre Oriental, are more diverse, but average about the same. Even though the average coloration has shifted so as to be closer to *phaeonotus* in passing south from Durango, it should be emphasized that no more change in character is involved than between this group and the southernmost group constituting typical *phaeonotus*.

The size characters are similar to those of the Chihuahua group. The curves for wing and tail (fig. 6) level out between Chihuahua and this region. Bill maintains a value similar to that in the Chihuahua and Chiricahua groups, and middle toe drops but slightly Tarsus is greatly increased over the peculiarly low Chihuahua average and is even higher than in the Chiricahua group. Thus in the size characters which show a gradient from one extreme to the other, there are no sharp breaks in the curve between this region and Chihuahua, and there is usually a tendency to remain constant. White in the tail remains about the same as in Chihuahua.

Region south of the Rio Grande de Santiago and the Rio Pánuco.—Within this large terminal area, which extends south to the limit of the rassenkreis in Oaxaca, no major changes in coloration or size occur. The rivers along the north of this area form convenient lines of demarcation and to a degree represent barriers in distribution to the northward. However, the break in distribution is no greater than between many mountains within this and other regions.

Color types 1, 2, and 3 are lacking. Type 4 occurs in only about 11 per cent, so that these populations are as uniform in color type as are the Arizona birds at the opposite extreme.

The curves representing size show distinct breaks between this region and the one next north with respect to wing, tail (at least of males), and bill lengths. Change in tarsus and middle toe are slight but continue trends of geographic variation already manifested farther north. The differences in wing, tail, and bill are no greater than between Arizona and Chihuahua or than between the Chiricahua and Huachuca mountains, but they are distinctly greater than between Chihuahua and Tamaulipas-Tepic. Tail white is not materially different.

Summary.—In the three intermediate geographic groups variation in color increases slightly compared with that at the geographic extremes. Variation in color of different parts of the body is not closely correlated in individuals, but is correlated as between populations. Change in average coloration is not abrupt between any two of the regions and constitutes an even gradient between extremes. Shift in average from a value closest to *palliatus* to one closest to *J. p. phaeonotus* occurs between Durango and Zacatecas (and neighboring states). The point marked by this shift, although possibly useful as a purely arbitrary boundary of races, is actually of no more biological significance than any of the other points along the gradient.

Length of wing, tail, and bill grade off to the southward with two principal breaks, one between Arizona and Chihuahua and the other between Tamaulipas-Tepic and the southernmost area. Tarsus shows peculiar local differentiation unrelated to the general geographic trend. It is therefore patent that although the color gradient is more or less uniform, the size gradient is more erratic, with subterminal breaks. The averages for tail markings show a single break, which, relative to the color gradient, is far north. Average changes along the cordillera are not effected in unison. Any nomenclatural treatment of these groups of intergrades would necessarily be extremely artificial.

Junco phaeonotus phaeonotus

Racial characters.—For characters distinguishing *J. p. phaeonotus* from *palliatus* see page 211.

A number of constant, strongly marked features distinguish *J. p. phaeo-*

notus from *Junco fulvescens* of central interior Chiapas. There is no point of contact between these forms. The southernmost point at which *J. p. phaeonotus* occurs in Oaxaca is 200 miles distant from the range of *fulvescens*, between which lie the extensive lowlands of the Isthmus of Tehauntepec. This is a greater interruption in the distribution of the genus than is found anywhere to the north on the mainland of North America. The back of *fulvescens* is raw umber rather than Kaiser brown, owing to a different type of pigment mixture (see p. 226). This color is confluent with the olive brown of the rump,

TABLE 9
MEASUREMENTS OF SOUTHERN MEXICAN J. P. PHAEONOTUS

Measurement	Sex	No. of specimens	Mean with prob. error	Standard deviation	Coeff. of var.
Wing............	♂	56	77.67 ± .20	2.27	2.92
	♀	25	73.97 ± .29	2.17	2.93
Tail.............	♂	49	72.66 ± .23	2.44	3.36
	♀	25	69.86 ± .38	2.79	3.99
Bill length.......	♂	58	8.10 ± .03	0.37	4.56
	♀	25	8.08 ± .04	0.33	4.04
Bill depth........	♂	50	6.19 ± .04	0.38	6.13
	♀	23	6.12 ± .04	0.30	4.98
Tarsus...........	♂	59	21.36 ± .05	0.57	2.66
	♀	26	21.25 ± .09	0.71	3.36
Middle toe.......	♂	52	11.78 ± .05	0.49	4.15
	♀	25	11.63 ± .06	0.43	3.71
Hind toe.........	♂	62	8.47 ± .03	0.41	4.84
	♀	26	8.36 ± .04	0.38	3.51

so that the reddish brown of the back is sharply defined only anteriorly. The sides are Sayal brown rather than buffy brown. Most striking is the heavy bill of *fulvescens*, which is 20 per cent deeper, with practically no overlap in individual variation with *phaeonotus*. Bill length is 6 to 9 per cent greater, tarsus 1 to 2 per cent greater, and the toes 4 to 10 per cent greater in *fulvescens*. Wing is 6 per cent shorter and tail 7 per cent shorter than in *J. p. phaeonotus*. Tail white is somewhat less extensive on the average on rectrices 5 and 6. The difference in proportions between *fulvescens* and the *J. phaeonotus* rassenkreis is more marked than between the rassenkreise *J. oreganus* and *J. caniceps*, or *J. caniceps* and *J. phaeonotus*, or *J. oreganus* and *J. hyemalis*.

Nongeographical variation.—There is a notable absence of striking plumage variations, compared with northern forms. Only one occurrence of red pileum and none of absence of red on wings or of white basal segment of culmen was found. One individual with whitish margins of the wing coverts was noted, but the margins hardly compare with the distinct white spots occasionally found in *J. caniceps* and other northern groups. Plumage variations, aside from the tail, are confined to lesser differences in tones of grays and hues of reddish browns and buffs.

Variation in the white of the tail is detailed on page 212.

Coefficients of variability are similar to those for *palliatus*. This means that the comparatively high variability of wing, bill depth, and middle toe is maintained in this member of the rassenkreis. No evidence of a different system of intraracial correlation of characters was detected.

Geographical variation.—No thoroughly conclusive evidence of variation of this sort in southern Mexico could be found. The series from restricted regions were not adequate to demonstrate colonial differentiation in dimensions. Material available from Vera Cruz suggests that birds from that region may be slightly shorter winged than those from the vicinity of Mexico City.

Van Rossem (1938, pp. 132–133) has described two races in southwestern Mexico from within the range of *J. p. phaeonotus* as here defined. These were based on very small series of adults in the British Museum. The same material had been studied by me and some notes taken on it in 1934. With our present knowledge of the types of variation in the genus *Junco,* it should be apparent that a conclusion concerning inherent differences in populations could not be safely reached on the basis of samples of from 5 to 7 adult individuals. Certainly if slight colonial differentiation is represented by these described forms, their equivalence to other races of junco is not apparent either in degree of differentiation or in certainty of the characters ascribed.

Junco phaeonotus colimae was described from 7 specimens from the Sierra Nevada de Colima, Jalisco. This is at the western end of the highlands of southern Mexico. The characters given are: "entire coloration decidedly darker and flanks browner." In studying this material I had recorded, as with all the *phaeonotus* complex, the general depth of coloration gained from inspection of several sections of the plumage (see p. 214). Particular attention was always paid to the flanks where in unworn plumage, as in this series of seven, the dark brown of *phaeonotus* shows to particular advantage. There is an appreciable seasonal alteration in the tones of the grays and browns in this species which had become apparent to me earlier in surveying the Mexican collections in American museums, and this was taken into account in classifying the types of coloration. Five of the 7 skins in question, including the type, were recorded by me as of average *J. p. phaeonotus* coloration, that is, of color type 5 (see p. 212); 2 were marked lighter than average *J. p. phaeonotus,* toward *palliatus,* or type 4. Thus I am unable to confirm even a minor differentiation in the Sierra Nevada de Colima or in specimens from Juanacatlan from the same section of the state of Jalisco.

Junco phaeonotus australis was described from the Sierra Madre of Guerrero, which is isolated from the main range of *J. p. phaeonotus* to the north by about 80 miles of unsuitable country. The characters ascribed are "upper parts brighter [than *J. p. phaeonotus*] and more extensively red, particularly on tertials and lower back; underparts more brownish-gray, particularly on flanks; bill decidedly larger and wing and tail shorter." Only 5 adults were available and but 3 of these were sexed. The birds are worn specimens taken in July and August. My own notations characterize all as typical *J. p. phaeonotus.* Duskier underparts and more prominent red on back and wings are features which become enhanced in worn plumage, and the extent of red is decidedly variable individually. Study of the distribution of the red would certainly be worth while in a larger series in fresh plumage. I was not impressed by any abnormal aspect of this pigment in my handling of the series. The 3 sexed birds were males, and 2 of these showed such wear in wing and tail that I felt measurement of the feathers would be misleading. My measurements of the one less-worn male are: wing, 76 mm.; tail, 72. This and van Rossem's measurements (wing, 70–77, and tail, 65–71) do not fall outside the curve of variation represented by the standard deviation for males and females of *J. p. phaeonotus* (see table 9, p. 218). (The 2 unsexed birds included by van Rossem may have been females.) They are lower than average, but a trend toward short wing and tail is in no sense proved statistically and is not even as clearly indicated as in the birds of Vera Cruz. Bill measurements, because not taken in comparable fashion by different workers, cannot be compared. My only measurement of a sexed bird was 8.6 for bill length; depth was not taken because the mandible was out of place. The length measurement is above average for *phaeonotus* but is exceeded by measurements of a number of individuals from the vicinity of the valley of Mexico. The only safe conclusion regarding dimensions is that differentiation in the Guerrero area is not yet demonstrable. We must have much better grounds than now available before differentiation of racial magnitude can be ascribed to the Guerrero population, even though the probabilities are good that there is a minor colonial differentiation there.

Distribution.—Permanent resident, in typical form, in the higher mountains on and surrounding the southern end of the Mexican Plateau, south of the Rio Grande de Santiago and Rio Pánuco, and including the highlands of Guerrero and Oaxaca. (For intergradient regions see p. 215.) The race occurs in pine or other coniferous forests of the Transition and Boreal zones, from 8000 feet to timber line (occasionally 14,000 feet). Some stations of record are apparently somewhat below 8000 feet. Prominent forest trees are *Pinus montezumae*, *Pinus liophylla*, *Abies religiosa*, and species of *Quercus*.

Known regions of occurrence include the mountains of central western Vera Cruz, central Puebla, Tlaxcala, southern Hidalgo, southern and eastern Mexico, central and western Michoacán, southern Jalisco, central Guerrero, and central Oaxaca to Zempoaltepec. Probably occurs in Morelos, extreme northern Hidalgo, Querétaro and Guanajuato.

Relationships with adjacent forms: J. p. palliatus (see p. 215). *J. fulvescens.*—No interbreeding with this form occurs, owing to wide separation geographically. There is no structural intergradation. For further discussion see pages 218 and 226.

Junco bairdi
Baird Junco

This species is fully isolated from all other juncos in the mountains of the Cape district of Lower California. Its affinities are not with the juncos of northern Lower California, but with the yellow-eyed forms of Mexico and Guatemala. Several important features point to this relation: lack of distinct hood demarcation; back color confluent with that of rump (*alticola* and *fulvescens*), inner secondaries, and coverts; proportions of wing and tail; relatively large bill; and reduced tail white. It possesses several peculiarities of color and proportion that sharply set it off from all others, yet, speaking broadly, it is a pale, dwarfed representative of the Central American juncos. Its range is probably as small as that of any other junco except *insularis* and possibly *vulcani*.

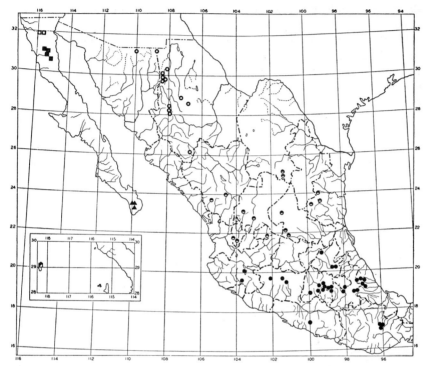

Fig. 7. Distribution of permanently resident juncos in Mexico northwest of the Gulf of Tehuantepec. Each symbol represents a locality from which specimens have been examined. Circles, *J. p. palliatus*; dots, *J. p. phaeonotus*; half dots, distinctly intermediate groups between *J. p. palliatus* and *J. p. phaeonotus* (see text); open squares, *J. o. pontilis*; solid squares, *J. o. townsendi*; solid triangle in square, *J. insularis*; triangles, *J. bairdi*.

Principal characters of the species.—Iris yellow; lower mandible pale yellow; upper mandible brown. Back cinnamon brown, the result of a pale yellowish phaeomelanin unaccompanied by tips of eumelanin on the barbules. This pigmentation might be related either to that of *J. phaeonotus*, which is similar but has a ruddier phaeomelanin, or to that of *fulvescens*, which has the same yellow phaeomelanin but also a peculiar scattered eumelanin which adds a faint olive hue. The cinnamon brown is almost the same as the color found in certain hybrids between *thurberi* and *caniceps* wherein eumelanin is lacking and the phaeomelanin is yellow (see p. 194). This likeness, although possibly identical genetically, does not mean that there is any immediate relationship.

Cinnamon brown confluent with similarly colored margins of inner secondaries and greater secondary coverts, and with buffy brown of rump. Lores dull black, but ocular region and pileum nearly uniform gray (light mouse to smoke gray). Throat and breast pale mouse gray confluent with gray of auricular region. Sides cinnamon, with color extending far forward onto breast, though without a definite line of demarcation with the gray; cinnamon sometimes meets across breast and spreads onto tips of feathers of throat. Color produced by a pure light yellowish phaeomelanin without any trace of the eumelanin found in *fulvescens*. Tail with two outer feathers partly, the outer often completely, white. A third feather (number 4) often with a white spot. Wing and tail short, wing of males averaging 71 mm., and tail 66 mm. Bill and feet moderately large, proportionately. Body size small, weight averaging 17.70 gr. in males, compared with 19.10 gr. in *palliatus* and 18.83 gr. in *townsendi*.

Sexual dimorphism is limited to size and tail pattern. Wing and tail show the usual 5-per cent differentiation. Bill and feet are differentiated to the extent of 1 to 2½ per cent, but none of these differences is entirely reliable statistically. The situation is comparable to that in *Junco phaeonotus*, where dimorphism in bill and feet is entirely lacking or is so slight as to be uncertain statistically. (See p. 224 for tail pattern.)

Wear lightens the cinnamon brown of the back, making it more yellowish, and may entirely remove the cinnamon edgings on the wing feathers. The cinnamon of the sides becomes yellower, less pinkish.

Differential characters.—Bairdi differs from *J. phaeonotus* subsp. most importantly in the extensive cinnamon sides that are devoid of eumelanin, in the lighter, duller cinnamon brown, rather than Kaiser brown or hazel of the back, in the confluency of back and rump colors rather than fairly sharp demarcation, and in shorter wing (9 to 11 per cent), tail (9 to 11 per cent), and tarsus (3 to 5 per cent). Other measurements are very close to the measurements of *J. p. phaeonotus*, while the grays of the head and breast are closely similar to those of *palliatus*. The white of the tail is perhaps closer to that of *palliatus* in total amount, but details of the pattern are, on the average, different (see p. 223).

In view of certain similarities with more southern forms, particularly *fulvescens*, points of difference with the latter may be noted: back of *bairdi* cinnamon brown rather than raw umber (fresh plumage); sides without eumelanin, hence pure cinnamon; pileum light mouse gray rather than dark olive gray; bill 5 to 7 per cent shorter, bill depth 20 per cent less, tarsus 7 per cent shorter; middle toe 5 to 8 per cent shorter; hind toe 7 to 11 per cent shorter. These differences in proportions I consider more important than the differences in color and in wing and tail length between *J. phaeonotus* and *bairdi*, since some of them show no overlapping. The resemblance in side color, throat color, and lower back of *fulvescens* and *bairdi* may be due to parallel lightening and brightening of pigments under comparable, though not identical, environmental influences (selection), and it does not certainly point to immediate relationships.

Bairdi differs so markedly from *insularis* and *townsendi* in the fundamental matters of eye and bill color and hood and back markings as to make detailed comparisons unnecessary. These northern forms are clearly related to the Oregon juncos (including *mearnsi*) and hence belong in a different section of the genus.

Nongeographical variation.—The plumage of *bairdi* is highly uniform. At least three-fourths of all birds have a buff patch on the nape suggestive of a

TABLE 10
COMPARISON OF OCCURRENCE OF RECTRIX PATTERN IN J. BAIRDI and J. PHAEONOTUS
(IN PERCENTAGES)

	Male			Female		
	bairdi	*palliatus*	*J.p.phaeonotus*	*bairdi*	*palliatus*	*J.p.phaeonotus*
Rectrix 4						
Inner web:						
pure black..................	71	33	83	85	61	80
(otherwise mixed or spotted)						
Outer web:						
pure black..................	82	95	100	100	97	100
(otherwise spotted)						
Rectrix 5						
Inner web:						
(always mixed)						
⅓ or less black.............	45	59	10	24	27	8
⅓ or less white.............	24	7	59	38	23	80
Outer web:						
pure black..................	26	47	98	15	69	100
nearly all black............	18	27	2	31	21	0
mixed......................	10	26	0	8	10	0
nearly all white............	19	0	0	31	0	0
pure white.................	27	0	0	15	0	0
Rectrix 6						
Inner web:						
pure white.................	43	16	7	27	14	4
Outer web:						
pure white.................	89	24	7	88	23	0
(otherwise mixed or with trace of black)						

comparable variant in *alticola* but not like the reddish crown of *J. phaeonotus*. This may be even more prevalent in *bairdi* than has been indicated, as some spring birds could have lost this buff patch through wear. About 25 per cent have traces of cinnamon across the middle of the breast. No pure white spots on the wing coverts were noted.

The same tail-pattern variants are found as in *palliatus;* additionally there are birds with white, and nearly pure white, outer webs of rectrix 5. On rectrix 4 the inner web averages between values for *palliatus* and *phaeonotus,* the outer web being less black than in *palliatus.* The inner web of rectrix 5 averages between values characteristic for the other two species, and the outer is markedly whiter than in *palliatus*. The sixth rectrix is somewhat whiter than that of *palliatus* on the inner web, but on the outer is much whiter. *Bairdi*, then, has a relatively great amount of white on the outer webs of all three feathers, though in total amount, both webs considered, it averages close to *palliatus*. This points to inherent difference in pattern despite similar amount of white. Sexual differentiation is seen chiefly in rectrices 4 and 5 as in *palliatus*, although there is also some in the inner vane of rectrix 6. The sexual dimorphism of the extremely variable outer vane of rectrix 5 takes a peculiar form. Males have more pure white and black variants, females more that are

TABLE 11
MEASUREMENTS OF J. BAIRDI

Measurement	Sex	No. of specimens	Mean with prob. error	Standard deviation	Coeff. of var.
Wing...............	♂	69	70.93 ± .15	1.87	2.63
	♀	32	67.52 ± .18	1.56	2.31
Tail................	♂	65	66.56 ± .20	2.41	3.62
	♀	28	63.96 ± .29	2.33	3.64
Bill.................	♂	65	8.30 ± .03	.35	4.22
	♀	30	8.18 ± .03	.26	3.17
Bill depth..........	♂	57	6.18 ± .02	.26	4.19
	♀	31	6.03 ± .03	.24	3.98
Tarsus..............	♂	71	20.29 ± .05	.61	3.00
	♀	33	19.99 ± .06	.53	2.66
Middle toe.........	♂	67	11.70 ± .04	.45	3.84
	♀	32	11.62 ± .05	.40	3.44
Hind toe...........	♂	70	8.34 ± .04	.53	6.36
	♀	34	8.12 ± .05	.41	5.04

not quite pure black or white. No tendency for more white in males is evident, the dimorphism being a matter of purity; and this is quite different from the condition in *palliatus*, where it is a difference in amount. Again, different controlling factors are indicated.

Variability in wing is slightly less than in *palliatus*, but the tail is as variable as in *alticola*. Instead of fitting into the mainland series of forms in which variabilities of wing and tail increase to the south more or less together (see p. 231), accompanied by lowered correlation, *bairdi* presents decrease in wing, and great increase in tail, variation. It is quite understandable that such an increase in one member only, whether genetic or somatic, would disrupt the correlation. In *bairdi* the coefficient of correlation is only 0.395 ± 0.07. It is not reasonable to suppose that the environment of *bairdi* is so different as to produce such marked differences in variation of wing and tail of a somatic kind. A higher degree of independence of genetic factors for wing and tail seems probable, with relative homozygosity for wing factors. Bill variation and that of the tarsus and hind toe are roughly comparable to those in *J. phaeonotus* and *J. c. dorsalis*. Middle toe variation is less, as in *J. caniceps*.

Although *bairdi* is not large footed like *fulvescens* and *alticola*, it shows a marked correlation of tarsus and middle toe ($.476 \pm .06$), which compares well with that (.534 ± .06) of *alticola*. (This correlation is totally lacking in *J. phaeonotus* and *J. caniceps*.) Tarsus and hind toe are also correlated (.452 ± .06). In *alticola*, both large size and correlation point to a special hereditary mechanism governing these parts. The evidence for *bairdi* points to a similar hereditary linkage, but involving factors for smaller size than in *alticola*, which must have different loci than those producing similar sizes in *J. phaeonotus*.

Bairdi shows a correlation of wing and bill depth at .240 ± .08, which, if reliable, is much greater than in any other forms except *alticola* and *mearnsi*, where it is .434 ± .08 and .427 ± .08, respectively. These correlations and resemblances in pigment weigh in the balance of relationships and make it problematical that affinities of *bairdi* are all with *J. phaeonotus*, although the resemblance still seems closer to that species than to *alticola* and *fulvescens* (see p. 220).

Distribution.—Resident of the Upper Sonoran Zone in the Victoria Mountains of the Cape district of Lower California. The region of the sierra in which it occurs is not more than 50 miles long. So far as known, there are no seasonal movements. Most specimens have been taken above 4000 feet in the section between Miraflores and La Laguna. Brewster's report (1902, p. 147) of a straggler at Triunfo, 1800 feet, is the only exception. The birds remain in winter in the highest part of their range at Laguna Valley, 6000 feet. Apparently there are no winter flocks of any proportion, the birds staying close to the nesting grounds, as is true of all the yellow-eyed forms. The region inhabited is forested predominantly with oak (*Quercus devia*), also with piñon (*Pinus cembroides*) and madroño (*Arbutus penninsularis*), and with cottonwoods (*Populus monticola*) in the canyons. Lamb (MS) took juncos repeatedly in black oak timber at 4000 feet at El Sauce in November and secured some individuals in mesquite bushes and in fan palms. In the spring there is surface water and a good covering of grass in the Laguna Valley.

Bairdi is one of a few juncos that breed under zonal conditions lower than Transition. Its tolerance for the environment in the Victoria Mountains very likely means a special physiological adjustment to it; other juncos usually require a forest of less Austral aspect. This is remindful of certain endemic avifaunas of the Andes, members of which have spread into different (higher) zones in the course of time and apparently have been modified in the process. If the mountains of the Cape once were more Boreal, the juncos may have been gradually forced to their present tolerance in course of gradual climatic change. Either that, or some small pioneering group successfully colonized this atypical habitat. The nature of the region compares most closely, as far as Junco-inhabited regions are concerned, with the Upper Sonoran oak belt of Arizona, which adjoins the habitat of *palliatus* in the Transition Zone and which at times is invaded by that race.

Relationships with other forms.—The nearest point where *J. phaeonotus* occurs is in the Sierra Madre of Mexico across the Gulf of California, roughly 200 miles distant. The isolation of *bairdi* is thus complete. The southern juncos, which are not given to wandering, would not be expected to immigrate to the Cape region today. Yet the connection of the Cape to the mainland of Mexico appears to have been remote geologically, if indeed it occurred at all in Tertiary times. Other members of the fauna do not reflect to any degree a connection in the immediate geologic past. Even if there were a connection when juncolike birds were in existence, it is unlikely that sufficient heights of land existed either north or east to provide continuous habitat. So far as the junco is concerned, the Cape district is now and probably always has been an island. How and when it was colonized must remain unanswered, but that the colonization was earlier than the Pleistocene and was from the south or east I am led to believe from the nature of the birds. The unrelated black-eyed juncos of northern Lower California are now isolated from *bairdi* by 550 miles of intervening desert.

Junco fulvescens
Chiapas Junco

This form is most closely related to *Junco alticola* of Guatemala. It does not interbreed with *alticola*, owing to its isolation. Certain of its characters are such that there is no overlapping, that is, no structural intergradation. Yet the resemblances between these two forms are greater than between *fulvescens* and *phaeonotus*, or *alticola* and *vulcani*, or *fulvescens* and *bairdi*. If its affinities and the degree of differentiation are appreciated, the arbitrary designation of *fulvescens* as race or species is of little moment. I have chosen the latter category in which to place it as more in keeping with treatment of other forms of the genus. *Fulvescens* is barely one step farther advanced in differentiation than a vicarious insular race which shows some overlapping in all characters with its nearest relative. Samples of the species are relatively uniform and the characters do not suggest hybrid origin through an ancient crossing of *phaeonotus* and *alticola*. The form probably has been locally evolved. I fail to see that it is a "variation towards *phaeonotus*," as stated by Dwight (1918, p. 303). Many of its characters indicate quite the opposite. Its pallor, compared with *alticola*, probably suggested this.

Principal characters of the species.—Iris probably yellowish; mandible yellow; upper mandible dark brown; feet straw color. Back raw umber, wearing to cinnamon brown and Dresden brown. Phaeomelanins of back comparatively yellow, rather than orange, with small amounts of eumalanin mixed throughout the barbules, although more concentrated on the tips. Mixture unlike the condition in northern forms, wherein the two pigments occupy different parts of the barbules. Mass color effect duller and more yellowish or greenish than in *J. phaeonotus*. Back area sharply delimited anteriorly, but confluent with cinnamon brown of inner secondaries and greater secondary coverts and with olive brown of rump. Lores and ocular region black, contrasting with dark olive gray of pileum and of auricular area. Throat and breast pale mouse gray, fairly sharply demarked from dark gray upper parts. Sides near Sayal brown. This color is produced by a scanty deposit of yellow phaeomelanin in the barbules, much more than in *J. p. phaeonotus*, together with eumelanin deposits in the tips of the barbules. Tail with two outer tail feathers part white and a third (number 4) often with a small white spot or tip. Wing and tail short, wings of males averaging 72 mm. Tail short relative to wing, more nearly as in *J. c. caniceps*. Bill very deep and stout, averaging the deepest of any member of the genus. Feet large, tarsus averaging 21.8 mm. in males.

Sexual difference in color is lacking. Nelson (1897, p. 62) describes the females as "rather duller colored." His series, which I have had for examination, is composed chiefly of molting birds, the females being markedly retarded in their molt so that fewer new, brightly colored flank feathers appear. Comparable plumages of the sexes appear exactly alike. Eleven specimens of each sex are all that are available for study and represent most if not all of the known skins of this species. Averages are, therefore, not as dependable as might be desired. Sexual difference in wing and tail averages about 5 per cent. The difference in bill length is about 4 per cent, but there is none in bill depth; tarsus is 1 per cent, and middle toe and hind toe are 3 and 7 per cent, respectively. An immature plumage is not distinguishable.

The effect of seasonal wearing and fading is to lighten the color of the back and make the green and yellow hues more prominent. The bright color of the sides is dulled and reduced in amount by wear. The white of the throat becomes darkened. Changes in the grays of the head are similar to those in *J. phaeonotus*.

Differential characters.—For comparisons with *J. p. phaeonotus* see page 218.

Fulvescens differs from *alticola* most importantly in the color of the sides, which are Sayal brown rather than buffy brown; the breast and underparts are much lighter—pale mouse gray with whitish areas, rather than deep olive gray. The pileum is lighter, not iron gray, and there is more contrast between auricular area and throat. Wing averages 5 per cent shorter; tail, 7 per cent shorter; bill length, 1 to 4 per cent shorter; bill depth, 3 per cent greater; tarsus, middle toe, and hind toe about 6 to 8 per cent shorter. Accordingly, *fulvescens* is smaller to equivalent degree in all measurements except for those of the bill. The bill is shorter but deeper. White is slightly more extensive on rectrix 6 in *fulvescens*.

Nongeographical variation.—Since the species is confined to one group of mountains, no geographically correlated variation is known. The small sample available precludes treatment of individual variation in fashion comparable to that for other forms. No peculiar plumage variations were detected. Tail markings show the following variations: inner vane of fourth rectrix all black (12) or white tipped or spotted (9); outer vane of fourth always black; inner vane of fifth rectrix no more than $\frac{1}{3}$ white except for one individual with half the vane white; outer vane of fifth always pure black; inner vane of sixth rectrix $\frac{2}{3}$ black (3) or chiefly white (3) but usually $\frac{1}{2}$ to $\frac{2}{3}$ white (15); outer vane of sixth pure white (1), pure black (1), chiefly black (4), chiefly white (8), or $\frac{1}{2}$ to $\frac{2}{3}$ white (7).

These variants are similar to those found in *J. phaeonotus*, but pure white inner webs of the sixth, and mixed (rather than pure black) outer webs of the fifth rectrix are absent. New types of variants, not found in *phaeonotus*, are pure black outer webs of the sixth, and inner webs of the sixth only ⅓ white. In the dropping out of extreme white variants and the addition of extreme black variants *fulvescens* shows a general reduction of white areas. Also, there are more individuals with greater proportions of black on the characteristically mixed vanes, but there is no decrease in white spotting of the fourth rectrix. The decrease of white is, therefore, not of great magnitude and is more of the order of difference found between the races of *J. phaeonotus*.

TABLE 12
MEASUREMENTS OF J. FULVESCENS

Measurement	Sex	No. of specimens	Mean with extremes
Wing	♂	12	72.56 (68.0–76.6)
	♀	11	67.85 (63.9–70.6)
Tail	♂	9	65.71 (63.0–69.6)
	♀	10	62.92 (57.0–66.0)
Bill	♂	11	8.93 (8.4– 9.5)
	♀	11	8.56 (7.9– 9.2)
Bill depth	♂	9	7.52 (7.2– 7.8)
	♀	10	7.50 (7.2– 8.0)
Tarsus	♂	12	21.83 (20.7–23.3)
	♀	11	21.52 (20.4–22.6)
Middle toe	♂	12	12.63 (12.1–13.0)
	♀	11	12.22 (11.3–13.1)
Hind toe	♂	12	9.41 (8.7–10.2)
	♀	11	8.74 (8.2– 9.3)

As nearly as may be ascertained without the aid of indices of variability, the variations and correlations are roughly comparable to those in adjacent forms, particularly *alticola*.

Distribution.—Limited to the pine forest area of the mountains of interior Chiapas that lie between the Rio Grande, a tributary of the Rio Grijaloa, and the Rio Jutate. The Sierra Madre south of the Rio Grande along the Pacific coast of Chiapas is inhabited by *alticola*, at least at its extreme eastern end.

Particulars relative to plant associations are not known and little information is at hand regarding climate. Nelson and Goldman (1926, p. 577) map the area as pine forest and refer to the rainfall in mountains of Oaxaca, Guerrero, and Chiapas, compared with most other parts of Mexico, as heavy. This impression was gained in part from their field observations of vegetation, some of which must have been made at stations where they took *Junco fulvescens* (San Cristóbal and Teopisca). Nevertheless, it is fairly certain that this area is not subject to the heavy rainfall and humid forest conditions described by Griscom (1932, p. 23) in Huehuetenango of Guatemala, where *alticola* occurs. The uplands of interior Chiapas do not exceed 8000 feet and this must mean more arid conditions than in the high forest habitat of *alticola*, which is chiefly above 8000 feet (Griscom, *op. cit.*, p. 362).

Relationships with adjacent forms.—For *J. phaeonotus* see pages 218 and 226.

J. alticola.—The ranges of *alticola* and *fulvescens* are separated by the lowlands (not in excess of 5000 feet) along the Chiapas-Guatemala border (northern Huehuetenango). The distance between suitable mountain terrains appears to be 40 to 50 miles. There is nothing to suggest any movement across this barrier. Colors characteristic of the two forms show no overlap, and the size characters, though overlapping, show some marked average differences (see p. 227). The two are thus fully isolated and all individuals are constantly and conspicuously differentiated with regard to certain characters.

Junco alticola
Guatemala Junco

This Guatemalan counterpart of *Junco fulvescens* is less restricted in distribution and better known than its relative of Chiapas. Part of its range lies near to that of *fulvescens*, but no evidence of intergradation has been found (see above). There is no intraracial geographic variation.

Principal characters of the species.—Iris variously reported as yellow and orange; lower mandible yellow; upper mandible dark brown; feet yellowish brown. Back raw umber to Prout's brown with pigments arranged in barbules as in back of *fulvescens* (see p. 226). Additional eumelanin seems to account for the slightly darker tone of red-brown. Brown of back sharply demarked anteriorly, but confluent with similarly colored areas on inner secondaries and greater secondary coverts and with olive brown of rump. Lores and ocular region black. Pileum, auricular area, and hind neck iron gray and deep mouse gray. Throat and breast deep olive gray, the chin distinctly lighter. Sides buffy brown, much as in *J. phaeonotus*. Color produced by eumelanins on tips of barbules with very small amounts of phaeomelanin proximally (not in granules). Tail with outer rectrix, and inner vane of next to outer, partly white. Fourth rectrix occasionally spotted. Wing and tail moderately long; wing averages 75 mm. in males. Tail long relative to wing as in *J. phaeonotus*. Bill deep and long. Feet very large, tarsus averaging 23.66 in males.

The plumages of the sexes average the same. Females appear no duller than males on the back and sides and do not possess more rufescent side color, as they do in *J. phaeonotus*. Sexual differentiation in wing and tail is 5 per cent. There are no statistically significant differences in the bill and toes. The tarsus is 2 per cent shorter in females. The sexual dimorphism in size, except for the tarsus, is therefore similar to that in *J. phaeonotus* in being limited to wing and tail. The sexual dimorphism in bill and feet of *fulvescens*, which occupies geographically intermediate territory, cannot be wholly accepted, owing to the impossibility of statistical treatment. Yet with respect to hind toe and bill length it seems certain that there is some true difference. In *alticola*, tail markings show slight sexual dimorphism only (see p. 231).

Seasonal alterations in plumage are comparable to those in *fulvescens*.

Differential characters.—For comparisons with *fulvescens* and *vulcani* see pages 227 and 234–236, respectively.

Nongeographical variation.—Some individuals have distinctly lighter gray breasts than normal, with the gray no darker than olive gray or light olive gray. Although this is a distinct approach toward *fulvescens,* there is never any real confusion of the two in coloration. These light variants are not restricted to the regions adjacent to *fulvescens.* Another variation found in nearly 30 per cent is buff tipping that extends forward from the red-brown of the back onto the hind neck and sometimes the crown. This probably always

TABLE 13

COMPARISON OF OCCURRENCE OF RECTRIX PATTERN IN J. ALTICOLA AND J. P. PHAEONOTUS
(IN PERCENTAGES)

	Male		Female	
	alticola	J. p. phaeonotus	alticola	J. p. phaeonotus
Rectrix 4				
Inner web:				
pure black	80	83	81	80
white spotted or tipped	20	17	19	20
Rectrix 5				
Inner web:				
black with white spot	26	59	44	80
⅔ black	68		56	
½ black	6	31	0	12
⅓ or less black	0	10	0	8
Outer web:				
pure black	96	98	100	100
(otherwise with trace of white)				
Rectrix 6				
Inner web:				
⅓ white	7	0	22	0
½ white	56	93	48	96
⅔ white	37		30	
pure white	0	7	0	4
Outer web:				
black	33	0	41	0
trace of white	31		41	
⅓ white	6		7	
½ white	11	93	4	100
⅔ white	6		4	
with trace of black	11		4	
white	2	7	0	0

disappears in late spring. It is not comparable to the reddish crown feathers found in some *J. phaeonotus* and *J. caniceps,* but is like variants of *bairdi.* The pileum and back are variable in color tone to a slight degree. The former may be faintly streaked with a darker tone of gray.

Rectrix patterns absent in *J. phaeonotus* are those with less than ⅓ of the inner vane of the sixth white, and those with pure black outer webs of the sixth; these already have appeared in *fulvescens. Alticola* lacks any with ⅓ or less of the inner vane of the fifth black, and any with pure white inner vanes of the sixth, which two types also are lacking in *fulvescens*. *Alticola* and *J. phaeonotus* are essentially identical with respect to the fourth rectrix and the outer web of the fifth. *Fulvescens* differs from *alticola* in the sixth rectrix, where there are only half as many that are black or black with only a trace of white. The greater amount of white spotting (40 per cent) on the fourth rectrix and the lack of an occasional individual with white in the outer web of the fifth are not significant in the sample of *fulvescens*. As to tail pattern, therefore, *alticola* and *fulvescens* are slightly differentiated, less so than the races of *J. phaeonotus* or of *J. caniceps*.

TABLE 14
MEASUREMENTS OF J. ALTICOLA

Measurement	Sex	No. of specimens	Mean with prob. error	Standard deviation	Coeff. of var.
Wing.............	♂	54	75.74 ± .21	2.28	3.01
	♀	30	72.05 ± .29	2.34	3.24
Tail..............	♂	50	71.53 ± .23	2.49	3.48
	♀	24	68.08 ± .41	2.98	4.37
Bill..............	♂	56	9.00 ± .04	0.48	5.33
	♀	29	8.95 ± .06	0.47	5.25
Bill depth.........	♂	54	7.35 ± .03	0.29	3.95
	♀	28	7.28 ± .05	0.41	5.63
Tarsus............	♂	60	23.66 ± .07	0.80	3.37
	♀	32	23.11 ± .07	0.63	2.72
Middle toe.........	♂	58	13.34 ± .05	0.51	3.82
	♀	31	13.18 ± .08	0.68	5.14
Hind toe...........	♂	62	9.87 ± .03	0.39	3.94
	♀	32	9.71 ± .05	0.45	4.63

Sexual differences in tail markings are not great. They are confined to rectrices 5 and 6, but chiefly the inner web of the former, much as in *J. phaeonotus phaeonotus*, although to a lesser degree.

It is appropriate here to review certain changes in variability found in passing from *J. caniceps* through *J. phaeonotus* to *J. alticola*. The wing is progressively more variable, the indices in males of *dorsalis*, *palliatus*, *J. p. phaeonotus*, and *alticola* increasing as follows: 2.70, 2.87, 2.92, 3.01; females, 1.96, 2.54, 2.93, 3.24. Tail in the same groups increases in variability, though less regularly and to much lesser degree in the males: 3.37, 3.17, 3.36, 3.48; females, 2.77, 3.07, 3.99, 4.37. Associated with this increase is an average diminution in wing and tail length. The tail diminishes less than the wing. Another associated phenomenon is the diminution of coefficients of correlation of the wing and tail. Coefficients for *dorsalis*, *palliatus*, *J. p. phaeonotus*, and *alticola* are: .721, .693, .573, .505.

Since in crosses of *caniceps* and *thurberi* correlations were found to be no higher than intraracial correlation (.71), it may be assumed that much of the intraracially correlated variation also is genetic. Two alternate explanations are then possible for the changes in correlation and variation in Central America. These are (1) that increased variability is of a somatic sort, not necessarily manifest to the same degree in wing and tail, and hence the correlated genetic variations of the two are partly obscured; and the alternative explanation (2), that the complex of multiple factors has different members, as it must to account for the general diminution in size, which are not in the same proportion common determiners for wing and tail or which are not linked.

Bill length in *alticola* is distinctly more variable than in *J. phaeonotus* and *J. caniceps*. This sharp increase accompanies increase in average length and

Fig. 8. Distribution of juncos in Central America. Each symbol represents a locality from which specimens have been examined. Dots, *J. p. phaeonotus*; open squares, *J. fulvescens*; solid squares, *J. alticola*; triangles, *J. vulcani*.

the frequency curves are skewed with modes higher than means. Length is comparable to that in *dorsalis*, but in the latter there is no such high variability. Correlation of bill depth with wing length is .434 ± .85. Tarsus and middle toe have a coefficient of .534 ± .063.

Distribution.—Resident of the high mountains of western Guatemala and extreme southeastern Chiapas (possibly also westward in the coastal mountains of Chiapas). Known from localities in the Department of Amatitlán west along the volcanic peaks and altos to southeastern Chiapas, just west of Volcán Tacaná. Also found in the interior in El Quiché and Huehuetenango almost to the Chiapas border. The birds inhabit the Boreal Zone (Griscom, p. 362, p. 65) in pine, mixed pine and oak, and humid cypress (*Cheirostemon*) and spruce forests usually above 8000 feet and most commonly at higher elevations up to 11,100 feet. Some localities, if they represent the actual point of capture, are as low as 6500 feet. Although Griscom compares the lower levels where *alticola* occurs to the southern Rocky Mountains of the United States, a majority of the population would appear to exist at elevations where the forest is very humid, with epiphytes abundant and sometimes with dense undergrowth. It is this upper humid forest that probably is lacking or restricted in the range of *fulvescens*.

Relationships with adjacent forms.—For *J. fulvescens*, see page 229.

J. vulcani.—This species, so different from other juncos that its generic assignment has at times been questioned, of course has no close relationship with *alticola*. It is separated geographically by 500 miles of country uninhabited by juncos. Although the highlands of Salvador and Honduras would seem to provide some limited areas of suitable habitat, no juncos have been discovered there, and there is no possibility of their occurrence in Nicaragua.

Junco vulcani
Irazú Junco

This terminal member of the series of Central American forms is the most aberrant species of the genus. Its distinctness has aroused suspicion concerning its generic affinities. Ridgway (1901, p. 272) comments as follows: "... even including *J. vulcani*, which is far out of place in any other recognized genus, *Junco* is a much more homogeneous group than *Spizella*, or indeed than most recognized genera containing an equal number of species." In some outstanding particulars such as the stripes on the back, the absence of pure white in the tail, and the prominently ridged culmen, *vulcani* is distinctly different. All these features are to some degree suggested in the rest of the genus. Not to be overlooked are the considerable resemblances to *alticola* in plumage coloration, exclusive of the stripes, as also in the color of the iris. Dimensions of *vulcani* are either similar to *alticola* or represent extremes of trends in differentiation that are in evidence in *phaeonotus, fulvescens,* and *alticola*. Too great emphasis should be avoided with respect to presence or absence of stripes in genera of the Emberizinae. Examination of details of feet, wing, and bill shows a reasonably close similarity in types. As genera of sparrows are now defined, I believe, in accordance with Ridgway, that *vulcani* definitely falls closest to *Junco*. Nothing would appear to benefit from designating a subgenus for it exclusively.

The question of including other groups of Central American or South American sparrows in the genus *Junco* has been studied only casually, as it is incidental to the main objectives of this investigation. *Phrygilus* of the Andean region, especially *P. alaudinus*, is as suggestive of *Junco* with respect to coloration as any Neotropical group known to me. However, the different type of tail markings in *alaudinus*, which consist of spots rather than stripes or longitudinal areas, is perhaps a greater departure from *Junco* than the obsolete white of *vulcani*. The bluish slate coloration of the head is not closely similar to that of *Junco*. The back is striped.

The characters of plumage, feet, and bill which are used throughout the sparrows to define genera are often unconvincing. It would be highly desirable to compare carefully the internal structure of *vulcani* with that of the other *Juncos*, as well as to compare related genera. It is not impossible that strong evidence might be disclosed which would not uphold the present allocation of *vulcani.* Other factors which would have some weight, in my judgment, are similarities in habit and, in a rough way, in habitat. The correspondence of *vulcani* with other juncos is fairly good in these respects, though more knowledge of details is desirable. The species is more terrestrial than are other juncos.

Principal characters of the species.—Iris variously reported as yellow, golden, or orange; mandible, maxilla and feet flesh color, resembling those of northern juncos. Bill exceptionally straight in outline on all sides, the width proportionately less than in other juncos; culmen distinctly ridged, the basal third with two grooves on either side parallel to ridge; relatively little inflation of maxilla. Back dark citrine, olive citrine, or buffy citrine, each feather with a large, somewhat guttate stripe of black on distal part of vane. Stripes fairly sharply limited anteriorly, becoming very faint, and dark citrine, variable in amount, confined to feather tips. Line of demarcation corresponds to margin of brightly colored area of other juncos. Back coloration and pattern extends with little modification to wing coverts and secondaries. Rump area sharply delimited, without stripes. Head smoke gray to light grayish olive, with olive citrine tipping. Lores and ocular region black as in *alticola.* Underparts light grayish olive, becoming lighter on belly and more tinged with olive buff. Sides buffy olive. Underparts generally similar to *alticola* except for more prevalent olive and yellowish hues of gray. Wings and tail blackish, with considerable edging of buffy citrine.

The citrine of the back is produced by eumelanin and a yellow phaeomelanin that has no trace of orange when viewed microscopically, the pigment being purer yellow than in *alticola* and *fulvescens.* The eumelanin is distributed in the barbules as in these species, that is, it is not confined to the tips of the barbules. The pigments of the flank feathers are comparable to those of the flanks of *alticola,* but the phaeomelanin is pure yellow and the eumelanin of the barbule tips is more extensive although well delimited. The phaeomelanin is nongranular.

Wing and tail moderately long, although possibly relatively short compared with body size. Wing of males averages 78 mm. Tail relatively long, less than 5 per cent shorter than wing. Bill deep and long. Feet extremely large, tarsus averaging 26.9 mm. in males.

Sexual dimorphism in plumage is absent. Wing and tail average 5 per cent shorter in females. No differences in bill or feet associated with sex can be proved statistically except for tarsus, which is 2 per cent shorter in females. The pattern of dimorphism is thus the same as that of *alticola.*

Wear decidedly alters the appearance of the head through the reduction of the olive or citrine overcast on the dorsal surface. Wear never completely removes this color, but it always accentuates the demarcation of head and back. The faint striping of the head is more obvious in worn plumage. The light-colored margins of rectrices and remiges disappear, especially the buff tips on the outer rectrices.

Nongeographical variation.—The newly grown outer rectrix probably always has a buff tip 1 to 2 mm. wide on the inner vane. Sometimes this is almost white, but often it is dark so as to contrast weakly with the remainder of the feather. Also on the inner vane of this rectrix the pigment may be pale distally to form blotched or clouded areas. This strongly suggests a blotching seen in other species of juncos on rectrices that are adjacent to white-bearing rectrices or that have white spots elsewhere on the vane of the same feather. The variation in color of the pileum, already noted, has a great range. Comparison of birds from different mountain summits reveals no colonial differentiation.

TABLE 15
MEASUREMENTS OF J. VULCANI

Measurement	Sex	No. of specimens	Mean with prob. error	Standard deviation	Coeff. of var.
Wing.............	♂	26	78.17 ± .24	1.84	2.36
	♀	18	74.10 ± .27	1.76	2.37
Tail..............	♂	23	74.43 ± .43	3.06	4.11
	♀	17	70.70 ± .47	2.85	4.03
Bill length........	♂	27	9.32 ± .06	0.44	4.68
	♀	18	9.53 ± .07	0.44	4.58
Bill depth.........	♂	19	7.21 ± .04	0.25	3.48
	♀	14	7.18 ± .07	0.39	5.50
Tarsus............	♂	23	26.96 ± .09	0.68	2.54
	♀	19	26.23 ± .12	0.79	3.00
Middle toe........	♂	23	14.66 ± .07	0.49	3.35
	♀	19	14.27 ± .07	0.47	3.32
Hind toe..........	♂	24	10.86 ± .10	0.79	7.26
	♀	19	10.76 ± .07	0.48	4.47

Variability in wing is materially decreased, being lower than in males of *dorsalis*. Thus, the southward increase in variability in the yellow-eyed juncos of the mainland does not hold for this species. Tail, on the other hand, is exceedingly variable, the most variable in the genus. As in *bairdi*, the coefficient of correlation of wing and tail is not especially high (.496 ± .108) where wing and tail each varies to a different degree. The correlation is not appreciably lower than in *alticola*, however.

Bill length does not vary so greatly as in *alticola*, but is comparable to that in *J. phaeonotus*. The frequency curve is skewed as in *alticola*. Bill depth is similar to that in *alticola*. Correlation between wing, tail, bill length, and depth is general in this species, whereas such relations involving the bill are found in only a few other juncos. The following are the coefficients: wing and bill length, .494 ± .104; wing and bill depth, .319 ± .129; tail and bill length, .354 ± .123; bill and bill depth, .599 ± .092. Tail and bill depth were not at all correlated. Unfortunately, the high probable errors detract from the significance of these coefficients. Those above .4 must certainly have some true value. The common factors that govern wing and tail would also seem to influence bill dimensions or to be linked to factors that do so. The lower correlation of tail with bill length and its absence in the case of tail and bill depth may be due to the unusually high variation in the tail, which either is controlled by independent factors connected solely with the tail or is somatic.

The only correlation involving the foot is that of tarsus and bill depth at .349 ± .129. There is no correlation of middle toe with tarsus such as in *alticola*, where the correlation suggested common or linked factors for long tarsus and long middle toe.

Distribution.—Resident of the highest peaks of central and southern Costa Rica, and of Volcán Chiriquí in extreme western Panama. The volcanoes of Irazú and Turrialba in the central cordillera are inhabited, as are also the highest peaks of the Dota Mountains and the paramo of Chirripo in the cordillera of Talamanca. It probably occurs at other points along this cordillera south to Chiriquí.

The juncos inhabit the scrub growth above timberline, never coming below 8500 feet. Austin Smith states in correspondence that they enter "to some extent the upper edge of the (oak) forest, but they are mostly found above timber line to the very crest, 11,400 ft. [of Irazú]. On ... Turrialba, it is not common—not more than 25 individuals. ... On the paramo [of Chirripo] above 11,000 ft., there are only 3 species of birds commonly met with," the junco being one of them. "*Junco vulcani* nests during April; mostly in tufts of grass; but sometimes in low bushes. It is essentially a terrestrial species; ... owing to more cover on the northern volcanoes it is there more skulking. ... This species as well as the one peculiar to Guatemala obtain a goodly part of their food supply from the seeds of various herbaceous Salvias. If my memory be not at fault its nuptial chant much resembles" that of the Arizona junco. The habitat of *vulcani* evidently is bleak and cold, but detailed comparison with the climate of the highlands of Guatemala cannot be made.

Comparison of vulcani and alticola.—The obvious differentiation of these two forms requires no special treatment for practical purposes of identification. There remain for detailed consideration comparative aspects that may bear on the problem of interrelationships. Throughout the genus the variation in tail white follows certain restricted lines. There is always a progressive diminution in white in the rectrices toward the center of the tail. No matter in what species, the rectrices, whether second, third, fourth, or fifth, that adjoin more lateral rectrices with appreciable amounts of white, display some variation in the form of white spots, or of diagonal or longitudinal areas of white, either terminally or separated from the tip. The major axes of such areas never lie across the feather, although in some forms they tend more that way than in others. In *vulcani*, the buff (occasionally whitish) tips sometimes reach fair proportions and, as they are placed diagonally along the margin of the inner web, are extremely similar to white tips and spots that appear on rectrices 4 and 5 in *alticola*. In *alticola* in such feathers that are predominately black the outer web rarely has white. The tail of *vulcani* is, then, quite conceivably merely an end type in the same series of variations found in other juncos. It is as though the last white or partly white tail feather were lost, leaving its faintly marked neighbor. The clouded effect on the outer rectrix, already mentioned, is contributory evidence.

The stripes of *vulcani* have no special counterpart in *alticola*, but in various members of the genus, in which there is eumelanin in the back area, occasionally there are weak, but quite apparent, dorsal streaks. In these the center of each feather vane is distinctly more blackish than the margins. Of course, the striped juvenal plumage in all young juncos is also suggestive, but this condition, although almost certainly primitive, is so prevalent in all sparrows as to have no particular significance here. Young of *vulcani* are heavily striped, with extension of the stripes to the head, rump, and underparts. The colors of juvenal *vulcani* and *alticola* are closely similar throughout.

Dr. M. Sassi of Vienna reports in correspondence that the iris of juveniles of *vulcani* is greenish yellow and that the bill and feet are like those of adults. The young he reports upon were taken on the 18th and 19th of July, 1930, and may have been acquiring adult coloration of the soft parts. The yellow eyes and flesh-colored bill of adults are a peculiar combination. The eyes agree well with the condition in other southern juncos, but the bill is like northern forms and possibly has been independently evolved. The characteristic bill shape of *vulcani* is not foreshadowed in *alticola*. Ratio of depth to length is more like that in *dorsalis*, and the shape of the bill is much closer to that in juncos with less inflated bills than to *alticola* and *fulvescens*. Also, in juncos with less inflated bills there are faint grooves paralleling the ridge of the culmen, but these never reach the development found in *vulcani*.

The wing of *vulcani* is 3 per cent longer than that of *alticola;* the tail, 4 per cent. Bill is 3 per cent longer, and because of its different proportions its depth is 1 per cent less. Tarsus is 14 per cent longer, middle toe and hind toe 10 per cent longer. The foot, then, throughout is large, without great alterations in proportion. Such an increase in foot size is not without parallel, for the tarsus of *alticola* is 10 per cent longer than that of *phaeonotus*. The proportions of *vulcani* are continuations of trends recognizable in *fulvescens* and *alticola* and the differences are in keeping with the degree of spatial separation and isolation of *vulcani*.

The pigments of *vulcani* are like those of *alticola* in the following respects: mixture of phaeomelanin and eumelanin of the back throughout each barbule, separation of these pigments in the flank feathers into separate parts of the barbules, and lack of granulation of phaeomelanin in the flank feathers. The phaeomelanins of the juncos south of *phaeonotus* are more nearly pure yellow and produce more olive and yellowish hues of brown than in *phaeonotus* and *caniceps*. *Vulcani* carries this to an extreme. The tones of grays are similar, though more mixed with faint yellow in *vulcani*. Facial mask, demarcation anteriorly of back area, and confluency of back color with that of the wings are characteristics of markings that *vulcani* shares with *alticola* and other southern juncos.

These facts all contribute to the view that *vulcani* is a member of the series of Central American juncos which, beside showing the extremes of certain trends, preserves some juvenal and probably ancient characteristics that may be like those of the prototype of the genus. Undeniably it has characters peculiarly its own, evolved probably in the course of its long isolation. Bill characteristics and tail coloration may fall in this category.

Rassenkreis Junco oreganus
Oregon Junco

This rassenkreis, the most extensive one in the genus in number of races, is composed of forms between which there is interbreeding or, at least, structural intergradation. Some of them are strongly differentiated, as for example *mearnsi* and *townsendi*, whereas others are faintly or very incompletely differentiated. *J. o. oreganus, shufeldti, montanus, mearnsi, thurberi, pinosus, pontilis,* and *townsendi* are included. In the breeding season the group occupies suitable areas on the Pacific coast from Yakutat Bay, Alaska, to the San Pedro Mártir Mountains, Lower California. In the north it is limited to the coast, but it extends farther inland in the Rocky Mountains of Montana and Wyoming, there culminating in *mearnsi*, which is found in many mountain ranges well within the Missouri River drainage. South of Oregon, Idaho, and northern Wyoming this *oreganus* type of junco is restricted to regions west of the Great Basin. Migrations vary from a complete type in the northern division of *montanus* to a state of absence in *pinosus* and *townsendi*.

Principal characters of the group.—In so extensive a rassenkreis few characters peculiar to the group remain constant throughout. The most important character is the presence, in the sides, of phaeomelanin which is sharply complementary to the black or gray of the hood. The phaeomelanin of the sides is granulated in the barbules, unlike that of the yellow-eyed juncos, which is uniformly distributed. The hood is convex posteriorly and is dark, although it may be as light as, or lighter than, that in *hyemalis*. Some phaeomelanin always is present in the middorsal area; eumelanin rarely is absent there. The two pigments are not mixed in the barbules, as in *fulvescens* and *alticola*, but occur in separate segments of the barbules. The mass effect produced by them is similar, although it is usually a redder, less greenish, hue of brown. Iris some hue of brown; bill flesh-colored; size moderately large to small.

Junco oreganus mearnsi

Racial characters.—*Mearnsi* may be distinguished from *montanus* by the following characters: slightly broader areas of phaeomelanin on sides, of pinkish cinnamon hue (not equal to any of Ridgway's colors) instead of vinaceous fawn, fawn color, or army brown; granular phaeomelanin of sides orange-yellow instead of yellow, and eumelanin absent from tips of barbules. Head neutral gray or slightly darker instead of dark neutral gray to black (males), or deep to dark neutral gray (females). Back usually less blackish, but phaeomelanin component often similar. Average measurements of *mearnsi* are greater than those of southern populations of *montanus:* wing, 3 per cent; tail, 5 per cent; bill depth, 3 per cent. White often more abundant on rectrices 3, 4, and 5. Sexual dimorphism in color less.

The major features that differentiate *mearnsi* from *J. caniceps* are essentially those of the entire *oreganus* group. Presence of phaeomelanin in the sides and absence of gray eumelanin, with the resultant hood marking, distinguish *mearnsi*. The back of *mearnsi* has a less abundant, yellower phaeomelanin and there is eumelanin in the tips of the barbules; the latter pigment is absent in *caniceps*. The brown area of the back is confluent with the brown and buff of the wing coverts and inner secondaries, instead of being sharply limited as is the reddish brown of *caniceps*. The head is a darker neutral gray (see p. 201). The lores and ocular area are less prominently black.

Wing and tail averages differ no more than 1 per cent in *mearnsi* and *caniceps*. These differences are not certainly reliable statistically. *Mearnsi* tends to be the smaller, particularly with regard to tail. The females of the two forms are more similar than the males in the samples at hand (see pp. 182, 241). Bill length of males is the same as in the shorter-billed populations of *caniceps*, but the females average 2 to 3 per cent shorter. Bill depth is 1 to 5 per cent less in both sexes of *mearnsi*, depending on the population of *caniceps* which is used for comparison. Bill depth approaches values for the Colorado and Utah populations of *caniceps*, but bill length is similar to that of the Colorado and Nevada groups. Thus it appears that size differences are all very small, many scarcely significant except that the trend in *mearnsi* in each difference is toward values characteristic of the main group of races of *J. oreganus*.

The white of the tail of *mearnsi*, so often reported to be great in amount, is remarkably similar to that of the Nevada populations of *caniceps* (see p. 240). *Mearnsi* does exceed the Nevada birds in amount of white on the third rectrix; 17 per cent of males and 5 of females have white on this feather, whereas such variants do not appear in *caniceps*. *Mearnsi*, nevertheless, is less differentiated from Nevada *caniceps* than the latter is from the other populations of *caniceps*. This still does not argue for a great degree of distinctness of the Nevada group, nor for special affinity with *mearnsi*, but merely emphasizes the incomplete differentiation of *mearnsi* with respect to tail pattern.

For comparisons with *aikeni* and *hyemalis* see pages 346 and 314–315, respectively.

Nongeographical variation.—*Mearnsi* is an especially variable race with regard to color, particularly when compared with *Junco caniceps*. This variability is seen in breeding areas remote from points of contact with *montanus* or *caniceps*. The range of variation in head color is fully as large as that represented in extremes of *caniceps* from Utah and Nevada. The pigment of the sides is variable in intensity, and sometimes is but slightly different from the vinaceous hue of *montanus*. Backs are uniform with respect to the eumelanin component, but some are much more ruddy, owing to phaeomelanins, than are others. The result is that certain individuals resemble the less blackish variants of *montanus*, being snuff brown, whereas others are drab so that the back is indistinctly contrasted with rump and wings and even with the head when buff tippings are present on the nape.

Reddish color on the pileum, a not uncommon variation in *caniceps*, was found in only one of 600 individuals. Reddish back feathers on birds from regions remote from points of *caniceps* contact were found in only three individuals (Montana and northern Wyoming). Whether these are new mutants, or whether they are the result of mixture with some stray *caniceps*, is not certain. In this connection should be recalled the significant occurrence

TABLE 16
Occurrence of Rectrix Pattern in J. c. mearnsi and J. c. caniceps (in Percentages)

	Male			Female		
	mearnsi	J. c. caniceps (Nevada)	J. c. caniceps (Colorado)	mearnsi	J. c. caniceps (Nevada)	J. c. caniceps (Colorado)
Rectrix 2						
trace of white	1	0	0	0	0	0
Rectrix 3						
Inner web:						
trace of white	17	0	0	5	0	0
Outer web:						
trace of white	7	0	0	0	0	0
Rectrix 4						
Inner web:						
pure black	1	0	7	12	6	10
trace of white	18			40		
mixed	64	100	93	47	94	90
trace of black	12			0		
pure white	4	0	0	0	0	0
Outer web:						
pure black	25	40	57	62	69	61
trace of white	21			15		
mixed	22	58	41	22	31	39
trace of black	21			0		
pure white	11	2	2	0	0	0
Rectrix 5						
Inner web:						
mixed	4	10	47	5	47	76
trace of black	26			52		
pure white	70	90	53	43	53	24
Outer web:						
mixed	14	38	70	32	66	67
trace of black	22			27		
pure white	64	62	30	41	34	33
Rectrix 6						
Inner web:						
pure white	100	100	100	98	100	90
Outer web:						
pure white	100	96	85	100	94	90

at the time of the spring migration, May 5, 1919, of a fully characterized *J. c. caniceps* at Glendive, Montana, far north of the range of the species.

White on the wings is somewhat more prevalent (1.82 per cent) than in *caniceps* (1.29 per cent), but it shows no greater incidence in regions nearest the range of the white-winged junco, *J. aikeni*. In addition to individuals with distinct white spots, there are many in which the coverts are edged with light gray. A single individual with a whitish throat patch has been noted.

Continuity of the areas of phaeomelanin across the breast is a frequent variation found in 8.10 per cent of all *mearnsi*. Its occurrence is much higher in females, 11.6 per cent, compared with males, 3.6 per cent. This difference exists in spite of small sexual difference in the hood pigment and no difference in the hue of the phaeomelanin. When a cross connection is present, it may extend irregularly across the margin of the hood, or it may form a line anterior to the margin, thus leaving a gray stripe posterior to it. It seems usually to represent a replacement of the gray hood pigment. This variant is com-

TABLE 17
MEASUREMENTS OF J. O. MEARNSI

Measurement	Sex	No. of specimens	Mean with prob. error	Standard deviation	Coeff. of var.
Wing............	♂	74	81.22 ± .17	2.22	2.73
	♀	46	77.16 ± .19	1.91	2.47
Tail.............	♂	73	73.15 ± .20	2.55	3.48
	♀	45	69.51 ± .23	2.30	3.30
Bill length.......	♂	64	8.22 ± .03	0.36	4.37
	♀	43	8.00 ± .04	0.38	4.75
Bill depth........	♂	54	6.17 ± .02	0.25	4.05
	♀	38	6.05 ± .03	0.24	3.96
Tarsus...........	♂	74	20.14 ± .04	0.55	2.73
	♀	44	19.75 ± .05	0.52	2.63
Middle toe	♂	62	11.52 ± .04	0.42	3.64
	♀	44	11.30 ± .05	0.47	4.15
Hind toe.........	♂	66	8.04 ± .03	0.36	4.47
	♀	44	7.99 ± .04	0.38	4.75

parable to that found in 25 per cent of the unrelated species *bairdi*, where the sides are likewise richly supplied with phaeomelanin. In *mearnsi* the phaeomelanin is granulated; in *bairdi* it is not.

Sexual dimorphism in color is restricted to the head, where the difference in gray is comparable to that in *caniceps*. An overcasting of buff on the hood is more frequent in females, but it is always limited to the extreme feather tips, so that the hood is demarked dorsally to some degree and is never obliterated as in some retarded plumages of *J. hyemalis*. Males may have buff tips on the hind neck. Probably such overcasting is more prevalent in first-year birds of both sexes, but accurately aged material taken in the fall is lacking and so this cannot be determined. Ages and sexes are less distinctly differentiated than in some other races of *J. oreganus*, though in all races the differentiation is less complete than is commonly supposed.

Wear alters the head but slightly. The back does not brighten with abrasion, even though the terminally situated eumelanin of the barbules is partly lost. Much of the phaeomelanin of the back feathers is situated in the more distal parts of the vanes, so that its loss probably compensates for the diminution of the dark tips of barbules. The color of the sides becomes less pinkish, more yellowish or cinnamon, as the season advances. The effect is comparable to the brightening of the phaeomelanin of the back of *caniceps*.

No intraracial geographic variation in the tail pattern of *mearnsi* can be found that grades either toward *J. aikeni*, with its greater amount of white, or toward *caniceps*, with less white. For discussion of resemblance to Nevada *caniceps* see page 239. Sexual dimorphism occurs chiefly in rectrices 3, 4, and 5.

Coefficients of variability for wing and tail are slightly greater than in *J. c. caniceps;* those for other measurements are similar. These differences are not at all comparable to those between *J. caniceps* and the species of Mexican and Central American juncos. The coefficient of correlation of wing and tail is $.865 \pm .02$, distinctly higher than in *caniceps* and much higher than in still more southern groups, where coefficients of variability are large. Tarsus and toes are not correlated. Wing and bill depth are definitely correlated, with a coefficient of $.427 \pm .07$. This peculiar relation, absent in *J. caniceps, J. phaeonotus,* and *J. o. montanus,* is all the more unexpected because there are no prominent differences in averages of these characters compared with adjacent forms. Certainly there must be a different arrangement of factors than in *J. caniceps,* where the coefficient (*dorsalis*) is $.074 \pm .07$. Sexual dimorphism is about 5 per cent in wing and tail, and 2 per cent in bill and feet, as in *J. caniceps*.

Geographical variation within the race.—Although no geographical variations can be clearly demonstrated at this time, it is not impossible that they may exist. The breeding material has not been representative of a variety of localities. Smaller size may occur in Montana north of the Yellowstone, but the sample is inadequate, and large size is again encountered in Saskatchewan. This possible diminution is unrelated to intergradation with *montanus* in western Montana. The birds of the Big Horn Mountains, Wyoming, are especially large. Populations of *mearnsi* are not isolated to the degree found in *J. caniceps,* except possibly in the region north of the Missouri River and in the Big Horn Mountains. The beginning of geographic variation in the direction of *montanus* is dealt with later (pp. 244–245).

Distribution in breeding season.—Populations of *mearnsi,* apart from those in which there is interbreeding with other forms, are found principally in the mountainous regions centering about the Yellowstone Plateau and Teton Range. Outlying populations, somewhat isolated from this fairly continuous inhabitable region, extend the breeding limits of the form into Saskatchewan, eastern Alberta, eastern Idaho, and north-central Wyoming (Big Horn Mountains). The form is to be thought of as one essentially endemic to that segment of the Rocky Mountains which lies north of the southern Wyoming deserts and east of the upper Salmon River of Idaho and the upper Missouri and Jefferson rivers of Montana. This mountainous section is moderately arid compared with northern Idaho and parts of interior British Columbia. Lodgepole pine (*Pinus contorta*) dominates the lower coniferous forests and there is junco habitat in aspen forests (*Populus tremuloides*). At higher altitudes Englemann spruce (*Picea englemanni*) forests predominate. *Mearnsi* apparently does not encounter the extreme aridity experienced by *caniceps* over great sections of its range, although it may tolerate relatively dry, scrubby growths of aspen locally and inhabit scanty yellow pine and Douglas fir forests on certain prairie ranges (see Saunders, 1921, p. 13). Nothing comparable to the *Cercocarpus* habitat of *caniceps* is occupied. Owing to the lower elevations of the Boreal areas, juncos may breed at altitudes as low as 5000 feet in parts of Montana and Idaho. The lower limits in central Wyoming are often 7000 feet.

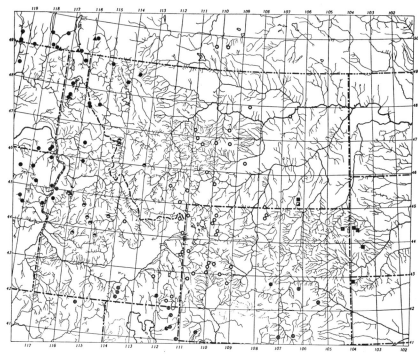

Fig. 9. Distribution of *Junco oreganus mearnsi* and neighboring forms in the breeding season. Northern Rocky Mountains of United States (Idaho, Montana and Wyoming, and parts of eastern Oregon and Washington, southern Canada, western South Dakota and Nebraska, and northern Nevada and Utah). Each symbol represents a locality from which specimens have been examined. Circles, *J. o. mearnsi*; dots, *J. o. montanus*; half dots, intermediates between *J. o. mearnsi* and *J. o. montanus*; circles enclosing crosses, hybrids of *J. c. caniceps* and *J. o. mearnsi*; triangles, hybrids of *J. o. montanus* and *J. hyemalis*; solid squares, *J. aikeni*; partly solid square, hybrids of *J. aikeni* and *J. o. mearnsi*.

Wyoming.—In Boreal Zone (Cary, 1917) in mountains of northeastern section from Salt River and Wind River ranges of Lincoln and Fremont counties northward and in Big Horn Mountains (see appendix for intermediates of central Wyoming).

Montana.—Chiefly in Boreal Zone in mountains west of about longitude 107° and south and east of the Missouri, Jefferson, and Beaverhead rivers. There is some evidence that populations just west of the last-named river represent the beginning of intergradation with *montanus*. Known definitely from Crazy, Little Belt, Big Belt, Big Snowy, and Big Horn mountains, from the mountains of extreme southern Musselshell County, and from those adjoining Yellowstone Park. Also occurs north of Missouri River in Little Rocky Mountains, Phillips County, and along Missouri River in Garfield County; probably also in other similar ranges in this section of the plains where there are limited areas of Transition Zone.

Saskatchewan.—Cypress Hills, in southwestern corner of province.

Alberta.—Known from Eagle Butte on southeastern border, adjacent to the Cypress Hills of Saskatchewan.

Idaho.—In eastern section of state north of Bear Lake, populations of pure *mearnsi* occur, but west of the lower Bear River and south of the Snake River all populations appear to be mixed. North of the Snake River on the Montana border pure populations are found west as far as the Lemhi Mountains. Westward in the Sawtooth and Salmon River mountains there is intergradation.

Distribution in winter.—*Mearnsi* winters commonly in the lower mountains of Arizona, New Mexico, and Colorado, the largest representations of winter birds coming from southeastern Arizona and the eastern flank of the mountains of Colorado. It is also known in winter from points on the plains as far east as central Nebraska, western Kansas, and the panhandle of Oklahoma. It occurs in western Texas (Presidio County), and in Chihuahua (south-central) and Sonora (northern). It migrates south through Utah and Nevada in small numbers; wintering records seem to be lacking for Nevada and are scarce for Utah (Presnall, 1935, p. 209). However, an intermediate between *mearnsi* and *montanus* is at hand from Belleview, Blaine County, Idaho, taken January 7, 1936. Of four record specimens for California, three are from the Colorado River region in Imperial County and one from western San Diego County. Apparently all birds leave the breeding grounds even though they remain in the fall until October or, sometimes, early November. Compared with *J. c. caniceps*, *mearnsi* leaves the breeding range more completely in winter and spreads out on the plains in greater number. The extreme limits of record are similar in the two forms; they are commonly associated in flocks in winter in Colorado and in the oak belt of New Mexico and Arizona.

Relationships with adjacent forms.—For *J. c. caniceps*, *J. h. hyemalis*, and *J. aikeni*, see pages 200, 326, and 351, respectively.

J. o. montanus.—The breeding Oregon juncos east of the Cascade and coast-range mountains from central Oregon to central British Columbia display considerable regional differentiation as well as much individual variation. Although all birds in this area may be considered to belong to a single race, *montanus* (formerly known as *shufeldti* or *couesi;* see Appendix), it is not possible to indicate any one phase of plumage out of the heterogeneous mixture as typical, nor is there a single center of racial differentiation. There are two main divisions of the group which are only very incompletely differentiated. The southern division occupies northern and western Idaho, eastern Oregon, and adjacent parts of Washington and Montana. It is fairly stable over this area and cannot be considered merely a hybrid or intergrade between the more northern division of *montanus* and *mearnsi*. It is essentially the same in such widely separated areas as Crook County, central Oregon, and Boundary County, extreme northern Idaho. It is only this division of *montanus* with which *mearnsi* interbreeds.

Study of the interrelations with *montanus* rests largely on series of breeding males taken by me at localities in Idaho beginning at the western border and extending east to the southeastern section of the Sawtooth Mountains at Alturas Lake. Juncos were secured at five localities spaced 20 to 25 miles apart. Those on the west were from normal populations of southern *montanus;* those at Alturas Lake were not all pure *mearnsi,* but the intergradation at that point had progressed more than halfway toward *mearnsi.* The principal points of contrast between *montanus* and *mearnsi* are those of coloration of the head, back, and sides. Size and tail pattern differ but slightly.

Head color effects the most complete differentiation of the two forms. In analyzing intensity of head color, ten successive stages were distinguished, ranging from a dark black (type 1), rare even in northern *montanus,* to the palest type of *mearnsi.* The ventral surface of the head was used chiefly. Southern *montanus* usually are of types 5, 6, and 7, with occasional darker individuals, the average being about 6. Pure populations of *mearnsi* are made up of types 9 and 10, the average falling about halfway between them. The graph (fig. 10) shows individual variation and change geographically in head color. The groups are spaced at intervals scaled to the actual spatial separation in an air line west to east. In males the uniformity across Oregon is evident. About 35 miles east of the Snake River the variability increases, significantly in the paler categories, and the average quickly drops. East of Alturas Lake the curve is projected on the basis of a small number of typical *mearnsi* from the region of the Lemhi Mountains, Idaho. A level doubtless is attained about here which is equal to that found in the Yellowstone region and which persists throughout the remainder of the range of *mearnsi.* Breeding range is more or less continuous across this entire area, certainly so in the region of intergradation in which I collected.

The females, similarly classified, form a more irregular curve, no doubt owing to the greater chances of variation from sampling. Several points of significance appear, however. Sexual dimorphism is twice as great in *montanus* as in *mearnsi.* The break of the curve at the region of intergradation is less abrupt. But also the curve shows that *montanus* influence first appears phenotypically at a point farther west than in males; similarly, typical *montanus* averages appear farther west.

Because back color is a complex of phaeomelanin and eumelanin mixed to produce a brown, the analysis of back variants is not easy. When the black component becomes especially large it may invade the basal portions of the vanes to produce solid blackish areas, whereupon the brown regions become more restricted distally. This redistribution, which is independent of tone of brown, is fairly certainly an effect of intermixture of *J. hyemalis.* The eumelanin component was classified as follows: *aa* and *a,* eumelanin of back extensive and intense, but usually not darkening the brown areas which may be distinctly bright (type *aa* is attributable to *hyemalis* without doubt); *b,* average extension of blackish areas and moderate component of eumelanin in brown areas; *c,* distinctly smaller black component in brown areas that results in lighter tones. The phaeomelanin component of the brown parts is classed with respect to intensity and, to some degree, amount: 1, clear reddish yellow; 2, less distinct yellow, the combination appearing duller brown; 3, dull yellow, with eumelanin producing a very dull gray-brown. *Mearnsi* all fall in classes 2c and 3c. *Montanus* of the southern area are of classes 2c, 3c, and 3b,

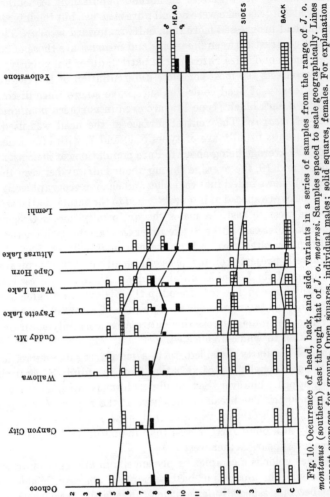

Fig. 10. Occurrence of head, back, and side variants in a series of samples from the range of *J. o. montanus* (southern) east through that of *J. o. mearnsi*. Samples spaced to scale geographically. Lines connect averages for groups. Open squares, individual males; solid squares, females. For explanation of variants see pages 245, 247, and 248.

with infrequent occurrences of 2b, and rarely others. The chief point to be noted in the transitional area is the disappearance of type b as *mearnsi* is approached (see fig. 10). The principal change takes place in the interval between Warm Lake and Alturas Lake, and so is restricted to a narrower zone than is that of the head. The change in the back is purely a shift of average, not an intergradation of completely different types as with the head.

The following list is a synopsis of symbols used in designating stages in reduction of intensity of pigments in the *oreganus* and *hyemalis* artenkreise. The symbols at the top of each column represent extreme saturation. Some members of the series of back types are not completely equivalent quantitatively, as explained in the text where new symbols are introduced.

BACK		HEAD
PHAEOMELANIN	EUMELANIN	
ooo	aaa	1
oo	aa	2
o or p	a	3
t	b	4
1	c	5
2	d	6
3	dd	7
4	y	8
5		9
6		10

The backs of females are distinctly poorer in blackish pigment than those of males, so that the phaeomelanins are more prominent and the general tone is lighter. All fall into class c in the southern part of *montanus* and in *mearnsi*. Phaeomelanin is either of type 2 or 3, rarely 1 in *montanus*. There is, therefore, no differentiation of the races on the basis of intensity of eumelanin as in the males.

This situation in the females and the fact that the differentiation in males is associated in general with head color, the latter more prominent and more sexually dimorphic in *montanus*, suggest some degree of correlation of head and back. That this is not perfect is obvious from the fact that distinctly dark-headed females of *montanus* are still of class c, as are the gray-headed *mearnsi*. Possibly it takes an extreme development of black, as in the heads of the darker males of *montanus*, to bring any concomitant extension or intensification of the eumelanin component in the back. To test this hypothesis, correlations of head type with back in male individuals of northern and southern *montanus* and in *montanus–mearnsi* intermediates have been determined. The general correlation for all groups combined was $.427 \pm .038$. It appeared possible that this correlation might be due in part to the diverse natures of the geographically different populations used in the correlation table. For example, if a population of pure *mearnsi* and one of *montanus* were combined, a high coefficient of correlation would result, indicative of the association of the characters in the races, but it would not necessarily be based on a fixed association in individuals. Consequently, separate correlation coefficients were determined for northern *montanus,* southern *montanus,* and the intermediates. The coefficients were: northern *montanus,* .402 ± .076; southern *montanus,* .435 ± .054; *montanus–mearnsi* intergrades (Warm Lake to Alturas Lake), .504 ± .086.

Inspection of all the correlation tables shows that types a and b are limited for the most part to birds with heads that are type 6 or darker. Only 8 per cent of type b are of head types pales than 6, as compared with 41 per cent of c that are paler than 6. In comparable fashion the pale back, c, occurs almost entirely with head types 5 or lighter. Only 5 per cent of c's constitute exceptions, while 35 per cent of b's are of head types darker than 5. Clearly there is some interdependency of eumelanins in these areas.

Any effect of head color on the back is seemingly suppressed in females. But, this may be because in southern *montanus* and in *mearnsi,* females rarely attain head color darker than 8 and never darker than 6, types which even in males are almost entirely associated with back c. There may, then, be no sex-limited situation, for certain dark-headed females of northern *montanus* attain back b. Probably there is some interaction of factors for back and head so that when a critical point in dark-headedness is reached the back cannot normally remain uninfluenced, and contrarily, when the head is extremely pale no great intensity of pigment in the back is possible.

A significant fact to be deduced from the coefficients is that intraracial correlation is not appreciably lower than in interracial hybrids. The racial characters and the intraracial variants are all part of the same complex. The intermediate populations are thoroughly intermixed, and extreme types that appear are probably chiefly Mendelian segregates. Variation is higher at a point like Warm Lake because of the contribution originally of parentage of diverse types from either side.

Side color is not completely differentiated in *montanus* and *mearnsi,* although it is much more so than is back color. Three categories were recognized: 1, extreme *montanus* type with colored areas moderate in extent and with nearly all feathers brownish or vinaceous fawn owing to small amounts of eumelanin additional to the dull pinkish phaeomelanin; 2, some feathers of the area brownish, but also some with conspicuous yellowish or cinnamon and no eumelanin; 3, all feathers yellowish or pinkish cinnamon, the area extending medially onto the breast farther than in types 1 and 2. All three types are found in *montanus,* but 3 is rare. *Mearnsi* are all of type 3, or they may show an even greater medial extension of the color (see p. 240). Type 3 might properly be subdivided into a yellowish type and a pinkish cinnamon one, both occurring regularly in *mearnsi,* but the latter solely in this race. However, wear and fading make it impractical to attempt such a distinction among worn plumages. Some difficulty is encountered in evaluating the width of the colored area because of variation in the make of skins.

The graph (fig. 10) indicates the intergradation and occurrence of types in *montanus* and *mearnsi.* Although it might appear that there is some definite change toward *mearnsi* in the Wallowa area, this actually is not the fact. The

Wallowa group is quite normal for *montanus* in proportions of types; the groups farther west happen to be lacking in type 3. Type 3 occurs occasionally north through Idaho to the northern limits of *montanus* in central British Columbia. The principal intergradation coincides with that of head color rather closely.

The sides of females are similar to those of males and intergrade in the same way. Type 3 is almost lacking west of Payette Lake and type 1 is lacking east of Warm Lake. The only sexual differences consist in slightly less dark brown on the lower flanks in type 1 and a somewhat paler yellow on the average in type 2. In *mearnsi* the yellowish type of 3 is a little more prevalent than the pinkish cinnamon type. Side color is not correlated individually or geographically with the phaeomelanin of the back or with eumelanin of back and head. Phaeomelanin increases in the sides toward *mearnsi*, whereas that of the back becomes reduced.

The wings of southern *montanus* and *mearnsi* differ on an average by only 2 per cent. Populations of *mearnsi*, as in the southern part of Montana adjoining the Yellowstone region, may measure 1 per cent less than *mearnsi* as a whole. The birds from Wyoming adjacent to Idaho are, however, not smaller. With such minor and rather unreliable differences in averages locally, it is almost impossible to indicate the region of intergradation with *montanus*, even though the general averages of the races are significantly different. As far east as Alturas Lake there is no sign of increase in wing length. Tail length, which differs by 5 per cent in southern *montanus* and *mearnsi*, shows no increase eastward until Alturas Lake is reached, where it rises definitely to an intermediate point, 3 per cent lower than in *mearnsi*.

The white of the third rectrix, which occurs irregularly in *mearnsi*, is about the only point of difference in the tail pattern that may be dealt with in the intergrades. Such variants appear in the Wallowa Mountains, but not with regularity except east of Payette Lake. The proportion of occurrence at Alturas Lake is fully as high as in *mearnsi*.

The situation north of the extensive intergradational belt in Idaho is not known in any detail. Saunders (1921, p. 128) reports breeding *mearnsi* from Jefferson, Silver Bow, Deer Lodge, and southern Powell counties. Apparently no critical examination of specimens was made, although the populations must have been essentially of *mearnsi* type. An early May bird from Deer Lodge County, which I have examined and which was taken after the migration dates given for the area, is somewhat atypical of *mearnsi*. Saunders tentatively gives the dividing line between *mearnsi* and *montanus* as the Northern Pacific Railroad between Helena and Missoula; thus, it is an east–west boundary line. From evidence at hand from scattered breeding birds, the Idaho intergrades, and from the knowledge of the northward extension of *mearnsi* in central Montana, I think that the belt of intergradation must extend more nearly north and south. From the Sawtooth Mountains, Idaho, it runs northeast through the southern Bitterroot Valley and northern Powell and Lewis and Clark counties. Two birds from Idaho City, Boise County, Idaho, suggest a population of a degree of intermediacy about equal to that at

Warm Lake. A series from the west side of the northern Bitterroot Valley, near Missoula, is *montanus* with only two individuals that suggest the beginning of intergradation. The series is about like that from Payette Lake, Idaho.

The populations of northwestern Montana, including that of the type locality of *montanus,* are in no way different from those of southern *montanus* in northern Idaho and eastern Oregon. Their considerable variability is encompassed in the normal range of variation found elsewhere in the race. The head colors of males are all of types 5, 6, and 7, average 5.8, and although the sides in some individuals look like *mearnsi,* such types are no more frequent than in Oregon and Idaho.

The meeting of *mearnsi* and *montanus* occurs in a region where there are no barriers to distribution. At least in Idaho and in parts of Montana they meet in areas of nearly continuous forest habitat. There is more discontinuity within the range of *mearnsi* in northeastern Idaho than in the region of principal intergradation. That there is some change in forest condition that reaches a critical point in the area of intergradation, it is difficult to conceive. In the range of *mearnsi* the forests are more open and arid than in parts of the range of *montanus.* Yet there is much diversity within the area of each race. *Mearnsi* frequently occupies arid lodgepole pine forest, but in parts of the Fraser River Valley *montanus* thrives under comparable forest conditions. If *montanus* were restricted to the humid forests of northern Idaho and British Columbia, one would be tempted to assume that such an environment was essential to this race and not to *mearnsi.* Adaptive value of the brightly colored sides of *mearnsi* or of its dull head is quite unlikely. The race appears to me to be possessed of characters of long standing which are not related to present climatic and vegetational conditions. In the zone of junction with *montanus* there is increased variability such as would arise through interbreeding of two diverse stocks. If the intermediates had differentiated *in situ* as the result of intermediate conditions, I believe the chances are that they would not show the same degree of variability, but only variability equivalent to that of either extreme population. If it were clear that the intermediates had to cope with two different phases of environment, corresponding to those of the racial extremes, highly variable intermediates might evolve, but in the absence of any indication that characters are associated with a particular environment this seems unlikely.

To speculate further, it is not probable that extreme types could have arisen and have become segregated into separate areas, as have *montanus* and *mearnsi,* if those areas were always in contact. One form surely would have swamped the other unless there was some distinct advantage in the new variant, for example, of the *montanus* sort. Isolation for a long period doubtless made possible the present differentiation, while at the same time ability to interbreed was not impaired. The present intergradation is to be viewed as a secondary junction similar to the interbreeding of *caniceps* and *mearnsi* and *caniceps* and *J. o. thurberi,* but not at all like the intergradation of *J. p. phaeonotus* and *palliatus.*

Most of the characters which separate *mearnsi* and *montanus* show blending, and multiple factor inheritance is the probable mechanism. Side color tends to segregate into two or three categories, but the separation is not perfect and there is some degree of blending. No evidence of dominance is at hand. It is to be noted that the chief differentiating characters of *mearnsi* and *montanus*, namely, the colors of head, back, and sides, involve the same pigments in both races. Proportions of phaeomelanins and eumelanins, with some modification of the hue of the former, alone account for the differences, great as they may appear to the eye. This was not true with *mearnsi* and *caniceps*, where the total absence of certain pigments was responsible for differences in back and sides. Thus, although in both these pairs of forms there is interbreeding, the differentiation of *caniceps* and *mearnsi* seems more fundamental, irrespective of modes of inheritance. Isolation of breeding ranges also is more complete in the case of *mearnsi* and *caniceps*.

Junco oreganus montanus

In areas north and west of the range of *Junco o. mearnsi*, the black-headed juncos vary with respect to color to a degree unparalleled in the forms of the southern Rocky Mountains. In addition, the birds of California, Oregon, Washington, Idaho, and British Columbia show geographic differentiations, some of which are fairly sharply outlined both as characters and range, and others of which are so incomplete as to make impractical a categorical separation into races. Within this mosaic of geographic variation, exclusive of the extremes in central and Lower California and central coastal British Columbia, it is doubtful if true racial units exist; instead, there are many independent geographic gradients in characters that run in different directions through the area. One can recognize some poorly defined breaks in the gradients that permit the mapping out of races that combine certain average values for various characters, but these races are in a sense artificial. This state of affairs should be fully appreciated by anyone attempting to identify birds pertaining to these regions.

The juncos of eastern Oregon, eastern Washington, interior southern British Columbia, northern and western Montana, and northwestern Idaho comprise a group which may be treated as a race. Within this race two divisions exist, a northern and a southern one. The extreme variants in each of the divisions are so different as to suggest well-defined races. But although these variants may be characteristic of an area, they are so frequently found elsewhere, or they are of such erratic occurrence, that they fail to differentiate the geographic divisions to a satisfactory degree. This composite race has had various names applied to it. It must now, by rule, bear the name *montanus* (see Appendix A).

Montanus as represented by the type series is not clearly an intergradient or hybridized group as supposed by Dwight and as I at one time thought. The critical fact not heretofore emphasized is that the race formerly known as *shufeldti* or as *couesi* is variable, enough so to include the original *montanus* series (see p. 392). The southern section of *montanus* as here considered is, then, like Ridgway's *montanus*. The type, unfortunately, is a light-headed variant, an extreme for the race.

Racial characters.—Compared with that of *J. o. oreganus*, the back of *montanus*, although quite variable, is grayer or less reddish; it never is Prout's brown, Brussels brown, or raw umber as in *J. o. oreganus*. This character effects a nearly perfect separation of the races outside of intermediate areas. Other characters that incompletely differentiate the two are lighter head color in *montanus*, averaging 4 to 6 (see p. 247) instead of 3.2 as in *J. o. oreganus*, and less rufescent, and often less dusky, sides. Wing and tail average 5 per cent longer; tarsus and toes 3 to 5 per cent shorter. The amount of white usually is greater on rectrices 4 and 5 (for details, see pp. 253 and 272).

For characters which differentiate *montanus* from *J. hyemalis* and from *J. o. shufeldti* see pages 314 and 264, respectively.

Nongeographic variation.—Color of head, back, and tail, and many of the measurements, are dealt with in a later section on geographic variation. For discussion of variation in side color see page 248. The variation in the side found in northern *montanus* is the same as that in southern *montanus*.

Reddish pileum, a mutation occasionally met with in *J. caniceps, J. phaeonotus*, and once in *J. o. mearnsi*, is unknown in *montanus*. However, brownish overcast of the head, which in extent may be very great, is extremely common. This rarely involves bases of the feathers to the extent that the red mutation does. It is a feature probably latent in all juncos of this and similar races, but it is most pronounced in this group; it finds expression in young and in females most often. Brownish overcast is, however, no certain indication of age or sex, but is properly viewed as a retarded phase of plumage (Mayr, 1933, p. 2), sometimes omitted entirely, but in other individuals found even after the first year. It is not highly correlated with the intensity of the black of the head. Undue concern with this feature may lead to confusion of the more significant characters of the race.

White on the wing coverts occurs in 1.17 per cent of the 342 breeding *montanus* that were examined. Odd color patterns in the ventral part of the hood and occasionally on the pileum are not uncommon. They appear to be due to incomplete molts with retention of some feathers of a previous retarded plumage. This is seen most frequently in the paler types of females and occasionally suggests the variants in *mearnsi* in which phaeomelanin is mixed along the ventral margin of the hood.

In parts of the range of the race adjacent to *J. hyemalis*, yet not in the regions where interbreeding occurs at all extensively, birds with some *hyemalis* characters appear as rare variants in otherwise typical populations of *montanus*. These so evidently have to do with the problem of hybridization with *hyemalis* that discussion of them will be deferred to a later section.

Sexual dimorphism of head color is about twice as great as in *mearnsi* (see fig. 10). Associated with this, the sides and back are often slightly lighter in females because of the lesser eumelanin component (p. 247). Sexual dimorphism occurs in rectrices 3, 4, and 5 about as in *mearnsi*.

The sides, and particularly the back, tend to brighten in the course of the year, becoming yellowish in those variants which have relatively large amounts of phaeomelanin. In the duller individuals seasonal change is not pronounced and is comparable to that in *mearnsi*.

TABLE 18

OCCURRENCE OF RECTRIX PATTERN IN J. O. SHUFELDTI, J. O. MONTANUS AND J. O. MEARNSI (IN PERCENTAGES)

	Male				Female	
	shufeldti	*montanus* N.	*montanus* S.	*mearnsi*	*montanus* S.	*mearnsi*
Rectrix 3						
Inner web:						
trace of white....	0	0	1	17	0	5
Outer web:						
trace of white....	0	0	1	7	0	0
Rectrix 4						
Inner web:						
pure black.......	17	23	4	1	20	12
trace of white....	44	60	50	18	61	40
mixed...........	40	15	46	64	20	47
trace of black....	0	2	1	12	0	0
pure white.......	0	0	0	4	0	0
Outer web:						
pure black.......	62	65	46	25	78	62
trace of white....	15	8	16	21	9	15
mixed...........	19	22	19	22	11	22
trace of black....	2	5	17	21	2	0
pure white.......	2	2	3	11	0	0
Rectrix 5						
Inner web:						
mixed...........	10	9	5	4	30	5
trace of black....	42	32	30	26	52	52
pure white.......	48	58	65	70	17	43
Outer web:						
mixed...........	17	23	9	14	13	32
trace of black....	17	26	25	22	52	27
pure white.......	67	51	66	64	33	41
Rectrix 6						
Inner web:						
pure white.......	100	100	100	100	95	98
Outer web:						
pure white.......	100	100	100	100	100	100

Geographic variations within the race.—*Montanus* is almost lacking in variants with white on the third rectrix and in those with pure white inner webs of the fourth rectrix. Otherwise the same variants are present as in *mearnsi*. The average amount of white on the fourth and fifth rectrices is less in *montanus*. In this respect the southern population stands about halfway between *mearnsi* and the northern group; the average for the fourth rectrix is closer to the northern, that for the fifth rectrix nearer *mearnsi*. Thus the interracial differentiation is no greater than intraracial differentiation with respect to these highly variable features.

Individual variability indicated by the coefficients is comparable to that in *mearnsi*. The coefficient of correlation for wing and tail is .646 ± .033 for males, .520 ± .069 for females; this is lower than in *mearnsi*, which has the highest correlation of any form of junco. The significance of wing and tail correlation is subject to the criticism that the correspondence may be due in part or entirely to a general variation in size which might affect all dimensions of the animal. F. B. Sumner has suggested the need of determining whether

TABLE 19
MEASUREMENTS OF J. O. MONTANUS, SOUTHERN DIVISION

Measurement	Sex	No. of specimens	Mean with prob. error	Standard deviation	Coeff. of var.
Wing.............	♂	141	79.25 ± .12	2.06	2.61
	♀	51	74.52 ± .15	1.58	2.14
Tail..............	♂	141	70.00 ± .13	2.28	3.28
	♀	51	65.69 ± .20	2.15	3.29
Bill length........	♂	125	8.25 ± .02	0.33	4.03
	♀	55	8.00 ± .03	0.28	3.57
Bill depth.........	♂	125	5.99 ± .02	0.25	4.23
	♀	52	5.93 ± .03	0.28	4.67
Tarsus............	♂	150	19.93 ± .03	0.53	2.67
	♀	58	19.96 ± .05	0.58	2.94
Middle toe........	♂	150	11.38 ± .02	0.39	3.42
	♀	59	11.21 ± .04	0.42	3.74
Hind toe..........	♂	150	8.02 ± .02	0.41	5.11
	♀	60	7.87 ± .03	0.35	4.39

there is a special correlation of these parts additional to general variability in size. He was able to do this in *Peromyscus* by determining the partial or net correlation, eliminating size. Unfortunately, in birds there is not available a good measurement of general size comparable to the measurement of total length in mammals. Even though new collections were to be made in which total length was ascertained at time of preparation, the extreme flexibility of the bird's neck would in some degree complicate the measurement. The best substitute that can be offered for total length is weight.

One hundred breeding male birds from within the range of southern *montanus* have been weighed (average 17.7 gr.). Wing and weight were found to be correlated to so low a degree (.237 ± .063) as to be of doubtful significance. This, together with the fact that tarsus and wing are not correlated, indicates the absence of any great correspondence between length of appendages and body size. In other words, there are no general size fluctuations which affect all parts; otherwise, variation in the tarsus and wing would correlate. To show more adequately that there is a net correlation over and above size, I have calculated the partial correlation (Fisher, 1934, p. 174) of wing and tail, eliminating weight. The coefficient is .623.

Despite the numerous factors which affect the weights of birds (see Linsdale, 1928, p. 312), the moderate variability of this measurement, compared with that of mammals, makes it fairly reliable. The breeding male juncos used in these calculations were taken over a period of six weeks. Practically all of them were devoid of obvious deposits of fat and rarely were their stomachs empty. The coefficient of variability was 5.2 per cent, which is about twice that of wing length but equal to that of some of the toe and bill measurements.

The unusual correlation of wing and bill depth found in *mearnsi* is lacking in *montanus*. Tarsus and hind toe show a low correlation of .356 ± .048.

Measurements of the northern division of *montanus* that occupies the Cariboo and interior Skeena districts of British Columbia show some minor differences compared with those of the southern group. Wing length is only 1 per cent shorter and tail is almost identical, so that no significant differentiations in these respects exist. Bill length and depth are 3 per cent less; tarsus is only 2 per cent longer, but the difference is significant statistically. Bill depth carries farther the trend toward small size seen in passing from *mearnsi* to southern *montanus*. Bill length, on the other hand, is the same in *mearnsi* and southern *montanus*. The degrees of difference in measurement between the divisions of *montanus* and between southern *montanus* and *mearnsi* are therefore comparable, though differing in detail. All differences are so purely matters of average as to be of no utility in attempts to segregate individuals into race groups.

Sexual dimorphism of wing and tail increases to 6 per cent compared with 5 in *J. caniceps* and *J. o. mearnsi*. Since tail is 12 per cent shorter than wing, compared with 10 per cent in *mearnsi*, an additional stage in tail shortening is manifest. Sexual dimorphism of bill and feet is almost lacking, except in bill length.

Head color varies geographically as well as individually. Normally, males of a population in the southern area are chiefly of categories 5, 6, and 7 (see p. 247), those of the northern area 3, 4, and 5. The average, as shown in figure 11, changes between the United States boundary and the southern Cariboo District (Clearwater River) of British Columbia. Just how abrupt this change is cannot be stated without more evidence from samples from the intermediate Kootenay District, but the rapid change in a twenty-mile interval between Meadow Creek and Creston suggests that most of the change may take place within sixty miles of the border. Increased variation at this point is well shown in the graph. The differentiation of the northern and southern groups on the basis of head is 69 per cent, but when it is considered that there is a certain amount of error in assignment to categories, identification of only about 60 per cent would be possible.

Back color, the components of which have already been fully analyzed in the racial hybrids, shows some average change northward through the Kootenay area. The phaeomelanin component continues to increase northward to the Skeena River. The graph which represents separately the alteration in average in the two pigments indicates that the principal change in the eumelanin occurs slightly south of the points where alteration in head pigment and phaeo-

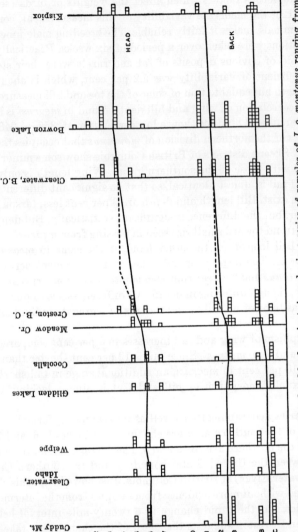

Fig. 11. Occurrence of head and back variants in males in a series of samples of *J. o. montanus* ranging from central western Idaho north and northeast to central interior British Columbia. Samples spaced to scale geographically. Lines connect averages for groups. Broken line represents probable curve projected on basis of differences in Meadow Creek and Creston groups. Two components of back coloration shown. For explanation of variants see pages 245 and 247.

melanin of the back is most pronounced. This situation exists despite the fact that in individual variation head and eumelanin of the back are correlated to a moderate degree (see p. 247). However, there is enough lack of conformity of head and back and enough variation in samples of the size dealt with to account for the imperfect geographical correlation of the changes.

Summary of intraracial geographic variation.—The incompletely differentiated northern and southern divisions of the race merge in the southern Kootenay area, where there is continuous forest habitat. Head color changes abruptly between Meadow Creek, Idaho, and Creston, B. C., as does the average for phaeomelanin of the back. The latter continues to change northward as far as the Skeena River. Eumelanin of the back changes principally south of the Canadian border. The average dimensions of bill and tarsus differ in the two geographic extremes, but the region of intergradation could not be localized on the basis of material at hand. The geographic intergradation of these various characters, because of their partial independence, does not appear to result from junction of types that once were more diverse, but may be an instance of differentiation *in situ* in which each character is affected independently.

Distribution in breeding season.—Populations of *montanus* occur in the mountains of Oregon east of the Cascades and north of the low desert areas of the southeast; in western Idaho north of the Snake River; in northwestern Montana; in eastern Washington; and in British Columbia north through the interior, west of the Selkirk Range and, more northerly, west of the continental divide and east of the coast ranges to the valley of the Skeena. It is especially difficult to define accurately the breeding range along its eastern limits in British Columbia because of the nature of the intermixture of *montanus* with *hyemalis*. Occasional stray *hyemalis*-like individuals may nest far west of the continental divide, as in the Skeena Valley and at Barkerville, but they are rare. Not until the continental divide is reached, as at Yellowhead Pass and Banff, is the mixture great. Even eastward of this point many *montanus*-like hybrids appear (for full discussion of zone of mixture see pp. 328 ff.), and some are known from points as far north as the interior of Alaska.

This breeding range is extremely diverse in character. In places the forest habitat is relatively arid, and is dominated by lodgepole pine (*Pinus contorta*) as in the range of *mearnsi;* but elsewhere, as in parts of northern Idaho, the habitat is remindful of coastal humid forests with rich growths of epiphytes and mosses and is dominated by hemlock (*Tsuga heterophylla*) and cedar (*Thuja plicata*). On the whole, the range does not include extensive forest habitats like those of the humid coastal archipelago of British Columbia or of the Puget Sound region. In Oregon there are large tracts of western larch (*Larix occidentalis*) and yellow pine (*Pinus ponderosa*) which, especially the latter, provide open parklike areas especially favored by juncos. Conditions here are similar to those within the range of *thurberi* in the Sierra Nevada. In parts of British Columbia, black spruce (*Picea mariana*) and timber-line associations are common habitats, as are also the hemlock and lodgepole pine forests. Probably if any environmental factor is correlated with the characteristics of these inland birds, it is not a particular forest habitat, but a general prevalence of fairly dry conditions. Certainly the tolerance of this junco for varied local environments is high.

British Columbia.—Occurs in the forested areas of the interior from elevations below 1000 feet in the north to timber line. It is not known north of the Skeena drainage basin or east of it in the Peace River Valley, except as hybrids. The race extends west to the headwaters of the coastal rivers in the coast range, where some intergradation becomes apparent. Southeasterly, relatively pure populations are restricted to areas west of the Selkirk Range and the Cariboo Range. The distribution is irregular and interrupted by Upper Sonoran areas in the Yale district, but is continuous in the Kootenay region southward into Idaho, eastern Washington, and northeastern Montana.

Washington.—Known from the mountains of eastern Okanagan County and above 3000 feet in the Blue Mountains of the southeast. Also occurs, beyond doubt, in the mountains of Ferry, Stevens, and Pend Oreille counties north of the Columbia and Spokane rivers.

Idaho.—Occurs in timbered areas over most of the northern part of the state south through Idaho, Adams, and Washington counties. Populations from western Valley County are but slightly intergradient. Lower limits of occurrence are at 1800 feet in the Clearwater River Valley and in more arid mountains, such as on Cuddy Mountain, at 4000 feet.

Montana.—Breeds in suitable forest areas northeast of the vicinity of Missoula and west of the continental divide, except locally as at Saint Mary's Lake.

Oregon.—Occupies the mountains east of the Cascades; these are chiefly the Blue and Wallowa mountains, which form an almost continuously habitable belt from Crook County eastward to Wallowa County south of the plains of the Columbia and north of the deserts of Malheur and Harney counties.

Distribution in winter.—Winters commonly in the lower mountains of Colorado, New Mexico, and Arizona. Also occurs generally in California, but more abundantly toward the east, where it is the predominant wintering form in suitable places in the desert areas. Rarely reaches the northwest coast of California. Extends southward to Chihuahua, and in Lower California to latitude 32° and possibly farther, though there are no definite records. There are records for Texas east to 97° longitude (Gainesville) and southeast to the vicinity of San Antonio (Ingram, Kerr County) and Fort Clark (Kinney County). In Oklahoma, Kansas, and Nebraska occasional individuals may appear, especially in the western regions, but also as far east as Lawrence, Kansas. Birds may remain as far north as southern British Columbia, in the Okanagan and lower Fraser valleys. Also winters in eastern Washington and Oregon, and Idaho, Nevada, and Utah. Occasional strays, which I have been able to verify by examination, have been taken far to the eastward, as for example at Waukegan, Illinois, and Branchport, New York. A well-characterized male has been examined that was taken November 29, 1936, near Monclova, Lucas County, Ohio. It is not unlikely, then, that *montanus* does actually occur in other states east of the Mississippi River, where it has been reported. Some of the record specimens have in the past been misidentified, however, and sight records are not to be relied upon.

Montanus and *shufeldti* on their wintering grounds cannot be identified with more than 80 per cent accuracy. Birds with 3b or 3c type of back may be identified as *montanus,* for these back types are found in 50 per cent of *montanus* but in only 4 per cent of *shufeldti.* Size criteria must be relied on largely in dealing with birds of back types 1 and 2. Back type *o*, or extreme ruddy individuals of type 1, may safely be considered *shufeldti*. Unfortunately, in the earlier stages of my work when great numbers of wintering *shufeldti* and *montanus* were examined, color criteria were thought to be adequate for these races and size was not considered; reëxamination and measurement of these specimens is not now feasible. At the present writing, juncos of back type 3 must be relied upon to a considerable degree to outline the range of *montanus* in winter. However, in the eastern limits of the winter range of Oregon juncos, the presumption, unless there is good evidence to the contrary, is that the birds are *montanus* and are not those from the breeding range of *shufeldti*, which as now known is restricted to coastal areas. Nevertheless, true *shufeldti* does occur in small numbers with *montanus* in the Great Basin and southern Rocky Mountains, where indeed the type was taken. It is regrettable, therefore, that certain wintering birds earlier identified cannot at present be considered as critically determined. The winter range of *montanus* outlined above rests on unquestionable specimens of this race.

Some direct evidence for the migration and wintering range of birds of the Cariboo district is afforded by a banding return provided me by Mr. T. T. McCabe. A breeding bird which was banded at Indianpoint Lake, near Barkerville, B. C., August 2, 1934, and which repeated August 11, was caught by Mr. E. D. McKee at Grand Canyon, Arizona, February 4, 1935, and repeated February 5, 6, 8, and March 22 and 24, 1935. Of different significance is the record of another bird whose breeding ground is not known. This bird, a migrant male, was banded April 28, 1927, at Indianpoint Lake. It repeated on April 30. In 1934 it was caught April 20, at Kitimat Mission at tidewater, west and only slightly north of Barkerville. It would appear that this bird in its 1934 migration was following a different route from that of 1927. Such variation in route might well lead to different breeding grounds and on occasion lead to intermixture of races. Although it might be argued that this individual could travel west from Barkerville, it is improbable that it would travel almost directly west across migration routes and the formidable coast ranges. This bird was at least eight years old at the time of its last capture, a fairly long life span under normal conditions for a small bird of this type.

Relationships with adjacent forms.—For *J. hyemalis hyemalis* and *J. hyemalis cismontanus*, see pages 328–343.

J. oreganus shufeldti.—The southern division of *montanus* approaches *shufeldti* only in central Oregon. There is a belt no less than 30 miles in width along the Deschutes and Crooked rivers and Bear Creek that is not suitable for breeding juncos. This barrier would seem to be quite effective, inasmuch as populations on either side of the Deschutes River at Bend (Ochoco Ranger Station, and Cascade Mountains 16 miles west of Bend) show no evidence of intermixture, but are entirely characteristic of their respective races.

Montanus as a whole is of course only about 80 per cent differentiated from *shufeldti* in wing and tail length, which are the best characters. But the southern division also has average color characters (pale head and type 3 back) which reach an extreme in east-central Oregon and maintain this with no diminution westward to the barrier. The head color, although it averages slightly darker in males at Ochoco Ranger Station, averages paler in females (see figs. 10 and 19). These fluctuations are best ascribed to sampling. With the addition of color characters, more than 90 per cent of the individuals of the populations on either side of the barrier at Bend can be distinguished. Nevertheless, the few atypical individuals in each race make it impossible to maintain that the two do not occasionally exchange individuals in this region.

The Boreal area of the Paulina Mountains, east of the Cascades and south of Bend, is completely connected with the Cascade region by suitable junco habitat. Although it is separated from the Blue Mountain area, across the Crooked River, it is faunally and spatially the region closest to the breeding range of *montanus*. The birds of the Paulina Mountains are essentially *shufeldti*, but they show some intermediate features.

The available sample consists of 17 males and 3 females. Head color averages 3.68 (3.91, west of Bend; 5.40 at Ochoco Ranger Station). Back varies from type *t* phaeomelanin to type 2, average 1.06 (west of Bend, *t*–3, average 1.16; Ochoco, 1–3, average 2.80). In color, then, the sample is typically *shufeldti* and shows no influence of *thurberi*, the characters of which appear not far to the southward.

Measurements in which the races differ in the southern region are those of wing, tail, and middle toe. The first two are much more decisive than the third. Paulina birds are definitely intermediate in wing length (see fig. 12), though closer to *shufeldti*. The variation is only slightly greater in the intermediates. Tail averages 67.5 mm.; 67.1 in birds from west of Bend, 70.7 in birds from Ochoco. Middle toe (11.34) is equivalent to the average for *montanus*. Bill measurements average distinctly small (length 7.98 mm., width 5.66), smaller than in either race or than in *thurberi–shufeldti* intermediates to the south. This feature can only be characterized as a local variation of the colony.

The intermediacy of the Paulina group is not one which clearly results from exchange of racial characters across the barrier. It is not even certain that there has been any significant mixture, for if there had been, the color characters should reflect it. Certainly the coloration of *montanus* would not be completely selected out in this comparatively arid forest habitat, which resembles that of the interior mountains quite as much as it does that of the coast. There is no evidence that *shufeldti* coloration is dominant over that of *montanus*. If the population resembles *shufeldti* because of the complete contact with that race and the continual influx of individuals from it, why does not wing length correspond equally? I conclude that the group has had a *shufeldti* ancestry and as a somewhat isolated group has been altered subsequently with respect to wing and tail, as so many interior forms are, in association with the more open habitats and the severe winter conditions which impose greater need of extensive migration. The intermediacy thus is the result not of interbreeding, but of an intermediate type of environment.

Montanus and *shufeldti* meet in southern British Columbia. Breeding habitat is continuous from interior localities such as Clinton and Lillooet southwest through the mountains to the coastal mainland at the mouth of the Fraser River. Measurements of birds from intermediate localities either are lacking or are too few to indicate the nature or exact locus of the intergradation. Observation and data from specimens indicate that probably there is intergradation in the region between Lillooet and Chilliwack. The birds of the Okanagan area that belong with the southern division of *montanus* show no diminution in wing suggestive of *shufeldti*. Southwestward into Washington there may be sufficient continuity of habitat to allow considerable intermingling.

The extent of intergradation northward in the coast ranges of southern British Columbia is not known, inasmuch as the exact northern limits of *shufeldti* have not been reliably established in this area. Intermediates between *J. o. oreganus* and *montanus* have much in common with *shufeldti* and *shufeldti–montanus* intergrades; in fact, they may be identical genetically even though they are of different history. Very probably, therefore, in the coast ranges there are continuous and similar populations in a north–south line which to the southward are *shufeldti* but which farther north one would view as intermediates of *oreganus* and *montanus*.

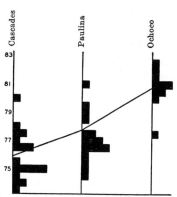

Fig. 12. Intergradation in wing length of males between *J. o. shufeldti* (Cascades) and *J. o. montanus* (southern division). Cascade sample from vicinity of Bend, Oregon. (Not spaced to scale geographically; see map, fig. 14.)

May not *shufeldti* have arisen from interbreeding of *montanus* and *oreganus*? I think this improbable in view of the range of *shufeldti*, which is situated too far to the southwestward. More likely, in view of the glacial history of the region, *shufeldti* of Oregon and Washington was the parent type, with *montanus* a larger interior counterpart. The small red *J. o. oreganus* differentiated in the northern coastal district. *Montanus*, in invading the interior with the glacial retreat, either through differentiation (see p. 257) or possibly through increments from *shufeldti*, gave rise to a slightly differentiated northern division. When the northern interior and northern coastal types met at the time of exposure of the high coastal mountains, they resynthesized *shufeldti*-like populations as intermediates geographically and structurally; there was rebuilt a similar genetic mosaic.

The high individual variability in color in *montanus* and *shufeldti* may be due to a crowding together during periods in the Pleistocene of stocks already somewhat diverse which failed to reassort upon expansion northward. The ruddier backed extremes of *shufeldti* seem associated to some degree with the humid coastal conditions on the Olympic Peninsula. Yet the mixture of this type with gray-backed birds both in the Puget Sound area and in the drier interior range of *montanus* suggests a shuffling of originally more uniformly differentiated races. Both long-time alterations in range through glaciation, and annual intermingling and straying in migration, have been factors.

J. o. oreganus.—Information is limited on the intergrading of *montanus* and *oreganus*. The sector of the coastal region from Queen Charlotte Sound to the Portland Canal is the region of contact. Exclusive of the Queen Charlotte Islands far off shore, breeding *oreganus* from this part of the coast are at hand from Calvert Island and from the vicinity of Princess Royal Island and the adjacent mainland. Many of the Calvert Island birds are somewhat intergradient toward *shufeldti*. A small series from Bella Coola may be taken as indicative of the intergradation of *J. oreganus* and *montanus* even if the former exists in submaximal form in this latitude. Calvert Island is somewhat south of Bella Coola, but a continuous waterway leads southwestward from Bella Coola to the island. The Princess Royal region is to the north and the mountains of the mainland are continuous with those of the interior at Bella Coola.

At the town of Bella Coola, McCabe found breeding juncos scarce; he encountered only a few in open timber above the valley floor and a few up the river to the eastward. These are not typical *J. o. oreganus*, although tidewater extends inland to this point. It appears, therefore, that *oreganus* is limited to the coastal lowlands west of the higher mountains, and does not in unmixed form extend to the head of such extensive inlets. Four birds, 2 males and 2 females, from localities ranging from Bella Coola to 18 miles east of Bella Coola, have the following back types: $oob(1)$, $ob(1)$, and $1b(2)$ (see pp. 245, 271 for explanation of types of back color). Type oob is a red-backed variant that never occurs in unmixed *montanus*, and ob is known only in one instance from *montanus*. This occurrence was at Hazelton in the Skeena Valley, possibly even there the result of intermixture. Type $1b$ never occurs in true *oreganus*. The oob bird of Bella Coola is barely within the limits of this color class, as it approaches ob. Its wing and tail are average for *oreganus*, but the tarsus is below the minimum of that race. The sides are heavily colored, clearly showing *oreganus* influence. The ob bird is inconclusive with respect to measurements, falling in the region of overlap for the curves of variation for the two races. One of the $1b$ birds was mated with the ob bird. This $1b$ bird is of average *montanus* wing length, though not beyond the range of variation of *oreganus*, but the tarsus is 21.3 mm., an extreme almost never attained in *montanus*. The other $1b$ bird is below the minimum of *montanus* wing and even below the average of *oreganus*. The tarsus is average for *montanus*.

At Bella Coola, then, characters of both races appear, in average, or often extreme form, but independently assorted in individual birds. The small sample has much in common with *shufeldti-oreganus* intergrades from Vancouver Island with respect to color, inasmuch as similar color extremes are intergrading.

About 30 miles east of Bella Coola in the transmontane by-pass and in the heart of the coast range, transition toward interior biotic conditions is found, and before the watershed is reached (east of the highest peaks, as this is an antecedent drainage system) typically interior conditions prevail. Four birds from 30 to 50 miles east of Bella Coola on the coastal drainage are *montanus* with respect both to color and size. These have been combined for the purposes of analysis with a group of birds from along the divide at Anahim Lake and from the Rainbow Mountains. Although these birds are essentially *montanus*, they do depart from average *montanus* in the direction of *oreganus* in an interesting fashion.

The wing average of 77.11 mm. (11 specimens) is decidedly intermediate between the 78.46 of northern *montanus* and the 75.00 of *J. o. oreganus*. The tail also is shorter than in *montanus*, but not to so great a degree. Upon studying individual wing lengths, it is seen that the lower average is not due to the presence of any very short-winged birds such as would suggest *J. o. oreganus* mixture, but that it results from a lack of the larger variants which contribute to the average of typical *montanus*. The types of back color occur: 1, 2, and 3 in the ratio of 9:0:2. The average is 1.3 compared with 1.9 in *montanus* to the eastward. The average is not much different from that of the Hazelton series, yet has fewer examples of types 2 and 3. Here, as in wing length, certain *montanus* variants (gray backs) have dropped out, but no *oreganus* variants have been added. This stage in intermediacy is thus very different from that along the lower Bella Coola Valley. An intermixture of *oreganus* and *montanus* types would probably increase variability in some characters after the first generation and something approaching the parental *oreganus* coloration should appear by segregation. Therefore this sample seems to be one representing a gradient in differentiation, not intermixture. This gradient may be merely a further manifestation of the dropping out of gray-backed types northwestward through *montanus*. But since it is coupled with shorter wing, one wonders if proximity to the coast may not in some manner influence the population and cause the dropping out of variants most unlike *oreganus*. This would be differentiation *in situ* in response, at least indirectly, to environment.

At Bella Coola there is interbreeding which results in high variation and in the appearance of extremes of character. The *montanus* variants found there do not surpass the extremes of the Anahim–Rainbow Mountain group. Thus, in this sector of the coast there are nicely demonstrated two kinds of intergradation, hybridization and intermediate differentiation. Earlier I indicated that the intergrades of the coast ranges have much in common with *shufeldti*. The back, for instance, averages 1.3 in both *shufeldti* and the Anahim Lake group. The fact that the wing is longer at Anahim, much as in *montanus*, shows that the history of the back in the two groups is not the same, but probably was arrived at independently.

At Swanson Bay on the mainland opposite Princess Royal Island, McCabe took five juncos near timber line, 4000–5000 feet. Four of these are of back type 1*b* or 1*a*, one is *ooa*. In size these birds correspond with *oreganus* of this vicinity from the outer coastal islands and from near sea level along the inlets. These birds are quite as intermediate as those of the Bella Coola area with respect to coloration. Although this small group does not conclusively show other intermediate features, it is difficult to explain the coloration except as diffusion westward of the interior type of back through the continuous mountains of the coast ranges.

This group of five is especially significant in comparison with birds of the very different lowland forest habitat at Swanson Bay. Eight birds taken from near tidewater, either here or on the outer islands, have among them but a single 1*a* variant. A zonal and associational difference correlates with back differentiation. The lowland habitat consists of muskegs in the timber, or the open parklike forest of the outer coast. Between the open outer coast and the balsam scrub of timber line is the heavy impenetrable forest at the base of high mountains unsuited to juncos and invaded by them only by means of occasional openings.

Junco oreganus shufeldti

This race occupies the coastal regions of Washington and Oregon and, even in extreme populations, is not fully differentiated from *montanus*. Yet, its divergence from this race is greater than the difference between the two divisions of *montanus*. *Shufeldti* is only partly migratory.

Racial characters.—*Shufeldti* differs from *montanus* in shorter wing and tail (average 4 per cent). Males of *shufeldti* are usually (75 per cent) less than 77 mm. in wing length; males of *montanus* usually (85 per cent) more than 76 mm. in wing length; females of *shufeldti* less than 73 mm. (66 per cent); *montanus* more than 72 mm. (85 per cent). By using these measurements, one can separate 75 to 80 per cent of the individuals of the two geographic groups. Similar limits could be set for tail length, but they would be less reliable because of greater individual variation (see table 20, p. 265). *Shufeldti* is similar to northern *montanus* in average head color, but the ruddier 1*b* type of back is more prevalent. Tail pattern, side color, and other measurements are similar.

Shufeldti differs from *J. o. oreganus*, as does *montanus*, in the grayer, browner, or less reddish hues of the back that never are the Prout's brown, Brussels brown, or raw umber of *J. o. oreganus*. Other average characters of *shufeldti* are less rufescent, often less dusky, sides, greater amount of white in the tail, especially in rectrices 4 and 5 (see p. 272), and shorter tarsus and toes (3 to 5 per cent).

For characters that differentiate *shufeldti* and *thurberi* see page 276.

Nongeographical variation.—The high variability of coloration in *shufeldti* is comparable to that in *montanus*. Only back color shows any average geographic modification within the race. Head color, classified as in *montanus*, averages 3.8 with variations from 1 to 6 (p. 247). Side color is of types 1 and 2 (p. 248). The brown overcast on feathers of the head, a phase of retarded plumage, is comparable in appearance to that found in northern *montanus*. There are no instances of white wing coverts. Mixed patterns in the ventral part of the hood occasionally occur in females. Sexual dimorphism and seasonal alterations are comparable to those in *montanus*.

Tail pattern for males is tabulated on page 253. It is similar to *montanus* in the kinds of variants present. Averages for the fourth rectrix and the inner web of the fifth fall closer to those of the northern division of *montanus;* those for the outer web of the fifth are closer to those of the southern division. Interracial differentiation is thus less than intraracial differentiation in *montanus.*

Variations and correlations are similar to those obtaining in southern *montanus* except for tarsal length, which is more variable. The high coefficient for tarsus seems to result from a peculiarly short-legged group from the Cascade Mountains near Bend, Oregon. This local variation may be without significance, as only 12 males were available. In any event, it bears no relation

TABLE 20

MEASUREMENTS OF J. O. SHUFELDTI

Measurement	Sex	No. of specimens	Mean with prob. error	Standard deviation	Coeff. of var.
Wing.............	♂	64	75.65 ± .14	1.71	2.26
	♀	24	71.83 ± .22	1.60	2.22
Tail..............	♂	56	67.14 ± .18	2.03	3.02
	♀	24	62.58 ± .29	2.12	3.38
Bill length........	♂	63	8.22 ± .03	0.36	4.42
	♀	23	8.03 ± .05	0.36	4.58
Bill depth.........	♂	57	5.89 ± .02	0.27	4.65
	♀	22	5.84 ± .04	0.28	4.75
Tarsus............	♂	67	19.82 ± .05	0.64	3.20
	♀	24	19.59 ± .09	0.68	3.49
Middle toe........	♂	65	11.50 ± .04	0.43	3.78
	♀	25	11.49 ± .06	0.42	3.67
Hind toe..........	♂	67	7.99 ± .03	0.38	4.79
	♀	25	8.02 ± .06	0.45	5.59

to any extensive geographic trends. Sexual dimorphism in wing and tail is 5 and 6 per cent, respectively, but it is lacking in bill and feet, except for bill length (2 per cent); this situation is similar to that in *montanus.* Tail is 12 per cent shorter than wing, as in *montanus.*

Geographical variations within the race.—Ruddier types of back are more common in the population of the Olympic Peninsula than in other sections of the race. That this is a result of exchange of breeding birds between the Olympic Peninsula and Vancouver Island, where a decidedly reddish *oreganus*-like population of intergrades exists, is doubtful. If such exchange is a common occurrence, why are not the populations of the adjacent mainland of southern British Columbia also involved? The increase in ruddiness is coincident with a more humid habitat which is not unlike that on parts of Vancouver Island. It is not impossible that the slight average difference between the birds of the Olympic Peninsula and those of the section of Washington lying immediately east of Puget Sound is due to the less frequent penetration into the extreme coastal regions of strays of *montanus* type from the interior. The graph (fig. 13) shows the contrast of southern *montanus* and *shufeldti* in Oregon and the slight intraracial variation in *shufeldti* with respect to depth and amount of phaeomelanin pigment. Male *shufeldti* are usually of type *b* back (eumelanin intensity) as in northern *montanus.* However, in the northern half of Oregon about 50 per cent are type *c*. This condition in combination with types 1 or 2 of phaeomelanin means the birds resemble the duller variants of *thurberi*. This tendency to diminish eumelanin is more and more apparent along the gradient toward *thurberi* in southern Oregon; it is not the influence of *montanus*.

The wing, tail, bill, and middle toe all average somewhat smaller throughout coastal Oregon than elsewhere in the race (fig. 18). This is not related to the gradient running south into *thurberi*, although *thurberi* may maintain

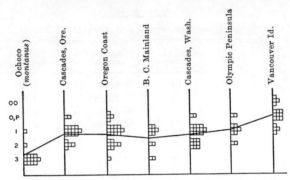

Fig. 13. Occurrence of back variants (phaeomelanin) in males in samples of *J. o. shufeldti*, in *montanus* adjacent to *shufeldti*, and in *shufeldti-oreganus* intergrades from Vancouver Island (not spaced to scale geographically). Line connects averages for samples. For explanation of variants see page 247.

some of these lower values along the coast. The coastal group of the Olympic Peninsula shows none of this diminution.

Distribution in the breeding season.—The race occurs in coniferous forests of the coastal mainland west of the more arid interior slopes of the coast range in British Columbia, and in Washington and Oregon west of the Columbia River plains and the Deschutes–Crooked River–Bear Creek drainage. It occurs south to Coos, northwestern Douglas, Lane and western Deschutes counties. Beyond these points the gradients in characters in the direction of *thurberi* become more pronounced. Distribution is nearly continuous in these areas, except in parts of the Willamette Valley.

The northern limit of *shufeldti* has not been precisely determined. There are specimens from Howe Sound just north of Vancouver, B. C. From Lund, about 80 miles northwest of Vancouver, a single male bird shows some approach to *oreganus*, as does also a bird from Johnstone Strait. It has been pointed out that *shufeldti*-like birds appear in the coast range farther north but that they are intergrades between *montanus* and *J. o. oreganus*.

The most common forest type in this region is the *Pseudotsuga–Thuja plicata–Tsuga heterophylla* combination, with local areas of mountain hemlock (*Tsuga mertensiana*) and fir and, on the coast, Sitka spruce. As a whole the forest habitat is much like the humid cedar-hemlock forests of northern Idaho that are occupied by populations of *montanus*. But climatic conditions are somewhat milder, and, perhaps as a result, there is less migration on the part of the juncos. The birds breed from sea level to timber line at 5000 to 7000 feet in the Cascade Mountains.

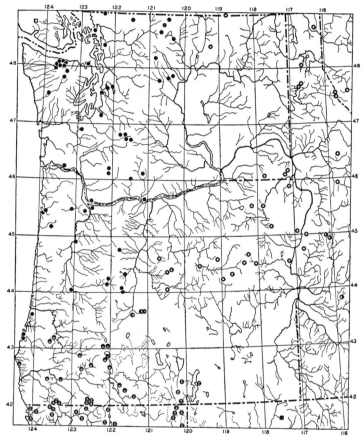

Fig. 14. Distribution of breeding juncos in Oregon and Washington, and portions of neighboring states. Each symbol represents a locality from which specimens have been examined. Dots, *J. o. shufeldti;* plain circles, *J. o. montanus;* half dots, left half solid, intermediates between *J. o. shufeldti* and *J. o. montanus;* half dots, upper half solid, distinctly intermediate groups between *J. o. shufeldti* and *J. o. thurberi;* half dots, right half solid, intermediates between *J. o. montanus* and *J. o. mearnsi;* circle with cross bar, *J. thurberi;* square enclosing white dot, intermediate between *J. o. shufeldti* and *J. o. oreganus;* solid square, *J. c. caniceps.*

Distribution in winter.—Occurs commonly at lower altitudes within the breeding range, north to the lower Fraser River, British Columbia, and throughout coastal Oregon and Washington. It also occurs fairly commonly in northern and central California, but rarely east of the Sierra Nevada or south to the San Diegan district and northern Lower California. The extent to which this race moves eastward to winter in the interior is difficult to determine. As explained on page 258, identification of individuals of *montanus* and *shufeldti* of back types 1 and 2 is not possible without data on measurements, which in some of my earlier work were not obtained. Study of large series of wintering birds from Arizona now at hand reveals that only a few fall below the minimum size limits of *montanus*. Others might be either small *montanus* or large *shufeldti*. Along with the type of *shufeldti* from New Mexico there are a few other specimens which on the basis of size or extreme coloration (type *o* back) must certainly have originated within the breeding range of *shufeldti*. These have been taken in Arizona, New Mexico, Nevada, Utah, and eastern Colorado. Strays may occur in the northern states of Mexico and in western Texas, but satisfactory determination of these has not as yet been made. The names *shufeldti* and *couesi* previously employed by me in studying certain collections apply, for all the latter and for most of the former, to the interior type of Oregon junco which now properly bears the name *montanus*. Inability to allocate all records of Oregon juncos from the southern Rocky Mountains and desert areas should not confuse the general situation, which can be stated with assurance, namely, that true *shufeldti* is a rarity east of the Cascade–Sierra Nevada mountain system, just as is *thurberi* in that area, and that the largest part of the *shufeldti* population winters within the breeding range of the race and in California.

Relationships with adjacent forms.—For *J. oreganus montanus* and *J. oreganus thurberi*, see pages 258 and 286, respectively.

J. oreganus oreganus.—It may be said that the juncos of the immediate coastal belt show a gradient from Oregon to Alaska in which the phaeomelanin of the back and sides and the size of tarsus and feet increase to the northward. Also there is some increase in eumelanin and diminution in white areas in the tail. The gradients in back color and tarsus, the most pronounced ones, do not coincide in detail; the chief increase in redness takes place farther south than does the elongation of the tarsus. Irregularities in the steepness of the gradients make it possible to fix racial limits arbitrarily (see locality lists and ranges), yet *shufeldti* and *oreganus* and their intermediates should really be viewed as a unit.

Figure 15 shows the gradient in redness of the back, with a distinct break at the Straits of Juan de Fuca between the Olympic Peninsula and southern Vancouver Island. Fully as pronounced a break occurs between the mainland of British Columbia and Vancouver Island (compare graph, fig. 13). The Vancouver Island group is predominantly of type *o* back, which is rare in *shufeldti, oreganus,* and *montanus*. Yet there are among the Vancouver Island males at least four that are indistinguishable from *shufeldti* and five that fall well within the limits of *oreganus*. Swarth (1912, p. 59) noted this resemblance to *oreganus*, but he rather overemphasized it because he considered extremely gray interior birds to be typical *shufeldti*. The next station that is at all well represented is Calvert Island. The island lies in the northern coastal faunal area, where conditions are much colder and rainfall heavier than on southern Vancouver Island. At this point the backs show another distinct advance in redness; they average nearly as red as in *oreganus* of Alaska and the Queen Charlotte Islands. Whether the increase occurs gradually or abruptly, perhaps at the north end of Vancouver Island where it is suspected that biotic conditions are more like those of Calvert Island, is not known. Two birds from Johnstone Strait are of back types *oo* and *o*.

The next series of birds to the northward is one from the vicinity of Princess Royal Island, British Columbia. Excluded from this group are some birds of intermediate character taken in the mountains of the mainland which show in-

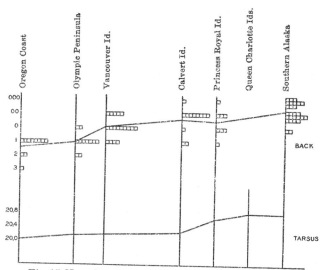

Fig. 15. Noncoördinated geographic variation in back color (phaeomelanin) and tarsal length in males along the Pacific coast from Oregon to southern Alaska. Groups spaced to scale geographically. Lines connect averages.

termediacy with *montanus*. Even without these, the group interrupts, slightly, the increase in ruddiness to the northward. This effect, if it is not the result of sampling, might well be associated with the more interior situations from which the birds come. Calvert Island is distinctly more coastal in aspect than the inner side of Princess Royal Island. This section of the coast has the maximum of rainfall found along the coasts of southern Alaska and British Columbia. Possibly the high occurrence of *a*-type back (rich in eumelanin) in this sector is associated with the heavy rainfall. This is a deviation quite independent of the degree of ruddiness.

From the southern boundary of Alaska northward, no intensification in *oreganus* characters occurs. The birds of the Queen Charlotte Islands are like those of Alaska, but not enough information is at hand to permit entering them on the graph.

The tarsal measurements show a very slight increase, uniform in rate, north to Calvert Island. Between there and Princess Royal Island a sudden increase occurs. The two gradients (color and tarsus) are thus independent and not susceptible to exactly the same geographic barriers and biotic changes, as, for example, at the Straits of Juan de Fuca. The reliability of the average for Calvert and Princess Royal birds is felt to be reasonably good. Females and males are combined by adding to the average of the females the normal sexual difference in Oregon juncos. The distribution of individual variants is such that the average appears to be reliable.

The Vancouver Island series appears to have sides not appreciably richer in color than *shufeldti*, but in the Calvert Island birds and those from farther north the color is that of true *oreganus*. The average amount of white in the tail of the Vancouver Island group is distinctly intermediate between *oreganus* and *shufeldti* (for details see table, p. 272). The Calvert Island and Princess Royal groups are not quite equal to *oreganus*. The average for head color of males from Vancouver Island is 3.6, which compares with 3.2 in *oreganus* and 3.8 in *shufeldti*.

The independence of gradients in character and the extended zone of intermediacy without increased variability suggest that the intermediacy results from progressive differentiation in association with geographic and environmental factors. There nowhere appears to be a group produced by intermingling and hybridizing of the racial extremes.

Junco oreganus oreganus

The humid coast belt from Yakutat Bay, Alaska, south to Queen Charlotte Sound, British Columbia, a region strongly characterized ecologically, is occupied in the breeding season by this race, which is the terminal differentiate of the species in the north. *J. o. oreganus* shows several features which seem related to conditions peculiar to this area and which parallel differentiations in other endemic birds of the region. The race is only partly migratory.

Racial characters.—Features which distinguish *J. o. oreganus* from *shufeldti* and *montanus* are given on pages 264 and 252. In summary, *J. o. oreganus* is short winged and short tailed like *shufeldti;* the phaeomelanins and eumelanins are intensified, especially the former, so that the back and sides are more ruddy and the head, sides, and back are darker on the average. The tarsus and toes are longer and heavier. Back color alone effects a 92-per cent segregation of *oreganus* and *shufeldti* and a 97-per cent segregation of *montanus*. Wing and tail length also permit an 80-per cent segregation of *montanus*, and tarsal length, to a nearly equal degree, sets *oreganus* off from both its conspecific neighbors. Other features are more purely matters of average.

Features which differentiate *J. o. oreganus* and *J. hyemalis cismontanus* are given on page 343.

Nongeographical variation.—Variability in coloration in *oreganus* is great, as in *montanus* and *shufeldti*. None of the colors shows intraracial geographic variation apart from the gradient in back coloration already discussed (see pp. 268 and 269).

Head color, classified as in other races, varies from type 2 to type 6, but averages 3.2 instead of 3.8 and 4.0 as in *shufeldti* and northern *montanus*. These classifications are made by inspection of the ventral surface of the hood. The pileum in *oreganus* seems to be relatively darker so that in the lighter categories, 4, 5, and 6, the pileum is more as in 3, 4, and 5 of *montanus*. Thus, the heads actually average darker than the indices show.

Back types *o*, *oo*, and *ooo* of *oreganus* are stages in phaeomelanin intensification and produce progressively more reddish color. Almost all *oreganus* are of types *oo* and *ooo*, neither of which occurs in adjacent forms. The eumelanin varies in intensity, producing variations in blackness or darkness of the red. Two categories, completely intergradient, were established. Type *a*, the blacker variant, is about as prevalent as *b* in males. In females type *a* is almost lacking. Categories *a* and *b* are comparable, but not equal, to types of eumelanin in *montanus*, in which the eumelanin is less completely mixed and, in the darker types, forms large gray basal areas on each feather vane. The eumelanin of *oreganus* is deposited throughout the vane in the barbule tips. The occurrence of categories *a* and *b* is not in any way correlated with types *oo* and *ooo*. The averages for the latter are the same in males and females. Intensity of head color and eumelanin of back are not correlated. The interrelation of pigments is thus decidedly different from that in *montanus*, in which back types *b* and *c* and head are correlated. The fact that eumelanin is largely in the basal parts of the back feathers of *montanus* and not so much mixed with phaeomelanin in the tips indicates that different mechanisms are involved; different correlations naturally result. However, there is some correlation between head color and phaeomelanin of the back. This also is true to some degree in northern *montanus* and probably in *shufeldti*. Why these two should be associated is not clear, since different pigments are involved. If the correlation were a matter of maleness, or retarded or progressive plumage, one would expect females to differ from males in the phaeomelanin of the back. Instead, they are deficient in eumelanin of the back, which in *oreganus* is not correlated with ruddiness. The association of head and phaeomelanin may represent a linkage of characters, and may not be brought about because of similar physiologic processes involved in the production of the pigments.

Side color is chiefly of two types: *a*, with relatively unmixed phaeomelanin, equivalent to type 2 in *montanus* but with extension of the area as in type 3 and with redder phaeomelanin (always a little eumelanin present); and *b*, with distinctly more eumelanin, comparable to extremes of type 1 of *montanus* but with phaeomelanin redder and hence total effect less vinaceous and closer to burnt umber. A third, distinctly sooty type, with the phaeomelanin materially obscured by eumelanin, was found in two breeding birds only. This variant almost suggests hybridization with *hyemalis*, but I believe it is independent of this, as one such bird came from Sitka. The chief intraracial variation consequently is one of the eumelanin component. Types *a* and *b* are about equally represented, type *a* being somewhat more common. There are no fewer type *b*'s in females. Eumelanin of the sides is in no way correlated with head pigmentation.

Buff feather tips are less prevalent than in *montanus* and *shufeldti*, yet some extremes occur. Their association with sex and age applies in the same fashion as in *montanus* (see p. 252). The buff tippings of the nape are usually more ruddy in this race. No instances of mixed phaeomelanin and eumelanin in the ventral part of the hood have been found. There are no instances of white on the wings or of red on the pileum deeper than the feather tips.

The only aspect of wear that is different from that in other Oregon juncos relates to the back; there the darker red brightens very materially from the Prout's brown of fresh feathers, but never becomes yellowish or whitish along the edges of the feathers as in *shufeldti, thurberi,* and *montanus.*

TABLE 21

OCCURRENCE OF RECTRIX PATTERN IN J. O. SHUFELDTI AND J. O. OREGANUS
(IN PERCENTAGES)

	Males			Females
	J. o. oreganus	Intermediates from Vancouver Is.	J. o. shufeldti	J. o. oreganus
Rectrix 4				
Inner web:				
pure black	56	35	17	86
trace of white	33	59	44	11
mixed	10	6	40	3
Outer web:				
pure black	91	88	62	97
trace of white	4	6	15	3
mixed	4	6	19	0
trace of black	0	0	2	0
pure white	0	0	2	0
Rectrix 5				
Inner web:				
mixed	37	26	10	78
trace of black	42	23	42	19
pure white	20	50	48	3
Outer web:				
pure black	11	0	0	31
trace of white	6	0	0	22
mixed	54	50	17	39
trace of black	22	18	17	8
pure white	7	32	67	0
Rectrix 6				
Inner web:				
mixed	0	0	0	3
trace of black	6	0	0	31
pure white	94	100	100	66
Outer web:				
mixed	0	0	0	6
trace of black	1	0	0	8
pure white	99	100	100	86

Sexual dimorphism in the color of the head is higher in *oreganus* than in southern *montanus,* in keeping with the trend of modification seen in *mearnsi* and *montanus,* in which the female lags behind in the increase of blackness. In southern *montanus* the average sexual difference of the ventral surface of the head is 2 (♂ 6, ♀ 8), in *oreganus* 3 (♂ 3.2, ♀ 6.2, both darker dorsally).

In *shufeldti* the dimorphism also appears to be high, but the representation of females is not entirely adequate. A slight sexual difference in average back color with less eumelanin in the back of females has been noted. This is comparable to that in *montanus*.

Oreganus lacks variants with white or nearly white outer webs of the fifth rectrix, but has new variants with pure black and nearly pure black outer webs of the fifth rectrix and with mixed webs of the sixth rectrix. Throughout the rectrices, average increase in black is to be noted relative to *shufeldti* and northern *montanus*. *Oreganus* is an extreme in this respect that parallels

TABLE 22
MEASUREMENTS OF J. O. OREGANUS

Measurement	Sex	No. of specimens	Mean with prob. error	Standard deviation	Coeff. of var.
Wing	♂	74	75.00 ± .14	1.81	2.41
	♀	38	71.02 ± .18	1.67	2.35
Tail	♂	74	66.13 ± .16	2.11	3.19
	♀	35	62.58 ± .27	2.43	3.88
Bill	♂	71	8.18 ± .03	0.32	3.93
	♀	33	8.21 ± .03	0.22	2.73
Bill depth	♂	64	5.77 ± .03	0.30	5.30
	♀	33	5.83 ± .03	0.27	4.67
Tarsus	♂	75	20.72 ± .04	0.57	2.74
	♀	38	20.35 ± .07	0.63	3.08
Middle toe	♂	65	11.98 ± .05	0.58	4.83
	♀	38	11.76 ± .06	0.54	4.56
Hind toe	♂	75	8.37 ± .03	0.39	4.66
	♀	39	8.15 ± .07	0.53	6.51

in an interesting fashion the southward increase in black among the yellow-eyed juncos of Central America. In both, the increase accompanies general increase in eumelanin in body plumage, though the latter is vastly different in the two series. The tail markings and the head or other body areas are not correlated in individual variation within a population, but the interracial correlation is striking. Sumner has found similar conditions involving color areas in *Peromyscus*. The difference in tail pattern between *oreganus* and *shufeldti* or between *oreganus* and *montanus* (northern) is greater than that between *shufeldti* and *montanus* or between the two divisions of *montanus*. It is about equal to the difference between *mearnsi* and northern *montanus*. Sexual dimorphism is apparent in all vanes of the outer three feathers.

Coefficients of variability are similar to those for *montanus* and *shufeldti*, except that those for middle toe are higher. The coefficients for tarsus are low as in *montanus*, not as in *shufeldti*. Sexual dimorphism is 5 to 6 per cent in wing and tail and, unlike the situation in *shufeldti* and *montanus*, is definitely present (2 per cent) in the feet; it is lacking in the bill. Tail is 12 per cent shorter than wing, as in *shufeldti* and *montanus*. Correlation of wing and tail is .717 ± .038, comparable to that in other northern juncos, though a little higher than in *montanus*. Thirteen males from the central British Columbian coast average 18.4 gr. in weight, compared with 17.7 gr. for 100 *montanus*. If the figure for this small sample should hold true for all *oreganus*, we would have a clear case of reduced wing and increased body size. Certainly from the evidence of weight and large feet it is difficult to conceive of *oreganus* as a smaller bird than *montanus*. It must be a less able flier. In this connection are to be noted the less extensive migration of *oreganus* and the less open character of the forests in its breeding and wintering ranges.

The coefficient of correlation for tarsus and hind toe is low and uncertain; tarsus and middle toe show a coefficient of $.407 \pm .066$, and middle toe and hind toe, $.301 \pm .069$. Although correlations of foot parts are erratic in all races of juncos, they are always slightly positive. It is perhaps significant that in several forms with large feet, compared with related races, tarsus and toe correlations increase materially. This is so in *oreganus*, and it is exactly paralleled in *alticola*, which is large footed in comparison with its Mexican relative *J. phaeonotus*.

Geographical variation within the race.—Once the full development of *oreganus* characters is attained in the Queen Charlotte Islands and in southern Alaska, no further change occurs to the northward. Birds from various divisions of the coastal archipelago, the outer islands, the mainland, the northern section around Glacier Bay, and the Queen Charlotte Islands were analyzed separately without positive results. The occasional duller variants from the coastal mainland, especially from the mouths of rivers draining from the interior, suggest intergradation and intermixture with *montanus* and *cismontanus* (see pp. 262 and 337).

Distribution in the breeding season.—The race breeds in the humid coastal district from Yakutat Bay south through the Alexander Archipelago to the Queen Charlotte Islands. In an easterly direction it ranges to the mainland coast, at least at low elevations, but extends inland no farther than the tidal inlets and lower river valleys; in some instances intergradation begins at these latter points. On the mainland it extends south to Princess Royal Island, British Columbia. At Calvert Island decided modification of certain characters is found, although the birds are still in most respects like the more northern populations; south of Queen Charlotte Sound intergradation is marked.

Juncos in much of this area are sparsely distributed, probably because of the prevailing density of the forest and undergrowth. They occur chiefly along the borders of muskegs, meadows, streams, and beaches and in the occasional tracts of parklike timber, especially on the outer islands. Cedar (*Thuja*), Sitka Spruce (*Picea sitchensis*), and western Hemlock (*Tsuga heterophylla*) dominate the forest. There is an abundance of epiphytic growth. No other junco experiences more overcast weather and precipitation during the breeding season. The sun may be obscured from view for many days in succession. Nearly all the occupied region is below 2000 feet in elevation. Some of the higher mountains of the islands may afford suitable junco habitat at timber line.

Distribution in winter.—The lowest mean monthly temperature at Sitka is 33°. This is indicative of the winter conditions which permit this race to winter, at least occasionally, as far north as this point. Little snow falls in the lower parts of the breeding area. Migration is therefore quite irregular. Yet, probably the majority of the birds move south, many wintering in western Washington and Oregon and intermingling with those members of *shufeldti* which fail to migrate. *Oreganus* ranges south in winter regularly to central

Fig. 16. Distribution of breeding juncos in Alaska and parts of northwestern Canada. Each symbol represents a locality from which specimens have been examined. Dots, *J. h. hyemalis*; circles, *J. o. oreganus*; solid squares, *J. h. cismontanus*; half solid square, intermediate between *J. h. hyemalis* and *J. h. cismontanus*; circles enclosing crosses, hybrids between *J. oreganus* subsp. and *J. hyemalis* subsp.

California, particularly the coastal district. Winter distribution has been based on unequivocal specimens only, those closely corresponding to the *oo* and the *ooo* types of back of the breeding season. There are substantiated records for New Mexico (Fort Bayard), Idaho (Nampa), Nevada (Carson and Ruby Lake), Arizona (Phoenix, Fort Verde, and the Huachuca Mountains), and Lower California (Santa Eulalia). These are the only certain records outside of western Washington, Oregon, and California, except for British Columbia and the breeding range. They represent less than 3 per cent of the winter-taken *oreganus* which I have examined. There are a few record stations along the Sierra Nevada, and in southern California on Santa Cruz Island and at Temescal. From the Monterey Peninsula northward along the Pacific coast, occurrence is much more regular. That *oreganus* migrates into the interior with regularity is evidenced by a number of specimens from the Okanagan region of British Columbia taken in midwinter. Winter stations in the breeding range include, beside Sitka, the Queen Charlotte Islands, B. C., and Wrangell, Alaska; at the last-named two places a number of examples have been taken.

The remarkable adherence of *oreganus* to humid forested areas in winter is highly suggestive of its preference and possible adaptation for comparable environments in summer. To a less degree this same tendency is seen in *shufeldti*. There is enough movement inland to indicate that it is not impossible, so far as migration routes and barriers for both these races are concerned, for them to spread to inland desert areas in winter. I believe it is preference for, or adjustment to, specific conditions in southerly latitudes which are most similar to those of their breeding habitat that largely determines the wintering range. The lesser power of flight of both compared with *montanus*—a condition frequently associated with denser floral habitat—may be an important factor.

Relationships with adjacent forms.—For *J. o. shufeldti, J. o. montanus,* and *J. hyemalis* subsp. see pages 268, 262, and 329, respectively.

Junco oreganus thurberi

This race, like *montanus,* is a complex one with incipient subdivision and numerous gradients in characters traceable within it. Several geographically variable characters grade south into and through it from *shufeldti.* Yet there is a mixture of pigments in the back that is found widely and almost exclusively in *thurberi.* This pigmentation becomes prevalent in the vicinity of the California-Oregon boundary and may be employed arbitrarily as a key character to delimit the races. Except for a few outlying areas, the breeding range comprises two tongues of the Transition and Boreal zones that extend southward in California. The larger of these is that of the southern Cascade–Sierra Nevada mountain system, which is extended discontinuously by the mountain-top Boreal "islands" of southern California. The smaller area includes the coastal regions of northern California south through Sonoma County. The two areas are extensively joined through the Shasta region, but the differentiation of the two groups of birds associated with them is greater than that within the long interior range from Lassen County to San Diego County.

Racial characters.—Thurberi is most completely differentiated from *shufeldti* by the lighter, more pinkish back. The color is Verona (fresh plumage), snuff, Sayal, or mikado brown, rather than the bister (fresh plumage), sepia, or Saccardo's umber of *shufeldti.* This difference permits 90-per cent segregation of individuals in the interior ranges, 75-per cent along the coast. The difference is the result of diminution of eumelanin in the barbule tips with unmasking, and increase of, phaeomelanin in *thurberi.* The latter pigment is of a pale yellowish type, different from that of *J. caniceps* but apparently not distinct qualitatively from that of the backs of other Oregon juncos. Sides are nearly lacking in eumelanin; phaeomelanin is restricted to extreme lateral regions; the color is near avellaneous and vinaceous buff instead of army brown, fawn color, or vinaceous fawn.

Other characters of *thurberi* such as slightly darker head and longer wing and tail (2 per cent) hold only for the Sierran division. The amount of white averages greater on rectrices 3, 4, and 5.

For characters which differentiate *thurberi* from *J. o. pinosus* and *J. o. pontilis* see pages 290 and 299, respectively. *Thurberi* does not meet *montanus*, a hiatus existing between their ranges in southeastern Oregon. The same color differences that distinguish *shufeldti* from *thurberi* distinguish *thurberi* from *montanus* even more sharply. Moreover, the head of southern *montanus* is a much duller black.

Nongeographical variation.—Most of the variable characters show some geographic correlation, so that all measurements and the colors of head, back, and tail are best discussed later. Color abnormalities are limited to white wing spots, which occur in about 1 per cent of the thousand birds examined. Pileum overcast with brown is found as in all Oregon juncos, but it is less prominent than in *montanus* and *shufeldti*. There are no instances of reddish pileum and few of interrupted hood margin. Individual variation in coloration of the sides is insignificant. Wear and fading may alter the back color decidedly. The dilute eumelanin of this race seems especially susceptible to fading so that the back becomes much brighter, and more yellow or tan in late summer. The exposure to intense light in parts of the breeding area, as compared with the exposure in more northern habitats, may accelerate fading of eumelanin and brightening of the phaeomelanin. The color of the sides becomes faint and is inconspicuous in badly worn plumages.

Sexual difference in head color is variable within *thurberi;* in the north it is equal to that in *oreganus* and in *shufeldti*, but it diminishes southward. The backs of females average somewhat lighter, that is, they have less eumelanin. The difference in sexes in other Oregon juncos is comparable with respect to eumelanin. The sides are similar in the sexes. Tail pattern differs in the average on rectrices 3, 4, and 5. The degree of difference is roughly comparable to that in *J. o. oreganus* and *montanus*. In *oreganus* rectrix 3 is not involved, owing to absence of white, but the variable sixth rectrix shows average difference.

Geographical variation within the race.—In *thurberi* white spotting appears occasionally on the third rectrix. This was not found in the coastal forms to the north and but rarely in *montanus;* the percentage in *thurberi* is not quite as great as in *mearnsi*. White or nearly white inner webs of the fourth rectrix appear as variants in this race. Aside from these, the classes of variants in *shufeldti* and *thurberi* are the same. Averages for white areas in rectrices 4 and 5 are higher in *thurberi* than in *shufeldti*. The differences are somewhat greater than those between southern *montanus* and *mearnsi*, or between the divisions of *montanus* and *shufeldti*, but not so marked as between *oreganus* and its neighboring races. A parallel between tail white and light coloring of sides and back with respect to the race as a whole is apparent, but, as in other such parallelisms, individual variation of the characters is not correlated. Unlike *mearnsi* and *J. o. oreganus*, in which similar parallelism is evident, *thurberi* has a darker head than *shufeldti* coincident with a tail that is whiter.

TABLE 23

OCCURRENCE OF RECTRIX PATTERN IN J. O. SHUFELDTI, THURBERI, AND PINOSUS
(IN PERCENTAGES)

	Male			Female	
	shufeldti	*thurberi* (Tahoe)	*pinosus*	*thurberi* (Tahoe)	*pinosus*
Rectrix 3					
Inner web:					
trace of white	0	11	0	0	0
Rectrix 4					
Inner web:					
pure black	17	3	43	4	54
trace of white	44	19	49	42	43
mixed	40	66	8	54	3
trace of black	0	6	0	0	0
pure white	0	4	0	0	0
Outer web:					
pure black	62	15	72	36	95
trace of white	15	12	18	12	0
mixed	19	27	7	30	5
trace of black	2	18	2	14	0
pure white	2	28	2	8	0
Rectrix 5					
Inner web:					
mixed	10	0	36	2	54
trace of black	42	3	38	18	35
pure white	48	97	26	80	11
Outer web:					
pure black	0	0	3	0	8
trace of white	0	0	8	0	16
mixed	17	1	26	6	43
trace of black	17	3	18	10	14
Outer web:					
pure white	67	96	44	84	19
Rectrix 6					
Both webs:					
pure white	100	100	95	100	86

The independence of these several characters, physiologically and genetically, suggests that such parallelism as is evident results from similar selective agencies working upon the different parts of the organism.

The values for tail white in the table are representative of the Sierran-southern Californian division of the race. In the coastal division certain parts of the tail pattern average the same as in *shufeldti* or *shufeldti–thurberi* intermediates of extreme northern California. The independence of the white spotting on different webs, even of the same feather, is manifest with respect to geographic gradients. Individually within a population there is correlation of markings in several of the feathers of the tail. For example, in 100 males from the Tahoe sector correlation of amount of white in the two vanes of the highly variable fourth rectrix is .531 ± .049; for white of the outer vanes of

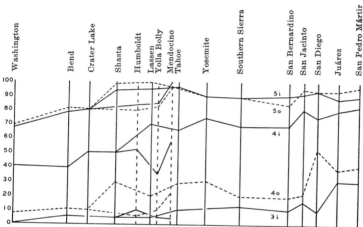

Fig. 17. Geographic variation in amount of white in tail along Pacific coast. Vertical lines represent samples spaced to scale geographically. Solid vertical lines mark positions of interior groups from the Cascades of Washington (*shufeldti*) to the Sierra San Pedro Mártir of Lower California (*townsendi*); broken vertical lines, groups of *thurberi* in coastal region of California. Figures at left represent percentage of occurrence of variants. 5i and 5o, inner and outer webs, respectively, of rectrix 5 pure white; 4i and 4o, webs of rectrix 4 mixed rather than black or chiefly black; 3i, inner web of rectrix 3 with trace of white rather than pure black. Males only.

rectrix 4 and 5, .620 ± .06. Correlation of inner web of rectrix 5, which has a small variation, and outer web of rectrix 4 is lower (.336 ± .06), yet it is probably significant. In working this latter correlation it was noted that black-spotted fifth rectrices occurred only with black or nearly pure black fourth rectrices.

The graph (fig. 17) shows in the upper two lines the percentage of occurrence of pure white in the two webs of the fifth rectrix in different geographic groups from Washington south to Lower California. The geographic gradients of the two are closely related. The *thurberi* level is attained in the Siskiyou-Shasta region with no significant alterations thereafter. The Crater Lake and Bend samples are intermediate. The correspondence of these trends is not a result of linkage of these characters. No correlation individually exists between the two vanes of this feather. (This does not mean that when greater amounts of black are present on this feather, as occurs in other forms of juncos, correlation individually does not exist.) Increase in white southward along the coast is shown by branch lines. The Mendocino and Yolla Bolly groups retain the intermediate values of the Crater Lake sample, and only in the terminal area of the coastal region is a level comparable to that of the Shasta region attained. The modification is thus retarded southward on the coast and a long region of nearly perfect uniformity is established at the level of the intermediate group of the interior. Southern coastal Oregon birds are similar to Washington *shufeldti*. The ultimate increase in white along the coast is not a result of *pinosus* influence.

The two vanes of the fourth rectrix show an interesting independence. The outer vane attains a level in the Shasta district that, with some irregularities, is maintained south to the San Jacinto Mountains. The extreme increase in San Diego County is a departure in the direction of the Lower California forms, but because other characters show no such resemblance this may be merely a local modification. The pronounced change in value in the north occurs between Crater Lake and Shasta. The Crater Lake group, unfortunately small, corresponds perfectly with *shufeldti* of the Cascades to the northward. On the inner web, white does not attain a *thurberi* level until the Lassen district is reached. Even if the Crater Lake group be ignored, the large Shasta group is decisive. Thus the average for one web changes north of Shasta entirely; the other accomplishes two-thirds of the total alteration south of Shasta. The outer web corresponds with rectrix 5.

In the coastal district the outer web of rectrix 4 again parallels rectrix 5; in that region it is about equal to *shufeldti*. The inner web is variable, but resembles in general that of the Crater Lake and Siskiyou intermediates. A small sample from Del Norte County, California, and from adjacent Oregon has a somewhat higher average for white, but this deviation is reasonably attributed to the size of the sample. In the group from Coos County, Oregon, amounts of white are typical of *shufeldti*. The differences in trends occur in spite of correlation individually of the two webs. This situation is comparable to that in the wing and tail, characters which may become altered independently, as for example in *J. c. caniceps* and *dorsalis*, and yet are correlated within a population. There appear to be common factors controlling both webs of the fourth rectrix, yet also separate factors that may affect one web or the other alone.

White on rectrix 3, a variation which is absent in *shufeldti*, appears occasionally from Bend south through the Crater Lake, Del Norte County, and Lassen groups. Between Lassen and Tahoe a higher frequency of occurrence is reached, a point of change which corresponds not at all with that for either rectrix 4 or 5. On the coast a diminution from Humboldt County southward is found that reverses the trend in the other tail feathers and runs counter to the gradient in the Sierra. In other features of tail pattern the affinity of the Humboldt group appears to be with populations farther north; in this instance it is with the Tahoe group.

The population in the Warner Mountains, Lake County, Oregon, and Modoc County, California, is intermediate between the Crater Lake group and the Shasta group with respect to tail pattern.

In general the indices of variability and the correlations of dimensions are extremely similar to those of *shufeldti*. Tail variability, however, is distinctly lower. The coefficient of correlation of wing and tail in males is .637 ± .043, similar to that in *montanus* (and probably also *shufeldti*). No measurements other than wing and tail are correlated. Weight apparently is no more related to linear measurements of limbs than it is in southern *montanus*, where the relation could be adequately analyzed. Average weights are available only for the coastal division of *thurberi*. Sexual dimorphism in wing and tail is between 5 and 6 per cent. Difference in wing and tail averages 12 per cent as in

TABLE 24
MEASUREMENTS OF J. O. THURBERI
(Tahoe sector of Sierra Nevada)

Measurement	Sex	No. of specimens	Mean with prob. error	Standard deviation	Coeff. of var.
Wing...............	♂	86	76.81 ± .12	1.64	2.13
	♀	48	72.59 ± .18	1.85	2.55
Tail...............	♂	86	68.82 ± .14	1.99	2.89
	♀	42	65.34 ± .18	1.80	2.75
Bill...............	♂	84	8.04 ± .02	0.29	3.69
	♀	42	7.82 ± .04	0.35	4.47
Bill depth..........	♂	81	5.59 ± .02	0.26	4.65
	♀	43	5.64 ± .03	0.28	5.07
Tarsus.............	♂	95	19.64 ± .04	0.57	2.90
	♀	49	19.39 ± .06	0.64	3.31
Middle toe.........	♂	90	11.29 ± .03	0.41	3.66
	♀	48	11.31 ± .04	0.42	4.24
Hind toe...........	♂	95	8.09 ± .03	0.39	4.89
	♀	49	8.03 ± .04	0.39	4.88

related races. Sexual difference in bill length is 2 per cent, but there is none in bill depth and it is uncertain in the foot. This pattern of sexual differentiation is exactly as in *shufeldti*.

Wing, tail, bill, bill depth, and middle toe show racial differences and intraracial trends that are comparable to those of the tail pattern in that there is lack of coincidence in some of the principal modifications. The graph (fig. 18) shows the average values plotted against miles in a general north–south direction for each adequately represented breeding population. Wing and tail lengths increase to the southward between central Oregon and Tahoe, and then maintain a fairly uniform level. Average wing length alters abruptly between Shasta and Lassen, the tail over a much greater distance. Nevertheless, the most important alteration in the tail lies in the same Shasta-Lassen interval. The continuation of the tail gradient upward from Lassen and Tahoe is probably only the effect of normal variation in samples. Likewise, the Bend group may not be significantly low; the Shasta average in reality is like more northern *shufeldti*. Therefore the difference in modification of wing and tail may be unduly emphasized by the graph.

Bill length shows an abrupt alteration between Tahoe and Yosemite. From Tahoe north, the value is essentially that of *shufeldti*. Bill depth declines between central Oregon and Shasta, then abruptly between Lassen and Tahoe.

The latter is an alteration not paralleled in other characters except middle toe, which changes chiefly here from a high northern to a low southern average. An extreme southern diminution also occurs in middle toe length.

One feature of the curves is the uniformity in the Sierra south of Yosemite and as far as the San Gabriel Mountains. The isolated populations still farther

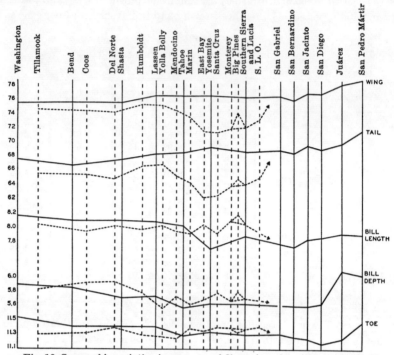

Fig. 18. Geographic variation in averages of dimensions (mm.) along the Pacific coast. Samples spaced to scale geographically. Solid lines, interior groups from the Cascades of Washington (*shufeldti*) to the Sierra San Pedro Mártir of Lower California (*townsendi*); broken lines, coastal groups from Oregon (*shufeldti*) to central California (*pinosus*). Males only.

southward give evidence of colonial variations, although they are no more isolated than the San Gabriel group. The three southernmost areas are all especially well represented, each having more than thirty breeding males.

Twenty-four males from the Warner Mountains of Oregon and California correspond with Lassen males in wing length, being large, but they also are long billed like populations of central Oregon.

Upon the coast, wing and tail lengths increase, as compared with the slightly smaller coastal groups of *shufeldti*, but the increase may be the result of the interior position of the Mendocino and Yolla Bolly groups, not their more southerly positions. A diminution evident in the Mendocino coastal area may be part of a gradient carried on into *pinosus*, or merely a return to a general level for the immediate coastal strip. As is true for most features of tail pattern, the coastal group is related to the southern Oregon populations, and in this instance to the Shasta population also.

Bill length in the coastal region again preserves a value typical of coastal *shufeldti*, though it is close also to the Shasta-Tahoe value. Bill depth south of Del Norte County is variable, but most like that of the Lassen group, except for the Yolla Bolly population, which drops to the level of the Sierran groups south of Lassen. Middle toe is short throughout as in coastal Oregon and the

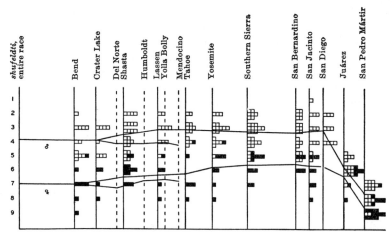

Fig. 19. Geographic variation in head color along the Pacific coast. Individuals represented by squares; open squares, males; solid squares, females. Samples spaced to scale geographically. Solid vertical lines interior groups from *shufeldti* of Washington to *townsendi* of Lower California; broken lines, groups of *thurberi* from coastal section of California. For explanation of variants see page 247.

Sierra Nevada. Except for bill depth the coast population clearly shows affinities with the birds of coastal Oregon or with those of northern interior California.

Weights of thirty-five breeding males are available from the Mendocino and Humboldt districts. The average is 17.58 gr., compared with 17.72 for southern *montanus*; coefficients of variability are 5.5 and 5.2, respectively. The difference in averages is but 0.7 per cent and is not significant. These same *thurberi* (also *shufeldti*) are 4 to 5 per cent shorter in wing and tail than *montanus*. The difference in the races is purely one of proportion of parts, therefore, and not a general difference in size.

Head pigmentation has been classified according to the same system as that used for *montanus*, only the ventral surface being considered. As in *oreganus* and to a less degree *shufeldti*, the categories are not, however, exactly the same as those of *montanus*. The intensities of melanin are the same, but the heads of *montanus* are more often ashy or bluish gray in the light gray variants, and in the coastal races they are sooty gray. Also, as in *oreganus*, the pileum normally associated with a given category is darker than in *montanus*.

Slight average increase in blackness takes place to the southward (see fig. 19). The principal change is in the Crater Lake–Shasta interval, with some increase thereafter to Tahoe. It should be noted that the range of individual variation is similar throughout, and that only a shift in proportion of types is involved. Extremes in *shufeldti* males are 1 and 6, in *thurberi* 1 and 5, except on the coast. The average is usually near 3.2; in *shufeldti*, 3.8.

On the coast, the *shufeldti* average is maintained southward to Sonoma County. The Warner Mountain group averages 3.4, the same as the Shasta birds; hence there is no intermediacy in the direction of *montanus*.

Sexual dimorphism in head color in *shufeldti* is about 3 in terms of head types. Females increase in blackness, as do males from Crater Lake to Shasta, with some increase on to Tahoe (see fig. 19). Unlike the situation in males, the increase continues from here through Yosemite so as to reduce the dimorphism to about 2½ south of this point. On the coast, dimorphism likewise diminishes from 3 to 2½, the females increasing in pigment, the males remaining constant.

In connection with other races of *oreganus*, it has been noted that dimorphism of the head appears to be correlated with progressive increase in black pigment. The females lag behind both in interracial differentiation (compare *montanus* and *oreganus*) and in hybrid intergradation. In a sense, then, females "catch up" with males in *thurberi*.

Analysis of back color is comparable to that made for other Oregon juncos. With regard to eumelanin, categories b and c occur which are quite comparable to those of other races; b is rare south of the intergradational area. Additionally, there is type d with even more reduced eumelanin. Beyond this, in the matter of pigment reduction, the variants lose eumelanin almost entirely from parts of the feather vanes. In worn plumage this produces a mottled appearance with light, bright areas dominating the back (type dd). This is the so-called mixed yellow of the *caniceps–thurberi* crosses. Its appearance rarely within the range of *thurberi* adjacent to the Great Basin would indicate some heterozygosity in the pair of factors, A A' (see p. 194). If the mechanism is correctly hypothesized, occasional A'A' or yellow (y) backs should appear where mixed yellows are moderately common. This is what happens in the San Jacinto Mountains, the only locality of occurrence outside of areas of obvious intermixture. Mixed yellow backs appear so much more ruddy or yellow than those with the d type of eumelanin that one is tempted to ascribe the effect to more phaeomelanin. Apparently, however, the color results merely from the complete unmasking of the phaeomelanin in certain areas. Mixed yellows are readily confused with badly worn backs of type d in which abrasion of the feathers causes resemblance to the normal areas of the mixed type that are without eumelanin.

Of the phaeomelanin variants, type 1 and occasionally 2 are found, but additional types with more phaeomelanin occur which are designated t and p. The series, 2, 1, t, p, represents in order simple increase in phaeomelanin. Class t may be essentially the same as class o of *shufeldti–oreganus* intergrades, but since it usually is combined with c or d eumelanin it produces a color much lighter, that is, more pinkish or tan. Because color under the microscope is difficult to evaluate, I hesitate to call p and o identical, although they would seem to be. Type p seems to be merely an additional stage of concentration and increase of phaeomelanin which, when combined with eumelanin of types c and d, appears ruddier than tc or td. Worn p types are often difficult to distinguish from *idd,* owing to breakdown of the feather margins. Type *p* is one that occurs in *pinosus,* but within *thurberi* it is not limited geographically to intergradational areas.

Figure 20 shows the increase southward of phaeomelanin and the diminution in eumelanin. The changes in the two pigments are well coördinated. Between Bend and Crater Lake and again between Crater Lake and the Siski-

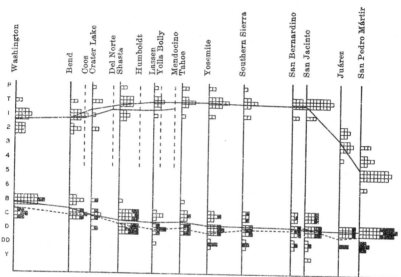

Fig. 20. Geographic variation in back color along the Pacific coast. Individuals represented by squares; open squares, males; solid squares, females. Vertical lines represent samples spaced to scale geographically. Solid vertical lines interior groups from *shufeldti* of Washington to *townsendi* of Lower California; broken vertical lines, groups from coastal Oregon and California. For explanation of variants see page 284. Upper section, phaeomelanin; branch line connects averages for coastal groups. Lower section, eumelanin; broken line connects averages for females.

you-Shasta area major alterations occur, slightly the larger one in the former region. Warner Mountains birds are similar to those of Shasta.

These characters of the back are the most conspicuous ones with the sharpest geographic breaks; they are best used if a definition of racial limits is desired. Type *c* eumelanin and 1 phaeomelanin are found at both geographic extremes, but they are not often combined. There is no correlation of eumelanin and phaeomelanin in individuals. Using back color, a practical separation of the northern and southern interior groups to the extent of about 90 per cent may be made. The same series has been handled that Dwight (1918, pp. 293–294) used in estimating that only 80 per cent of El Dorado County breeding males, and less of females, are distinguishable from northern races. The higher percentage now attained is, I believe, attributable to better understanding of the variations of northern races and to the weighing of both phases of back pigmentation in the analysis. Females have been found to be as readily separable as males.

The sexual difference in the back is confined to the eumelanin. The broken line in figure 20 is the average value for females, with individual females shown by solid squares. The averages are generally, but only slightly, lower. A curve for phaeomelanin of females coincided so closely with that for males as to be impractical of representation.

On the coast the averages for back color change less abruptly to the southward and do not attain values equal to those of the Sierran division. They are similar to those of the Crater Lake intermediates. A 75-per cent segregation of coastal *thurberi* from *shufeldti* is possible on the basis of back coloration. (Individual variation is shown in fig. 23.) Types such as 2c and 2d are fairly common on the coast. These are almost lacking in *shufeldti*, so that there results an appearance of incipient differentiation of a sort not found elsewhere. But when it is considered that such a type as 2d is merely a combination of *shufeldti* and *thurberi* pigment, one hesitates to view the coastal tendency as a new departure.

In the southern section of the interior division the occurrence of types dd and y, which possibly have something to do remotely with *caniceps*, depress the eumelanin curve. The Modoc group averages the same as the Lassen sample.

Summary of geographic variation within the race and relationships with shufeldti.—Thurberi and *shufeldti* and their intergrades form a complex, with more or less continuous breeding range, through which run geographic gradients in characters which are not always correlated and which sometimes run counter to one another. The gradients are of differing steepness and regularity. Each population or colony may have an average genetic pattern more or less peculiar to it. The entire mosaic is at best only artificially subdivided into so-called races which have no high degree of unity. Races could be mapped out on the basis of one group of characters which would not coincide with races based on other characters. The selection of characters is arbitrary, and hence the races are man-made, not natural. If a point could be found where there was maximum change in several coördinated and well-marked gradients, this might be accepted as a natural dividing line. A satisfactory point of this kind has not been found, however.

No great discontinuity of breeding range occurs except south of the Sierra Nevada. Biotic differences are recognizable between Bend and Crater Lake, Crater Lake and Shasta, Shasta and Lassen, Lassen and the Sierra Nevada, and even within the Sierra. The tabulation shown on the following page is an attempt to outline the principal breaks in the curves for characters of male juncos in terms of faunal breaks or discontinuities in distribution (see map, fig. 21).

The contrast between birds of Mount Lassen and the Yolla Bolly Mountains on either side of the Sacramento Valley is considerable, and it is sharper than that between any other adjacent groups. Intervention of the noninhabitable valley and different environmental conditions in the two regions are associated with this contrast. Even so, this differentiation is not nearly so great as the changes occurring between Bend and Shasta, or Bend and Lassen. The changes in these intervals are of greater magnitude, are cumulative, and result in two fairly well-defined extremes.

OUTLINE OF BREAKS IN CURVES FOR CHARACTERS OF MALE JUNCOS IN TERMS OF FAUNAL BREAKS

Interval	Number of characters showing important change	Characters
Successive intervals in south to southeasterly direction:		
Washington–Bend, Oregon	2	Rectrix 5, eumelanin
Bend–Crater Lake	2	Eumelanin, phaeomelanin
Crater Lake–Siskiyou Mountains and Shasta	5	Rectrices 4o, 5, eumelanin, phaeomelanin, head
Shasta–Lassen	3	Rectrix 4i, wing, tail
Lassen–Tahoe	3	Rectrix 3i, bill depth, middle toe
Tahoe–Yosemite	2	Bill, eumelanin
Yosemite–southern Sierra Nevada	0	None
Southern Sierra–San Gabriel (discontinuity)	0	None
San Gabriel–San Bernardino (discontinuity)	1	Wing
San Bernardino–San Jacinto (discontinuity)	3	Rectrix 3i, wing, tail
San Jacinto–San Diego (discontinuity)	2	Rectrices 3i, 4o
Intervals in general southerly direction on coast:		
Coos County–interior Humboldt Co. (intervening area in Oregon and Del Norte Co. poorly represented)	6	Rectrix 5, rectrix 4i, wing, tail, bill depth, phaeomelanin
Humboldt–Yolla Bolly Mountains	3	Rectrices 3i, 4i, bill depth
Yolla Bolly–Mendocino and Sonoma counties (coasts)	6	Rectrices 4o, 4i, 5, wing, tail, bill depth
Intervals in general east–west direction		
Bend–Coos Co. (southwest)	5	Rectrix 5, 4i, wing, tail, eumelanin
Shasta–interior Humboldt Co.	7	Rectrices 3i, 4o, 5, middle toe, head, eumelanin, phaeomelanin
Lassen–Yolla Bolly Mountains (discontinuity)	8	Rectrices 4i, 5, wing, tail, bill depth, middle toe, head, phaeomelanin

It can be concluded that the principal modifications occur at several points in southern Oregon and northern California in regions where there is nearly continuous breeding distribution. Color characters as a group change north of the point where size characters change. The coastal division of *thurberi* resembles the populations of southern Oregon in many respects, either because of continuity, or because of similarity of environment, or both. Regional differentiation in the whole complex appears in the absence of physical barriers to breeding distribution, but where there are environmental differences. Such differentiation is sometimes absent where there is isolation. The barrier of the lower Sacramento Valley doubtless is important in maintaining extremes on

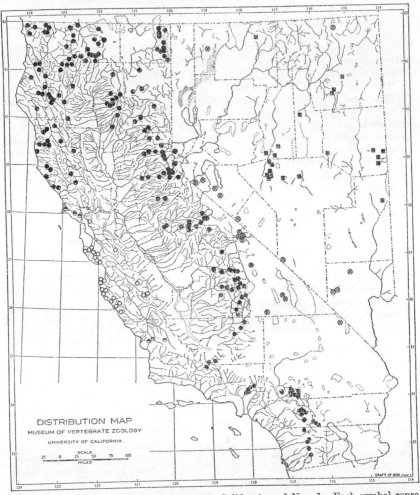

Fig. 21. Distribution of breeding juncos in California and Nevada. Each symbol represents a locality from which individuals have been examined. Dots, *J. o. thurberi;* circles enclosing crosses, hybrid groups involving *J. o. thurberi* and *J. c. caniceps*, and in northeastern Nevada, *J. o. mearnsi* and *J. c. caniceps;* squares, *J. c. caniceps;* circles, *J. o. pinosus;* half dots, intermediates between *J. o. thurberi* and *J. o. pinosus.*

either side that are developed in the course of the extensive regions of diverse but habitable nature that indirectly connect them.

Distribution in the breeding season.—Intermediates are distributed continuously south from Coos, northern Douglas, Lane, and Deschutes counties in Oregon to the California line. If for cataloguing purposes it is essential to name intermediates, the Crater Lake district and the area just south of the Rogue River may be taken as the northern regions for application of the name *thurberi*. Distribution in this sector extends from the coast to the Warner Mountains, with local discontinuity in Klamath and Lake counties. The Warner Mountain colony is isolated to the southwest from the Shasta and Lassen districts.

In California, distribution of intermediate and fully differentiated populations is coextensive with the Transition and Boreal forests of the Cascade-Sierran system, south to Walker Pass and the Kern River. From there south, *thurberi* is found in isolated areas of coniferous timber as follows: Piute Mountains, Kern County (a specimen from Fort Tejon, Kern County, may actually have been taken in the higher mountains near there); Mount Pinos, Fraser Mountain, and probably Cobblestone and Pine mountains, Ventura County; San Gabriel, San Bernardino, and San Jacinto mountain systems (including Santa Rosa Mountain); Santa Ana Mountains, Riverside and Orange counties; and in San Diego County from Palomar Mountain to the Laguna Mountains. On the coast in the north, distribution is continuous from Del Norte County to central Sonoma County and in the interior south along the mountains to northern Napa County. Marin County birds and those from San Luis Obispo County show affinities with *pinosus* of the central coastal district. Birds east of the Sierra Nevada show the result of mixture with *caniceps*, although they are principally of *thurberi* character.

Coniferous forests are the usual habitat throughout the range. Moist places, especially meadows, are preferred but are not required. Suitable ground cover consisting of grass and low herbage is most often found in such places, or in open or down timber, or along a forest edge. In places on the coast, madrone and oak timber, usually along a stream, may be inhabited, and in southern California and in the interior nesting pairs are occasionally found in the deep canyons among golden-cup oaks and alders. The dense redwood forests of the north coast are uninhabited, but the edges of such timber are used. On the coast there are great areas where the population is sparse.

A marked contrast is apparent between the forests of the interior and southern ranges and those of the coast of California. The former are relatively arid in summer but support a Boreal flora that is dependent on the winter snow pack. The coastal district, for the most part of lower elevation, is more humid in summer, more subject to summer rains, and more nearly uniform in daily range of temperature. Only certain higher parts of the coastal region are snow-covered in winter. Much difference in the herbaceous plant cover results which might be of significance to breeding juncos. The altitudinal range of *thurberi* is great, extending from sea level to timber line at 9000 to 11,500 feet. The lower limits rise in the interior and to the southward to about 4000 feet with the retreat upward of the Transition Zone.

Distribution in winter.—Migratory movements are variable, some individuals staying on or near the breeding grounds, as in the north coast district, whereas others move considerable distances into Arizona and northern Lower California. The majority winter in the foothill and Upper Sonoran areas of the Pacific slope of California. Unquestionable specimens of *thurberi* have been examined from Siskiyou County that were taken in midwinter, and intermediates like those breeding in southern Oregon occur in the Klamath and Rogue river drainage basins of that state.

East of the Sierra Nevada the proportion of *thurberi* in winter flocks is not high. There are a few records from north of Owens Valley, as at Eagle Lake, Lassen County. Along the bases of the mountains they are often plentiful, but they are rare in the desert ranges farther east. In Arizona, specimens have been examined from the vicinities of Flagstaff and Prescott and from the Chiricahua and Huachuca mountains. Strongly characterized birds from the latter range demonstrate that at least some of the Arizona records of *thurberi* are not based merely on aberrant members of more northern races. *Thurberi* is to be expected as an occasional visitor throughout the parts of Arizona suitable for wintering juncos.

From Colorado (Golden) and northern New Mexico (Charma) I have examined one bird each that might be referred to *thurberi*. In view of the overlapping of racial characters, these are not considered sufficient to warrant the conclusion that these areas constitute part of the wintering territory of the race. Southwestern New Mexico is undoubtedly visited occasionally by *thurberi* (one record, Big Burro Mountains, Grant County).

In Lower California, *thurberi* ranges southward on the west coast and in the San Pedro Mártir district to about latitude 30°. There are a few records for the Colorado Delta region. The occurrence on Guadalupe Island of some Oregon junco (Bryant, 1887, p. 299) can never be substantiated, owing to loss of the specimen; therefore it seems hazardous to assign the record to any particular race. *Thurberi* has been taken on several of the islands off southern California.

Relationships with adjacent forms.—For *shufeldti* see the foregoing discussion, pages 277–286. For *pinosus* see page 295.

Junco oreganus pinosus

Pinosus is a well-differentiated race that is sharply defined geographically. Discontinuities in distribution between it and *thurberi* are not greater, nor are they attended by greater environmental changes, than those that occur within the range of *thurberi*. The characters of the race are not the culmination of gradients that extend southward along the coast, but often represent reversals of these trends. The area occupied is the coastal region from San Francisco Bay south through Monterey and San Benito counties to San Luis Obispo County. The race is not limited to the true coastal forests of the Monterey region, but inhabits a great diversity of biotic situations, displaying a tolerance for low zonal conditions not found in *thurberi*. The race is nonmigratory.

Racial characters.—*Pinosus* is differentiated from *thurberi* by its ruddier back and sides. Back russet (fresh plumage) or cinnamon brown (worn plumage) instead of the Verona (fresh plumage), snuff, Sayal, and mikado browns of *thurberi*. Exceptions to this characterization amount to less than 10 per cent in *thurberi*, 25 per cent in *pinosus*. Sides near Sayal brown and cinnamon, although more dilute and not identical with these colors. Sides resemble less vivid types of *mearnsi*. Such color is almost never found in *thurberi* and is lacking in less than 5 per cent of *pinosus*. Head color of males lighter in *pinosus* (average 6, instead of 3 to 4), but there is 33 per cent of overlapping with coastal *thurberi*. Reduction in sexual dimorphism of head in *pinosus* involves maintenance of pigment intensity in females nearly equivalent to that of *thurberi*. Wing and tail much shorter in *pinosus* (6 and 7 per cent), but even so there is some overlapping with *thurberi*. Bill length, bill depth, and middle toe length average slightly (1 to 2 per cent), but consistently, greater. Tail pattern with less white, the averages approaching, but not quite equaling, values for *oreganus*. The averages are closer to *oreganus* than are those of *shufeldti*.

Nongeographical variation.—Head types of males are 5, 6, and 7. These are identical with similarly designated variants in the western races of *Junco oreganus*. The range of variation is less than in *thurberi* (1–5). The females vary more than males, ranging from 5 to 9. The average is about 7.5, compared with 6 for males. The net dimorphism thus is 1½, whereas it is 3 in *shufeldti* and 2½ in southern *thurberi*. Dimorphism follows the general rule according to which less sexual difference occurs in lighter-headed races. *Montanus*, with a head average of 6 in males, shows a sexual difference of 2.

Back color with respect to phaeomelanin is either of type t or p, but more commonly the latter. Both these categories occur in *thurberi*. The difference in *pinosus* is one of relative abundance of the two types (see fig. 23). There are some *pinosus*, however, in which the intensity is more extreme than in any *thurberi* of type p; but the difference is not such as to warrant establishing an additional category. Under the microscope the greater amounts of phaeomelanin in the barbules in type p may readily be seen. As in other Oregon juncos, females show no higher frequency of the more intense types of phaeomelanin. For this reason males and females are combined for graphical representation. *Pinosus* owes much of its brilliance to a more prevalent unmasking of the phaeomelanins by the eumelanins than occurs in *thurberi*, so that the type p phaeomelanin shows to full advantage. Stages in reduction of eumelanin are c, d, dd, and y as in *thurberi*. Type dd is slightly less distinctly spotted than in *thurberi*, the areas with and without eumelanin being less sharply contrasted. Microscopic examination of the y type reveals essentially no eumelanin. The slight sexual difference in eumelanin in other Oregon juncos holds true in this race. The sexual difference in eumelanin is indirectly the cause of the ruddier appearance of females.

Enough variation in eumelanin of the back is found in *pinosus* to make feasible the calculation of correlation with head color. The coefficient is $.274 \pm .08$. This is decidedly lower than in *montanus*, but it doubtless represents a similar relation of the pigments of the head and back.

Side color varies but little. The few individuals with *thurberi*-like sides do not have other characters of that race.

Variations in tail patterns are shown in the table on page 279. In *pinosus* new types of variants (compared with *thurberi*) are: mixed inner webs of the fifth rectrix, black and partly black outer webs of the fifth, and sixth rectrices with traces of black. Variants that disappear in *pinosus* are: inner web of fourth pure white or nearly white, and third rectrix with trace of white. The variants that occur have average values similar to those for *oreganus*. Average differences between *pinosus* and *thurberi* involve all vanes of the outer four rectrices, and in magnitude they are as great as those between *oreganus* and its neighboring races. The differentiation of *pinosus* and coastal *thurberi* is not

TABLE 25

MEASUREMENTS OF J. O. PINOSUS

Measurement	Sex	No. of specimens	Mean with prob. error	Standard deviation	Coeff. of var.
Wing	♂	62	72.05 ± .13	1.50	2.09
	♀	40	69.20 ± .22	2.03	2.93
Tail	♂	62	64.03 ± .19	2.27	3.55
	♀	41	61.26 ± .33	2.15	3.50
Bill	♂	58	8.13 ± .03	.29	3.68
	♀	36	8.07 ± .03	.29	3.66
Bill depth	♂	58	5.67 ± .02	.27	4.88
	♀	32	5.76 ± .03	.25	4.42
Tarsus	♂	67	19.76 ± .05	.55	2.79
	♀	41	19.62 ± .07	.62	3.18
Middle toe	♂	64	11.41 ± .03	.48	4.23
	♀	41	11.27 ± .05	.43	3.91
Hind toe	♂	67	7.99 ± .03	.39	4.84
	♀	39	7.83 ± .04	.37	4.74

quite so great as that of *pinosus* and the interior group, because the coastal birds retain some of the low values for tail white of *thurberi–shufeldti* intermediates. The values for north-coast birds are not the result of *pinosus* influence, for the Mendocino and Sonoma county populations nearest *pinosus* have the greatest amount of white. *Pinosus* controverts the rule of increase in white in tail correlative with paling of the head.

Indices of variability do not differ importantly from those of *thurberi* except for that of the tail, which is much more variable. The coefficient of correlation of wing and tail is .534 ± .061, somewhat lower than in related races. Other correlations are uncertain; that of tarsus and hind toe, .301 ± .075, possibly has significance. Difference in wing and tail is 11 per cent. Sexual dimorphism of wing and tail is only 4 per cent instead of the customary 5 or 6 per cent, and it is uncertain in other measurements. In general the measurements show less sexual difference than in related races. This is in keeping with a similar situation with respect to head color.

Available weights of breeding males of typical populations number only 17. The average, 17.44 gr., is about 2 per cent less than in coastal *thurberi*. Wing and tail are about 4 per cent shorter than in this division of *thurberi*. The reduction in the planing surface of the wing and tail that supports the weight is therefore disproportionately great.

Color variants such as white-spotted wing coverts have not been found. The usual variation in overcasting in retarded plumages is found. Heavily overcast heads are especially common in females in association with their general light pigmentation. The whole race in a broad sense is one characterized by retarded plumage and low sexual dimorphism. Wearing of the back results in brilliantly colored breeding birds, the brightening effect being comparable to that in *Junco caniceps*. Some of the populations in more arid situations show greater fading than those occupying the fog belt. The sides become more yellowish in the breeding season, but are always prominently colored.

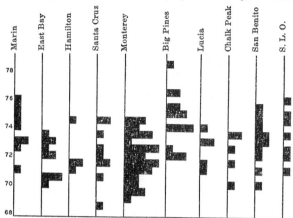

Fig. 22. Occurrence of wing variants in males in populations of *J. o. pinosus* and groups intergradient toward *J. o. thurberi*. Note especially the Big Pines colony.

Geographical variation within the race.—There is little variation locally in the relatively small range of this form that is not a part of peripheral intergradation. The graphs showing intergradation indicate the uniformity within the confines of the race. The various characters show their extreme development at different places, but most often in the Santa Cruz and Monterey districts.

One population displays a most suggestive colonial differentiation. This group is found in the yellow pine and black oak forests of the Monterey Mountains (Big Pines) between the Carmel and Little Sur rivers. The area lies at an elevation of 2700 to 3700 feet, above the fog belt and inland 5 to 9 miles. At lower elevations are large tracts of chaparral that isolate this region from breeding areas in the coastal redwood canyons and lower oak belt. These birds (16 males and 10 females) are distinctly longer winged (see fig. 22) and longer tailed. The females bear out the differences revealed in the males so that there is little doubt that the character of this colony is accurately indicated by the sample. The weights of these birds also are greater, but no features of color are different. The bill averages appreciably longer. Specimens from the coastal canyons a few miles west are small throughout.

One might suppose that this population represented an interior phase of *pinosus,* or in fact a weak race in itself, isolated altitudinally and faunally from the coastal groups, were it not for the fact that samples from the Santa Lucia Mountains to the south do not show like tendencies. Santa Lucia Peak is higher and as much isolated from the coast as the more northern pine areas occupied by the group in question. Across the Salinas Valley to the eastward in the arid mountains about San Benito the juncos are not as distinctly large.

It must be concluded that this population at the headwaters of the Little Sur is differentiated independently of other adjacent populations that occupy comparable habitats. Its isolation is effected by unfavorable tracts of Upper Sonoran chaparral and perhaps by some discontinuity of habitat along the crest of the mountains to the south. The birds may not descend from this region in winter. When one encounters trenchant differentiation in the middle of the range of a race, with only moderately extensive uninhabitable tracts isolating it, his imagination is aroused with respect to the possibility of race formation. This colony could, given the opportunities of continued isolation and at the same time expansion of range, readily lead to a distinct race. Here are the seeds of a race.

Of what does this potential race really consist? The characters concerned presumably are determined by multiple factors. The distribution of sizes shown in figure 22 indicates strikingly the long-winged variants and the absence of short-winged variants that are normal in *pinosus*. But in the females, besides long-winged variants, one small bird, far below the average of *pinosus* females, occurs. Probably occasional small males also occur, but they are not represented in the sample. It is likely, then, that the complex of multiple factors has had a new factor or factors for large size added to it that have spread extensively in the population. If the number of factors concerned is at all large, the variation as a result of this mixture would not in a small sample appear much greater than that of the original racial stock. The occurrence of minimum *pinosus* variants would be extremely rare even if all the *pinosus* factors for small size still were extant in the population.

Distribution.—There are no records of *pinosus* in the fall or winter for localities more than a few miles from the habitats occupied in the breeding season. Definition of breeding areas thus indicates geographic distribution throughout the year. Scattered breeding populations occur in west-central California in the Upper Sonoran and Transition zones from the Golden Gate and Carquinez Straits southward in the hills and mountains of the interior ranges and in coastal wooded areas. The race now breeds in many residential districts of cities in the San Francisco Bay region where formerly it was absent. It breeds occasionally in Contra Costa County (Las Trampas Creek; Mount Diablo), and fairly plentifully in the hills of northwestern Alameda County. There are small numbers in the hills south of Livermore and in the Hamilton Range of Santa Clara County. In San Mateo County, distribution is continuous south along the Santa Cruz Mountains through Santa Cruz County. South of Monterey Bay, *pinosus* breeds in the coastal forests and along the crests of the mountains to the extreme southern part of Monterey County. In San Luis Obispo County intergradation becomes apparent. Isolated colonies occur in San Benito County.

Associations that are inhabited are various and include the following: moist redwood canyons; heavy, but arid, live oak timber (*Quercus agrifolia*); Monterey pine forest consisting chiefly of *Pinus radiata* and *Pinus muricata;* yellow pine (*Pinus ponderosa*) and Douglas fir (*Pseudotsuga taxifolia*) forests; black oak (*Quercus kelloggii*), golden-cup oak (*Quercus chrysolepis*), and madroño (*Arbutus menziesii*) associations; and digger pine (*Pinus sabiniana*) and Coulter pine (*Pinus coulteri*) forests. Almost any shaded forest land with some form of ground cover that remains green throughout the summer can support nesting birds. Usually, perhaps always, surface water is available. The variation in rainfall and humidity is almost as great as in the range of *thurberi*. *Pinosus* is tolerant of low zonal conditions to a degree found only in *bairdi*. Accordingly, probably a larger proportion of the race than of *thurberi* exists in regions of low rainfall. Yet a large part of the range of *pinosus* is in the fog belt, an environment met by *thurberi* only on the coast of northern California. Few places in the range of *pinosus* have severe winter climate with significant snowfall, a circumstance which appears to explain absence of migration. In winter, flocks may wander into open country where there is limited cover in the form of isolated clumps of trees. They occur often in orchards in the open valleys. Many are known to stay on the breeding territory in small flocks, possibly family groups.

Relationships with adjacent forms.—*Pinosus* intergrades with *thurberi* in Marin County, and in San Luis Obispo and probably Santa Barbara counties. *Thurberi* at the southern limits of habitable forest along the Russian River in Sonoma County gives little indication of *pinosus* influence. Wing and tail here diminish in average value to begin the downward trend toward *pinosus*, but possibly this is a level characteristic of the immediate coastal regions even farther north. The general decrease in bill depth southward in *thurberi* stops at this point. The colors of head, back, and sides are unchanged. There is, nevertheless, a single male with the *py* type of back; in this one feature the bird is typical of *pinosus*. Tail white actually increases, instead of diminishing toward *pinosus*. Whether there is any mixture of *pinosus* in this group is therefore somewhat doubtful.

In Marin County the situation is different. All characters show intermediate averages, or even attain *pinosus* values. The Marin woodlands are isolated from those of Sonoma County by about 20 miles of country unsuitable for breeding. A similar gap separates the breeding areas north and south of San Francisco Bay. The isolation afforded does not appear great, but in resident populations it means much more than in migratory races. The Marin County representation that has been studied is not as large as might be desired (9 males and 5 females). Averages for wing and tail are about halfway between those of coastal *thurberi* and *pinosus*. The sample is too small to determine whether there is increased variability associated with this intermediate average. In other measurements the margin of differentiation is too small to permit any statement with respect to the average of this group. Tail pattern is in general intermediate (fig. 24). Head color averages as light as in *pinosus*, but, significantly, there is an exceptionally great range of variation for so small a sample (fig. 23). With respect to back, the numbers of *thurberi* and *pinosus* types are about equal. The backs of the group do not show a uniform blend of the two racial types, but can be classified quite as readily into the *t* and *p* categories as can those of members of either race.

The side color of Marin birds is likewise of two distinct types, rarely a blend. *Pinosus* and *thurberi* coloration occurred in about the ratio of 2 to 1. The

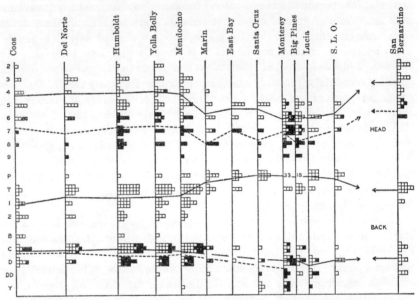

Fig. 23. Geographic variation in head and back color (two pigments) in the immediate vicinity of the coast from Coos County, Oregon (*shufeldti*), south through *J. o. pinosus* to southern California. Connection with group from San Bernardino Mountains (*thurberi*) suggested by arrows through inadequately represented Mount Pinos and San Gabriel Mountain regions. Individuals represented by squares: open squares, males, except upper line for back (phaeomelanin), both males and females; solid squares, females. Samples spaced to scale geographically. For explanation of variants see pages 247 and 284.

Marin group, therefore, shows increased variability with respect to characters that clearly segregate as though they might be based on single factors. This variability is not evident in dimensions, probably because of the many factors involved and the limited size of the sample.

The intergradation is like that of *mearnsi* and *montanus* in that all characters become altered in the same region; the zone of intergradation is narrower, however. It appears that the Marin area is inhabited by birds of dual origin and that the group may continue to receive increments from either side. The situation is very different from the intergradation of *thurberi* and *shufeldti*, where modification in the various characters does not correspond geographically and where the intermediates are probably not the result of any immediate racial hybridization.

In San Luis Obispo County occasional pairs of juncos breed among the oaks and digger pines on shaded hill slopes. A group of 7 adults and 3 young from the vicinity of Santa Margarita resemble *pinosus* but show fairly conclusively a trend toward *thurberi*. This is seen in wing and tail length and in tail pattern. In these characters the birds are roughly equal to the Marin County population. With respect to head and back color, the proportion of types is about as in *pinosus*. Two adults, however, have typical *thurberi* sides; the other 5 have *pinosus* coloration. As far as may be surmised, the nature of the intergradation is like that in Marin County, but with a higher percentage of

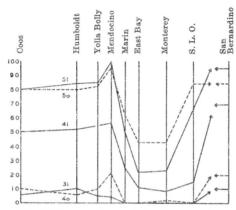

Fig. 24. Geographic variation in amount of white in tail in immediate coastal belt from Coos County, Oregon (*shufeldti*), south through *pinosus* to southern California. Vertical lines represent samples spaced to scale geographically. Figures at left represent percentage of occurrence of variants. 5*i* and 5*o*, inner and outer webs, respectively, of rectrix 5 pure white; 4*i* and 4*o*, webs of rectrix 4 mixed rather than black or chiefly black; 3*i*, inner web of rectrix 3 with trace of white rather than pure black. Connection with group from San Bernardino Mountains (*thurberi*) suggested by arrows through inadequately represented Mount Pinos and San Gabriel Mountain regions. Males only.

pinosus parentage. It is not known how continuous the distribution may be between Santa Margarita and the mountains of southern Monterey County. Because birds successfully breed in the live oak belt, even in such arid interior regions as near La Panza, San Luis Obispo County, it is probable that occasional pairs are scattered throughout the coast range north and south from Santa Margarita. Instead of there being sharply defined barriers as in the northern intergradational zone, in the south there are extremely sparse, scattered populations over a large area. This situation may be as effective a deterrent to intermixture of the main populations of the races as are absolute discontinuities in distribution. Juncos apparently occur in Santa Barbara County (Willett, 1933, p. 174) in a habitat similar to that about Santa Margarita and in scattered forests of yellow pine eastward to Mount Pinos, Ventura County, where typical *thurberi* is present.

Pinosus, even though tolerant of a considerable variety of environmental conditions in the central coast district, may in its origin have been intimately related to the ancient coastal closed-cone pine forest of this district. The diminution of this forest type since the Pleistocene may have left *pinosus* in areas that now are much more arid than formerly.

Some suggestive points of general resemblance can be found between *Junco insularis* and *pinosus,* the former inhabiting the southern outpost of the coastal closed-coned forest on Guadalupe Island. *Pinosus* of course most resembles *thurberi,* with which it undoubtedly has long been in occasional contact, but its resemblances to *insularis* are nonetheless real.

Pinosus with respect to ruddy coloration and tail markings strongly suggests *J. o. oreganus* and the resemblance may be phyletic. Absence of a ruddy coastal type in northern California is indeed peculiar, since many vertebrate animals show this kind of modification there. Can *pinosus* be a southern residuum of an ancient junco stock somewhat like *J. o. oreganus,* which once crowded southward? It could have been preserved in central California by isolation while lost on the northern California and Oregon coasts where an interior type may have crowded westward. Against this view is the distinct lack of resemblance to *oreganus* in the pale head and the diminution of blackish pigment generally.

Probably the prototypes of the Oregon juncos, which must have had relatively pale heads, early gave rise to ruddier coastal and paler interior types which have since become modified and partly lost, and which remain chiefly in isolated peripheral regions. In final judgment, *pinosus* seems to me to be a remnant of such an early differentiation. It stands in relation to the pale forms, *mearnsi, townsendi,* and *insularis,* as *oreganus* does to *montanus, shufeldti,* and *thurberi.* At the same time it has probably been modified by *oreganus* and *thurberi–shufeldti* contact. If *pinosus* were merely a local coastal differentiate of a *shufeldti–thurberi* stock, why is the head so pale? Southern populations of *thurberi* in arid regions tend to be fully as black as the northern groups. The origin of *pinosus,* as with many other races, seems to have involved not merely adaptation to the present environment, but also historical factors such as interpenetration of races.

Junco oreganus pontilis

This annectent form, which occupies a region between *thurberi* of San Diego County, California, and *townsendi* of the Sierra San Pedro Mártir of Lower California, proves to be relatively uniform in character. Isolation in the Sierra Juárez of extreme northern Lower California is an important factor in maintaining it as a discrete racial entity. Were the region connected by breeding habitat to the north and south, permitting infiltration of related races, the population probably would have taken on the aspects of a hybrid or intergradational complex. *Pontilis* shows a remarkable resemblance to the southern division of *montanus.* But its seemingly independent origin, its geographic range, and its lack of migration, which probably entails other physiologic adjustments to its habitat, preclude grouping it with *montanus.* Such a grouping certainly would obscure the true history of the population.

Racial characters.—*Pontilis* is differentiated from *thurberi* by grayer head (types 5, 6, and 7 in males; 6, 7, and 8 in females). Type 5 in males equivalent ventrally to 5 of *thurberi*, but usually duller on the dorsal surface. Types 6, 7, and 8 of *pontilis* entirely equivalent to similarly designated types of *montanus*. Type 5 of *pontilis* usually grayer above than 5 of *montanus*. Only one male *thurberi* from southern California matches perfectly the heads of the two darkest *pontilis*.

Back of *pontilis* duller brown—olive brown, or nearly drab rather than Verona (fresh), snuff, Sayal, or mikado brown. Back color, classified according to types found in other Oregon juncos, of types 3c, 2c, 3d, and 2d. Types 3c and 3d do not occur in *thurberi*, and 2c and 2d are found only in the coastal division of this race.

Sides of *pontilis* pinker than in *thurberi*, more nearly vinaceous fawn, hence like the less brown and less yellow variants of *shufeldti* and *montanus*. Color not the cinnamon of *mearnsi* or *insularis*. Contrary to Oberholser's (1919, p. 120) observations, which were based on four birds, I find that the colored area of the sides is slightly more extensive than in *thurberi*.

Tail of *pontilis* with more white on the average on rectrix 3 (30 per cent with some white) and on outer web of rectrix 4. San Diego County *thurberi* equivalent to *pontilis* with respect to rectrix 4, but not 3. Wing 2 per cent longer; tail slightly less than 2 per cent longer, but not significantly longer than in southern California populations; bill, 1½ per cent longer than Tahoe *thurberi*, but equivalent to southern California populations; bill depth about 8 per cent greater; feet similar, but tarsus tending to be longer and hind toe shorter.

For characters differentiating *pontilis* and *townsendi*, see page 303.

Because *pontilis* parallels closely the southern division of *montanus*, it may be useful to summarize such few differences as exist between these two. Color of head, back, and sides in *pontilis* all duplicate common variants in *montanus*. *Pontilis* more often has white on rectrix 3 (30 per cent as against 1); rectrices 4 and 5 have more white on the average, but same variants present in both. Bill 4 per cent shorter and 2 per cent deeper. In northern *montanus* bill-length similar, but depth even more different; there is of course some difference in head color in northern *montanus*. All these differences are so purely matters of average that there is no feasible means of assorting individuals of the two races.

Nongeographical variation.—A total of 15 breeding males and 7 females has been examined. Additionally, 6 October birds have been studied which almost certainly are *pontilis*, not *montanus*. Six fully grown juveniles also have been used in compiling certain of the average measurements. In such an assemblage the principal color variations in head, back, and sides may be expected to appear. Rare color abnormalities would not be encountered. Average values for tail white and for measurements have high probable errors, yet the main trends probably are indicated.

Head color averages 5.73 in males, varying from 5 to 7, and 6.42 in females, varying from 6 to 8. (For frequency of occurrence of types see fig. 19.)

The range of variation (3) is less than in *thurberi* (4) and the same as in *townsendi*. No greater degree of heterogeneity occurs in this intermediate form, therefore. The reduction of sexual difference to less than 1 is striking; it is equivalent to the situation in *mearnsi*.

Back color is fairly uniform. Phaeomelanin of types 2, 3, and 4 (*townsendi*) occur (see fig. 20). None of these is found in *thurberi* of southern California, but 3 and 4 occur in *townsendi*, although 5 is predominant there. Categories 4 to 6, which represent much reduced phaeomelanin, are not found in more

TABLE 26
AVERAGE AND EXTREME MEASUREMENTS OF J. O. PONTILIS

Measurement	Sex	No. of specimens	Average and extreme measurements
Wing	♂	16	78.48 (75.0–81.2)
	♀	12	73.93 (71.6–76.0)
Tail	♂	16	70.01 (66.6–73.8)
	♀	11	66.58 (61.3–68.7)
Bill	♂	14	7.93 (7.4– 8.6)
	♀	7	7.70 (7.3– 8.2)
Bill depth	♂	14	6.12 (5.8– 6.6)
	♀	7	5.97 (5.8– 6.1)
Tarsus	♂	16	19.91 (19.0–20.5)
	♀	12	19.51 (18.8–20.8)
Middle toe	♂	16	11.24 (10.8–11.8)
	♀	12	11.02 (10.5–11.5)
Hind toe	♂	16	7.87 (7.4– 8.3)
	♀	12	7.64 (7.1– 8.1)

northern members of the rassenkreis. The eumelanin is either of type c or d. The range of variation is like that in *townsendi* and is less than in *thurberi*, where types b, c, and d occur. Again, no evidence of increased heterogeneity is seen. Oberholser (*op. cit.*, p. 120) commented on the diversity among his four birds and the close approach of some to *thurberi* and of others to *townsendi*, particularly with reference to back and head coloration. It should be emphasized that this diversity does not now appear to be so prominent as suggested by him. The original four birds, it so happens, represent extremes of the race. Sexual dimorphism of the back is comparable to that in *thurberi*, but the scanty evidence suggests that there is also a difference in the phaeomelanin; the females possess on the average more of this pigment as well as less eumelanin. Such a difference is of significance only because a comparable situation is met with in the large series of *townsendi*.

Some variation in side coloration exists, but the state of wear obtaining in the series precludes classification of variants. All but about three or four are visibly more pinkish than average *thurberi*. The width of color area is usually distinctly greater than in *thurberi*, but not, in mass effect, so much greater as is that of *townsendi*. There are a considerable number of *pontilis* that are indistinguishable from certain *thurberi* variants. Nearly all might be confused with certain *townsendi* variants. Doubtless in unworn plumage *pontilis* would prove to be relatively uniform in color, with few *thurberi*-like sides appearing in the population.

The table on page 304 shows the variants in tail pattern present in *pontilis*. Significant new variants, compared to *turberi*, are found in the third rectrix where mixed outer webs appear; we cannot conclude concerning absence of variants of the blacker types in *pontilis*, because the sample is too small. Nevertheless, the addition of white variants and the dropping out of certain black variants in the transition from *thurberi* to *townsendi* is significant.

Indices of variability for *pontilis* cannot be determined, but comparison of range of variation and of histograms indicates no tendency to high variability.

Sexual dimorphism in wing and tail appears to be of the usual magnitude, that is, 5 to 6 per cent. In the other measurements the differences indicated are not to be relied on, but the difference in bill length is like that in *thurberi*. Slight differences in the feet are like those in *townsendi*. Difference in wing and tail is 12 per cent of tail length.

Distribution.—Resident of the Transition Zone forests of the Sierra Juárez, northern Lower California in the vicinity of Laguna Hanson near latitude 30°. Mr. Laurence M. Huey estimates that there is about 20 to 25 miles of scattered Transition Zone forest along this section of the Sierra at elevations between 5000 and 6000 feet. Granitic outcrops are frequent and, in combination with exposure, are causes of the interdigitation of tracts of Upper Sonoran forest. The junco habitat in summer is the parklike yellow pine forest. The fall-taken specimens all come from these mountains. One bird was taken at an elevation of 4200 feet at Los Pozos about 30 miles north of Laguna Hanson, October 31. There is no evidence of migration, therefore. There may be some tendency, as in *townsendi*, to descend to lower levels in winter.

Relationships with adjacent forms.—The spatial separation of breeding ranges of *thurberi* and *pontilis* appears to be about 40 miles. The intervening region, although occupied by the low northward extension of the Sierra Juárez, is not inhabitable in the breeding season. *Thurberi* winters in the range of *pontilis*. To the southward the isolation from breeding areas of *townsendi* is also 40 miles, as far as known, but the distance may be somewhat less. Low passes across the mountains separate the higher zonal areas of the Sierra Juárez and the Sierra San Pedro Mártir.

In the comparisons of *pontilis* with other races and in the analysis of variation within the race, the degree of resemblance to *thurberi* and *townsendi* is set forth in detail. In summary, average coloration of head and back is almost exactly intermediate, but there is a little more overlapping with *townsendi* than with *thurberi*. With respect to these colors, there is little "structural intergradation" or "transgressive variation" with either of the neighboring races. Side color is much as in *townsendi*. Tail white averages close to that in *townsendi*, but there is a distinct indication of intermediacy. The various average measurements are not equally intermediate and they more often are closer to those of *townsendi* than of *thurberi*.

As to the history of *pontilis*, certain possibilities may at least be ruled out. The population is not like the hybrids between *caniceps* and *thurberi*, or like the hybrids between *caniceps* and *J. o. mearnsi*. From these hybrids it differs in being relatively homogeneous, with no greater variability than is found in the adjacent forms. It does not contain members that are entirely equal to one or the other of its relatives. If the group has arisen by hybridization, such hybridization does not take place now, nor has it probably occurred in the recent history of the group, for if it had, more diversity would be manifested, even where multiple factors are concerned. If hybridization entered into the origin, certain of the factors of the parents have by now been dropped out of the complex.

Townsendi does not fit well into the scheme of differentiation correlated with environment that prevails in the Sierra San Pedro Mártir. The trend in most local races of vertebrates of that area is toward darker, more leaden hues. In *townsendi* the head is comparatively light and the sides extensively pink, so that one is led to suspect that its characters are not the product of an environmental response to conditions now affecting other animals there. To be sure, the dull color of the back may parallel the modifications in other animals. It is probable that the characters of *townsendi* date back in part to an earlier period of adjustment of the species to environment, and that its present racial characters are more matters of history than adjustments to special conditions now extant. Of course its characters must not be detrimental in its present environment.

Elsewhere I have suggested that *townsendi* and other pale-headed Oregon juncos are peripheral remnants of a lighter-headed, more ancient group of Oregon juncos. Is *pontilis*, then, (1) a remnant of an intermediate stage in the evolution of a darker-headed type, or (2) a result of a fairly old junction of a dark with a light type that has become stabilized and more homogeneous since the time of origin, or (3) an intermediate differentiate, corresponding in some way with the intermediate geographic region it occupies, assuming that these mountain environments affect juncos differently from other vertebrates? I can see no conclusive evidence to support any one of these suppositions, but there is least in favor of the last. Possibly *pontilis* is a direct descendant of a remnant of an ancient intermediate to which there have been additions from time to time by interbreeding with neighboring relatives. It might be thought that *montanus* has established *pontilis* as a remote, separate breeding center through colonization by birds failing to return northward in spring migration. But this seems unlikely on physiologic grounds, and the coincidence of *montanus* colonizing a region where it happened to fit perfectly as a connecting form in a chain of races overtaxes the law of probability.

Junco oreganus townsendi

Townsendi, a strongly characterized race, is the southernmost member of the *oreganus* rassenkreis. It is resident of the Sierra San Pedro Mártir of northern Lower California. Although *pontilis* serves to connect it with the remainder of the rassenkreis, geographic isolation prohibits interbreeding of these two. Even the strikingly differentiated *J. o. mearnsi* interbreeds freely with its neighboring race. *Townsendi* in many respects parallels *mearnsi;* the parallel is not nearly so striking as that between *pontilis* and *montanus*, however.

Racial characters.—*Townsendi* differs from *pontilis* in grayer head; males average 7.6 (7–9), females 8.7 (7–10). Only one male *pontilis* was examined that was as light as type 7. One female *pontilis* was type 7 and one type 8, whereas 1 *townsendi* was type 7 and 5 were type 8, the others being chiefly 9. Head types 9 and 10 in *townsendi* are usually darker dorsally than the same types in *mearnsi*, although equal ventrally. The back of *townsendi* is usually duller, with less phaeomelanin; color hair brown or drab instead of olive brown (*pontilis*); or, of types 3, 4, 5, or 6 instead of 2, 3, and rarely 4. Sides similar, though somewhat more uniformly and extensively vinaceous fawn. White of tail more abundant on the average on rectrix 4 (see table, p. 304). Wing and tail 1 and 2 per cent longer, respectively (the difference cannot be established statistically, but the correspondence of both sexes is confirmatory); tarsus and middle toe about 2 per cent longer.

Townsendi has been compared to *mearnsi*, Dwight having overstressed the similarity by grouping them with *insularis* in a species separate from other Oregon juncos. The important differential characters of these geographically remote races are, briefly: head darker in *townsendi* (average 7.6) than in *mearnsi* (9.5); back duller, usually not with phaeomelanin of type 3 as in *mearnsi*; sides pinkish in both, but *townsendi* not cinnamon and color area narrow. Tail white more or less comparable, but white on the average more extensive in *townsendi*, especially on outer web of rectrix 4 and on rectrix 5. Wing and tail in *townsendi* 2 per cent shorter; bill 3 per cent shorter in males only.

For comparison with *J. insularis*, see page 307.

Nongeographical variation.—The range of variation in head color is shown in figure 19. Not only is the range small, but average deviation from the mean is slight. Sexual dimorphism is slight as in *pontilis* and *mearnsi* and all pale-headed juncos. Back coloration is relatively uniform with respect to eumelanin, especially in males (fig. 20). Sexual dimorphism occurs in phaeomelanin as well as in eumelanin. Side color shows no variations of significant magnitude. Females in fresh plumage appear to average slightly yellower on the sides; this may be related to the dimorphism in comparable pigments of the back.

Because the average occurrence of tail patterns in *pontilis* is not based on a thoroughly adequate series, it is well at this point to note the total alteration that has taken place south of *thurberi*. For this purpose the San Diego County population (34 males) of *thurberi* is used, but it is to be noted that already in this population and in the San Jacinto group an increase in white on rectrix 4 has taken place, compared with Sierra Nevadan *thurberi*. In *townsendi* the following new variants appear: rectrix 3, inner web mixed; outer web with merely a trace of black (mixed type in *pontilis*). A pure black inner web of rectrix 4 fails to appear. A general increase in percentage of occurrence of white types throughout rectrices 3, 4, and 5 is evident. Average sexual difference occurs in rectrices 3, 4, and 5 in *townsendi* as in *thurberi*.

Wing variability is similar to that in *thurberi, shufeldti, pinosus,* and the majority of Oregon juncos, including *mearnsi*. Tail variability is about as great as in *shufeldti* and *mearnsi*, but is higher than in *thurberi* and lower

TABLE 27

OCCURRENCE OF TYPES OF TAIL PATTERN IN J. O. PONTILIS AND J. O. TOWNSENDI (IN PERCENTAGES)

	Males		Females	
	pontilis	*townsendi*	*pontilis* (only 11 individuals)	*townsendi*
Rectrix 3				
Inner web:				
trace of white	31	30	18	18
mixed	0	1	0	0
(otherwise pure black)				
Outer web:				
trace of white	6	6	0	0
mixed	6	0	0	0
trace of black	0	3	0	0
(otherwise pure black)				
Rectrix 4				
Inner web:				
pure black	0	0	9	0
trace of white	19	3	36	16
mixed	81	83	55	77
trace of black	0	11	0	6
pure white	0	3	0	0
Outer web:				
pure black	6	0	27	10
trace of white	13	6	0	22
mixed	13	28	45	29
trace of black	31	25	18	18
pure white	38	41	9	20
Rectrix 5				
Inner web:				
mixed	6	0	9	0
trace of black	6	10	55	12
pure white	88	90	36	88
Outer web:				
mixed	0	0	0	2
trace of black	6	3	45	16
pure white	94	97	55	82
Rectrix 6				
pure white throughout				

than in *pinosus* and *insularis;* variability of bill and foot is comparable to that in *thurberi* and related races, except that the coefficients for hind toe are especially high. Wing and tail correlation is .741 ± .037. The peculiar wing–

bill-depth correlation of *mearnsi* is entirely lacking. Foot measurements show the usual indecisive but always positive correlations. The highest is between middle toe and hind toe (.256 ± .074).

Weights of 24 male *townsendi* average 18.83 gr. Sixteen of these are of breeding birds, 9 of fall birds, but the averages for the two groups are insignificantly different. The average weights of *pinosus, thurberi,* and southern *montanus* are 17.44, 17.58, and 17.77, respectively. *Townsendi,* 7 per cent heavier than *thurberi,* would appear to be a larger bird throughout. In other words, the long wing, tail, and tarsus of *townsendi* merely reflect a large-sized

TABLE 28
MEASUREMENTS OF J. O. TOWNSENDI

Measurement	Sex	No. of specimens	Mean with prob. error	Standard deviation	Coeff. of var.
Wing............	♂	70	79.31 ± .17	2.16	2.72
	♀	55	74.68 ± .16	1.81	2.42
Tail.............	♂	66	71.20 ± .20	2.43	3.38
	♀	52	67.90 ± .18	1.95	2.87
Bill.............	♂	70	7.94 ± .03	0.34	4.21
	♀	52	7.95 ± .03	0.31	3.88
Bill depth........	♂	57	6.07 ± .02	0.25	4.20
	♀	47	6.08 ± .03	0.30	4.95
Tarsus...........	♂	70	20.10 ± .05	0.63	3.16
	♀	55	19.87 ± .05	0.58	2.94
Middle toe........	♂	70	11.49 ± .03	0.39	3.40
	♀	52	11.23 ± .03	0.35	3.14
Hind toe.........	♂	70	7.90 ± .04	0.44	5.57
	♀	53	7.87 ± .05	0.54	6.95

bird, whereas they do not in *montanus.* The wing is actually 5 per cent longer in *townsendi* than in *thurberi.* The correlated increase of weight and wing certainly is more to be expected in a sedentary race than in a migratory form like *montanus.*

The wing is 10 to 11 per cent longer than the tail, hence slightly shorter relatively (or the tail longer) than in *thurberi.* This reduction in difference of wing and tail length parallels the reduction in *J. c. caniceps, J. c. dorsalis,* and *J. p. palliatus,* groups in which, in the Rocky Mountain area at similar latitudes, migratory habit disappears.

Sexual dimorphism in wing and tail is 5 to 6 per cent. The bill is identical in the sexes. Tarsus differs by 1 per cent, which is not significant, and the middle toe by 2½ per cent. The dimorphism is thus slightly different from that in *thurberi,* but there is no general reduction in dimorphism to parallel the situation in head coloration.

No instances of white wing spots have been noted. Other aberrations are limited: one spotted albino has been recorded which had a small number of white feathers about the head. *Townsendi* has no retarded plumage marked by overcast neck and pileum. This seems to disappear with reduced head pigmentation much as in some other pale-headed forms. The color of the back is not so much altered by wearing and fading as in *thurberi,* as there are no bright, lightly pigmented regions in the feather vanes to be exposed or to be broken down. The side patches become reduced with wear and often become brighter; they are never abraded to the inconspicuous state common in worn *thurberi.*

Distribution.—Permanent resident of the Sierra San Pedro Mártir of northern Lower California from about latitude 30° 36' to 31° 10' (fig. 7). The birds breed in the Transition and Canadian zone forests above 6000 feet, but may descend the mountains to some extent in winter, as at Valladares (2700 feet). Huey (1931, p. 128) records one bird taken October 29, 1930, at San Agustín, at latitude 30°, south of the Sierra [specimen atypical, the back being ruddier than in *townsendi; montanus*?]. The Transition yellow pine forest (*Pinus ponderosa*) is relatively open and arid and is interrupted by granitic outcrops. Ground vegetation is sparse. The juncos are probably concentrated in the parks and grassy basins. In the higher parts of the range are Jeffrey pines (*Pinus ponderosa* var. *jeffreyi*), lodgepole pines (*Pinus contorta*), and small tracts of aspens (*Populus tremuloides*). These forest types doubtless provide favorable breeding areas. On the whole, the mountains do not appear to differ greatly from some of the junco-inhabited ranges of southern California. They are perhaps slightly more arid and the forests more open, but no data are available with which to make a meteorological comparison.

Relationships with other juncos.—In the accounts of *pontilis* and *insularis* the isolation and relationships of *townsendi* are outlined. In these discussions the close relationship with *mearnsi* suggested by Dwight is criticized. The nature of its environment in the Sierra San Pedro Mártir does not suggest reasons for the considerable differentiation of *townsendi* from California forms, nor is the differentiation closely parallel to the alterations in other animals in this area. The racial characters may represent an earlier adjustment to environment, and hence are now probably largely "historical" or paleotelic; the characters may never have been developed in response to a local climatic regime. *Townsendi*, with the other light-headed members of the *oreganus* artenkreis, appears to be a relic of an early *oreganus* stock. In this sense it is related to *mearnsi,* but at present it maintains, or retains, connection with the black-headed races by overlapping of characters.

Junco insularis
Guadalupe Junco

This species, the only truly insular form in the genus, is confined to Guadalupe Island off the coast of Lower California. It possesses nearly all the characters of the rassenkreis *oreganus,* briefly summarized as follows: phaeomelanin on sides, the area sharply demarked from gray of hood; hood convex; both eumelanin and phaeomelanin in back, making this area contrast distinctly with head; iris brown; moniliform pigment in side feathers; some segregation of eumelanin and phaeomelanin pigments in barbules of back feathers. These similarities strongly indicate affinities with the Oregon juncos, but there also exist peculiarities of bill, dimensions, tail pattern, and combinations of colors that strongly differentiate *insularis* from *Junco oreganus*. The relationships with *J. oreganus* are comparable in many ways to the relationships of *bairdi* to the yellow-eyed juncos of the mainland. *Insularis* and the *oreganus* rassenkreis *J. oreganus* may be thought of as the *oreganus* artenkreis.

Principal characters of the species.—Iris brown; lower mandible dull flesh color, parts of maxilla similar, but other areas, especially basally, horn color (maxilla not uniformly horn color as in *bairdi*, though basally it is as dark). Feet dusky flesh color (dried specimens), definitely darker on the average than in other juncos, and never yellowish. Back bister, warm sepia, or (worn plumage) sepia, confluent with similar colors on the wing coverts; back fairly sharply demarked from gray of head. Head brownish gray above, equivalent to type 9 of *montanus–mearnsi* series; chin and throat distinctly lighter than dorsal and auricular areas, contrasting somewhat as in *J. phaeonotus* but tones of gray darker throughout; lores and ocular region blackish as in *J. caniceps*; ventral portion of hood gray, type 11; posterior hood margin distinctly convex. Sides with fairly broad area of pinkish cinnamon, closely similar to *J. o. mearnsi*, but also like the less yellowish variants of *J. o. pinosus*. Tail with two, often three, feathers with some white. Pattern shows less contrast on different rectrices than in other juncos. Pure white outer tail feathers uncommon. Wing and tail extremely short relative to feet and bill; measurements for feet moderately high compared with Oregon juncos. Bill deep and exceptionally long, longer than in *aikeni*, the latter a species obviously much larger throughout. Bill the longest in the genus except for *vulcani*; shape more like the average junco bill than is that of *vulcani*; ratio of bill length to depth equivalent to that in certain races of *J. oreganus*, such as *pinosus* and *thurberi*.

Sexual dimorphism in coloration is essentially lacking. The dorsal surface of the head is a little more frequently mixed with brown in females than in males; the dimorphism is less than in *J. o. mearnsi* and *J. c. caniceps*. Wing and tail show a 5-per cent average difference. Difference in bill length, 1½; bill depth, 3; hind toe, 2 (these latter three differences are barely significant statistically, and others involving the foot are even smaller). Average difference in tail white is found in the outer three tail feathers, especially where there is considerable latitude of individual variation; it is most pronounced on the outer web of the fifth rectrix and on the sixth.

Wear and fading makes the color of the sides more yellowish and brighter. The head changes but little with the season. The back becomes somewhat lighter, but there is no brightening of the brown.

Differential characters.—*Insularis* differs from *J. o. townsendi*, its closest neighbor, in the sepia or bister, instead of hair brown or drab of the back, and in the less pure gray head with black mask; the hood is less contrasted with belly and sides. Sides pinkish cinnamon instead of near light vinaceous-cinnamon to vinaceous-fawn. Wing 12 per cent shorter (not longer than 73 mm. in males, 69 in females; *townsendi* longer than 74 and 72); tail 13 per cent shorter; bill 16 per cent longer (not shorter than 8.5 mm. in *insularis*; not longer than 8.5 in *townsendi*); bill depth 5 to 8 per cent greater; hind toe 5 to 6 per cent longer. White on tail averages less on rectrices 4, 5, and 6. Fifth rectrix rarely without black areas, whereas in *townsendi* usually pure white. Maxilla partly dusky.

Insularis differs from *pinosus* in head and back color, and somewhat in side color. The contrasts are as great as between *insularis* and *townsendi* but different in detail with respect to back, *pinosus* being ruddier than *insularis*. Wing of *insularis* 4 per cent shorter; tail 2 to 3 per cent shorter; bill 14 per cent longer; bill 11 to 16 per cent deeper; tarsus 2 per cent longer; hind toe 7 per cent longer. Total amount of white in tail similar, but *insularis* with less white on outer webs and on outer tail feathers, *pinosus* the converse. Maxilla partly dusky.

Insularis differs from *bairdi* in many of the fundamental points which distinguish the black-eyed and yellow-eyed groups of juncos. Briefly the important differences are: presence (in *insularis*) of eumelanin in back and sides; flesh-colored instead of yellowish lower mandible; eye color brown; convex hood mark; back sepia instead of cinnamon brown and not confluent with color of rump. Wing and tail of both species are small and feet are of comparable size; these are probably parallelisms and perhaps are related to comparable degrees of insularity. The bill of *insularis* is much longer (12 per cent) and more slender than that of *bairdi*. Tail patterns differ in details, though total amount of white similar. Outer webs of *insularis* average blacker and inner webs, especially of inner rectrices, whiter.

Variation.—Coloration is relatively uniform except for the back, which is appreciably ruddier in some individuals than in others; the range is represented by the colors sepia and warm sepia. There is nothing comparable to the retarded and advanced plumage types of the darker-headed races of Oregon juncos. No instances of white wing bars or other comparable plumage abnormalities have been noted.

Variations in tail white are here compared with *J. o. pinosus*, as the latter is as close as any Oregon junco to *insularis* in these respects and may have a special affinity with this form not shared by other members of the rassenkreis (see p. 298). The peculiar tendency toward relatively dark outer webs and reduced contrast between rectrices 4, 5, and 6 is as important a departure from the usual condition in the genus as are great increases or decreases in total amount of white (*aikeni*, for example). Interestingly, *bairdi* shows a disproportionate increase in white on the outer webs compared with its continental relatives. In both, a fundamental change in the factors for tail pattern is indicated.

Wing and tail variability is similar to that in *pinosus*, both races having a fairly high tail coefficient that is not equaled in *thurberi* and *townsendi*. Bill length variation is a little higher than in *pinosus*, but it is equaled in *thurberi;* bill depth and foot measurements are similar. The variation in dimensions reflects no tendency to uniformity such as might be expected as a result of inbreeding in a small insular population; coloration, however, is comparatively uniform.

The coefficient of correlation of wing and tail is $.571 \pm .049$; of tarsus and middle toe, $.459 \pm .047$. The peculiarly long hind toe is not well correlated with the tarsus. A correlation of .261 ± .066 for bill and bill depth may have significance. It is associated with a general increase in the bill, in which a ratio of depth to length equal to that in *pinosus* and *thurberi* is maintained. The

TABLE 29

OCCURRENCE OF RECTRIX PATTERN IN J. INSULARIS AND J. O. PINOSUS
(IN PERCENTAGES)

	Males		Females	
	insularis	*pinosus*	*insularis*	*pinosus*
Rectrix 4				
Inner web:				
pure black	3	43	14	54
trace of white	32	49	50	43
mixed	66	0	36	3
Outer web:				
pure black	90	72	95	95
trace of white	8	18	5	0
mixed	2	7	0	5
trace of black	0	2	0	0
pure white	0	2	0	0
Rectrix 5				
Inner web:				
trace of white	2	0	0	0
mixed	40	36	68	54
trace of black	45	38	27	35
pure white	13	26	5	11
Outer web:				
pure black	20	3	68	8
trace of white	21	8	18	16
mixed	49	26	14	43
trace of black	10	18	0	14
pure white	1	14	0	19
Rectrix 6				
Inner web:				
mixed	1	0	5	0
trace of black	29	5	50	14
pure white	70	95	45	86
Outer web:				
trace of white	3	0	0	0
mixed	20	0	23	0
trace of black	45	5	73	14
pure white	32	95	4	86

factors for bill size that are in part the cause of the greater dimensions may be common factors, thus giving rise to this correlation.

Distribution.—*J. insularis* occurs only on Guadalupe Island, Lower California, Mexico. The exact distribution on the island is not fully known. I am only acquainted with records from the northern part of the island; because the southern section is barren and arid, it is unlikely that juncos regularly occur there. At the northern end the species occurred originally chiefly in the pines (*Pinus radiata*) and cypresses (*Cupressus guadalupensis*) at elevations of from 3000 to 4500 feet. But Bryant (1887, p. 300) reported it in the palms (*Erythea edulis*) and oaks at lower elevations. In spite of reduction of the forests, the junco has persisted. Even in the earliest report, Ridgway (1876, p. 188) stated that birds came on board ship for food, thus indicating that they inhabited the island at points near sea level. Several exploring parties have found them nesting along the beach, and Swarth (1933, p. 40) speaks of the abundance of juncos at the northeast cove, where young barely able to

TABLE 30
MEASUREMENTS OF J. INSULARIS

Measurement	Sex	No. of specimens	Mean with prob. error	Standard deviation	Coeff. of var.
Wing	♂	84	69.30 ± .12	1.70	2.45
	♀	25	65.70 ± .24	1.82	2.77
Tail	♂	85	62.65 ± .17	2.34	3.73
	♀	22	59.27 ± .30	2.09	3.52
Bill	♂	94	9.30 ± .03	0.37	4.01
	♀	28	9.15 ± .05	0.42	4.62
Bill depth	♂	95	6.59 ± .02	0.31	4.77
	♀	29	6.39 ± .04	0.28	4.49
Tarsus	♂	98	20.20 ± .04	0.62	3.08
	♀	29	20.06 ± .07	0.60	2.97
Middle toe	♂	96	11.54 ± .03	0.42	3.61
	♀	28	11.43 ± .06	0.46	4.04
Hind toe	♂	98	8.55 ± .03	0.40	4.73
	♀	30	8.38 ± .04	0.45	5.38

fly were found. He says: "At the time of our visit [March 16, 1932] the ground was covered thickly with green grass and clovers. The juncos were in the grass or skulking under huge boulders, like Rock Wrens." Thus there is tolerance for a wide range of biotic conditions, fostered by an absence of predators (except the introduced cats) and absence of competitors in the unusual habitats invaded by this species. The breeding of *insularis* in rocky cliffs, grasslands, and beaches devoid of trees is a remarkable departure from the norm for the genus.

The lower parts of the island are classed as Upper Sonoran Zone and the forests as Transition. Nelson (1921, p. 94) says: "Guadalupe has the same wet and dry seasons as the California coast. Owing to its situation in the midst of cool ocean currents and the prevalence of the northwesterly winds, it is much cooler than any part of the coast of Lower California at the same altitudes.... Dense fogs are more abundant in winter but prevail also about higher parts of the island throughout most of the year.... During May and June, 1906, W. W. Brown ... found the temperature so low that it was necessary to keep a roaring fire in camp.... For 10 days at a time the sun was hidden and everything was ... saturated with moisture ..."

The climate and the affinity of the flora to that of the closed-cone pine forest of the coast of California point to considerable similarity of the island habitat to that of *pinosus* in the Monterey district of central California. The ability to breed in localities zonally lower than the Transition is common to these two forms, though more pronounced in *insularis*. In this connection the parallel adjustment to Upper Sonoran conditions in the well-isolated *bairdi* is noteworthy (see p. 225).

Relationships with other Juncos.—Guadalupe Island is about 200 miles from the range of *J. o. townsendi* in the Sierra San Pedro Mártir of Lower California. It is more than twice that far from the range of *bairdi* in the Cape district (fig. 7). The range of *pinosus* is 500 miles to the north.

There is no evidence of interbreeding of *insularis* with other juncos today as the result of vagrants crossing the water. Guadalupe is a volcanic island and the intervening depths between the island and the mainland are as great as 2000 fathoms. There is general unanimity of opinion from studies of the biota and geology that the island was never connected with the mainland. Grinnell (1928, p. 10) says that "all of its land birds must have reached it as vagrants, the eight differentiates in more or less remote times. . . . Despite its location 135 miles from the nearest mainland, Guadalupe Island . . . is receiving from year to year vagrant delegates from a wide range of mainland species. Only now and then do the ecologic conditions there, in coincidence with sufficient numbers of arriving vagrants of a given species, plus other critical factors of persistence, permit of colonial establishment." This is a very satisfactory statement relating to the mode of origin of *insularis*.

There is a record of Bryant's (1887, p. 299) of another form of junco on the island. The bird, which was collected, was reported as *J. o. oreganus*, but it could have been one of several races of the *oreganus* group; the specimen probably was destroyed in the San Francisco fire of 1906. Bryant remarks that the bird was attacked by the local juncos. Possibly this action meant no more than that local birds were beginning to breed (February 16), for most juncos are extremely tolerant of related species in winter. Even though juncos can cross to the island today, *insularis* as it exists now, with reduced wings (see Lucas, 1891, p. 220), would probably not be able to fly to the mainland.

The source of the ancestral vagrant that established *insularis* remains an open question. There is nothing favoring an idea of origin from the south. The resemblances of *insularis* are with the *oreganus* rassenkreis; furthermore, southern juncos are relatively sedentary. The most likely ancestor would be a boreal form which, at least in the past, was to some degree migratory.

Ridgway was struck by the resemblance of *insularis* and *mearnsi* (*annectens* of Ridgway, 1877, pp. 60–64). This was occasioned by the similarity of their gray heads and pink sides. At the time of his writing, the pale-headed *pinosus* was not known and *mearnsi* was thought of as more distinct from *J. oreganus* than has since proved true. There is a remarkably close approach of the head color of female *pinosus* to *insularis*. Dwight (1918) pursued the idea of relationship of *mearnsi* and *insularis* further, even placing them in the same species. This again was done because of similar head color and pinkish tones of the sides. These colors, in his opinion, are qualitative characters and are indicative of specific unity. Elsewhere I have indicated that gray and black are colors dependent on different quantities of the same eumelanin pigment. The slight distinction between the more ruddy phaeomelanin of the sides of *mearnsi* and the yellow phaeomelanin of other Oregon juncos is not impressive. Indeed, some *insularis* and *pinosus* are not certainly distinguishable in this regard. Wherein, then, are there grounds for stressing "qualitative differences"? The back of *insularis*, rather than being close to *mearnsi*, is the same as that of certain variants in *shufeldti* and *montanus*. Dwight's concept of *mearnsi* (or *insularis*) as a dispersed species consisting of the races *mearnsi*, *townsendi*, and *insularis* was destroyed by the discovery of an intermediate form *pontilis*, which links *townsendi* with *thurberi*. Thus *townsendi* with its "qualitatively different" pink sides is associated with yellow-sided races of the *oreganus* rassenkreis.

Despite the overstress on similarities of *mearnsi* and *insularis*, a certain fundamental, rather remote, relation still is to be conceded. Around the southern, eastern, and western periphery of the *oreganus* artenkreis are more or less isolated, light-headed forms with richly colored sides. These are *mearnsi*, *pinosus*, *townsendi*, and *insularis* (see discussion, p. 298). In all, the sexual dimorphism is relatively low. It is likely that these forms are to some degree relics of an original *oreganus*-like stock. Their paleness far surpasses that which might be expected to appear in association with the aridity of their environments. Perhaps the fully black-headed junco is a newer type, bearing the same general pattern of its paler relations, which has originated in and come to dominate the center of the area of the artenkreis, and which has carried sexual differentiation to greater extremes.

Accordingly, *insularis* may date back to an early period in the evolution of the *oreganus* group, quite possibly retaining in its isolation much of the ancestral coloration. The other relics have had comparable histories, but each has been subject to different influences and to the intermixing of new stocks (especially *pinosus* and *townsendi*). Peculiar tail pattern, short wing and tail, large bill, and moderately large feet are specializations developed in *insularis* and, except the first, are features often associated with insularity. Comparable trends are seen in lesser degree in *pinosus* and *bairdi*. The most primitive features of *insularis*, which may even hark back to a period of closer approach of *oreganus* to *caniceps* and *phaeonotus*, are the black loral area and the dusky maxilla.

Primitive *oreganus* migrants from the range of the rassenkreis to the north may have established *insularis*. Considering present migration routes and similarity of forest habitat, we may suppose that these vagrants probably came from the coast of California, where *pinosus* now breeds, rather than from the Rocky Mountain district. In the Pleistocene, the Monterey coastal forest was a larger and more southerly extensive biotic area which may have brought the ancestral stock into the San Diegan district or to the islands of southern California. However, establishment of *insularis* may have been pre-Pleistocene.

Rassenkreis Junco hyemalis
Slate-colored Junco

This rassenkreis is small in point of numbers of forms, but it occupies an area greater than that of any other rassenkreis. The central and principal race, *hyemalis*, ranges from the northern coast of Alaska to the Atlantic seaboard. The race *carolinensis* is a differentiate in the southern Appalachian Mountains. Another form, not obviously a local differentiate, and probably a stabilized hybrid, is found in the Cassiar district of British Columbia. The *hyemalis* rassenkreis is the most boreal one of the genus; it extends north to the limit of trees and reaches such latitudes as 71° at Point Barrow, Alaska. The southern limits coincide roughly with the occurrence of Oregon juncos in southern coastal Alaska, in coastal British Columbia, and in central interior British Columbia. In the interior of the continent the plains regions limit the breeding area in the south.

Junco aikeni, a conspicuously differentiated, isolated species, obviously is related to this rassenkreis and is geographically complementary to it. With *Junco hyemalis* it may be said to constitute an artenkreis of *hyemalis*-like juncos. Except for *carolinensis*, members of this artenkreis migrate, often to great distances south of the breeding area. The movements of *carolinensis* in winter appear to be chiefly altitudinal.

Principal characters of the rassenkreis.—Sides with some feathers, if not all, fairly pure gray or slate (eumelanin pigment). Slate may be obscured by extensive buff or cinnamon, but, even so, it is visible beneath tips, at least far anteriorly. Ventral hood margin concave posteriorly, owing to confluency of gray of breast and sides. Lores and ocular region not distinctly darker than remainder of head. Back area indistinctly or not at all differentiated from dorsal surface of head. Back chiefly gray or slate, never deep black; phaeomelanin, if present, uniformly distributed on dorsum. Phaeomelanin occurs chiefly on distal parts of feathers and never produces a bright or ruddy brown. Eumelanin is uniformly distributed in barbules, except at extreme bases. When phaeomelanin is present, it is not restricted to clear basal parts of barbules, as it usually is in *Junco oreganus*, but is suffused throughout as though eumelanin were partly altered. Thus, in the arrangement of pigments there is some resemblance to *Junco alticola*, except that in *hyemalis* eumelanin predominates. Pigments of sides, when other than black or gray, also show mixture in barbules, but there is some tendency to concentration of eumelanin distally. Phaeomelanin or eumelanin is less distinctly moniliform than in *Junco oreganus*. Phaeomelanins appear chiefly in distal parts of vanes in feathers of sides.

Iris dark brown; bill and feet flesh color. Tail with two outer feathers chiefly white; usually some white on fourth rectrix. Size moderately large.

Junco hyemalis hyemalis

This race is associated with the transcontinental boreal forests. Except in certain populations in the southeast which are intermediate toward *carolinensis*, migration is extensive and is participated in by all individuals. In view of the size of the breeding range, the degree of geographic differentiation within it is exceptionally slight, much less than within the relatively small areas occupied by *J. o. thurberi* or *J. o. shufeldti*. However, significant discontinuities in distribution north of central New York are lacking and the environment is comparatively uniform in the summer season. The areas of intergradation between *J. h. hyemalis* and the other two members of the rassenkreis are extensive.

An exceedingly great amount of individual variation is characteristic; the prevailing slaty pigmentation is strongly modified or replaced by buffs and browns (phaeomelanins) to produce striking retarded plumages of varying degrees. Such plumages are not entirely matters of sex or immaturity, and so they cannot properly be named as such. The retarded plumages are comparable to those in *Junco oreganus*, but they affect more areas of the body and lead to greater diversity in appearance. They involve suffusion of the entire plumage and not so much just the feather tips.

Differential characters.—Features of *hyemalis* that distinguish it from *Junco oreganus oreganus*, *J. o. montanus*, and *J. o. mearnsi* are, most importantly, those already given as characteristic of the rassenkreis *hyemalis*: gray of varying degree in the sides which precludes appearance of uniformly colored buff, brown, or pinkish cinnamon sides, as in *Junco oreganus;* ventral hood line never convex posteriorly; head never sharply set off from back.

In addition to noting these fundamental differences, it is well to compare other details with races of *oreganus* which border the range of *hyemalis*. *J. h. hyemalis* differs from *J. o. oreganus*: back dark mouse gray, or dark olive gray to olive brown instead of Prout's brown, Brussels brown, or raw umber; head slate, not darker than type 6 in males, a type that is rare in *oreganus,* or else head overcast with less brilliant ruddy buff and brown; tail with fifth rectrix much more often pure white; average amount of white on fourth rectrix greater; wing and tail, 4 per cent longer; bill depth, 3 per cent greater; middle toe, 3 per cent shorter.

Contrast in coloration between *hyemalis* and *J. o. montanus* is similar to that between *J. h. hyemalis* and *oreganus*. The head of *hyemalis* is lighter, although the contrast is not so strong as with *oreganus,* especially so far as southern *montanus* is concerned; the back of *hyemalis,* when suffused with brown, may be the same as that of *J. o. montanus,* but basally the feathers are gray or slate, not brown. *Montanus,* even in decidedly retarded plumage, usually shows some trace of head demarcation dorsally. The heads of females in retarded plumage and with similar degrees of buff suffusion are nearly always paler slate in *hyemalis,* but an extremely pale buffy female *montanus* may be no darker slate than a less retarded female *hyemalis*. The tail markings, on the average, are nearly the same. Significant differences in average measurements between northern *montanus* and *hyemalis* are: tail, 1 per cent shorter in *hyemalis;* bill depth, 3 per cent greater (equal to southern *montanus*); tarsus, 1 per cent longer; toes, 2 per cent longer.

Hyemalis differs from *mearnsi* much as it does from *montanus* with respect to sides and back, except that the sides of *mearnsi*, which lack phaeomelanin and are of great extent, are even more strikingly different. The back of *mearnsi* is always slightly lighter brown (nearer drab) and the pileum never more than lightly overcast, the head demarcation being preserved dorsally. Male *hyemalis* are darker headed than *mearnsi*, which latter have head types 9 and 10 instead of 6 or 7 (relatively lighter dorsally). Female *hyemalis* may match male or even female *mearnsi* in general intensity of gray, but the gray is always less pure or ashy in tone because of associated buff tips. *Hyemalis* shows less white in the tail, particularly on the third and fourth rectrices. Average differences in measurements are: wing, 3 per cent shorter; tail, 5 per cent shorter; bill, 2 per cent shorter; bill depth, 3 per cent less; tarsus, 2 per cent longer; hind toe, 3 per cent longer.

For characters differentiating *hyemalis* and *aikeni* see page 346; for *carolinensis*, page 326.

Nongeographical variation.—White tips on the wing coverts occur more frequently in *hyemalis* than in any other form except *aikeni*. White is never found on the secondaries as in some *aikeni*, but it may be fully as prominent on the secondary coverts. Some pure white areas occur in 2.66 per cent of the 1556 birds that were examined. No geographic localization of occurrence is apparent. With so large a sample it is possible to state the relative abundance in the sexes: 3.50 per cent in 970 males, 1.19 per cent in 586 females. The highest occurrence in other juncos, exclusive of *aikeni*, is 1.82 (in *mearnsi* both sexes combined). These occurrences probably represent mutations that have occurred a number of times in different places and have to varying degrees been spread through the population. The different occurrence of the character in different races may be as much a matter of success in spreading as of rate of mutation.

Variations in the suffusion of phaeomelanin in the plumage have been described in the foregoing sections. There is no doubt that this occurs more frequently and more extensively in females than in males, but it is also clear that no complete sexual difference exists. Females probably never attain a pure slate-colored back, but neither do all males. Without large numbers of specimens at hand at one time, it is impossible to classify different stages of retarded plumage and to indicate their occurrence within each sex. Much the same holds true for age. The age of relatively few fall birds has been reliably determined from the skull, but a few have shown conclusively that adult and immatures of the same sex and season may on occasion be identical in coloration. Adult males in fresh plumage always have some buff tipping.

Where buff tipping is prevalent, wear decidedly alters the mass color effect. Wear always tends to purify the gray and slate of the hood and to make it slightly darker, for the bases of the vanes are heavily pigmented. In females that are heavily suffused with phaeomelanin, little alteration may occur by the time of the breeding season. The faintly brownish backs of females usually are accentuated by wear, but do not become sharply defined. Prenuptial molt, as Dwight has stated (1918, p. 287), is not an important factor in these changes.

Head color of males varies with respect to the gray of the dorsal surface of forehead and pileum. Certain of the series of eastern breeding birds were classified as follows: light slate (dorsal surface equals type 8 of *montanus*, relatively lighter ventrally); slate (equals 7 dorsally); dark slate (equals 6 dorsally); dull black (darker than 6 dorsally, but not equal to 5). A certain amount of variation in the proportion of occurrence of these types was found, but only in areas where there is a gradient toward *carolinensis* or where *Junco oreganus* or *J. hyemalis cismontanus* is approached. The percentages of occurrence of types in eastern Canada is: 10 (1), 60 (2), 17 (3), 13 (4). In the Mackenzie-Athabasca area and in Alaska the proportions are similar, with about 50 per cent of type 2. Heads of females, when gray, vary from slate through light slate to still lighter categories equivalent dorsally to types 9 and 10 of *mearnsi*. Females are most frequently light slate, or type 9.

Inner webs of the fourth rectrix with only a trace of black occur rarely in other samples of *J. h. hyemalis* than the one here tabulated. In *carolinensis* there are a few variants with pure white outer webs on the fourth that do not occur in *hyemalis*; and there is none with black on the sixth. No well-represented class of variants is present in *carolinensis* that is not present in *hyemalis*. Sexual dimorphism in *hyemalis* is remarkably slight; only on the inner web of the fifth rectrix is it significant. In *carolinensis* dimorphism is definite on both webs of rectrices 4 and 5.

The chief difference in pattern between the races *carolinensis* and *hyemalis* lies in the greater average amount of white in *carolinensis*, but because of the difference in sexual dimorphism in the two, females of the two are not strongly differentiated. In other words, the sexual dimorphism in *carolinensis*, which is of normal amount, is about equal to the racial difference in males. The racial difference is not great compared with that obtaining between some forms of *Junco oreganus*. Also to be emphasized is the near identity of pattern in *J. h. hyemalis* and *J. o. montanus*, though in other respects they seem fundamentally differentiated.

Coefficients of variability are similar to those in Oregon juncos. There are no significant departures from the values for *J. o. montanus* except in wing length, which is lower, more as in *J. o. oreganus*. The coefficients for *carolinensis* are extremely similar to those of *J. h. hyemalis*. In computing averages for different subgroups within *hyemalis*, coefficients of variability also were determined. These usually depart from the values here tabulated by little more than the rather large probable errors of the coefficients. Wing and tail are much less variable (1.63 and 2.70) in the group from the maritime provinces of the Northeast. Tail also is low in the New England area. Tarsus is exceptionally low (1.80) in Alaska. Differences in coefficients in the other more variable characters are probably not to be trusted. Sexual difference in wing is 6 per cent; tail, 5 per cent; bill depth and middle toe, 2 per cent; tarsus, 1½ per cent; otherwise it is absent. Difference in wing and tail is 12 per cent in males, 10 in females.

Wing and tail are well correlated, as usual: $\male$ .676 ± .036. No correlations involving wing and tail with bill or feet occur. Among the foot parts there are the following correlations: tarsus and middle toe, .316 ± .056; tarsus and hind toe, .218 ± .060; middle toe and hind toe .499 ± .048.

Geographical variation within the race.—Attention has been directed to

TABLE 31

OCCURRENCE OF RECTRIX PATTERN IN J. H. HYEMALIS (QUEBEC) AND J. H. CAROLINENSIS (IN PERCENTAGES)

	Males		Females	
	J. h. hyemalis	carolinensis	J. h. hyemalis	carolinensis
Rectrix 3				
Inner web:				
trace of white.................	2	9	0	0
Outer web:				
trace of white.................	2	1	0	0
Rectrix 4				
Inner web:				
pure black....................	16	1	22	6
trace of white.................	50	22	48	44
mixed........................	34	71	30	47
trace of black.................	0	5	0	3
pure white....................	0	1	0	0
Outer web:				
pure black....................	54	14	48	44
trace of white.................	10	21	26	22
mixed........................	20	36	13	22
trace of black.................	12	15	9	8
pure white....................	4	14	4	3
Rectrix 5				
Inner web:				
mixed........................	14	1	39	14
trace of black.................	18	8	13	33
pure white....................	68	91	48	53
Outer web:				
trace of white.................	0	0	4	3
mixed........................	36	9	35	22
trace of black.................	20	18	13	22
pure white....................	44	73	48	53
Rectrix 6				
Inner web:				
pure white....................	100	100	96	100
Outer web:				
pure white....................	100	100	96	100

the remarkable uniformity of populations within the extensive range of *hyemalis*. The few minor differences that occur are shown by means of graphs that illustrate the gradients in characters toward *carolinensis*. In measurements, the population breeding in the maritime provinces of Canada is peculiar with respect to the bill, which is long. In New England, wing and tail diminish to the lowest level for the race despite the fact that this group in certain characters begins to grade toward the large *carolinensis*. The northeastern maritime group again differs in tail white. White on the fourth rectrix is materially greater on the average, approaching the situation in *carolinensis*. The fifth rectrix does not show similar modification. The Alaskan population has some reduction of white.

TABLE 32

MEASUREMENTS OF JUNCO HYEMALIS HYEMALIS
(Groups from Alaska to Quebec north of the St. Lawrence River)

Measurement	Sex	No. of specimens	Mean with prob. error	Standard deviation	Coeff. of var.
Wing	♂	104	78.17 ± .12	1.79	2.29
	♀	50	73.50 ± .12	1.32	1.79
Tail	♂	104	68.98 ± .14	2.22	3.22
	♀	48	65.62 ± .20	2.06	3.14
Bill	♂	92	8.02 ± .02	0.34	4.23
	♀	53	8.03 ± .03	0.34	4.20
Bill depth	♂	89	5.98 ± .02	0.27	4.65
	♀	41	5.87 ± .03	0.29	4.89
Tarsus	♂	116	20.56 ± .04	0.65	3.16
	♀	57	20.26 ± .05	0.58	2.85
Middle toe	♂	116	11.65 ± .03	0.53	4.77
	♀	57	11.38 ± .03	0.38	3.34
Hind toe	♂	113	8.29 ± .03	0.38	4.61
	♀	57	8.24 ± .04	0.39	4.82

Distribution in the breeding season.—Generally distributed from the Arctic coast in Alaska southeastward in the transcontinental forest to Labrador, Newfoundland, Nova Scotia, New England, and New York; in the latter two regions intergradation is perceptible.

Alaska.—From vicinity of Point Barrow (Bailey, Brower, and Bishop, 1933, p. 38) and probably other parts of Arctic coast, south to base of Alaska Peninsula, Kenai Peninsula, and Prince William Sound, and from Bering Strait eastward into Yukon. Not known to breed on the Alaska Peninsula or the Aleutian Islands (see fig. 16).

Yukon.—Probably north nearly to Arctic coast. Distributed throughout drainage system of Yukon River south to about 61°, beyond which change in character begins.

Northwest Territories.—North along Mackenzie River probably to delta region, certainly as far as Fort Good Hope. Known from upper Mackenzie Valley, and Great Slave Lake district (fig. 28). Northern limits probably coincide closely with forest margin that runs from mouth of Mackenzie River southeast toward Churchill on Hudson Bay.

British Columbia.—May occur in pure form in northeast corner of province from Fort Nelson northeastward. No specimens examined.

Alberta.—South nearly to Edmonton in central and eastern section. On plains just east of mountains and west of Edmonton intermediate populations occur. In western half of province pure populations probably occur southwesterly at least to Peace River Landing and Lesser Slave Lake.

Saskatchewan.—South to about latitude of Prince Albert, except along eastern boundary, where it follows edge of prairies southward to a point slightly north of Canadian Northern Railway.

Manitoba.—From shores of Hudson Bay south nearly to Winnipeg; east of Winnipeg extends south to Lake of the Woods.

Ontario.—Throughout the province as far south as Toronto, but sparsely and irregularly distributed in vicinity of Lake Ontario and upper St. Lawrence River.

Quebec.—Throughout province north to Richmond Gulf on Hudson Bay; also on Magdalen Islands.

Newfoundland Labrador.—North to Tikkoatokok Bay, but probably not beyond the Kiglapait Mountains just north of there.

Newfoundland and Maritime Provinces of Canada.—Occurs throughout, including Sable Island.

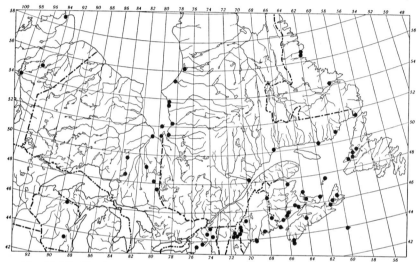

Fig. 25. Distribution of breeding *J. h. hyemalis* in eastern Canada and the northern United States. Each symbol represents a locality from which specimens of this junco have been examined.

Minnesota.—Roberts (1932, p. 406) reports breeding "in the coniferous forests of the northern counties, as far south as central Pine County (Surber) and as far west as Itasca Park and eastern Marshall County. The Junco, as a nesting bird, is confined well within the typical *Canadian Zone*, not reaching the boundaries of this region either on the south or west."

Wisconsin.—Northern half, south to about latitude of Green Bay, Brown County. A bird taken May 25, 1913, at Beaver Dam, Dodge County, may be a late migrant (see Roberts 1932, p. 406, for dates).

Michigan.—Northern peninsula and southern peninsula, south to Saginaw Bay, but also sporadically to Kent and Ingham counties (Barrows, 1912, p. 513).

New England.—Throughout Maine, Vermont, New Hampshire, and Massachusetts, and sporadically in Connecticut, principally in north and west (see Forbush, 1929, p. 86).

New York.—Northern section, especially the Adirondack Mountain area, and Catskill Mountains of southeastern New York. Sporadically in western section from Tompkins County to Cattaraugus County and east of Hudson River from Rensselaer County to Dutchess County.

Birds breeding in Pennsylvania are considered intermediate toward *carolinensis*. This probably is true also for those reported from northeastern Ohio (Campbell, 1940, p. 173).

Any attempt to describe the habitat of *hyemalis* cannot but fail to cover many local situations within the extensive geographic range. Further, information relative to the exact associations in which slate-colored juncos breed is not so plentiful as might be wished. The author has only a limited field acquaintance with this form and must therefore compile information from various sources, credit for which can be given only in part. The chief requirements of *hyemalis* are like those of Oregon juncos. Bush and tree cover is essential, with moist, somewhat open, plant-covered ground in which to forage. These conditions are encountered widely in the northern forests, except where the trees become too dense or the ground too much saturated with water. *Hyemalis* has a fairly high tolerance for dense forests. Grinnell (1900, p. 52) says that in the Kowak River Valley, Alaska, "they were always met with in the deep spruce woods." Places where I have seen the species in Pennsylvania in the broad-leaf forest seemed densely grown to vegetation compared with the habitat of southern races of *J. oreganus*. Probably the habitat of *Junco o. oreganus* is as dense as that of any junco, however. Swarth (1926, p. 132) speaks of the race *"connectens"* (= *J. h. cismontanus*), which undoubtedly is similar in habitat preference and tolerance, "singing from tops of bushes and small trees" in the Carcross region, Yukon. Nests "were all in fairly open bottom land, on the ground, and well concealed in sheltering grass and other vegetation." Thus is indicated a considerable range of tolerance in openness of habitat that seems to hold throughout the range. Broadly speaking, absence of trees is the limiting factor in the north, and prairie or austral forests the limiting features southward except where other species and races replace *hyemalis* in this direction. On the coast of Alaska, the mountains and glaciers between Prince William Sound and Yakutat Bay appear to serve as a barrier between *hyemalis* and *J. o. oreganus*. How complete a barrier they are must be answered by further exploration. In Alaska the more southern mountains may be inhabited up to 3500 feet, but in the north the junco is restricted to the forests of the valleys, the only places, in fact, where forests occur. Precipitation varies from 130 and 190 inches in the Prince William Sound district to only about one-tenth as much in parts of the interior.

Trees which dominate the forests occupied by *hyemalis* are various. In coastal Alaska, Sitka spruce (*Picea sitchensis*), giant cedar (*Thuja*), and alpine hemlock (*Tsuga mertensiana*) are the prevalent conifers. In the interior of Alaska, white spruce (*Picea canadensis*) and black spruce (*Picea mariana*) are the important forest types, together with balsam poplar (*Populus balsamifera*) and aspen (*Populus tremuloides*). These four comprise much of the forest of the interior of the continent, with the addition of larch (*Larix americana*), which becomes more prevalent east of Alaska, balsam fir (*Abies balsamea*), and Banksian pine (*Pinus banksiana*). This forest extends east of Hudson Bay to Labrador. In southern Quebec, Ontario, and the Maritime Provinces, beside the coniferous forests, deciduous forest types constitute in part the habitat of juncos. Some of the trees here concerned are beech (*Fagus grandifolia*), sugar maple (*Acer saccharum*), and yellow birch (*Betula lutea*). White pine (*Pinus strobus*) at lower elevations and red spruce (*Picea rubra*) at higher elevations together with these hardwood forest trees constitute junco habitat in New England and New York.

Distribution in winter (November 1 to March 1).—Winters chiefly south of the breeding range, except in the eastern United States, where certain populations may have only a vertical migration (Forbush, 1929, p. 87). Most of the wintering population is found in the United States east of the Rocky Mountains, south of latitude 45° and north of latitude 33°. West of the Rocky Mountains a great many, perhaps half, of the *hyemalis*-like juncos are *cismontanus* or hybrids with *J. oreganus*, and west of the Cascade–Sierra Nevada system typical *J. h. hyemalis* is exceedingly rare.

The exact limits of known winter occurrence based on specimens examined by me or on records reasonably attributed to *hyemalis* rather than *cismontanus* are as follows: northward as far as southern British Columbia (Fraser and Okanagan valleys), probably northern Idaho, Montana, and North Dakota (no definite records north of southern Idaho, Wyoming, and Nebraska), Minnesota except the extreme north, Wisconsin, the southern peninsula of Michigan, Ontario north at least to Toronto, southern New York north to Oswego and Warren counties, New England north to southern Vermont, New Hampshire, and Maine (*fide* Forbush). Winter specimens from central Maine (Moosehead Lake) and Nova Scotia (no specific locality) have been examined. In the west there is a winter-taken specimen from Juneau, Alaska. A record (Palmen, 1887, p. 284) for eastern Siberia (June 4, 1879) probably represents lost or strayed birds that might, however, have wintered on the Asiatic side. Southward, through California, exclusive of the coastal district of northern California, to about latitude 32° in northern Lower California; southern Arizona, northern Chihuahua (Chihuahua City), Texas to Beeville, the gulf coast of Louisiana, Mississippi, and Florida south to Orange County. There is a record (Howell, 1932, p. 467) from Chokoloskee, Collier County, southern Florida.

Relationships with adjacent forms. Junco h. carolinensis.—Beginning in New England and continuing southwestward, gradation in averages of characters takes place which culminates in a terminal population in the Carolinian mountain region that is fairly well differentiated from *hyemalis* of Canada. No new characters appear, except for certain extremes in size and in amount of white in the tail. The differentiation consists of different average occurrence of variants.

The accompanying tabulation (see p. 322) shows the occurrence, in percentages, of the head types of males along this gradient. The geographic areas there represented are: (1) eastern Canada, including Ontario, Quebec, and the Maritime Provinces; (2) New England, including also the Adirondacks of northern New York; (3) southern New York, the Catskill Mountains south of the Mohawk River; (4) Pennsylvania, all parts of the state where juncos breed, from the Delaware River south, and including western Maryland; (5) North Carolina, chiefly, but including adjacent parts of Tennessee, Georgia, and Virginia north to the latitude of the James River. At least 25 males from each area were used in computing averages, and from Canada and North Carolina the number exceeded 100.

To be noted particularly is the shift of the mode from slate to light slate in passing southward and the disappearance of dark slate and dull black types. At least 10 per cent of the Canadian population cannot be distinguished from those of the southernmost area on the basis of this series of characters. Some of the types occur in all the groups, but, the proportion being different, the patterns of the group mosaics differ accordingly.

The degree of discontinuity between these intergrading populations is variable. The breeding populations of Canada and New England are continuous. The southern New York group from the Catskill Mountains is isolated to some extent. There are regions to the north and east along the Mohawk and Hudson rivers and to the southwest in the Delaware Valley that are not occupied by breeding juncos. The discontinuity between the Catskills and the Berkshire Hills is not more than 30 miles, and that between the Catskills and the moun-

Head type	Eastern Canada	New England	Southern New York	Pennsylvania	North Carolina
Light slate	10	25	55	60	98
Slate	60	50	30	40	2
Dark slate	17	16	15	0	0
Dull black	13	8	0	0	0

tains of northeastern Pennsylvania is not more than 60 miles. Breeding juncos are distributed fairly continuously in the mountains of Pennsylvania from Luzerne and possibly Lackawanna and Wayne counties in the northeast and Tioga, Potter, McKean, Warren, and Erie counties in the north, south into Maryland. The juncos of western New York are not isolated from those of Pennsylvania; no representations from the former region have been examined. The material from Pennsylvania may be subdivided into a small northeastern group (7 males), northeast of the west branch of the Susquehanna River, a northern group (11), north of latitude 41° and the west branch of the Susquehanna, and a southern group (21), south of the west branch of the Susquehanna (fig. 26). With respect to head color, the three groups seem to form a unit. No higher percentage of light slate heads occurs in the southern group than in the northern and northeastern groups. Breeding range appears to be fairly continuous south through western Maryland and eastern West Virginia. No adequate samples from these states or from Virginia north of the James River have been examined. Probably this area, which connects Pennsylvania and southern Virginia, has a population less extreme in *carolinensis* character than those farther south.

The graph (fig. 27) shows the increase in all measurements to the southward. This increase is most pronounced in wing, tail, bill, bill depth, and middle toe. Wing and tail begin to increase south of New England and continue at a nearly uniform rate; the gradient for tail length is steeper, thus establishing a different ratio of wing to tail in *carolinensis*. Averages for both these measurements change within the Pennsylvanian region, as shown when the subgroups (see above) are analyzed, so that the gradient is uniform and unbroken through this sector. The northern Pennsylvania subgroup is intermediate between the northeastern and the southern.

Bill length reaches a low point for *hyemalis* in southern New York. From here south it increases steadily through Pennsylvania. Bill depth is exceptional in that the value typical of *carolinensis* is attained in Pennsylvania. The southern subgroup in Pennsylvania exceeds slightly the average for *carolinensis*, whereas the northern and northeastern subgroups are nearly

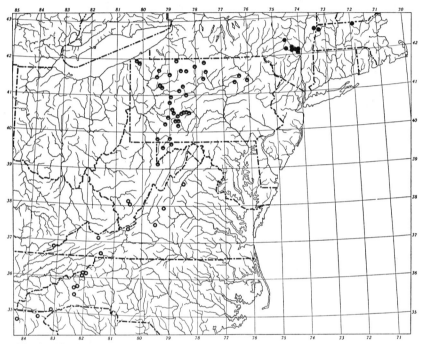

Fig. 26. Distribution of breeding juncos along the Atlantic coast. Each symbol represents a locality from which specimens have been examined. Dots, *J. h. hyemalis*; half dots, distinctly intermediate groups between *J. h. hyemalis* and *J. h. carolinensis*; circles, *J. h. carolinensis*.

identical with those of southern New York. Thus, the curve flattens out when the New York level is reached, and begins to rise again only in southern Pennsylvania.

Tarsus and middle toe increase fairly evenly from New England southward, with the steepness of the curve slightly greater between New York and Pennsylvania. Differences among the Pennsylvanian subgroups are not significant trends. Middle toe increases from the Maritime Provinces southward in a uniform curve of considerable steepness. Southern and northeastern subgroups in Pennsylvania are different, corresponding to the line of the general gradient.

In an earlier paper (Miller, 1938b), I indicated that the principal changes in averages in passing from *hyemalis* to *carolinensis* were in some instances localized more than the gradients that are here presented show. This resulted from failure in the earlier work to take into account the actual geographic spacing of the groups used. Although some irregularity exists in the gradient for bill depth, the evenness of the other gradients is striking. The present

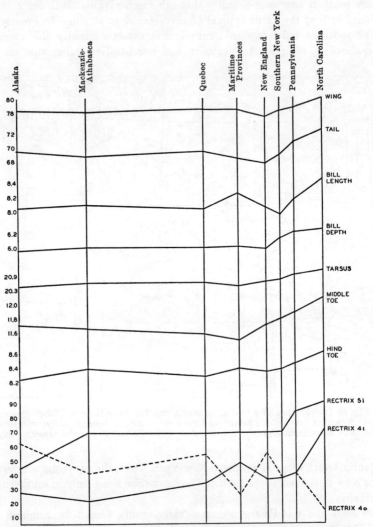

Fig. 27. Geographic variation in measurements and amount of white in tail in the transcontinental range of *J. h. hyemalis* and southward in the Appalachian Mountains to *J. h. carolinensis*. Vertical lines represent samples spaced to scale geographically. Measurements in millimeters. Figures for tail white represent percentage of occurrence of variants. 5o, outer web of rectrix 5 white; 4i, inner web rectrix 4 mixed rather than black or partly black; 4o, outer web of rectrix 4 entirely black rather than partly black. Males only.

analysis of the situation still shows that the gradients or changes are not perfectly coördinated. Thus wing, tail, bill depth, tarsus, and hind toe all begin to change south of New England; middle toe begins south of New Brunswick, that is, in New England; bill length begins south of New York. Gradients in measurements do not culminate north of the Carolinian area, except that of bill depth, which definitely reaches a level in southern Pennsylvania. Accordingly, the view previously presented (p. 281) that each character intergrades independently of the others is in general borne out. Some characters such as wing, tail, tarsus, hind toe, and head color may be thought of as related in their changes. But these characters are not linked or associated in individual variation, except for wing and tail, and tarsus and hind toe.

Figure 27 illustrates the increase in amount of white in rectrices 4 and 5. South of New York the remarkable uniformity in average for the inner web of the fifth is abruptly broken. There is some change between Pennsylvania and North Carolina, but much more between New York and Pennsylvania. The inconstant values for the two webs of the fourth rectrix within *hyemalis* make it apparent that the only substantial intergradational change is between Pennsylvania and North Carolina. Here, as in *shufeldti–thurberi* intergrades, the different rectrices are not coördinated in their geographic variation, even though they may be correlated to some degree in individual variation.

Summary.—Characters of *J. h. hyemalis* are remarkably uniform west to east across the continent, but southward in the mountains of the east coast there is extended gradation toward *carolinensis*. Most of the changes involved in this intergradation occur at a uniform rate geographically, despite some discontinuity of breeding populations in New York. There are three characters—bill depth, white on rectrix 5, and white on rectrix 4—that change more or less abruptly. These abrupt changes occur: one between New England and New York, one between New York and Pennsylvania, and one between Pennsylvania and North Carolina. The characters with uniform gradients begin to change: south of the Maritime Provinces (2), or south of New England (5), or south of New York (1). The only character which reaches full *carolinensis* value north of North Carolina is bill depth.

Some points of comparison with the intergradation of *J. o. shufeldti* and *J. o. thurberi* are significant. In *hyemalis* there are no gradients that run throughout *hyemalis* and on into *carolinensis,* as there are in the *shufeldti–thurberi* complex. There are more characters of *hyemalis* and *carolinensis* that grade evenly through three or four regions in the area of intermediacy. The degree of continuity of populations is about the same in the two situations. In both, not all gradients and breaks in gradients of different characters are coördinated.

How may we delimit the races *hyemalis* and *carolinensis?* As for *shufeldti* and *thurberi,* delimitation must be arbitrary if it is to be attempted. Intergradations in all characters neither begin nor end at a single point. One character does not seem more important than another. Sudden changes in steepness of gradients, infrequent in this instance, occur at different places. How can anyone name the birds from Pennsylvania? The cataloguer desires definiteness and simplicity. Unfortunately these qualities do not exist.

Because birds of New England fall so much closer to *hyemalis* than to *carolinensis,* they may be included in the former. The same may be done with those of southern New York, though they are near the halfway point between the races. Their inclusion in *hyemalis* seriously distorts the conception of that race. The Pennsylvania birds are intermediate, and little more can be said. I cannot see that they fall closer to one race than to the other by a sufficient margin, all characters considered, to warrant a rational decision.

For interrelationships with *J. hyemalis cismontanus*, *J. o. oreganus*, and *J. o. montanus*, see pages 329–343; for *J. aikeni*, page 353.

J. hyemalis hyemalis does not meet *mearnsi*. Possibly there is a near approach, or actual junction, of hybrids of *montanus* and *J. hyemalis* with *mearnsi* in the plains of southern Alberta, but as yet there is no evidence to suggest this. The area between the Cypress Hills in southeastern Alberta, where *mearnsi* breeds, and the east slopes of the Rocky Mountains directly to the west, where *montanus* occurs, probably supports no breeding juncos.

Junco hyemalis carolinensis

This race is the southern culmination of geographic variation in the mountains of the Atlantic coast. Its nature and status are to a considerable extent dealt with in analyzing its relationships with *hyemalis*. The populations from southern Virginia southward display no perceptible intensification of character and thus in this terminal area have the qualifications of a race as ideally conceived. The breeding range is restricted to the higher Appalachian Mountains, corresponding to the higher position of favorable zonal conditions to the southward. Migration, so far as is known, is limited to altitudinal movements.

Racial characters.—The color of the hood most completely differentiates *carolinensis* from *hyemalis*. A uniform light slate (type 8, see p. 247) head is nearly universal in males of *carolinensis*, whereas it is found in only 10 per cent of fully differentiated *hyemalis*. Heads of females are usually type 9, but often types 8 and 10. Thus many are difficult to segregate from *hyemalis*. Usually the gray of females is less mixed with buff or brown than in *hyemalis*. It is reported that *carolinensis* in life usually has a dark basal spot on the maxilla and that the lighter parts of the bill are light bluish horn (Ridgway, 1901, p. 282). In dried skins dark basal spots are not a constant feature, occurring in about 75 per cent. They are like those dark areas which usually border the nostril of *J. caniceps* (types 2 and 3), a feature uncommon in *J. oreganus* and *J. h. hyemalis*. For practical identification of study skins the bill color, in this instance, is of little value. Average differences in *carolinensis* and typical samples of *hyemalis* are as follows: wing, 2 per cent longer in *carolinensis*; tail and bill, 5 per cent longer; bill depth, 4 per cent greater; tarsus, 2 per cent longer; middle toe and hind toe, 4 per cent longer. There is more white in the tail of *carolinensis*, particularly among males, in rectrices 3, 4, and 5, but average differences are not great. (See p. 317 for details.)

Nongeographical variation.—But one instance of white on the wing coverts has come to my attention. Dr. Alexander Wetmore has described to me a male identified by him as *carolinensis* which was taken March 15, 1936, in the Blue Ridge Mountains of Virginia. White is present on both middle and greater coverts and even on the inner tertials [= secondaries] and, faintly, on the distal upper tail coverts. The latter feature suggests an albinistic tendency not found associated with the white-winged variants of *hyemalis*. Since *carolinensis* is in some respects a southern counterpart of *hyemalis*, as is *aikeni* of the middle west, it is noteworthy that the white-winged variation is especially rare and in occurrence parallels in no way the situation in *aikeni*.

Variation in head color is much less than in *hyemalis*, both with respect to tone of gray and to suffusion of brown and buff. The latter rarely obscures completely the gray of the sides, even in females, and is lighter colored. *Carolinensis* has a less pronounced retarded plumage with respect to buff suffusion, even though it has a characteristically pale hood, which in *hyemalis* often is a manifestation of the retarded state. The occurrence of buff with respect to age and sex presents a situation like that in *hyemalis*, and indeed other juncos, but the degree of mixture is so much less that average age and sex differences are slight.

The uniformity of the slate tone of the head is shown in the table on page 322. The exact average of tone of gray in females cannot be given, but the sexual difference is evidently somewhat less than in *hyemalis*. A situation like that in *Junco oreganus* exists, in which darker-headed races as a rule show more dimorphism in blackish pigment. Seasonal alteration of plumage is less than in *hyemalis*, owing to the lesser amounts of buff and the greater uniformity of gray both upon the individual feathers and in the different parts of the hood and mantle.

Variations in tail pattern are given on page 317.

Sexual difference in wing is 6 per cent; tail, 5 per cent; bill length and middle toe, 2 per cent. Average difference in wing and tail 10 per cent in males, 9 per cent in females.

Coefficients of variability are remarkably similar to those of *hyemalis* with respect to males. In females the correspondence is not so perfect, probably owing to the smaller sample. Coefficients of the Pennsylvania intermediates (40 birds) show no uniform tendency to higher values. The coefficients depart in varying degrees from the values typical of *carolinensis* and *hyemalis*, but apparently without regard to the degree of intermediacy of the character concerned. For certain characters the group appears to be more completely homozygous than the parental races, whereas for others it is more heterozygous, and preserves and combines the parental extremes. Among the distinctly more variable characters are wing length and bill depth, but tail, bill, tarsus, and the toes are less variable. It does not appear, therefore, that the Pennsylvania group is of immediate hybrid origin, although the apparent absence of increased variability in hybrids where many multiple factors are involved is not conclusive evidence. Diminished variability would hardly be expected, however. The genes governing dimensions in *hyemalis* and *carolinensis* must differ and it is remarkable that the same degree of variability is found in the two, especially in view of the situation in the intermediates. Is it possible that the racial extremes have reached a comparable degree of unification and adjustment to environment while the intermediates with respect to some characters have not attained this and with respect to others have surpassed it? The explanation is not obvious; the whole situation is bound up with unknown details of the history of the intermediates that are requisite to complete understanding.

Coefficients of correlation in *carolinensis* are: wing and tail, .633 ± .045; wing and bill depth, .315 ± .073; wing and tarsus, .261 ± 0.67; tail and bill depth, .378 ± .074; bill and bill depth, .364 ± .067; bill depth and tarsus, .419 ± .062; tarsus and middle toe, .258 ± .068; tarsus and hind toe, ± .171 (not significant, but compare *hyemalis*); middle toe and hind toe, .422 ± .06.

TABLE 33
MEASUREMENTS OF J. H. CAROLINENSIS
(Areas south of the latitude of the James River, Virginia)

Measurement	Sex	No. of specimens	Mean with prob. error	Standard deviation	Coeff. of var.
Wing	♂	79	80.25 ± .14	1.83	2.29
	♀	42	75.55 ± .21	2.06	2.72
Tail	♂	79	72.48 ± .17	2.30	3.18
	♀	40	68.95 ± .19	1.81	2.63
Bill length	♂	76	8.43 ± .03	.35	4.21
	♀	38	8.26 ± .04	.34	4.11
Bill depth	♂	76	6.24 ± .02	.29	4.73
	♀	31	6.21 ± .03	.28	4.53
Tarsus	♂	93	21.11 ± .05	.66	3.13
	♀	37	20.93 ± .07	.66	3.17
Middle toe	♂	86	12.08 ± .04	.51	4.20
	♀	43	11.89 ± .04	.40	3.39
Hind toe	♂	93	8.60 ± .03	.44	5.11
	♀	47	8.46 ± .04	.43	5.12

Some important contrasts with *hyemalis* exist. Wing and tail are comparable in the two races and the relation of parts within the foot is extremely similar. Association of bill depth and bill length, entirely lacking in *hyemalis*, suggests new common factors responsible for the large size of the bill. Some of these factors may be size-determining for many parts of the body, thus perhaps accounting for some of the correlation of bill depth with tarsus and for correlation of wing, tail, and bill depth. Wing and tail each correlate with bill length to a degree (.209 and .207) that, although not statistically reliable, seems real in light of bill depth correlations. Certain factors entering into the usual correlation of wing and tail, to the extent of a correlation value of about .2, apparently affect all dimensions of the bird. Such general factors for size are not found in *hyemalis*. It is reasonable to suppose that these factors are peculiar to *carolinensis* and that they account in part for the averages of that race.

Distribution in the breeding season.—Occurs in the Appalachian Mountain system in essentially typical form from the Potomac River to extreme northern Georgia. The lower limits of occurrence are about 3000 feet in North Carolina in the oak and chestnut forests of the mountain slopes. The limits possibly are as low as 2500 feet in Virginia, but they do not extend below 3500 feet in Georgia. Most of the population is found between 4000 feet and the summits of the mountains, somewhat over 6500 feet.

Virginia.—Blue Ridge Mountains irregularly north as far as Page County and in the Allegheny and Shenandoah mountains; generally distributed in higher mountains of the southwest.

West Virginia.—Allegheny and Shenandoah mountains from Hardy and Randolph counties south to Mercer County.

Kentucky.—Cumberland Mountains (Big Black Mountain) of the extreme southeast.

North Carolina.—Entire extent of Blue Ridge Mountain area of the far western section, above about 3000 feet elevation.

Tennessee.—Blue Ridge Mountain system along North Carolina boundary.

South Carolina.—Northwestern boundary adjacent to North Carolina.

Georgia.—Rabun, Towns, Union, Gilmer and, probably Fannin counties in the extreme north.

The forest habitat in Alleghenian and Boreal zones consists, in the lower levels, of certain of the southern hardwoods, chestnut (*Castanea dentata*), hickories, and yellow poplar (*Liriodendron tulipifera*), mixed with the northern hardwood forest associates such as beech, birch, and sugar maple, as in Pennsylvania. Evergreens of the Boreal region are chiefly red spruce (*Picea rubra*), fir (*Abies fraseri*), and hemlock (*Tsuga canadensis*). The region is one of heavy summer precipitation, the annual rainfall amounting to 50 to 70 inches. The forest floor inhabited by the juncos is arid and open compared with the habitats of *hyemalis*, with more relatively barren ground and exposed rock.

Distribution in winter.—The movements of this race in winter are extremely limited and the majority of the birds would seem to be essentially nonmigratory. It is reported that they move down the mountains to winter in the lowlands. There are, however, midwinter records in abundance from such mountain localities as Black Mountain and Craggy Mountain, North Carolina. The danger of confusion with misidentified wintering *hyemalis* is great. Among winter birds, personally studied, I have seen none that was certainly *carolinensis* that was taken in the piedmont region more than a few miles from the bases of the mountains. Numbers of lowland specimens considered to be *carolinensis* prove to be *hyemalis* or of doubtful origin with respect to breeding area.

Junco hyemalis cismontanus and the relationships of Junco hyemalis and Junco oreganus

The relationships of *Junco hyemalis* and *Junco oreganus* need to be fully elucidated to explain the complicated plumages that apparently result from interbreeding of these species. The status of the race formerly called *J. hyemalis connectens* by Swarth (1922, p. 244), which now must go by the name *cismontanus* Dwight (see Appendix, p. 403), is a part of this problem. To anticipate conclusions, it may be stated that *cismontanus* is a fairly uniform group of birds, doubtless of hybrid origin to begin with, which now perpetuates itself independently of further hybridization of the parental types, *J. oreganus* and *J. hyemalis*. *Cismontanus* on the margins of its breeding range breeds freely with *J. h. hyemalis*, which it most resembles. In certain regions it does not interbreed freely with *J. oreganus*, whereas at other localities hybridization is evidently frequent. As pointed out elsewhere in connection with possible origin of forms through hybridization, the degree of uniformity, geographic segregation, and the autonomy from the related or parental forms are the criteria for judging of the existence of a race; the mode of origin is not significant. The probable genetic identity of members of a race such as *cismontanus* with newly produced hybrids of the parental types is not denied. The logical mode of analysis of *cismontanus* is first to deal with all *oreganus–hyemalis* hybrids and then proceed to the formal, practical circumscription of the race, even though this procedure may not pattern after the treatment accorded other races.

Characters which are practicable for analysis in *oreganus–hyemalis* hybrids are those of color. Dimensions and the white areas of the tail are similar in *J. h. hyemalis* and *J. oreganus montanus* so that in treating of small groups of individuals little can be said concerning them. When *J. hyemalis* and *J. o. oreganus* interbreed, dimensions are worth considering. The color areas concerned are the head, sides, and back. Each of these shows interesting complications in hybridization. Lines of demarcation, as at the junction of the hood with the sides and with the back, are to some degree the result of combinations of conditions in the separate areas. The relation of hood and sides is comparable to that found in *J. c. caniceps–J. o. thurberi* hybrids.

The most striking difference in color is in the sides, which in *J. hyemalis* are slate and in *J. oreganus* are wood brown or pinkish brown. Although the latter species does not have distinct gray areas in the side feathers (for rare exceptions in *J. o. oreganus* see p. 271), a considerable amount of eumelanin may be mixed with phaeomelanin to produce dark brown. In *hyemalis,* in the retarded plumage, the slate of the sides may be partly or completely replaced by brown so that they are indistinguishable from the sides of *oreganus,* although the distribution of pigment granules in the barbules is somewhat different (see p. 313). If the plumage of *hyemalis* is thus retarded, the hood is equally affected and the color areas are confluent. In retarded plumages of *J. oreganus* the demarcation is visible beneath the buff feather tips. The part of the side immediately adjoining the hood constitutes the critical area where differentiation between the species in all plumages may be sought. The close approach of many female *hyemalis* to *oreganus* makes it much more difficult to allocate females than males, even though genotypically there probably is as much difference in the females as in the males of the two species.

In males, obvious hybrids show nearly every degree of mixture of slate and brown. The two colors are segregated into different areas of the feather vanes. The anterior part of the side area is more prone to slate color, the posterior region to buff and brown. No occurrence has been found of an anterior section browner or buffier than the posterior section.

Such differentiation within the side area was not found in *caniceps–oreganus* hybrids. The *hyemalis* condition seems to involve a gradient of slate color from the head to the sides; in *oreganus* this gradient is broken and the sides are autonomous. To modify these regions in the direction of *oreganus* (1) the gradient must be interrupted and (2) phaeomelanin must replace (or nearly replace) the slate-producing eumelanin. In hybrids of *hyemalis* and *oreganus*, then, the degree of contrast of head and sides may not be solely the summation effect of different colors in the two areas. The independence of the gradient and pigmentation factors is shown by hybrids in which the sides are pure gray yet distinctly marked off from the dark slate of the hood, and by others with buff or brown sides which anteriorly have some slate that grades into the hood. There are also birds with mixed gray sides that do not grade evenly into the hood.

An additional interrelation of hood and sides is shown in hybrids in which the sides are to a varying extent gray and the typical *hyemalis* gradient is manifest, but in which the *oreganus*-like darkening of the eumelanin of the hood is also reflected in the sides. A comparable intensification effect was seen in the eumelanin of *caniceps–oreganus* hybrids uncomplicated by the gradient effect of *hyemalis*. To summarize, three sets of controls are operative in the hood-side complex: determination of slate or buff-brown in the sides; presence or absence of a tendency to a gradient of color between head and sides; and a common intensification of the eumelanin of head and sides. A particular situation with respect to one set of factors may modify or mask the expression of another set.

In the *hyemalis–oreganus* hybrids there is nowhere a fully isolated hybrid population that appears to have been set up in the simple fashion of the *caniceps–thurberi* hybrids of southern Nevada. One cannot assume, as in the *caniceps–mearnsi* hybridization in Utah, a bilateral infiltration of parental types into the zone of hybridization, because the obvious hybrids are more irregular in distribution and there is some restraint of hybridization. Consequently, it is not safe to draw positive conclusions from evidence of phenotypic combinations concerning the dominance of characters or the degree of simplicity of the factorial mechanism. In at least two regions where populations occur that are predominately of *hyemalis* character, but with some evidence of mixture, the proportion of *oreganus* characters in the sides to *oreganus* characters in the back is 7:13 and 4:8. These populations are from the Cassiar (Stikine to Carcross) and Jasper Park sections, respectively. They do not mix freely with *oreganus*, and perhaps they were established from a particular hybrid group which through segregation had more *oreganus* back characters than side characters. But it is remarkable that two samples from well-separated areas should show such similar ratios. It would appear possible that in males there is some degree of dominance of *hyemalis* sides over *oreganus* sides. In the scattered hybrids closest geographically and phenotypically to *oreganus* there is an apparent preponderance of *hyemalis* sides relative to *hyemalis* back, so that the suggestion of dominance is borne out.

The sides of females present a different situation. Buff- or brown-sided hybrids may owe their resemblance to *oreganus* to a retarded *hyemalis* state or to *oreganus* factors. The tendency already present in *hyemalis* to show buff sides seems to lead to *oreganus* coloration if any *oreganus* factors are present.

Thus slate sides are not dominant or at least do not appear with great frequency, because of interaction of sex factors or hormones in females. As Swarth (1922, p. 244) has pointed out, the females from the Cassiar area resemble *oreganus* much more than do the males. Out of 41 males of the Cassiar area, 7 showed some buff in the sides, but out of 32 females from the same region 26 showed some buff. The confluency of head and sides is another matter, however. Occurrence of an even gradient in side and hood color is essentially as frequent in females as in males in groups of dual affinity. Still, the prevalence of buff sides in the females gives a mass effect much more like that of *oreganus* and tends to obscure, and draw attention away from, the gradation of the color areas.

The heads of males of both species vary considerably and the range of intensity of eumelanin overlaps in northern *montanus* and *hyemalis*, both forms possessing variants classed as type 6 (dull black or dark slate). The extent of the dark areas, if identical in tone, is never the same. The darker *hyemalis* variants do not, with rare exceptions, show extension of the black over the nape to a definite transverse line in front of the interscapular area. In hybrids there are, therefore, extensions and intensifications of varying degrees, increasingly toward *oreganus*. The intensification shows a finely graded series of intermediate stages, as in other similar situations within the genus, and presumably there is a series of multiple factors operative. The extension posteriorly is independent of intensity, although of course if the pileum is light slate there is no contrast with the rest of the head and no extension can be seen. Intermediate stages in extension are numerous, but they do not appear as progressive changes in position of a sharp line of black, but as progressive suffusion of darker color back over the neck to a definite line which, with greater intensity of black, becomes more and more apparent. In females comparable changes take place, complicated by buff tipping and lighter tones of gray, slate, and black. The buff tipping is more uniform in *hyemalis* and is complementary in part to the dark slate of the forehead, so that in a sense a uniform brown neck and pileum is equivalent to a uniform slate dorsum. Retreat posteriorly of the heavy brown tipping is the counterpart of posterior extension of the dark slate or black in male, or advanced female, plumages.

Pigments of the back consist solely of eumelanin in advanced male plumages of *hyemalis*, except for the extreme feather tips, which rapidly wear away. Just proximal to these tips is a zone of pale gray, and farther proximally is the true slate area that is the dominant color in worn plumage. In hybrids with *oreganus*, phaeomelanin, not unlike that of the extreme tips, extends into the outer gray zone, modifying the light gray to produce light brown. This gives a spotted appearance to the back when the basal slate parts are exposed through wear. A further influence of *oreganus* is manifested in progressive reduction of the basal slate or blackish areas until a fairly uniform brown feather results (type 3b of *montanus*). Some reduction of the basal slate areas may occur without the addition of phaeomelanin, but the area is never fully replaced merely by light gray. Some brown feathers appear to have an unusually large amount of eumelanin mixed with the phaeomelanin (type 3a) throughout the vane, but usually any increment of eumelanin in *oreganus*-like feathers appears basally without toning down the brown beyond 3b. Most of the *montanus* that are classed as 3a have small basal slate spots attributable to *hyemalis*. Increased black or slate basal spots are classified as a, aa, and aaa. The last two types are found in typical *J. h. hyemalis*. Sometimes phaeomelanin of intensity 1 or 2, as well as 3, is combined with aa, and rarely with aaa.

It has been noted that hybrids with aa or aaa pattern and with black *oreganus*-like heads may "reflect" the dark head color in these basal feather spots. The general intensification of eumelanin is thus comparable to that which is seen in the slate of the sides. Invasion of the back by phaeomelanin always takes place over the whole area, and as this pigment increases the back becomes sharply contrasted with the head.

The line of demarcation of head and back is then the result of phaeomelanin of the back and the intensification and extension to a definite neck line of the black of the dorsal part of the head. A line may be distinct as a result of contrast of black neck and gray back (slate, with light gray margins), or by contrast of slate neck with brown backs 3b, 3aa, or 3aaa. There is no gradient factor, as there is between hood and sides. The most striking contrast naturally results from combined black neck and 3b, or, even better, oob (*J. o. oreganus*) or 1b.

In fresh overcast plumage, much of the contrast of back and head may be obscured, but it never disappears in pure-blooded *J. oreganus*. In female *hyemalis* the extension of buff far basally on the back feathers often results through wear in a brown back patch in the breeding season. Such a patch usually is not sharply defined; yet it is much like a 3b or 3a *montanus* back. In the Cassiar group, brown-backed females of this sort plus those of *oreganus* blood make up a large group of females that resemble *oreganus*, both with respect to the back itself and the contrast of the head. Thirteen of 41 males have no type 3 phaeomelanin, whereas 25 out of 32 females possess it. This difference is not truly a matter of dominance of *hyemalis* character in males that is lacking in females, but results merely from the phenotypic identity of retarded plumages of *hyemalis* and hybrid back types. With respect to sides, the even greater sexual difference (♂, 7 with *oreganus* character out of 41, and ♀, 26 out of 32) is probably a summation effect of differences in dominance in the two sexes and similarity of *oreganus* and its hybrids to the retarded female *hyemalis* plumage.

The colors of head, back, and sides appear to assort independently in hybrids except that when eumelanin is present in two or more areas intensification may be correlated. Probably few hybrids of the F_2 generation from pure parental types exist, so that the full amount of assortment is not often seen. Most interbreeding is probably between types one or both of which already is heterozygous. This situation, together with multiple factors and intensification factors, obscures the typical aspects of assortment.

With respect to distribution, intermediate birds fall chiefly into two classifications, namely, (1) those populations either out of contact, or not freely mingling, with *J. oreganus* which nevertheless are of hybrid ancestry, and (2) scattered individuals of more immediate hybrid origin either in the zone of contact between *J. oreganus* subsp. and the birds of class 1, or at localities distant from such regions. Group 1 differs in uniformity in different regions,

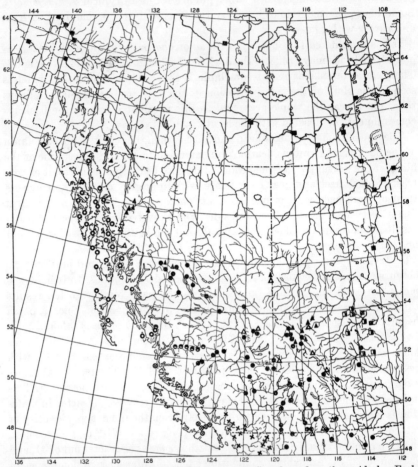

Fig. 28. Distribution of breeding juncos in western Canada and southern Alaska. Each symbol represents a locality from which specimens have been examined. Dots, *J. o. montanus*; half dots, intermediates between *J. o. montanus* and *J. o. oreganus*; plain circles, *J. o. oreganus*; circles enclosing crosses, intermediates between *J. o. oreganus* and *J. o. shufeldti*; crosses, *J. o. shufeldti*; open triangles, hybrids between *J. oreganus* subsp. and *J. hyemalis* subsp.; solid triangles, *J. h. cismontanus*; half-open squares, intermediates between *J. h. cismontanus* and *J. h. hyemalis*; solid squares, *J. h. hyemalis*.

but its characters partake more of *hyemalis* than of *oreganus*. In general it occupies the eastern foothills of the Rocky Mountains and the edge of the plains from the vicinity of Jasper Park northwestward, extending into the Stikine and Taku drainages east of the coast range to the Atlin area. Along its western margins it meets *J. oreganus* geographically, but there is evidence to show that it does not freely interbreed. There is nothing comparable in the way of a group that is chiefly of *oreganus* type but with all individuals to some extent intermediate. The sharpest division between *hyemalis* and *oreganus*, then, is one between *J. o. oreganus* or *J. o. montanus* and *hyemalis*-like birds already to some degree possessed of *oreganus* blood.

In southern interior Alaska occasional evidence of mixture (usually reported as *montanus*) is found in populations otherwise of pure *hyemalis* type; the same is true of the Slave River Valley and probably of parts of Yukon Territory. Localities of such occurrences which I have verified are: mountains near Eagle, Alaska; Fort McMurray, northern Alberta. The bird from the latter locality (no. 231616 Mus. Comp. Zoöl.) was taken May 8, 1920; the ovaries were "scarcely enlarged," so that this may be another instance (see p. 240) of a stray migrant. It again suggests how intermingling might take place at localities remote from the regions of hybridization or intergradation. Dwight (1918, p. 295) mentions *"montanus"* occurring far north in the Mackenzie Valley. No details are given, and I have not been able to verify such an occurrence although essentially all the material that Dwight studied has passed through my hands.

In south-central Yukon Territory, south of Lake Marsh, *oreganus* characters in the *hyemalis* population become frequent. From Carcross, Yukon, and Bennett City and Atlin, British Columbia, a considerable series is available. Out of 54 summer resident birds only 8 are without some *oreganus* character; none is typical of *J. o. oreganus*, which breeds in the coastal belt just west of the coast range. Among males of this area, the sides of 31 are pure slate or blackish slate; the remaining 4 are mixed with brown, but slate predominates. The ventral hood color ranges from type 4 to 8 intensity, with an average of 6.4 compared to 4.0 in northern *montanus* and about 7 in *J. h. hyemalis*. The intensity of black on the pileum is slightly greater than in *hyemalis*, the modes of occurrence being dark slate and slate, instead of slate; light slate (2) and black (4), deeper than the dull black of *J. h. hyemalis*, occur. The black, irrespective of intensity, extends to a neck line to contrast distinctly with the backs in all but 7 of the males; 4 of these are fall birds in which overcasting obscures what might otherwise be a fairly sharp contrast. With respect to the back, only 1 shows type *a* eumelanin; the rest are *aa* or *aaa*. Type *aa* is not common in *J. h. hyemalis*, but it is frequently found in these populations. Type 3 phaeomelanin occurs in 8 out of 31 males. In summary, all characters of males of this region are distinctly closer to *hyemalis* than to *oreganus*. In each color area, not more than one-fourth are really different from *J. h. hyemalis*. But with respect to extension of black of neck posteriorly to a definite line, 80 per cent are distinct from *hyemalis*. This is probably the situation because any one of three *oreganus* characters (extension of black on nape, brown back, or especially dark head) will set up some degree of *oreganus*-like contrast.

The only evidence for interbreeding of this adulterated *hyemalis* stock of the Atlin sector with *J. o. oreganus* is supplied by a specimen (no. 165749 U. S. N. M.), from Glacier, White Pass, Alaska, north-northeast of Skagway in the pass leading to Lake Bennett in the interior. The locality is in the center of the coast range. This bird has sides that are mixed with gray only anteriorly, the head is type 7, the pileum is dull slate and extended to a neck line, and the *ooa* back is sharply contrasted. Type *oo* phaeomelanin never seems to appear except close to the range of *oreganus*, and of course it would never be derived from intermixture with *J. o. montanus*. This hybrid has a wing 76.0 mm. long, which is below average for *hyemalis* and above average for *J. o. oreganus*. The foot is not conspicuously large as in *oreganus*, but the tail pattern is that of average *oreganus*. The abundance of such obvious hybrids in this region is unknown. Probably the population is sparse but largely hybrid, in view of our knowledge of similar situations that obtain in the lower Stikine River to the southward. It is suspected, but not known, that hybrids like this also occur in the somewhat limited suitable territory between the interior and coast in the Alsek and Taku river valleys to the north and south. In the north there might be a direct meeting of *J. o. oreganus* and typical *J. h. hyemalis*.

One hundred and fifty miles south of Atlin in the interior Stikine River basin another group of breeding juncos has been taken. Swarth (1922) based his concept of the Cassiar junco on this group and gave a rather complete description of the nature of the population. However, it is necessary to review the group in terms of the analysis of variants employed in this work. Juncos were taken at Telegraph Creek, Glenora, and Dock-da-on (*op. cit.*, p. 129), all places well within the interior biotic province. There were 18 males and 15 females.

This group closely resembles the Atlin series, but upon careful analysis it proves somewhat more homogeneous. It includes only one individual, a female, that might be confused with *J. h. hyemalis*. Yet among the males there is no individual that could be confused with *J. oreganus* subsp.; there are only two such females and they probably possess *hyemalis* blood. Certainly this group is not one that has arisen from immediate hybridization of parental types, or else these parental types would be present in some quantity in the population.

Of the 18 males, 4 have some brown in the sides. The hood varies from 5 to 7, average 6.1. The intensity of black of the pileum varies from dull black to slate, much as in the Atlin group, but 10 individuals are dull black, so that there is less average variation and a single mode of higher value. Black or dark slate extends to the neck line to contrast with the back in all but 1. There is only 1 type *b* back, no type *a*, and about half and half *aa* and *aaa*. There are 9 with type 3 phaeomelanin, a higher proportion than at Atlin (8 out of 31). The principal differences in the Stikine group are, then, more uniform head color, above and below, more constant neck line, and absence of *hyemalis*-like types, which latter factor increases slightly the percentage occurrence of *oreganus*-like back and side characters. The heterogeneity, though easily perceived, and more prominent than Swarth indicated, is no greater than in *J. h. hyemalis, montanus, shufeldti,* or *thurberi*. This similarity to situations in other forms of juncos led Swarth appropriately to conceive of the group as a race, even though he could not determine its geographic limits.

Westward, down the Stikine River, Swarth endeavored to obtain breeding birds at Flood Glacier and Great Glacier, where, however, they were scarce. The former locality is in the area of transition toward coastal conditions. Eight adults from this place show distinct mixture of characters. The combination of characters in these and in two birds from Great Glacier (nos. 40002, 40004), which is farther down stream where there are coastal conditions, is as follows.

M.V.Z. no.	Sex	Sides	Lower surface of head	Upper surface of head	Back	Wing	Bill depth	Middle toe
39988	♂	slightly slate	6	dull black	2b	77.0	6.3	11.2
39990	♂	buff-brown	6	black	1b	79.8	5.7	11.1
39998	♂	slate	5	black	1a	77.0	6.4	10.8
39993	♂	slate	6	dull black	2a	77.5	6.6	11.5
39987	♂	buff-brown	5	black	1a		5.7	11.1
39999	♂	slate	6	slate	aa	77.8	6.1	11.6
39994	♀	buff-brown	6	slate	oob	69.0	6.2	11.3
39991	♀	buff-brown	7	slate	oob	73.0	5.8	11.4
40002	♂	buff-brown	5	black	1b	78.3		11.7
40004	♀	buff-brown	6	slate	oob	71.6	5.9	11.5

Some of the Flood Glacier birds are truly hybrids. Both parental types are present, and also birds with various mixtures and assortments of characters. Side color in males seems to segregate more than back color, which tends to be intermediate (1b, 1a, 2b, 2a) between the extremes (aa and oob). Dominance of slate sides in males is suggested even by this small sample. Only 1 typical aa back occurs to 3 slate sides. Most of the wing lengths are variously intermediate (multiple factors); bill depth and middle toe show a rather great variation for so small a sample.

Swarth was much concerned with the resemblance of the backs of these hybrids to *montanus* (= *shufeldti* of Swarth). He felt it to be impossible for *montanus* to extend northward between his *connectens* (= *cismontanus*) and *J. o. oreganus* on the coast, and for that reason discounted the possibility that *montanus* contributed to these birds. I believe Swarth overlooked the tendency for juncos to inhabit timber-line and subarctic terrain. McCabe and others have found juncos nesting above timber line in the coastal mountains. A more cogent reason for believing that these *montanus*-like birds are *J. o. oreganus* × *J. h. cismontanus* is the fact that both parental types are present at Flood Glacier. The *montanus* type of phaeomelanin, No. 3, already is common in *cismontanus*. If now additional *oreganus* factors, irrespective of race, are added that reduce the eumelanin, 3b or *montanus* type of back results. Types 1 and 2 of phaeomelanin that occur in *montanus*, but not in *cismontanus*, are certainly to be considered as intermediates from the breeding of type 3 with oo of *oreganus*. One of two birds from Great Glacier has a 1b back, again probably because of hybrid ancestry. A *montanus*-like bird, with type 1*b* back, from Bradfield Canal, Alaska, probably is of similar origin.

Two hundred miles south of the Stikine River in the upper Skeena Valley, *montanus* is the summer resident junco, so that somewhere in the interval *montanus* must push north between *J. o. oreganus* and *J. h. cismontanus*.

Of 54 *montanus* taken in the Skeena region (some fall birds, but probably all near their breeding grounds) only 1 has *hyemalis* character and this consists only of a small amount of gray anteriorly on the sides. However, two pairs of *cismontanus* were seen by Swarth, and a male and female, not necessarily a pair, were taken. Both these are completely slate sided; 5 and 7 (♀) with respect to head below, and above, dull black and light slate (♀); the backs are 3*b* and *aa* (♀). Swarth in his notebook speaks of the *"connectens"* being in pairs and implies that members of the pairs were equally well-characterized slate-colored juncos. The implication that the two types do not interbreed is borne out by the rarity of birds of decidedly mixed characters such as the one mentioned which had a small amount of gray in the sides. Surely more of this type should appear if the *cismontanus* that were there interbred freely with *montanus*. Interbreeding should result in a complex something like that in the Flood Glacier district of the Stikine Valley.

The next region to the south that is well represented by birds comparable to those of the Stikine is the Jasper Park area of the Rocky Mountains. Here a different geographic situation prevails. *Cismontanus* occupies the east slope of the mountains primarily, while *montanus* is on the west slope. There is no isolated intermontane basin for this group of *cismontanus*-like birds to occupy.

From a transection from Edmonton through the continental divide at Jasper Park to the headwaters of the Fraser River there are 43 breeding males and 17 females available. Those in the plains area, from Edmonton south and west about 80 miles, constitute a unit (11 males, 5 females). In this group only 1 male has a mixture of brown in the sides. The hood varies from 6 to 8 and averages 6.3. The color of the pileum varies from dull black to slate, with the mode at dull black (5). Black or dark slate extends from the back to the neck line to contrast with the back in 6 only. There is but 1 type *b* back and no *a*; the others are chiefly *aaa*; there are only 2 with type 3 phaeomelanin.

This group has a lower percentage of *oreganus* characters than does the Atlin series. The females likewise bear this out. Only in the intensity of head color do the males equal or (dorsally) surpass the average of the Atlin series. The more fundamental characters of sides and back are more uniformly like *J. h. hyemalis*. It should be recalled that north and east of Edmonton birds appear to be typical of *hyemalis*, except for sporadic *montanus*-like birds, as at Fort McMurray (see p. 335).

Along the base of the Rocky Mountains, and extending up nearly to the divide, is a piedmont region from which there are 19 males and 4 females. This zone is about 130 miles wide, but most of the specimens are from the westerly half of this sector. As Riley (1912, p. 68) has pointed out in reporting on the juncos of this area in the collection of the United States National Museum, most of the birds of these eastern slopes are like *hyemalis*, that is, in present terminology, *cismontanus*. In comparison with the Edmonton group, 6 out of the 19 males have some brown in the sides. Two of these have no gray in the sides and no other *hyemalis* characters whatsoever. Two others have more brown than gray in the sides, but obviously are of mixed parentage. The 2 birds that are seemingly *montanus* in all respects may be left out of the comparison, for it is possible that they were more or less isolated sexually from the *cismontanus*-like group. In the remaining 17 the hood averages 6.3, as in the Edmonton district, although type 5 variants appear. Dorsally the situation is comparable, the mode is dull black, as at Edmonton, but black variants are added. Only 3 show indistinct contrast of head and neck. Back type *b* is absent, but 3 are of type *a* and the rest are divided about equally between *aa* and *aaa*. Type 3 phaeomelanin occurs in 6.

This group shows no average increase in head pigmentation compared with the Edmonton group, but a few darker variants appear. However, *montanus*-like backs and sides are much more common, reaching a frequency of occurrence roughly equivalent to the Atlin series. The uniformity of character lies somewhere between that of the Atlin and Stikine series. *Montanus* characters are no more prevalent than at Atlin, but pure *hyemalis* are less frequent. There are only 3 males that could be confused with *J. h. hyemalis*. Among the females, 2 are in all respects *montanus*. The other 2 are of mixed character.

The third group (13 males, 8 females) in this transection come from Yellowhead and Moose passes along the continental divide, or from points in the mountains not more than 10 miles west of the divide. All the females and 10 of the males are indistinguishable from *J. o. montanus*. The other 3 males are possessed of full slate sides, 6 or 7 type of head, dull black to slate pileum, *aa*, 3*aa* and 3*a* backs. All three come from the passes, not west of them. They are as strongly *cismontanus* as are any birds from the eastern slopes of the mountains.

This abrupt appearance of *J. o. montanus* in essentially unmixed populations as soon as the high mountains are entered is remarkable. Nowhere else in the genus do two such distinct forms come into such contact and fail to interbreed freely (compare *J. c. caniceps* and *J. o. mearnsi*, or *J. c. caniceps* and *J. o. thurberi*). The localities in the passes where the three males of *cismontanus* were taken and the two localities (Henry House and 15 miles south of Henry House) where *montanus* occurred east of the divide do not produce birds of the decidedly mixed type such as are found in the lower Stikine River, where *cismontanus* and *J. o. oreganus* hybridize. However, there must be some hybridization, else it is difficult to explain the increase in *montanus* characters westward in the foothills from Edmonton. The situation may be similar to that in the Skeena region (Kispiox Valley) where Oregon juncos segregate with their own kind, yet occasionally cross with *cismontanus*. In the Jasper region the sharp ecological differentiation of the plains and mountains, in addition to the probable sexual intolerance of the species, may aid in keeping the two types separate.

Some speculation concerning the nature of the intolerance may be in order, even though little critical evidence can be brought to bear. The similarity of females of the two species suggests that if plumage is important, it would be easier for females to distinguish males than for males to distinguish females. In captivity, males of different species seem to be responsive to any kind of female junco, but several females have shown no interest in males of different species; there may, however, have been other factors than preference and recognition of species that inhibited the females. Nevertheless, selection or recognition by females is possibly the controlling factor. A female *montanus* thrown in the company of *J. h. hyemalis* exclusively might be receptive, as indeed they sometimes appear to be in captivity, but if some *montanus* males were present they might be preferred. The same might hold for females of *hyemalis*, although it is possible that the females of one species are more discriminating than those of the other.

One hundred and sixty miles northwest of Jasper Park in the Peace River district in the Tupper Creek region, latitude 55° 30′, longitude 120°, a small sample taken by Cowan (1939, p. 59) is about equivalent to the birds of the piedmont area of the Jasper Park transection. One of 4 males is close to *montanus* and one of 4 females is identical with that race in back, side, and hood pattern. The other 6 correspond well with the Atlin series. Backs are *a* to *aaa*, and sides slate or predominately slate. It is not certainly known whether the *montanus*-like birds in this sector remained somewhat segregated from the slate-colored birds. Indications on the labels with respect to mated pairs suggest that they may.

One hundred miles west of Jasper Park in the Barkerville region *montanus* is the summer resident form. Yet here, within a series of 26 breeding male birds, there are 4 with some *hyemalis* character. These are: no. 448 McCabe collection, with slate sides, head 5, pileum black, neck line distinct and 3*a* back; no. 65708 Mus. Vert. Zoöl., with brown sides, head 3, pileum black, neck line distinct, and 2*aa* back; no. 186 McCabe, with slate sides, head 2, pileum black, back 3*aaa*; no. 65711 Mus. Vert. Zoöl., with sides mixed, but predominantly brown, head 6, pileum black, neck line distinct, and 3*b* back. Two of these birds, which I took, did not possess enough *hyemalis* character to be recognized as such in the field. Evidently these two arose from hybridization. Individuals of *cismontanus*, common in migration in this area, if they lagged behind to breed, would not find their own kind among the great numbers of *montanus* and consequently would hybridize, whereas such birds would not usually be "forced" to do this in the Jasper region, nor perhaps in the Skeena Valley.

Although mountains are rarely barriers to juncos, they may influence dispersal in the breeding season in the north. Spring migratory movements seem to be along the lowlands, and thence upward as the snow retreats to timber line. Thus, in the Jasper area *cismontanus* moves up the eastern slopes in spring and abruptly meets *montanus* that has come up through the lower parts of the western mountains. Along the Fraser Valley and in the Skeena area there is a northward migratory flow of both types, each going to its own breeding grounds. This movement along the same line may facilitate mixture.

South of the Jasper Park area, along the eastern slopes of the Rocky Mountains, the slate-colored types do not dominate. This might be expected because *J. h. hyemalis* is not found eastward of this region. Few specimens are extant, and it is probable that juncos are not widely distributed or abundant in the foothills about Calgary. One female from Didsbury is indistinguishable from *montanus*, but in this sex this means little. From Banff, just east of the divide, 10 males and 2 females have been examined. Only one male has mixed sides; 1 has slate sides; the others have *montanus* sides. Heads of males average 5.3 (4 to 7), somewhat paler than in northern *montanus* (4), but distinctly darker than in *cismontanus* of Jasper (6.3). All heads and backs are well contrasted. All backs but one are type *b* and phaeomelanin is for the most part type 3, with variants 1 and 2 present. Thus the group is chiefly *montanus*, nearly as much so as is the Barkerville group, and it is equivalent to the group found in and west of Yellowhead Pass.

West and south of Banff in the East and West Kootenay districts, scanty evidence indicates that the populations are similar to the one at Banff. They are essentially *montanus*, with occasional hybrids. Since there are no populations of *hyemalis* in the vicinity, these hybrids must be the result of interbreeding of stray *hyemalis* or *cismontanus* with *montanus*. As at Barkerville, far from populations of pure *hyemalis* or *cismontanus*, nothing seems to deter free intermixture. Male birds, entirely of *montanus* character, have been examined from Revelstoke, Summit, Glacier, and Field. A bird with partly gray sides and 3*aaa* back from Cranbrook, and one each with some gray in the sides (one with 3*aa* back) from Clearwater River and Rossland, B. C., have been seen. At the last two localities named, other birds are all *montanus*. Two birds from Creston near the Idaho line have partly gray sides, and a third has a 1*aa* back; 14 breeding birds from this area have been examined. South of the Canadian border in Montana and Idaho no conclusive evidence of mixture has come to my attention. One bird from the Clearwater River, Idaho (no. 61682 Mus. Vert. Zoöl.), has type *a* back, but the back is not spotted with distinctly outlined dark areas. It may be a variant unrelated to *hyemalis*.

Summary.—*Hyemalis* and *J. oreganus* hybridize when individuals of one type, for various reasons, occur among populations of the other. This results in the widely scattered occurrence of hybrids, for each species may wander far into the range of the other. When the two species both occur in numbers in a region, they tend to segregate and mate like with like, although even here some hybridization may take place. Through occasional hybridization (involving probably the race *montanus*), modified *hyemalis*-like populations (*J. h. cismontanus*) have been established adjacent to *J. oreganus* subsp. Certain *montanus* characters are spread fairly uniformly through it. The modified groups behave toward *J. oreganus montanus* as indicated above, that is, they do not interbreed freely at points of junction. The groups differ in uniformity. The group in the inner Stikine Valley is especially stable where it is partly isolated from *J. oreganus* and from pure populations of *J. h. hyemalis*. Quite probably these *cismontanus* groups have arisen at different times. In some, new intermixture of pure *oreganus* and *hyemalis* may still be taking place; others may have been established much earlier. With the retreat of the glaciers and the opening up of an intermontane area in the north, some modified *hyemalis* individuals that arose from interbreeding farther south may have moved in to take possession to the exclusion of the parental types. Being isolated, the characters of the original stock remained undiluted and their average occurrence about the same. The heterozygosity of the original stock would, through segregation, occasionally give rise to birds that resembled the parent types in a number of characters.

This probable course of events in the history of the Stikine and Atlin groups is what was predicated as necessary for the establishment of a race in the *caniceps–thurberi* zone of intermixture. In this zone the conditions were not quite right for isolation of a yellow-backed hybrid. In *cismontanus* a further step was made, and a race unit of appreciable extent and fair uniformity has resulted. In *dorsalis,* a race thought to be of hybrid origin, the uniformity and isolation are more nearly perfect. No longer does hybridization occur; the initial formative stage is passed, and the race may continue to evolve independently.

The striking thing about *cismontanus* is that in different areas one can see the genetic type produced in nature by hybridization and at the same time in other places see the same type (phenotypically at least) isolated to a fair degree and expanded into fairly uniform populations which compare favorably in all essential qualities with other races of juncos. Rare indeed are such opportunities of seeing the origin of a race and the "finished" product at the same time.

It is worth noting again that an *oreganus*-like group with some *hyemalis* characters mixed uniformly through it has not appeared. Several reasons may be suggested for this. *Hyemalis* may not have moved into the territory of *oreganus* so much as *oreganus* has moved into *hyemalis* territory. Or, hybrids may compete more successfully with *hyemalis* than with *oreganus*. On the other hand, *cismontanus* may be more exactly intermediate genetically than would appear. In the male the sides of *hyemalis* type show evidence of dominance. The partial intolerance of the species might rest on this feature, so that despite its hybrid nature *cismontanus* would on the average breed with *hyemalis* more than with *oreganus*. This would lead to accumulation of more *hyemalis* characters and make *cismontanus* tend even more toward *hyemalis*. The greater importance of specific recognition marks in the males is a premise in this line of reasoning. If this latter explanation is correct, it is not surprising that populations of *oreganus*-like intermediates (*transmontanus* of Dwight) do not exist.

As a practical matter it seems wise to continue Swarth's practice of according the Cassiar populations (*"connectens"*) definite nomenclatural status as a race. It seems that this group, though clearly of hybrid origin, has passed the formative stages and now in certain districts is a race. The naming of the large group of birds of this type would be difficult unless this procedure were followed. The reasons for the use of the name *cismontanus* in place of *connectens* are given in the appendix on nomenclature. Since genetically, and certainly phenotypically, birds of immediate hybrid origin are identical with those of relatively uniform breeding populations of *cismontanus*, all birds of this appearance must be designated by this name when on the wintering grounds or in migration. Generally I do not favor the practice of considering a bird apart from the breeding population of which it is a member. For example, it does not seem reasonable in the chickadees for intergrades of *P. atricapillus septentrionalis* and *P. a. occidentalis* to be considered *P. a. atricapillus* because they look like the latter, when they arise as intergrades in an area geographically remote from *P. a. atricapillus*. But the situation in the juncos is different. Here the hybrids (comparable to the intergrades) have seemingly at one time produced a race. The use of the name for all *cismontanus*-like hybrids as well as for the true race is advocated as a practical measure. This does not apply to sporadic isolated hybrids on the breeding grounds. The Jasper Park slate-colored juncos may for convenience be called *cismontanus*, even though by themselves they could be thought of as hybrids, or perhaps intergrades, rather than as part of a definite race.

The gradual mergence of *cismontanus* with *hyemalis* in the Edmonton and southern Yukon areas has been indicated. The intergradation does not consist of intermediate differentiation, but is a diffusion by interbreeding of characters of *cismontanus* (the *oreganus* elements) northeastward through continuous populations until such characters to all purposes disappear in pure *hyemalis*.

A summary of information useful to the taxonomist follows.

Junco hyemalis cismontanus

Racial characters.—*Cismontanus* is differentiated from *J. h. hyemalis* by extension posteriorly of dark area of pileum to a definite neck line in 80 to 95 per cent. Head averages darker, in males 6.1–6.4, instead of 7. Back rarely *a* or *b*, usually *aa* and *aaa* as in *J. h. hyemalis*, but about one-third to one-half have type 3 phaeomelanin even in advanced plumages. Females seldom entirely gray on sides. In both sexes sides frequently set off from hood, even though gray. No useful differences in size and tail markings exist.

Cismontanus differs from *J. o. montanus* in the presence of pure gray or slate sides in males or in gray or partly gray sides anteriorly in females (80 to 85 per cent). Head averages lighter; males 6.1–6.4 instead of 4.0. Contrast of head and back less pronounced, though distinct. Back darker in 95 per cent of males; *hyemalis* backs of retarded type (females) and *montanus* backs often identical. A male by reason of the presence of at least some gray in the sides together with other *hyemalis*-like features is never to be confused with pure *montanus*. Tail markings similar. Toes average 2 per cent longer.

Cismontanus differs from *J. o. oreganus* in the same way as from *montanus*. Additionally, there is a greater contrast in head color, and the back of *cismontanus*, which never is brighter than type 3, is never to be confused with types *oo* or *ooo* of *oreganus*. *Cismontanus* averages 5 per cent longer in wing and tail, 1 per cent shorter in tarsus and toes. Tail white on the average is much more extensive on rectrices 4 and 5.

Nongeographical variation (see also pp. 329–341).—Averages for the Stikine and Atlin groups correspond closely, except for those for bill depth, thus making possible, with the one exception, the combination of the groups for statistical treatment. Wing and tail are about 1 per cent longer than in *hyemalis* (Alaska and Mackenzie). The difference, though slight, is probably reliable statistically. Thus *cismontanus* surpasses slightly its presumed parents, northern *montanus* and *J. h. hyemalis*. Certainly there is no evidence from measurements that the short-winged *J. o. oreganus* was involved in the origin

TABLE 34
MEASUREMENTS OF J. H. CISMONTANUS
(Populations from Atlin and interior Stikine River regions)

Measurement	Sex	No. of specimens	Mean with prob. error	Standard deviation	Coeff. of var.
Wing............	♂	46	79.08 ± .18	1.88	2.37
	♀	40	74.33 ± .19	1.80	2.42
Tail.............	♂	44	70.09 ± .22	2.17	3.09
	♀	38	65.45 ± .20	1.82	2.78
Bill.............	♂	34	8.11 ± .03	0.29	3.67
	♀	32	8.10 ± .05	0.39	4.77
Tarsus...........	♂	51	20.43 ± .04	0.45	2.21
	♀	40	20.25 ± .06	0.53	2.59
Middle toe........	♂	51	11.74 ± .04	0.39	3.40
	♀	40	11.51 ± .04	0.37	3.20
Hind toe..........	♂	51	8.22 ± .03	0.34	4.20
	♀	40	8.05 ± .04	0.40	4.92
Bill depth (Atlin)...	♂	19	6.02	—	—
	♀	17	6.03	—	—
Bill depth (Stikine)	♂	17	5.78	—	—
	♀	13	5.70	—	—

of *cismontanus*. Bill length is similar to the parental types, which are nearly identical. Bill depth in the Atlin group averages about the same as in *hyemalis*, the averages for females being a little higher; the Stikine group averages about the same as in northern *montanus*. This curious difference between groups that otherwise appear to be essentially the same stock emphasizes the possibility of separate but comparable derivation in which one group retained the deep *hyemalis*-like bill and the other the shallow bill of northern *montanus*. Foot measurements agree closely with those of *hyemalis*.

Coefficients of variability compare closely with those of *hyemalis*. There is no increased variability. Again, as in populations of immediate hybrid origin, increased variation in size characters is not manifest.

Coefficients of correlation are: wing and tail, .762 ± .043; wing and tarsus, .375 ± .085; middle toe and hind toe, .360 ± .083. The last two of these correlations are somewhat doubtful because of the large probable errors. Sexual dimorphism of wing and tail is 6 per cent. It is absent in the bill. Slight dimorphism (1 to 2 per cent) occurs in the feet. The dimorphism of wing and tail is comparable to that in the parent races; that of middle toe and tarsus is like that in *hyemalis;* dimorphism in bill depth also is absent in *montanus*.

The difference in the hind toe may not have significance, but it is like that found in *montanus*.

Tail patterns are of exactly the same types as found in *J. h. hyemalis* and the average occurrence of types is essentially identical (see table, p. 317). There are no significant differences in *montanus*, *cismontanus*, and *hyemalis* pattern. The absence of any tendency in *cismontanus* to show reduction of white is important as further evidence that *J. o. oreganus*, with its reduced tail white, was not concerned in its origin.

Distribution in breeding season.—Occurs from extreme south-central Yukon Territory (Carcross) south through interior British Columbia east of the coast ranges into the interior Stikine River basin. The eastern limits in the northern Cassiar district are not known; probably the race extends eastward into the headwaters of the Dease and Liard rivers. South of the Stikine region it occurs east of the continental divide in the upper Peace River drainage south to Jasper Park. (For detailed occurrence of hybrids and of populations of *cismontanus* see foregoing discussions, pp. 333–341.)

Distribution in winter.—The impossibility of distinguishing hybrids of *hyemalis* and *oreganus* from the stabilized hybrid race *cismontanus* precludes accurate definition of the wintering range of the latter. The following account covers the distribution of "*cismontanus*-like" birds, hybrids included.

Winters most commonly in the southern Rocky Mountain region, together with *montanus* and *hyemalis*, and with fair regularity in California. In the north, winter-taken specimens of certain identity are known from southern British Columbia (Sumas); Minnesota (Minneapolis, Lanesboro); Wisconsin (Beaver Dam, Milton); Michigan (Ann Arbor); and eastern New York (Hastings). Southern records are from Lower California (Laguna Hanson), southern Arizona, and New Mexico (many localities), and Texas to Kendall and Kerr counties in the San Antonio district. There are records from Oklahoma, Arkansas (Van Buren, Delight, Winslow), Missouri (Shannon County), Illinois, and Indiana in a southeasterly direction. East of the Mississippi, *cismontanus* is rare and even in Kansas it constitutes but a small fraction of the wintering juncos. Records of occurrence on the eastern seaboard are for the most part doubtful, the birds probably representing variants of *hyemalis* (Massachusetts, New York, District of Columbia). A bird from Vanceville, North Carolina, seems indistinguishable from this race. There is some question, however, of whether it actually arose in the breeding area of *cismontanus*. Yet banding records reported by Lincoln (1933, p. 338) show that juncos (*J. h. hyemalis ?*) may move eastward from Minnesota to New Jersey, and from North Dakota to Pennsylvania. On the Pacific coast, *cismontanus* occurs rarely in western Oregon and Washington. In California, records for the coastal districts are chiefly from localities south of the northern humid coast belt.

Junco aikeni
White-winged Junco

This strongly marked member of the *hyemalis* artenkreis is geographically one of the most isolated species occurring on the continent north of Mexico.

Not only is the breeding range in the Black Hills district of the plains region of small extent, but the true wintering range is small and well defined. There are almost no instances of migration out of the plains and Rocky Mountain regions that are based on critically determined specimens.

Principal characters of the species.—Sides fairly pure light gray (deep to light mouse gray); the color may be obscured somewhat by drab and smoke gray tipping. Ventral hood margin concave posteriorly owing to confluency of gray of sides and breast. Lores and ocular region not distinctly darker than rest of head. Back area indistinctly differentiated from head and of same color when drab streaks and tipping are absent. In advanced plumages dorsum clear light gray, same as ventral surface of hood. Arrangement of pigments in feathers as in *J. hyemalis* (see p. 313).

Middle and greater wing coverts usually with some white. Tail with three outer tail feathers largely white, and white usually present on third rectrix. Dimensions distinctly large throughout. Wing and tail longer than in any other junco (some overlapping); bill tumid, about as large as in *alticola;* feet relatively small, only slightly larger than in *J. h. hyemalis* and *J. o. oreganus*. Weight of males (average of 19) 22.0 gr. Iris brown; bill and feet flesh color.

Differential characters.—*Aikeni* differs from *J. h. hyemalis* in its lighter slate, almost ashy dorsal coloration; only in decidedly retarded plumages might there be confusion of the two forms. *Aikeni* usually has white spots on middle and greater secondary coverts; when these are absent, margins of feathers light gray, unlike *J. h. hyemalis;* white spots occur in 3 per cent of *hyemalis*. White on third rectrix occurs in 94 per cent of *aikeni*, 2 per cent of *J. h. hyemalis*. Bill heavier, usually somewhat tumid basally, and culmen not so distinctly ridged. Wing averages 12 to 13 per cent longer; tail, 15 to 16 per cent longer; bill, 9 to 11 per cent longer; bill depth, 20 per cent greater; tarsus, 2 to 3 per cent longer; middle toe, 2 per cent longer.

There is essentially no overlapping in measurements of wing, tail and bill depth. The absolute minimum (1 out of 50) of *aikeni* may be equaled by the maximum (1 out of 100 or more) of *hyemalis*. Wing length of 82 mm. or greater, tail of 75 or greater, and bill depth of 6.8 or greater in males may be taken as reliable diagnostic measurements of *aikeni*. Comparable minima of females are: wing, 79; tail, 72; bill depth, 6.6.

In spite of the approach of *J. h. hyemalis* and *aikeni* in all the essential points of difference, the absence of any consequential overlap in a large number of characters precludes the possibility of confusion of the two forms when all of them are considered. I have yet to examine an equivocal specimen, so far as these two forms are concerned. Their status as distinct species seems entirely satisfactory.

Aikeni differs from *mearnsi*, its nearest neighbor in the breeding season, in size to nearly the same degree that it does from *hyemalis*. In coloration it differs in the same way that the entire artenkreis of slate-colored juncos differ (see p. 314) from *mearnsi* and the other Oregon juncos. There is no approach in side coloration (slate or gray, as against pinkish cinnamon) or in the color and demarcation of the back (slate or buffish slate as against brown, definitely delimited anteriorly from the head). *Mearnsi* does have white in the third rectrix in 17 per cent of males (94 in *aikeni*) and occasional white spots on the secondary coverts (1.82 per cent).

Nongeographical variation.—The amount of white present on the wing is exceedingly variable. The percentage of occurrence of pure white areas on different groups of feathers is:

Sex	No. of specimens examined	No white	White on middle coverts but not on greater coverts	White on middle and greater coverts	White on inner secondaries
♂	86	5	4	91	22
♀	134	22	8	70	2

No account is taken in the foregoing tabulation of the number of feathers in each series marked with white. On the inner secondaries the amount of white is highly variable. Usually it is confined to the tips and inner webs of the three innermost secondaries (so-called tertials), but it may extend to five or six of the secondaries and on some it occupies a terminal area that extends 3 or 4 millimeters along the vane. The pronounced sexual difference is similar to the condition in *hyemalis* (see p. 315). Association of larger amounts of white with the advanced plumages of the male does not support the suggestion that the wing bars of *aikeni* are a primitive character. The obviously immature birds of both sexes most often have inconspicuous white areas. The areas are, however, the same in position and outline as the buff wing spots of juveniles of all juncos. Juvenal *aikeni* may not have pure white spots, but they average whiter than in other species. The best view concerning this character in *aikeni* is, then, that a juvenal, presumably archaic, feature has been accentuated to some degree in this species.

Variations in the amount of suffusion of the light slate with buff are much less pronounced than in *hyemalis*. The gray of the sides is never fully replaced or obscured by buff. The relation of retarded plumage to sex and age is comparable to that in other slate-colored juncos, particularly to that in *J. h. carolinensis*. Variation of the gray or slate of the head ranges in males from types 8 to 10, chiefly 9, and in females from 10 to 11.

All variants involving white on the first and second rectrices are peculiar to the species *aikeni*. (There is one *mearnsi* with a white spot on the second rectrix.) Likewise, mixed white, predominantly white and pure white patterns on the third rectrix are peculiar to *aikeni*. Absence of pure black fourth rectrices, of black and mixed black fifth rectrices, and of impure white sixth rectrices are departures from *J. h. hyemalis* in accord with the general trend of increased whiteness. There is no other junco except *vulcani* so strongly set off from adjacent forms in amount of tail white. Sexual dimorphism as usual is present in all feathers that show appreciable variation in pattern (rectrices 3, 4, and 5). The sexual dimorphism is much greater than in *J. h. hyemalis*, in which, abnormally for the genus, it is almost lacking.

The coefficients of variability are relatively high, compared with those of *J.*

TABLE 35
OCCURRENCE OF RECTRIX PATTERN IN J. AIKENI AND J. H. HYEMALIS (IN PERCENTAGES)

	Males		Females	
	aikeni	*J. h. hyemalis*	*aikeni*	*J. h. hyemalis*
Rectrix 1 (innermost)				
Inner web:				
trace of white................	0	0	3	0
Rectrix 2				
Inner web:				
trace of white................	22	0	8	0
Outer web:				
trace of white................	6	0	3	0
Rectrix 3				
Inner web:				
pure black....................	6	98	8	100
trace of white................	29	2	64	0
mixed........................	57	0	22	0
trace of black................	6	0	6	0
pure white....................	2	0	0	0
Outer web:				
pure black....................	27	98	50	100
trace of white................	8	2	19	0
mixed........................	37	0	25	0
trace of black................	20	0	6	0
pure white....................	8	0	0	0
Rectrix 4				
Inner web:				
pure black....................	0	16	0	22
trace of white................	0	50	3	48
mixed........................	2	34	14	30
trace of black................	16	0	31	0
pure white....................	82	0	52	0
Outer web:				
pure black....................	0	54	0	48
trace of white................	0	10	2	26
mixed........................	25	20	50	13
trace of black................	12	12	17	9
pure white....................	63	4	31	4
Rectrix 5				
Inner web:				
mixed........................	0	14	0	39
trace of black................	2	18	0	13
pure white....................	98	68	100	48
Outer web:				
trace of white................	0	0	0	4
mixed........................	0	36	0	35
trace of black................	6	20	6	13
pure white....................	94	44	94	38
Rectrix 6				
Inner web:				
pure white....................	100	100	100	96
Outer web:				
pure white....................	100	100	100	96

hyemalis and *J. oreganus*. Wing length is as highly variable as in any other species of the genus (equivalent to *alticola*). The tail reflects none of this great variability. Bill length is slightly less variable and bill depth is much less variable than in *hyemalis*. The hind toe is more variable. In such a restricted form it is indeed interesting to find instances of high variability. The relative uniformity of bill is important in a form where this member is of such distinctive size.

Correlations are as follows: wing and tail, .758 ± .041; wing and bill length, .552 ± .066; wing and tarsus, .317 ± .088; tail and bill length, .411 ± .079;

TABLE 36
MEASUREMENTS OF JUNCO AIKENI

Measurement	Sex	No. of specimens	Mean with prob. error	Standard deviation	Coeff. of var.
Wing.............	♂	50	87.80 ± .27	2.81	3.19
	♀	38	83.08 ± .26	2.36	2.84
Tail..............	♂	50	79.60 ± .23	2.51	3.14
	♀	35	76.61 ± .28	2.51	3.27
Bill..............	♂	50	8.89 ± .03	0.31	3.51
	♀	35	8.76 ± .04	0.34	3.91
Bill depth.........	♂	49	7.17 ± .03	0.27	3.82
	♀	35	7.16 ± .03	0.28	3.97
Tarsus............	♂	50	21.13 ± .05	0.55	2.59
	♀	36	20.76 ± .07	0.66	3.21
Middle toe........	♂	47	11.89 ± .05	0.52	4.39
	♀	37	11.62 ± .06	0.55	4.69
Hind toe..........	♂	47	8.44 ± .05	0.52	6.15
	♀	36	8.11 ± .05	0.41	5.03

tail and tarsus, .280 ± .088; bill length and tarsus, .516 ± .072; bill length and bill depth, .281 ± .088; tarsus and middle toe, .415 ± .081. The correlations above .4 certainly are significant, as also probably all those here listed. Other correlations are below .25. A common correlation of wing, tail, bill, and tarsus suggests that there are some general factors for size in this giant species. In *hyemalis* only wing and tail were correlated and, separately, the parts of the foot. In *carolinensis*, which is somewhat larger than *hyemalis*, extensive correlations exist that are similar to those of *aikeni*, except that bill depth rather than bill length is concerned. The situations are therefore comparable but not identical. In *carolinensis* bill and bill depth are correlated at .364; in *aikeni* the correlation is .281; there is no correlation in *hyemalis*. As in other systems of correlated parts, it appears that certain common factors are present but that such factors are not equally important in determining the adult size in the several parts. Each part has its own independent set of controls. Certainly bill depth, so distinctive in *aikeni*, must have a set of factors not associated with other size characters. Possibly some of these affect bill length, or are linked with those that do. Middle toe and tarsus are correlated as usual, but hind toe definitely is not; perhaps significant in this connection is the fact that the hind toe is no larger than that in other northern juncos. Middle toe is not correlated with wing, tail, or bill, so that its association with the tarsus rests on different factors than the general size factors that have been hypothesized.

Significant sexual dimorphism is as follows: wing, 5 per cent; tail, 4; tarsus, 2; hind toe, 4. This situation differs from that in *hyemalis* in the lesser dimorphism in wing and tail and in the absence of significant differences in the bill. Difference in wing and tail in *aikeni* is 10 per cent in males, 9 per cent in females. Both figures are lower than in *hyemalis* and equal to those in the nonmigratory *carolinensis*.

Distribution in the breeding season.—Breeds in the Black Hills of South Dakota and in similarly pine-forested areas in southeastern Montana, northeastern Wyoming, and northwestern Nebraska. The regions lie chiefly within areas mapped by Cary (1917) as Canadian Zone, but some are Transition in character (Saunders, 1921, p. 13). Forests are relatively arid, with yellow pine (*Pinus ponderosa scopulorum*) dominant. Second-growth pine forests and pine forests mixed with scrubby growth of oaks may be inhabited. The forests lie in regions that receive from 15 to 25 inches of rainfall annually. This environment is comparable to that in the lower arid regions of the breeding ranges of *caniceps* and *mearnsi*. Altitudinal range is from below 4000 (Long Pine Hills, Montana) to 7200 feet. (See fig. 9.)

Montana.—Known to breed in the Long Pine Hills of Carter County (Visher, 1911, p. 14) and in western Powder River County near Otter. Birds taken in April at Miles City, Custer County (Saunders, 1921, p. 124), probably were not on their breeding grounds. Professor Joseph Kittredge (MS) reports seeing the species in May in the forests of Powder River County, south of Stacey. Probably occurs in essentially unmixed populations in various of the pine hills in Fallon, Powder River, Rosebud and eastern Big Horn counties.

Wyoming.—Breeds in the Bear Lodge Mountains (Warren Peaks), in the vicinity of Sundance, and in the Black Hills of eastern Crook and Weston counties; possibly also in extreme eastern Niobrara County.

South Dakota.—Black Hills of Lawrence, and western Meade, Pennington, and Custer counties. Probably occurs also in hills of Harding County.

Nebraska.—Breeds in northern Sioux County and probably also in pine areas at the headwaters of the White River in Dawes County.

The record of breeding in Colorado (Cary, 1909, p. 182) is not substantiated by specimens. It is extremely unlikely that *aikeni* nests in that state. I have examined a specimen taken June 26, 1917, at Alice, Clear Creek County, by R. S. Niedrach. There is no certainty that this bird was breeding; it was probably a retarded migrant. Tate's record of the breeding of *aikeni* in Cimarron County, Oklahoma, needs further confirmation. The region is one that does not appear to offer suitable breeding habitat, even though a spur of the Rocky Mountains extends eastward at this point, bringing with it scattered low-zone coniferous trees. If juncos should breed there, the probability is great that they would prove to be *caniceps* rather than *aikeni*, because the former occurs about 100 miles west of this point. The nearest proven breeding grounds of *aikeni* are 600 miles to the northward.

Distribution in winter (November 1 to March 15).—Remains in the vicinity of breeding areas as at the Long Pine Hills, Montana (Saunders, p. 124), Hat Creek and Newcastle, eastern Wyoming, and Crawford, Nebraska. Also migrates southward to winter along eastern slopes of the Rocky Mountains and the western part of the plains as far as Oklahoma (Fort Reno) and New Mexico (Las Vegas and vicinity of Santa Fe). Probably occurs in northern Texas. Occasional in Arizona in the vicinity of Flagstaff and in the White Mountains in 1937 (Hargrave, Phillips, and Jenks, 1937, p. 258).

The majority of known specimens are from the bases of the mountains in central and northern Colorado. Birds have been examined from Wallace and Morton counties, Kansas. There is little reason to doubt that the record (Cooke) from Ellis, Ellis County, pertains to this species, but supposed *aikeni* from eastern Kansas that I have examined have proved to be *J. h. hyemalis*. Cooke's record from Caddo, Bryan County, southern Oklahoma, is based on a skin; he presents additional sight records. The specimen has not been available to me, but his statement that two of the tail feathers are all white, and a third partly so, leads me to suspect that the bird was not *aikeni*. His record for Wisconsin is likewise to be doubted, for he states that the specimen was exactly like the one from Caddo. It is entirely possible that *aikeni* ranges east of the vicinity of Oklahoma City, for there are several sight records east as far as Tulsa, but, as in eastern Kansas, it would seem best that specimens critically compared with *aikeni* with regard to all characters be at hand before definite conclusions are drawn.

Relationships with other species. J. o. mearnsi.—Evidence of hybridization of these species was first found in a single specimen from the University of Colorado collection at Boulder. This bird, a male, no. 99 of Dille, the collector, was taken at Altona, Boulder County, February 6, 1902. Of a number of supposed hybrids taken in winter, this is the only one that is not merely a *mearnsi* with the white wing spots which may be considered a "normal" variation of that species. Dille's specimen is certainly a hybrid. The head is about type 9, which is an intensity of color common to the two species. The coloration of the back is *mearnsi* but the back is not very sharply set off from the head. The hood line below is concave as in *aikeni*. The sides are mixed bright pinkish cinnamon (*mearnsi*) and light gray (*aikeni*). The manner of mixing is comparable to that found in *J. o. montanus–J. h. cismontanus* hybrids or in *J. c. caniceps* × *J. o. mearnsi*. There is no trace of white on the wing coverts, but gray edgings are prominent. Wing and tail measurements fall between the averages of the two forms, at a point about equal to the minimum of one species and the maximum of the other. The bill is large, but it is closer to that of *mearnsi* than that of *aikeni;* it is not so distinctly tumid as in *aikeni*. White on the tail extends to the third rectrix as it does occasionally in *mearnsi* and nearly always in *aikeni*. The amount of black on rectrices 4 and 5 is greater than in *aikeni*.

The only alternative to considering this bird a hybrid of *mearnsi* and *aikeni* would be to look upon it as a hybrid of *mearnsi* and *hyemalis*. But all the *hyemalis* elements in it could equally well have come from *aikeni*. Besides, the wing and tail length could never have been derived from *hyemalis*, and there is little chance that the light gray in the sides would have come from that source.

Subsequent to the study of this bird a field party from the Museum of Vertebrate Zoölogy visited a forest area in western Powder River County, Montana, which seemed the most likely point of origin of such a winter-taken hybrid. At 4 miles west of Fort Howe Ranger Station a breeding sample of 30 adult juncos, with 9 juveniles, was taken. The population was dominantly *aikeni*. Only 3 birds showed evidence of *mearnsi* blood, but this evidence was conclusive.

Apart from these 3, the *aikeni* population seemed essentially typical. Average measurements except for bill length were a little lower than in the main sample of winter-taken *aikeni* from Colorado that form the basis of the tables presented for this species. Most of the differences are not reliable statistically and even if valid may well represent a faint colonial differentiation independent of *mearnsi* influence. This seems all the more likely because the amount of tail white in the sample is prevailingly a little greater than in other *aikeni*, a trend opposite from that which would result from *mearnsi* mixture. As regards white on the wing, a larger percentage ($\male$ 16, $\female$ 33) than normal were without it, but also a larger percentage ($\male$ 50, $\female$ 22) had extreme development of white that extended to the inner secondaries. The color of the hood is apparently normal for breeding *aikeni*, although it might prove to be a little darker than normal if a large breeding series from the Black Hills were compared.

The 3 hybrids from this population, 2 males and a female, were all in full breeding condition. The nature of the mates of none of them is known. One male (no. 7328 W. C. Russell) is in most respects *mearnsi*. The head is type 8. The back is brown as in *mearnsi* on many of the distal parts of the feather vanes, but unlike the coloring in *mearnsi* there is much of each feather that is slaty. The mass effect is one of intermediacy. I have seen no *mearnsi* in which the back is anything but pure brown. The hood line is convex as in the *oreganus* group and the sides are pure pinkish cinnamon, although the area of color is narrower than normal for *mearnsi*. Wing is near maximum for *mearnsi* and minimum for *aikeni*; tail is below average for *mearnsi*; bill length is almost average for *aikeni*, but bill depth is minimum for *aikeni* and maximum for *mearnsi*. White in the tail extends inward to the third rectrix, but the black on rectrix 4 is more than in males of *aikeni*. There is no white in the wings.

The second hybrid male (no. 1025 F. H. Test) is chiefly *aikeni* in character. The head is type 9 and the back without trace of brown. The hood line is, however, convex as in *mearnsi*. The sides are not pinkish cinnamon, but are smoke gray and pale slate so that they contrast with the hood. Wing and tail are minimum and average for *aikeni*, respectively; bill measurements are above average for *aikeni*, approaching maximum. Tail white is normal for *aikeni*. There is no white in the wing.

The female (no. 7325 W. C. Russell) is decidedly mixed in character. The head is type 9 and the back is without trace of brown. The hood is convex and the sides are mixed light slate and bright pinkish cinnamon, the former predominating. The wing has white spots on the middle coverts, larger than in the variants that occur normally in *mearnsi*. Wing and tail are above average for *aikeni*; bill measurements are near maximum for *aikeni*. The white of the tail extends inward to the second rectrix, a coloration detail that is never found in female *mearnsi*.

These 3 hybrids from the breeding range of *aikeni* prove that occasional *mearnsi* wander eastward from their breeding range in the Big Horn Mountains of Montana and Wyoming to breed with *aikeni*. The rarity of such hybrids in the large wintering flocks of Colorado would indicate that hybridization of the two is rare indeed. *Mearnsi* and *caniceps* hybrids are encountered with fair regularity, whereas their breeding ranges are but little more closely connected than those of *mearnsi* and *aikeni*. Between the Fort Howe area and the Big Horn Mountains there are scattered areas of yellow pine forest. The greatest interval between junco habitats is probably no more than 20 miles and probably much less, and the total interval is about 70 miles. In view of this fact it is remarkable that more hybridization does not occur, although it is granted that populations of more mixed character than now known may be found in Rosebud and eastern Big Horn counties. Here are two forms, each representative of an extreme modification of a distinct rassenkreis, that remain peculiarly apart, considering their migratory habit, when in close proximity geographically in the breeding season. They must have had a long history of divergent evolution. Through some factors of habitat preference and specific intolerance, or both, they remain essentially separate, as do species rather than races. This is all the more remarkable in view of their physiological capability of hybridization. The ability to hybridize becomes particularly dubious as a sole criterion of specific distinction when situations like this are encountered.

The breeding ranges of *aikeni* and *hyemalis* are separated by at least 500 miles of uninhabited territory. One may suspect that during glacial periods *hyemalis* was forced southward in the breeding season, possibly once to the vicinity of South Dakota. The Black Hills remained free of glaciation in the Pleistocene. *Aikeni* may then have been a local southern race of *hyemalis*, as *carolinensis* is today. Many of its characters correspond to modifications in other forms in arid regions. Its large size and pale color parallel, and even surpass, *carolinensis*. Subsequent isolation with the retreat of northern populations could have aided in the preservation and propagation of new variants within it.

Alternative views concerning origin cannot be overlooked. *Aikeni* might represent an archaic remnant of the original *hyemalis* stock. But the large amount of white in the tail is far from primitive, viewing the genus as a whole. This and the accentuated juvenal wing bars appear to correspond with the prevailing arid environment. An ancestral junco might be expected to be of a size more like that of the average of the genus, not a giant form. *Aikeni* may have been pre-Pleistocene in origin, dating back to the arid period in the plains in the Pliocene. In any event, it resembles a geographic race, although it is now fully dissociated from its one-time racial relatives as a species. It is not clearly a primitive form in the sense that *vulcani* may be so considered. It may well have diverged contemporaneously with *hyemalis* from a common stock, each retaining unmodified certain of the original characters of that stock.

DISCUSSION AND SUMMARY

Intraracial Correlations

The most striking series of correlations has to do with wing and tail length. In no race is correlation of these two absent, but it may differ greatly in value (see graph, fig. 32). Two explanations for the association of wing and tail length have been offered. The first is that the same series of factors determines the growth of the similar major feathers of wing and tail. These determiners might be effective in controlling size of body as a whole and might control wing and tail length only incidentally. Therefore, the net correlation of wing and tail additional to the correlation of each with body size should be known. This relation is difficult to determine, because there is not much information that can be obtained from the material at hand to indicate body size. Lengths of skins mean nothing. Weights are the most useful measure and are more constant than is usually supposed (see p. 254). In one race (*montanus*), in which an adequate number of weights is available, the net correlation of wing and tail is .623, whereas the gross correlation is .646; the coefficient for wing and tarsus is .237. Tarsus is not correlated with wing and tail to a significant degree, except in a few races. We can conclude, then, that wing and tail correlations are in large measure independent of body size.

The second explanation of correlation rests on the linkage of factors that might independently affect wing and tail. It is assumed from study of naturally occurring hybrids that wing and tail lengths are governed by multiple factors (see p. 195); this is in no sense to be considered as proved. Multiple factors, that could be distributed in a number of chromosomes, would have little chance of being so linked as to effect a fairly constant and close association of a given wing and tail length. Especially is it unlikely that such a linkage system would hold in a number of distantly related forms where new factors must come into the complex (perhaps, however, only as multiple allelomorphs) to produce different wing and tail lengths characteristic of these forms. There is little likelihood, therefore, that genetic linkage is the cause of correlation.

The variation in wing and tail that is not correlated, although partly somatic, must to some degree be determined by factors that affect these parts independently. If such specific factors did not exist, it would be difficult to explain the difference in wing and tail ratio in two closely related races, each of which possesses an equivalent coefficient of correlation of wing and tail. Further evidence along the same line is found in the fact that some races show an increased coefficient of variability in the tail while there is a decreased variation in the wing, or vice versa (see graph, fig. 32). In general, where there is a close correspondence in the coefficients of variability of wing and tail, these parts tend to be more highly correlated, and where variability is distinctly different, the correlation drops. Thus, in the latter situation it is supposed that some new factors have appeared to control independently one or the other member. Associated with this, there is often a difference in absolute value of wing or tail length as compared to that of the most nearly related races. The consistently higher variation of the tail is most probably due to somatic variation, although there may be more independent factors concerned with this member.

Consideration of the foregoing factors indicates a mechanism of variation which shapes the result of selection pressure and mutation pressure and which doubtless has been important in the origin of the characteristic average dimensions of the races as they exist today. In summary, this mechanism would seem to consist of (1) a set of multiple factors which to a high degree determines the lengths of both remiges and rectrices, (2) factors for general body size, which have only a very slight control over length of the feathers, (3) some factors which may control wing or tail independently, or at least disturb the balance of the two, and which effect alterations in relative length, and (4) somatic variation. Theories concerning the trends in evolution of wing and tail length and the method of such evolution should take into account this fundamental system of control.

No other intraracial correlations of large magnitude run through all forms of juncos with a constancy approaching that of the wing and tail. Bill and bill depth in a few scattered instances are correlated. These two dimensions depend on somewhat independent growth processes. Both would of course reflect a general massiveness of bill, but the length depends partly on growth of the horn structure terminally, whereas the depth depends on the size of the bones and to some degree the massiveness of the whole head. The only juncos that display significant correlations are the relatively large-billed forms, *vulcani, carolinensis, aikeni,* and *insularis,* with coefficients of .599 ± .192, .364 ± .067, .281 ± .088, and .261 ± .066, respectively. Other distinctly large-billed types, *alticola* and *dorsalis,* do not show this. In each of the forms showing correlation the bill is larger, to varying degrees, than in the forms which obviously are most nearly related. One is tempted to suppose that the new factors responsible for larger bill are those that act on both bill length and depth. Sometimes they may be general size factors, for there is some correlation of a great many parts in these forms, excepting *insularis*. But because weights are not available, this cannot be proved. In *insularis,* the correlation clearly is confined to the bill, and the species obviously is a small form. The factors responsible for its strikingly great bill size would therefore seem to be partly factors that affect this member exclusively and cause some measure of correlation in the two bill dimensions.

Bill length is correlated with both wing and tail in *vulcani* and *aikeni,* and slightly so in *carolinensis*. Bill depth is correlated with wing in *carolinensis* and *vulcani,* but not in *aikeni*. Despite this latter exception, this evidence supports the idea of common size factors in these forms. In the bill low correlations may be obscured by variations due to wear and to displacement of the mandibles in the course of preparation of the skins. These factors are partly responsible for the prevalently high coefficients of variability.

In a different category are the correlations of wing and bill depth in *alticola* (.434 ± .08), *bairdi* (.240 ± .08), and *mearnsi* (.427 ± .07). In these forms there is no widespread correlation of parts and none of significance of bill and bill depth or of bill length and wing. *Bairdi* and *mearnsi,* at least, are not importantly different from closely related forms in absolute value of bill depth. These correlations suggest to me linkage, which of course must mean either that there are different factors involved with different loci, compared with related forms, or that some translocation of parts of chromosomes has occurred to establish close linkage of the factors.

To summarize, bill length and depth are not often correlated, and if they are, the coefficients usually are not high. In a few instances general size factors would seem to play a part in bringing about correlation. In one race common factors for bill and bill depth are suggested. Otherwise the determination of the two bill dimensions is more or less independent, but bill depth occasionally shows strong association with wing length (probably wing and tail), an association which suggests a linkage phenomenon. Judging from what now exists, variations and associations of several kinds may appear with respect to the bill, thus leaving the way open for evolution in bill proportion of any sort and degree, and for change in general size of bill with or without corresponding modification in other parts.

Tarsal and toe length would seem to be determined by a number of factors. The usual picture of quantitative inheritance is shown in such natural hybridizations as have been studied. The correlation of foot parts seldom reaches .5. Usually it is .2 or .3, or lower, so that its significance is doubtful in small samples. Correlation of foot parts is erratic in the genus as a whole. Some forms show correlation of tarsus and middle toe only, others of tarsus and hind toe only. The correlation is always slightly positive, but it may be no larger than the probable error. In those forms which have general systems of correlation it is not always present. It seems to be quite absent in *vulcani,* for example. Many of the relatively large-footed forms, such as *alticola, insularis,* and *J. o. oreganus,* show much correlation, but others do not, notably *vulcani.* The indices for variability in toe length run high. More than with the wing, errors in measurement enter in that partly account for the high variability and that tend to obscure correlations of low or moderate magnitude.

Probably there are some general size factors that affect the feet to a small degree. Additionally there probably are some races with linked factors, or common factors, operative only on the foot, that have some effect. But there is almost always much independent variation of the parts concerned, which, if it is in any great measure hereditary, must be controlled in each instance by separate factors.

Correlations of characteristics other than those of dimensions have not been investigated extensively. Often the difficulty of classifying color in the available material has hindered statistical treatment. Amount of white on separate rectrices (4 and 5) was found to be well correlated ($.620 \pm .06$) in *thurberi.* Such an association probably is generally prevalent in the genus. Apparently, extent of the wing bars and tail white are correlated in *aikeni,* although this could not be determined quantitatively. The amount of white in the tail is not correlated with coloration in other parts of the plumage; that is, lighter-headed birds do not have more white in the tail, even though such characters may show parallel evolution in a series of races.

In *montanus*, it was found that eumelanin of the head and of the back were correlated (.427 ± .038). Most probably there is some interaction of the factors for this pigment in the two areas, so that when a critical degree of dark-headedness occurs the back is influenced, and contrarily, when the head is pale to an extreme degree no great intensity of pigment in the back is possible. In *J. o. oreganus* a comparable correlation is lacking, but there is some association of dark head with intensity of phaeomelanin in the back. The color of the sides is not correlated with coloration of back or head in Oregon juncos, nor to any degree in *J. p. phaeonotus*. There seems to be no prevalent interrelation of different color areas. Parts of the body that are essentially monochromatic, like the head and sides in the *hyemalis* artenkreis and in *J. caniceps*, show correlation.

INTERRACIAL CORRELATION

It has been found in studies of geographic variation, for example by Sumner (1932, pp. 52–53), that characters that show correlative differentiation in a series of geographic forms may not be correlated among individuals of a given race. Presumably, selective factors have then been operative which have shaped the parallel modification of two features which are unrelated within the organism. Or else (Bubnoff, 1919, p. 166; Sumner, *loc. cit.*), the racial differences are genetic whereas the intraracial variation is nongenetic. Conversely, there are individually correlated characters which do not differentiate in races in any associated way. However, if two factors are associated because of common factorial control or related physiology of development, they might be expected in most instances to show some degree of parallel racial modification.

In the preceding section on intraracial correlation, it was brought out that the best example of prevalent association of characters is that of wing and tail length. The graph (fig. 29) demonstrates the interracial and interspecific correlation of these two. The parallelism is remarkable throughout, irrespective of whether the forms are weakly differentiated races or strikingly different species (*vulcani*). An increase in tail relative to wing is to be noted, especially in the yellow-eyed group of nonmigratory forms that ranges from Arizona southward. This is also seen in the southern members of the *hyemalis* and *oreganus* artenkreise. However, these independent alterations are insignificant in magnitude as compared with the correlated modification. The common genetic control of wing and tail must have had much to do with the correlative differentiation of the two in the genus.

Average weights for a few races (fig. 29) throw some light on general correlation of size of body parts. The wing and tail are but poorly correlated with weight, just as is true in intraracial variation. Complete reversal of the expected association of weight and length of wing and tail is seen in comparing *J. o. oreganus* and *J. o. montanus*. The small *bairdi* and *pinosus* are, however, short winged, and the giant *aikeni* is of course correspondingly long winged. Lack of correlation is very definite in *townsendi* and *montanus* and in *caniceps* and *palliatus*.

Tarsus correlates but slightly with wing and tail. Extreme differentiation in wing length is usually reflected in the tarsus, but not in equivalent degree (see the southernmost forms, fig. 29). A negative correlation is seen in the series of forms beginning with *dorsalis* and running through *J. p. phaeonotus*,

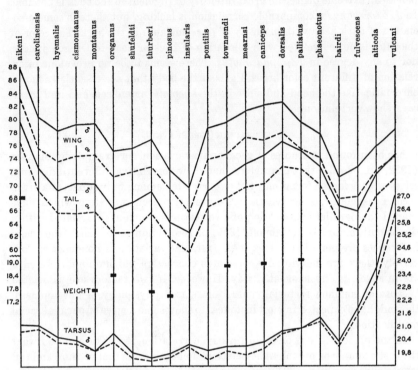

Fig. 29. Comparison of average wing, tail, and tarsal lengths in the 21 forms of the genus *Junco*. Measurements in millimeters: numbers at left for wing and tail; at right, for tarsus. Average weights in grams (numbers at lower left) for breeding males of certain races are shown by spots.

and in *J. o. oreganus*, and, slightly, in *insularis*. Weights correlate with tarsus, and might, with more data, show a moderately strong correspondence throughout.

The middle and hind toes and the tarsus show a fair degree of correspondence throughout the genus (fig. 30). Only in magnitude of differentiation are there striking differences, as in *insularis*. Minor discrepancies in correlation are seen in comparing *aikeni* and *carolinensis*, and *palliatus* and *J. p. phaeonotus*. Modification must always preserve a functional balance, and hence radical changes in one part of the foot alone are not likely to appear.

Bill depth and length show some measure of correspondence, but there is a marked negative correlation of differentiation in the three southernmost forms (fig. 31). There is no prevalent correspondence of the bill with wing, tail, and foot. *Aikeni*, *alticola*, and *vulcani* are in a sense giant forms and accordingly have large bills, but this is most certainly not true of *insularis*. The bill, because of a usual lack of common factorial control of its dimensions and because there is no necessity for correspondence with body size for the sake of efficient action, becomes modified independently of other parts. Preservation of a certain balance of proportions within the bill may be expected even

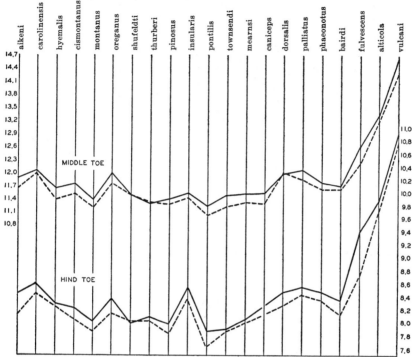

Fig. 30. Comparison of average middle-toe and hind-toe lengths in juncos. Measurements in millimeters: numbers at left for middle toe; at right for hind toe.

without developmental correlation, but even this can be, and is in some measure, broken down.

In tracing correlation of colors, forms should be considered according to geographic groups. Each rassenkreis or artenkreis may show its own system of interracial correlation, but between major groups there is so much difference in color because of long separate evolutionary history that correlation may be obscured. Color areas, when correlated, seem to correspond to some environmental factor that would appear to have a molding influence (probably indirect).

To exemplify, *J. hyemalis carolinensis* has a blacker head than *J. phaeonotus phaeonotus*, but it has a much whiter tail. But within the *phaeonotus* rassenkreis and within the *hyemalis* rassenkreis the correlation is quite clear. We might say that there is interracial correlation but not interspecific correlation. In measurements, both interspecific and interracial correlation occur, and they fit into one system.

It is difficult to represent the whiteness of the tail in simple fashion since the modifications of the pattern are complex. However, I have attempted to show in table 37 the gross differences in amounts of white.

It should be understood that tail white in *dorsalis* and *pinosus*, although roughly similar in amount, differs importantly in pattern. In *J. caniceps* and in the yellow-eyed groups the white is more widely distributed on several

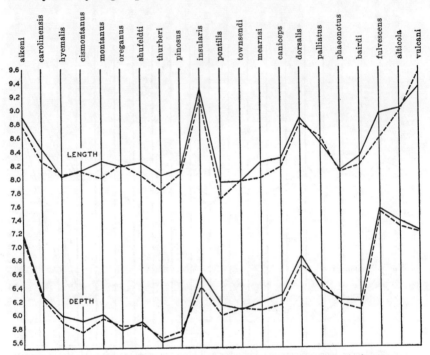

Fig. 31. Comparison of average bill length and depth, in millimeters, in juncos.

feathers (a primitive feature?), and less concentrated in large areas on a few rectrices. *Bairdi* and *insularis*, compared to each other and to their nearest relatives, have peculiarities in pattern. Using the seriation in the table, comparison with relative head color reveals the following associations. In the *hyemalis* artenkreis, head color and tail white correlate, *aikeni* being lightest and *hyemalis* and *cismontanus* darkest. *Cismontanus* possesses a distinctly darker head than *J. h. hyemalis*, yet its tail is no less white. Here another factor enters in, namely, the fairly recent derivation of the dark head of *cismontanus* by hybridization with *J. oreganus*. *J. h. hyemalis* and *J. o. montanus*, the presumptive parents, have the same amount of tail white. Further, there is no reasonable association of head color of *cismontanus* with conditions of aridity or moisture, as there is in *aikeni* and *carolinensis*.

In the *oreganus* artenkreis the correlation of head and tail color is uncertain. Related races can be grouped which show correlation, but other groupings, quite as natural, reverse the correlation completely. Thus, *mearnsi, montanus, shufeldti*, and *J. o. oreganus* show a positive correlation which extends even to intraracial geographic differentiation. Also, in the *thurberi–pontilis–townsendi* group there is correlation. But the *shufeldti–thurberi* series is entirely out of keeping with this plan, as are also *pinosus* and *insularis;* it cannot be asserted that *shufeldti* and *pinosus* are unrelated to *thurberi*. They form an essentially continuous series of variant populations in which there is a nega-

TABLE 37
TAIL WHITE

Stages of increase in white	1 *vulcani*	2 yellow-eyed, plain-backed species	3 *caniceps* artenkreis	4 *oreganus* artenkreis	5 *hyemalis* artenkreis
1	*vulcani*				
2					
3		*alticola*			
4		*fulvescens*			
5		*J. p. phaeonotus*			
6		*palliatus, bairdi*			
7				*J. o. oreganus insularis*	
8			*dorsalis*	*pinosus*	
9			*caniceps*		
10				*montanus, shufeldti*	*J. h. hyemalis, cismontanus*
11				*thurberi*	*carolinensis*
12				*mearnsi*	
13				*pontilis townsendi*	
14					*aikeni*

tive correlation of the features in question. If the forms are grouped into primitive or light-headed (*insularis, townsendi, pontilis, pinosus,* and *mearnsi*) and advanced (*J. o. oreganus, montanus, shufeldti,* and *thurberi*) groups, the correlation within each is far from perfect.

In the *caniceps* rassenkreis, correlation of head color with tail white is negative. *Dorsalis,* however, is probably of hybrid origin, which would explain its relatively light-colored head and dark tail. Also, it is thought that the Nevada division of *caniceps* owes its slightly darker head and whiter tail to a small amount of hybridization with *J. o. thurberi*. In the yellow-eyed juncos there is a fair general correlation of head and tail color, except for *vulcani,* and except for some discrepancy when the *phaeonotus* rassenkreis and the *alticola* artenkreis are compared. In general, in this division of the genus, the differentiation of head is slight compared with that of the tail.

In those forms in which there is a eumelanin component in the back, the correlation of this with head color requires consideration. In *J. o. montanus* it was found that among individuals these areas were correlated, although in other races this was not necessarily true. To some degree they are associated interracially in *montanus* and *mearnsi*, but the back of *thurberi* has much less eumelanin than that of *oreganus*, although the two races are equally black headed. In *oreganus* the great amount of phaeomelanin masks the eumelanin or replaces it, yet the saturation of the eumelanin in the back of this blackheaded race is certainly high as compared with the variants of *shufeldti* and *montanus* that have comparable extension of phaeomelanin. Among the southern Oregon juncos a general but imperfect correlation of dark pigment in back and head prevails. The chief exception to correlation occurs in *thurberi* because of its especially black head.

In the *hyemalis* artenkreis, head and back color, which are similar in tone, are closely correlated interracially. Correlation is also evident in the *alticola* artenkreis, where some dusky melanin enters into the back complex.

It follows from the foregoing discussion that eumelanin of the back and tail white also must show some correlation in certain closely related races. This is distinctly not true of *insularis* and *pinosus* in comparison with their relatives.

In groups of forms where the side coloration is grayish or has a eumelanistic component and is *confluent* with the hood area, there is interracial correlation of the color of these areas. This is manifest in the *hyemalis, caniceps,* and *phaeonotus* groups. In the Oregon juncos, the sides and head are separately controlled and the former is almost lacking in eumelanin; consequently no correlation is found. Presence and intensity of phaeomelanin in the sides are not correlated with the modifications in eumelanin. Phaeomelanins of the back and sides are not correlated.

In summary, correlated modifications in different color areas involve only the dark pigments (eumelanin), and are often imperfect even within groups of closely related races. In general, a light-colored form will to a varying degree show correlated differentiation in independently controlled parts of the plumage.

Environmental Correlation

In matters of environmental correlation of color, the practice of treating each rassenkreis or artenkreis separately must again be followed. Intensity of pigmentation, especially of the head or hood, and the partly correlated (interracially) pattern of the tail, might by analogy with situations in other higher vertebrates (Gloger's Law) be expected to show association with rainfall and humidity. Through the northern members of the *oreganus* group this is clearly true, at least in a rough way. Just as Sumner has pointed out in *Peromyscus,* environmentally correlated characters in *Junco* by no means coincide in their occurrence with the precise geographic area over which the particular environmental conditions prevail, probably because of the effect of other factors such as barriers and population pressures. *Mearnsi, montanus, shufeldti,* and *J. o. oreganus* show the expected differences in intensity of pigment in correspondence with the environments which each encounters. In *mearnsi,* the degree of difference is unaccountably great. But southward along the Pacific coast in California the climatic gradients and racial differentiation do not correspond. Coastal *thurberi* of northern California is lighter headed than the black-headed interior bird of the more arid Sierran system. *Pinosus* of the coast is lighter headed than *thurberi.* The correlation of tail white is better than that of the head. There is no reversal of the general law, except that *pinosus* in the southern coastal district has less white than coastal *thurberi* and *shufeldti* in the more humid coastal regions to the north. In the *thurberi–pontilis–townsendi* series, correlation of head color with climate prevails, but it is surprising that a great deal of difference should exist in the birds of mountain ranges that experience relatively similar climates. The suggested archaic group of pale-headed Oregon juncos should not be forgotten in this connection. In the group comprising *thurberi, pinosus, pontilis, townsendi,* and *insularis* the correlation of tail white and climate is positive.

In the *hyemalis* artenkreis there is correlation of pale head and white tail. However, within the large range of *hyemalis,* which is extremely variable climatically, no local differentiation can be detected. If climatic correlation comes about through selection, the selection cannot be very rigid and it must be so slow that intermixture within this vast race keeps the whole at a uniform average level (see Dobzhansky, 1937, pp. 186–188).

In *Junco caniceps caniceps* the local differentiation in the Nevadan area is thought to be the result of occasional hybridization and there is, therefore, the anomalous occurrence of slightly darker-headed populations in the most arid sections of the Rocky Mountain–Great Basin area. The whiter tail of the Nevada birds might be thought of as associated with climate, yet it probably results from intermixture with *thurberi.* The summer habitat of *dorsalis,* to judge from the flora, is possibly more moist than that of *caniceps* as a whole. The head of *dorsalis* is of course lighter and the tail darker, both of which features may be attributed to the hybrid origin of the race.

In the yellow-eyed juncos both head and tail color correlate well with the degree of moisture of the habitat (see discussions under each form).

With respect to back, obviously there is environmental correlation where head and eumelanin of the back are closely associated as in the *hyemalis* group. The changes in the back in *oreganus* really comprise two phases in the seeming response to the climatic gradient of increasing moisture and humidity: (1) replacement of eumelanin by phaeomelanin, and (2) intensification of such eumelanin as persists. The back color increases from dull brown toward richer and warmer browns progressively in the following series: *mearnsi, montanus* (south), *montanus* (north), *shufeldti,* and *oreganus.* Again, there is no correlation in passing from this series to the southern *thurberi–pinosus–pontilis–townsendi–insularis* group. But within the latter there is an association with humidity. *Thurberi* and *pinosus* are ruddier than their northern counterpart, *shufeldti,* but they do have less eumelanin. Thus one phase of the correlation (eumelanin) with humidity is carried out, whereas the other is reversed at this one point. Coastal *thurberi* may owe its slightly duller coloration to extensive infiltration of *shufeldti* blood. Whatever the cause, there is a reversal of the expected trend of differentiation of coastal and interior groups in this race.

The occurrence of negative correlations in the transition from the northern to the southern group of races of Oregon juncos suggests that we are dealing with two units within each of which the association of color and climate has arisen somewhat independently. There may be some significance in the fact that the northern group is chiefly migratory, the southern more or less sedentary, although I cannot suggest a causal relation. It has seemed probable that the type of intergradation that exists between *thurberi* and *shufeldti* has not come about merely by the junction of two diverse stocks, but instead is one that results from the independent and noncoördinated change in various characters. The causative factors in the trends of differentiation in this region remain obscure. That there is no reversal of correlation in tail white and eumelanin of the back at this point should not be lost sight of.

Again it is well to stress that environmental correlation within a race is often strikingly absent. An extreme example of this is found in *pinosus*: some characters of the race appear to correspond to conditions in the humid coastal belt of central California, where the center of abundance and perhaps the center of origin of the race lie, but the scattered populations that occur in the interior arid coast ranges encounter extremes of aridity and low zonal conditions unequaled elsewhere in California in the range of any junco. Yet these birds of the interior are essentially as ruddy as the coastal populations.

Back color in the *caniceps* and yellow-eyed groups is chiefly phaeomelanin; it is of a type different from that of the Oregon juncos. It is not subject to much fluctuation, but it is slightly brighter in tone in forms that occupy especially arid regions (*bairdi* and *palliatus*).

The eumelanin of the sides, which is associated interracially with head color in all but the Oregon juncos, corresponds with climate in about the same fashion as does the head, and shows the same imperfections. The essentially phaeomelanistic sides of Oregon juncos are more ruddy in hue and are more extensive in the presumed archaic peripheral group (*mearnsi, townsendi, pontilis, insularis,* and *pinosus*). Within this group no correlation with climate is apparent. In the other Oregon juncos (*montanus, shufeldti, oreganus,* and *thurberi*), intensity, extent, and ruddiness (versus yellow) of the sides increase with increased humidity.

Among the yellow-eyed juncos, the most brilliantly colored sides occur in *bairdi* of the arid Cape district of Lower California and in *fulvescens* of the comparatively (see p. 228) arid interior of Chiapas.

In general, Gloger's Law of increased phaeomelanin and reduced eumelanin in dry hot climates finds unsatisfactory confirmation in this genus. With respect to the back area, although the eumelanin is reduced in arid regions, the phaeomelanin is also diminished; phaeomelanin is increased in some humid regions. With regard to the sides, some humid coastal races have quite as much phaeomelanin as do arid warm interior types.

The evidence for size differentiation in accordance with Bergmann's Law is unsatisfactory or wanting. In the genus as a whole, the largest species are either southern or central in location. Within the *hyemalis* group the large forms occur in the warmer southern regions. The same is true within the yellow-eyed juncos, although not perhaps within the *phaeonotus* rassenkreis (no weights for the southern member are available). Here, however, the differences in average temperature in the ranges of juncos are not known and possibly they are not great. The temperature in the habitat of the small *bairdi* should average rather high. Among Oregon juncos, the smallest forms are *pinosus, thurberi,* and possibly also *insularis* of the south; but one northern form (*montanus*) is scarcely any larger, and the largest form is *townsendi* (perhaps equaled by *mearnsi*) of Lower California.

If the principles of Bergmann's Law are valid, the low temperatures of winter with which a junco must contend may be critical factors. The small northern forms migrate from the breeding grounds and encounter no greater cold than do certain of their southern relatives. The average temperature in the wintering grounds of *hyemalis* may be less than in that of *aikeni* in the high cold plains region east of the Rocky Mountains. But still no general adherence to the Law can be discerned in the Oregon juncos. *J. o. oreganus* may be large because a large proportion of the race winters in a cool northern coastal region. But why is *townsendi* large and *thurberi* moderately small?

Relatively increased tail, foot, and bill lengths (Allen's Law) in southern members of the *hyemalis* artenkreise cannot be proved, for lack of specific data on general body size, which also increases to the south. In tail length there is, relative to wing, a definite increase; this and other changes in this ratio have already been shown to have some relation to migration (p. 357). In *oreganus*, except for wing-tail ratios, there is no consistent manifestation of Allen's Law. In *J. caniceps* and in the yellow-eyed group a definite relative increase in the foot occurs. Tail and wing appear to decrease relative to body size. Many of the races that are short winged and short tailed in comparison with their nearest relatives inhabit restricted isolated areas or veritable islands (*insularis, bairdi, fulvescens*); not all of these may be relatively short winged, as they also are small throughout—as for example, *bairdi*. Other short-winged forms are found in regions of comparatively dense forest habitat (*J. o. oreganus* and *J. o. shufeldti*) where daily flight requirements may be less (see Miller, 1931, p. 102).

Degree of Variability

The accompanying graph (fig. 32) shows that variability of tail is greater than that of wing; variabilities of bill length and of bill depth are successively greater than that of tail length. The tarsus in general is similar to wing in variability, though usually slightly higher, and variabilities of middle toe and hind toe are successively, though irregularly, higher than that of tarsus (see fig. 33). Hind toe shows the highest variability of any of the mensurable characters. In general the characters with high coefficients show the most diversity in coefficients as between different races and species.

Races that are highly variable in some characters may be relatively low in others. This of course merely reflects the genetic independence of the characters. In the insular forms, *insularis, bairdi,* and *vulcani,* the coefficients are not low, even though one might suppose that these species (the first two consisting, each, of but a single population or colony) would be relatively homogeneous because of inbreeding and restocking from small residua. But certain characters are especially variable in these insular species. Probably mutation pressure is not counteracted by selection pressure to the degree that it is in

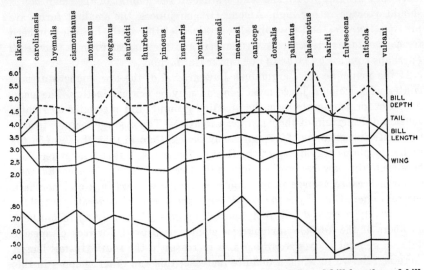

Fig. 32. Comparison of coefficients of variability of wing, tail, and bill lengths and bill depths in males of the 21 forms of the genus *Junco.* Lowest line represents coefficient of correlation of wing and tail.

other juncos. *Insularis* particularly is relatively free from predators and from competing species. It occupies ecologic situations of marked variety, and of types not ordinarily invaded by the genus.

With respect to measurements, certain of the races constitute mere segments of extensive geographic gradients. Values for each of these races (*montanus, thurberi, caniceps, phaeonotus*) in the graphs represent one restricted part of the race. Inclusive variability for the whole group of populations that are comprised in such a race would certainly be great. The variation in any one of the populations of the race is of "normal" magnitude. In hybrid populations it has been found that the variation usually is not greater than in the parent populations. This may be because there are involved a large number of factors, which would result in variability in generations subsequent to the F_1 that might not be noticeably higher than that of the parents. Bill and tail dimensions in certain hybrids of *caniceps* show segregation and increased variability, which suggests that a small number of factors is concerned.

Variability in coloration has not lent itself to statistical expression. Variability of measurements was not found to be uniformly high in some forms and low in others, but in coloration some are distinctly uniform throughout, whereas others are variable in almost all parts of the plumage. The northern migratory races, *J. h. hyemalis, J. o. oreganus, montanus, shufeldti,* and *thurberi,* are especially variable. However, some permanently resident forms (*phaeonotus*) in any one local province and some insular species (*vulcani*) are also distinctly variable. *Caniceps, dorsalis, palliatus, townsendi,* and *carolinensis* are relatively uniform. The most variable races with respect to color may not show high variation in any of the dimensions. Nearly all races are highly, though unequally, variable in tail pattern.

The dimensional and color characters thus far considered are all more or less quantitative in nature. They appear to provide ample material in the

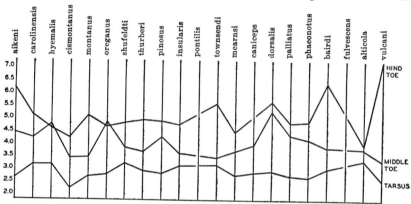

Fig. 33. Comparison of coefficients of variability of tarsal, middle toe, and hind toe lengths in male juncos.

way of variations for further evolution of races and species. But it is clear that certain features of color are variable whereas others are not, and the variable ones have limited lines of variation which they follow.

Certain other color variations appear that resemble the larger mutations familiar in laboratory genetics. Some are like characters of other species, but others are not. They occur more frequently in some forms than in others. Because they are not merely deviations in the quantitative expression of a character, but are highly distinctive features, they serve to give an idea of the rate of occurrence and spread of new characters in a race. Probably there is no fundamental difference between this type of variant and the variants in a quantitative series. They merely are more readily detected as new departures; probably more often they depend on single factors.

The following characters are never predominant in any race: reddish brown crown feathers (not merely feather tips); phaeomelanin of areas in midventral part of hood; spotted albinism of various patterns. Others which are characteristic of some forms appear as anomalies in other races or species. These are: white wing-covert spots; reddish brown margins on wing coverts and secondaries; absence of normal reddish brown on wing; white base of culmen on an otherwise black maxilla. (Characters that almost certainly represent segregation following a more or less recent hybridization are not included.)

The percentage of occurrence of these characters is usually much too high to represent actual mutation rate. Many of the characters doubtless have been propagated through a race, following the initial mutation, so that they now are rather frequently met with. Some are probably rare recessive expressions resulting from a complex heterozygous situation that is widespread. The varying frequency of appearance may be due to factors that influence dominance in different races. The characters which are typical of certain species but which occur as anomalies in another species may be characters once common to the two, or they may have resulted from interbreeding sometime in the past. Reddish inner secondaries and white base of culmen are variants in *dorsalis*, probably of long standing, that are best explained in this way.

White wing bars occur in *hyemalis, carolinensis, caniceps, dorsalis, palliatus, mearnsi, montanus*, and *thurberi*. It is unbelievable that interbreeding with *aikeni* could have taken place to produce these variants. Also, in the races only remotely related to *aikeni* it is difficult, though not impossible, to conceive of this character as one retained from some common ancestor. The appearance of it in such diverse species and its nonlocalized occurrence in *hyemalis* favor the idea of recurrent mutation. The occurrence varies from 2.66 per cent in *hyemalis* to 0.5 in *palliatus*. The difference in occurrence in the sexes in *hyemalis* suggests that the frequency of expression is definitely modified by other factors. In *aikeni* a comparable difference obtains.

The phaeomelanistic median hood margin, which occurs more often in females than in males and which appears in 8 per cent of *mearnsi* and 25 per cent of *bairdi*, obviously is a feature that has become widely propagated in these unrelated forms following mutation. The red-crowned variant appears in *caniceps, dorsalis, palliatus, phaeonotus*, and *mearnsi*. Its occurrence varies from 6.2 per cent in *caniceps* to 0.2 in *mearnsi*. A single occurrence in the southern Mexican highlands, a region far removed from the range of *caniceps*, probably represents a separate mutation of the same kind. In *caniceps* this feature must have become considerably multiplied in the population.

The most significant aspect of these variations is the potentiality which they reveal for the formation of races or species of distinctly different plumage. The white-winged junco may be a form that has arisen from the establishment of variants of this type, although the wing bar in *aikeni* is not the most fundamental character of the species. However, the processes leading to the segregation and establishment of a character in a population, plus eventual association of characters and hence race formation, seem to be the crucial steps in speciation.

KINDS OF INHERITANCE

Nothing can be stated with finality regarding inheritance without evidence from extensive breeding experiments. Nevertheless, a large amount of circumstantial evidence from the study of naturally occurring hybrids, correlations, and types of variability permits certain provisional statements. Even presumptions regarding heredity cannot be overlooked in considering the methods by which races and species have evolved. They may throw light on the cogency of certain hypotheses concerning origin. A summary of types of heredity will be given here without repetition of the reasons given in earlier sections for my tentative conclusions.

The most prevalent method of inheritance is apparently of the multiple-factor type. The quantitative characters of dimension, as usual, fall in this category. But there is some evidence of certain exceptions. The number of factors for any character is, as a rule, probably large. The color of the head seems to be inherited in this way, as also many of the differences in pigmentation of the back and sides.

Smaller numbers of factors, sometimes just one pair, appear to be involved (1) in presence or complete absence of eumelanin in the back, (2) in certain types of phaeomelanin of back and sides (crosses with *caniceps*), (3) in presence or absence of phaeomelanin in the sides, (4) in demarcation of the side area (crosses of *J. hyemalis* and *J. oreganus*), (5) in black base of the culmen (*dorsalis*), (6) in a certain degree of bill depth and length in *caniceps*, (7) in tail length of *dorsalis*, and (8) in the infrequent qualitative variations of plumage (see p. 367). There is nothing to indicate that characters inherited in this manner are any more fundamental than those determined by multiple factors. The characters of races in contrast to those of species cannot be distinguished by mode of inheritance or by any other means. Nevertheless, characters that depend on few factors could more readily be segregated into sharply defined species.

Dominance, sometimes variable and sex-modified, is shown by the phaeomelanin of the back of *mearnsi* over that of *caniceps;* by phaeomelanin of *caniceps* over that of *thurberi* (experimentally demonstrated; Miller, 1939, p. 212); by phaeomelanin of the sides of *mearnsi* over absence of the pigment in *caniceps;* by flesh-colored maxilla; especially the base of culmen of *caniceps* over the black maxilla of *dorsalis* (experimentally demonstrated); by bill depth of *caniceps* over that of *thurberi;* by bill length of *caniceps* over that of *dorsalis;* by tail length of *caniceps* over that of *dorsalis;* and by sides of *hyemalis* over those of *oreganus* in males.

Evidence for the interaction of sex factors (possibly through the endocrine mechanism) is found in reduction or replacement of eumelanins in females; the lesser concentration of eumelanin of the head and back in this sex; the less frequent white spots and less extensive tail spots in females; the more frequent interruption of the ventral hood margin; smaller dimensions throughout, of constant and considerable magnitude only in the wing and tail, however.

Interaction not concerned with sex has been noted between some factors for different pigments in the back and between factors for demarcation of sides and hood regions. Many of the low correlations of dimensions of parts might be thought of as resulting from interaction of factors instead of from common factors.

SEXUAL DIFFERENTIATION

The average difference in measurements of wing and tail in the sexes is fairly constant in the different juncos (see graphs). The difference usually is close to 5 per cent. The *hyemalis* and *oreganus* artenkreise are roughly similar, whereas there is slightly more sexual difference in the *caniceps* rassenkreis and slightly less in the yellow-eyed group. There is a small, almost uniform average difference in length of tarsus, which amounts to 1 or 2 per cent. Sexual difference in bill dimensions is seldom statistically reliable, but, the genus being viewed as a whole, the averages for males in the majority of cases exceed those for females. Quite as positive is the general tendency toward sexual difference in toe length.

Indices of variability and the coefficients of correlation are usually not significantly different in the sexes. No general trends are obvious in the genus.

The most obvious sexual dimorphism involves color, but this is true only of certain species. Dimorphism of the body plumage tends in the direction of reduced eumelanin and increased phaeomelanin in females. This may consist of replacement of one by the other, or extension of the phaeomelanin, or mere dilution of eumelanin (see also p. 369). In the rectrices, conversely, eumelanin is more extensive in females of all forms, except *vulcani*. The interspecific and interracial differences in dimorphism of tail pattern, unlike those of body coloration, follow no appreciable trend in the genus. Dimorphism of tail pattern is peculiarly low in *J. h. hyemalis*, whereas that of the body plumage is great. But it is not low in the related *J. h. carolinensis* and is low in unrelated forms, some of which have body plumages that are identical in males and females. It is not feasible to express color dimorphism quantitatively except with respect to intensity of head color, which displays a considerable range of sexual difference. The following list summarizes the degree of color dimorphism, chiefly of the head (for scale of numbers representing head pigmentation see p. 247):

Lacking: *vulcani, alticola, fulvescens, bairdi*
Nearly lacking: *phaeonotus, palliatus, dorsalis, insularis*
Slight: *caniceps*
Differential in head pigment (average) 1: *mearnsi, pontilis, townsendi, carolinensis*
1½: *pinosus, aikeni*
2: *montanus, cismontanus, hyemalis*
2½: *thurberi*
3: *shufeldti, oreganus*

The association of dimorphism in plumage with intense head pigmentation is frequent, but it is not consistent. The *hyemalis* group shows dimorphism equal to that in many members of the *oreganus* group, but the intensity of black or gray is generally less. No increased dimorphism appears in *J. p. phaeonotus*, nor in the dark-hooded *alticola*, compared with the lighter *J. p. palliatus*. In *thurberi* there is less sexual difference than in some related lighter-headed races; *aikeni* shows more difference than does the darker-headed *carolinensis*.

Because dimorphism of the head is associated to some degree with intensity of pigmentation, the sexual difference in color may be thought of as an interaction of certain intensity factors with sex factors (or hormones). If in the course of evolution these factors for intensification were lost, a species would revert to a less dimorphic state. Forms with dimorphism probably are not primitive, at least with regard to pigmentation, but those with low dimorphism might have secondarily attained this condition (*carolinensis?*). Reasons based on distribution already have been advanced for believing that the darkest-headed Oregon juncos are the most recently evolved forms in this artenkreis.

Retarded and female plumages often resemble fully developed plumages of duller-headed, less dimorphic forms. They do not closely resemble the rather different pattern of juvenal plumages, but they are less radically different from these than are those of adult males.

PHYLOGENY

Phylogeny within a single genus often is a matter of much uncertainty when all facts are taken into consideration. Like the problem in the crossbills (Griscom, 1937, p. 208), we do not know in *Junco* the characters of the ancestral species, and the antiquity of the characters with which we are now confronted cannot be soundly judged. Only a few speculations on the general course of phylogeny deserve attention here. Discussions of the probable origin and evolution of each form are included in the earlier parts of this paper.

The genus *Junco* seems to constitute a natural unit and, being confined to North and Middle America today, it may be assumed to have arisen in this area. Of the Emberizinae of Europe and South America, which in some measure resemble *Junco*, none has qualities which suggest that it is ancestral to the genus.

Where within the Nearctic area was the center of origin, and which juncos retain in greatest measure the ancestral condition? A few characters seem definitely to be primitive in the light of the prevalent patterns and juvenal plumages in other Emberizinae. Such are dorsal stripes, undifferentiated head region, slight sexual dimorphism, and absence or slight amount (not reduced in correlation with environment) of tail white, dark-colored maxilla, and dark iris. White wing bars are of doubtful significance in this connection. It is clear that these features, except for eye color, are more prevalent in the southern members of the genus.

Boulton (1926) has postulated, without especially convincing evidence, that the Zonotrichiae of Ridgway originated in the southern highlands of Mexico. Was this so in the genus *Junco?* The usual question is presented: Did the southern, perhaps truly primitive, juncos become crowded south along the cordillera, or did the other juncos spread out from the southern highlands as modified northward-invading forms? I fail to see grounds for a conclusive answer. The isolated southern areas may properly be viewed as an asylum for antiquated types, but are they centers of origin? Juncos as a group are adapted to Boreal conditions. It is not likely that they would have arisen on the fringes of such regions and would have come eventually to dominate them. Probably far back into the Tertiary they were adapted to Boreal situations and enjoyed a wide distribution before, or at the time, the color characters arose that now appear primitive. There must have been early geographic differentiates, some of which may have led to the basic stocks of the present rassenkreise and to isolated species; then later the races of the present rassenkreise arose as subsequent evolutionary inflorescences. Probably no form has been at a standstill with respect to evolution. The greatest environmental changes in recent geologic history have been in a broad zone bordering the glacial areas. The areas of greatest environmental stability would preserve the primitive types, and these in general must have been the southern and insular areas.

Vulcani appears to represent a primitive, possibly ancestral, stage, irrespective of point of origin. The characters common to the yellow-eyed juncos may have characterized the next stage in advance in some Tertiary species. Contemporaneously, perhaps, a more northern representative of the genus arose on the west coast that was not unlike the present *insularis* in coloration. Also there may have been *caniceps*-like and *hyemalis*-like forms in the interior and in the East. The isolation of these basic types must have been fairly complete to have permitted marked differentiation. In the glacial periods, perhaps even earlier, environmentally correlated differentiation within these Tertiary types occurred and some of them were thrown together as ranges were rejoined. *Aikeni* was established and in interglacial times it became isolated. *Caniceps* and the yellow-eyed juncos hybridized to give rise to *dorsalis*. In the Northwest a new dark-eyed group of races within the *oreganus* sphere arose which crowded out, or spread and modified, many of the ancestral paler-headed types. *Phaeonotus* and *hyemalis* proper differentiated into races, with characters that were environmentally correlated. Finally, the later glacial changes in British Columbia influenced the composition of races immediately to the south, and in the last phase of retreat permitted junction of *hyemalis* and *oreganus*, certain of whose hybrids, with favorable isolation, set up the form *cismontanus*.

INTERRELATIONSHIPS OF FORMS IN NATURE

Different forms of juncos retain a remarkable ability to interbreed, in view of the degree of structural differentiation which they display. Their readiness to accept a mate of another species is paralleled in other avian groups, yet certainly it is more pronounced than in most genera of fringillids. Little is known of the fertility of hybrids of distinct forms, but one may suppose from the successful nesting of hybrid generations in the wild that it is fairly high.

Experimental breeding in the aviary has not progressed far. A male *J. o. dorsalis* and a female *J. o. pinosus* accepted each other and mated. Three sets of eggs that were laid have been fertile, but embryos of two of the sets died shortly before hatching. The cause of death is unknown. Two normal young hatched from the third set and lived two days, when they succumbed because of dietary deficiencies (Miller, 1938a, pp. 92–93). In the spring of 1938, with riboflavine added to the parents' diet, two vigorous young hatched and, by providing insects, were kept alive. One young was attacked by ants when it scrambled out of the nest at eight days. The other was hand-raised from that age. In its adult plumage it served to show dominance of red back and flesh-colored bill and intermediacy (blending) of head color and size characters generally. No success was had in breeding this bird back to *pinosus* in 1939, but in 1940 a female *pinosus*, not its parent, mated and laid three sets of eggs. The first set of three hatched only one young, which seemed weak and died the same day. The other eggs had embryos that died at hatching and at about 5 days' incubation. The second set of four hatched only one young, which died at 30 hours. One of the eggs was cracked and may or may not have been fertile; the other two had embryos, one of which died at about 5 days. No riboflavine was fed to the parents in 1940 and the eggs were especially small. The eggs of the third set were somewhat larger and two of the three hatched and the young were successfully reared. The F_1 male obviously is fertile, but there may be some weakness in the cross that leads to low hatching. Again, as with the initial hybridization, the nutritional factor cannot as yet be evaluated.

Mearnsi and *aikeni* have interbred in the wild to produce at least the first generation and at least some birds that must be of later generations (see p. 351). *Caniceps* and *mearnsi* hybridize extensively when thrown together, as do *thurberi* and *caniceps* (see p. 205). A hybrid male *caniceps* × *mearnsi* has been observed behaving as though mated with a female *J. o. thurberi* (Miller, 1935, p. 468). *J. o. montanus* and *J. h. hyemalis* hybridize, but show some evidence of preference for their own kind (see p. 341).

We may conclude that a psychologic barrier to interbreeding in the wild is usually absent or is not insurmountable. Whether such an aberrant form as *vulcani* would hybridize with other juncos is doubtful. If there is no certain sterility of hybrids and if psychologic factors and differences in season of breeding in adjacent forms are of limited importance, the usual controlling factor in hybridization is geographic isolation.

Data from experimental breeding do not precisely indicate specific relationships (see Robson, 1928, pp. 14–15). Our concern is with natural processes of evolution and with the natural relations of species. Developmental failure in hybridization is but one cause of isolation; psychic, ecologic, and geographic barriers are others. Natural isolation of specific units may result from the action of several incomplete barriers of different kinds. Because the genus *Junco* somewhat resembles a group of insular forms, the members of which have the potentiality to interbreed in some measure, it might be suggested that it constitutes a single rassenkreis. Certainly the members of it are almost without exception geographically complementary. But even the most ardent exponent of the formenkreis principle would hesitate to include *vulcani* in the same *species geographicum* with the other juncos. As Rensch has stated (1934, p. 23), "systematic judgement" must be resorted to in perplexing problems where taxonomic decisions are after all purely arbitrary and involve somewhat artificially defined units.

Our problem here is to summarize in synoptic form the degrees to which the different juncos merge in nature. The series demonstrates how completely the natural relations fail to assort into a few simple categories.

a) Complete isolation at all seasons, consequently no hybridization; complete differentiation (in at least a few characters): *vulcani; alticola; fulvescens; bairdi*.

b) Complete isolation in breeding season, but junction in winter; complete differentiation: *dorsalis* and *palliatus; aikeni* and *hyemalis; aikeni* and *caniceps; mearnsi* and *hyemalis* (so far as present evidence shows).

c) Complete isolation at all seasons; some overlapping in all characters: *pontilis* and *townsendi*.
d) Complete isolation in breeding season, but junction in winter; some overlapping in all characters; no known hybridization: *pontilis* and *thurberi*.
e) Complete isolation in breeding season, but junction in winter; complete differentiation; four instances of hybridization probably arising from exchange in migration, that is, return of a bird to the breeding ground of the other species: *aikeni* and *mearnsi*.
f) Complete isolation in breeding season, but junction in winter; complete differentiation; hybrid populations in few small areas, well isolated from one or both parents: *caniceps* and *thurberi*.
g) Essentially complete isolation in breeding season, but junction in winter; complete differentiation; hybrid populations in a number of somewhat discontinuous or greatly restricted intervening breeding areas: *caniceps* and *mearnsi*; *caniceps* and *dorsalis*.
h) Partial isolation; complete differentiation; at least occasional (free?) hybridization in annectent areas of sparse population: *hyemalis* and *oreganus*; *cismontanus* and *oreganus*.
i) No physical isolation; complete differentiation; hybridization not free where both parental types are present; exchange in migration considerable: *hyemalis* and *montanus*; *cismontanus* and *montanus*.
j) No physical isolation; complete differentiation; free hybridization over broad area, giving appearance of "intergradation"; all characters grade and mix essentially at the same point and variability is locally increased: *mearnsi* and *montanus*.
k) No physical isolation; incomplete differentiation; intergradation chiefly through hybridization, with increased variability locally; all characters grade and mix at same point: *thurberi* and *pinosus*; *montanus* and *oreganus* (one phase as in *l*).
l) No physical isolation; incomplete differentiation; intergradation not the result of any immediate hybridization; some characters alter independently at different geographic points: *phaeonotus* and *palliatus*; *shufeldti* and *oreganus*; *shufeldti* and *thurberi*; *shufeldti* and *montanus* (southern); *montanus* and *oreganus* (one phase as in *k*); *hyemalis* and *carolinensis*.

The foregoing outline inevitably distorts the picture of the relations of any two forms. Reference should be made to the sections on relationships in the accounts of races. Particularly to be stressed is the fact that isolation is meant to indicate very considerable spatial separation. The intangible barriers (see Grinnell, 1915) that exist may not be absolute, but they are quite real, I believe, in their influence on infiltration of stocks and hence upon the type of intergradation encountered in the latter categories of the list. Particularly difficult is a brief characterization of the situation in *caniceps*, as also the relationship of the *hyemalis* and *oreganus* groups. The characterization of forms as completely or incompletely differentiated is rather arbitrary.

THE FORMATION OF SPECIES

The process of evolution is one of differentiation and segregation. The differentiation consists of separate hereditary changes, either minute or large. Evolution proceeds through the accumulation and segregation of these changes in separate populations. This study cannot touch upon the causes of mutations, nor can it produce certain types of desirable data on the selective value of characters. However, the kinds of mutations that appear and the way in which they are interrelated have in some measure been discerned. Their nature may fix the channels of future evolution, as it seems to have done in the past. The ultimate evolutionary structure is in part molded by the kinds and quantity of the building stones.

The segregating process has many phases. Some of these are demonstrable through analysis of individuals, colonies, populations, races, and species. Segregation begins with differential average occurrence of phenotypes in populations and becomes progressively more perfect through the action of a variety of isolating agencies, including genetic isolation sooner or later. Modification of populations by natural selection is a form of segregation. When full differentiation and absolute isolation are attained, further change merely increases the magnitude of differentiation; extinction of annectent groups emphasizes morphologic and physiologic separation and obscures the evolutionary pathway that has been followed. The genus *Junco* contributes a rather complete exemplification of the stages and processes that lead to the first milepost in the evolutionary path, the full species.*

In wild populations the variations in measurement, in some degree hereditary, are of roughly comparable magnitude in different forms. There is always enough latitude of variation so that some selection might take place. Racial differentiation in dimensions is rarely absolute, because with multiple-factor inheritance the restriction of variation in each race necessary for elimination of intermediates would require incredibly rigid selection.

Because wing and tail lengths are in large measure determined by common factors, their modifications can be expected to proceed in parallel more often than independently. Indeed, this is what seems to have taken place. Variations of other parts are fairly independent and permit independent differentiation. Much the same situation holds true for small differences in tone of color; but here some races and populations are much less uniform than others, permitting more modification through selection or other types of segregation. In many instances it is difficult to imagine, however, that limited variation has been brought about by more rigid selection. *Caniceps* is less variable than *J. o. thurberi* and undoubtedly it is more homozygous with respect to back color. This difference may result from some intermingling of stocks in *thurberi*, but it also may reflect a higher mutation rate, which I suspect obtains in all Oregon juncos.

Color characters of considerable magnitude which are controlled by few factors apparently spread through a population or, with comparative ease, become established in small isolated colonies. Basic mutation rate may have something to do with this, but the nature of the characters and their inheritance, especially with dominance, promote it. White wing spots, reddish brown on the crown, and other similar characters seem to arise independently from the germ line of several forms. Accordingly, it is more likely that white-winged juncos would have evolved than juncos with black spots on the breast. The paths of evolution are guided, so to speak, by the available variations, by

* In this discussion of speciation a fundamental similarity may be noted to many of the views expressed in Dobzhansky's (1937) excellent review of the subject, which appeared when preparation of this paper was completed. Although it has been possible to insert a few references to his work, this cannot be done for all places where our opinions coincide.

the points of weakness (or strength) in the germ plasm where changes may take place. Much of the adherence to certain trends of modification, especially where proportion is concerned, may be necessitated by retention of functional practicability, but not so with regard to some features of color pattern. Differentiation of predominantly eumelanistic sides has occurred independently in distantly related species; this is also true of the invasion of the gray or black hood by phaeomelanin (*bairdi* and *mearnsi*). Tail white has differentiated in parallel fashion in a number of different forms that are not closely related. It may be conceded that the white has some sort of value, perhaps as a recognition mark. But the waxing or waning always follows a nearly identical series of intermediate stages. This is suggestive of the vg series of allelomorphs of *Drosophila* and their developmental interrelations as outlined by Goldschmidt (1938, pp. 229–230). Never are there predominantly transverse boundaries between dark and light areas, never a pattern with more white on a given rectrix than on its neighbor next lateral. Available variation (mutations) is of a restricted type. This means that there is a kind of orthogenesis or nomogenesis. To summarize, juncos have differentiated in characters which usually are prominent as intraracial variants, and the variants are in some measure restricted in type by the nature of the germ plasm.

What now are the stages in segregation of variants? To begin with, colonies and populations show differences in frequency of occurrence of phenotypes (Dobzhansky, 1937, p. 63). Two phases are recognizable: (1) spotted local deviations in average occurrence, and (2) deviations as part of slight, but fairly general and often evenly graded, geographic trends. The two types may in part be coincident. The magnitude of the differences in averages ranges from the lowest limit of statistical reliability upward. Type 1, when entirely unrelated to widespread geographic trends, appears not to be correlated with local environmental differences; such segregation is found in small colonies that have some measure of isolation from neighboring populations. Examples are the weakly differentiated colony within the race *pinosus* and the Huachuca and Chiricahua populations of *palliatus*. Usually the number of characters involved is small and often there is but one character or one group of related characters (wing and tail length). It has been suggested that the fortuitous occurrence of variants within small isolated populations, or the establishment of the colony from small stocks not representative of the entire range of variability of the race, are factors in colonial differentiation. (Important and related factors are random fixation and loss of genes, Dobzhansky, *loc. cit.*, p. 131).

Differentiated colonies could subsequently spread by reason of alteration in barriers, and perhaps might supplant other juncos because of physiological or structural superiority attained in significant degree during the period of isolation. The colonial nature of the inception of such evolution would be obscured. Also, through continued isolation and further differentiation colonies might lead to such distinct forms as *aikeni*, *bairdi*, and *insularis*. The characters possessed by some of these species do show some correspondence with widespread geographic correlations. But such correlated modification of an isolated form may arise, if there is any molding force (indirect) in the environment, without there ever having been a continuous geographic gradient. In the genus there are no examples showing complete transition from the isolated colony that is not a part of a geographic gradient to a fully differentiated form. *Townsendi* and *pontilis* which might seem to exemplify this, are better explained as members of a broken-up gradient system (*pontilis* may have arisen from the junction of distinct types).

Segregation in the form of geographic gradients (type 2) can be traced from incipiency through a complete series of stages to the establishment of fully differentiated units. The causative agents in the origin of an extended geographic gradient in a character are not certainly known. The gradient results partly from the dispersal of variants (genes) from one center toward another with lowered frequency of occurrence. The extreme groups in a gradient are, so to speak, partly isolated by their own population pressures. But the gradient also results from many local intermediate frequencies established in balance with intermediate environments. Both factors are in varying measure concerned. Often gradients for two or more characters are not coincident. This suggests that diffusion started at different times and places, so that the degree of dispersal in two characters is not the same. But these characters may also be affected by different critical limits in the environmental gradient.

The stage in segregation in which there is a number of independent gradients is illustrated by conditions in *thurberi*. This form is a group of more or less continuous breeding populations through which run geographic gradients which are not all correlated and which sometimes take distinctly different directions (north to south as against east to west). The gradients differ in steepness and uniformity. Each colony or small population may have an average genetic pattern more or less peculiar to it.

How does such a complex break up into more definite units? Goldschmidt (1934) in dealing with groups of like nature in *Lymantria* found that units defined on one set of physiologic characters did not correspond to those based on another or upon structural characters. This is a good illustration of lack of correspondence in geographic or environmental trends. Nevertheless, at times we do see arising from such complexes some units the members of which have a number of characters in common that are rarely found in other units. So in *montanus*, instead of a mosaic consisting of many slightly different populations there is a polarized arrangement, with northern and southern divisions. Fifty to 60 per cent of the individuals from the two divisions can be differentiated if all characters are considered. The gradients run more or less in parallel and most of them show breaks at nearly the same point. There is a fairly marked change in average environment at this point. It would appear that the distribution of critical environments and their effectiveness as molding agents determine whether or not the mosaic will be shaped in the direction of two subraces or, as elsewhere, many variously definable subraces.

In *phaeonotus* the breaks in the parallel gradients do not coincide as well as in *montanus*, but the gradients are so extensive that the polar populations are to all purposes completely differentiated; we call the extremes races. The situation in *hyemalis* and *carolinensis* is similar, but the terminal groups are only 90 per cent differentiated. *Shufeldti* and *montanus* are about 80 per cent differentiated. The geographic differentiation in the genus tends to form race chains (Rassenketten) rather than spheres of races (Rassenkreise), because distribution is chiefly along cordilleran systems. In the horned larks (*Otocoris*), with distribution in open valleys and lake basins, the situation is reversed.

Mosaics lead to subraces and races, then, by the development of parallel gradients in different characters and by the development of coincident breaks in the gradients. The environment is basic in all this, working indirectly, we may believe. In establishing the initial gradients, slightly different rates of selection may set up different equilibria at the extremes of a geographic axis. The character of the intervening regions largely determines the gradation that results from diffusion of characters and from intermediate equilibria. If the environment changes abruptly, there is more chance that several characters will show a shift in equilibrium at this point, and there may be retardation of diffusion in one or both directions because of the change. When moreover the intermediate area is at some point less favorable to the species as a whole, a sparse breeding population results. The infiltration of characters will tend to be checked at this point, and contrasting populations will build up on either side. The few intermediates that occupy the region then become relatively more heterozygous and variable and are in fact hybrids. Of course in some instances they probably pass out of existence in certain seasons—to be replaced by individuals from either side which then hybridize. The barrier to infiltration of this type may not coincide with the point of critical environmental change that causes different equilibria. Because of population pressure and diffusion, the alterations in characters, neverthless, usually occur at the barrier. Some phases of the intergradations of *montanus* and *shufeldti*, and of *montanus* and *oreganus* (see pp. 260 and 263), constitute exceptions.

More complete isolation is exemplified by *thurberi* and *pinosus*. Areas between their ranges support small isolated breeding groups which are intermediate. This discontinuous system rather completely isolates the racial populations. Loss of such intermediate groups between *thurberi, pontilis,* and *townsendi* is the next stage in isolation. Here the only chance of interbreeding is by failure of birds to migrate in the spring (*thurberi*) and their staying to breed with members of another form. In other groups of races there is some evidence that this occasionally takes place.

The most complete isolation is that of a species such as *bairdi* in which there is no chance of mixture through stray migrants. Any further isolation must be genetic and physiologic. Prolonged geographic separation, with accumulation of adjustments to different situations, would lead to this sooner or later.

Complete isolation naturally does not bring full structural differentiation, but full differentiation cannot come about without some degree of isolation unless we assume a much more rigid selection than we have reason to think exists. Complete differentiation following isolation depends on occurrence of new characters which through selection or mutation pressure, or both, spread in the isolated population or become established by chance fixation.

The usual contributory isolation in the process of segregation is geographic. Ecologic isolation is not apparent, though ecologic differences may, as incomplete barriers, aid in determinng breaks in gradients. There is no conclusive evidence for isolation of adjacent forms through infertility. The partial isolation of *J. hyemalis* and *J. oreganus* is thought to be psychologic. This probably became manifest upon recombination of the forms after they were strongly differentiated and following complete geographic isolation.

The question may be asked, Is differentiation and segregation accomplished entirely by this essentially geographic and ecologic process? Is it possible for new variants to appear and spread through all parts of a wide-ranging species? It seems incredible that such variants could become uniformly established over a large diversified area on a more or less common time front. Inevitably, irregularities would arise, corresponding in some measure to the geographically separated units. Only through such changes as involved sterility with the parental stock could isolation be secured that would lead to a new form which occupied the same grounds as its predecessor. This is of course a method of origin for which there is evidence in various plants and animals. It does not seem to have taken place in the genus *Junco*. If it had, we might expect that two species would occupy the same range in part. Without this kind of sudden origin, the genus has progressed far in differentiation and one can hardly deny that species, even in the most conservative sense, have been formed, when we consider such types as *vulcani*, *alticola*, *bairdi*, and *insularis*.

Until segregation has proceeded to the point of adding infertility, there is always the possibility of its being broken down when forms are again thrown together in the same region. Even with infertility, unless it be quite absolute, there is some possibility. The tolerance for mates of a different type has resulted in many breakdowns in segregation in *Junco*. These have complicated the history of certain races and species. The hybridization that may result from junction of species leads to nothing completely new, but new combinations may constitute the beginning of distinct races or species, which through further change give rise to distinct phyletic lines.

What chance is there of a hybrid's becoming established over a definite area as a relatively uniform type? The possibilities would seem to lie in the possession of some advantageous quality resulting from the interaction of factors and new combinations (see Dobzhansky, 1937, p. 187), and in isolation. Isolation is the only method clearly suggested in juncos. Three instructive cases have been considered: (1) that of *caniceps* and *thurberi*, in which hybrid populations came near, so to speak, to establishing an annectent race (p. 197); (2) that of *hyemalis* and *montanus*, in which certain types of hybrids became established in a somewhat isolated area that probably was rather recently made available to juncos; and (3) that of *dorsalis*, a fully established form which gives evidence in its peculiar combination of characters of origin by the hybridization of *J. c. caniceps* and *J. p. palliatus*.

Aside from the matter of origin of a race from hybrids, there is the problem of influence of occasional hybridization upon the composition of sharply defined races already extant. The problem is difficult to solve because characters added in this fashion and thoroughly mixed through a race are difficult to recognize as such. It is thought that intermixture of this kind has had an effect on *shufeldti* (see p. 261), on a subdivision of *caniceps,* and possibly on *pontilis,* if hybridization was not its actual mode of origin. Intermixture of races has contributed some of the so-called historic features of races, features that do not accord with prevailing environmental and interracial correlations.

The origin of races and species involves an interplay of many factors which, although subject to analysis, withstand resolution into a simple picture. Emphasis on a few phases of the process is certain to yield a faulty concept. It is not unlikely that in this present attempt to consider the numerous ramifications of the process as applied to a particular genus limitations will be evident to persons working with other organisms where different phases seem more apparent.

APPENDIX A

NOMENCLATURE OF RECOGNIZABLE FORMS

Junco caniceps caniceps (Woodhouse)

Struthus caniceps Woodhouse (1853, p. 202), original description
Junco caniceps, Baird (1858, p. 468)
Junco annectens Baird (1870, p. 564), part (hybrid)
[*Junco*] *cinereus* var. *caniceps*, Coues (1872, p. 141), part
[*Junco hyemalis*] var. *caniceps*, Ridgway (1873a, p. 613)
Junco hyemalis var. *annectens*, Ridgway (1873b, p. 182), part (cf. Baird)
J[*unco*]. *hyemalis annectens*, Ridgway (1874, p. 174), part
Hybrid between *oreganus* and *caniceps*, Baird, Brewer, and Ridgway (1874, p. 579), part (cf. *annectens* Baird)
[*Junco cinereus* var. *caniceps*] b. *caniceps*, Coues (1874, p. 143), part
Junco oregonus annectens, Drew (1881, p. 90), part (cf. Baird)
Junco cinereus caniceps, Drew (1881, p. 90)
Junco hiemalis annectens, Coues (1882, p. 56), part (cf. Baird)
Junco hiemalis caniceps, Coues (1882, p. 56)
Junco annecteus Morrison (1890, p. 38), part (cf. Baird)
Junco canceps Morrison (1890, p. 38)
Junco ridgwayi Mearns (1890, p. 243), part (hybrid)
Junco phaeonotus caniceps, A. O. U. Committee (1908, p. 378)
Junco oreganus mutabilis van Rossem (1931, p. 329), original description, part (hybrid)
J[*unco*]. c[*aniceps*]. *caniceps*, Miller (1934, p. 167)
Junco oreganus caniceps, van Rossem (1936, p. 58)

Type.—No holotype was designated by Woodhouse. With reference to the material on which the description was based, he says (p. 203) : "My attention was first called to this bird by ... Cassin, who ... suggested an examination of several specimens of males in the collection of the Academy in connection with another in his possession, and a female in the collection made by me when attached to the Exploring party under the command of Capt. Sitgreaves, ... in the San Francisco Mountain, New Mexico. One of these specimens in the collection of the Academy is from Mexico, the others are from Texas. My specimen is from New Mexico." These specimens may be considered to constitute cotypes. Search in the collection of the Philadelphia Academy of Sciences revealed no *Junco caniceps*, or other juncos at all readily confused with this species dating back to 1852. The only bird apparently now extant that represents the type material is No. 9281, United States National Museum. This bird evidently is the single female referred to by Woodhouse which he took on San Francisco Mountain. He speaks (1853, p. 84) as though but one junco of this new kind was taken there, and (p. 37) that at some point between Camps 15 and 18 (October 10–15, 1851; see Sitgreaves, pp. 10–12). No. 9281 bears the date, October 14, 1851. There is no reasonable doubt, therefore, that it is a cotype, and as the only known type remaining may be designated a lectotype and used as *the type*.

Type (= lectotype), no. 9281 U.S. Nat. Mus., examined by me, July, 1933.

First label, apparently original:

Fringilla oregana? Towns. | ♀ Struthus caniceps Woodhouse! | San Francisco Mt. N. M. 9281 | Oct. 14th 1851 S. W. Woodhouse MD; [reverse side] Struthus caniceps Woodhouse.

U.S. Nat. Mus. printed label, blue edged, with following data entered:

9281 | Junco caniceps (Woodh.) ♀ | San Francisco Mt. Ariz. Dr. S. W. Woodhouse; [reverse side] Oct. 14, 1851 | This is the "type" of ♀ ad | The ♂♂ mentioned in orig. descr. are in the Phila. Acad.

The skin is flattened and somewhat stained. The hood is of the gray type characteristic of the breeding birds of eastern Nevada, central and southern Utah and Colorado. It would be impossible to relate it definitely to the paler or darker variants of *caniceps* as defined in this revision (p. 182). Sides are faintly buffy (probably not merely soiling), as is frequently the condition in females of this race. Back mahogany red; secondaries edged with buff; greater secondary coverts tipped with white (for occurrence of this variation see p. 182). Bill with tip broken; base black; central region somewhat darkened, probably due to blood stain, but not black.

Type locality.—The most specific locality mentioned by Woodhouse is San Francisco Mountain, New Mexico (now Arizona). This fortunately is the place represented by the only existing type; it may be considered as the type locality, as already has generally been done. More specifically, the locality may be restricted on the basis of the known movements of the field party to their Camp 17 (October 14) on the west side of the mountain, probably Hart Spring. The locality and type of course constitute a winter or fall migration occurrence, as did also, undoubtedly, all the cotypes.

Remarks.—The description given by Woodhouse gives no good evidence of being based on more than one form; it clearly applies to the species represented by the existent type. The only doubtful point is his reference to the upper mandible as "dark brown, almost black." The frequent discoloration of the flesh-colored mandibles of *J. c. caniceps* and the presence of a discolored, injured bill in the type satisfactorily explain his statement. *J. c. dorsalis* is the only form that might be confused, and certainly its bill would ordinarily be described as black without qualification.

Junco caniceps dorsalis Henry

Junco dorsalis Henry (1858, p. 117), original description
[*Junco*] *cinereus* var. *caniceps*, Coues (1872, p. 141), part
[*Junco caniceps*] var. *dorsalis*, Henshaw (1874, p. 113)
[*Junco cinereus* var. *caniceps*] b. *caniceps*, Coues (1874, p. 143), part
Hybrid between *caniceps* and *cinereus*, Baird, Brewer, and Ridgway (1874, p. 579)
Junco cinereus var. *dorsalis*, Henshaw (1875b, p. 270)
Junco cinereus dorsalis, Brewster (1882, p. 195)
Junco hiemalis dorsalis, Coues (1882, p. 56)
Junco cinereous dorsalis, Morrison (1890, p. 38) (probably based on *J. c. caniceps*)
Junco phaeonotus dorsalis, Ridgway (1895, p. 391)
Hybrid between *caniceps* and *phaeonotus*, Dwight (1918, p. 299)
Junco caniceps dorsalis, Miller (1932, p. 99)

Type.—No holotype was designated by Henry. The status of the existing type, no. 9272, in the United States National Museum is not entirely satisfactory. In the museum catalogue for no. 9271 is the entry: "type, 'so stated by Baird in 1860.'" This explanation is in Richmond's handwriting. I find that there is no definite statement in the writings of Baird for this year concerning the selection of a lectotype from among Dr. Henry's material upon which the description was based. Baird does indicate that more than one specimen was used by Henry by the following statement (1858, p. 467): "The only specimens yet known of this species are those collected at Fort Thorn by Dr. Henry." What Richmond's basis was for his notation about Baird's statement of the type is not known. In Baird's reports of 1858 and 1860 the only *Junco dorsalis* listed is no. 9270. In a copy of Baird's work Richmond has changed 9270 to 9271. No. 9270 is the number of a dickcissel in the catalogue. Further information is found on the wooden block now in the main collection which refers to the type as no. 9272; on this block in Richmond's handwriting is the statement, "9271 is the type." Apparently no. 9271 was erroneously mentioned by Baird as 9270. In any event, 9271 does not now exist, so far as is known, and it may never have been formally selected as a lectotype.

No. 9272 has had the following history as written in the museum catalogue: "returned to Dr. Henry in 1859"; then later in another hand, "now in N. M." Irrespective of whether through some mix-up in cataloguing or relabeling no. 9271 became 9272, the present 9272, from the information associated with it, evidently constituted part of the material on which the description was based. It is, then, a cotype, along with the supposed no. 9271. Unless other evidence can be brought to bear concerning selection of 9271 as a lectotype, no. 9272 may be definitely selected as the type—a selection already implied by the type label which it bears. No. 9272 agrees with the original description in all matters of importance, if not in details.

Type (lectotype or cotype), no. 9272 U.S. Nat. Mus., examined by me, July, 1933. The first label, a printed Smithsonian Institution label with "type" printed on it, hence not an original label, has the following entries:

(No. 9272) Junco dorsalis. Henry | Γ.' Thorn. N. M. Dr. T. C. Henry; [reverse side] hardly distinguishable from J. | caniceps | except black mandible | [in pencil] throat paler as in cinereus I t

Second label, a U.S. Nat. Mus. printed red type label with the following:

9272 | Junco dorsalis | Ft. Thorn N. M. Dr. T. C. Henry: [reverse side] I.

The skin is in poor condition with neck badly twisted and feathers ruffled. The upper mandible is blackish with only a very small area slightly lighter at base of culmen (= type 8, p. 188). Bill open. Head somewhat discolored, brownish. Head and throat equal in coloration to those of birds from Mogollon Mountains, Arizona. Back dark reddish brown, hence not the bright worn red of a late summer bird. Measurements: wing, 83.3 mm.; tail, 79.5; bill length, 8.9; tarsus, 21.6; middle toe, 12.1; hind toe, 8.4. The wing measurement, which is the only one likely to afford a comparison with Henry's figures, appears not to correspond with Henry's 3.05 inches; it would be about 3.25 inches. The tail of 9272 has the following pattern; rectrix 6 pure white; outer web of fifth white, inner web $\frac{1}{5}$ black; outer web of fourth entirely black, inner web with small white spot. Henry speaks of both 5 and 6 as white, but he might have disregarded the small amount of black on the fifth; it is not clear what was intended by his "third [from outside, = 4th] with brown on the inner edge." Measurements and tail white suggest, but do not prove, that another bird than 9272 entered into the description and may have been used more specifically in the writing of the diagnosis. The tail pattern is a common one for the race; the measurements exceed, for the most part, the averages for males, but do not represent extremes.

Type locality.—W. W. Cooke, in Bailey's Birds of New Mexico, gives the following discussion (p. 740) : "The Red-backed Junco was described in 1858 by Henry, and the explicit statement is made in the original description that the type locality is Fort Thorn. This has been accepted without question as the actual type locality, but in a publication made in 1859 giving a summary of all his bird observations in New Mexico, Henry says of *Junco dorsalis* 'Found only near Fort Stanton, among the mountains, where I should judge they nested. Never observed during winter.' The species is now known to be a common breeder near Fort Stanton, while it does not breed at Fort Thorn and occurs there only in migration or in winter, at which season Henry says he has never seen it. Rereading the original description with these facts in mind it is evident that the locality Fort Thorn belongs under the preceding species, *Toxostoma dorsalis* = *Toxostoma crissalis*, which has no type locality ascribed to it, and by a printer's error was placed under *Junco dorsalis*. The real type locality of the Red-backed Junco is therefore to be considered as Fort Stanton." As further aspects of this problem I would direct attention to the fact that the original label is lacking from no. 9272 and that no date is given. From 1854 to 1858 Henry had headquarters at Fort Thorn, yet during this period he visited other localities, Fort Stanton among them, in 1855, and perhaps also at other times. Probably specimens were sent in from Fort Thorn without details concerning locality, or else in Washington an error was made in providing the label now on the specimen, which may have been made to correspond to the faulty (?) locality in the original description. Although no. 9272 is not a worn breeding bird, it could easily be an early spring bird taken on the breeding grounds. As further substantiation of the opinion that Henry found the bird on the breeding grounds, and hence not at Fort Thorn, is his statement (1859, p. 107) : "This species is an excellent songster." I concur, therefore, with Cooke in considering Fort Stanton, Lincoln County, New Mexico, as the type locality.

Remarks.—With doubt remaining concerning the exact status of the type, there is, nevertheless, nothing about the description or existing type to suggest that the name *dorsalis* applies to any other than the black-billed (maxilla), pale-headed race of *Junco caniceps* for which it is currently used.

Junco phaeonotus palliatus Ridgway

[*Junco caniceps*] var. *cinereus*, Henshaw (1874, p. 113), part
Junco cinereus, Henshaw (1875a, p. 328), part
Junco hiemalis cinereus, Coues (1882, p. 56), part

J[unco]. *cinereus palliatus* Ridgway (1885, p. 364), original description
Junco palliatus, Sharpe (1888, p. 655)
Junco phaeonotus palliatus Ridgway (1895, p. 391)

Type.—Ridgway in the original description designated no. 68817 U.S. Nat. Mus. as the type. This bird examined by me, July, 1933. First label a printed National Museum label with the following entries:

68817 | Junco cinenus ♂ ad Sep. 19–1874 | Mt. Graham Ariz — H W. Henshaw; [reverse side] I.

Red, printed type label:

68817 | Junco cinereus palliatus Ridgw. | Mt. Graham, Arizona. H. W. Henshaw; [reverse side blank].

Skin only slightly dirty. Reddish brown extensive on secondaries and coverts. Bill black above, yellow below. Tone of grays as in resident birds of Chiricahua Mountains, with only a trace of pale smoke gray on sides. Measurements: wing, 74.0 mm.; tail, 70.4; bill length, 8.3; bill depth, 6.1; tarsus, 20.7; middle toe, 11.6; hind toe, 8.9. Tail pattern: outer web of sixth rectrix nearly pure white, inner web ¾ white; outer web of fifth rectrix black, inner web ½ white; outer web of fourth rectrix black, inner web with white spot. The tail pattern is a common one, about average for *palliatus* of Arizona; measurements except for hind toe are below average for both males and females. Those for wing and tail are so small as to suggest strongly that the bird, which was taken in September, was missexed and really was a female.

Type locality.—Mount Graham, Graham or Pinalino Mountains, Graham County, Arizona. Henshaw collected in the "pineries" (1875, p. 140). These mountains support a permanently resident population that breeds in the coniferous forests. The type almost certainly was a bird that had bred locally. Some of Henshaw's September specimens were juveniles. No migratory movements are known and the only other breeding grounds for this form north of the Graham Mountains is Pinal Mountain, which has a very small population.

Remarks.—No question attaches to the application of the name *palliatus*. Fortunately the type is from a part of the range of the race where there is none of the nomenclaturally confusing modification in the direction of *Junco phaeonotus phaeonotus*. Although *dorsalis* breeds to the northeast of the Graham Range, across the Gila Valley, there is no known hybridization to cause complications.

Junco phaeonotus phaeonotus Wagler

Fringilla cinerea, Swainson (1827, p. 435), preoccupied by *Fringilla cinerea* Gmelin 1788
J[unco]. *phaeonotus* Wagler (1831, p. 526), original description
[*Junco*] *cinerea*, Bonaparte (1850, p. 486)
Junco cinereus, Cabanis (1851, p. 134)
[*Junco*] *phaeonotus* Bonaparte (1853, p. 918)
[*Junco hyemalis*] var. *cinereus*, Ridgway (1873a, p. 613)
[*Junco caniceps*] var. *cinereus*, Henshaw (1874, p. 113), part
[*Junco cinereus* var. *caniceps*] a. *cinereus*, Coues (1874, p. 143)
Junco hiemalis cinereus, Coues (1882, p. 56), part

Junco phaeonotus phaeonotus, Ridgway (1901, p. 275)
Junco phaeonotus colimae van Rossem (1938, p. 132), original description
Junco phaeonotus australis van Rossem (1938, p. 133), original description

Type.—Wagler did not specify a type, but there is a specimen in the Munich Museum (Zoologische Sammlung des Bayerischen Staats) upon which the description was based. Someone has designated it as the type and it is probably the only specimen of Keerl's that Wagler had. The description is contained in a report on the Keerl collection. Wagler may have made cursory examination of two Deppe-taken specimens, nos. 6179 and 6178, from Real Arriba and Chico, Mexico, in the Berlin Museum. The type at Munich was examined by van Rossem, August 21, 1933. He very kindly took full notes on the specimen for me and wrote on that date, giving the following particulars: "Wagler's type of *Junco phaeonotus.* No number. No sex. Wing, right 78.0 [mm.], left 77.5; tail, 69.5 [not equivalent to my tail measurements]; exp. cul., 11.1 [not equal to bill length]; tar., right, 21.6, left, broken; middle toe—claw, right, 14.1, left 14.0 [not equivalent to my measurements]. Skin taken down from mount and in bad condition. Mandible missing; maxilla blackish horn color. Plumage abraded as though collected in mid-summer—abrasion not excessive, however. Greater coverts (outer webs), 'tertials' and scapulars reddish like back, but duller. Outer tail feather white for 47 mm. along vane on outer web (terminal 17 mm., however, dusky) and for 42 mm. along vane on inner web; second lateral tail feather entirely dusky on outer web and white, terminally for 34 mm., on inner web along the vane and for 23 mm. on edge of inner web; remaining tail feathers with no white whatever. Rump inclining to olive gray.

"Three tags are attached: an old (original?) green one reading 'Junco cinereus Sw [later crossed out] Junco phaeonotus Wagler. Mexico. Keerl leg'; a new, white museum tag, 'Junco phaeonotus Wagler. Mexico. Keerl coll.'; and finally the red type tags in Hellmayr's handwriting."

Type locality.—Mexico is all that is indicated on the type and by the context of Wagler's article that contains the description. It is not known where Keerl obtained specimens, but from knowledge of the distribution of other species which he secured the locality was apparently somewhere in the southern Mexican Plateau, and certainly not anywhere near the range of *palliatus.*

Remarks.—The tail pattern of the Keerl specimen is the type that is most prevalent in populations of *J. p. phaeonotus* from the southern Mexican Plateau, although such a pattern is not unknown in *palliatus.* Measurements, where comparable, are close to the average for males of *J. p. phaeonotus.* Significant is van Rossem's description of the rump as inclining to olive gray, thus indicating the darker more olive rump of the southern race. Wagler's description clearly represents the species *phaeonotus,* but there is nothing in it that would indicate racial characters.

Junco bairdi Ridgway

Junco bairdi Ridgway (1883, p. 155), original description
Junco hiemalis bairdi, Coues (1887, p. 875)
Junco oreganus bairdi, Hellmayr (1938, p. 551)

Type.—Ridgway designated as types nos. 89811 ♂ ad. and 89810 (sex not determined) of the U.S. Nat. Mus. Subsequently someone has selected 89810 as *the type* (lectotype), even though it was mentioned after 89811 in the original description. Ridgway may have made this selection himself. Indication of this selection appears in penciled notations on the labels, apparently in Richmond's handwriting.

Type, no. 89810 U.S. Nat. Mus., examined by me, July, 1933. Skin taken down from mount, in good condition. Two labels attached. The first is a Smithsonian Institution, National Museum label with "Explorations in Lower California" printed across the top; it has the following entries:

89810 | Junco ? new species? ♂ [sex mark in different hand?; mark has been made over original; according to description, sex not determined at time of preparation] | Laguna feb 2 1883 L. Belding [name in print]; [reverse side] Junco bairdi Belding | Iris bright yellow | Type I [in blue pencil].

Red type label of U.S. Nat. Mus. with the following:

89810 | type of [printed] Junco bairdi ♂; [reverse side blank].

Cotype, no. 89811 U.S. Nat. Mus. Skin somewhat flattened. Two labels attached. First label of same kind as on type with respect to printed matter, with following entries:

89811 | } iris color of ripe oranges in | all shot about a dozen | Junco ? iris bright yellow | Laguna feb 2 1883 | under mandible bright yellow; [reverse side] Junco bairdi Belding | Type I [purple stamp] I [blue pencil].

Second label a printed type label with the following:

89811 | not the [inserted in pencil before the word type] type of [printed] Junco bairdi Belding | Laguna L. Cal. { Feb 2 / 1883 } L. Belding; [reverse side] not the real | type! [in pencil, I believe Richmond's handwriting] I.

The type, no. 89810, shows coloration characteristic of the species. Plumage appears slightly stained, hence pileum relatively dark, and reddish brown of back slightly darker. Pale gray of throat tinged with buff. Maxilla light horn; mandible yellow. Measurements: wing, 71.1 mm.; tail, 69.4; bill length, 8.0; bill depth, 5.8; tarsus, 19.6; middle toe, 11.3; hind toe, 6.9. Tail pattern: outer web of sixth rectrix white, inner web with trace of black; outer web of fifth rectrix white, inner web half white; outer web of fourth rectrix black, inner web with small white spot. Wing and tail measurements are above average for males, hence the sex as subsequently added to the label probably is correct. Measurements of bill and feet are distinctly below average and that for the hind toe is a minimum extreme.

Cotype, no. 89811 U.S. Nat. Mus., coloration characteristic; plumage not so much stained as in 89810, and reddish brown of back relatively light. Pileum partly overcast; and throat and breast buff edged. Measurements: wing, 67.8 mm.; tail, 66.3; bill length, 8.4; bill depth, 5.9; tarsus, 19.5; middle toe, 10.8; hind toe, 7.5 mm. Tail pattern: outer web of sixth rectrix white, inner web nearly pure white; outer web of sixth nearly pure black, inner web $\frac{1}{3}$ white; fourth rectrix all black. No sex mark appears on the labels, although Ridgway (*loc. cit.*) listed 89811 as a male. The wing and tail measurements are inconclusive, falling between the averages for males and females.

Type locality.—Laguna [=La Laguna], Sierra de la Laguna, Cape district, Lower California, Mexico.

Junco fulvescens Nelson

Junco fulvescens Nelson (1897, p. 61), original description
Junco alticola fulvescens, Dwight (1918, p. 302)
Junco phaeonotus fulvescens, Hellmayr (1938, p. 554)

Type.—Nelson designated no. 14906 U.S. Nat. Mus., Biological Survey Collection, the type; original number, 3079. This bird examined by me, July, 1933. Original printed label headed "Biological Explorations, U.S. Dept. Agl.," with the following entries:

143906 [vertical line at left] | ♂ | San Cristobal Sept 2 1895 [year printed] Chiapas, Mexico | Nelson & Goldman; [reverse side, vertical entries] 143906 [at left] | 3079 [at right].

Red type label with printed caption as on first label with the following:

143906 [vertical] | Junco fulvescens, Nelson ♂ | San Cristobal. Sept 21, [remainder printed] 1895, Chiapas, Mexico | Nelson & Goldman; [reverse side] 143906 [diagonal] | *Junco fulvescens* | 3079 [vertical].

Type comparable to others of the original series taken by Nelson and Goldman. Skin in good condition. Measurements: wing, 74.8 mm.; tail, 69.6; bill length, 9.5; bill depth, 7.6; tarsus, 22.8; middle toe, 12.9; hind toe, 9.7. Tail pattern: outer web of sixth rectrix, ½ white, inner web ¾ white; outer web of fifth rectrix black, inner web ⅓ white; fourth rectrix all black. Measurements above average for males throughout, but extreme only with respect to bill length. Tail pattern more or less average for the series. Type was in process of molt, the back plumage still consisting of worn reddish feathers.

Type locality.—San Cristóbal, Chiapas, Mexico.

Junco alticola Salvin

Junco alticola Salvin (1863, p. 189), original description
[*Junco hyemalis*] var. *alticola,* Ridgway (1873, p. 613)
[*Junco caniceps*] var. *alticola,* Henshaw (1874, p. 113)
[*Junco cinereus*] var. *alticola,* Baird, Brewer, and Ridgway (1874, p. 580)
Junco alticola alticola, Dwight (1918, p. 302)
Junco phaeonotus alticola, Hellmayr (1938, p. 554)

Type.—Salvin designated no type. Sharp (1888, p. 657) lists the four specimens in the British Museum that were taken by Salvin prior to the publication of the description. Two of these, the birds first taken and the only ones of 1861, are designated types of the species. The first-named bird is unsexed; it is the only one now bearing a type label in the British Museum. The other bird taken in 1861, a female, is also present. The unsexed bird, no. 85.12.14.984, appears to be a legitimate type (lectotype) selected from the original series used by the describer.

Type no. 85.12.14.984 Brit. Mus. examined by me, July, 1934. Skin in good condition, though slightly flattened. Bill open. Three labels attached. The first, apparently the original:

Junco alticola | Pine Forest Volcan de Fuego | November 1861   763; [reverse side] iris 88 white [underline in pencil].

Second label:

[Left-hand section] E. Mus. [printed] | Nov. | 1861 | O. S. and F. B. G.; [right-hand section] Junco alticola Salvin | Pine forest Volcan de Fuego, Guatem. | Fig •BCA O. Salvin   763; [reverse side] 10 [on a blue paster] A [in red] | 85.12.14.984.

Third label, red:

[Vertical printing on left margin] British Museum | Type [printed] Junco alticola Salv. | Salvin, P. Z. S. 1863, p. 189 | Loc. Pine forest Volcan de Fuego, Guat; [reverse side] Brit Mus. Reg [printed] | 1885.12.14.984 | collected by O. Salvin in Nov. 1861 | Salvin-Godman Collection.

Type comparable in coloration to series in British Museum; lower mandible black basally, unlike normal *alticola*, tip yellowish; maxilla black. Measurements: wing, left, 78.1 mm., right, 78.5; tail, 74.7; bill length, 9.3; tarsus, 23.5; middle toe, 13.5; hind toe, 10.4. Tail: outer web of sixth rectrix nearly pure black, inner web $\frac{2}{3}$ white distally; outer web of fifth all black, inner web $\frac{1}{5}$ white distally; outer web of fourth all black, inner web with minute white tip. Measurements for wing and tail are well above average for males; they are above the limits of the standard deviation, but are not extreme. The bird is, therefore, almost certainly a male. Bill and hind toe are distinctly above average; tarsus and middle toe are near average for males. The tail pattern is made up of feather types that are well represented in the population.

Type locality.—Pine Forest, Volcán de Fuego, Department of Sacatepéquez, Guatemala. Salvin indicates that the species occurs at 8000 feet, but this does not necessarily mean that the type was not taken higher on Volcán de Fuego; in the light of the statements made by Salvin and Godman (1886, p. 374), it probably was taken between 10,000 and 12,000 feet.

Junco vulcani (Boucard)

Zonotrichia vulcani Boucard (1878, pl. 4, p. 57), original description
Junco vulcani, Ridgway (1878, p. 255)

Type.—Boucard designated no type; he mentions (p. 57) "Several specimens from the Volcano of Irazu, obtained at the altitude of 10,000 feet." I am greatly obliged to Mr. A. J. van Rossem for examining the two Boucard birds, male and female, which are marked as types in the Paris Museum. The birds would appear to be valid types since they have been selected by someone from the author's original material. They should be regarded as cotypes. It might be appropriate at this time to designate the male as *the type*, a lectotype, as there is no indication of previous selection of one in preference to the other. Van Rossem, at the Paris Museum on July 25, 1933, made the following notations: "Two birds in the Boucard Collection. . . . Unmounted skins in fair condition though in somewhat abraded plumage. Each still bears the original field tag in (I presume) Boucard's writing | 'Volcan d' Irazu 5/77' | one is marked '♂'—the other '♀'. On the face of the labels is written, just above the locality, | Zonotrichia volcani | type |. Later tags with the well known Boucard Museum label read: Museum [vertical at left] | Zonotrichia vulcani ♂ | Col. [printed] Boucard Volcan Irazu | Costa Rica | 1877 | Type of species ♂ [In English as written] | Boucard [vertical at right]. The new tag on the second specimen is identical with the first save that it bears the ♀ sex mark."

Type locality.—Volcán de Irazú, 10,000 feet, Costa Rica.

Remarks.—Van Rossem makes no comment on the characters of the types. The birds are obviously the same as those subsequently taken from Volcán de Irazú and called *vulcani*. The original description, and especially the figure, fully identify the name with series which have been examined.

Junco oreganus mearnsi Ridgway

Junco annectens Baird (1870, p. 564), original description; part (hybrid *mearnsi* × *caniceps*); all of some authors
Junco oregonus, Merriam (1873, p. 681), part
Junco hyemalis var. annectens, Ridgway (1873b, p. 182), part (cf. Baird)
[Junco oregonus] var. annectens, Henshaw (1874, p. 113)
J[unco]. hyemalis annectens, Ridgway (1874, p. 174), part
Hybrid between *oregonus* and *caniceps*, Baird, Brewer, and Ridgway (1874, p. 579), part (cf. *annectens* Baird)
[Junco cinereus var. caniceps] b. caniceps, Coues (1874, p. 143), part
Junco oregonus annectens, Drew (1881, p. 90), part (cf. Baird)
Junco hiemalis annectens, Coues (1882, p. 56), part
Junco ridgwayi Mearns (1890, p. 243), original description; part (hybrid)
Junco mearnsi Ridgway (1897, p. 94), original description
Junco montanus, Macoun (1904, p. 505), part
Junco hyemalis mearnsi, American Ornithologists' Union Committee (1908, p. 378)
Junco mearnsi mearnsi, Dwight (1918, p. 296)
Junco insularis mearnsi, Dwight (1919, p. 287)
Junco (oreganus ?) mearnsi, Hellmayr (1938, p. 551)

Type.—Ridgway designated no. 11164 U.S. Nat. Mus. as the type. This bird examined by me, July, 1933. Skin taken down from mount. Three labels. Original, a printed Smithsonian Institution–South Pass Wagon Road Exp. label with the following entries (parts in italics printed):

No. 11164 Junco oreganus ♂ 177 | *Fort Bridger Utah*, (*Camp Scott*) apr. 12, *1858*. C. Drexler.

Second, a printed Smithsonian Institution–National Museum label with the following:

11164 | Junco oreganus | Ft. Bridger, Utah Drexler; [on reverse side] April 12, 1858.

Third, a Smithsonian Institution–National Museum type label bearing the following:

Type of Junco mearnsi, Ridgway.

Type specimen characteristic of the species. Depth of head color of type 10 (see p. 245). Back of ruddier type, 2c (see p. 247); there is none of the distinctly reddish brown color of *caniceps*, of certain *mearnsi* × *caniceps* hybrids, and of the types of *annectens* and *ridgwayi*. Measurements: wing, 82.6 mm.;

tail, 77.4; bill length, 7.9; tarsus, 21.1; middle toe, 11.7; measurements of feet possibly affected by former mounting of skin. Tail pattern: rectrices five and six pure white; outer web of fourth rectrix nearly pure black, inner web ⅓ white. Wing and tail slightly above average for males; bill distinctly below average; foot parts large, but measurements probably not to be relied upon. Tail pattern of most prevalent type in all respects.

Type locality.—Camp Scott, Fort Bridger, Uintah County, Wyoming. This locality is one at which juncos may breed. The breeding population consists, however, of *caniceps* or *caniceps* × *mearnsi* hybrids. There is no evidence of "pure-blooded" *mearnsi* breeding there, though this may possibly occur; it is probable that the *mearnsi* characters of hybrids were introduced by a pure *mearnsi* breeding there at some time. The type specimen, taken April 12, may well have been a migrant.

Remarks.—The name *annectens* was used extensively for this race, which now is properly called *mearnsi*. This name and *ridgwayi* are based on types that are hybrids of *mearnsi* and *caniceps*, as Ridgway (1897, p. 94 and 1901, p. 276) has shown. Both types are red backed (*caniceps*) and pink sided (*mearnsi*), hence with equal affinities to the two parental types. Therefore, they cannot arbitrarily be assigned to either of the parental forms and the names cannot be applied to a "pure" population. For details about types of *annectens* and *ridgwayi* see pages 405 and 406.

Junco oreganus montanus Ridgway

[*Struthus*] *oregona*, Bonaparte (1850, p. 475), part
Niphoea oregona, Baird (1853, p. 316), part
Struthus oregonus, Woodhouse (1853, p. 83), part
Junco oregonus, Baird (1858, p. 466), part
[*Junco hyemalis*] var. *oregonus*, Ridgway (1873a, p. 613), part
Junco hyemalis oregonus, Ridgway (1875, p. 19), part
[*Junco hyemalis*] c. *oregonus*, Trippe (1874, p. 145), part
Junco hiemalis oregonus, Coues (1882, p. 56), part
Junco hyemalis shufeldti Coale (1887, p. 330), original description, part
Junco shufeldti, Sharpe (1888, p. 840), part (cf. Coale)
Junco oregonus shufeldti, Mearns (1890, p. 244), part
Junco hiemalis shufeldtii Rhoads (1893, p. 63)
J[*unco*]. *h*[*yemalis*]. *connectens* Coues (1897, p. 94), part
Junco montanus Ridgway (1898, p. 321), original description
Junco oreganus shufeldti, Ridgway (1898, p. 321), part
Junco hyemalis montanus, American Ornithologists' Union Committee (1908, p. 378)
J. shufeldti × *J. mearnsi*, Sclater (1912, p. 381)
Junco oregonus couesi Dwight (1918, p. 291), original description, part
[*Junco*] "*transmontanus*" Dwight (1918, p. 295), original description, part (hybrid)
Hybrid between *mearnsi* and *oregonus*, Dwight (1918, p. 297)
Junco oreganus montanus, Oberholser (1918, p. 211)
J[*unco*]. *o*[*reganus*]. *couesi*, Stone (1918, p. 488), part

Type.—No. 133253 U.S. Nat. Mus. was selected by Ridgway as the type; examined by me, July, 1933. Three labels are attached. The first has the following:

Junco hyemalis | May 7, '94 Columbia Falls, Mont. RSW; [reverse side] 133253.

Second, a Smithsonian Institution–National Museum label:

133253 | Junco [cancelled: hyemalis shufeldti; interlineated:] montanus Coale | ♂ ad | Columbia Falls, Montana. R. S. Williams.

Third, a red label of the National Museum:

133253 | Junco montanus, Ridgw.; [reverse side blank].

Type specimen fairly characteristic of the southern division of the race. Head of type 7, paler than average for males (6) and palest variant normally occurring in the race; type 7 occurs in moderate numbers in all populations of southern *montanus* and occasionally in northern division. Pileum faintly streaked with different tones of black. Back, type 1c; thus the back is light (eumelanin), of a type frequent in southern *montanus;* at the same time it is fairly ruddy (phaeomelanin) to a degree rare in southern *montanus* but frequent in northern *montanus* of British Columbia. Sides of type 2, that is, characteristic of the race, and not suffused with eumelanin as in some; not extensively cinnamon or yellowish as in rare type 3 that resembles *mearnsi*. Measurements: wing, 76.4 mm.; tail, 69.3; bill length, 8.1; bill depth, 5.6; tarsus, 20.3; middle toe, 11.5; hind toe, 7.7. Tail pattern: sixth rectrix all white; outer web of fifth rectrix nearly pure white, inner web ⅔ white; outer web of fourth rectrix ¾ white, inner web with large white spot. Wing, and to less degree tail, distinctly shorter than average of southern division; bill less than average; feet, except middle toe, above average. Fifth rectrix blacker than average, outer vane of fourth rectrix whiter than average; but tail pattern falls within the limits of variability.

Type locality.—Columbia Falls, Flathead County, Montana. Juncos like the type breed in the vicinity of Columbia Falls. The date of collection of the type does not give assurance that the bird was breeding there, although the probability is strong that it was.

Remarks.—On page 250 will be found a discussion of the population of *montanus* from which the type was selected. The birds from the northwestern section of Montana are no more variable and no more like *mearnsi* than any of the populations from eastern Oregon, eastern Washington, or western Idaho. The type series, therefore, fits perfectly into the race, even though representing the southern division. The type itself is unfortunately a pale-headed, but not rare, extreme. Confusion in the allocation of the name *montanus* has resulted from the fact that the type series consists almost entirely of females, which naturally are pale headed, and from a general failure to appreciate the degree of paleness of the head in the females that breed throughout the interior section of the northwest. Furthermore, Ridgway selected the pale-headed variants from northwestern Montana for naming and allocated certain dark-headed breeding birds from the same localities to what was then called *shufeldti*, that is, the darker-headed groups of interior British Columbia. It is possible that Ridgway at one time thought that two species (*montanus* was described as a full species) bred side by side in this region.

In the account of *montanus* (pp. 251–257) the great range of variation of the race is stressed, as also the impracticability of subdividing it into two races. Three older names have been applied to *montanus: oreganus, shufeldti,* and *connectens.* All these are now restricted to other forms, either as usable names or as synonyms (see pp. 396, 394, and 407). *Shufeldti,* a name which recently has been used for *montanus,* is properly applied to the bird of coastal Washington and Oregon. The type is more ruddy on the back than is any interior breeding junco. Dwight (1918, p. 291) is to be credited with ascertaining this fact. What he failed to see was that the type represents at least one phase of the short-winged coastal form of Oregon and Washington. Dwight named the present *montanus* as *couesi* in the belief that Ridgway's *montanus* was a hybrid, or that at least it did not fall within the limits of his *couesi.* The name *connectens* cannot apply to *montanus,* as Ridgway has stated (1901, p. 276). The type is a female *hyemalis* in retarded plumage. Swarth's efforts to prove it was not were hampered by his inability personally to examine the type.

Junco oreganus shufeldti Coale

Fringilla oregona Audubon (1839, p. 68), part
Niphaea oregona Audubon (1839, p. 107), part
Niphoea oregona, Baird (1853, p. 316), part
Struthus oregonus, Woodhouse (1853, p. 83), part
Junco oregonus, Baird (1858, p. 466), part
[*Junco hyemalis*] var. *oregonus,* Ridgway (1873a, p. 613), part
Junco hiemalis oregonus, Coues (1882, p. 56), part
Junco hyemalis oregonus, American Ornithologists' Union Committee (1886, p. 275), part
Junco hyemalis shufeldti Coale (1887, p. 330), original description, part
Junco shufeldti, Sharpe (1888, p. 840), part (cf. Coale)
Junco oreganus shufeldti, Ridgway (1898, p. 321), part
Junco oregonus shufeldti, Kaeding (1909, p. 23), part
Junco hyemalis connectens, American Ornithologists' Union Committee (1910, p. 266), part
Junco oregonus couesi Dwight (1918, p. 291), original description, part
Junco oreganus couesi, Stone (1918, p. 488), part

Type.—Coale selected no. 106035 U.S. Nat. Mus. as the type; examined by me, July, 1933. Two labels are attached. The first is a printed Smithsonian Institution–Shufeldt Donation label; the words "Shufeldt Donation" have been stricken out; the label bears the following:

13 Oct. 85 | 106035 | Junco oreganus shufeldti [last name in different handwriting] ♂ | Ft. Wingate   Dr. R. W. Shufeldt, U.S.A. | N. Mexico; [on reverse side] W 78   3.15 × 2.72 | 106035.

Second, a U.S. National Museum–Smithsonian Institution label:

[Cancelled: 106033] | 106035 | type of [printed] Junco hyemalis shufeldti | Coale; [reverse side blank].

Type specimen a relatively ruddy-backed Oregon junco, of class *ob* (see p. 271); the head is of about type 4 (p. 247), with some buff on the nape; sides army brown, thus with noticeable eumelanin component. Measurements: wing, 79.2 mm.; tail, 82.5; bill length, 8.2; bill depth, 5.5; tarsus, 20.2; middle toe, 11.6; hind toe, 8.2. Tail pattern: rectrices 5 and 6 all white; outer web of rectrix 4 nearly pure black, inner web with white spot.

Type locality.—Fort Wingate, McKinley County, New Mexico. This locality is, of course, remote from the breeding range of *J. oreganus*. *Montanus* and *mearnsi* are the most abundant races of Oregon juncos in this region in the winter.

Remarks.—The application of the name *shufeldti*, on the basis of the type, to one of the three dark-headed races of the Northwest presents a problem. Unfortunately the type has coloration of a kind found only in the coastal forms, and wing and tail length more like the average of interior birds. The intention of the describer, and certainly of Ridgway as quoted by him, was that the name should apply to the interior bird (now *montanus*) with a dull back and long wing. Most of the specimens other than the type are *montanus*. Two forms are therefore included in the description. There is no alternative but to determine which breeding population the type most closely resembles, since it is not typical of *J. o. oreganus* or of the coastal bird of Oregon and Washington, or of the interior bird of these states and British Columbia. The evidence is as follows. The type has a moderately ruddy brown back, type *o*, which has not been found in interior birds. (There is a single exception in the Hazelton breeding population which may be intergradient toward the coastal *J. o. oreganus*.) Type *o* is an intermediate type between *J. o. oreganus* and the coastal bird of Oregon and Washington, but it occurs in about 10 per cent of *J. o. oreganus* and in 10 per cent of Washington birds from the Cascade Mountains; it is the dominant type in the Vancouver Island intermediates (see p. 268). These considerations would seem to rule out the interior bird, which, therefore (see p. 392), must be called *montanus*. The wing and tail of the type, although long and equal to those of average *montanus*, do not exceed the maxima for the coastal forms; the measurements fall within the highest 5 per cent. Other measurements are not diagnostic individually so far as *montanus* is concerned, nor is the tail pattern.

There remains the problem of whether the type of *shufeldti* is *J. o. oreganus* (as Dwight thought, 1918, p. 291) or a member of the coastal Oregon-Washington race. The back is undoubtedly intermediate and is not more characteristic of one than of the other. The tail pattern is of an average type for coastal Oregon and Washington, but it is a type rare in *J. o. oreganus*. A pure white fifth rectrix, for example, occurs only in 7 per cent of *J. o. oreganus* but in 67 per cent of the other race. The type does not show excessively ruddy sides such as are found in many *J. o. oreganus*. The measurements of the feet are intermediate between the averages of the races. It may be concluded that the type is an intermediate but that the sum total of characters relates it more closely to the birds of coastal Oregon or Washington. It could conceivably have arisen either in the breeding range of *J. o. oreganus* or in that of the more southern race; the probabilities favor the latter. The best policy in nomenclature is to conserve familiar names when this is reasonably justified. The name *shufeldti* cannot be established as a synonym of *J. o. oreganus*. If the type is an intermediate, it is not closest to *J. o. oreganus*. Therefore *shufeldti* should be used for the race of western Oregon–western Washington. This usage has an advantage in that this name is currently applied to this race as well as to much of *montanus* as I have defined it.

The occurrence of a bird, such as the type of *shufeldti*, in New Mexico, far southeast of its breeding ground, is unusual. Nevertheless, some other *shufeldti*, and occasionally *J. o. oreganus*, cross east of the coastal mountains into the Great Basin and Rocky Mountain areas.

Junco oreganus oreganus (Townsend)

Fringilla oregana Townsend (1837, p. 188), original description
Struthus oreganus, Bonaparte (1838, p. 31)
Fringilla oregona Audubon (1839, p. 68), part
Niphaea oregona Audubon (1839, p. 107), part
Fringilla atrata Brandt (Icon. Ross. t. 2. 8, *fide* Bonaparte, 1850, p. 475)
Niphoea oregona, Baird (1853, p. 316), part
J[unco]. oregonus, Sclater (1857, p. 7)
Struthus oregonus, Newberry (1857, p. 88), part
Fringilla (Zonotrichia) atrata, Kittlitz (1858, p. 199)
[*Junco hyemalis*] var. *oregonus*, Ridgway (1873a, p. 613), part
Junco hyemalis oregonus, Ridgway (1875, p. 19), part
Junco hiemalis oregonus, Coues (1882, p. 56), part
Iunco oregonus, Hartlaub (1883, p. 272)
Junco oreganus oreganus, Ridgway (1901, p. 273)
Junco h[yemalis]. oreganus, American Ornithologists' Union Committee (1902, p. 323)
Junco oregonus oregonus, Dwight (1918, p. 291)

Type.—No type was selected by Townsend at the time of the description of the species. Two of his specimens now bear type labels, one, no. 1947 U.S. Nat. Mus., the other no. 24048 Acad. Nat. Sci. Phila. Which of these was first selected as a lectotype is not entirely certain, but the National Museum specimen has been accepted by Ridgway and Dwight as the type. As early as 1887 in the description of *shufeldti*, Ridgway, as there quoted, indicates that Townsend's original specimen was in the National Museum collection. The Philadelphia specimen bears no date, so that it is not certain that it was one of the original series used by Townsend; but undoubtedly it was collected in the course of Townsend's stay (fall of 1834 and April, 1835, to June, 1836) in the vicinity of the Columbia River. Therefore, I will continue to consider no. 1947 U.S. Nat. Mus. as the type; it apparently was the first specimen taken by Townsend. Fortunately, the Philadelphia specimen matches the type closely.

Type, no. 1947 U.S. Nat. Mus., examined by me, July, 1933. Two labels attached. The first bears the following:

Niphoca oregona Towns. | 1947 ♂ I | Columbia River J R.T. 5 Oct. 1834; [reverse side] Junco | oregonus | Type | of species | I [on a blue paster].

Second, a printed National Museum–Smithsonian Institution type label: 1947 | Type of [printed] Fringilla oregona Towns ♂ | Columbia River — Oct. 5–1834. J K Townsend; [reverse side] 2.98 × 250 | I.

Coloration characteristic of the most completely differentiated populations of the race. Back rich bright red-brown, type *ooob*, and head dark black, about type 3. There is some buff on the nape, as is usual in fall plumage. The sides are of average intensity. Measurements: wing, 75.0 mm.; tail, 66.8; bill, 8.1; bill depth, 5.9; tarsus, 20.3; middle toe, 13.3; hind toe, 8.6. Tail pattern: sixth rectrix pure white; outer web of fifth ½ white, inner web ⅕ white; fourth rectrix (?) black. Wing and tail length and bill length average for males; bill depth slightly greater than average; tarsus somewhat below average; toes, especially middle toe, approaching maximum, hence representing an extreme of this large-footed race. Tail pattern close to the average type throughout.

No. 24048 Phila. Acad. Sci. examined by me, July, 1933. Only the red type label is attached. Skin in good condition, taken down from a mount. Label with "Academy Nat. Sciences Philadelphia" printed on it and with the following entries:

24048 | Type [printed] of Fringilla oregana Towns. | Jour. A. N. S. Phila. VII p. 188; [reverse side] "Columbia R. | "JK.T"

Coloration essentially identical with *the type;* back color *ooob.* Fairly certainly a male on the basis of the head color and wing length. Measurements: wing, 75.2 mm.; tail, 68.6; bill length, 7.7; bill depth, 5.8; tarsus, 20.9; middle toe, 11.7; hind toe, 9.3. Tail pattern: sixth rectrix pure white; outer web of fifth ⅘ white, inner web all white; fourth rectrix all black.

Type locality.—Forests near the Columbia River, more specifically the vicinity of Fort Vancouver, Vancouver, Clark County, Washington.

During October, 1834, Townsend apparently made his headquarters at Fort Vancouver, having arrived only a few days before from the interior. He moved twenty miles down the Columbia River on the first of November. The two juncos the labels of which bear dates were taken on October 5 and 16.

Remarks.—*J. o. oreganus* winters commonly in coastal Washington and Oregon. The type and other Townsend-taken specimens are all identifiable with the breeding race of southeastern Alaska and parts of coastal British Columbia; they are clearly birds that have migrated south into the breeding range of *shufeldti.* Townsend (1839, p. 346) states: "Common on the Columbia river in winter."

Junco oreganus thurberi Anthony

Struthus oregona, Heermann (1853, p. 265)
Niphoea oregona, Baird (1853, p. 316), part
Struthus oregonus, Newberry (1857, p. 88), part
Junco oregonus, Baird (1858, p. 466), part
[*Junco hyemalis*] var. *oregonus,* Ridgway (1873a, p. 613), part
Junco hyemalis oregonus, Ridgway (1874, p. 173)
Junco hiemalis oregonus, Coues (1882, p. 56), part
Junco hyemalis thurberi Anthony (1890, p. 238), original description
Junco thurberi, Swarth (1900, p. 39)
Junco oreganus thurberi, Ridgway (1901, p. 273)
Junco oregonus thurberi, Stone (1904, p. 583)
Junco oreganus mutabilis van Rossem (1931, p. 329), original description, part (hybrid)

Type.—A. W. Anthony designated no. 3072 of his collection as the type. This is now no. 14814 of the Carnegie Museum, Pittsburgh. Examined by me there in June, 1933. Skin in good condition, although the neck is bent. Two labels attached. First with the following (printed words in italics) :

3072 | *Collection of* A. W. Anthony; *Junco oreganus thurberi* | Junco hyemalis thurberi ♂ ad | *Collector* E. C. Thurber 5-24-90; [reverse side] Wilsons Peak, Calif. | Type 14814.

Second a blue Carnegie Museum label:

14814 [vertically] *Carnegie Museum, Pittsburgh, Pennsylvania* | Type: Junco hyemalis thurberi | Anthony: Zoe, I, 1890, 238; [reverse side] = Junco oreganus thurberi | W.E.C.T.

The type is an average example of the race with respect to coloration. The back is of the common *td* type (p. 284) and the head of type 3. The sides are a light vinaceous buff. Measurements: wing, 73.0 mm.; tail, 66.2; bill length, 7.5; bill depth, 6.0; tarsus, 20.7; middle toe, 12.2; hind toe, 8.4. Tail pattern: fifth and sixth rectrices pure white; outer web of fourth rectrix ¾ white, inner web ½ white. Wing and tail are extremely short, being about equal to the minimum for males. Bill length is well below average, even for the southern populations of *thurberi;* bill depth is somewhat above average. Tarsus is especially long, but the middle toe is near average. Thus in dimensions the type is a poor representative of the race. The tail pattern is normal.

Type locality.—Wilson's Peak (= Mount Wilson), about 5500 feet altitude, San Gabriel Mountains, Los Angeles County, California.

Remarks.—The type is a bird taken on the breeding grounds. No question attaches to the application of the name to the breeding populations of interior California, of which the type, at least with respect to coloration, is characteristic.

Junco oreganus pinosus Loomis

Fringilla hudsonia, Lichtenstein (1839, p. 432), probably referred to this race, principally or entirely
Fringilla hyemalis, Vigors (1839, p. 20), probably referred to this race in part
S[truthus]. oregonus, Gambel (1847, p. 49), part
Junco oregonus, Cooper (1870, p. 199), part
Junco hiemalis oregonus, Coues (1882, p. 56), part
Junco hyemalis oregonus, American Ornithologists' Union Committee (1886, p. 275), part
Junco pinosus Loomis (1893, p. 47), original description
Junco hyemalis pinosus, American Ornithologists' Union Committee (1894, p. 47)
Junco oreganus pinosus, Ridgway (1901, p. 273)
Junco hyemalis thurberi, Fisher (1904, p. 108), part
Junco oreganus thurberi, Squires (1916, p. 202), part
Junco oregonus pinosus, Dwight (1918, p. 291)

Types.—Loomis refers to nos. 278 and 281 of the Leland Stanford Junior University Museum as types. The male, no. 278, is the first one designated and may be considered as *the type*. Types examined by me on March 8, 1935. Both in worn plumage. The type, no. 278, bears two labels. First, a Stanford Museum label with the following (italics indicate printed matter) :

Leland Stanford Junior University | Type of | *No.* 278 Junco pinosus Loomis. ♂ | *Vicinity of Monterey, Cal., July 4, 1892;* [reverse side] 5.8 | *Collector,* Leverett M. Loomis (935).

Second, a red label:

Leland Stanford Jr. University | *Type of ♂* Junco pinosus Loomis.; [reverse side] *described in* "The Auk," X, p. 47. | *Figured in* "The Auk," XI, pl. vii.

Cotype, no. 281, bears two comparable labels. First:

Leland Stanford Junior University | Type of | *No.* 281. Junco pinosus Loomis ♀ | *Vicinity of Monterey, Cal., July 4, 1892.;* [reverse side] 5.7 | *Collector, Leverett M. Loomis* (930).

Second, type label identical with that of male except for ♀ sign.

Type, no. 278, of characteristic coloration; head about type 6; back cinnamon brown, type *pdd* (see p. 291); sides ruddy, though badly worn. Measurements: wing, 69.9 mm.; tail, 58.7; bill length, 8.1; bill depth, 5.5; tarsus, 19.9; middle toe, 11.4; hind toe, 8.0. Tail pattern: rectrices 5 and 6 pure white; outer web of rectrix 4, ½ white, inner web ¼ white. Wing and tail lengths are distinctly below average, near the minimum for males; other measurements do not depart greatly from the averages. The tail is whiter than the average, especially on rectrix 4, but the pattern is not uncommon in the race.

Cotype, no. 281, characteristically colored except for back, which represents the dull extreme of the race, type *tc;* head about type 8 and sides ruddy. Measurements: wing, 66.8 mm.; tail, 59.7; bill length, 8.0; bill depth, 5.8; tarsus, 18.7; middle toe, 10.4; hind toe, 8.0. Tail pattern: sixth rectrix pure white; outer web of fifth nearly pure black, inner web nearly pure white; outer web of fourth black, inner web with white spot. Wing, tail, tarsus, and middle-toe lengths are well below averages; the other measurements and the tail pattern are near the averages.

Type locality.—Vicinity of Monterey; more exactly, woods close to Point Pinos, near Pacific Grove (see Grinnell, 1932, p. 309), Monterey County, California. The types are, of course, breeding birds, taken at a place where the race breeds abundantly.

Junco oreganus pontilis Oberholser

Junco hyemalis oregonus, Bryant (1889, p. 301), part
Junco oreganus pontilis Oberholser (1919, p. 119), original description

Type.—Oberholser designated no. 196964, U.S. Nat. Mus., Biol. Surv. coll., as the type. Examined by me in July, 1933. Plumage worn. Type bears two labels. First, a Biological Survey label with the following (italics indicate printed matter):

196964 [vertically at left] | *Biological Survey, U. S. Dept. Agriculture* | Junco o. [*cancelled:* thurberi] pontilis | *Mexico:* El Rayo *Lower Calif. Nelson & Goldman* | June 4, 1905; [reverse side] 196964 [vertically at left] 11276 [vertically at right; original number] | Hansen Laguna Mts. | type.

Second, red Biological Survey label:

196964 [vertically at left] *Biological Survey, U. S. Dept. Agriculture* | Junco o pontilis Oberholser | *Mexico:* El Rayo *Lower Calif.* | *Nelson & Goldman* | June 4, 1905; [reverse side] Hansen Laguna Mts | 11276 [vertically at right].

Type an average specimen of the race with respect to coloration; head type 6, back about type 2c or 3c, hence duller than *thurberi* of adjacent parts of California, although some tips of feathers worn and hence light colored. Measurements: wing, 75.0 mm.; tail, 67.2 (worn); bill length, 8.0; bill depth, 6.3; tarsus, 19.0; middle toe, 10.9; hind toe, 7.7. Tail pattern: fifth and sixth rectrices all white; outer web of fourth rectrix black, inner web with white spot. Wing, tail, tarsus, and middle toe minimum for males of the race; others near average. Tail blacker than average with respect to fourth rectrix.

Type locality.—El Rayo, 4700 feet altitude, Hanson Laguna Mountains (Sierra Juárez), Lower California, Mexico. El Rayo is a ranch on the western flank of the Sierra a little south of lat. 32°, 7 miles west of Laguna Hanson (Grinnell, 1928, p. 23).

Junco oreganus townsendi Anthony

Junco townsendi Anthony (1889, p. 76), original description
Junco hiemalis townsendi, Coues (1894, p. 900)
Junco h[yemalis]. townsendi Anthony (1895, p. 183)
Junco oreganus townsendi, Oberholser (1918, p. 211)
Junco mearnsi townsendi, Dwight (1918, p. 296)
Junco insularis townsendi, Dwight (1919, p. 287)

Types.—Anthony specified two types: first, male no. 2539 of his collection (now no. 14910 of the Carnegie Museum) and second, female no. 2538 (now no. 14909, Carnegie Museum). The male, the first named, may be considered as *the type*. Both the type and the cotype examined by me at Carnegie Museum, Pittsburgh, in June, 1933; skins in good condition. The type carries two labels. The first is an Anthony collection label bearing the following (italics indicate printed matter):

Collection A. W. Anthony 2539 | *Junco townsendi* | Junco townsendi 14910 ♂ | San Pedro, L. Calif. 4/28 *1889;* [reverse side] (*Type*).

Second, a blue type label:

14910 [vertically at left] | *Carnegie Museum, Pittsburgh, Pennsylvania* | *Type:* Junco townsendi | Anthony: Proc. Calif. Acad. Sci., (2) II, 1889, 76; [reverse side blank].

Cotype, no. 14909, bears two comparable labels. First:

Collection A. W. Anthony 2538 | *Junco townsendi 14909* | Junco townsendi | San Pedro. L. Calif. 4/29 *1889;* [reverse side] type.

Second, a blue label same as that of the type except for the number 14909.

Type characteristically colored, but back extremely dull; head about type 8; back, 4c. Cotype comparable, with back somewhat browner than average; head about type 9. Measurements (of type): wing, 78.2 mm.; tail, 72.2; bill length, 7.8; bill depth, 6.0; tarsus, 19.7; middle toe, 11.6; hind toe, 7.8. Tail pattern: rectrices 5 and 6 pure white; outer web of fourth with trace of black, inner web ⅓ black. Measurements and pattern do not depart greatly from averages for males. Measurements of cotype: wing, 74.1; tail, 68.3; bill length, 7.8; bill depth, 5.4; tarsus, 18.8; middle toe, 11.1; hind toe, 7.7. Tail pattern: rectrices 5 and 6 pure white; outer web of fourth rectrix ½ white, inner web

⅓ white. Foot measurements and bill depth distinctly below averages for females; other measurements and tail pattern close to average.

Type locality.—"San Pedro Mountain" (San Pedro on labels), Sierra San Pedro Mártir, northern Lower California.

Remarks.—The types, although taken in April, probably were on their breeding grounds. There is no doubt that they represent the breeding population of the Sierra San Pedro Mártir.

Junco insularis Ridgway

Junco insularis Ridgway (1876, p. 188), original description
Junco mearnsi insularis, Dwight (1918, p. 296)
Junco insularis insularis, Dwight (1919, p. 287)
Junco oreganus insularis, Hellmayr (1938, p. 550)

Type.—Ridgway mentioned no type in the original description. Specimens nos. 70015–70018 and 70020–70027 in the United States National Museum were listed by him. The first of these, no. 70015, was at some time selected as *the type*. This bird examined by me in July, 1933; skin in good condition. Two labels attached. First, a printed Smithsonian Institution–National Museum label with the following written entries:

70015 | Junco   10 | Guadalupe Id ♂ | E. Palmer; [reverse side] Feb. 12, 1875 JLR | I Type [in pencil].

Second, a Smithsonian Institution–National Museum red type label:

70015 | Type of [printed] Junco insularis Ridgw ♂ | Guadalupe Isld. Dr. E. Palmer; [reverse side] I.

Type, of normal coloration. Bill dark horn color at tip, along margins and at base of maxilla. Measurements: wing, 68.4 mm.; tail, 61.8; bill length, 9.6; bill depth, 6.2; tarsus, 20.7; middle toe, 11.6; hind toe, 8.4. Tail pattern: outer web of sixth rectrix ⅔ white, inner web all white; outer web of fifth with trace of white, inner web with trace of black; outer web of fourth rectrix black, inner web ½ white. Wing, tail, and bill-depth measurements slightly below average; bill length, tarsus, and middle toe above average. Tail normal, but slightly darker than average on outer webs.

No. 70023 U.S. Nat. Mus., has "'type of ♀'" marked on back of label in pencil.

Type locality.—Northern end of Guadalupe Island, Lower California, Mexico. In the description, the occurrence of birds on board ship at the island and on the summit of Mount Augusta, 3500 feet, is mentioned. It is not certain precisely where the type was taken.

Junco hyemalis hyemalis (Linnaeus)

[*Fringilla*] *hyemalis* Linnaeus (1758, p. 183), description based on *Passer nivalis* Catesby (1731, p. 36, pl. 36)
[*Emberiza*] *hyemalis* Linnaeus (1766, p. 308)
Fringilla hudsonias Forster (1772, p. 406)
Fringilla hudsonia Gmelin (1788, p. 926)

P[*asser*]. *nivalis* Bartram (1791, p. 291, *fide* Coues, 1875, p. 351; p. 289 of reprint of 1792, which seen)

Fringilla nivalis, Wilson (1810, pp. xii and 129, pl. 16, fig. 6), part, this name used only in contents, p. xii.

Struthus hyemalis, Bonaparte (1838, p. 31)

Niphaea hyemalis, Audubon (1839, p. 106)

Euspiza [*Niphoea*] *hyemalis*, Blyth (1849, Cat. Birds Mus. Asiat. Soc. Bengal, p. 130, *fide* Ridgway, 1901, p. 280)

Niphaea . . . hudsonia, Lichtenstein (1854, Nomencl. av. Mus. Berol., p. 43, *fide* Ridgway, 1901, p. 282

J[*unco*]. *hyemalis*, Sclater (1857, p. 7)

[*Junco hyemalis*] var. *hyemalis*, Ridgway (1873a, p. 613)

[*Junco hyemalis*] a. *hyemalis*, Coues (1874, p. 141)

[*Junco hyemalis*] b. *hyemalis*, Trippe (1874, p. 145)

Junco hiemalis Coues (1882, p. 55), part

J[*unco*]. h[*iemalis*]. *connectens* Coues (1884, p. 378), original description, part

J[*unco*]. *hyemalis connectens*, Ridgway *in* Beckham (1887, p. 122), part

Junco hyemalis hyemalis, Ridgway (1901, p. 273)

There is no type. Linnaeus based his description on Catesby's account. Linnaeus characterizes the species as follows (p. 183): "*hyemalis*. 30 F[ringilla]. nigra, ventre albo. *Passer nivalis. Catesb. car.* I. *p. 36. t. 36. Habitat in* America." Although this would appear to refer to *Junco hyemalis* as now known, Catesby's more explicit account leaves no doubt. Catesby says, p. 36: "The bill of this Bird is white: The Breast and Belly white. All the rest of the Body black; but in some places dusky, inclining to Lead-colour. In *Virginia*, and *Carolina* they appear only in Winter: and in Snow they appear most. In Summer none are seen. Whether they retire and breed in the North (which is most probable) or where they go, when they leave these Countries in the Spring, is to me unknown." As Brewster (1886a, p. 109) has stated, this description clearly applies to the northern race of slate-colored junco. The fact that part, and not all, of the dark parts are "inclining to Lead-colour" may be taken as eliminating *carolinensis*. Also the winter occurrence suggests the prevalent flocks of the northern form rather than *carolinensis*, which, I believe, does not commonly leave the vicinity of the mountains (see p. 329).

Type locality.—In the fourth edition of the A.O.U. Check-list (1931, p. 345) the type locality is restricted to South Carolina. Catesby mentions both Virginia and Carolina, in the order given. It is not particularly significant which region is specified, as all this section of the Atlantic coast is in the winter range of the race and there is no more chance of intermixture of other races in the winter flocks in one area than in another.

Junco hyemalis carolinensis Brewster

Fringilla nivalis, Wilson (1810, pp. xii and 129, pl. 16, fig. 6), part; this name used only in contents, p. xii

Fringilla hudsonia, Wilson (1810, p. 129, pl. 16, fig. 6), part

Fringilla hyemalis, Jardine ed. Wilson's Am. Ornith., 1832, p. 272, pl. 16, fig. 6, *fide* Ridgway, 1901, p. 280), part

Junco hyemalis, Baird, Brewer, and Ridgway (1874, p. 580), part

Junco hiemalis Coues (1882, p. 55), part

Junco hyemalis carolinensis Brewster (1886a, p. 108), original description
Junco carolinensis Brewster (1886b, p. 277)
J[unco]. h[iemalis]. *carolinensis*, Coues (1887, p. 874)
[*Junco hiemalis*] Subsp. a. *Junco carolinensis*, Sharpe (1888, p. 649)

Type.—Brewster designated nos. 10597 and 10567 as types. The first one mentioned, the male, no. 10597, may be considered as *the type*. It is now no. 210597, Mus. Comp. Zoöl. It and the cotype examined by me in Cambridge in July, 1933. Type bears two labels. First, a Brewster Collection label with the following (italics indicate printed matter):

Junco h. carolinensis ♂ ad. Type ♂ | North Carolina (Black Mts.) Jun 2 | 1885 | [vertically at right] *No.* | 10597; [reverse side] *Collection of William Brewster* | *Type* | *Collection* William Brewster.

Second, a red Museum of Comparative Zoölogy label:

No. 210597 *Museum Comparative Zoology* | Cotype | Junco hyemalis carolinensis | [cancelled: *Loc.*] Brewster   *coll.;* [reverse side] Auk III. Jun. 1886. p. 108.

The type is a characteristic light slate bird, with tail with maximum amount of white. Measurements: wing, 80.5 mm.; tail, 69.5; bill length, 8.4; bill depth, 6.3; tarsus, 21.3; middle toe, 13.3; hind toe, 8.8. Measurements close to average except tail and middle toe, which are below average.

Cotype, no. 210567 Mus. Comp. Zoöl. Two labels attached comparable to those of type. First:

Junco hiemalis carolinensis type ♀ ad ♀ | North Carolina (Highlands) May 28, 1885. | [vertically at right] *No.* 10567; [reverse side] *Collection of William Brewster* | Type of ♀ | *Collection* Wm. Brewster.

Second, red label:

No. 210567 *Museum Comparative Zoology* | Junco hyemalis carolinensis | [cancelled: *Loc.*] Brewster | Cotype   *coll.;* [reverse side] Auk III Jan. 1886 p. 108.

Cotype pale slate, about type 9 or 10, with sides and back washed with smoke gray. Measurements: wing, 74.9 mm.; tail, 68.5; bill length, 8.5; bill depth, 5.9; tarsus, 20.5; middle toe, 11.7; hind toe, 8.7.

Type locality.—Black Mountain, Buncombe County, North Carolina. Evidently the type was taken at a higher elevation than Black Mountain Post Office, probably to the north in the Black Mountains (see label) at 5000 feet altitude or more, certainly above 4300 feet (Brewster, 1886a, pp. 95 and 109). The types clearly are breeding birds. The cotype was taken at Highlands, Macon County, North Carolina.

Junco hyemalis cismontanus Dwight

This synonymy includes names applied to hybrids of *J. oreganus montanus* or of *J. oreganus oreganus* with *J. hyemalis*, as it is not possible to distinguish certain hybrids from the race *cismontanus* that breeds in interior northern British Columbia (see p. 345).

Hybrid between *hyemalis* and *oregonus*, Baird, Brewer, and Ridgway (1874, p. 579)
J[unco]. h[iemalis]. *connectens* Coues (1884, p. 378), original description, part

J[unco]. h[yemalis]. *connectens* Coues (1897, p. 94), part
Junco montanus Ridgway (1898, p. 321), original description, part
Junco hyemalis × *Junco oreganus shufeldti*, Ridgway (1901, p. 276)
Junco hyemalis, Macoun (1904, p. 499), part
Junco oreganus shufeldti, Macoun (1904, p. 504), part
Junco hyemalis shufeldti, MacFarlane (1908, p. 411), part
Junco hyemalis montanus, Osgood (1909, p. 41), part
Junco h[yemalis]. hyemalis, Riley (1912, p. 68), part
[*Junco*] "*cismontanus*" Dwight (1918, p. 295), original description
[*Junco*] "*transmontanus*" Dwight (1918, p. 295), original description, part
Junco hyemalis × *oregonus*, Dwight (1918, p. 295)
Junco oregonus × *hyemalis*, Dwight (1918, p. 295)

Winter-taken or migrant hybrids and members of the race *cismontanus* have had various forms of the names *oreganus, shufeldti, montanus,* or *hyemalis* mistakenly or loosely applied to them. These names have been used apparently without intention of expanding their application to hybrids or to *cismontanus*. Only when they have been used for breeding *cismontanus*, or in reference to breeding range that clearly includes areas of prevalent hybridization, have they been considered by me in compiling the above synonymy.

No type was specified by Dwight, for it was his intention to suggest a name for a certain kind of hybrid without recommending its use. The name "*cismontanus*" is valid because Dwight gave an "indication" which beyond reasonable doubt fits the race to which Swarth (1922, p. 244) endeavored to apply the name *connectens* (see p. 407). This indication is contained in the following statement (1918, p. 295) : "... east of the Rocky Mountains we have a *hyemalis* darkened or blackened by the *oregonus* strain that, for convenience, might be called '*cismontanus*.' . . . If these names [referring also to *transmontanus*] could be restricted to definite geographical areas there would be some grounds for admitting the existence of three races." This latter condition with respect to *cismontanus* is fulfilled, as shown by Swarth, and substantiation has been offered in the present study. As pointed out earlier, *cismontanus* is a group of hybrid origin, now stabilized, which occupies a geographic area in the breeding season to the exclusion of other forms. In other regions birds of identical appearance may be produced by hybridization. Thus, although Dwight coined the name solely for hybrids, it has since been found that a stabilized group indistinguishable from the hybrids exists, and the name must be used. In similar fashion the name *connectens* could have been applied, except that a type identical with *hyemalis* was selected.

Dwight's indication describes perfectly the males of the race in question. Part of the race breeds east of the Rocky Mountains, as he states. Further, there is a series of specimens in Dwight's collection that bear his penciled identification: " 'cismontanus' D." These are comparable to breeding birds from the Atlin, Carcross, and interior Stikine districts of British Columbia and southern Yukon.

Selection of type.—Since Dwight's own series was before him at the time he coined the name *cismontanus*, it is appropriate to select a lectotype from it. Accordingly, no. 402559 Am. Mus. Nat. Hist. (12281, Dwight collection) is chosen. It bears two labels (*fide* John T. Zimmer, March 24, 1938). Original label:

Junco hyemalis | 13 Feb. 05 ♂ Sumas, B. C. | Allan Brooks col.

Dwight label:

Junco o. shufeldti ♂ | B. C. Sumas. Feb. 13, 1905 | 12281 | hyemalis × oreganus D | = cismontanus | AMNH 402559.

Because Dwight is not known to have made use of any particular breeding specimens in formulating his concept of *cismontanus*, a winter-taken *cismontanus* unfortunately must be selected.

The type was examined by me in July, 1933, at the American Museum. It was compared directly with no. 44868 Mus. Vert. Zoöl., which is a typical breeding male of the race, from Carcross, Yukon Territory, taken May 26, 1924. The type, although in less worn plumage, would have been equivalent in appearance by breeding time. The only difference is that the type has a somewhat blacker head; size was not compared. The coloration of no. 44868 is as follows: sides completely slate; hood no. 6; pileum dull black; head well contrasted with back; back *aaa* (see pp. 330–333).

Type locality.—Sumas, New Westminster District (lower Fraser Valley), British Columbia. The local breeding bird in this region is *Junco oreganus shufeldti*.

Junco aikeni Ridgway

Junco hyemalis var. *aikenii* Aiken (1872, p. 201), *nomen nudum*
[Junco hyemalis] var. *aikeni* Ridgway (1873a, pp. 613, 615), original description
J[unco]. *aikeni*, Baird, Brewer, and Ridgway (1874, pl. 26, fig. 6)
[Junco hyemalis] b. *aikeni*, Coues (1874, p. 141)
[Junco hyemalis] a. *aikeni*, Trippe (1874, p. 145)
Junco hyemalis *aikeni*, Drew (1881, p. 90)
Junco hiemalis *aikeni*, Coues (1882, p. 55)
Junco hyemalis *danbyi* Coues (1895, p. 14), original description
Junco *danbyi*, American Ornithologists' Union Committee (1897, p. 133)

Type.—Ridgway based his description on no. 61302 of the United States National Museum. This, although not specifically mentioned as the type, is the only specimen that can be so regarded; others were not listed in the formal description. Type examined by me in July, 1933, at the United States National Museum. Two labels are attached. First, an Aiken collection label:

Collection of C. E. Aiken, [printed] ♂ 135 Peoria St., Chicago. | 1053 Junco [*cancelled:* hyemalis] var. aikeni, Ridgw. | Collected [printed] El Paso Co., Col. T. Dec. 11. 71 — Aiken [printed]; [reverse side] Type of Ridgway's description 61302 | I.

Second, a red, printed Smithsonian Institution–National Museum label:

61302 Type of [printed] Junco aikeni 1053 ♂ | El Paso Co., Col. $\left\{ \begin{array}{c} \text{Dec. 11} \\ 1871 \end{array} \right\}$ C. E. Aiken.

The type is a light slate-colored bird, brownish as are females and immatures. There is white on the middle secondary coverts. Measurements: wing, 84.0 mm.; tail, 78.6; bill, 8.7; bill depth, 7.3; tarsus, 21.2; middle toe, 11.6; hind toe, 9.4. Tail pattern: fifth and sixth rectrices white; outer web of fourth rectrix ⅔ white, inner web all white; outer web of third rectrix black, inner web ⅓ white. Wing and tail are below average for males, the former nearly as short as the average for females; the hind toe is especially long; the other measurements do not depart importantly from the averages. The tail pattern is normal, but it is closer to the average for females than it is to that for males. This and other features of the type suggest the possibility that the bird may have been missexed.

Type locality.—El Paso County, near Fountain, Colorado. More specifically, Turkey Creek, "some eighteen miles southwest from Colorado Springs" (Warren, 1936, p. 235).

Remarks.—Although the type and the original specimens are winter taken, they all clearly pertain to the species that nests in the Black Hills of South Dakota.

Type Specimens upon Which Discarded Names Were Based

It is well to place on record information relative to types that formed the basis for certain invalid names. For some of the early names, such as *Fringilla cinerea, atrata, nivalis,* and *hudsonias,* that were relegated to synonymy, there are no types. All types connected with the more recent names have been examined. *Transmontanus,* for which there is no type, was proposed by Dwight (1918, p. 295) for a certain phase of hybrid of *J. oreganus* and *J. hyemalis*. No stable group of birds of the kind he describes exists. Hybrids with the essential characters of *oreganus,* but with partly slate-colored backs, occur within parts of the range of *montanus* adjacent to *J. h. hyemalis* or *J. h. cismontanus.* These correspond to Dwight's *"transmontanus."* The name remains that of a hybrid, then, as Dwight intended it should be.

Junco annectens Baird

Type.—When Baird (1870, p. 564) named *annectens,* he had several specimens before him, all but one of which were *mearnsi*. Ridgway (1901, p. 276) says: " . . . it is nevertheless easy, in view of the characters most prominently mentioned in the diagnosis, to determine which should be considered as the type. For instance, the phrase 'whole interscapular region . . . light chestnut rufous,' found in the description, applies only to no. 11164." With this opinion I am in agreement, but there is an unfortunate error regarding the number, which probably was incorrectly transcribed along with the data, and should be no. 10701. In Ridgway's original description (1897, p. 94) of *mearnsi,* in which he develops a similar argument for the restriction of the name *annectens* to a particular specimen, he designates no. 10701 as Baird's type and 11164 as the type of *mearnsi*. My recent examination of the types confirms this. No. 10701 is the red-backed bird, and 11164 and others are brown backed. No. 10701 is therefore a properly selected type of *annectens*.

The type bears two labels. First, a printed Smithsonian Institution–South Pass Wagon Road Exp. label:

No. [printed] 10701 [*cancelled with blue pencil:* Hybrid—J. oregonus & caniceps; *interlineated:*] annectens ♀ | 474 | Fort Bridger, Utah, (Camp Scott) [printed] May 28, 1858.

C. Drexler [year and name printed]; [reverse side, in blue pencil:] "Hybrid" [in ink:] junco ridgway Mearns. [in blue pencil:] connectens × caniceps | R.R.

Second, red Smithsonian Institution–National Museum type label:
Type of [printed] Junco annectens Baird; [reverse blank].

The back is the full reddish brown of *caniceps*, and the sides are broadly pinkish cinnamon as in *mearnsi*. Some of the secondary coverts are tipped with reddish brown. The bird is exceptionally worn for the date, May 28. It undoubtedly was breeding, unlike the type of *mearnsi* taken earlier in the season at the same locality. Measurements: wing, 78.5 mm.; tail, 69.4; bill, 8.0; bill depth, 6.2; tarsus, 20.4; middle toe, 11.7; hind toe, 7.5. Tail pattern: sixth rectrix all white; outer web of fifth ½ white, inner web with small amount of black; outer web of fourth black, inner web with white spot.

Type locality.—Camp Scott, Fort Bridger, Uintah County, Wyoming.

Identification.—Hybrid of *J. c. caniceps* × *J. o. mearnsi* (see synonymies of these).

Junco ridgwayi Mearns

Type.—Mearns (1890, p. 243) named *ridgwayi* from a type in his collection, no. 2770, now no. 52902 Am. Mus. Nat. Hist. Examined by me in July, 1933. Three labels attached. First (italics indicate printed matter):

Collected By Edgar A. Mearns M D | *No.* 2770 ♂ ad. April 22, 1884 | *Locality,* Whipple Barracks, Arizona | caniceps × mearnsi [in pencil]; [reverse side] Junco ridgwayi Mearns | 163; 257; 80; 77. | Culmen 12; tarsus 20.5; middle toe & claw 20.

Second:

Mearns | *Collection* [vertically at ends of label] | *American Museum* 52902 *of Natural History* | April 22, 1884. ♂ ad | Junco ridgwayi | Arizona. Whipple Barracks Dr E. A. Mearns; [reverse side]; No. 2770 Mearns Coll | Type of Junco Ridgwayi, Mearns | Auk, vol. p.

Third, a red printed type label of the American Museum with no entries upon it.

The type is of the same coloration as the type of *annectens;* back pure reddish brown, sides pinkish cinnamon. Measurements: wing, 79.5 mm.; tail, 74.0; bill, 8.0; bill depth, 6.4; tarsus, 20.0; middle toe, 11.5; hind toe, 8.4. Tail pattern: sixth rectrix all white; outer web of fifth white, inner web with trace of black; outer web of fourth with trace of white, inner web ¼ white.

Type locality.—Whipple Barracks (Fort Whipple at Prescott), Yavapai County, Arizona. The locality represents a winter or migration station.

Identification.—Hybrid *J. c. caniceps* × *J. o. mearnsi*. Mearns identified the type with No. 10701, U.S. Nat. Mus., the bird that later proved to be the type of *annectens. Ridgwayi* is, therefore, a synonym for the hybrid *annectens* (see p. 405).

Junco oreganus mutabilis van Rossem

Type.—No. 31,126, collection of Donald R. Dickey at the University of California, Los Angeles. Two labels attached. First:

Dickey | Col. No. 31,126 [vertically at left] Junco oreganus mutabilis Type van Rossem ♂ im | Charleston Mts., Clark Co., Nevada | Col. by A J. van Rossem | Sept. 14, 1930; [reverse side] col. no. 31,126 [vertically at left] Collection of Donald R. Dickey ♂ im | alt. 8200 ft. Assoc. Yellow Pine | Lee Cañon AJvR. #13,221.

Second, a red type label:

Collection of Donald R. Dickey No. 31,126 | Type of Junco oreganus mutabilis van Rossem | Described in Trans. San Diego Soc. Nat. Hist., 6, no. 22, June 5, 1931, p. 329.

Back of pure *caniceps* type, sharply limited. Hood not strongly set off from sides, which are gray anteriorly. Sides posteriorly partly buff, with phaeomelanin. Head as in lighter types of female *thurberi*. Measurements: wing, 78.4 mm.; tail, 70.9; bill length, 8.3; bill depth, 6.2; tarsus, 20.4; middle toe, 12.0; hind toe, 8.8. Tail pattern: sixth rectrix all white; outer web of fifth white, inner web ¾ white; outer web of fourth black, inner web with white spot.

Type locality.—Lee Cañon, yellow pine association, 8200 feet, Charleston Mountains, Clark County, Nevada. The bird, although taken in September, undoubtedly was on the breeding grounds where it was raised.

Identification.—Hybrid *J. c. caniceps* × *J. o. thurberi*. The hybrid nature of the population breeding on the Charleston Mountains has been fully discussed (see p. 189 ff.). The type is a fair representative of this heterogeneous population; it combines characters of the parent forms. The population is highly varied and there are many birds that are distinctly different from the type.

Junco hiemalis connectens Coues

Type.—Coues (1897, p. 94) states that types were selected from the Brewster Collection when he characterized *connectens* (Coues, 1884, p. 378). Allen and Brewster (1883, p. 189) mention two specimens taken by Brewster in 1882 shortly before Coues named *connectens*. The first of these two similar specimens, no. 7046, apparently was chosen as *the type*. Certainly it was one of those used by Coues, and it probably was selected by him.

Type, no. 207046 Mus. Comp. Zoöl., examined by me in July, 1933. Two labels attached. First (italics indicate print):

Collection of William Brewster | Junco h. connectens Type ♀ | Colorado (Colo. Springs) Apr. 26, 1882 *W. B.* | *No.* 7046 [vertically at right]; [reverse side] ♀ Type of Junco | connectens | Coues.

Second:

No. 207046 Museum of Comparative Zoology | Type | Junco h. connectens Coues | [*Cancelled: Loc*] (= Junco h. hyemalis (Linn).); [reverse side] Key N. Am. Birds 2nd Ed | 1884. p. 378.

The type was identified by Ridgway and by Dwight as an immature female *J. h. hyemalis*. Coues' description also might in part refer to such an individual of this race. Swarth (1922, pp. 244–245) in resurrecting the name for application to the Cassiar breeding bird was unable to examine the type personally. He sent some sample specimens east to Outram Bangs for comparison with the type. Bangs reported to Swarth that the type was the counterpart of no. 39957 Mus. Vert. Zoöl. from Telegraph Creek, British Columbia, May

29, 1919: " 'Indeed you would have difficulty in telling the two apart, except that the type . . . is . . . in a little more worn plumage. I can't find spring females from the east just like these, but on the other hand, autumnal females much resembling them . . . are common in our series' " (Swarth, 1922, p. 245).

It was with much interest that I compared this same no. 39957 with the type at the Museum of Comparative Zoölogy. In color values and detailed distribution of buff overcasting the similarity is very great. The overcasting on the breast of the type is a little more extensive and, as a result, the hood line less distinct. The less worn feather tips of no. 39957 conceal a slightly deeper black than exists in the type. The most important character of the type is, however, the extension of the hood mark onto the sides beneath the feather tips. Consequently the hood line is concave, not convex as in no. 39957. This was overlooked by Bangs, probably because of concentration upon the superficial buff color, but it is the crucial point in the determination. The type is, then, a fairly typical female *hyemalis* which can be matched by eastern birds. The buff overcast is a little less uniform on the breast than in the majority of female *hyemalis*.

Measurements: wing, 75.0 mm.; tail, 63.2; bill, 7.9; bill depth, 5.5; tarsus, 19.8; middle toe, 11.7; hind toe, 8.4. Tail pattern: fifth and sixth rectrices pure white; outer web of fourth rectrix nearly pure white, inner web ⅔ white; outer web of third rectrix black; inner web with small white spot.

Type locality.—Colorado Springs, El Paso County, Colorado. The locality is of course a winter or migration station.

Identification.—*Junco hyemalis hyemalis.*

Junco oregonus couesi Dwight

Type.—No. 16969, collection of J. Dwight. Examined by me July 20, 1933. Two labels attached; a third type label was added on this date by J. T. Zimmer. First, a small round label:

16969 | A B/$_D$ | Breeding | D; [reverse side] J. h. shufeldti | 14.5.06 ♂ | Okanagan.

Second (italics indicate print):

Collection of J. Dwight, Jr. | Junco hyemalis [*cancelled:* shufeldti$_D$] connectens ♂ | B. C., Okanagan May 14, 1906 | 16969 [vertically at right]; [reverse side] Like L. B. ♀ 4129 but paler | back   blacker head. | A.M.N.H. | oreg.$_D$ [all in Dwight's hand, except A.M.N.H.].

Third, a pink American Museum type label:

Type of Junco oregonus couesi | Dwight.

Dwight (1918, p. 291) provided this name for birds that breed in the interior of British Columbia and in parts of both coastal and interior Washington and Oregon. He eliminated the name *shufeldti* by synonymizing it, incorrectly, with *oreganus,* whereas it can be applied to the coastal division of his *couesi.* He ruled out the name *montanus* because he believed it to represent a hybrid. An insufficient amount of breeding material made it impossible for Dwight to appreciate the variability of this race and hence the fact that the type and topotypes of *montanus* fit into the pattern of variation (see p.

392). *Couesi* as described, therefore, is a synonym partly of *shufeldti* and partly of *montanus*, but the type is an example of *montanus*.

The type has a back 2b, and head of about No. 4 intensity (see p. 247).

Type locality.—Okanagan, British Columbia.

Identification.—*Junco oreganus montanus.*

Junco hyemalis danbyi Coues

Coues did not name a type in the original description (1895, p. 14). Later he (1897, p. 94) states that the type was deposited in the United States National Museum, but does not give the number. This presumably is no. 153188, now labeled as the type, which I examined in July, 1933. Two labels are attached. First, a National Museum label from which the top part has been cut off:

Type specimen | No. [printed] Junco danbyi (=aikeni Sex. [printed] | Locality [printed] Custer City, So. Dak. Date [printed] Sept. 1895 Dr. E. Coues; [reverse blank].

Second, a National Museum–Smithsonian Institution type label:

53188 | Type of [printed] Junco danbyi Coues; [reverse blank].

The bird is typical of *aikeni;* the overcast plumage suggests that it is a female. The last primary is still in its sheath, indicating a primary molt which is just being completed. This means the bird is an adult, not a juvenile, or immature as Coues indicated. The absence of well-defined wing bars is not a constant sign of age; juveniles may show white spots on the coverts (see p. 347).

Measurements: wing, 87.7 mm.; tail, 79.0; bill length, 8.7; bill depth, 6.8; tarsus, 20.7; middle toe, 12.1; hind toe, 8.0. Tail pattern: rectrices five and six all white; outer web of fourth rectrix ¼ white, inner web ½ white; outer web of third black, inner web with white spot. The wing length is great for a female.

Type locality.—Custer City (Custer), Custer County, South Dakota. This is a breeding locality.

Identification.—*Junco aikeni.*

Junco phaeonotus colimae van Rossem

Type.—No. 99.2.1.2255, British Museum. Examined by me in 1934 before it was made a type. Sierra Nevada de Colima, Jalisco, Mexico, April 9, 1889; collected by W. B. Richardson.

Noted by me as possessed of characters average for *J. p. phaeonotus*.

Type locality.—Sierra Nevada de Colima, Jalisco, Mexico. The type probably was a breeding bird and almost certainly was on its breeding grounds.

Identification.—*Junco phaeonotus phaeonotus* (see p. 219). Original description by van Rossem (1938, p. 132).

Junco phaeonotus australis van Rossem

Type.—No. 99.2.1.2169, British Museum. Examined by me in 1934 before it was made a type. Omilteme, 8000 ft., Guerrero, Mexico, August 1, 1888; collected by Mrs. H. H. Smith.

Noted by me as possessed of characters average for *J. p. phaeonotus*.

Type locality.—Omilteme, 8000 ft., Guerrero, Mexico. This is a breeding station.

Identification.—*Junco phaeonotus phaeonotus* (see pp. 219–220). Original description by van Rossem (1938, p. 133).

APPENDIX B

Breeding Localities

The localities listed in this appendix are those from which breeding or permanently resident birds have been examined. Only typical or unmixed populations are represented under a given name, unless otherwise stated.

Junco caniceps caniceps

COLORADO

Larimer County: Medicine Bow Range; Estes Park; Longs Peak.
Jackson County: Mount Zirkel; Hill Creek.
Routt County: Williams Range.
Boulder County: Boulder.
Grand County: Grandby.
Jefferson County: Golden; Wigwam Creek, near Deckers.
Gilpin County: Mount McClellan, Steven's Mill.
Clear Creek County: Alice; Berthoud's Pass.
Summit County: Breckenridge.
Eagle County: west slope Gore Range.
El Paso County: Pikes Peak; Mount Bross, near Alma; Montgomery; Fairplay; Bailey; South Park; Deer Creek; Elk Creek; Mount Ptarmigan.
Pueblo County: Beulah.
Fremont County: Sangre de Cristo Range.
Gunnison County: Cebolla.
Saguache County: upper Saguache River; Cochetopa Pass.
San Juan County: Silverton.
San Miguel County: Trout Lake; south fork of Rio San Miguel.
Dolores County: Rico.
Costello County: Fort Garland.
Conejos County: Osier.
Huerfano County: Wet Mountains.
Archuleta County: Little Navajo River; Pagosa.

NEW MEXICO

Colfax County: Taos Peak; Elizabethtown.
Taos County: 5 miles south of Twining.
Rio Arriba County: 8 to 12 miles north of El Rito.
San Miguel County: Pecos; Willis.
Santa Fe County: Pecos Forest Reservation; Pecos Baldy.
Sandoval County: James Mountains, 25 miles west of Española.
McKinley County: Chuska Mountains, Long Lake, 8800 feet.

UTAH

Summit County: 10 miles east of Kamas; Mount Baldy.
Sanpete County: Great Basin Experiment Station east of Ephraim.
Sevier County: 10 miles north of Fish Lake.
Garfield County: 7 miles east of Widtsoe.
Iron County: Cedar Breaks.
San Juan County: Navajo Mountain.

NEVADA

Humboldt County: Martin Creek Ranger Station, 7000 feet, Santa Rosa Mountains.
Elko County: Secret Pass; Three Lakes; south fork of Long Creek (all in Ruby Mountains).
White Pine County: Lehman Creek and Baker Creek, Snake Mountains; Cleve Creek,

Shell Creek Range; 3 miles west of Hamilton, White Pine Mountains; Hendry Creek, 9100 feet, 1½ miles east of Mount Moriah.

Eureka County: 4 miles south of Tonkin, Denay Creek, Roberts Mountains.

Lander County: Birch Creek and Kingston Creek, Toyabe Mountains; Peterson Creek, Shoshone Mountains.

Nye County: Green Monster Cañon, Monitor Range; north slope of Toquima Peak; head of Reese River, Arc Dome, Mohawk Ranger Station, Wisconsin Creek, north of Twin River and South Twin River in Toyabe Mountains.

Lincoln County: Wilson Peak [Cedar Range], 8500 feet.

Intermediates (hybrids) between *J. c. caniceps* and *J. c. dorsalis*

ARIZONA

Coconino County: Bright Angel; 1 mile north of National Park Boundary, Kaibab Forest; 2–3 miles west of V. T. Ranger Station; 3 miles south of Dry Park Ranger Station; 4 miles north of Jacob's Lake (all on Kaibab Plateau).

NEW MEXICO

Valencia County: Bear Ridge, Zuni Mountains; San Mateo Mountains (probably hybrid population).

McKinley County: Nutria, Zuni Mountains.

Intermediates (hybrids) between *J. c. caniceps* and *J. o. thurberi*

NEVADA

Clark County: Lee Cañon, 8200 to 9200 feet; Clark Cañon, Charleston Park, and north base of Charleston Peak, Charleston Mountains; Hidden Forest, 8500 feet, Sheep Mountains.

Nye County: Grapevine Peak, 8000–8500 feet; 2½ miles east, 1 mile south of Grapevine Peak, 6700 feet.

Esmeralda County: Chiatovich Creek, 8000 to 8200 feet, and Quinn Mine, 10,000 feet, White Mountains.

Mineral County: 2 miles southwest of Pine Grove, 7250 feet; Cottonwood Creek, 7900 feet; Lapon Cañon, 8900 feet; south fork of Cat Creek, 8500 feet.

Douglas County: Desert Creek, 7100 feet, Sweetwater Range.

Humboldt County: Duffer Peak, 8400 feet, Pine Forest Mountains.

CALIFORNIA

San Bernardino County: northwest side Clark Mountain, 7300 feet.

Inyo County: Mountain Spring Canyon, 6600 feet, Argus Mountains; Hanaupah Canyon, 9300 feet, and "Coalkilns" [Wildrose Canyon], Panamint Mountains; head of Lead Cañon, 10,000 feet, Inyo Mountains; Weyman Creek, 8200 feet, White Mountains.

Mono County: McAfee Creek, 10,000 feet; near Blanco Mountain, 10,500 feet; near Big Prospector Meadow, 10,300 feet; Poison Creek, 9600 feet; McCloud Camp on Cottonwood Creek, 9200 feet (all in White Mountains).

Intermediates (hybrids) between *J. c. caniceps* and *J. o. mearnsi*

WYOMING

Albany County: Laramie Peak, 10,000 feet.

Carbon County: south base of Bridger Peak, 8800 feet, Sierra Madre Mountains; Black Hills [=Medicine Bow Range].

Natrona County: Rattlesnake Mountains, 7500 feet; Casper Mountains.

Uintah County: Camp Scott, Fort Bridger.

UTAH

Cache County: 12 miles west of Garden City.
Rich County: 8 miles west of Randolph; 14 miles southwest of Woodruff.
Weber County: Beaver Canyon, 7500 feet [Beaver River].
Summit County: Porcupine Ridge near Upton.

IDAHO

Bannock County: Swan Lake.
Cassia County: Albion; Mount Harrison, 10 miles south of Albion; 8 miles southwest of Elba.

NEVADA

Elko County: summit between Copper and Coon creeks, Jarbidge Mountains.
Humboldt County: Duffer Peak, 8400 feet, Pine Forest Mountains.

Junco caniceps dorsalis

TEXAS

Culberson County: Guadalupe Mountains.

NEW MEXICO

Lincoln County: Alto; southwest slope of Capitan Mountains; south branch of Carrizo Creek, near foot of Sierra Blanca Peak, 7600 feet.
Otero County: Tularosa Canyon, 6850 feet, 4 miles above Mescalero; Cloudcroft.
Socorro County: head of Water Canyon, 9300 feet, Magdalena Mountains.
Catron County: 8 miles southeast of Mogollon.
Grant County: Head of Mimbres [River]; Big Rocky Creek, 8000 feet.

ARIZONA

Greenlee County: Hannagan Meadow.
Apache County: 7 miles north of Big Lake; Marsh Lake [Big Lake]; Horseshoe Cienega; Willow Springs; summit of White Mountains; Cooley [McNary]; 6 miles south-southwest of Greer; Odart Mountain, 8300 feet; PS Ranch, 28 miles south of Springerville.
Navajo County: Apache [Fort Apache].
Gila County: Aztec Peak, Sierra Ancha.
Coconino County: Baker's Butte; Pivot Rock Spring; 3 miles southwest of Long Valley; Little Spring, north side of San Francisco Mountain; San Francisco Mountain; 8 miles north of Maine [Sitgreaves Mountain]; Williams.
Yavapai County: Mingus Mountain; Mount Union, Bradshaw Mountains; Mount Francis, 6500 feet, Prescott.

Junco phaeonotus palliatus

NEW MEXICO

Hidalgo [formerly Grant] County: Animas Peak, 8000 feet; Summit of San Luis Mountains; Big Hachita Mountain.

ARIZONA

Gila County: Pinal Mountain, 7500 feet.
Graham County: Graham [Pinalino Mountains], 8400–9200 feet; Mount Graham.
Pinal County: Santa Catalina Mountains: Catalina Mill; [San] Pedro slope; 4000 feet.
Pima County: Santa Catalina Mountains: Soldier's Camp, 7700 feet; Mount Bigelow, 8300 feet.
Cochise County: Chiricahua Mountains: Pinery Canyon, 6000–7000 feet; Ida Peak, 8000 feet; Rustler Park, 8700 feet; Bar Foot Park, 8300 feet; Paradise, 6000 feet; Flys Peak, 9500 feet; Monte Vista, 8000 feet. Whetstone Mountains. Huachuca Mountains: 7000–8500 feet; Palmerlee; head of Miller Canyon, 8400 feet; Ramsay Canyon, 9000 feet.
Santa Cruz County: Santa Rita Mountains: Madera Canyon, 4700 feet; Stone Cabin Canyon, 8500 feet.

MEXICO

Sonora: San José Mountains.
Chihuahua: San Luis Mountains; Colonia Juárez; near Colonia Garcia; Water Cañon, 7200 feet, 3 miles south of Colonia Garcia; 9 miles southeast of Colonia Garcia, 8200 feet; Pachaco [= Pacheco]; Chuchuichupa [= Chuhuichupa and Chuichupa]; Bustillos; 30 miles west of Miñaca; Jesús María; Pinos Altos; El Carmén; Rio Verde, 8000 feet.

Localities in intergradient areas in central Mexico between *J. p. palliatus* and *J. p. phaeonotus*

MEXICO

Coahuila: Sierra Encarnación [20 miles southwest of Carneros]; Sierra Guadalupe [20 miles southwest of Saltillo].
Durango: Arroyo de Buey, 7500 feet [northwest Durango]; El Salto; Ciudad Durango.
Tamaulipas: Miquihuana; Ciudad Victoria; Galindo.
Nuevo León (no specific locality).
San Luis Potosí: Sierra de San Luis Potosí; mountains 25 miles west of Charcas, 7000–8000 feet; mountains near Jesús María.
Zacatecas: Jérez; Sierra Valparaíso.
Aguascalientes: Sierra de Aguascalientes.
Jalisco (northern): Bolaños; Sierra Bolaños; Sierra Nayarit, 8000 feet.
Nayarit (Tepic): Sierra Madre.

Junco phaeonotus phaeonotus

MEXICO

Vera Cruz: Las Vigas, 8000 feet; Coatepec; Jalapa; Jico [= Xico]; Perote; Mirador.
Pueblo: Chalchicomula, 8700 feet; Pine Forest, 10,000 feet, Mount Orizaba; Chachapa; Pinal.
Tlaxcala: Maquey Verde, 7100 feet, 8½ miles northeast of Zimapan; Mount Malinchi.
Hidalgo: Real del Monte; El Chico.
Mexico and Federal District: Rio Frio, Ixtaccihuatl; Amecameca; Popocatepetl, 12,000 feet; Tenango; City of Mexico [probably vicinity of]; Coajimalpa and Chimalpu, vicinity of Tacubaya; Culhuacán, Mexicalcingo, and Hacienda Eclava, vicinity of Tlalpam; Salazar; Las Cruces, Valley of Mexico; Tetalco, vicinity of Xochimilco; Ajusco; north slope of Volcán Toluca.
Michoacán: Pátzcuaro; Nahuatzen; Patambán.
Jalisco (southern): Sierra Nevada de Colima; Volcán de Nieve, 10,000–14,000 feet, Sierra Nevada de Colima; Juanacatlán.
Guerrero: Omilteme, 8000 feet.
Oaxaca: Villa Alta; Tonagnia; Totontepec; Mount Zempoaltepec.

Junco bairdi

MEXICO

Lower California. Cape Region: Sierra de la Laguna (Sierra Laguna); Victoria Mountains; La Laguna; Laguna Valley; El Sáuz (also El Sauce); 14 miles northwest of San Bernardo; Mount Miraflores.

Junco fulvescens

MEXICO

Chiapas: San Cristóbal; Teopisca.

Junco alticola

GUATEMALA

Amatitlán (Department of): Volcán de Agua.
Sacatepéquez: Volcán de Fuego, 10,200–12,000 feet, pine forests.

Chimaltenango: Tecpam, 9500 feet; Santa Llana, 10,000 feet (about six miles from Tecpam).
Solola: San José Solola, 8000 feet.
Quezaltenango: Quezaltenango [7700 feet]; Chuipaché; San Martín; Volcán Santa María; Calel.
San Marcos: San Marcos [8000 feet]; Tajumulco, 10,500 feet; Zanjón (9500 feet).
Totonicapán: Momostenango [6500 feet]; Totonicapán [8000 feet].
El Quiché: Chichicastenango [6500 feet].
Huehuetenango: San Mateo [9000 feet]; Hacienda Chancol, 11,000 feet [=Chancol]; Todos Santos.

MEXICO
Chiapas: Volcán Tacaná, 11,000 feet; Niquivil [or Niquihuil]; Chiquihuite, 8200 feet.

Junco vulcani
COSTA RICA
Volcán Irazú [various slopes of mountain]; Volcán Turrialba, 8500–11,000 feet; Las Vueltas [Dota Mountains].
PANAMA
Volcán Chiriquí, 11,000–11,200 feet.

Junco oreganus mearnsi
SASKATCHEWAN
Maple Creek District: Cypress Hills near Maple Creek.
ALBERTA
Medicine Hat District: Eagle Butte.
MONTANA
Garfield County: 20 miles west of Lismas [possibly in Valley County].
Phillips County: Ruby Creek, Zortman, Little Rocky Mountains.
Fergus County: 15 miles south of Heath, Big Snowy Mountains; 5 miles northwest of Hilger.
Judith Basin County: Buffalo Canyon, 13 miles west of Buffalo; Dry Wolf Creek, 20 miles southwest of Stanford, Little Belt River Mountains.
Cascade County: Belt River Canyon [Belt Creek].
Meagher County: Sheep Creek, 16 miles north of White Sulphur Springs, Little Belt Mountains.
Musselshell County: 16 miles south of Roundup.
Sweetgrass County: 14 miles south of Big Timber; head of Timber Creek, Crazy Mountains.
Gallatin County: Mystic Lake [20 miles south of Bozeman]; west fork of West Gallatin River, 6500 feet.
Madison County: Ward Peak, Washington Creek, 6500 feet.
Big Horn County: Lodgegrass Creek, 8000 feet, Big Horn Mountains.
WYOMING
Sheridan County: Big Horn Mountains near Sheridan.
Big Horn County: head of Trapper's Creek, west slope of Big Horn Mountains, 5500–8500 feet.
Park County: northeast base of Black Mountain; Whirlwind Peak; Grinnett Creek near Pahaska Tepee; Pahaska [=Pahaska Tepee]; 12 miles north of Wapiti; Valley; Needle Mountain.
Yellowstone Park: Lower Geyser Basin.
Fremont County: Whisky Mountain, 5 miles south of Dubois, 10,000 feet; Bull Lake.
Teton County: south of Moose Creek, 10,000 feet, Teton Mountains.
Sublette County: Kendall; 12 miles north of Kendall, 7700 feet; 7 miles south of Fre-

mont Peak, 10,400 feet; 5 miles northwest of Merna, 8500 feet; 1 mile north of Green River Lakes, 8300 feet.

Lincoln County: Head of Dry Creek, Salt River Mountains, 9200 feet.

IDAHO

Custer County: head of Pahsimeroi River.

Fremont County: Mack's Inn near Trude; 17 miles east, 4 miles north of Ashton, 6275 feet.

Teton County: 3 miles southwest of Victor.

Bonneville County: Big Hole Mountains, 8 miles northeast of Swan Valley.

Bannock County: Inkom.

Bear Lake County: west rim of Copenhagen Basin, 8400 feet, Wasatch Mountains.

Franklin County: 20 miles northeast of Preston, Strawberry Creek, 6700 feet.

Intermediates between *J. o. mearnsi* and *J. o. montanus*

MONTANA

Deer Lodge County.

IDAHO

Custer County: Mill Creek, 14 miles west-southwest of Challis, 8370 feet; Stanley Lake; 5 miles west of Cape Horn, Sawtooth Mountains.

Blaine County: Alturas Lake.

Valley County: 5 miles east of Warm Lake; Lardo (nearly typical *montanus* population).

Adams County: 3 miles west of Payette Lake (nearly typical *montanus*).

Boise County: Idaho City.

Junco oreganus montanus

Breeding localities at which there are relatively unmixed populations. Where mixture with *J. hyemalis* is known, the locality is marked with an asterisk.

BRITISH COLUMBIA (districts as indicated in Rand McNally Commercial Atlas of 1934)

Skeena District: Kispiox Valley,* 23 miles north of Hazelton; Nine-Mile Mountain,* 20 miles northeast of Hazelton, 4000 feet; Hazelton, 959 feet; New Hazelton; Driftwood River, 15 miles northwest of Tacla Lake; 10 miles west of Tacla Lake, Babine Trail; 15 miles east of Babine; Telkwa.

Cariboo District: Stuart Lake; Vanderhoof; mouth of Big Salmon River [= McGregor River]; Cottonwood;* Indianpoint Lake;* Isaac Lake; 3 miles east of Moose Lake (probably mixture); north fork, and mouth, of Moose River (probably mixture); Yellowhead Lake;* Yellowhead Pass;* Moose Pass;* Kleena Kleene; 15 miles southwest of Kleena Kleene; 5 miles northwest of Redstone; Alexis Lake; 20 miles south of Chezacut; 150 Mile House; Lac la Hache; Clearwater; 12 miles north of Clearwater;* Grizzly Peak, 10 miles west of Clearwater; Sicamous; Ducks; Revelstoke; Ashcroft; 12 miles north of Clinton; Lillooet.

West Kootenay District: Summit [= Summit Lake?]; Deer Park, Lower Arrow Lake; Nelson; Rossland;* Trail; Creston.*

East Kootenay: Glacier; Field;* Cranbrook.*

Yale District: Okanagan; Okanagan Landing; Shuswap Falls; Vaseaux Lake, Okanagan Valley; Inkaneep Creek, Okanagan Valley; Midway.

Alberta (all localities in regions of considerable intermixture): Henry House; 15 miles south of Henry House; Fork of Blindman and Red Deer rivers; Red Deer River at Didsbury; Banff.

WASHINGTON

Okanagan County: Tunk Mountain.

Columbia County: 21 miles southeast of Dayton, east fork of Touchet River, Blue Mountains.

Asotin County: Anatone, Blue Mountains, 3500 feet; 15 miles north of Paradise [Oregon] at Horse Creek, 2000 feet.

IDAHO
Boundary County: 4 miles west of Meadow Creek, 3000 feet.
Bonner County: 5 miles west of Cocolalla, 3000 feet.
Kootenai County: 5 miles north of Coeur d'Alene; Fernan Hill, Coeur d'Alene.
Shoshone County: Thompson Pass; Glidden Lakes, 5700 feet.
Clearwater County: 2 miles northeast of Weippe, 3000 feet.
Idaho County: Castle Creek Ranger Station, south fork of Clearwater River, 1800 feet.
Adams County: 1 mile north of Bear Ranger Station, southwest slope of Smith Mountain, 5400 feet; Thorn Creek Ranger Station, 3 miles west of Payette Lake, 5400 feet (nearly typical *montanus*).
Valley County: Lardo (nearly typical *montanus*).
Washington County: 1 mile northeast of Heath, southwest slope of Cuddy Mountain, 4000–4600 feet.

MONTANA
Lincoln County: Tobacco Plains [=Tobacco].
Flathead County: Columbia Falls; Flathead Lake.
Glacier County: Saint Mary's Lake.
Sanders County: Thompson Falls.
Missoula County: Lolo Creek, 3470 feet, 6½ miles west of Lolo (nearly typical *montanus*).
Ravalli County: Florence, west of, 7300 feet (nearly typical *montanus*).

OREGON
Jefferson County: Haycreek, 12 miles east of Foley Creek.
Crook County: Ochoco Ranger Station, 4000 feet; Howard; Maury Mountains.
Grant County: John Day River; Beech Creek; Strawberry Mountains; 12 miles south of Canyon City, 5500 feet; Cold Spring, 4900 feet, 8 miles east of Austin.
Umatilla County: Meacham.
Wallowa County: Billy Meadows Ranger Station.
Union County: Elgin; Catherine Creek, 7 miles northeast of Telocaset, 3500 feet.
Baker County: Homestead, 3500 feet; Bourne; McEwen; Home, 3000 feet.

Intermediates between *J. o. shufeldti* and *J. o. montanus*

OREGON
Deschutes County: Lapine; Paulina Lake; 3 miles west Paulina Lake; 1 mile south East Lake, Paulina Mountains.

Junco oreganus shufeldti

BRITISH COLUMBIA (extreme southwestern mainland)
Desolation Lake, Howe Sound; Brackendale, Howe Sound; Seymour Mountain; New Westminster; Port Moody; Mount Lehman; Chilliwack; Agassiz; Sumas.

WASHINGTON
Whatcom County: 35 miles east of Glacier, 3500 feet; Bonita Mine, Barron, 6000 feet; Lummi Island.
Okanagan County: Hidden Lakes, 4100 feet; Mazama, 2100 feet; west fork of Pasayten River, 4700 feet, at mouth of Holman Creek.
Skagit County: Mount Vernon.
Chelan County: Railroad Creek, 3900 feet, Hart Lake; Tyee Peak, 3000–5000 feet, Entiat; Stormy Peak, 5500 feet, near Entiat; Wenatchee Lake, 1870 feet.
King County: Kirkland; Seattle; Redmond.
Pierce County: Tacoma; Indian Henry's Ranger Station, 5300 feet, Mount Rainier; above Paradise Park, 6500 feet, Mount Rainier.
Lewis County: Reflection Lakes, 4900 feet, Mount Rainier; Stevens Creek, 3000 feet, Mount Rainier.
Clallam County: 2 miles southwest of Mount Angeles, 6000 feet; head of Little Creek, 5000 feet, Mount Angeles; Cañon Creek; 3 miles south of Soleduck [Sol Duc] River, 3550

feet; Lake Sutherland, 500 feet; mouth of Boulder Creek, 560 feet, Elwha River, Olympic Mountains; Happy Lake, 4900 feet; Frazier Creek, 800 feet, 4 miles southwest of Port Angeles.

Jefferson County: Port Townsend; Elwha Basin, Elwha River, 2750 feet; upper Quinault.
Grays Harbor County: Quinault Lake.
Thurston County: Tenino.
Yakima County: Gotchen Creek, 4000 feet, Mount Adams; Signal Peak.
Wahkiakum County: 4 miles east of Skamokawa.
Clark County: Vancouver.
Skamania County: Mount Saint Helens; 60 miles east by trail from Toledo [Lewis County].
Klickitat County: Grand Dalles.

OREGON

Tillamook County: Netarts; Mount Hebo, 3165 feet; Tillamook.
Washington County: Forest Grove; Beaverton.
Multnomah County: Portland.
Marion County: Salem.
Jefferson County: Mill Creek, 20 miles west of Warm Springs.
Linn County: Lookout Mountain.
Lane County: Eugene; McKenzie Bridge.
Deschutes County: Sisters; 16 miles west of Bend, 6500 feet.
Coos County: Lakeside; 3 miles south of Cape Arago; 2 miles north of Bandon.

Intermediates between *J. o. oreganus* and *J. o. shufeldti*

BRITISH COLUMBIA

Calvert Island (nearest *oreganus*): Johnstone Strait; Lund. Vancouver Island: Mount Arrowsmith; Halls Ranch, Alberni Valley; Errington; Golden Eagle Mine, 18 miles south of Alberni; Great Central Lake; Little Qualicum River.

Junco oreganus oreganus

ALASKA

Yakutat; Glacier Bay; Gustavus Point, Glacier Bay; Haines; Port Frederick; Kelp Bay [Baranof Island]; Red Bluff Bay, Baranof Island; Hooniah Sound; Sitka; Old Sitka; Taku River; Port Snettisham; Windfall Harbor; Mole Harbor; Farragut Bay; Thomas Bay; Petersburg, Mitkof Island; St. John Harbor, Zarembo Island; Kasaan Bay, Prince of Wales Island; Three Mile Arm, Kuiu Island; east end of Heceta Island; east side of Warren Island; Rocky Bay, northwest Dall Island; Wrangell; Etolin Island; Portage Cove, Revillagigedo Island; Ketchikan.

BRITISH COLUMBIA

Fort [Port] Simpson. Queen Charlotte Islands: McIntosh Meadows near Massett; Massett; Langara Island; Cumshewa Inlet. Yule Lake, Swanson Bay; "Soda Creek," 52° 47' N, 128° 33' W, Princess Royal Island; Borrowman Bay, Aristazabl Island.

Intermediates between *J. o. oreganus* and *J. o. montanus*

BRITISH COLUMBIA

Bella Coola, 5000 feet; 13 miles east of Bella Coola; 18 miles east of Bella Coola (intermediates). Thirty miles east of Bella Coola; Cariboo Mountain, 5000 feet, Stuie [Bella Coola Valley]; Plateau above Hotnarko River, 2–4 miles east of Precipice; Anahim Lake; Rainbow Mountains (all nearest *montanus*). Peak above Swanson Bay (near *oreganus*).

Intermediates (hybrids) between *J. o. oreganus* and *J. h. cismontanus*

ALASKA

Glacier, White Pass; Bradfield Canal.

BRITISH COLUMBIA

Flood Glacier, Stikine River; Great Glacier (chiefly *oreganus*), Stikine River.

Intermediates between *J. o. shufeldti* and *J. o. thurberi*

OREGON (Principal regions of intergradation)

Curry County: Port Orford.

Douglas County: Reston; Lookingglass; Drew; Minnehaha Creek, Diamond Lake Road, Crater Lake National Forest; 5 miles south Mount Thielsen.

Josephine County: Grants Pass; Onion Mountain Lookout.

Jackson County: Trail; Prospect.

Klamath County: Crater Lake; Anna Springs [Crater Lake National Park]; Fort Klamath; Algoma; Swan Lake Valley.

Junco oreganus thurberi

OREGON (many somewhat intermediate toward *shufeldti*)

Josephine County: Lind Home, 5 miles north-northeast of Holland; junction of Grayback and Sucker creeks; 5 miles northwest of Bolan Lake.

Jackson County: Siskiyou; Siskiyou Mountains; Colestin; north base of Ashland Peak, 4000 feet.

Klamath County: Bonanza.

Lake County: 3 miles east of Lakeview, 5400 feet; north base of Crook Peak, Warner Mountains; Fort Warner, Hart Mountains; Barley Camp, 14 miles southwest of Adel.

CALIFORNIA

Del Norte County: 7 miles east-northeast of Smith River, 1800 feet; Bald Hill, 9 miles east of Crescent City; Patrick's Creek; east fork of Illinois River, 1900 feet, ¼ mile south of Oregon line.

Siskiyou County: Poker Flat, 5000 feet, 12 miles northwest, and 9 miles west, of Happy Camp; Bald Mountain; head of Doggett Creek, 5800 feet, Siskiyou Mountains; west fork of Cottonwood Creek, 4000 feet, 4½ miles southwest of Hilt; Old Piney Mine, 4 miles southwest of Greenview; Salmon Mountains near Greenview; Fort Jones; Forest House Mountain, 8 miles west of Yreka; Vernon Mine near Yreka; Jackson Lake, 5900 feet; south fork of Salmon River, 5000 feet; head of Rush Creek, 6400 feet; Kangaroo Creek; Castle Lake, 5434 feet; Bray; Goose Nest Mountain; Mount Shasta, 7500 feet; head of Squaw Creek, 7600 feet, Mount Shasta; timber line, Mud Creek, Mount Shasta; Horse Camp, 7200 feet, Mount Shasta; Sisson; Bear Creek; Weed; Stewart Springs near Weed; T. H. Benton estate, Butte Creek; McCloud.

Modoc County: Camp [Fort] Bidwell; Lake City; Cedarville; head of north fork of Parker Creek, 7300 feet, Parker Creek 5500 feet, middle fork of Parker Creek, 7500 feet, and head of Dry Creek, 7000 feet, Warner Mountains; east face of Warner Peak; Eagleville; near Willow Ranch; Sugar Hill, 5000–6000 feet; near Davis Creek, Sugar Hill, 5500 feet.

Humboldt County: 8–12 miles east of Arcata, 1200 feet; Hoopa Valley; Coyote Peak, 3000 feet; Horse Mountain, 4700–5200 feet; South Fork Mountain, 5500–5700 feet, near Blake Lookout; 4 miles northeast of Bridgeville, 1800 feet, McLellan Ranch; forks of Van Deuzen River, 8 miles east of Bridgeville; 10 miles north of Garberville.

Trinity County: Norgaar's Ranch, 2000 feet, South Fork, Trinity River; summit of South Fork Mountains, 3 miles east of Forest Glen; 4–10 miles northwest of Forest Glen, 3100–3600 feet; Salyer; Peanut; Mad River Ford, 2700 feet, above Ruth; Bully Choop Mountains, 20 miles east of Weaverville; head of Grizzly Creek, 6000 feet; north fork of Coffee

Creek, 4500 feet; head of Bear Creek, 6400 feet; divide, 12 miles north of North Yolla Bolly Mountain.

Shasta County: 3 miles west of Knob; Lake Helen, 8500 feet, south base of Lassen Peak; Manzanita Lake, 6000 feet; Burney.

Tehama County: ½ mile to 2 miles south of South Yolla Bolly Mountain; Lyonsville; 5 miles northwest of Lyonsville; 2 miles west of Black Butte, 6800 feet; Round Valley Ranger Station; Mineral, 6000 feet.

Lassen County: Hayden Hill; Mount Lassen; southeast side of Lassen Peak; Mill Creek, south base of Mount Lassen; Bogard Ranger Station; Honey Lake; Eagle Lake.

Mendocino County: 3 miles south of Covelo; Cahto; Laytonville; Sherwoods; 14 and 22 miles west [by road] of Willets; Mount Anthony; 4500 feet, 6000 feet, and 3 miles west of summit of Mount Sanhedrin; Mendocino City, 75 feet; Ornbaun Post Office; 5 miles west of Ornbaun Spring.

Lake County: Glenbrook, Cobb Post Office.

Colusa County: Fout's Springs; Snow Mountain.

Napa County: Mount Saint Helena.

Sonoma County: Gualala River; Gualala, 50 feet; Sea View; 7 miles west of Cazadero.

Butte County: Jonesville; Magalia.

Plumas County: Mohawk; Quincy; Meadow Valley; Johnsville; Lights Cañon.

Sierra County: Sierra City, 6800 feet; Sierraville, 5300–5500 feet.

Nevada County: Independence Lake.

Placer County: Blue Cañon; Emigrant Gap; Cisco; Donner, 7900 feet; Squaw Valley near Truckee; Tahoe Tavern.

El Dorado County: Placerville; Fyffe, 3500–4000 feet; Kyburz; Pyramid Peak; Glen Alpine; Mount Tallac; Echo, 7000 feet; Summit of Tahoe-Placerville Road; Meyers, Lake Tahoe.

Alpine County: 3 miles south-southwest of Woodfords, 7800 feet.

Calaveras County: Big Trees.

Tuolumne County: Stanislaus River, 4600–5000 feet; head of Lyell Canyon, 9700 feet.

Mariposa County: near Porcupine Flat, 8100 feet; east fork of Indian Canyon, 7300 feet; 1 mile east of Merced Lake; near Mono Meadow, 7300 feet; Yosemite Valley; Merced Grove of Big Trees, 5500 feet; Dudley, 3000 feet; 5 miles northeast of Coulterville, 3200 feet.

Mono County: Warren Mountain, 10,000 feet, near Mono Lake; Parker Creek, 7500 feet; near Walker Lake, 8000 feet; Mammoth Lakes.

Fresno County: south fork of San Joaquin River at Evolution Creek junction, 8500 feet; Bubb's Creek Cañon; Hume; Bullfrog Lake, 10,600 feet; Horse Corral Meadows.

Tulare County: Whitney Meadow, 9800 feet; Taylor Meadow, 7000 feet; Whitney Creek, 11,000 feet; Mount Whitney; Monache Meadow, 8000 feet; Jackass Meadow, 7750 feet; Sirretta Meadow, 9000 feet; Olancha Peak, 10,000 feet; Coyote Creek, east side of summit of Western Divide; east fork of Kaweah River, 9700 feet; Sequoia Park; North Tule; Weisher's Mill [6000 feet, east fork of Kaweah River].

Inyo County: Owens Valley; Onion Valley, 8500 feet, and Flower Lake [= Heart Lake], 10,500 feet, Kearsarge Pass; Cottonwood Lakes, 11,000 feet; Little Cottonwood Creek, 9800 feet.

Kern County: Kiavah Mountain, 6000–7000 feet near Walker Pass; Mount Breckenridge, 6000 feet, 12 miles below Bodfish; Piute Mountains; Fort Tejon; northeast side of Mount Pinos.

Ventura County: Mount Pinos, 8500 feet.

Los Angeles County: Arroyo Seco Canyon, near Pasadena; Buckhorn; Pine Flats; Chileo; Wilson's Peak [= Mount Wilson]; Strain's Camp; West Fork [San Gabriel River] near [?] Pasadena (all in San Gabriel Mountains).

San Bernardino County: Bear Valley; Bluff Lake, 7400 feet; Fish Creek, 6500 feet; south fork of Santa Ana River, 6200–8500 feet; Sugarloaf, 8000–9000 feet; Santa Ana

[River], 5500–6600 feet; San Bernardino Peak; Lost Creek, 6400 feet; Oak Glen; San Gorgonio Peak, 11,500 feet (all in San Bernardino Mountains).

Riverside County: near Beaumont; Round Valley, 9000 feet; Fuller's Mill, 5900–6000 feet; New Mill Site, near Fuller's Mill, 5300 feet; Tahquitz Valley, 8000 feet; Thomas Mountain, 6800 feet; Strawberry Valley, 6000 feet (all in the San Jacinto Mountains). Garnet Queen Mine, 6000 feet; Toro Peak, 8000 feet; Santa Rosa Peak, 7500 feet (all in the Santa Rosa Mountains).

Orange County: Los Pinos Peak, 4000 feet, Santa Ana Mountains.

San Diego County: Julian; Campbell's Ranch, Laguna Mountains; Laguna; Cuyamaca Mountain; Volcan Mountain; Volcan; Smith Mountain [Palomar Mountain].

NEVADA (typical populations only)

Washoe County: Galena Creek, 7000–8000 feet.
Douglas County: Glenbrook.

Intermediates between *J. o. thurberi* and *J. o. pinosus*

CALIFORNIA

Marin County: Inverness; San Geronimo; San Rafael.

San Luis Obispo County: Santa Margarita, 996 feet; Salinas River, 1000 feet, 3 miles northeast of Santa Margarita (closer to *pinosus* than to *thurberi*); 4 miles west of La Panza.

Junco oreganus pinosus

CALIFORNIA

Alameda County: Strawberry Canyon and University of California campus, Berkeley.
Santa Clara County: east side of Mount Hamilton; Alma.
San Mateo County: King Mountain; Pescadero; Pescadero Creek Basin; Butano Basin.
Santa Cruz County: Santa Cruz Mountains, 2300 feet; summit between Los Gatos and Boulder, Santa Cruz Mountains; Big Trees; Boulder Creek; Santa Cruz; 2½ miles east, 1 mile north-northeast of Santa Cruz, 300 feet.
San Benito County: Big Oak Flat, 3300 feet, 4 miles northeast of San Benito; Butts Ranch, 3000 feet, 5 miles north-northeast of San Benito; summit, 5250 feet, and 1 mile southeast of summit, 4400 feet, San Benito Mountain; Laguna Ranch, 4000 feet, 4 miles south of Hernandez.
Monterey County: Monterey; Pacific Grove; Point Pinos; Garrapatos Creek; Little Sur River; head of Turner Creek, 2700 feet; Big Pines, 3500–3700 feet [=The Pines and Government Camp, Danish Creek]; Big Sur River; Lucia P. O., headwaters of Big Creek; Santa Lucia Peak, 5600 feet; 6 miles below Tassajara Springs, Tassajara Creek; Partington Point, 1¼–2 miles south of Chalk Peak, 3000 feet.

Junco oreganus pontilis

MEXICO

Lower California. Sierra Juárez: Laguna Hanson; El Rayo.

Junco oreganus townsendi

MEXICO

Lower California. Sierra San Pedro Mártir: San Pedro Mártir Mountain, 7500–8500 feet; Concepción, 6000 feet (November); La Joya, 6200 feet (October); Vallecitos, 7500–8500 feet; La Grulla, 7200–7500 feet; Santa Rosa Flats (6000 feet).

Junco insularis

MEXICO

Lower California. Guadalupe Island, sea level to 3500 feet.

Junco hyemalis hyemalis

ALASKA

Point Barrow; Wainwright; Kowak River; Cape Prince of Wales; Wales; Cairn, Norton Sound; north fork of Kuskokwim River, base of Mount Sischu; Dry Creek, 2600 feet, and Savage River, 2800 feet, Mount McKinley district; Sheep Creek [of Talkeetna Mountains?]; Lake Clark; Iliamna Village, Lake Iliamna; Knik River, Cook Inlet; Hope; Kenai Mountains; Seward; Homer; Seldovia; head of Port Nell [Nellie] Juan and head of Cordova Bay, Prince William Sound; Circle; Yukon River, 20 miles above Circle; Charlie Creek [= Kandik River]; mountains near Eagle; Rapids; Tanana Crossing.

YUKON

Forty-mile; Chandindu River, Yukon River; Dawson; White River, Yukon River; Russell Mountains, near forks of Macmillan River.

MACKENZIE

Fort Good Hope, Mackenzie River; Great Bear River; Fort Simpson; Willow River near Fort Providence; Hay River; Great Slave Lake; Fort Resolution; Great Slave River; Maufelly Point, Taltheilei Narrows and Belle Isle, Charlton Bay, Great Slave Lake.

ALBERTA

Bon Accord; Hastings Lake; Athabasca River; Fort Smith; Fort Chipewyan; Cypress Point and 10 miles northeast of Sand Point, Lake Athabasca.

SASKATCHEWAN

Lake Athabasca: 8 miles northeast of Moose Island, mouth of Beaver River, and north shore.

MANITOBA

Churchill; Norway House; Oxford.

ONTARIO

Moose Factory; East Point, James Bay; Cochrane; Dane; New Liskeard; Brunswick House [New Brunswick]; G. T. P. Crossing [= Mattice]; Coboconk.

QUEBEC

Richmond Gulf; Cairn Island, Richmond Gulf; James Bay; Great Whale River, James Bay; Aquatuk Bay; East Main, James Bay; Fort George, James Bay; Charlton Island, James Bay; Rupert House; Tadousac; Point Natashquan; mouth of Necatina [= Mekattina] River; Ste. Margaret [Margarite] River; Fauriel, Grosse Isle, Magdalen Islands.

NEWFOUNDLAND LABRADOR

Tikkoatokok Bay; Nain; Groswater Bay [= Lake Melville]; Lanc au Loup [= L'Anse au Loup, Loup Bay].

NEWFOUNDLAND

Bay of Islands; Port-au-Port; Spruce Brook; Nicholasville.

PRINCE EDWARD ISLAND

Tignish; Summerside; Souris.

NEW BRUNSWICK

Bathurst; Salisbury; Hillsborough; Hampton; Rothsay; Point Lepreaux; Grand Manan.

NOVA SCOTIA

Kelly's Cove, Cape Breton Island; Plaster Cove, Strait of Canso; Whycocomagh; Sable Island; New Port; Barrington; Cape Sable Island.

WISCONSIN

Oneida County: Woodruff.
Dodge County: Beaver Dam (May 25, breeding?).

NEW YORK (somewhat intermediate)
   Franklin County: Chateaugay; St. Regis Lake.
   Lewis County: Locustgrove.
   Herkimer County: Fulton Chain Lakes.
   Essex County: Keene.
   Rensselaer County: Berlin.
   Greene County: Hunter Mountain; High Peak, 3800 [3600] feet; Stony Clove Notch; Lanesville; Plateau Mountain (all in Catskill Mountains).
   Ulster County: Overlook Mountain; Summit [Mountain].
   Delaware County: Stamford.

NEW HAMPSHIRE
   Coos County: Lancaster; Shelburne; 2 miles by road from summit, and Luckerman's Ravine, Mount Washington.
   Grafton County: Franconia.

MAINE
   Aroostook County: Fort Fairfield
   Somerset County: King and Bartlett Lake.
   Washington County: Long Lake.
   Hancock County: Bar Harbor.
   Knox County: Allen's Island, George's Harbor.
   Oxford County: Lake Umbagog; Upton.
   York County.

MASSACHUSETTS
   Berkshire County: Mount Greylock.
   Worcester County: Mount Watatic.

Intermediates between *J. h. hyemalis* and *J. h. carolinensis*

PENNSYLVANIA
   Erie County: Tamarack Swamp [8½ miles north-northeast of Edinboro]; Edinboro.
   Warren County: Tidioute.
   McKean County: Bradford; Katrine Swamp; Norwich; and 3 miles west of Norwich.
   Potter County: Newfield Junction; Coudersport; Cherry Spring.
   Forest County: Tionesta; Sheffield Junction.
   Elk County: 1 mile north of Weedville.
   Clinton County: Tamarack Swamp [= Tamarack, head of Drury Run].
   Clarion County: Clarion Junction; ½ mile west of Clarion.
   Jefferson County: Punxsutawney.
   Clearfield County: Winterburn; Woodland.
   Lycoming County: North Mountain.
   Sullivan County: Lopez.
   Luzerne County: Harvey's Lake.
   Indiana County: Rochester Mills; Pine Flats.
   Cambria County: Belsano; Gallitzin; Cresson; Summit; Portage, 4 miles up Trout River; Dunlo.
   Blair County: Allegrippus; Altoona.
   Westmorland County: Conemaugh Furnace; Laughlintown.
   Fayette County: Sugar Loaf Mountain [= Sugarloaf Knob].
   Somerset County: Crumb; Keystone Junction.

MARYLAND
   Garrett County: Finzel; Bittinger.

WEST VIRGINIA
   Tucker County: Davis.

Junco hyemalis carolinensis

WEST VIRGINIA
  Greenbrier County: Cold Knob; Job's Knob.

VIRGINIA
  Page County: Stony Man.
  Rockbridge County: Davis.
  Bedford County: White Top, Peaks of Otter.
  Giles County: Mountain Lake.
  Tazewell County: Tazewell.
  Grayson County: Mount Rogers; Trout Dale.

KENTUCKY
  Harlan County: Big Black Mountain.

TENNESSEE
  Carter County: top Roan Mountain.

NORTH CAROLINA
  Avery County: Cranberry.
  Mitchell County: 5500–6300 feet and summit of Roan Mountain.
  Yancey County: Mount Mitchell, 6000–6300 feet.
  Buncombe County: Black Mountain; Craggy Mountain, 6300 feet.
  Macon County: Highlands.

GEORGIA
  Towns County: Brasstown Bald, 4500 feet.
  Gilmer County: Rich Mountain, 4000 feet.

Intermediates between *J. h. cismontanus* and *J. h. hyemalis*

YUKON
  Lake Marsh.

ALBERTA
  Edmonton region; 25 miles northwest of Edmonton; Lac La Nonne; 12 miles west of St. Ann's; Edmonton; Camrose; Buck Lake; near Red Deer; Red Deer River at Alta; Red Deer River, 7 miles below Nevis.

Junco hyemalis cismontanus

(Asterisk denotes known presence of *J. o. montanus* or hybrids in same region.)

YUKON
  Carcross; Caribou Crossing.

BRITISH COLUMBIA
  Skeena District: Bennett City; Atlin; Monarch Mountain, 4500 feet, Atlin; Telegraph Creek; 4–12 miles north of Telegraph Creek; second south fork, 50 miles east Telegraph Creek; Glenora, Stikine River; Doch-da-on Creek, Stikine River; Kispiox Valley,* 23 miles north Hazelton; Nine-Mile Mountain,* 4000 feet, northeast Hazelton.
  Cariboo District: Tupper Creek;* Moose Pass;* Yellowhead Pass.*

ALBERTA
  Peace River District: Henry House;* Jasper House, and 5 miles south of Jasper House; Smoky River, Moose Branch; Jasper Park* [= east side of divide]; Brule Lake;* Prairie Creek; 140 miles west Edmonton.

Hybrids between *J. h. cismontanus* (or *J. h. hyemalis*) and *J. o. montanus*

ALASKA
Mountains near Eagle.

BRITISH COLUMBIA
Kispiox Valley, 23 miles north of Hazelton; Nine-Mile Mountain, 4000 feet, northeast of Hazelton.
Cariboo District: Tupper Creek; Cottonwood; Indianpoint Lake; 12 miles north of Clearwater.
West Kootenay District: Rossland; Creston.
East Kootenay District: Field; Cranbrook.

ALBERTA
Peace River District: Jasper Park [= east side of divide]; Brule Lake; Brazeau Lake.
Athabasca District: Fort McMurray (essentially *montanus*; breeding?).

Junco aikeni

SOUTH DAKOTA
Lawrence County: Spearfish, 4600 feet; Deadwood (October 6).
Custer County: Custer.

WYOMING
Crook County: Sundance.

NEBRASKA
Sioux County: Monroe Canyon [Hat Creek] (August 18; molting wing).

MONTANA
Powder River County: 4 miles west of Fort Howe Ranger Station, 4000 feet (also 3 *aikeni* × *mearnsi*).

LITERATURE CITED

AIKEN, C. E.
  1872. Notes on the birds of Wyoming and Colorado territories. By C. H. Holden, Jr., with additional memoranda by C. E. Aiken. Proc. Bost. Soc. Nat. Hist., 15: 193–210.

ALLEN, J. A.
  1879. Rare birds in Michigan. Bull. Nutt. Ornith. Club, 4:123.
  1893. List of mammals and birds collected in northeastern Sonora and northwestern Chihuahua, Mexico, on the Lumholtz Archaeological Expedition, 1890–92. Bull. Am. Mus. Nat. Hist., 5:27–42.

ALLEN, J. A., and BREWSTER, W.
  1883. List of birds observed in the vicinity of Colorado Springs, Colorado, during March, April, and May, 1882. Bull. Nutt. Ornith. Club, 8:189–198.

AMERICAN ORNITHOLOGISTS' UNION COMMITTEE
  1886. The code of nomenclature and check-list of North American birds adopted by the American Ornithologists' Union (New York, Am. Ornith. Union), viii + 392 pp.
  1894. Sixth supplement to the American Ornithologists' Union check-list of North American birds. Auk, 11:46–51.
  1897. Eighth supplement to the American Ornithologists' Union check-list of North American birds. Auk, 14:117–135.
  1902. Eleventh supplement to the American Ornithologists' Union check-list of North American birds. Auk, 19:315–342.
  1908. Fourteenth supplement to the American Ornithologists' Union check-list of North American birds. Auk, 25:343–399.
  1910. Check-list of North American birds. Third edition (New York, Am. Ornith. Union), 430 pp., 2 maps.
  1931. Check-list of North American birds. Fourth Edition (Lancaster, Pa., Am. Ornith. Union), xix + 526 pp.

ANTHONY, A. W.
  1889. New birds from Lower California, Mexico. Proc. Calif. Acad. Sci., ser. 2, 2:73–82.
  1890. A new junco from California (*Junco hyemalis thurberi*). Zoe, 1:238–239.
  1895. Junco hyemalis shufeldti in Lower California. Auk, 12:183.

AUDUBON, J. J.
  1839. Ornithological biography, or an account of the habits of the birds of the United States of America (Edinburgh, Adam and Charles Black), vol. 5, xxxix + 664 pp.
  1839. A synopsis of the birds of North America (Edinburgh, Adam and Charles Black), xii + 359 pp.

BABCOCK, E. B., and CLAUSEN, R. E.
  1918. Genetics in relation to agriculture (New York, McGraw-Hill Book Company, Inc.), xx + 675 pp., 4 pls., 239 figs. in text.

BAILEY, A. M., BROWER, C. D., and BISHOP, L. B.
  1933. Birds of the region of Point Barrow, Alaska. Prog. Activities. Chicago Acad. Sci., 4:15–40.

BAILEY, F. M.
  1928. Birds of New Mexico (New Mexico Department of Game and Fish), xxiv + 807 pp., 79 pls., 136 figs. in text.

BAIRD, S. F.
  1853. Birds. *In* Appendix C of H. Stansbury's exploration and survey of the valley of the Great Salt Lake of Utah, pp. 314–335.
  1858. Birds. Pacific Railroad Reports, pt. 2, vol. 9, lvi + 1005 pp.
  1870. Appendix. *In* J. G. Cooper, Ornithology, Geol. Surv. Calif., vol. 1, land birds, pp. 563–592.

BAIRD, S. F., BREWER, T. M., and RIDGWAY, R.
1874. A history of North American birds (Boston, Little, Brown, and Company), land birds, vol. 1, xxviii + 596 + vi pp., 26 pls., many figs. in text.

BARROWS, W. B.
1912. Michigan bird life. Special Bull., Dept. Zoöl. Physiol., Mich. Agr. College, xiv + 822 pp., 70 pls., 151 figs. in text.

BARTRAM, W.
1792. Travels through North and South Carolina, Georgia, east and west Florida, the extensive territories of the Muscogulges, or Creek confederacy, and the country of the Chactaws (Philadelphia, James and Johnson, 1791; reprinted, London, J. Johnson, 1792), xxiv + 520 pp. + index, 6 pls., map, frontispiece.

BECKHAM, C. W.
1887. Additional notes on the birds of Pueblo County, Colorado. Auk, 4:120–125.

BONAPARTE, C. L.
1838. A geographical and comparative list of the birds of Europe and North America (London, John van Voorst), vii + 67 pp.
1850. Conspectus generum avium (Lugduni Batavorum, apud E. J. Brill), vol. 1, 543 pp.
1853. Notes sur les collections rapportées en 1853 par M. A. Delattre, de son voyage en Californie et dans le Nicaragua. Troisième Communication. Compte Rendu des Séances de l'Académie des Sciences, 37:913–925.

BOUCARD, A.
1878. On birds collected in Costa Rica. Proc. Zoöl. Soc. London, 1878, pp. 37–71, pl. IV.

BOULTON, R.
1926. Remarks on the origin and distribution of the Zonotrichiae. Auk, 43:326–332.

BREWSTER, W.
1882. On a collection of birds lately made by Mr. F. Stephens in Arizona. Bull. Nutt. Ornith. Club, 7:193–212.
1886a. An ornithological reconnaissance in western North Carolina. Auk, 3:94–112.
1886b. Junco hyemalis nesting in a bush. Auk, 3:277–288.
1902. Birds of the cape region of Lower California. Bull. Mus. Comp. Zoöl., 41:1–241, map.

BRYANT, W. E.
1887. Additions to the ornithology of Guadalupe Island. Bull. Calif. Acad. Sci., 2:269–318.
1889. A catalogue of the birds of Lower California, Mexico. Proc. Calif. Acad. Sci., ser. 2, 2:237–320.

BUBNOFF, S. VON
1919. Ueber einige grundlegende Principien der paläontologischen Systematik. Zeit. für induktive Abstammungs- und Vererbungslehre, 21:158–168.

CABANIS, J.
1851. Museum Heineanum. Verzeichniss der ornithologischen Sammlung des Oberamtmann Ferdinand Heine, auf Gut St. Burchard vor Halberstadt. Mit kritischen Anmerkungen und Beschreibung der neuen Arten, systematish bearbeitet von Dr. Jean Cabanis. I thiel, Singvögel (Halberstadt, R. Frantz), viii + 176 pp.

CAMPBELL, L. W.
1940. Birds of Lucas County. Toledo Mus. Sci. Bull., 1:1–225.

CARY, M.
1909. New records and important range extensions of Colorado birds. Auk, 26:180–185.
1911. A biological survey of Colorado. No. Am. Fauna, 33:1–256, 12 pls., 39 figs. in text, map.
1917. Life zone investigations in Wyoming. No. Am. Fauna, 42:1–95, 15 pls., 17 figs. in text, map.

CATESBY, M.
1731. The natural history of Carolina, Florida and the Bahama Islands (London, printed at the expense of the author), vol. 1, xii + 100 pp., 100 pls.

COALE, H. K.
1887. Description of a new subspecies of junco from New Mexico. Auk, 4:330–331.

COOKE, W. W.
1828. *In* F. M. Bailey, Birds of New Mexico (New Mexico Department of Game and Fish), pp. 740–741.

COOPER, J. G.
1870. Ornithology of California. Geol. Surv. Calif., Ornithology. Vol. I. Land birds (published by authority of the Legislature), xi + 592 pp., many ills.

COUES, E.
1872. Key to North American birds (Salem, Naturalists' Agency), 361 pp., 6 pls., 238 figs. in text.
1874. Birds of the northwest. U. S. Geol. Surv. Terr., Misc. Publ. No. 3, xi + 791 pp.
1875. Fasti ornithologiae redivivi.—No. I. Bartram's 'Travels.' Proc. Acad. Nat. Sci. Phila., 1875, pp. 338–358.
1882. The Coues check list of North American birds. Second edition (Boston, Estes and Lauriat), 165 pp.
1884. Key to North American birds. Second edition (Boston, Estes and Lauriat), xxx + 863 pp., 561 figs. in text.
1887. Key to North American birds. Third edition (Boston, Estes and Lauriat), xxx + 895 pp., 561 figs. in text.
1892. Key to North American birds. Fourth edition (Boston, Estes and Lauriat), xxx + 907 pp., 561 figs. in text.
1895. Letter from Sylvan Lake, So. Dak. Nidiologist, 3:14–15.
1897. Rectifications of synonymy in the genus *Junco*. Auk, 14:94–95.

COWAN, I. M.
1939. The vertebrate fauna of the Peace River district of British Columbia. Occ. Papers Brit. Columbia Prov. Mus., No. 1, 102 pp., 5 pls., 2 maps.

DOBZHANSKY, T.
1937. Genetics and the origin of species (New York, Columbia University Press), xvi + 364 pp., 22 figs. in text.

DREW, F. M.
1881. Field notes on the birds of San Juan County, Colorado. Bull. Nutt. Ornith. Club, 6:85–91.

DWIGHT, J.
1918. The geographic distribution of color and of other variable characters in the genus Junco: a new aspect of specific and subspecific values. Bull. Am. Mus. Nat. Hist., 38:269–309, pls. 11–13, 5 maps.
1919. A correction involving some juncos. Auk, 36:287.

FISHER, R. A.
1934. Statistical methods for research workers. Fifth edition (Edinburgh, Oliver and Boyd), xii + 391 pp.

FISHER, W. K.
1904. Two unusual birds at Stanford University, California. Condor, 6:108–109.

FORBUSH, E. W.
1929. Birds of Massachusetts and other New England states. Pt. III. Land birds from sparrows to thrushes (Commonwealth of Massachusetts), xlviii + 466 pp., pls. 63–93, many figs.

FORSTER, J. R.
1772. An account of the birds sent from Hudson's Bay; with observations relative to their natural history; and Latin descriptions of some of the most uncommon. Philos. Trans., 62:382–433.

GAMBEL, W.
1847. Remarks on the birds observed in upper California, with descriptions of new species. Jour. Acad. Nat. Sci. Phila., n. s., 1:25–56, pls. 8–9.

GMELIN, J. F.
1788. Caroli A Linné, Systema naturae per regna tria naturae, secundum classes, ordines, genera, species, cum characteribus, differentiis, synonymis, locis. Ed. 13 (Lipsiae, impensis Georg. Emanuel. Beer), 1: pt. 2, 501–1032.

GOLDSCHMIDT, R.
1934. Lymantria. Bibliographia Genetica, 11:1–186, 74 figs.
1938. Physiological genetics (New York, McGraw-Hill), ix + 375 pp., 54 figs. in text.

GÖRNITZ, K.
1923. Ueber die Wirkung klimatischer Faktoren auf die Pigmentfarben der Vogelfedern. Jour. für Ornith., 71:456–511, pl. 7.

GRINNELL, J.
1900. Birds of the Kotzebue Sound region. Pacific Coast Avifauna, 1:1–80, 1 map.
1928. A distributional summation of the ornithology of Lower California. Univ. Calif. Publ. Zoöl., 32:1–300, 24 figs. in text.
1932. Type localities of birds described from California. Univ. Calif. Publ. Zoöl., 38: 243–324, 1 map.

GRISCOM, L.
1932. The distribution of bird-life in Guatemala, a contribution to a study of the origin of Central American bird-life. Bull. Am. Mus. Nat. Hist., vol. 64, vi + 439 pp., 11 figs., 2 maps.
1937. A monographic study of the red crossbill. Proc. Bost. Soc. Nat. Hist., 41:77–210.

HARGRAVE, L. L., PHILLIPS, A. R., and JENKS, R.
1937. The white-winged junco in Arizona. Condor, 39:258–259.

HARTLAUB, G.
1883. Beitrag zur Ornithologie von Alaska. Jour. für Ornith., 31:257–286.

HEERMANN, A. L.
1853. Notes on the birds of California, observed during a residence of three years in that country. Jour. Acad. Nat. Sci. Phila., n. s., 2: pt. 3, 259–272.

HELLMAYR, C. E.
1938. Catalogue of birds of the Americas. Part XI. Zoöl. Ser. Field Mus. Nat. Hist., vol. 8, vi + 662 pp.

HENRY, T. C.
1858. Descriptions of new birds from Fort Thorn, New Mexico. Proc. Acad. Nat. Sci. Phila., 1858, pp. 117–118.
1859. Catalogue of the birds of New Mexico as compiled from notes and observations made while in that territory, during a residence of six years. Proc. Acad. Nat. Sci. Phila., 1859, pp. 104–109.

HENSHAW, H. W.
1874. Report upon and list of birds collected by the expedition for geographical and geological explorations and surveys west of the one hundredth meridian in 1873, Lieut. G. M. Wheeler, Corps of Engineers, in charge. In Report upon ornithological specimens collected in the years 1871, 1872, and 1873 (Washington, Government Printing Office), pp. 55–148.
1875a. Am. Sportsman, Feb. 20, 1875, 328 (fide Ridgway, 1901, p. 301).
1875b. Report upon the ornithological collections made in portions of Nevada, Utah, California, New Mexico and Arizona, during the years 1871, 1872, 1873, and 1874. Rep. Geog. and Geol. Explor. and Surv. west of One Hundredth Meridian, 5: chap. iii, 131–989, 15 pls.

HOWELL, A. H.
1932. Florida bird life (Florida Department of Game and Fresh Water Fish), xxiv + 579 pp., 58 pls., 72 figs. in text.

HUEY, L. M.
1931. Two new birds and other records for Lower California, Mexico. Condor, 33:127–128.

KAEDING, H. B.
1909. Index to the Bulletin of the Cooper Ornithological Club, Volume I—1899, and its continuation The Condor, Volumes II–X—1900–1908. Pacific Coast Avifauna, vol. 6, iv + 48 pp.

KITTLITZ, F. H. VON
1858. Denkwürdigkeiten einer Reise nach dem russischen Amerika, nach Mikronesien und durch Kamtschatka (Gotha, Perthes), vol. 1, xvi + 383 pp., many ills.

LICHTENSTEIN, H.
1839. Beitrag zur ornithologischen Fauna von Californien nebst Bemerkungen uber die Artkennzeichen der Pelicane und über einige Vögel von den Sandwich-Inseln. Abhandl. König. Akad. Wiss. Berlin, 1838, pp. 417–451, 5 pls.

LINCOLN, F. C.
1933. A decade of bird banding in America: a review. Smithsonian Report for 1932, pp. 327–351, 5 pls.

LINNAEUS, C. (CAROLI LINNAEI)
1758. Systema naturae per regna tria naturae, secundum classes, ordines, genera, species, cum characteribus, differentiis, synonymis, locis. Ed. 10 (Holmiae, impensis direct. Laurentii Salvii), 1:1–823.
1766. Systema naturae per regna tria naturae, secundum classes, ordines, genera, species, cum characteribus, differentiis, synonymis, locis. Ed. 12 (Holmiae, impensis direct. Laurentii Salvii), 1:1–532.

LINSDALE, J. M.
1928. Variations in the fox sparrow (*Passerella iliaca*) with reference to natural history and osteology. Univ. Calif. Publ. Zoöl., 30: 251–292, pls. 16–20, 38 figs. in text.
1932. Frequency of occurrence of birds in Yosemite Valley, California. Condor, 34: 221–226, 2 figs.

LOOMIS, L. M.
1893. Description of a new junco from California. Auk, 10:47–48.

LUCAS, F. A.
1891. Some bird skeletons from Guadalupe Island. Auk, 8:218–222.

MACFARLANE, R.
1908. List of birds and eggs observed and collected in the north-west territories of Canada, between 1880 and 1894. *In* Mair, Through the Mackenzie Basin (Toronto, William Briggs), pp. 287–470, 2 pls.

MACOUN, J.
1904. Catalogue of Canadian birds (Geol. Surv. Canada). Pt. III, pp. iv+415–733+xxiii.

MAYR, E.
1933. Birds collected during the Whitney South Sea Expedition. XXVII. Notes on the variation of immature and adult plumages in birds and a physiological explanation of abnormal plumages. Am. Mus. Novitates, No. 666, pp. 1–10.

MEARNS, E. A.
1890. Descriptions of a new species and three new subspecies of birds from Arizona. Auk, 7:243–251.

MERRIAM, C. H.
1873. Report on the mammals and birds of the expedition. *In* F. V. Hayden, Sixth Annual Report, U. S. Geol. Surv. Terr., 1872, pp. 661–715.

MILLER, A. H.
1931. Systematic revision and natural history of the American shrikes (*Lanius*). Univ. Calif. Publ. Zoöl., 38:11–242, 65 figs. in text.
1932. The summer distribution of certain birds in central and northern Arizona. Condor, 34:96–99.
1934. Field experiences with mountain-dwelling birds of southern Utah. Wilson Bull., 46:156–158.
1935. Some breeding birds of the Pine Forest Mountains, Nevada. Auk, 52:467–468.
1938*a*. Hybridization of juncos in captivity. Condor, 40:92–93.
1938*b*. Problems of speciation in the genus *Junco*. Proc. Eighth Internat. Ornith. Congress, Oxford, 1934, pp. 277–284.
1939. Analysis of some hybrid populations of juncos. Condor, 41:211–214.
1940. A Transition island in the Mohave Desert. Condor, 42:161–163.

MORRISON, C. F.
1890. A list of the birds of Colorado. Ornithol. and Oöl., 15:36–38.

NELSON, E. W.
1897. Preliminary descriptions of new birds from Mexico and Guatemala in the collection of the United States Department of Agriculture. Auk, 14:42–76.
1921. Lower California and its natural resources. Nat. Acad. Sci., vol. 16, First Memoir, 194 pp., 35 pls.

NELSON, E. W., and GOLDMAN, E. A.
1926. Mexico. *In* Shelford, Naturalist's guide to the Americas (Baltimore, The Williams and Wilkins Company), pp. 574–596.

NEWBERRY, J. S.
1857. Report upon the zoology of the route. Report of explorations and surveys, to ascertain the most practicable and economical route for a railroad from the Mississippi River to the Pacific Ocean, vol. 6, pt. iv, no. 2, pp. 35–110, 5 pls.

NICE, M. M.
1931. The birds of Oklahoma. Rev. ed. Biol. Surv., Univ. Oklahoma, 3:1–224, 11 figs., frontispiece.

OBERHOLSER, H. C.
1918. Third annual list of proposed changes in the A. O. U. check-list of North American birds. Auk, 35:200–217.
1919. Description of an interesting new junco from Lower California. Condor, 21:119–120.

OSGOOD, W. H.
1909. Biological investigations in Alaska and Yukon Territory. No. Am. Fauna, 30:1–96, 5 pls., 2 figs. in text.

PALMEN, J. A.
1887. Bidrag till Kännedomen orn Sibiriska Ishafskustens Fogel fauna enligt Vega-Expeditionens iaktagelser och samlingar. Vega-Expeditionens Vetenskapliga Iaktagelser, 5:241–511.

PRESNALL, C. C.
1935. The birds of Zion National Park. Proc. Utah Acad. Sci., Arts, Letters, 12:196–210.

RENSCH, B.
1934. Kurze Anweisung für zoologisch-systematische Studien (Leipzig), Akademische Verlagsgesellschaft (M. H. B.), 116 pp., 22 figs. in text.

RHOADS, S. N.
1893. The birds observed in British Columbia and Washington during spring and summer, 1892. Proc. Acad. Nat. Sci. Phila., 1893, pp. 21–65.

RIDGWAY, R.
1873a. On some new forms of American birds. Am. Nat., 7: 602–619.
1873b. The birds of Colorado. Bull. Essex Inst., 5:174–195.
1874–1875. Lists of birds observed at various localities contiguous to the Central Pacific Railroad, from Sacramento City, California, to Salt Lake City, Utah. Bull. Essex Inst., 6:169–174; 7:10–24.
1876. Ornithology of Guadeloupe Island, based on notes and collections made by Dr. Edward Palmer. Bull. U. S. Geol. Geog. Surv. Terr., 2:183–195.
1877. The birds of Guadalupe Island, discussed with reference to the present genesis of species. Bull. Nutt. Ornith. Club, 2:58–66.
1878. Descriptions of two new species of birds from Costa Rica, and notes on other rare species from that country. Proc. U. S. Nat. Mus., 1:252–255.
1883. Descriptions of some new birds from Lower California, collected by Mr. L. Belding. Proc. U. S. Nat. Mus., 6:154–156.
1885. On *Junco cinereus* (Swains.) and its geographical races. Auk, 2:363–364.
1895. *Junco phaeonotus* Wagler, not *J. cinereus* (Swainson). Auk, 12:391.
1897. Notes on *Junco annectens* Baird and *J. ridgwayi* Mearns. Auk, 14:94.
1898. New species, etc., of American birds. II. Fringillidae (continued). Auk, 15:319–324.
1901. The birds of North and Middle America. U. S. Nat. Mus. Bull., vol. 50, pt. 1, xxx + 715 pp., 20 pls.

RILEY, J. H.
1912. Birds collected or observed on the expedition of the Alpine Club of Canada to Jasper Park, Yellowhead Pass, and Mount Robson region. Canadian Alpine Journal (Alpine Club of Canada), special number, pp. 47–75, 2 pls.

ROBERTS, T. S.
1932. The birds of Minnesota (Minneapolis, University of Minnesota Press), vol. 2, xv + 821 pp., pls. 50–90, figs. 299–606, frontis.

ROBSON, G. C.
1928. The species problem, an introduction to the study of evolutionary divergence in natural populations (Edinburgh, Oliver and Boyd), xvii + 283 pp.

SALVIN, O.
1863. Descriptions of thirteen new species of birds discovered in Central America by Frederick Godman and Osbert Salvin. Proc. Zoöl. Soc. London, 1863, pp. 186–192, pls. XXIII, XXIV, 1 fig.

SALVIN, O., and GODMAN, F. D.
1879–1904. Biologia Centrali-Americana. Aves, vol. 1, xliv + 512 pp.

SAUNDERS, A. A.
1921. A distributional list of the birds of Montana. Pacific Coast Avifauna, 14:1–194, 35 figs.

SCLATER, P. L.
1857. Notes on the birds in the museum of the Academy of Natural Sciences of Philadelphia, and other collections in the United States of America. Proc. Zoöl. Soc. London, 1857, pp. 1–8.

SCLATER, W. L.
1912. A history of the birds of Colorado (London, Witherby and Co.), xxiv + 576 pp., 17 pls., 1 map, frontis.

SHARPE, R. B.
1888. Fringilliformes: Pt. III. Fringillidae. Cat. Birds Brit. Mus., vol. 12, xv + 871 pp., 16 pls.

SITGREAVES, L.
1854. Report of an expedition down the Zuni and Colorado rivers (Washington, Beverly Tucker, senate printer), 198 pp., many pls., map.

SQUIRES, W. A.
1916. Sierra junco in Golden Gate Park. Condor, 18:203.
STONE, W.
1904. On a collection of birds and mammals from Mount Sanhedrin, California. Proc. Acad. Nat. Sci. Phila., 56:576–585.
1918. Dwight's review of the juncos. Auk, 35:486–489.
SUMNER, F. B.
1929. The analysis of a concrete case of intergradation between two subspecies. II. Additional data and interpretations. Proc. Nat. Acad. Sci., 15:481–493.
1932. Genetic, distributional, and evolutionary studies of the subspecies of deer mice (*Peromyscus*). Reprint from Bibliographia Genetica (The Hague, Martinus Nijhoff), vol. 9, 106 pp., 24 figs. in text.
SWAINSON, W.
1827. A synopsis of the birds discovered in Mexico by W. Bullock, F. L. S. and H. S., and Mr. William Bullock, jun. Philos. Mag., n.s., 1:364–369, 433–442.
SWARTH, H. S.
1900. Avifauna of a 100-acre ranch. Condor, 2:37–41.
1912. Report on a collection of birds and mammals from Vancouver Island. Univ. Calif. Publ. Zoöl., 10:1–124, pls. 1–4.
1914. A distributional list of the birds of Arizona. Pacific Coast Avifauna, 10:1–133.
1922. Birds and mammals of the Stikine River region of northern British Columbia and southeastern Alaska. Univ. Calif. Publ. Zoöl., 24:125–134, pl. 8, 34 figs. in text.
1926. Report on a collection of birds and mammals from the Atlin region, northern British Columbia. Univ. Calif. Publ. Zoöl., 30:51–162, pls. 4–8, 11 figs. in text.
1933. Off-shore migrants over the Pacific. Condor, 35:39–41.
TANNER, V. M., and HAYWARD, C. L.
1934. A biological survey of the La Sal Mountains, Utah. Report No. 1 (Ecology). Proc. Utah Acad. Sci., Arts, Letters, 11:209–234, pl. 10.
TAYLOR, W. P.
1912. Field notes on amphibians, reptiles and birds of northern Humboldt County, Nevada, with a discussion of some of the faunal features of the region. Univ. Calif. Publ. Zoöl., 7:319–346, pls. 7–12.
TOWNSEND, J. K.
1837. Description of twelve new species of birds, chiefly from the vicinity of the Columbia River. Jour. Acad. Nat. Sci. Phila., 7:187–192.
1839. Narrative of a journey across the Rocky Mountains, to the Columbia River and visit to the Sandwich Islands, Chili, &c. with a scientific appendix (Philadelphia, Henry Perkins), viii + 352 pp.
TRIPPE, T. M.
1874. *In* Coues, Birds of the northwest, pp. 144–145.
VAN ROSSEM, A. J.
1931. Descriptions of new birds from the mountains of southern Nevada. Trans. San Diego Soc. Nat. Hist., 6:325–332.
1936. Birds of the Charleston Mountains, Nevada. Pacific Coast Avifauna, 24:1–65.
1938. [. . . descriptions of twenty-one new races of Fringillidae and Icteridae from Mexico and Guatemala. . . .] Bull. Brit. Ornith. Club, 58:124–138.
VIGORS, N. A.
1839. The zoology of Captain Beechey's voyage . . . in his Majesty's Ship Blossom. Ornithology (London, Henry G. Bohn), pp. 13–40, pls. 3–14.
VISHER, S. S.
1911. Annotated list of the birds of Harding County, northwestern South Dakota, Auk, 28:5–16.
WAGLER.
1831. Einige Mittheilungen über Thiere Mexicos. Isis von Oken, 1831, pp. 510–535.

WARREN, E. R.
  1936. Charles Edward Howard Aiken. Condor, 38:235–238, fig. 47.
WILLETT, G.
  1933. A revised list of the birds of southwestern California. Pacific Coast avifauna, 21:1–204.
WILSON, A.
  1810. American Ornithology; or, the natural history of the birds of the United States (Philadelphia, Bradford and Inskeep), vol. 2, xii + 167 pp., pls. 10–18.
WOODHOUSE, S. W.
  1853. Description of a new snow finch of the genus *Struthus*, Boie. Proc. Acad. Nat. Sci. Phila., 6:202–203.
  1854. Report on the natural history of the country passed over by the exploring expedition under the command of Brev. Capt. L. Sitgreaves, United States Topographical Engineers, during the year 1851. *In* Sitgreaves, Report on an expedition down the Zuni and Colorado rivers, pp. 33–105.

Transmitted by author May 19, 1938; but corrections and some additions made up to November 15, 1940.—A. H. M.

CRITERIA OF SUBSPECIES, SPECIES AND GENERA IN ORNITHOLOGY*

By

Ernst Mayr

The American Museum of Natural History, New York, N. Y.

It is an impossible task to give an adequate discussion of the criteria of subspecies, species and genera in the short time of 15 minutes. So let us lose no time in preliminaries but start immediately with the discussion of *species*, which after all is the most important of all taxonomic categories. The species concept has become increasingly confused in recent years by the application of rigid and arbitrary criteria. Some authors determine the status of a natural population with the help of *genetic* criteria; others, through *morphological* criteria, such as lack of intergradation or degree of morphological difference; still others, on the basis of diminishing *fertility*. In determining whether or not such criteria are valid, we must go back to the fundamentals of the species concept. Let us first ask: *What is a species?*

The species concept is almost as old as mankind itself. I once spent several months with a primitive tribe of Papuans in the interior of New Guinea and found that these people had a different vernacular name for nearly every species of bird occurring in their territory. What these natives distinguish as different kinds of birds corresponds exactly to the species of the taxonomist. The same is true for our local birds. The field naturalist recognizes, for example, five kinds of thrushes of the genus *Hylocichla* in the American Northeast: the wood-thrush, veery, hermit thrush, gray-cheeked thrush and olive-backed thrush. Their ranges overlap broadly; in fact, up to three of these kinds of birds may nest in the same woods, but no intermediates or hybrids are known. Furthermore, these five species differ in their songs, courtship habits, nests, migratory habits and about every other attribute that has been studied carefully. Local populations within any of these five species are *interbreeding*, but each of the five units is completely separated from the others; it is *reproductively isolated*, in the terminology of the modern biologist. And thus we have arrived at two of the basic criteria, in fact, at *the* two basic criteria of species: first, interbreeding of the local populations belonging to the species; second, reproductive isolation of those populations which do not belong to the same species.

---

*A more detailed survey of this field is presented in the author's book, "Systematics and the Origin of Species" (Columbia University Press).

After the museum taxonomist obtained specimens of what the field naturalist recognized as species, he discovered that each of these interbreeding units, called species, showed certain morphological characteristics which permitted the easy identification of dead specimens. These characteristics, the so-called taxonomic characters, gained an ever increasing importance, particularly when it concerned the classification of specimens from countries or systematic groups about which field naturalists were unable to provide pertinent data. Eventually most taxonomists forgot that species were aggregates of living organisms; they forgot that in the museum they were dealing merely with dead samples of the true species of nature, and one of the consequences of this working technique of the taxonomist was that the taxonomic character advanced from a convenient handle in practical work to the level of an absolute criterion. The conclusion, "the level of reproductive isolation—that is, the species level—is associated with certain morphological differences," was reversed to read: "a certain degree of morphological difference proves reproductive isolation—that is, it proves specific difference." This conclusion is one of the most famous fallacies of logic. The ancient Greeks illustrated it in the form of a so-called "vicious syllogism":

(1) Birds have two legs.
(2) Man has two legs.
(3) Therefore, man is a bird.

In our case the fallacious syllogism reads:
(1) Forms that are known to be good species are separated by morphological gaps.
(2) Forms A and B, whose specific status is in doubt, are separated by a morphological gap.
(3) Therefore, A and B are good species.

This, of course, is true only in some cases, not in others. The absurdity of the strictly morphological criterion would become perfectly evident if we were to apply it to males and females in sexually dimorphic species, or if we were to recognize larval or immature stages as separate species or give specific rank to the various forms of a polymorphic species. As a matter of fact many "species" of the taxonomic literature have turned out to be nothing but such intraspecific variants. However, as soon as it was proven that they interbreed with other variants, they were deprived of their specific rank, even though they remained separated from each other by clear-cut, unbridged gaps in morphological characters. To repeat once more, in a given locality the criterion of interbreeding versus reproductive isolation will permit us to determine, in nearly all cases, which individuals belong to one species and which to another. Morphological differences are of practical convenience, but not a primary criterion.

What is true for contiguous individuals—Poulton calls them *sympatric* individuals or species—is equally true for those that do not occur in the same geographical districts, the so-called *allopatric* forms. No degree of morphological distinctness is in itself proof of specific distinctness. The probability of reproductive isolation is the primary criterion. To emphasize this point is important, because the degree of reproductive isolation of a geographically isolated form is not necessarily correlated with the degree of morphological distinctness, nor is the degree of morphological distinctness necessarily a good index of genetic distinctness. A single gene difference may produce a phenotypic difference which might cause some taxonomists to call the population carrying it a different species. This consideration again leads us to the inevitable conclusion that a species concept, based primarily on the criterion of morphological distinctness or of a gap in morphological characters, is invalid.

Please do not misunderstand me. I am not demolishing the morphological species concept merely because I enjoy being an iconoclast. No, I attack it because I consider it contrary to the fundamental concept of species. But, you will ask me, how should we treat geographically isolated populations? Should we follow Kleinschmidt and call *all* of them subspecies? Such a procedure would have the advantage of consistency, but actually it is just as unscientific as calling all morphologically separated populations species, because there is little doubt that many of these isolated populations have already reached the species level. But how shall we decide whether or not these allopatric forms are species?

A complete analysis, including experiments on mating preference and a cytological examination of possible hybrids, is required in order to reach a satisfactory decision. To make such a painstaking analysis is out of the question in practically all the cases which interest the taxonomist. The taxonomist is forced, in most cases, to determine the status of a geographically isolated population by indirect methods. Reproductive isolation is correlated with a certain degree of morphological difference, which is rather typical for any given genus. It is very small in the flycatcher genus *Empidonax* or in the mosquito genus *Anopheles*, but extraordinarily large among the birds of paradise. If we want to determine the taxonomic status of a geographically isolated population, we must first study all the species of the genus to which this population belongs, and the species of related genera. In this manner we can work out a scale of differences between unquestionably valid species and between unquestionable subspecies. The status of the doubtful population will have to be decided by inference.

In conclusion I might say that the species concept, as just set forth by me, is by no means endorsed by all ornithologists. Many of them consider, for example, the yellow-shafted and the red-shafted flicker, or the Oregon and the eastern junco as separate species, even though individuals of the respective forms seem to interbreed indiscriminately wherever they come into geographical contact. A proponent of a biological species concept has no choice but to consider such completely interbreeding forms as conspecific. So much for the species.

The *subspecies* is composed of a group of local populations and can be distinguished from other such groups by one or several taxonomic characters. There are three principal difficulties involved in the recognition of subspecies.

(1) *How can it be determined whether an isolated population is a subspecies or a species?* This question was just discussed in regard to the species and no further comments are needed.

(2) *Is the geographical race the only kind of subspecies?* Invertebrate and plant taxonomists recognize ecological races. Recent research indicates that some of these so-called "races" are purely phenotypical, while others are microgeographical races. No such ecological races of birds have been described. It might be emphasized, however, that every geographical race owes a greater or smaller proportion of its characteristics to the selective influences of the particular local environment. Every geographical race is, thus, to a greater or lesser degree also an ecological race. In some cases, as for example on black lava flows, geographical races develop which owe their most conspicuous (and often only) taxonomic character to the selective qualities of this particular habitat. I see little advantage in calling such forms ecological races rather than geographical races, because each of these races has many characteristics which are not the result of local selective factors. It will be the task of the student of plants and invertebrates to determine whether or not it is possible to recognize the ecological race as something quite distinct from the geographical race. One point should not be overlooked in such an analysis. Whenever two ecologically differentiated races overlap in the same locality, a careful analysis of the situation usually shows that this is a secondary condition and that the original ecological difference developed during a preceding geographical isolation;

that is, the race was a geographical race before it became an ecological race.

(3) The third difficulty in recognition of the subspecies can be expressed in the question: *How can subspecies be delimited from each other in continuous populations?* Subspecies borders are generally drawn where there is a distributional gap, or a change of environment, or a significant change in the taxonomic characters of the continuous populations. A detailed analysis of the populations near subspecies borders shows in many cases that the change from one subspecies into the next is so gradual that the placing of these borders is left to an arbitrary decision. In other cases equally arbitrary decisions must be made concerning the degree of difference between two groups of populations which is to be considered sufficient for separation of subspecies. There are splitters and lumpers. In general it is stated, in the ornithological literature, that 75 per cent of the individuals of one race must be clearly separable from all the individuals of the other race. This is a very unsatisfactory method of handling the problem, since the status of many subspecies changes with the increase in number of collected specimens. A way should be found to express the necessary degree of distinctness in more absolute terms, such as standard deviations. Beginnings are being made along these lines, but they have not yet found their way into the taxonomy of birds.

And now for the *genus!* Recent historic research has proven conclusively that the genus of Linnaeus and his forerunners goes back to an equivalent concept of folk-lore. There was a concept "oak," before there was a genus *Quercus,* and a concept "finch" or "thrush," before there were any such genera as *Fringilla* and *Turdus.* Up to this point the genus presents a parallel case to the species. However, whereas the species of Linnaeus, in the great majority of the cases, still corresponds to the species of today, the genus of Linnaeus only rarely does so. In most cases it is now equivalent to a subfamily, a family, or even an order. There is little argument among contemporary ornithologists concerning the delimitation of species, but the divergence of opinion in regard to the genus is tremendous. The genus of such splitters as Mathews, Roberts and Oberholser has little in common with the genus of lumpers as represented by Stresemann, Peters and Mayr, to mention some of the contemporary ornithologists. Let us illustrate the difference between splitters and lumpers by examining a part of a phylogenetic tree.

Let us assume we have two phylogenetic branches, A and B. Branch A again will break up into four or five twigs and twiglets. Both lumper and splitter agree that all the twigs on branch A are derived from a common ancestor; they also agree that there is a definite gap between A and B, and that all the species of A have certain characters that distinguish them from all the species on B. However, the splitter will emphasize the reality of the little gaps between the twigs on branch A, while the lumper will contend that the recognition in nomenclature of all minor gaps will obscure rather than clarify the true relationships and the basic pattern of classification.

We now must ask, what are generic criteria?

It is perhaps not always realized or at least not kept in mind that subspecies, species and genus differ from all other systematic categories by being the only ones that are mentioned in the scientific name of an individual animal. The significance of this nomenclature must therefore be examined if we want to consider the status of the categories which they represent. We can eliminate the subspecies from our discussion, since what is true for the species is also true for the subspecies, and concentrate on genus and species. There is a fundamental difference between these two categories. The species is an individual unit; the species name therefore emphasizes distinctness.

The genus, on the other hand, is a collective unit and the joint application of the same generic name to a number of species indicates their similarity or relationship. The functions of the two components of the scientific name as proposed by Linnaeus are therefore diametrically opposite. The species name signifies singularity and distinctness, the generic name implies the existence of a group of similar or related units. This difference in the functions of species and genus names is completely ignored by many recent taxonomists, particularly the so-called generic splitters. It is their aim to express difference not only in the specific, but also in the generic name. This tendency, if carried to its logical extreme, leads to uninomialism, and some of the leading generic splitters have openly or in a veiled form endorsed this principle of nomenclature. To me it seems to indicate a complete misunderstanding of the principle of binomial nomenclature, if somebody uses the generic name primarily to express difference. This is the function of the species name. The generic name was introduced by Linnaeus into nomenclature in order to relieve the memory and this should remain its principal function. Wherever the genus becomes too small, it loses its usefulness.

The splitting fever has played havoc in bird taxonomy. We have now more than 10,000 generic names for an estimated 8500 species of birds. Even conservative authors admit 2600 genera, which amounts to only 3.27 species per genus of birds. An average of five species per genus would bring the number of bird genera down to about 1700, which would be within the capacity of the memory of a single human individual. The trend toward larger genera has been unmistakable in ornithology and it has been accelerating in recent years. I do not know of a single younger author who could be classed as a generic splitter.

The genus should contain groups of similar species—that is, species which we consider related—but so far as I know nobody has ever found an *objective* criterion of the genus. Personally, I like best a genus definition which is based on honest admission of the subjective nature of this unit. It may not be possible to go much beyond the definition of the entomologist Thorpe who said: "The genus, to be a convenient category in taxonomy, must in general be neither too large nor too small." A more comprehensive definition would be: "A genus is a systematic unit including one species or a group of species, presumably of common phylogenetic origin, which is separated from other similar units by a decided gap. It is demanded for practical reasons that the size of the gap be in inverse relation to the size of the unit." This latter qualification will prevent the recognition of many monotypic genera.

We can summarize this discussion as follows: The taxonomic category of the genus is based on the fact that species are not evenly distinct from one another, but are arranged in smaller or larger groups, separated by smaller or larger gaps. Recognition of the genus is, therefore, based on a natural phenomenon. How many of such groups are to be included in one genus and how the genus should be delimited from other genera are matters of convenience left to the judgment of the individual systematist. Taxonomic characters that prove generic distinctness do not exist. Taxonomic literature could have been spared many unnecessary generic names if the taxonomists had kept in mind Linnaeus' warning: "The characters do not make the genus, but the genus gives the characters."

One more word in conclusion—all taxonomic categories are collective units. The subspecies, the species and the genus all consist of groups of unequal components: the subspecies of local populations, the species of local populations and subspecies, the genus of more or less distinct species. Those authors who try to obtain homogeneous units by splitting them down to their last elements are not only bound to fail in their endeavor, but they also obscure the basic relationships.

HISTORY OF THE NORTH AMERICAN BIRD FAUNA

BY ERNST MAYR

THE bird student cannot help becoming envious on observing with what accuracy and amazing detail the student of mammals reconstructs the history of that class. Rich finds of fossils have enabled the paleomammalogist to determine the probable region of origin not only of families but also of genera, sometimes even of species, and to trace past modifications in their ranges. The student of birds is far less fortunate. Bird bones, being small, brittle, and often pneumatic, are comparatively scarce in fossil collections. The majority of Tertiary species of birds described from North America belong to zoogeographically unimportant families of water birds. Even fewer fossil birds are known from South America. The absence of certain families or orders from the fossil record of either North or South America proves nothing as far as birds are concerned. Furthermore, the history of birds is more difficult to reconstruct than that of mammals for two other reasons. Birds seem to be a more ancient group than the mammals, many or most of the Recent families having been in existence at the beginning of the Tertiary. And secondly, since birds cross water gaps more easily than mammals, the isolation of a land mass does not necessarily result in the isolation of its bird fauna. It would seem on these premises that it would be almost impossible to trace the history of the components of a local bird fauna, but this is by no means the case. Indirect methods of faunal analysis lead to fairly reliable results, since most families of birds are rich in genera and species. A quantitative analysis is, of course, impossible in small families, and their place of origin (as, for example, that of the limpkins) can be determined only with the help of fossils. In a paper read in 1926 before the International Ornithological Congress at Copenhagen, Lönnberg (1927) demonstrated the productivity of the indirect method by applying it in an investigation of the origin of the present North American bird fauna. Although most of Lönnberg's conclusions are still valid today, so much additional knowledge has accumulated during the past 20 years that a fresh analysis seems timely.

Faunal and Regional Zoogeography

There have been trends and fashions in the science of zoogeography as in any other science. The zoogeography of the nineteenth century—the classical zoogeography of Schmarda (1853), Sclater (1858), and Wallace (1876)—was merely descriptive, essentially regional, and nondynamic. It was based on the premise that different parts of the world are inhabited by different kinds of animals; and each of these major areas was called a zoogeographical region. This method seemed successful while knowledge of the distribution of animals was still incomplete. As far as the boundaries between these regions were concerned, it was recognized that they "depend upon climatic conditions, which are in a measure determined or modified by features of topography" (Allen, 1893:120). However, as the various parts of the world became better known, it became evident that the various regions proposed were of unequal value. This led to the proposal of new regions or to the fusion of previously separated regions into larger units. It is impossible to give here the history of the never-ending attempts to find a "perfect" zoogeographical classification. For example, it was soon found that the fauna of North America was somewhat intermediate between that of Asia and that of South America, which resulted in conflicting proposals concerning the zoogeographic position, or rank, of North America.

According to one school, North America was only part of a larger region combining North America, Europe, and north Asia. Gill (1875:254) called this region the Arctogaean, while Heilprin (at the suggestion of Newton) called it the Holarctic (Heilprin, 1883:270). This region (with the Palearctic and Nearctic as subregions) is perhaps even today the most frequently adopted zoogeographical classification of the northern hemisphere. Reichenow (1888:673 ff.) took emphatic exception to this classification. He showed that, as far as birds were concerned, North America was much closer to the "Neotropical" than to the Old World, and that North and South America should be combined in a "Western Zone" or "New World Region." This point is well substantiated by his statistics. J. A. Allen (1893:115) showed that the Old World element in the warm temperate parts of North America amounted to only 23 to 37 per cent of the genera, but he did not draw any conclusions from these figures. Subsequent writers almost completely ignored Reichenow's conclusions. Heilprin (1883) went to the opposite extreme. He refused to recognize the Nearctic even as a subregion. He drew a zoogeographic boundary right across North America, putting the northern half into the "Holarctic Region," the southern half in the "Neotropical Region." Wallace himself thought (1876:66) that it was a question "whether the Nearctic Region should be kept separate, or whether it should form part of the Palaearctic or of the Neotropical regions." The literature, particularly of the 1880's and 1890's, was filled with discussions of this question.

Eventually it was realized that the whole method of approach—*Fragestellung*—of this essentially static zoogeography was wrong. Instead of thinking of fixed regions, it is necessary to think of fluid faunas. As early as 1894, Carpenter said: "No zoological region can be mapped with the hard and fast line of a political frontier, and the zoologist must always think more of faunas than of geographical boundaries" (1894:57). The faunal approach made slow but steady progress in Europe and in America. In Europe it has led to such excellent studies as those of Stegmann (1938a) on the birds of the Palearctic and of Stresemann (1939) on the birds of the Celebes. In America it was E. R. Dunn who was the pioneer of this concept. In a spirited attack on the older, static, regional zoogeography, he stated (1922:336):

> There has been a constant search for some sort of scheme whereby ranges of animals might be reduced to a common denominator. . . .
> By far the most generally used of these philosophical methods is that of Realms, Regions and Zones. These are all based on the idea that large numbers of species have the same range, and that by picking out some of the conspicuous forms and mapping their ranges one has *ipso facto* a set of regions, to which other ranges may be referred, and with which other ranges should agree.
> This is, in some degree, true, but in nearly every case in which the ranges of any two species agree, the agreement is due to the geographic factors and not to the zoologic factors.
> It is obvious that the zoogeographical realms are nothing save and except the great land masses with lines drawn to corespond to the physiographic barriers. There is a great philosophical difference between such terms as Holarctic Fauna and Holarctic Region. In the first case we speak of zoological matters in terms of zoology, in the second of geographical matters in terms of mythology.
> The Palearctic fauna is an aggregate of species and may invade (in fact *has* invaded) Australia without forfeiting its name.

Following up these thoughts, Dunn (1931:107) analyzed the reptile fauna of North America and found that it could be classified into the following three groups:

> (1) A northern, circumpolar, modern element. This would be truly *Holarctic*.
> (2) A more southern, older element, which I shall call *Old Northern*. . . .
> (3) A still more southern, still older element, the original fauna of South America, with its analogues in the Australian or Ethiopian regions. This I shall call *South American*, as I wish to avoid the term Neotropical. . . .

I have attempted in the following sections to classify the North American bird fauna in a similar manner. This classification, tentative as it is under the circumstances, is very useful as a test of the various arrangements proposed by regional zoogeographers. It provides at least provisional answers to such questions as: "Is it justifiable to recognize a neotropical fauna and a nearctic fauna?" "Is the nearctic fauna, if it exists, part of a New World or of a holarctic fauna?" "Does North America have a fauna of its own, or is it merely an area of intergradation between the Eurasian and the South American faunas?" "Are the faunas of given geographical areas sufficiently homogeneous to justify the recognition of zoogeographic regions, or does the delimitation of zoogeographic regions convey an erroneous impression?"

RECENT ADVANCES

We are in a much better position today to answer these questions than was Lönnberg 20 years ago. First, there has been a general advance in the whole field of zoogeography—a complete change in the concept of the functions of the science—signalized by the important publications of Simpson, Stegmann, and Stresemann. Classical zoogeography asked: What are the zoogeographic regions of the earth, and what animals are found in each region? The modern zoogeographer asks when and how a given fauna reached its present range and where it originally came from; that is, he is interested in faunas rather than in regions. In the light of this new concept of the science, such familiar terms as holarctic, nearctic, and neotropical acquire completely new meaning. Secondly, there have been many very specific recent additions to our knowledge, contributed partly by the paleontologist and partly by the taxonomist, which permit a more accurate analysis than Lönnberg could give.

Recent contributions of the paleontologist. The number of important discoveries of fossil birds has been greatly augmented in recent years, the Californian school and Alexander Wetmore having made the most valuable contributions. Finds of particular zoogeographic significance concern the following groups (Wetmore, 1940): 1. The Aramidae. The limpkin (*Aramus*) is the only living representative of this family; and, as Lönnberg said (1927:24), "if one has to judge only from the present distribution, [it] would certainly be regarded as South American"; but the fact that there are two extinct Tertiary genera (*Badistornis* and *Aramornis*) in North America favors a North American origin for the family. 2. The Old World vultures (Aegypiinae), which are now restricted to the Old World. Nobody would suspect the former occurrence in the New World of this subfamily of the Accipitridae if fossil remains of three extinct genera had not been found in the Miocene (*Palaeoborus*), Pliocene (*Palaeoborus, Neophrontops*), and Pleistocene (*Neogyps, Neophrontops*) of North America. No conclusion can be drawn, however, as to the origin of the family. 3. The New World vultures (Cathartidae), which Lönnberg (1927:22) listed as a South American family. The fact that Wetmore (1940 and 1944) has found several striking genera in the early Tertiary of North America indicates either a North American or pre-Tertiary origin for the family. 4. The Cracidae (curassows and guans), whose present center of distribution is in South America, where the vast majority of the species occur and where most of the genera are endemic. Even though seven Recent species occur in Central America and two genera are endemic there (*Penelopina* and *Oreophasis*), this family would surely be considered a comparatively recent arrival in North America, were it not for the occurrence of two species in the Tertiary of North America (*Ortalis tantala* in the lower Miocene; *O. phengites* in the lower Pliocene) and for the occurrence in the Wyoming Eocene of the related (fossil) family Gallinuloididae.

Recent contributions of the taxonomist. Unsound classifications have caused much confusion in zoogeography, as ably pointed out by Simpson (1940b) in a discussion of the so-called evidence for an antarctic land bridge. Of particular zoogeographic significance are the following recent changes in the classification of birds.

"New World Insect Eaters." From a study of a number of South American genera it would seem that the tanagers (Thraupidae)—including the South American swallow-tanagers (Tersinidae), honeycreepers (Coerebidae), wood warblers (Parulidae—formerly "Compsothlypidae"), vireos (Vireonidae) — including the shrike-vireos (Vireolaniidae) and the pepper-shrikes (Cyclarhidae), blackbirds and troupials (Icteridae), and some of the finches (the subfamily Emberizinae) are closely related, constituting a single superfamily, perhaps the New World equivalent of the Old World family Muscicapidae of recent authors (J. T. Zimmer, verbal information).

Troglodytidae. Sharpe's Hand-list (vol. 4, 1903) and other older taxonomic works included among the wrens a considerable number of south Asiatic genera (*Pnoëpyga, Elachura, Spelaeornis, Sphenocichla,* and sometimes *Tesia*). Lönnberg (1927:9–10) consequently had considerable difficulty in proving an American origin for this family. Recent taxonomic work has clearly established the fact that none of the listed Asiatic genera (superficially wren-like babbling thrushes and Old World warblers) belongs to the Troglodytidae and that *Troglodytes troglodytes* is the only wren that occurs in the Old World. The strictly American character of the wren family is now beyond dispute.

"Chamaeidae." The Wren-tit (*Chamaea*) is not the sole representative of a separate family, but a member of the Paradoxornithinae (parrot bills and suthoras), and possibly not even generically separable from *Moupinia* of southwest China.

Fringillidae. The so-called finches are an assemblage (probably highly artificial) of seed-eating birds with cone-shaped bills. Three major groups can be distinguished within the fringillids that are established in North America: (a) Carduelinae—the carduleine finches; (b) Emberizinae—certain buntings and American sparrows; and (c) Richmondeninae—the cardinals, or South American finches. (See Sushkin, 1924 or 1925.) There is little doubt that the Carduelinae are Old World in origin; the Emberizinae North American, although some species are found in the Old World; the Richmondeninae South American, although some genera have become thoroughly established in

North America. (It should be noted that no final decision can be reached on the last two groups until it has been determined whether certain South American genera belong to the Emberizinae or to the Richmondeninae. A discussion of the characters of the fringillid subdivisions, as well as an incomplete listing of the genera, will be found in Sushkin.)

THE GEOLOGICAL HISTORY OF NORTH AMERICA

The North America of today is connected with South America by an isthmus and is separated from Asia only by a narrow oceanic strait. These connections with the two adjoining faunal areas are of the greatest importance, and a study of their history, both geologically and climatically, is a prerequisite to full understanding of the faunal history of North America. There is also a loose connection directly with Europe through the arctic islands of the North Atlantic (Greenland, Iceland), but it is doubtful whether it ever played a greater role for land birds than it does today. The Wheatear (*Oenanthe oenanthe*) is one of the few birds that has come to us via this bridge.

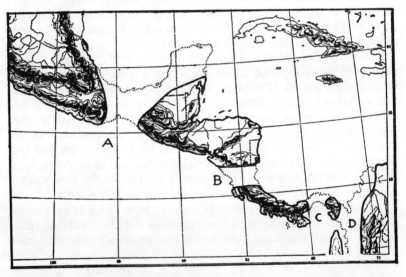

Figure 1. Tertiary water gaps between North and South America. A = Tehuantepec gap (late Miocene to middle Pliocene), B = Nicaraguan gap (late Eocene to middle Miocene), C = Panamanian gap (late Eocene to ? late Oligocene), D = Colombian gap (middle Eocene to late Miocene). (*Free reconstruction from various geological sources.*)

The coast line of North America in former geological periods was not always where it is today. There is, for example, good evidence for a former land connection across Bering Strait, as well as for oceanic gaps across what is now Central America (Figure 1). The extent of these changes in the outlines of land areas is being debated rather vigorously by the geologists and paleogeographers, who tend to interpret the available evidence to fit the concepts of one of the following three schools. The oldest concept is that of a continuous large-scale change in the surface of the earth. Some land masses sink to the bottom of the ocean while others arise by buckling up. Old continents break to pieces as new ones are being formed. Today few authors believe in such violent upheavals. The prevailing theory today is perhaps that of "permanence of continents and oceans." The continents, as well as the major oceanic basins, are relatively stable according to this school of thought. "Sea bottoms" that dry up and lands that become submerged are merely the shallow "amphibious" zones on the continental shelves. The relative position of continents and oceanic basins has not changed materially, according to this theory, since Mesozoic times or even before. The third theory includes elements of the other two, but combines them in a very original way. It agrees with the second theory that continents will always remain continents and ocean bottoms will stay ocean bottoms, but denies that their relative positions are fixed. Rather it holds that the continents are floating on the magma of the earth like ice floes in the arctic sea and that they are continuously shifting their position (Wegener's theory of continental drift). As Simpson (1943a) and others have pointed out, the zoogeographical evidence is on the whole opposed to the theory of continental drift, at least for the Mesozoic and Tertiary periods.

Although some points are still controversial, the following facts seem to be well established:

(1) South America was separated from North America for the greater part of the Tertiary. The isthmus between Colombia and central Mexico was broken into a series of islands by several ocean channels between the Pacific and the Caribbean (Figure 1). A complete land connection between South and North America probably did not exist between the lower Eocene (50 to 70 million years ago) and upper Pliocene (about 2 million years ago).

(2) Asia and North America were repeatedly connected by dry land across Bering Strait during the Tertiary. There is no evidence that this bridge was ever much more extensive than the present shelf, nor is there any evidence for a complete land bridge to Asia across the Aleutians. The Bering Strait bridge may have existed as recently as the last ice age.

A few more words about the nature of these land bridges before we examine what faunal elements have reached or left North America on them. The ocean gaps between North and South America must have been considerable (perhaps even wider than shown in Figure 1), since they almost completely prevented an interchange of the mammals of

North and South America. Ground sloths were apparently the only South American mammals to reach North America during the period of separation; only raccoons (procyonids), with possibly also monkeys and opossums, crossed from North to South America (Simpson, 1940a:158). For birds, these ocean channels were much less of a hindrance, as will be shown below.

Most important for an understanding of the origin of the North American fauna is the fact, emphasized by Lönnberg (1927), Dunn (1931), and Simpson (1943b), that the whole southern half of North America was subtropical or tropical during most of the Tertiary, when it was separated from South America by oceanic gaps. Even in the later Tertiary, a tropical climate prevailed in the southernmost section of North America. This means that (with the exception of those animals that cross water gaps easily) there was not merely one tropical American fauna, the "Neotropical," but two quite distinct ones: one south of the ocean gaps, the other north of them. F. M. Chapman (1923) showed that the motmots (Momotidae), usually referred to as a "typically Neotropical" family, had actually originated in Middle America "where the ancestral forms of the existing genera were possibly developed during the Oligocene when this region consisted of scattered islands which would afford the isolation favorable to differentiation" (p. 58). Lönnberg (1927:12) states correctly that the same would probably be found to be true, if other families were examined as "thoroughly and masterfully" as the Momotidae were by Chapman. In the meantime, Dunn (1931), Simpson (1943b:428), and Hubbs (1944:271) have emphasized the importance of this Middle American (i.e. tropical North American) element among reptiles and fishes.

The mid-Tertiary fauna of North America was probably not only highly peculiar but also rather homogeneous. To visualize its composition, one must look at the South America of today. The temperate zone of South America, which admittedly is rather small because of the continent's triangular shape, does not have a fauna which is basically different from that of the tropical areas. It has its share of endemic species and even genera, but its fauna (although poorer) is composed more or less of the same families as that of the warmer portion. A similar faunal homogeneity was perhaps true for North America during Tertiary times, the faunas of the tropical, of the subtropical, and of the warm-temperate zones being very much alike in composition. The present-day contrast between the fauna of tropical-subtropical Central America and that of temperate North America, has two causes: (1) the climatic deterioration in the late Tertiary and Pleistocene, which eliminated all tropical elements then existing in North America, (2) the invasion (from South to North America) of a new tropical element after the closing of the Central American water gaps. This faunal mixing during the late Pliocene and the Pleistocene led to a complete reshuffling of faunal elements. As far as birds are concerned, we can see only the final result of the opposing processes of range expansion on the one hand and extinction on the other. Simpson (1940a:158) has shown in detail what happened to the mammalian faunas. "Just before the two continents were united, South America had about 29 families of land mammals and North America about 27. With two doubtful exceptions [Didelphidae and Procyonidae], they did not then have any families in common. Shortly after the union of the continents, in the Pleistocene, they had 22 families in common, 7 of South American origin, 14 North American, and 1 doubtful." Considerable extinction and further migration have resulted in the Recent fauna, which consists of 38 families of land mammals, of which 14 are common to both continents, 15 confined to South America, and 9 confined to North America. Four North American families (tapirs, camels, peccaries, and short-faced bears) have become extinct in all or nearly all of their original home country, but are surviving in South America. Obviously it would be a zoogeographical error to classify such families, which were originally North American, with the truly autochthonous* South American families. Yet, nearly all the older zoogeographical treatises classify as "Neotropical" what is really a mixture of North and South American faunal elements. An effort has been made in the following classification to avoid this error. (In this paper zoogeographical North America is considered to extend southward to the edge of the tropical rain-forest.)

CLASSIFICATION OF THE FAUNAL ELEMENTS OF THE AMERICAS

Three Tertiary land masses are the primary contributors to the present fauna of the Americas: South America, North America, and Eurasia. It would therefore appear that the simplest classification of faunal elements would be into the same categories: South American, North American, and Eurasian (or "Old World"). These three classes undoubtedly must be recognized, but they are not sufficient to cover all families and genera of birds. First, an additional category must be recognized for groups that cannot be analyzed for one reason or another (to be stated below). Second, there are certain groups ("holarctic," or "panboreal," elements) which have moved back and forth across Bering Strait so freely that they cannot be assigned with certainty to either continent. Others ("pan-American") crossed the Central American water gaps sufficiently freely to obscure their center of origin. Finally, there is an old tropical element ("pantropical") which is of such similar composition in the Old World and New World tropics that it is impossible at the present time to determine the original home.

---

* In this paper I have used the terms "endemic" and "autochthonous" as follows: Endemic = restricted to a given region; not found elsewhere. Autochthonous = having originated in a given region; now sometimes found beyond the borders of that region.

It is into these categories (Figure 2) that I have tried to classify all the families of birds known to occur in the Americas, whenever possible carrying the analysis even further: to subfamilies, genera, and occasionally to species. This is particularly necessary in the case of families that originated outside of North America, for parts of which North America became a secondary center of evolution (e.g. quails, jays, thrushes), and of those other families that reached North America repeatedly at different geologic periods (e.g. the swallows).

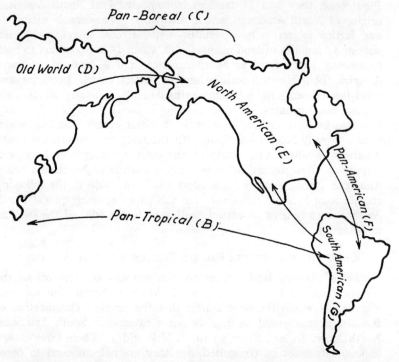

Figure 2. Diagram of the faunal elements of North America. The unanalyzed Element (A), whose geographical origin cannot be determined is, of course, omitted from the map.

Criteria

Unfortunately the bird geographer has, as stated above, relatively few fossils to guide him in his analysis. He is therefore forced to utilize indirect evidence, which is often difficult to evaluate. For example, both the Ruby-throated. Hummingbird (*Archilochus colubris*) and the Horned Lark (*Otocoris alpestris*) are widespread North American birds. But the Horned Lark is obviously only a recent arrival in the New World; it is the only member of the Alaudidae, a typical Old World family, to occur in North America and is not even an endemic species; whereas the hummingbird is clearly South American in origin. These cases indicate what evidence can be used. The larks are a family of more than 70 species and are represented in all parts of the Old World. Only certain subspecies of a single species occur in the New World. There can be no shadow of doubt concerning the family's Old World origin. Sometimes the distribution of the nearest relatives can be used as a clue. The gnatcatchers (Polioptilinae), for example, seem to be a branch of the rich Old World group of Insect Eaters (Muscicapidae) and they are without near relatives in the New World; these facts indicate an Old World origin for the subfamily.

These indirect methods are fully reliable only in richly developed families. The value of the evidence is uncertain in regard to families consisting of only one or merely a few species. Mammalogists like to cite in this connection the present distribution of the llamas (relatives of the camels) and the tapirs, two groups formerly widespread in North America but now surviving only in tropical or South America and (the tapir) in southeast Asia. However, both these groups would probably be considered northern elements, even without fossil evidence, because of the distribution of their relatives.

A. The Unanalyzed Element

The separation of land masses, which is responsible for the divergent development of terrestrial faunas, has little bearing on the evolution of sea bird faunas. Roughly, the oceanic birds can be classified into (1) a southern group: penguins (Spheniscidae) and sheath-bills (Chionidae); (2) a tropical group: tropic-birds (Phaëthontidae), boobies and gannets (Sulidae), frigate-birds (Fregatidae); (3) a northern group: skuas and jaegers (Stercorariidae); (4) a world-wide group: albatrosses, shearwaters, fulmars, and petrels (Tubinares), gulls and terns (Laridae). A further analysis and determination of the point of origin of these sea birds is outside the scope of this paper.

Equally obscure is the place of origin of the partly oceanic, partly fresh-water, families of the pelicans (Pelecanidae) and the cormorants (Phalacrocoracidae). Among the true fresh-water groups, a number of families are so evenly distributed in the Old and New World as to make determination of their centers of origin impossible. These include the grebes (Colymbidae), herons and bitterns (Ardeidae), storks and jabirus (Ciconiidae), ibises and spoonbills (Threskiornithidae), flamingos (Phoenicopteridae), the ducks, geese, and swans (Anatidae), and the rails, coots, and gallinules (Rallidae). With most of these, it is not simply the family as a whole that is widespread, but also the subfamilies, many of the genera, and frequently even the individual species. This point is well illustrated by the duck family, of which an up-to-date classification is available (Delacour and Mayr, 1945). Of the nine recognized tribes (or "subfamilies"), only the monotypic torrent duck tribe (Merganettini) is restricted to a single continent. Of the 40 genera, no less than 18 are found on two or more continents. Many species are circumtropical or at least very widespread. For example, the White-faced Whistling Duck (*Dendrocygna viduata*): South America, Africa, Madagascar; the Fulvous Whistling Duck (*Dendrocygna bicolor*): America, Africa, India; the superspecies *Tadorna ferruginea* (which includes the four species formerly separated as *"Casarca"*): Europe, Asia, South Africa, Australia, New Zealand; the black duck–mallard group of river ducks (*Anas platyrhynchos–fulvigula*): spread over most of the world except South America; the superspecies *Aythya nyroca* (white-eyed ducks): Madagascar, Eurasia, east Asia, Australia, and New Zealand; the Muscovy Duck group (*Cairina*, including *"Pteronetta"* and *"Asarcornis"*): America, Africa, India; the mergansers (*Mergus*, including *"Mergellus"* and *"Lophodytes"*): Holarctic region, Brazil, Auckland Islands; the southern ruddy ducks (*Oxyura australis*, including *maccoa*, *ferruginea*, and *vittata*): South America, Africa, Australia.

Widespread genera and species are typical also of other families of fresh-water birds. A few examples are: the grebes (*Colymbus* [*Podiceps*]), which occur on all continents; the gray heron group (*Ardea cinerea–herodias*), the green heron group (*Butorides virescens –striatus*), the Egret (*Egretta alba*), the night heron group (*Nycticorax nycticorax–caledonicus*), and the bitterns (*Ixobrychus* and *Botaurus*), all of which are world-wide. Many additional examples could be cited from other fresh-water families, particularly from the rails.

Most of the families of shore birds also are so widespread as to make it impossible to trace their origin. This is particularly true for the oyster-catchers (Haematopodidae), the plover family (Charadriidae), avocets and stilts (Recurvirostridae), and thick-knees (Burhinidae). In the case of the snipes, woodcock, and sandpipers (Scolopacidae) an origin in the northern hemisphere appears probable.

Though all these families of fresh-water and shore birds cannot be analyzed at the present time, it seems certain that new evidence may bring us a good deal further. Most of them are composed of medium-sized and large forms, which we find represented in fossil recoveries to an ever-increasing extent. Furthermore, certain subdivisions within these families are sometimes clearly Old World, New World, or even more specifically South American. Finally, a study of their parasites might facilitate the finding of the center of origin, as Szidat (1940) has suggested.

Among the strictly terrestrial birds, there are eight families that are so widespread or so evenly distributed as to make analysis difficult at the present time. These families are the hawks and eagles (Accipitridae), the osprey (Pandionidae), falcons and caracaras (Falconidae), nightjars (Caprimulgidae), swifts (Apodidae), woodpeckers (Picidae), and swallows (Hirundinidae). The evidence indicates that all of these families originated at such an early date (Eocene or Cretaceous) that subsequent shifts in distribution have obliterated most of the clues.

Indirect clues, however, permit a guess for two of these families. The Caprimulgidae may well be of New World origin, since this is the home not only of the entire subfamily nighthawks (Chordeilinae), but also of 10 of the 15 genera of goatsuckers (Caprimulginae). However, a comparison of the numbers of genera in the two regions does not give an entirely accurate picture, since the American birds are more finely split by the taxonomists. Students of New World Caprimulgidae employ 14 genera for 29 species, while Old World ornithologists recognize only 6 genera for 37 species. The woodpeckers (Picidae) are represented about equally well in the Americas and the Oriental regions. They are rather poorly developed in Eurasia and Africa and are absent from the Australian region and from Madagascar. This pattern of distribution suggests a New World (but very early) origin for the family, although the fact that their nearest relatives, the wrynecks (Jyngidae), are exclusively Old World would seem to indicate the opposite.

The swallows are also a very ancient family; it is particularly rich in species in South America and Africa but also extends to Madagascar and Australia. The place of origin of the family as a whole is uncertain, but it is fairly easy to determine where each of the (approximately) seven major subdivisions (Mayr and Bond, 1943) of the family first developed. The specialized mud-nest builders, *Hirundo* and *"Petrochelidon,"* as well as *Riparia,* are of Old World origin, being recent arrivals in America from the Palearctic. It is uncertain whether the family originated in South America, and retained one primitive branch in the Americas (*Progne–Atticora–Stelgidopteryx*), sending another branch to the Old World (*Psalidoprocne,* etc.) that gave rise to the specialized mud-nest builders and other Recent Old World forms, or whether the "old-American" swallows are descendants of early invaders from Asia. Parallel cases in other animal groups favor the second alternative.

B. The Pantropical Element

While representatives of the hawks, owls, and swifts are found in several climatic zones, there are certain other families which are also widespread but only within the tropical belt. For five families of freshwater birds (in some cases, partly marine), the area of origin is difficult to fix because each of them is found both in the Old World and New World tropics, though represented only by a single, or merely a few, species. These families are the snake-birds (Anhingidae), sungrebes (Heliornithidae), jacanas (Jacanidae), painted snipes (Rostratulidae), and the skimmers (Rynchopidae). All of them now have widely disrupted ranges, as can be easily seen from the map of the sun-grebes (Figure 3). It is also remarkable that the Recent Old World and New World representatives are often the members of a single species or superspecies (*Anhinga*, *Rostratula benghalensis*, *Rynchops*). This would indicate either extremely slow evolution or an enormous capacity for transoceanic dispersal.

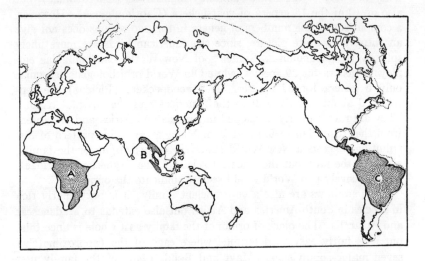

Figure 3. Present distribution of the sun-grebes (Heliornithidae), a typical family of the pantropical group. A = *Podica*, B = *Heliopais*, C = *Heliornis*.

Among the land birds, three families are pantropical. The barbets (Capitonidae) and the trogons (Trogonidae) have a notably similar distributional pattern. The ranges of both families are restricted to the humid tropics, and are bounded in the east by Wallace's Line. Fossil trogons have been found in the Eocene of France, and this fact, together with the scarcity of trogons in South America, has led most authors to assume an Old World origin for the family. On the other hand, trogons are much more diversified in Central America than in the Old World tropics; in fact, all the African and Indian species could be included in a single genus. Tropical North America or the Oriental region is the most likely place of origin. The barbets, with a similar distributional picture, are so much more richly developed in the Old World tropics than in the New that an Old World origin is probable (cf. Ripley, 1945:543–544).

The distribution of the parrots (Psittacidae) is considerably more extensive than that of the barbets and trogons. The parrots, with about

315 species, are one of the richest of all bird families, but about an equal number are found in the Old and the New World. However, most of the more aberrant types, such as the lories (Loriinae), cockatoos (Cacatuinae), and pigmy parrots (Micropsittinae), are found in the Old World, more specifically in the Australian region. It is, therefore, probable that the Psittacidae originated in the Old World, but the great number of endemic genera and species in America indicates a very early arrival in the New World. This might well have taken place before the Eocene separation of South America from North America.

The present ranges of these circumtropical families are widely disrupted, and they have therefore been used as "evidence" of former transatlantic or transpacific land connections by the advocates of such land bridges. We shall investigate in a later section how well founded their argument is.

C. The Panboreal Element

The loons (Gaviidae) among the fresh-water birds, the phalaropes (Phalaropodidae) among the shore birds, and the auk family (Alcidae) among the sea birds are typical of a large class of circumboreal birds. All three families are distributed in the arctic or in the north temperate zone and are about equally well represented in the Old and the New World. The auk family and the loons are known from the Tertiary of both North America and Europe. The temperate zones of Eurasia and America were in such direct contact for a good part of the Tertiary (by means of the Bering bridge) that it will be very hard to determine which of the two land masses was the giver and which the taker of the members of this temperate zone group. Among genera and species, this circumboreal element is much stronger than among families. Well over 80 per cent of the species of the circumboreal tundra zone belong to it, and it is impossible to determine their ultimate source. Stegmann (1938a) believes that Asia, more particularly Siberia, has probably made the greatest contribution to the group because it is the largest land mass in the temperate zone.

D. The Old World Element

It is generally admitted that the connection between Asia and North America across Bering Strait is very ancient (pre-Tertiary). As far as birds are concerned, a more or less active faunal exchange probably took place right through the Tertiary, even during periods when the two land masses were separated by water. This long-standing accessibility of North America to Old World immigrants is reflected in the taxonomic composition of the Old World element in America. According to the date of their immigration, these birds have either (1) not changed at all, e.g., the Alaska Yellow Wagtail (*Motacilla flava alascensis*), the Red-spotted Bluethroat (*Luscinia* ["*Cyanosylvia*"] *suecica robusta*), and the Wheatear (*Oenanthe oe. oenanthe*); (2) they have become subspecifically distinct, e.g., Kennicott's Willow Warbler (*Phylloscopus* ["*Acanthopneuste*"] *borealis kennicotti*), the Northern Shrike (*Lanius excubitor borealis*), Brown Creeper (*Certhia familiaris americana*); or (3), if they arrived very early, they have evolved into separate species, genera, or even subfamilies—that is, America has become for them a secondary center of evolution.

The third case is true of the Old World pheasant family (Phasianidae), which has produced the American quails (subfamily Odontophorinae). And it is probably true of the cuckoos (Cuculidae). In this family, Peters (Check-list, vol. 4, 1940) recognizes six subfamilies. Three of these, the Cuculinae, the Couinae (Madagascar), and the Centropodinae, are restricted to the Old World; the Crotophaginae are American; the Neomorphinae have five genera in the New World, one in the Old; and the Phaenicophaeinae have nine in the Old World, three in the New. The evidence points toward an Old World origin of the family, and to tropical North America as a secondary center of evolution for three subfamilies.

It is highly probable that the typical owls (Strigidae) originally came from the Old World, since the closely related family Tytonidae is clearly of Old World origin (only one of its species occurring in the New World) and since in the Old World there are twice as many endemic genera of Strigidae as in the New World. However, this must have been a very early invasion, since there are now six endemic genera in the New World, and since four fossil species of the extinct family Protostrigidae are known from the Eocene of North America (Wetmore, 1940:66–67).

The gnatcatchers (subfamily Polioptilinae, comprising the three genera *Polioptila, Microbates,* and *Ramphocaenus*) offer a puzzling problem both to the taxonomist and the zoogeographer. They are usually treated as a subfamily of the Old World warblers ("Sylviidae"), but there seems little beyond the fine bill to support such a classification. They are surely one of the branches of the Old World Insect Eaters (Muscicapidae), but what their nearest relatives are is still obscure. Although more species of Polioptilinae are found in South than in Central America, it seems probable that tropical North America was the secondary evolutionary center of this group after its arrival from the Old World. Lönnberg (1927:17) expressed a similar opinion.

The pigeons (Columbidae) are world-wide in distribution—which indicates their great age. However, the rich development of the family in the Australian region, where the most aberrant members of the family occur (e.g., *Caloenas, Goüra, Otidiphaps,* and *Didunculus*), and the fact that most American species belong to just a few phyletic lines, prove an Old World origin. It seems probable that some species reached

South America as early as the middle Tertiary and established a second evolutionary center.

Both the crow family (Corvidae) and the thrushes ("Turdidae") are examples of Old World groups which have established minor secondary evolutionary centers in North America, particularly in the tropical part. For the Corvidae, Amadon (1944:16–20) has presented detailed evidence. The blue jay group (*Cyanocitta*) developed in America, but since there is not a single endemic genus in South America, it is obvious that the jays reached there only after the closing of the Central American water gaps in the late Tertiary. The genera *Corvus*, *Nucifraga*, and *Perisoreus* represent separate later invasions of the Corvidae into North America. In view of the early arrival of the jay group, it seems conceivable that some of the palearctic genera (*Perisoreus, Nucifraga, ? Garrulus*) evolved in America and crossed back to Asia by Bering Strait, but it would be impossible to prove this.

The thrush subfamily Turdinae (see Mayr, 1941:106) presents a very similar distributional pattern and probably had a similar history. Thrushes are rich in species in South America (where there are no less than 20 full species of *Turdus*), but all the genera (even the solitaires, *Myadestes*, and the nightingale-thrushes, *Catharus*) belong to a single natural group; and even with the two (not very pronounced) West Indian genera (*Mimocichla* and *Cichlherminia*), there are only a total of 12 genera in the New World—excluding the recent immigrants, *Oenanthe* (Wheatear) and *Luscinia* (the Bluethroat, "*Cyanosylvia*"). This compares with several dozen widely divergent genera of thrushes in the Old World, such as the Old World nightingales, redstarts, robins, and chats. There are about 244 Old World and 60 New World species. Since also all of the closer relatives of the Turdinae—babbling thrushes (Timaliinae) and Old World flycatchers (Muscicapinae)—are Old World in origin, there can be no question of the Old World origin of the subfamily. The interesting aspect of the American thrushes is, however, that they demonstrate very graphically the effect of the continuous availability of the Bering bridge. There was an early immigration of a *Turdus*-like stock which produced some of the endemic South and Central American genera; there was the later arrival of another group which gave rise to the solitaire, nightingale-thrush, and hermit-thrush groups (*Myadestes, Catharus, Hylocichla*); then the immigration that resulted in the bluebird genus *Sialia;* then additional members of the genus *Turdus,* which changed specifically but not generically; and finally the most recent immigrants, the Bluethroat (Alaska) and the Wheatear (Alaska and Labrador), in which not even subspecific differences have developed.

The cranes (Gruidae) are known from North America as far back as the middle Pliocene—perhaps even earlier (see Wetmore, 1940). However, they would seem to be an unquestionably Old World family on the basis of their present distribution. There are 13 species (4 genera) in the Old World as compared with 2 species (one genus) in the New World.

The kingfishers (Alcedinidae) are a rich Old World family of which only one branch (Cerylinae) has reached the New World. This colonization cannot have been very recent, since a few species (the neotropical group *Chloroceryle*) are sufficiently distinct from their nearest Old World relatives to be considered by most authors a separate genus.

The cardueline subfamily of the Fringillidae is an Old World group, but one of the lines seems to have arrived in America rather early, since it has produced a number of endemic South American species (*"Spinus"*) and an endemic West Indian genus, *Loximitris* (Hispaniolan Siskin), which is closely related to *"Spinus."* *Hesperiphona* (Abeillé's and Evening Grosbeaks) is the only endemic North American genus, but it is closely related to the Himalayan *Mycerobas*—if at all separable from it. The purple and house finches (*Carpodacus*), pine grosbeaks (*Pinicola*), crossbills (*Loxia*), and rosy finches (*Leucosticte*) are even more recent arrivals from the Old World.

The Paridae (titmice) are a mainly Eurasian family, which has repeatedly invaded North America, where it has even developed two endemic genera, verdins (*Auriparus*) and bush-tits (*Psaltriparus*). But the latter genus seems closely related to the Asiatic genera *Aegithaliscus* and *Psaltria,* while the other American titmice are still more closely related to Asiatic species; some are even conspecific. They must have crossed Bering Strait during or after the late Pleistocene.

As stated above, the genus *Chamaea* (wren-tit) of the west coast of North America is not the sole representative of a separate family, but a member of the Paradoxornithinae (parrot-bills and suthoras) and probably congeneric with *Moupinia* of China. All the other genera of the Paradoxornithinae are palearctic, as are those groups of babbling thrushes (Timaliinae) which are the closest relatives of this subfamily.*

The wagtails and pipits (Motacillidae) are a definitely Old World family, about equally well represented in Africa and Asia. The family is a rather recent arrival in America but has developed six endemic species in North and South America.

Six additional Old World families (or subfamilies) have colonized the Americas so recently, and the New World representatives are still so similar to the Old World forms (congeneric or even conspecific), that North America cannot be considered, for them, a secondary evolutionary center. These are: barn owls (Tytonidae), larks (Alaudidae), nuthatches (Sittidae), creepers (Certhiidae), Old World warblers and

---

* As J. T. Zimmer has pointed out to me, it may be necessary to call the subfamily "Chamaeinae," a name first used by Baird in 1863. The name Paradoxornithidae seems to have been used first by Oates about 20 years later. However, I have not made a thorough investigation of this nomenclatural complication. Furthermore, it may not be possible to separate the group from the Timaliinae.

kinglets (Sylviinae), and shrikes (Laniidae). The Old World origin of most of these groups has been discussed by Lönnberg (1927) and earlier authors. Only two of them (the larks and barn owls) have reached South America, and that so recently that the South American representatives are no more than subspecifically distinct.

E. The North American Element

The fauna that developed in North America during the Tertiary, while this continent was separated from South America and connected with Asia only by the Bering Strait bridge, is of great zoogeographical importance. It was much neglected in the past, when some of its components were labelled "Holarctic," others "Neotropical." The greater part of the Tertiary North American continent had a subtropical or tropical climate, as mentioned above, and it is therefore not surprising that tropical families and genera are well represented in this North American element.

The reasons have already been stated why the New World vultures (Cathartidae) and the limpkins (Aramidae) have to be considered North American in origin. Lönnberg (1927:7–12) considered that the thrashers and mockingbirds (Mimidae), vireos (Vireonidae), wood warblers (Parulidae), the waxwings (Bombycillidae) with their relatives the silky flycatchers (Ptilogonatidae), the wrens (Troglodytidae), and motmots (Momotidae) are also North American in origin. The monotypic family palm-chats (Dulidae) also belongs to this group. In all these cases there are so many more endemic genera in North than in South America that no fault can be found with Lönnberg's conclusions. Among the Mimidae, for example, only two genera have reached South America, one of which, the mocking-thrush (*Donacobius*), is endemic. Five genera (three endemic) occur in Central America, five genera (four endemic) on the islands of the Caribbean, and four genera (two endemic) in North America. The tropical origin of the family is indicated by the fact that none of the United States species has entered the Canadian zone.

The vireos, shrike-vireos, and pepper-shrikes have six genera (two endemic—*Neochloe* and *Vireolanius*) in Mexico and Central America, as compared with four genera (none endemic) in South America. The single genus occurring in North America is rich in species (11), of which 2 (*solitarius* and *philadelphicus*) are at home in the Canadian zone. There are 7 endemic species in the Caribbean. Even though no less than 20 species are found in South America, the combined weight of the other facts favors a North American origin for the family.

The wood warblers (Parulidae) present a very similar picture. There are 16 genera in North America (many endemic) and only 6 in South America (none endemic). However, the genera *Myioborus* and

Basileuterus have respectively 6 and 17 endemic South American species. In the genus *Dendroica* alone there are about 20 endemic North American species, a good many of which are restricted to the Canadian zone coniferous forest. All the facts combined indicate a North American origin for the family.

A North American origin may also be postulated for the turkeys (Meleagrididae), grouse (Tetraonidae), dippers (Cinclidae), and the subfamily Emberizinae.

The evidence is unequivocal as far as the turkeys are concerned. The two Recent genera and the only known extinct one (*Parapavo*) have been found only in North America.

The grouse family presents a more difficult case. It has a wide distribution in the northern hemisphere, from Spain to Kamchatka, and from Alaska to Newfoundland and southward almost to Mexico. Absent from the subtropical and tropical belts of the Old and New World, the grouse show the typical distributional picture of a holarctic family. As both Lönnberg (1927:12) and Stegmann (1938a) have pointed out, there is much that favors an American origin for the family. Only three genera are endemic to the Old World (*Tetrao, Lyrurus,* and *Tetrastes*), all three being more or less Siberian taiga (moist coniferous forest) elements which have apparently radiated only quite recently into the western palearctic (Stegmann, 1932:396–397). The Old World has no equivalent of the American grassland genera *Tympanuchus, Pedioecetes,* and *Centrocercus*. Extinct genera of grouse have been reported from the Miocene and Eocene of North America.

The dippers (Cinclidae) are a family with only a single genus and too few species for a reliable analysis. There are three closely related species in the New World and two in the Old; one of the latter (*Cinclus pallasii*) is restricted to the eastern Palearctic. Relationship to the wrens (Troglodytidae), which is assumed by most authors, would indicate a North American origin.

The subfamily Emberizinae is apparently of North American origin, though (as mentioned above) no final decision can be reached without first determining which of the South American genera actually belong to the Emberizinae. Perhaps there was a continuous faunal exchange with South America throughout the Tertiary. One single branch of the Emberizinae, consisting of closely related forms, has reached the Old World. Even though more than 30 species are now found there, they all belong either to the genus *Emberiza* or to *Fringillaria, Miliaria,* and *Melophus,* which hardly deserve to be called more than subgenera. It can therefore be assumed that the invasion of the Old World by the Emberizinae must have taken place rather late in the Tertiary.

As stated in the preceding section, on the Old World element, North America became a secondary center of evolution for several Old World groups: American quails (Odontophorinae), the blue jay (*Cyanocitta*) group of the family Corvidae, the *Myadestes–Catharus–Hylocichla* group of thrushes, and some others. In particular, the Odontophorinae, a whole subfamily restricted to North America, and known there as far back as the Miocene, well deserve to be included among the typically North American fauna. Part of the pan-American element (certain Icteridae), discussed below, has also now become sufficiently well established in North America to be considered part of the North American element.

F. The Pan-American Element

The water gaps that existed between North and South America from the lower Eocene to the late Pliocene produced an almost complete separation of the mammalian faunas of the two continents (Simpson, 1940a:157–163). The intervening chain of islands (Figure 1) permitted colonization by only a few groups especially adapted to "island hopping." On the whole, the geographical picture of this line of islands was apparently very similar to that of the Malay Archipelago, where colonization by mammals was almost completely prevented, even though the islands were more numerous and the water gaps comparatively small. For birds, these inter-island straits of the Malay Archipelago were much less of a barrier, as I have recently pointed out (Mayr, 1944a:171–194). The same is true for the inter-American island belt. It explains many of the difficulties of the bird geographer. There are quite a number of American families that are so rich, both in North and South America, in endemic genera and species that it is impossible to determine their primary country of origin without fossil evidence. It is rather obvious that these are the families able to utilize islands as stepping stones from one continent to the other. During the greater part of the Tertiary, the whole southern part of North America was apparently more humid, and certainly warmer, than it is today. It would have been more difficult for many of the species that developed in this climatic zone to enter the more temperate parts of North America than to cross into tropical South America. In the reverse direction, the same was true for species of tropical South America. This is one of the reasons that the contrast between the North and the South American Tertiary faunas is much less pronounced in birds than in mammals, and much less than one would expect on the basis of the length of separation of the two continents. On the other hand, the factor of age should not be left out of consideration. In the Eocene, when North and South America were connected, there were more bird families than mammal families with representatives on both continents.

Families almost certainly South American in origin, known to be successful transoceanic colonizers (West Indian fauna!), and rich in elements endemic to Central and North America, are the hummingbirds

(Trochilidae), the tyrant flycatchers (Tyrannidae), the tanagers (Thraupidae), and the blackbird–troupial family (Icteridae).

It is significant that not one of these families has crossed Bering Strait into the Old World although all four are rich in species and all four have at least a few species in temperate North America, some extending even as far as Alaska.

Among South American families of the suborder Mesomyodi, only the aggressive tyrant flycatchers (Tyrannidae) have penetrated far into North America. But many of these have reached the Canadian zone, and they were undoubtedly the first birds of this group to become established north of South America. There is every reason to believe that the invasion took place prior to the connection of the two continents in the late Pliocene. Nevertheless, their arrival must be considered comparatively recent. Of the 117 currently recognized genera of this family, only 10 are not indigenous to South America, and none of these is particularly distinctive; in every case the relationship to South American genera is more or less obvious, viz *Tolmarchus* (related to *Tyrannus*); *Hylonax, Deltarhynchus, Eribates,* and *Nesotriccus* (related to *Myiarchus*); *Blacicus* and *Nuttallornis* (related to *Contopus*); *Aechmolophus, Xenotriccus,* and *Aphanotriccus* (related to *Praedo*)—according to James Bond (*in litt.*).

The tanagers are more poorly represented in North America. There are a few genera in Central America; there are 5 endemic genera and 11 endemic species in the West Indies, but only one genus (*Piranga*) reaches the United States (with 4 species).

The blackbirds and troupials include 35 genera, of which no less than 16 are endemic to South America. There are two endemic genera in Central America, two in the West Indies (11 endemic species) and three in North America. (See also Lönnberg, 1927:10.) The family is well established in the temperate zone of North America with such hardy birds as the Bronzed Grackle (*Quiscalus quiscula*), Cowbird, (*Molothrus ater*), Meadowlark (*Sturnella*), Rusty Blackbird (*Euphagus carolinus*), and Red-wing (*Agelaius*). These species are so thoroughly at home in North America that a very early immigration is indicated.

Elements of the pan-American fauna that were perhaps originally North American are the curassow (Cracidae) and the cuckoo (Cuculidae) families. Both families are now richer in South, than in North, America, but both have relatives in the Old World (the mound-builders, family Megapodiidae, are at least distant relatives of the Cracidae). In the Cracidae, 5 out of 11 genera, 38 out of 46 species, are restricted to South America. On the other hand, the chachalaca *Ortalis* is known from the Pliocene and lower Miocene (Wetmore, 1940:42) of North America. The case of the Cuculidae has been discussed above in the section on the Old World element.

All of the families listed in this section have endemic genera or species in both North and South America. These are sufficiently peculiar to make it exceedingly unlikely that they could have developed in the short time since the re-establishment of the Panamanian land connections at the end of the Tertiary. They must have had as ancestors birds with the faculty of transoceanic colonization. On the other hand, there is not sufficient difference between the North American and the South American groups of genera to force us to assume an Eocene split of any of these families (by the separation of the two continents) into a northern and a southern section.

For the sake of completeness it will be useful to mention here those groups of Old World birds, discussed above, that arrived in North America at an early date and then crossed to South America with the help of the insular stepping stones. This includes, apparently, the pigeons (Columbidae), gnatcatchers (Polioptilinae), some thrushes (Turdinae), and some cardueline finches.

G. The South American Element

Certain families are very richly developed in all parts of South America, relatively scarce in Central America, even in the tropical parts, and extremely rare, or completely lacking, north of the tropics; and with these families, there can be no doubt about their South American origin. This is true for the tinamous (Tinamidae), potoos (Nyctibiidae), jacamars (Galbulidae), puff-birds (Bucconidae), toucans (Ramphastidae), oven-birds (Furnariidae), wood-hewers (Dendrocolaptidae), antbirds (Formicariidae) and two small related families, the ant-pipits (Conopophagidae) and tapaculos (Rhinocryptidae), the cotingas (Cotingidae), manakins (Pipridae), honey-creepers (Coerebidae), and the cardinal group (Richmondeninae). A South American origin is very probable also for the following families (though each contains less than five species, and some caution is therefore advised): rheas (Rheidae), screamers (Anhimidae), hoatzins (Opisthocomidae), trumpeters (Psophiidae), sun-bitterns (Eurypygidae), cariamas (Cariamidae), seed-snipe (Thinocoridae), oil-birds (Steatornithidae), sharp-bills (Oxyruncidae), and plant-cutters (Phytotomidae).

The cotingas (Cotingidae) may be cited to illustrate the distribution pattern characteristic of a typical South American family. Of the 31 genera of the family, only 12 reach Central America, and only one the United States; 19 genera are restricted to South America, not a single one to Central or North America; only one species (*Platypsaris niger*) has reached the West Indies (Jamaica). The oven-birds, wood-hewers, and antbirds are even more closely restricted to South America, and none of them has reached the West Indies.

The cardinals (Richmondeninae) apparently belong to the South American element, but, as already stated, nothing final can be said about this subfamily without first determining which genera belong to it.

As stated above, some of the families listed with the pan-American element are also of primary South American origin. This is reasonably certain for the hummingbirds (Trochilidae), tyrant flycatchers (Tyrannidae), tanagers (Thraupidae), and the blackbird–troupial family (Icteridae).

It is most remarkable that none of the families that are clearly South American in origin has developed any species that have crossed into the Old World. Old World families, on the other hand, have sent many branches into South America. Perhaps this means that a temperate zone family can more easily become adapted to the tropics than a tropical family to a temperate climate.

The above analysis is summarized in Table 1.

TABLE 1

ANALYSIS BY ORIGIN OF AMERICAN BIRD FAUNA

A. UNANALYZED ELEMENT

OCEANIC BIRDS
  Spheniscidae, *penguins*
  Procellariiformes, *tubinares*
  Chionidae, *sheath-bills*
  Sulidae, *boobies, gannets*
  Fregatidae, *frigate-birds*
  Phaëthontidae, *tropic-birds*
  Stercorariidae, *skuas, jaegers*
  Laridae, *gulls, terns*

SHORE BIRDS
  Haematopodidae, *oyster-catchers*
  Charadriidae, *plovers*
  Scolopacidae, *snipes, woodcock, sandpipers*
  Recurvirostridae, *avocets, stilts*
  Burhinidae, *thick-knees*

FRESH-WATER BIRDS (partly marine)
  Colymbidae, *grebes*
  Pelecanidae, *pelicans*
  Phalacrocoracidae, *cormorants*
  Ardeidae, *herons*
  Ciconiidae, *storks*
  Threskiornithidae, *ibises*
  Phoenicopteridae, *flamingos*
  Anatidae, *ducks, geese, swans*
  Rallidae, *rails*

LAND BIRDS
  Accipitridae, *hawks, eagles*
  Pandionidae, *osprey*
  Falconidae, *falcons, caracaras*
  N Caprimulgidae, *nightjars*
  Apodidae, *swifts*
  N Picidae, *woodpeckers*
  o Hirundinidae, *swallows*

B. PANTROPICAL ELEMENT

FRESH-WATER BIRDS (partly marine)
  Anhingidae, *snake-birds*
  Heliornithidae, *sun-grebes*
  Jacanidae, *jacanas*
  Rostratulidae, *painted snipes*
  Rynchopidae, *skimmers*

LAND BIRDS
  o Psittacidae, *parrots*
  N Trogonidae, *trogons*
  o Capitonidae, *barbets*

C. PANBOREAL ELEMENT

  Gaviidae, *loons*
  Alcidae, *auks, murres, puffins*
  Phalaropodidae, *phalaropes* (and many other groups of shore birds)

D. OLD WORLD ELEMENT

EARLY IMMIGRANTS
  Gruidae, *cranes*
  Columbidae, *pigeons*
  Cuculidae, *cuckoos*
  Strigidae, *typical owls*
  Corvidae, *crows, jays* (part)
  Turdinae, *thrushes* (part)

FAIRLY EARLY
  Alcedinidae, *kingfishers*
  Corvidae, *crows, jays* (part)
  Paridae, *titmice*
  Sittidae, *nuthatches*
  "Chamaeidae," *wren-tit*
  Motacillidae, *wagtails, pipits*
  Carduelinae, *cardueline finches* (part)

N = Probably originated in the New World.   o = Probably originated in the Old World.

RECENT

Tytonidae, *barn owls*
Alaudidae, *larks*
Hirundinidae, *swallows* (part)
Certhiidae, *creepers*
  Turdinae, *thrushes* (part)
  Sylviinae, *Old World warblers, kinglets*
Laniidae, *shrikes*
  Carduelinae, *cardueline finches* (part)

[Also of Old World origin are the Phasianidae, represented in the Americas by the quail (subfamily Odontophorinae); and the Muscicapidae, to which the American subfamily gnatcatchers (Polioptilinae) is undoubtedly related.]

E. NORTH AMERICAN ELEMENT

Cathartidae, *New World vultures*
Tetraonidae, *grouse*
  Odontophorinae, *American quail*
Meleagrididae, *turkeys*
Aramidae, *limpkins*
Todidae, *todies*
Momotidae, *motmots*
Cinclidae, *dippers*
Troglodytidae, *wrens*
Mimidae, *mockingbirds*
  Polioptilinae, *gnatcatchers*
Bombycillidae, *waxwings*
Ptilogonatidae, *silky flycatchers*
Dulidae, *palm-chats*
Vireonidae, *vireos, shrike-vireos, pepper-shrikes*
Parulidae, *wood warblers*
  Emberizinae, *typical buntings*

[Some genera and species belonging to families listed under; A. (hawks, nightjars, woodpeckers, swallows); B. (trogons, barbets); D. (cuckoos, typical owls, pigeons, jays, thrushes, titmice, wren-tit, cardueline finches); are distinct enough to require mention under this heading.]

F. PAN-AMERICAN ELEMENT

APPARENTLY ORIGINALLY NORTHERN

Cracidae, *curassows, guans*

PROBABLY ORIGINALLY SOUTH AMERICAN

Trochilidae, *hummingbirds*
Tyrannidae, *tyrant flycatchers*
Thraupidae, *tanagers*
? Icteridae, *blackbirds, troupials*

[The cardinals (Richmondeninae) may have to be transferred from the South American group to this class.]

G. SOUTH AMERICAN ELEMENT

*Rheidae, *rheas*
Tinamidae, *tinamous*
*Anhimidae, *screamers*
*Opisthocomidae, *hoatzins*
*Psophiidae, *trumpeters*
*Eurypygidae, *sun-bitterns*
*Cariamidae, *cariamas*
*Thinocoridae, *seed-snipe*
*Steatornithidae, *oil-birds*
Nyctibiidae, *potoos*
Galbulidae, *jacamars*
Bucconidae, *puff-birds*
Ramphastidae, *toucans*
Dendrocolaptidae, *wood-hewers*
Furnariidae, *oven-birds*
Formicariidae, *antbirds*
Conopophagidae, *ant-pipits*
Rhinocryptidae, *tapaculos*
Cotingidae, *cotingas*
Pipridae, *manakins*
*Oxyruncidae, *sharp-bills*
*Phytotomidae, *plant-cutters*
Coerebidae, *honey-creepers*
Richmondeninae, *cardinals*

[Families marked with an asterisk contain less than five species, and their allocation is consequently somewhat doubtful. In most cases it is well supported by circumstantial evidence.]

Conclusion

The results of this analysis of the North American fauna can be summarized as follows: Most North American families and subfamilies are clearly either Old World in origin, South American in origin, or members of an autochthonous North American element that developed during the partial isolation of North America in the course of the Tertiary. Although many details of this analysis are still questionable, its major outlines are established facts. These facts are, however, merely descriptive raw material. It is only by correlating them with established concepts in related fields that their full significance becomes apparent. Such a correlation will be attempted in the following sections.

AN ANALYSIS OF NORTH AMERICAN BIRD POPULATIONS

In Table 2, the song birds of various areas in North America are analyzed according to their point of origin. The endemic North American genera among the swallows (Hirundinidae) and the blackbird–troupial group (Icteridae) were included with the North American element. It would have been most desirable to extend the type of analysis used in Table 2 to all the families of birds, but I failed in an attempt to do so. Many species of non-passerines were in the doubtful categories, *A*, *B*, and *C*, of Table 1; others belonged to the difficult families of cuckoos (Cuculidae), owls (Strigidae), and pigeons (Columbidae).

TABLE 2

ANALYSIS BY GEOGRAPHICAL ORIGIN OF THE BREEDING PASSERINE SPECIES OF SEVERAL DISTRICTS OF NORTH AMERICA

	South American	North American	Old World
Yakutat Bay, southeast Alaska (Hudsonian Zone)[1]	3%	39%	58%
Oregon[2]	14	47	39
Nipissing area, southern Ontario, 46° N (Canadian Zone)[3]	13	57	30
New Jersey[4]	14	63	23
Florida[5]	20	59	21
Sonora, Mexico[6]	27	52	21

[1] Shortt, T. M. 1939. The summer birds of Yakutat Bay, Alaska. *Roy. Ont. Mus. Zool. Contr. No.* 17.
[2] Gabrielson, I. N., and S. G. Jewett. 1940. Birds of Oregon. Corvallis, Ore.
[3] Ricker, W. E., and C. H. D. Clarke. 1939. The birds of the vicinity of Lake Nipissing, Ontario. *Roy. Ont. Mus. Zool. Contr. No.* 16.
[4] Original data.
[5] Howell, A. H. 1932. Florida bird life. Tallahassee, Fla.
[6] van Rossem, A. J. 1945. A distributional survey of the birds of Sonora, Mexico. *La. State Univ. Mus. Zool. Occ. Paper No.* 21.

It might be claimed that the neglect of the non-passerines introduces so great a degree of uncertainty as to jeopardize the validity of the figures as indices of the composition of the North American fauna as a whole. This argument is not well founded for two reasons. One is that the families of Group A are composed of essentially the same mixture of South American, North American, and Old World elements, in essentially the same proportions, as are the analyzed families as a whole. This is quite obvious from a cursory study of the hawks and rails, for example. The second reason is that most of the families of Group A (composed chiefly of large birds and other non-passerines) are comparatively rare. In faunal lists in which the species have equal value, these birds may constitute a significant percentage. But they are negligible if each species is weighed on the basis of numerical frequency. To determine the faunal composition of the bird population of a given type of forest, it would be necessary to analyze the total number of pairs instead of the total number of species. I suggested (Mayr, 1944b) that this should be done to test the validity of Wallace's Line, but no data were available for such an analysis. Fortunately, however, good census data are available for North American birds in the Audubon breeding-bird censuses initiated by William Vogt (Hickey, 1937–1944). Table 3 shows that the unanalyzed element is negligible. It becomes important only in aquatic habitats.

TABLE 3

ANALYSIS BY GEOGRAPHICAL ORIGIN OF THE BREEDING PAIRS REPORTED [1] FROM FIVE NORTH AMERICAN HABITATS

	South American	North American	Old World	Un-analyzed	Total Number of Pairs
Red and White Spruce in Maine (No. 27, 1941 [1938 data])	0.0%	73.0%	25.9%	1.1%	85
Northern Forest in Idaho (No. 27, 1944)	12.5	62.5	25.0	0.0	56
Beech-Maple in Ohio (No. 20, 1941)	23.0	52.5	23.0	1.5	131
Southern Hardwood in Alabama (No. 21, 1944)	25.8	54.8	16.2	3.2	62
Desert in southern California (No. 5, 1941)	37.1	48.6	14.3	0.0	35

[1] Audubon breeding-bird censuses (Hickey, 1937–1944).

If Table 2 (species analysis) is compared with Table 3 (pair analysis), a few interesting facts are apparent. One is the basic similarity of the figures. In both cases, the North American element makes up a large proportion of the total (47 to 63 per cent * in the species analysis, 48 to 73 per cent in the pair). The South American and the

---

* Unless one includes the marginal Yakutat Bay area (39 per cent).

Old World elements share the rest. However, the Old World element, largely consisting of permanent residents, is significantly lower in the pair, than in the species, tabulation, indicating a lower density. The South American element, on the other hand, composed mainly of hummingbirds (Trochilidae), tyrant flycatchers (Tyrannidae), tanagers (Thraupidae), and cardinals (Richmondeninae), is higher in the pair than in the species list.

A number of additional facts become obvious from a study of these tabulations. There is a decrease of the Old World element from the north to the south, but even as far south as Florida or Sonora, one-fifth of the species, or one-sixth of the pairs, are still of Old World origin. In mountainous western North America there is, naturally, a higher percentage of Old World elements than in a similar latitude in the lowlands of the eastern states. It is not justifiable, as far as birds are concerned, to include North America either in a "Neotropical" or in a "Holarctic" region, since the autochthonous North American element comprises up to 50 per cent, or even more, of the North American fauna in all habitats except the arctic. As is to be expected, from north to south, there is an increase of the South American element. However, even as far south as Sonora, only 27 per cent of the species are South American. Finally, it appears, again as is to be expected, that the faunal change from north to south is quite gradual—there are no "step clines" anywhere. Since each of the approximately 200 species involved in these analyses has different ecological requirements and a different distribution-pattern, it is not surprising that there is no sharp change in the gradient. The most rapid faunal change appears to occur near the northern tree limit.

The exact line, north of which more than 50 per cent of the bird species belong to the Panboreal and Old World element, has never been accurately drawn, but it runs somewhere through the middle of the Canadian coniferous forest. This 50:50 line does not by any means coincide with any major physiographic feature. There is, however, as stated above, a sharp drop in the percentage of American elements along timber line. Those who want zoogeographic regions may do well to follow the lead of the zoogeographers who recognize an Arctic (circumpolar) region as distinct from the Palearctic region. This was, I believe, first proposed by Schmarda (1853:225–226), later adopted by J. A. Allen (1871:381–382), by Reichenow (1888:673), and by the recent Russian zoogeographers (Stegmann, 1938a). Similarly, it will be advisable to include all the wooded parts of North America in the "North American region," even though the North American element might be slightly in the minority along the northern fringe. Since the only major avifaunal break occurs along the tree limit, it seems legitimate to accept the tree limit as a regional border.

The Arctic or tundra zone is inhabited by few land birds. The bird fauna consists almost entirely of sea birds, fresh-water birds, and shore birds. This fauna is strikingly different from that of the wooded parts of the continent, but it is practically identical on the two sides of Bering Strait. There are 104 species of birds that now breed in the arctic regions. Of these, only the following species seem to be restricted to the American continent: Canada Goose (*Branta canadensis*), Ross's Goose (*Anser rossi*), Bald Eagle (*Haliaeëtus leucocephalus*), Eskimo Curlew (*Numenius borealis*), Bristle-thighed Curlew (*Numenius tahitiensis*), White-rumped Sandpiper (*Ereunetes fuscicollis*), Stilt Sandpiper (*Micropalama himantopus*), Buff-breasted Sandpiper (*Tryngites subruficollis*), and the Surf-bird (*Aphriza virgata*). (Certain additional species usually considered exclusively North American I would include in superspecies that occur in both North America and Siberia.)

The same small number (nine species) are restricted to the Old World: Lesser White-fronted Goose (*Anser erythropus*), Red-breasted Goose (*Branta ruficollis*), Dotterel (*"Eudromias" morinellus*), Temminck's Stint (*Ereunetes temminckii*), Siberian Pectoral Sandpiper (*Ereunetes acuminatus*), Curlew Sandpiper (*Ereunetes ferrugineus*), Eastern Asiatic Knot (*Calidris tenuirostris*), Spoonbill Sandpiper (*Eurynorhynchus pygmeus*), and the Red-throated Pipit (*Anthus cervinus*). Thus, except for 18 species (of which 12 are shore birds), the arctic bird faunas of Asia and America are practically identical in composition. Furthermore, the arctic fauna is remarkable in that more than 50 per cent of its species are restricted to the Arctic zone, and in its almost complete difference from the fauna of the coniferous zone. The northern tree limit is, so far as birds are concerned, one of the clearest faunal boundaries on the earth.

I shall refrain from drawing any zoogeographical boundaries south of the timber line. Simpson (1943b:427–429) distinguishes five regions in America: Boreal, Middle, and Southern, in North America (including Mexico and Central America); Equatorial and Austral, in South America. It seems to me that this attempt to reconcile the historico-faunistic findings with descriptive-regional zoogeography is not entirely successful. As far as birds are concerned, none of the five regions mentioned by Simpson is well characterized by its present faunal contents, nor are the boundaries between the regions clear. Distinctive faunas develop only in isolation, and zoogeographic regions can retain their faunistic integrity only if they are separated from other regions by geographical or ecological barriers. The union of the North American and the South American tropical zones at the end of the Pliocene has resulted in such a mingling of the respective faunas that it seems futile to draw a line through Panama separating a tropical "Southern North America" from an "Equatorial South America." The faunas of the two "regions" are today essentially identical. If one wants zoogeographic regions, one may have to go back to the solution of the classical zoogeographers, who looked for a physiographic border line and found it in Mexico along the northern edge of the tropical rain-forest belt. This is where Wallace (1876:79) placed the border between his Neotropical and Nearctic regions. So far as I can see, it is along this line that the only major faunal break occurs in the warmer parts of North America. However, I agree with Dunn (1931) and Simpson (1943b) that the term Nearctic is misleading. To call the region north of the tropics (i.e. north of the tropical rain-forest) simply the North American region is probably the best solution.

COMPARISON OF BIRDS WITH OTHER ANIMALS AND WITH PLANTS

On a walk through the woods in temperate North America, one encounters flowers and trees which differ but little from species found in temperate Asia. The admixture of tropical South American elements is negligible. The same is true for mammals. The porcupine and the armadillos are apparently the only South American elements in the present North American mammal fauna, compared with a 13 to 20 per cent South American element in the bird fauna, except at the northern fringe (Table 2). I do not know of any exact published figures, but I gather from the writings of mammalogists that more than 50 per cent of the temperate North American mammals are of Old World origin. (Is the percentage even higher in plants?) In birds (again excepting the northern fringe), it is only a third or less.

There are mainly two reasons why the Old World element is so much weaker among North American birds than among most other animal groups—or perhaps I should better say: why the South American and warm North American element in temperate America is so much stronger in birds than in other animal groups. One of these reasons is the ability of birds to cross water gaps. Thus, while the indigenous mammals were imprisoned in South America during the Tertiary separation of the two continents, several groups of South American birds crossed the water gap into the northern continent. Among the invading groups that became thoroughly established in North America are the blackbirds and troupials (Icteridae), tyrant flycatchers (Tyrannidae), and cardinals (*Richmondena, Hedymeles, Passerina*, etc.). Some of these genera and generic groups must have arrived in North America at a very early date. Pre-empting many ecological niches, the 40 or 50 species of these originally South American groups have helped stem the influx of Old World species.

A second and more important factor is bird migration. It enables many tropical or semitropical birds to include in their breeding range the areas of the temperate zone that have a hot summer season and move back into their tropical home when the cool season begins. An analysis of the mid-winter avifauna of temperate eastern North America shows that it is composed almost entirely of Old World elements. The difference in migratory behavior between the autochthonous and the Old World elements is illustrated in the following statistics. Among the 28 species of permanent residents (excluding water birds and unanalyzable species) listed by Cruickshank (1942:25–26) for the New York region, no less than 23 (82.1 per cent) are of Old World origin. On the other hand, among 67 analyzable species of summer residents (which migrate south in the fall) only 8 (11.9 per cent) are of Old World origin. If the 95 species of the two categories are combined, it is found that of the 12 species of the South American element only one (8.3 per cent) is a permanent resident, of the 52 species of the North American element only 4 (8.3 per cent) are permanent residents, while of the 31 species of the Old World element no less than 23 (76.7 per cent) are permanent residents.* The Old World element, which, as Stegmann (1938a) has shown, developed for the most part in the always cold land mass of northern Siberia, is so thoroughly adapted to the cold that it can survive in this latitude without migration, whereas the autochthonous American element, most of which developed in a warm zone, survives the winter by avoiding it.

The combination of these two factors has resulted in the peculiar composition of the contemporary North American bird fauna. It is, therefore, obvious that no general zoogeographic scheme can be based on the distribution of birds, and that the ornithologist will find zoogeographical classifications inapplicable that are based on the distribution of mammals or reptiles. This difference between birds and other

---

* I present these analyses of Cruickshank's data merely as an illustration of a trend. Because the classification by origin of the birds of such populations (with different migratory status) involves weighing evidence and probabilities, such an analysis inevitably varies somewhat with the individual. For the benefit of students who may wish to make similar analyses of other populations and compare results, I give the following outline of my classification of the populations.

List of Permanent Residents. South American: Cardinal; North American: Ruffed Grouse, Bob-white, Carolina Wren, Song Sparrow; Old World: Sharp-shinned Hawk, Red-tailed Hawk, Bald Eagle, Marsh Hawk, Duck Hawk, Sparrow Hawk, Barn Owl, Screech Owl, Great Horned Owl, Barred Owl, Long-eared Owl, Short-eared Owl, Pileated Woodpecker, Hairy Woodpecker, Downy Woodpecker, Prairie Horned Lark, Blue Jay, Crow, Black-capped Chickadee, Carolina Chickadee, Tufted Titmouse, White-breasted Nuthatch, Goldfinch. (Not analyzed: Cooper's Hawk, Red-shouldered Hawk, Red-headed Woodpecker, Water birds; not considered truly permanent residents: Flicker, Meadowlark, Fish Crow, Swamp Sparrow, Field Sparrow.)

List of Summer Residents. South American (11 = 16.4%): Hummingbird, Kingbird, Crested Flycatcher, Phoebe, Acadian Flycatcher, Alder Flycatcher, Least Flycatcher, Wood Pewee, Scarlet Tanager, Rose-breasted Grosbeak, Indigo Bunting; North American (48 = 71.7%): Flicker, Tree Swallow, Rough-winged Swallow, Purple Martin, Short-billed Marsh Wren, Long-billed Marsh Wren, House Wren, Catbird, Brown Thrasher, Cedar Waxwing, White-eyed Vireo, Yellow-throated Vireo, Red-eyed Vireo, Warbling Vireo, Black and White Warbler, Worm-eating Warbler, Golden-winged Warbler, Blue-winged Warbler, Nashville Warbler, Yellow Warbler, Black-throated Green Warbler, Chestnut-sided Warbler, Pine Warbler, Prairie Warbler, Oven-bird, Louisiana Water-thrush, Kentucky Warbler, Yellow-throat, Yellow-breasted Chat, Hooded Warbler, Redstart, Meadowlark, Bobolink, Red-wing, Orchard Oriole, Baltimore Oriole, Purple Grackle, Cowbird, Towhee, Savannah Sparrow, Swamp Sparrow, Field Sparrow, Grasshopper Sparrow, Henslow's Sparrow, Sharp-tailed Sparrow, Seaside Sparrow, Vesper Sparrow, Chipping Sparrow; Old World (8 = 11.9%): Kingfisher, Bank Swallow, Barn Swallow, Fish Crow, Robin, Wood Thrush, Veery, Bluebird. (Not analyzed: First 31 species listed; added: 5 species from permanent-resident list.)

animal groups is the reason for much of the "New World" versus "Holarctic" controversy. Those who wanted to unite North and South America into a single "New World" based their conclusion mainly on a study of birds. Those who wanted to include North America with Eurasia in a "Holarctic" region based their conclusions on mammals or reptiles.

The History of the Pantropical Element

In a previous section I discussed a number of families which are more or less restricted to the tropics, but are found in the Old as well as in the New World. A similar distribution has been documented for various families and subfamilies of turtles (Simpson, 1943b), and other reptiles (Dunn, 1931), as well as for mammals (e.g. tapirs) and other groups. Various explanations have been advanced to account for this type of distribution. In a few exceptional cases, for example, the White-faced and Fulvous Whistling Ducks (*Dendrocygna viduata* and *D. bicolor*) and the Southern Pochard (*Netta erythrophthalma*), it is reasonably certain that transoceanic colonization is the answer. This explanation is, however, exceedingly improbable for most of the other groups, which have closely related representatives in the tropics of both the Old and the New World, for example, some of the snake-birds (Anhingidae), the sun-grebes (Heliornithidae), jacanas (Jacanidae), barbets (Capitonidae), trogons (Trogonidae), and parrots (Psittacidae) among the birds that I have classified with the Pantropical element; as well as some of the storks (Ciconiidae), ibises (Threskiornithidae), flamingos (Phoenicopteridae), nightjars (Caprimulgidae), woodpeckers (Picidae), and hawks (Accipitridae and Falconidae). A different explanation must be found for their movement from one continent to another.

The "land-bridge builders" considered this pattern of distribution as evidence of a former land connection across the Atlantic and Pacific. The objections to their theories were summarized by Matthew (1915), who showed that fossil finds indicate that many of these families formerly had much wider ranges (probably continuous across the Bering Strait bridge) in the temperate zones. A faunal agreement is particularly close between tropical-subtropical North America and the Old World tropics. It indicates that the present separation of the faunas is of comparatively recent date and that it must have been preceded by a long period of faunal exchange. Matthew (1915), Simpson (1943a:9), and others have postulated that the Bering Strait bridge was the pathway of this faunal exchange, which continued until late in the Tertiary (and, as far as non-tropical elements are concerned, down to the present). Stegmann (1938b) objects to this solution. He quotes considerable evidence from the field of paleobotany and paleoclimatology which indicates (p. 485): "that the climate in the region of Bering

Strait was at times warmer than it is now, but never reached tropical temperatures. Indeed it is quite certain that in northwestern America and in nearly all of Siberia the climate was never tropical or even subtropical during the entire Cenozoic and Cretaceous. . . . The Bering region was thus far outside the tropics during the entire period that needs to be taken into consideration for the evolution of Recent birds, so that it is without the slightest significance as a 'land bridge' for tropical groups." The records of American plant paleontologists support this contention. Chaney (1940) shows that as far back as the Eocene only a temperate climate existed in the countries east and west of the Bering Strait bridge. (See Figure 4.) One has to go as far south as the State of Washington on the American side, and to China on the Asiatic side, to find fossil plants that indicate even a subtropical climate.

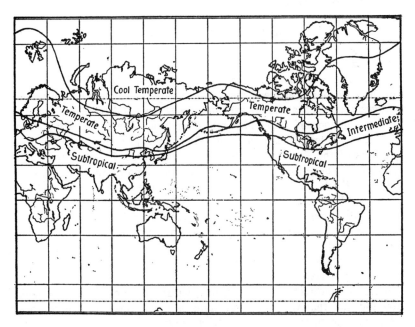

Figure 4. Eocene climatic zones as indicated by fossil plants. (*Based on Chaney, 1940.*)

A generation ago the opinion was widespread among paleogeographers that there were past periods during which a uniformly tropical climate prevailed all over the world. Reputed finds of Tertiary palms in Greenland seemed to strengthen this theory. However, these botanical reports have since been found to be erroneous; furthermore, certain geophysicists have made it abundantly clear that climatic zones must have always existed on the earth. This is a corollary of the earth's curvature. Less radiated heat from the sun will reach a given area in the higher latitudes than will reach the equatorial districts, where at noon the sun is nearly overhead during the greater part of the year. Furthermore—and this is a factor strangely neglected in books on past climates—the axis of the earth is inclined at an angle of $23\frac{1}{2}°$ to the perpendicular to the plane of the ecliptic. This inclination causes our seasons. The northern hemisphere is turned away from the sun during the winter and turned toward the sun during the summer. Geophysicists believe that this angle of the ecliptic has not changed significantly during the geological past. This means that north of the Arctic Circle an Arctic winter night has existed at all times, including the so-called "warm periods" of the earth. The Arctic Circle goes exactly through Bering Strait, and there can be little doubt that an Arctic "winter" (in terms of daily sunlight) must have existed at least as far south as the Aleutians, in other words beyond the southern edge of the Bering shelf. Surely this would not be a favorable condition for tropical faunas and floras to pass freely back and forth between Asia and America.

Yet the close relationship between the Old and New World members of the Pantropical element, whose ranges are now widely discontinuous, proves that such a faunal exchange must have taken place, and this places the zoogeographer in a real quandary. The customary solution for the problem is to ignore it. Stegmann (1938b:492) and other authors of the Russian school (e.g. Wulff, 1943:173–196) attempt to solve it by suggesting a modified Wegenerian land connection across the North Atlantic lasting at least until the middle of the Tertiary. Simpson (1943a:20–22), however, objects to this proposal on the basis of the small number of early Tertiary mammalian forms that were common to Europe and North America. A similar objection comes from the field of botany. The Eocene floras of Europe and North America "were remarkably different" according to Reid and Chandler (1933:70–88). There could have been no direct land connection between the two areas. Additional indirect evidence against a transatlantic bridge is provided by the fact that the American fauna is much closer to the southeast Asiatic than to the European-African fauna.

In view of the improbability of a North Atlantic land connection, various attempts have been made to find new routes for the transpacific migration. I shall refrain from a discussion of the various proposed transpacific land bridges. They are faunistically possible, but find no geological support. There is, however, some evidence for considerable recent tectonic activity in and south of the Aleutian island region, as well as for a pronounced lowering of the floor of the Pacific as a whole. Malaise (1945) and other authors have therefore made the assumption that the Bering Strait bridge was formerly very much wider than it is now, wide enough, in fact, to reach southward into a tropical climate.

Another assumption sometimes made is that there was, during the Tertiary, a much stronger contrast than now on the Bering bridge between the warm climate of its southern shore and the temperate climate of the interior, owing to the shutting off of the Arctic Ocean and the stronger influence of the warm Japan Current. This theory can account for the strictly temperate climate character of all fossil plants found in the Bering bridge area only by assuming that they have come exclusively from inland stations. Also this theory necessarily minimizes the effects of the arctic winter season.

Strict adherents of the theory of permanence of oceans and continents will look for a different explanation of the intercontinental migration of tropical faunas. Perhaps the common ancestors of the tropical faunas in the Old and New Worlds were not so narrowly tropical as are their living descendants. Furthermore, many representatives of tropical families are not nearly so heat-loving as is generally assumed— although they live in equatorial latitudes, their habitat is not tropical. In the characteristically "tropical" family of trogons, for example, *Harpactes wardi* (Burma, Indochina) lives in the mountains between 2,500 and 3,000 meters; *Trogon personatus* and other South American species reach even higher altitudes. The climate at these altitudes is distinctly temperate. Most other "tropical" families of birds, particularly the parrots, have some members that live in an equable humid temperate climate. Species with similar ecological requirements might have been able to exist in the warm temperate parts of Bering Strait bridge, even during the rather dark winter days. It must not be overlooked that the tropical regions were apparently more arid at earlier geological periods than they are today. Perhaps the warm temperate zone was in the late Mesozoic to early Tertiary a refuge for species with a preference for an equable humid climate, just as the tropics are today.

These comments may suffice to indicate that the problems of the faunal exchange between Old and New World are by no means solved. However, the questions that need to be asked are beginning to crystallize, and the information needed to answer them is beginning to accumulate. We have advanced beyond the stage of pure speculation.

Faunal Zoogeography and Ecology

We are all familiar with the fact that among the birds of the northern coniferous woods there is a high percentage of recently immigrated palearctic species. The South American element, on the other hand, is almost non-existent in these forests. It would be a rewarding task to analyze the bird life of all the major North American habitats and determine their faunistic composition from the point of view of origin. To do this in detail would require much more space than can be given in this paper; furthermore, there are not enough reliable published tabulations of the characteristic species of the various habitats to provide the material for such a study. For example, I have looked in vain for a good tabulation of the typical birds of the chaparral or of some of the more specialized habitats in the Southwest. No comprehensive account of the breeding birds of the various types of prairie is available.

One of the striking features of North American faunal history is that not a single species of the originally South American fauna has crossed the Bering Strait bridge into the Old World. On the other hand, numerous Old World birds have been able to invade South America. Some became adapted to life in the tropics, for example, certain jays, thrushes, kingfishers, and carduline finches. Others—the Short-eared Owl (*Asio flammeus*) and Horned Lark (*Otocoris alpestris*)—simply jumped the tropical gaps.

It would be tempting to reconstruct the climate on Bering Strait bridge throughout the Tertiary by analyzing the ecological requirements of the birds that passed this bridge at a given period. At present, for example, the bridge is passable only for birds of the tundra and of the coniferous belt (taiga = "Hudsonian"). Stegmann (1938b) lists the birds that could pass Bering Strait under climatic conditions similar to or slightly warmer than the present. But as we go further back in time, the analysis becomes more difficult. Again it seems that the Old World contributed more than the New. The only birds of North American origin that have spread into the Old World are the grouse (Tetraonidae), the finches of the subfamily Emberizinae, one species of wren (*Troglodytes troglodytes*), and—if these are indeed North American—two species of dippers (*Cinclus cinclus, C. pallasii*), and two species of waxwings (*Bombycilla garrula, B. japonica*). Even such richly developed North American families as the mockingbirds (Mimidae), vireos (Vireonidae), and wood warblers (Parulidae) * have not crossed for reasons that are difficult to understand. On the other hand, nearly every family of temperate Eurasia has entered North America, and most of them have sent at least one representative as far as South America.

It is conceivable that the fauna of each of the major habitats or ecological formations of North America would have its peculiar composition from the point of view of origin. However, a glance at Table 3 shows that there are no major differences, at least as far as forest habitats are concerned. What differences there are can be attributed mainly to latitude. Also there seems to be no striking difference from the point of view of origin between the faunas of climax and second growth. Among 159 breeding pairs listed in two years (1932, 1934) on a study area in a climax Maple-Beech-Hemlock forest Saunders (1938:32–33) records 10.0 per cent South American, 71.1 per cent North American, and 18.9 per cent Old World pairs. Among 104 pairs

---

* The Myrtle Warbler (*Dendroica coronata*) and the Northern Water-thrush (*Seiurus noveboracensis*) have recently crossed into Anadyrland.

(listed in 1932, 1933) in near-by second growth Cherry-Aspen there were 6.8 per cent South American, 71.1 per cent North American, and 22.1 per cent Old World pairs. The figures were thus almost identical.

In specialized habitats there are sometimes significant deviations from the faunal composition exemplified in Tables 2 and 3. For example, all of the species usually listed as typical for the mid-western prairie are of North American origin: Prairie Chicken (*Tympanuchus cupido*), Upland Plover (*Bartramia longicauda*), Burrowing Owl (*Speotyto cunicularia*), Western Meadowlark (*Sturnella neglecta*), Bobolink (*Dolichonyx oryzivorus*), Grasshopper Sparrow (*Ammodramus savannarum*), and Savannah Sparrow (*Passerculus sandwichensis*). This may mean that the great humidity of both the Bering and the Panama bridges prevented an influx of the faunas of the more arid habitat of Eurasia and South America. The ecological niche of the North American grasslands thus could be filled by the autochthonous North American element. The land birds of marshes also tend to be prevailingly (80 to 100 per cent) North American. For example, the Long-billed Marsh Wren (*Telmatodytes palustris*), Short-billed Marsh Wren (*Cistothorus stellaris*), Swamp Sparrow (*Melospiza georgiana*), Sharp-tailed Sparrow (*Ammospiza caudacuta*), Seaside Sparrow (*A. maritima*), Red-wing (*Agelaius phoeniceus*), and Yellow-headed Blackbird (*Xanthocephalus xanthocephalus*). The Old World element, on the other hand, is, as a rule, comparatively strong at the higher altitudes in the mountains.

It would be interesting to analyze in a similar manner other specialized habitats, such as the Californian chaparral, the creosote bush–mesquite thickets of the Southwest and the Caribbean mangroves, but adequate census data are not available. This brief discussion is to be considered merely as a hint at the interesting relationship between ecology and faunal history, which constitutes a fertile field for future investigators.

LITERATURE CITED

ALLEN, J. A.
- 1871 On the mammals and winter birds of east Florida, with an examination of certain assumed specific characters in birds, and a sketch of the bird faunae of eastern North America. *Bull. Mus. Comp. Zool.*, 2, No. 3, pp. 161–450.
- 1893 The geographical origin and distribution of North American birds, considered in relation to faunal areas of North America. *Auk*, 10:97–150.

AMADON, DEAN
- 1944 The genera of Corvidae and their relationships. *Amer. Mus. Novit.* No. 1251.

CARPENTER, GEO. H.
- 1894 Nearctic or Sonoran? *Nat. Sci.*, 5:53–57.

CHANEY, RALPH W.
1940 Tertiary forests and continental history. *Bull. Geol. Soc. Amer.*, 51:469–488.

CHAPMAN, FRANK M.
1923 The distribution of the motmots of the genus Momotus. *Bull. Amer. Mus. Nat. Hist.*, 48:27–59.

CRUICKSHANK, ALLAN D.
1942 Birds around New York City. *Amer. Mus. Nat. Hist. Handbook Ser.*, No. 13.

DELACOUR, JEAN, and ERNST MAYR
1945 The family Anatidae. *Wils. Bull.*, 57:3–55.

DUNN, EMMETT REID
1922 A suggestion to zoogeographers. *Science*, 56:336–338.
1931 The herpetological fauna of the Americas. *Copeia*, 1931:106–119.

GILL, THEODORE
1875 On the geographical distribution of fishes. *Ann. and Mag. Nat. Hist.* (Ser. 4), 15:251–255.

HEILPRIN, ANGELO
1883 On the value of the "Nearctic" as one of the primary zoological regions. Replies to criticisms by Mr. Alfred Russel Wallace and Prof. Theodore Gill. *Proc. Acad. Nat. Sci. Phila.*, 1883:266–275.

HICKEY, JOSEPH J.; MARGARET B. HICKEY
1937–1944 Bird-Lore's [Audubon Magazine's] breeding bird-census. 1–8. *Aud. Mag.* [formerly *Bird-Lore*], 39–46.

HUBBS, CARL L.
1944 [Review of "Studies in the genetics of Drosophila. III"]. *Amer. Nat.*, 78:270–271.

LÖNNBERG, EINAR
1927 Some speculations on the origin of the North American ornithic fauna. *Kungl. Svenska Vetenskapsakad. Handl.* Ser. 3, vol. 4, no. 6, pp. 1–24.

MALAISE, RENÉ
1945 Tenthredinoidea of southeastern Asia. *Opusc. Entom.*, suppl. 4.

MATTHEW, W. D.
1915 Climate and evolution. *Ann. N.Y. Acad. Sci.*, 24:171–318.

MAYR, ERNST
1941 List of New Guinea birds. Amer. Mus. Nat. Hist., New York.
1944a The birds of Timor and Sumba. *Bull. Amer. Mus. Nat. Hist.*, 83:123–194.
1944b Wallace's Line in the light of recent zoogeographic studies. *Quart. Rev. Biol.*, 19:1–14.

MAYR, ERNST, and JAMES BOND
1943 Notes on the generic classification of the swallows, Hirundinidae. *Ibis*, 85:334–341.

PETERS, JAMES LEE
1940 Check-list of birds of the world. vol. 4. Harv. Univ. Press, Cambridge, Mass.

REICHENOW, A.
1888 Die Begrenzung zoogeographischer Regionen vom ornithologischen Standpunkt. *Zool. Jahrb., Abtheilung für Systematik, Geographie und Biologie der Thiere*, 3:671–704.

REID, ELEANOR MARY, and MARJORIE ELIZABETH JANE CHANDLER
1933 The London clay flora. Brit. Mus. (Nat. Hist.), London.

RIPLEY, S. DILLON
1945 The barbets. *Auk*, 62:542–563.

SAUNDERS, ARETAS A.
1938 Studies of breeding birds in the Allegany State Park. *N. Y. State Mus. Bull. No.* 318.

SCHMARDA, LUDWIG K.
1853 Die geographische Verbreitung der Thiere. vol. 2. Vienna.

SCLATER, PHILIP LUTLEY
1858 On the general geographical distribution of the members of the class Aves. *Jour. Proc. Linn. Soc. London for Feb.* 1858, pp. 130–145.

SHARPE, R. BOWDLER
1903 A hand-list of the genera and species of birds. vol. 4. Brit. Mus. (Nat. Hist.), London.

SIMPSON, GEORGE GAYLORD
1940a Mammals and land bridges. *Jour. Wash. Acad. Sci.,* 30:137–163.
1940b Antarctica as a faunal migration route. *Proc. Sixth Pac. Sci. Congr.* (1939), vol. 2, pp. 755–768.
1943a Mammals and the nature of continents. *Amer. Jour. Sci.,* 241:1–31.
1943b Turtles and the origin of the fauna of Latin America. *Amer. Jour. Sci.,* 241:413–429.

STEGMANN, B.
1932 Die Herkunft der paläarktischen Taiga-Vögel. *Arch. f. Naturg.* (N.F.), 1:355–398. [Leipzig]
1938a Principes généraux des subdivisions ornithogéographiques de la région paléarctique. *Faune de l'URSS* (N.S. 19), *L'Oiseaux,* vol. 1, no. 2.
1938b Das Problem der atlantischen Landverbindung in ornithogeographischer Beleuchtung. *Proc. Eighth Internatl. Ornith. Congr.* (Oxford), 1934: 476–500.

STRESEMANN, ERWIN
1939 Die Vögel von Celebes. Teil 2. Zoogeographie. *Jour. f. Ornith.,* 87:312–425.

SUSHKIN, PETER P.
1924 [Résumé of the taxonomical results of morphological studies of the Fringillidae and allied groups]. *Bull. Brit. Ornith. Club,* 45:36–39.
1925 A preliminary arrangement of North American genera of Fringillidae and allied groups. *Auk,* 42:259–261.

SZIDAT, L.
1940 Die Parasitenfauna des Weissen Storches und ihre Beziehungen zu Fragen der Oekologie, Phylogenie und der Urheimat der Störche. *Z. Parasitenkunde,* 11:563–592.

WALLACE, ALFRED RUSSEL
1876 The geographical distribution of animals. vol. 1. Harper & Bros., New York.

WETMORE, ALEXANDER
1940 A check-list of the fossil birds of North America. *Smiths. Misc. Coll.,* 99, no. 4.
1944 A new terrestrial vulture from the upper Eocene deposits of Wyoming. *Ann. Carnegie Mus.,* 30:57–69.

WULFF, E. V.
1943 An introduction to historical plant geography. Chronica Botanica Co., Waltham, Mass.

THE AMERICAN MUSEUM OF NATURAL HISTORY, NEW YORK 24, N. Y.

Speciation in Birds

Progress Report on the Years 1938—1950

Ernst Mayr

The American Museum of Natural History, New York

To discuss species and speciation in Uppsala seems particularly appropriate for the name Linnaeus is in the mind of every biologist inseparably linked with the words species and Uppsala. It was Linnaeus who placed taxonomy on a solid foundation through the introduction of binomial nomenclature. On this foundation it has prospered and grown into the large field it is today. However, there is a great difference between the systematics of 1758 and that of 1950. Whereas Linnaeus knew only 564 species of birds, we now know 8 600. But, more important, the emphasis in the days of Linnaeus was on the discovery of new species, their description, and formal classification. The emphasis today is on the nature of species, the origin of species, and the relation of species to their environment.

Twelve long years have passed by since the last International Ornithological Congress, long years in more than one sense. But in spite of all the misery and destruction which the past unholy war has brought upon the world, it has not reduced scientific productivity to a very marked extent. More than 18 000 ornithological papers have been published since the Rouen Congress and nearly 200 of them touch upon the problem of speciation. Only a cursory survey of this vast literature can be attempted in the forty-five minutes allotted to me. Fuller details can be found also in a number of recent summaries that were published in book form (Mayr, 1942; Huxley, 1942; Lack, 1947a). The genetic basis of speciation and other more general aspects of evolution have been discussed in other volumes (Dobzhansky, 1941; Bauer and Timoféeff-Ressovsky, 1943; Simpson, 1944; Rensch, 1947; Jepsen, Simpson, and Mayr, 1949). My own discussion will concentrate on the controversial issues and unsolved problems. We ornithologists can be proud of our contributions to this field. The present biological concept of the species, as well as the theory of geographical speciation, was to a large extent developed by ornithologists on the basis of ornithological material.

No rational theory of the origin of species was possible until the concepts of the nature and structure of species had matured. Let us therefore say a few words on these two topics.

The nature of species. The biologist of today considers species to have the following essential characteristics: A species is an aggregate of interbreeding natural populations which are not only reproductively isolated from other such aggregates but also ecologically specialized sufficiently so as not to compete with other such species. The problem of the origin of species then resolves itself in the question: How can *one* group of interbreeding populations break up into two or more which are not only reproductively isolated from each other but also ecologically compatible?

A study of the population *structure of species* has yielded the clues for the solution of this problem. As stated, species are usually polytypic, that is, they are composed of lesser units, subspecies and local populations, which deviate more or less from each other. Furthermore, through their geographical extension species are multidimensional systems. Species, as understood by us today, are a far cry from the necessarily monotypic, typological species of Linnaeus and his contemporaries.

The study of the structure of species is the chief occupation of the bird taxonomist of today. It is impossible to refer to more than merely a fraction of the recently published monographic studies on the geographical variation of species (Aldrich, 1943, 1946b; Behle, 1942; Bond, 1943; Chapin, 1948, 1949; Chapman, 1940; Dunajewski, 1939; Hawbecker, 1948; Jany, 1948; Johansen, 1945, 1946, 1947; Lack, 1945, 1947a, 1947b; Mayr, 1941, 1942a, 1949a, 1949b, 1949c, 1950b; Mayr and Amadon, 1947; Mayr and Moynihan, 1946; Miller 1941; Moreau, 1948b; Pateff, 1947; Pitelka, 1945a; Ripley and Birckhead, 1942; Ripley, 1941; Salomonsen, 1944; Snyder and Shortt, 1946; Stresemann, 1940; Vaurie, 1949a, 1949b; Voous, 1947, 1949; Voous and Van Marle, 1949). All these studies show that most species of birds are polytypic, that is, they are composed of several subspecies. In New Guinea, for instance, 80 per cent of all species are polytypic (Mayr, 1942) and among temperate zone birds about 60–80 per cent. Accurate comparative studies on the percentage of monotypic species in various parts of the world have not yet been made. They can be undertaken only where reliable faunal lists are available.

Subspecies

Though not strictly part of our subject, I would like to say a few words about *subspecies.* Unlike the effective local population (Sewall Wright) or the species, which can be defined biologically, the subspecies is primarily a taxonomic concept. It is a population or aggregate of populations which differs taxonomically from other such aggregates within a species. A subspecies cannot clearly be delimited from local populations on one hand and species on the other. In the past, certain authors have tried to name every population that differs in average characters. That this policy is nonsense has been made clear by the population geneticists who have shown that no two populations of sexually reproducing animals are exactly alike in the frequencies of polymorph genes and the means of multifactorial characters. To be recognized as different subspecies it is therefore not sufficient that two populations merely differ; they must differ "taxonomically." The standard of this "taxonomic" difference, in the opinion of some leading contemporary ornithologists, is best expressed by the so-called "75 per cent rule." A population can be recognized as a valid subspecies if 75 per cent of it differs from 100 per cent of the population with which it is compared. Amadon (1949) shows how this is best calculated from the available sample. In view of the slanting of the population curve a difference of 75 per cent from 98 per cent would seem to express current standards more accurately than the standard proposed by Amadon. The various standards were discussed by Rand and Traylor (1950).

The difference must be a "taxonomic" one, that is, it must be recognizable in museum specimens. It is important to emphasize this in view of the many slight differences among intrasubspecific populations and the physiological differences between populations and geographical races. In the White-crowned Sparrow (*Zonotrichia leucophrys*) the resident subspecies *nuttali* of the San Francisco Bay region differs in many ways from the migratory *pugetensis* (southern British Columbia to northern California). This includes differences in regard to beginning and length of breeding season, development of gonads and endocrine glands, song, and morphology (Blanchard, 1941). There is a broad transition zone, and some populations in northern California, though morphologically like *pugetensis*, are sedentary like *nuttali*. A similar physiological difference exists between the British and continental populations of the Starling (*Sturnus vulgaris*) (Bullough, 1942), but since there is no taxonomic difference, it is not admissible to recognize a separate subspecies for the British Starling (Thomson, 1943). Populations that differ with respect to migratory behavior and the physiology of the breeding cycle occur in many widespread subspecies of the temperate zone. It would lead to chaos to recognize all these population differences taxonomically.

Even so, bird taxonomy is loaded down with subspecies, which often obscure rather than clarify trends of geographical variation. On continents most subspecies intergrade with each other rather imperceptibly and also show a good deal of intrasubspecific variability (Behle, 1942; Lack, 1946). If there are reversals of the trends of variation, far distant populations may be indistinguishable and will have to bear the same subspecific name. There are many situations which provoke the feeling that trinomials have outlived their usefulness (Lack, 1947b).

Instead of expending their energy on the describing and naming of trifling subspecies, bird taxonomists might well devote more attention to the evaluation of trends in variation. Three basic rules are now quite evident, but require more quantitative documentation:

(1) Much geographical variation, particularly on continents, is clinal. These character clines are correlated with gradients in the selective factors of the environment.

(2) Peripheral or otherwise isolated subspecies of a species are the most distinct (Mayr and Vaurie, 1948).

(3) All taxonomic characters are in the last analysis controlled by selective factors of the physical and biotic environment.

More on (1) and (3) will be said in the next section.

The often striking divergence of peripheral subspecies or semispecies has been commented upon by many recent authors. Instances are *Junco vulcani* (Miller, 1941), *Ptilinopus huttoni* (Ripley and Birckhead, 1942), *Ducula galeata* (Mayr, 1942a), *Dicrurus megarhynchus* and four other groups of drongos (Mayr and Vaurie, 1948), *Dicaeum tristrami* (Mayr and Amadon, 1947), *Erithacus rubecula hyrcanus* and *superbus* (Lack, 1946), and many others. In fact, there is hardly a group of polytypic species which does not have some illustrations of this rule. Many of these peripheral populations are in the twilight zone between subspecies and species, and they are particularly important to the student of speciation as indicating the course of species formation.

The Process of Speciation

The fact that species are not amorphous-homogeneous but that instead they have a definite structure consisting of populations and subspecies greatly facilitates the understanding of speciation. But, what do we mean by speciation? The term has been used in two different meanings. First it has been used to indicate a transformation in time, the phyletic evolution of the paleontologist. With his scanty fossil material, the ornithologist cannot make much of a contribution to this problem, although a promising beginning has been made for the Pleistocene of California (Howard, 1947a, 1947b; Fisher, 1947). When speaking of speciation, the neozoologist usually means the multiplication of species in space rather than their transformation in time. Excellent progress in this field has been made in recent years, in fact, the problem can be considered as solved in its major outlines. Many details remain to be filled in. The solution was the result of a painstaking analysis of many species, genera, and families of birds, such as those by Rensch of the birds of the Lesser Sunda Islands, Miller of Junco, Mayr of South Sea birds, Lack of the Galapagos Finches, and others quoted above. It is most gratifying to state that these authors are in almost complete agreement as to their results. These studies will therefore be summarized as a whole.

Speciation, in the sense of multiplication of species, proceeds in two steps. During the first step, a population (or subspecies or subspecies group) becomes isolated by extrinsic barriers. Gene interchange with the parental population is completely or almost completely interrupted. During this period of separation the isolated population becomes genetically reconstructed. If this change of the genetic makeup includes factors for reproductive isolation (isolating mechanisms), as well as for ecological compatibility, the isolated population may reach species level and can re-invade the range of the parental population. This process is called geographic speciation since geographical isolation (by barriers or by distance) seems to be needed to permit the genetic reorganization (Mayr, 1947).

Both the ecological compatibility as well as the reproductive isolation are controlled by hundreds of genes, and we can express the genetical difference between two sympatric (co-existing in the same geographical area) species as follows:

$$GD_{A-B} = IM_n + IM_p + IM_s + EC_n + EC_p + EC_s.$$

These letters mean:

GD_{A-B} = Genetic difference between species A and B
IM = Isolating mechanisms
EC = Ecological compatibility factors
n = Neutral
p = Preadapted
s = Acquired by selection

This formula means that the genetic differences between two species are composed of three sets of factors: those that are neutral with respect to isolating mechanisms (IM_n) and ecological compatibility (EC_n); those which the species acquired during their geographical isolation and which preadapted them for a subsequent overlap of the ranges (IM_p and EC_p); and finally those differences which were favored by selection after the species had become sympatric (IM_s and EC_s). It is logical (1) that no overlap between two species can occur unless IM_p and EC_p have been acquired during isolation, and (2) that there will be a high selective premium on the acquisition of factors that improve reproductive isolation and minimize ecological competition. Hence, there will be a continued accumulation of such factors (IM_s and EC_s). The role of selection before and after the establishment of overlap was discussed by Mayr (1949d).

The Proof for Geographical Speciation

The course of speciation through population divergence, as established by the taxonomist, is a vastly different concept of speciation than that held by the early Mendelians (De Vries, Bateson), who believed in instantaneous speciation by major mutations. Modern geneticists are nearly unanimous in endorsing the speciation concept of the naturalists, although there are still a few dissenters (e.g. Goldschmidt). A second line of attack on the concept of geographic speciation came from the proponents of sympatric speciation (see Thorpe, 1945 vs. Mayr, 1947). In view of these criticisms it may not be out of place to summarize once more the proofs for geographic speciation.

Geographic variation of species characters

If isolating mechanisms (IM_p) and ecological compatibility (EC_p) evolve during geographical isolation, as postulated by the theory of geographical speciation, then one should find intermediate stages. Particularly one should find that isolating mechanisms, as well as ecological characteristics, are geographically variable. This is indeed the case. Owing to technical difficulties of keeping birds there is not as much proof for the geographical variability of isolating mechanisms in birds as in insects, mice, or fish. Fertility between geographical races seems to be occasionally reduced. Song in birds is an important component of the ethological isolating complex. Many cases of geographical variation of song have been described in the literature. Benson (1948) found it in thirty-three among 209 widespread species or groups of allopatric species (i. e. species with the geographical ranges not overlapping) of African birds. No voice difference was detected by him among thirteen pairs of sympatric species. I know of no pair of European, North American, or Papuan birds with identical songs or call notes. Perhaps Benson sets a higher standard for "difference." All sixteen species of birds that occur both in Britain and Teneriffe, Canary Islands, differ in their voices. In four species the song was strikingly different (Lack and Southern, 1949). The cases of circular overlap (see below) are good indirect proof for the geographical variation of isolating mechanisms.

Borderline cases

There are many bird populations that are on the border between the subspecies and species level. No less than 94 of the 755 species of North American birds listed in the last A.O.U. Checklist are considered subspecies by some authors (Mayr, 1942). Twenty-five forms of Anatidae, considered subspecies by Delacour and Mayr (1945), are treated as species by other authors. Many of the examples of pseudo-conspecificity, listed below (Appendix), are also such borderline cases.

Some of these cases have received special attention in recent years. Among the African love birds (*Agapornis*) none of the eight forms widely overlaps the range of any other. The ranges of two forms, *fischeri* and *personata*, approach each other within twenty miles without signs of intergradation in spite of ecological similarity (Moreau, 1948 b). The woodpecker *Dendrocopos leucopterus* coexists geographically with *D. major* at various localities in central Asia, but it always inhabits broad-leaved trees (mostly poplars) in the lowlands, while *major* occurs at higher altitudes in coniferous forests. Some authors treat these forms as conspecific, others as separate species (Voous, 1947). A somewhat similar situation occurs in the *bokharensis* group of *Parus major* (Vaurie, 1950a).

The geese with their singular population structure are rich in borderline cases. The bean geese (*Anser fabalis*) include a southern group of large, long-billed subspecies in the taiga and a northern group of small short-billed subspecies in the tundra. (Johansen, 1945). In the Canada Geese (*Branta canadensis* group) it occasionally happens that neighboring nesting colonies differ in body size, voice (honking vs. cackling), nesting site (seacoast vs. inland), nest (mound vs. scrape), migration season, and other features. Intergradation between adjacent colonies may be entirely lacking and the temptation is great to assign these populations to two or three different species (e.g. Aldrich, 1946b). Yet, it has been impossible to arrange the subspecies satisfactorily into species. In fact, Bailey (1948) shows almost conclusively that *leucopareia* and *parvipes* form a perfect link between the small cackling geese on one side and the large honkers on the other. By secondary range expansion, however, large and small subspecies have come into secondary contact at various places without interbreeding. Thus, we have here another interesting borderline case between subspecies and species. Others occur in the *Rhipidura rufifrons* group (Mayr and Moynihan, 1946) and in *Pyrrhula* (Voous, 1949).

Circular overlaps

The ideal proof for geographical speciation is presented by the cases of "circular overlap." The end links of a continuous chain of interbreeding subspecies overlap in such a case without interbreeding. Some examples were cited by Mayr (1942). Others have been described more recently. According to Johansen (1947) the chiff-chaff (*Phylloscopus collybita*) is another such case with the Caucasus being the area of overlap. There the northwestern *abietinus* occurs in the forests of the lower slopes while the southeastern *lorenzii* (of the *neglectus* group) is restricted to the sub-alpine forest zone.

Recent work indicates that in the most celebrated case of overlap matters are not as simple as once believed. I refer to the Herring Gull (*argentatus*)—Lesser Black-backed Gull (*fuscus*) overlap in Europe. This group is not composed of clinally intergrading populations, but rather of groups of subspecies which in certain zones of contact show no signs of intergradation and are ecologically segregated (Stresemann, 1943a). These gulls were apparently divided during the Pleistocene into a number of isolated populations which reached various levels of reproductive isolation and ecological compatibility (Stresemann and Timoféeff-Ressovsky, 1947). A similar situation exists in the American Herring Gulls, where *kumlieni* and *smithsonianus* overlap in northern Labrador and southern Baffin Land without interbreeding (Rand, 1943). The relation of these two forms to *thayeri* remains to be cleared up.

Recently Completed Speciation

Recently completed speciation is indicated by three types of phenomena: (a) double invasions, (b) superspecies, and (c) marginal overlaps. The geographic basis of these three phenomena is apparent and they can be cited as proof for geographic speciation.

(a) Double invasions. If an island is repeatedly invaded by colonists from a distant mainland, it may happen that the descendants of the first wave of colonists have changed so much that they are reproductively isolated from the new arrivals. Stresemann was the first to recognize the importance of this phenomenon and Mayr (1942) listed many cases. Others have since been described: *Phaeornis* and *Loxops* on Kauai (Amadon, 1947); *Dicrurus* on Celebes and the Andamans (Vaurie, 1949a); *Pernis* in the Philippines (Stresemann, 1940); *Cinnyris jugularis* and *solaris* on Sumbawa and Flores (Mayr, 1944); *Philemon moluccensis* in Australia (Mayr, 1944); and six cases of completed double invasions in Ceylon and south India (Ripley, 1949).

(b) Superspecies. A monophyletic group of closely related allopatric species may be called a superspecies (Mayr, 1942). Superspecies are exceedingly common in island regions. They are 17 superspecies among 130 species of Solomon Islands birds. There are at least 11 superspecies (with 26 species) among the 145 species of Anatidae (Delacour and Mayr, 1945). Superspecies are found in the *Vini palmarum* and *V. peruviana* group, in *Ptilinopus luteovirens*, in *Ducula pacifica* and *Ducula latrans* (Amadon, 1943). They are common in the Celebes region: *Cyornis hoevelli* and *sanfordi*, *Eudynamis scolopacea* and *melanorhyncha*, *Hypothymis azurea* and *puella* (Stresemann, 1939), and in the Moluccas (*Gerygone fusca* group, *Oriolus*, *Sphecotheres*, *Philemon*, *Macropygia*) (Mayr, 1944). *Dicrurus hottentottus* and *D. adsimilis* belong to widespread superspecies (Vaurie, 1949a) and so does *Halcyon chloris* (Mayr, 1949c). Superspecies occur also in the *Dicaeidae* (Mayr and Amadon, 1947) and in *Agapornis* (Moreau, 1948b).

In all these cases the differences between allopatric forms are so striking that they appear to have reached species level during geographical isolation.

(c) Marginal overlaps. Occasionally one of the species of a superspecies expands into the range of another member of the superspecies with the result that there will be a smaller or greater overlap along the margins of their respective ranges. Several such cases were described by Mayr (1942); others are listed below in the list of pseudo-conspecific species pairs (Appendix). They occur in the *Treron curvirostra* group, in the *Anthreptes malaccensis* group, and in *Geokichla* (Mayr, 1944), in the *Rhipidura rufifrons* group (Mayr and Moynihan, 1946), in the *Ptilinopus purpuratus* group (Ripley and Birckhead, 1942), in *Anaimos* and *Dicaeum* (Mayr and Amadon, 1947), in fact in most actively speciating genera. Ripley (1945) has proposed the term interspecies for such ex-superspecies, but Moreau (1948a) considers this term superfluous. They have often been referred to simply as species groups.

The final step in this speciation process is the production of groups of full species that are more or less extensively sympatric. That they originated by geographical speciation is most evident in the case of species flocks that evolved on archipelagos. This has been worked out brilliantly by Lack (1947a) for the finches of the Galapagos Islands and by Amadon (1950) for the Drepaniidae of the Hawaiian Islands. On islands that are on or at the edge of continental shelves differentiation may not rise above the subspecies level, as is characteristic for most of the forms on the west Sumatran Islands (Ripley, 1944). This is due to the recency and insufficient isolation of these islands.

Speciation on Continents

That speciation proceeds via geographical isolation is now universally admitted for insular species. At first sight, however, this explanation appears to meet difficulties when applied to continental species flocks. Genera rich in widespread species, such as *Parus* and *Sylvia* in Europe, *Emberiza* and *Carpodacus* in Asia, *Ploceus* in Africa, and *Dendroica* in America seem to defy the concept of geographic speciation. Closer analysis indicates, however, that speciation on continents obeys the same rules as on islands. The extrinsic isolating barriers in this case are not water gaps, but stretches of terrain that are uninhabitable because of ecological unsuitability. The best proof for this process of speciation is given by the many instances of incipient speciation in continental birds. These cases have frequently been obscured in the past by the bad habit of taxonomists to treat slightly and strikingly different subspecies alike, and either not to indicate subspecies groups at all or to raise them to species level.

The subspecies group indicates an important level of speciation as was pointed out by Stresemann (1940, Ornith. Monatsb., 48: 103–104) for the Jay *(Garrulus glandarius)*. Well-defined subspecies groups occur in many species of the northern hemisphere. It can usually be shown that the Pleistocene glaciation was responsible for the development of these intraspecific groups. When the species were pushed southward they had to retreat into wooded refuges separated by cold steppes. This concept was first applied to European birds by Stresemann, but Rand (1948) has shown that it applies to North American birds equally well. Voous traces the history of subspecies groups in *Dendrocopos* (1947) and *Pyrrhula* (1949) with the help of this concept. Among the brown-eyed Juncos *(Junco)* of North America, the four continental subspecies groups also seem to fit the picture of refuges: *hyemalis* (Appalachians), *aikeni* (Black Hills), *caniceps* (Rocky Mountains), and *oreganus* (coast ranges). The three subspecies groups of the Ruffed Grouse *(Bonasa umbellus)*, distinguished by Snyder and Shortt (1946), seem to have become differentiated in the Appalachians, Rocky Mountains, and coast range refuges. It would no doubt yield interesting results if somebody would systematically test all species pairs and subspecies groups in North America for possible origin in these Pleistocene refuges.

Speciation is incomplete in nearly all the mentioned cases and they do not, therefore, completely elucidate the problem of continental speciation. The isolated populations have often diverged so far morphologically that the subspecies groups are considered as species by some authors as, for instance, the four groups of brown-eyed Juncos by Miller (1941). Still they have apparently not reached either complete reproductive isolation or ecological compatibility. The period of isolation was apparently not long enough to permit the perfection of isolating mechanisms and the rapid post-Pleistocene northward movement of the vegetation has favored a mixing-up of the previously isolated populations.

A better example of the amount of isolation that may permit the completion of speciation is perhaps given by tropical forest birds that live in isolated patches of forest (particularly in the mountains). Numerous examples of this occur in Africa (although never properly tabulated and evaluated) and on the tabletop mountains near the Brazilian-Venezuelan frontier as shown by the splendid explorations of the Phelps, father and son.

Australia shows a different manifestation of the same process. This island continent has an arid interior surrounded by a more humid marginal zone. This humid belt is broad in humid periods, but contracts into a few coastal pockets at the height of an arid cycle. These pockets serve then as refuges (Gentilli, 1949) for the humidity-loving fauna and flora, and distinct populations evolve in each refuge. In the case of *Neositta*, the populations isolated during the last aridity cycle have not yet reached species level (Mayr, 1950). Pseudo-conspecific pairs of largely allopatric species also demonstrate geographic speciation on continents (Appendix).

Allopatric Hybridization as Evidence of Incomplete Speciation

According to the theory of geographic speciation isolating mechanisms and ecological compatibility evolve during the geographical isolation of populations. It is to be expected that such geographical barriers sometimes break down before the speciation process is completed. Two things may happen in such a case. Either ecological compatibility has not yet developed, then we will get a peculiar mutual exclusion of the species ranges without interbreeding in the zone of contact. Such cases are listed in the Appendix.

If, on the other hand, the isolating mechanisms have not yet evolved, hybridization will take place. There is little agreement in the literature on how to define hybridization. The term was coined to describe a relationship of individuals and difficulties are caused by extending it to populations. Hybridization in nature can be classified into two categories: sympatric and allopatric hybridization. Under sympatric hybridization I understand the occasional interbreeding of two otherwise well-delimited sympatric species. There is no conceptual difficulty in regard to this type of hybridization.

The real difficulty concerns allopatric hybridization, the interbreeding of allopatric populations in their zone of contact. If the two temporarily isolated populations diverged only very slightly we are loath to call such interbreeding hybridization, we call it intergradation. I have referred to it as secondary intergradation (Mayr, 1942) to distinguish it from the intergradation of populations that have always been in continuous contact and owe their difference to opposing selection pressures. All secondary intergradation indicates incomplete reproductive isolation. Such cases are extremely common, they probably occur in every but the most localized species. Miller (1941) has shown for *Junco* how many races "hybridize" in zones of contact, and Pateff (1947) has pointed out how many names have been given to hybrid populations in *Sturnus vulgaris*. Occasionally even strikingly different forms hybridize in a zone of contact as Meise has shown (Mayr, 1942). Such a case in northern Persia was described by Paludan (1940) and illustrated in a beautiful color plate. It involves two buntings of the genus *Emberiza* (*icterica* and *melanocephala*) which seem to interbreed indiscriminately in a narrow hybrid zone within which all individuals show signs of hybridism. This recalls a color plate by Sutton (1938) depicting a similar hybridization in American orioles. Those who accept a species definition based on reproductive isolation will call such freely interbreeding forms subspecies.

Incomplete speciation in southeastern Asia is indicated by the striking difference of subspecies in *Lanius schach* (Dunajewski, 1939) and *Microscelis madagascariensis* (Mayr, 1941).

Hybrid populations are normally characterized by pronounced variability but hybrid populations may become stabilized and the variability may drop to the "normal" level. Miller (1941) has shown this for *Junco cismontanus*. Whether or not *Pardalotus ornatus* is a species that owes its origin to a stabilized hybrid population is not certain (Hindwood and Mayr, 1946).

There is one specially interesting class of hybridizations which does not fit into the above categories. A most instructive case of it was described by Chapin (1948) for the African species of the flycatcher genus *Terpsiphone*. Here, certain species are essentially allopatric because habitat differences keep them segregated. Two species, *rufocinerea* and *rufiventer*, live in the rain forest of equatorial Africa, a third (*viridis*) in second growth and in wooded savannas. There are three areas in Africa (northwest Angola, Port. Guinea, and Uganda) where hybrid flocks have developed and individuals of the parental species are either present or absent. Chapin not only unmasked the polymorphism in these regions as hybridism but also provided an explanation. He showed that ecological conditions are responsible for it. Hybridism in *Terpsiphone* occurs where through the destruction of the rain forest by man (white or native) the forest and the savanna species are brought into exceptionally close contact. The hybridization between *Passer domesticus* and *P. hispaniolensis* also has an ecological explanation. It occurs where the Willow Sparrow has shifted from river bottom into human settlements. Ecological isolation is in birds as in the lower animals the isolating mechanism that is most easily broken down.

Rate of Speciation

Numerous attempts have been made in recent years to fix the exact date at which a certain subspecies or species originated. Students of mammals base such calculations on an analysis of the known age of fossils. The avian record is rarely substantial enough to permit this method. The fossil Golden Eagles (*Aquila chrysaëtos*) from the Pleistocene of Rancho la Brea (California) differ from the living ones by a relatively low, flat, and broad skull with a heavy, broad beak and strong jaw. This fossil subspecies, which intergrades completely with modern Golden Eagles, is evidently ancestral to the modern form (Howard, 1947a). The exact dating of the tar pit remains has unfortunately not been possible. Presumably they are between 50 000 and 200 000 years old. The change in the California Condor (*Gymnogyps californianus*) from the Pleistocene to the recent subspecies was analyzed by Fisher (1947). There has been a very pronounced faunal change since the Pleistocene, for 33–49 per cent of the individuals among the fossils in the tar pits of Rancho la Brea belong to extinct species (Miller, 1940).

Since an adequate fossil record is available only for a few large-sized birds, calculations on the rate of speciation are usually based on indirect evidence, such as distributional data and amount of geographical variation. For instance, if a bird population is uniform in the wide area from eastern Siberia to northern Europe, the assumption is made that it has spread through this range post-glacially. This is assumed, for instance, by Voous (1947) for *Perisoreus*, *Dryocopus*, and *Dendrocopos leucotos*. Although there is much validity in such reasoning, rates of speciation and dispersal depend on many cryptic factors that are often overlooked. For instance, Zeuner (1943) tries to reconstruct the history of the *Troides* (*Ornithoptera*) section of the butterfly genus *Papilio* in the Malay Archipelago by correlating the ranges of the species and subspecies with the shifts of sea level in the wake of the various Pleistocene glaciatons. In order to simplify the picture, he has to make the assumption that evolution is equally rapid on small and on large islands, and also that there is no transoceanic dispersal. Since it is evident that neither of these assumptions is valid, the analysis loses most of its value. The non-biologist is apt to forget that species are not all alike. Each has its own particular evolutionary behavior. Not only are the inherent rates of evolution different in every species but also the physical and biological factors of the environment and their changes, degree of isolation, and size of population. The North American species that were split into separate populations in the various glacial refuges reached very different levels of speciation (Rand, 1948). Most water and shore birds reached species level (*Gavia immer* and *adamsi; Calidris semipalmata* and *maurii; Limnodromus griseus* and *scolopaceus; Nyroca affinis* and *marila;* and perhaps also *Branta bernicla* and *nigricans* and *Somateria mollissima* and *v-nigra*). A similar separation in many birds of the forest belt, such as in the genera *Parus, Canachites, Dendroica, Junco, Perisoreus, Sphyrapicus,* and *Zonotrichia,* has in most cases not led to completed speciation.

The geological data on which most of the dating is based are very insecure. Some geologists, for instance, state that the Hawaiian Islands originated in the Pliocene (Amadon, 1947). This would allow only an unbelievably short time for the evolution of the bizarre animal and plant life of these islands. As a matter of fact, the Hawaiian Islands are the tops of a chain of volcanic pinnacles rising from the ocean bottom. What the geologist can see is only the uppermost "skin" that covers up the earlier geological history. If the ocean bottom has dropped in the Hawaiian region as much as it has near Bikini (Mayr, 1951), it is quite possible that the history of these islands reaches back to the Mesozoic.

The exact dates of the recent pluvial and arid periods in the tropics are equally uncertain. Zoogeographers (Gentilli, 1949; Mayr, 1950) and geologists agree that there must have been a period of extreme drought in the recent past of Australia, but the geologists wonder whether it was as recently as 6 000 years ago or as early as 150 000 years ago. The avian data suggest that an earlier date (at least 50 000 years) is more likely.

The history of the last million years has been exceedingly complex. There were four or five glaciations with long interglacials; there were wet and dry cycles in the unglaciated areas as far south as the tropics; there were extensive shifts of sea level on all continental shelves; and in connection with this there were extensive shifts of the vegetational belts. It is not surprising therefore that students of speciation disagree occasionally on the dating of certain speciation events. Stresemann established the thesis, in a series of papers in 1919 and 1920, that the Scandinavian and Alpine ice caps had separated the eastern and western populations of many European bird species and thus caused subspeciation or speciation. As examples of subspeciation he listed the eastern and western races of *Sitta europaea, Aegithalos caudatus, Pyrrhula pyrrhula,* and *Corvus corone,* as cases of speciation the two creepers *Certhia familiaris* and *brachydactyla.* Salomonsen later extended this list to include quite a few western Palearctic species pairs. Steinbacher (1943, 1948) denies the idea of two European forest refuges and believes in a post-Pleistocene origin of the differences between west and east European subspecies. He makes the assumption that the forest belt was continuous south of the Alps and that there were no essential climatic differences between eastern and western Europe. The cited evidence for these assertions is not convincing. For instance, the fact that some loess occurs in northern France does not in the least prove that the climate in the western European forest belt near the Atlantic was the same as, let us say, in the Balkans or in Asia Minor. Also, there is no evidence for the complete continuity of the forest belt in the region south of the Alps. Stresemann's thesis of the origin of the more pronounced subspecies pairs during the Pleistocene still seems valid.

As far as the species pairs of forest birds are concerned that were listed by Salomonsen (including *Certhia familiaris* and *brachydactyla*), it seems probable that they originated in drought refuges either during interglacials or during the late Pliocene.

On the basis of the available knowledge it thus may seem very difficult to arrive at an exact timing of the speciation process. This is an over-pessimistic attitude. A few accurate timings are available, as for instance, for the age of the races of larks in the Nile delta (Moreau). It is also possible for the populations and subspecies that live on the Anamba and Natuna Islands and other small islands on the Sunda Shelf (Stresemann, 1939; Mayr and Vaurie, 1948). These populations became separated from those of the adjacent mainlands only when the Sunda Shelf was drowned at the end of the Pleistocene and are therefore less than 20 000, perhaps even less than 10 000 years old.

Claims have been made repeatedly for rapid recent changes in the genetic composition of polymorph populations of birds. The occurrence of such changes is highly probable in view of similar changes in insect populations, both free-living and in laboratories. However, for birds it is not well substantiated. It has been questioned by Mayr (1942) for *Rhipidura fuliginosa* and *Coereba flaveola* and has been refuted by Benson (1946) for *Lybius zombae*. There is some evidence for a rapid increase of the Blue Goose (*Anser coerulescens*) at the cost of the Lesser Snow Goose ("*hyberboreus*"), but it is not certain how far this has been due to a relaxation of hunting pressure.

Factors of Speciation

The factors that cause and guide speciation have been discussed in a number of recent books (Mayr, 1942; Huxley, 1942) so that it is unnecessary to discuss them in detail.

As far as the genetic factors are concerned, it is now realized that species differ by very many genes, certainly hundreds, perhaps even thousands. This has been established for animals other than birds since birds are particularly unsuitable for genetic research. The few thorough genetic examinations of birds are, however, fully consistent with this hypothesis. For instance, merely the genes that control the antigenic characters of the red blood cells of three species of *Streptopelia* are situated on no less than nine or ten chromosomes and the evidence indicates the presence of multiple characters on several of these chromosomes (see Irwin, 1949, for the latest summary of the work of Irwin and his associates). Those interested in a more complete discussion of the genetic aspects of speciation should consult Dobzhansky (1941), Bauer and Timoféeff-Ressovsky (1943), or Jepsen, Simpson, and Mayr (1949).

The deflation of the role of single mutations goes even further. The De Vriesian mutation concept became the basis of Cuénot's preadaptation principle. According to this typological concept the sudden emergence of new genetic types forces them into new niches for which they are preadapted. Modern genetics, thinking in terms of populations, make the unlikeliness of this process apparent. In fact, with a good deal of justification, some authors now stress the importance of habit formation as a forerunner of genetic change. Once established through conditioning, such a habit favors the selection of facilitating structures. One might call this a process of genetic post-adaptation. In a population of falcons that has learned to feed on bats, there will be selection pressure in favor of individuals that have the best equipment for the job. As a result of this selection pressure, a gradual genetic modification of the population will take place (Cushing,

1944). In regard to epigamic structures (crests, plumes) the movement behavior precedes the development of the structure, as Lorenz has emphasized correctly. Conditioning alone, however, can never lead to the existence of two sympatric populations (Mayr, 1947).

Of more immediate interest to the naturalist than the genetic factors of speciation are the ecological ones. Since speciation is dependent on the genetic reconstruction of isolated populations, the nature of the isolating barriers is of primary importance. Gentilli (1949) has given a summary of such barriers. The effect of water gaps in the speciation of island birds is too obvious to require further discussion (Lack, 1947a). Ecologically unsuitable terrain is the usual barrier on continents. The role of ice and correlated vegetational changes for the speciation of North American birds is stressed by Rand (1948). Droughts which restrict the vegetation to small pockets or refuges are an important agent of speciation in arid countries as Australia (Gentilli, 1949; Mayr, 1950), India (Ripley, 1949), and Africa (Moreau, 1948b).

Equally important as the extrinsic barriers themselves are the intrinsic factors, such as habitat selection and homing that re-enforce them (Mayr, 1949a). Much attention is paid in population genetics to effective population size, dispersal, in short, to all the factors that contribute to the factor m in Sewall Wright's formulae ($m=$ rate of change in the genetic composition of a population due to immigration). Miller (1947) and Haartman (1949) have shown how different these factors are in different species and even in different parts of the range of the same species. We are only at the beginning of this important branch of ornithological research which is based on the banding and careful analysis of local bird populations. Dispersal facility may be very different in closely related species. *Zosterops rendovae* respects water gaps of only one and one-half miles in width while *Z. lateralis* crossed in 1856 the 1 200-mile wide gap between Australia and New Zealand. The eleven species and superspecies of drongos (Dicruridae) can be divided into four groups according to their dispersal faculties. Only two superspecies (*adsimilis* and *hottentottus*) are freely capable of colonizing oceanic islands. It is not surprising that these two superspecies have developed the greatest number of species and subspecies in the family (Mayr and Vaurie, 1948). The total size of a species population is also of evolutionary importance. Palmgren (1942) has calculated it for some of the forest-inhabiting species in southern Finland and Miller (1947) for some insular species.

Speciation and Higher Categories

There are quite a few biologists, particularly paleontologists, who insist that the origin of higher categories is a process that differs basically from the origin of new species. That this opinion is wrong has been demonstrated conclusively in a number of recent works (Simpson, 1944; Rensch, 1947; Jepsen, Simpson, and Mayr, 1949) so that it is superfluous to go once more over the entire ground. However, the bird speciation literature has shed considerable light on the subject and some of this may be summarized here.

The first step in the origin of higher categories is the origin of new genera. If this could be explained, much would be gained toward an understanding of the origin of the still higher categories. No progress can be made in this until we know what a genus is. It is now evident that the genus cannot be defined objectively as can the species (Mayr, 1943). Most genera have been based, in the past, on "generic characters." If we study such characters as for instance the formerly accepted generic characters in the drongos (crest, shape of tail feathers), we find that they vary geographically (Mayr and Vaurie, 1948). The same is true for plumes in the birds of paradise (*Astrapia, Parotia, Diphyllodes*) and for the bill structure in the Dicaeidae (Mayr and Amadon, 1947). In all these cases, the so-called genera obviously originated by geographical variation, but it must be admitted that most of these genera are based on rather slender foundations. The so-called generic characters are in many of these cases epigamic characters of local populations of males.

During the past twenty years there has been a steady trend away from the narrow, purely morphologically defined genus. This movement has led not only toward a great simplification of nomenclature by reducing the number of genera from about 4 000 to about 1 500–1 700, it has also given the genus a new meaning, a biological one. The genus is now defined not merely as a group of related species but as a group of species with similar ecology. When a species shifts into a major new ecological niche, the stage is set for the origin of a new genus. I say major niche because even every species is, of course, in a different niche. Unfortunately it cannot be defined exactly what a major niche is, and this is the reason why genera cannot be delimited as accurately as species.

This process of the entering of new niches has been discussed by Lack (1947a) and by Miller (1949). It occurs most frequently and most rapidly where there are the greatest number of empty niches, namely on well isolated ancient islands. Examples of rapid adaptive radiation and hence the rapid origin of higher categories by geographic speciation are presented by the Geospizinae in the Galapagos Islands, by the Vangidae and Philepittidae of Madagascar, and especially by the Drepaniidae of Hawaii. A particularly instructive example in the genus *Hemignathus* has been analyzed in detail by Amadon (1947). This process of shifting into a new niche is also often evident in continental genera of birds. Miller (1949) shows how the thrashers (*Toxostoma*) evolved from mockingbird (*Mimus*)-like ancestors through a reconstruction of the bill (and other parts of the body) in correlation with a shift into a new niche, namely, a more terrestrial mode of life with digging as the major method of food getting.

The entering of new ecological niches can be followed step by step in these and many other cases and there is no need to postulate evolution by major saltations.

Open Problems

Much has been learned about the process and the causes of speciation, but much still remains to be learned. No sooner has one problem been solved than several new ones become apparent. What then are the problems that lie ahead of us? It would lead too far to give a complete inventory, but I would like to single out three groups of problems that are of immediate interest. Work on them has already started actively.

Comparative systematics

The first is the problem of comparative systematics. By this I mean the study of the differences in speciation patterns in different groups of birds, in birds of different regions, and in birds with different ecologies. Beginnings in this field have been made by Rensch, Mayr, and others, but no comprehensive analysis is as yet available. Nobody has yet studied differences in speciation patterns between predators and prey, between Passeres and non-Passeres, or compared insect-, nectar-, and seed-eating birds. In particular, the question should be studied whether or not there are inherent rates of speciation in different orders and families of birds. On the whole, one has the impression that genetic heritage is rather unimportant in the matter of subspecies and species formation as compared with such factors as population structure and dispersal facilities.

Rensch compared subspeciation in Palearctic birds and found that the number of subspecies per polytypic species was about twice as high in sedentary species as in migratory ones (Mayr, 1942). A similar analysis for North America and other parts of the world would be very desirable.

No correlation between size of species range and environmental factors has so far been worked out. However, water birds and other birds of large size usually have much larger ranges than the small song birds.

Topography is by far the most important element affecting species structure. In island regions many localized species or strikingly different subspecies occur; in continental regions most species are widespread with slight or no geographical variation. In the Solomon Islands, for instance, 52 per cent of the species are in the former category and only 24 per cent in the latter; in continental Manchuria only 3 per cent of the species are localized or with striking subspecies, while 69 per cent are widespread with slight or no geographic variation (Mayr, 1942).

Additional work in comparative systematics would permit us to state the relative importance of the various speciation factors more precisely.

Taxonomic characters

The analysis of the significance of taxonomic characters has made significant progress in recent years. Not so many years ago it was believed that only the characters of genera and higher categories had adaptive value and that the characters of species and lower categories were without selective significance. It was thought that they were merely accidents of nature. More and more evidence is now accumulating to indicate that most, if not all, species and subspecies characters have positive selective values.

The first line evidence consists of the ecological rules. As Rensch and others have shown (Mayr, 1942) many taxonomic characters, such as size, proportion, relative size of extremities and other appendages, color, and others, vary geographically in a most regular manner, parallel to the determining factors of the environment. The work of Jany (1948) on the manifestations of Gloger's Rule in North African shrikes and of Lack (1947c) on the geographical variation of clutch size are among the more recent contributions to this field.

If the change of character is very regular, it may form a geographical gradient or *cline*, as Huxley has called it. Such clines have been frequently discussed in the recent ornithological literature, as for *Pyrrhula* by Voous (1949), for several species of *Dicrurus* by Mayr and Vaurie (1948), and for the white-breasted nuthatch (*Sitta carolinensis*) by Hawbecker (1948). Contrary to expectations the southernmost populations of this nuthatch have the largest measurements. The author did not investigate whether this unexpected trend was due to competition with a smaller species or due to geographical variation of favorite food or due to some other reason.

After we eliminate all taxonomic characters on whose variation the ecological rules are based, we are still left with a large residue of characters that are not under the obvious selective control of the environment. Let us take, for example, the differences between the North American warblers of the genus *Dendroica* or the titmice of the genus *Parus*. It has been suggested that these species characters are epigamic characters and may serve as isolating mechanisms between closely related species. Is this really their entire significance?

Unsurmountable difficulties may seem to prevent a solution of this problem. How should one determine whether or not a black cap or a pectoral band have selective values? Fortunately, some very convincing evidence has turned up as byproduct of the study of polymorphism. A short survey of this field would seem appropriate in view of its new significance.

Polymorphism. Polymorphism is the occurrence in the same population of two or more distinct forms of a species in such proportions that the rarest of them cannot be maintained by recurrent mutation (after Ford).

An earlier summary of the occurrence of polymorphism in birds was presented by Stresemann (1926) and a later one by Mayr (1942). The polymorphism of many additional species has been studied in recent years as, for instance, in *Falco rusticolus* (Todd and Friedmann, 1947), *Anser coerulescens* (Manning, 1942), *Accipiter novaehollandiae* (Southern and Serventy, 1947), *Macronectes giganteus* (Serventy, 1943), *Uria aalge* (Southern and Reeve, 1942), *Stercorarius parasiticus* (Southern, 1943), *Macropygia mackinlayi* (Amadon, 1943), *Halcyon saurophaga* (Mayr, 1949c), *Cracticus quoyi* (Stresemann, 1943b), *Poëphila gouldiae* (Southern, 1946), *Oenanthe* (Vaurie, 1949b), and *Chlorophoneus* (Chapin, 1947).

The fact that there is geographical variation in the frequency of the alternative forms and that this frequency often changes in the form of a rather regular cline had long suggested that the alleles in polymorphic populations differ in selective values. On this basis, however, it was hard to understand how the balance between the alleles should be maintained at a given locality. This puzzle was solved by R. A. Fisher, who postulated a condition of balanced polymorphism due to a superiority of the heterozygotes over both classes of homozygotes. The frequency of the polymorph genes is then determined by the ratio of the viability of the homozygotes to each other and to the heterozygotes. This brilliant hypothesis has been substantiated in the meantime by experimental work (references in Mayr and Stresemann, 1950).

The selective value of the alleles should be different in different environments and this is indeed indicated by the fact that there is geographical variation of the frequency of the alleles in nearly all known cases of polymorphism. Presumably, there are also some shifts from year to year but the counts are rarely accurate enough to substantiate this possibility. In birds there is the case of the black mutant of the Rock Dove (*Columba livia*) of the Faeroe Islands which increases in years that are favorable for the species until it comprises about 14 per cent of the population. This same mutant is more vulnerable to severe conditions and decreases in frequency to about 5 per cent during a hard winter (Petersen and Williamson, 1949). Here it is clearly established that the black mutant is superior during favorable, warm periods, and the "wild type" during severe winters.

The bearing of this observation on the interpretation of subspecies and species characters is evident. Indeed, we find that in many polymorph species an allele has become fixed in parts of the range and has become *the* subspecies character as, for instance, in *Oenanthe* and *Cinclus* (Mayr and Stresemann, 1950). Such difficult taxonomic situations, as that found in *Motacilla flava*, are best interpreted on this basis, as was done correctly by Grote (1937). Although "pure" subspecies of this bird are known from many areas, it is true that exceedingly polymorph populations have been described from others. Gray-headed, yellow-headed, and black-headed birds breed together in such areas. To call each of these plumage types a different species is a simple, but biologically incorrect solution. Not only are there no differences except such of coloration, but all sorts of intermediate types have been found and there is much evidence for indiscriminate interbreeding. On the basis of the information presented by Johansen (1946) it appears that *Motacilla flava* includes several strikingly distinct subspecies groups and that instead of blending intergradation in the hybrid belts where they meet there is Mendelian segregation leading to polymorph populations.

Adaptive characters. Why a yellow-headed bird should be superior in one area and a gray-headed in another is not apparent. Still, we must never lose sight of the fact that it is not necessarily the taxonomic character that is adaptive but rather the underlying gene. This gene produces not only the visible phenotype but it also controls all sorts of invisible physiological processes which may have very pronounced selective properties. This seems to be true particularly for those pigmentation characters that follow Gloger's Rule. There is a definite correlation between the deposition of eumelanin and phaeomelanin in the feathers and humidity or temperature in the local climate (see Jany, 1948, and Frank, 1938, for special cases). Snyder and Shortt (1946) find that the Appalachian group of races of the Ruffed Grouse (*Bonasa umbellus*) are characterized by darkness and high amounts of phaeomelanin, while the Rocky Mountain group of the interior tends to paler and grayer coloration. They report a similar distribution of pigment characters in the Song Sparrow (*Melospiza melodia*), the House Wren (*Troglodytes aedon*), and the Horned Owl (*Bubo virginianus*). A systematic study of such distribution patterns of coloration is badly needed.

The phenotype may also have a direct adaptive value. The best known case is the cryptic coloration of desert birds, particularly larks. This was discussed exhaustively in the earlier literature, most recently by Hoesch and Niethammer (1940) and by Meinertzhagen (1950). In the American Horned Larks (*Eremophila alpestris*) there is in several races a fairly close correlation with soil color: reddish *rubea* lives on the red soils of Sacramento Valley, dark *merrilli* on the dark soils of northwestern California, and pallid *utahensis* on the whitish alkaline soils of the Great Salt Lake region (Behle, 1942).

Such cryptic coloration occurs not only as subspecies, but also as species character. Most desert birds have a sandy coloration and most birds of dry savannas are inconspicuously or disruptively colored. It is interesting that most of the ant-eating woodpeckers have barred upperparts, such as the flicker (*Colaptes*) and the ant-eating species in the genus *Dendrocopos* (*macei, atratus, analis*). Sitting on the ground, while feeding on ants, they are exposed to heavy predation, and this disruptive coloration is undoubtedly of considerable survival value.

There are some exceptions to this rule of the cryptic or disruptive coloration of birds of the open country which at first sight are very puzzling. For instance, many of the desert chats (*Oenanthe*) are of a conspicuous black and white color. Cott (1947) made the remarkable discovery that this conspicuousness is, in fact, a warning coloration! He found that the flesh of such conspicuously colored birds has a very disagreeable taste. Meinertzhagen (1950) confirmed that the taste of a conspicuously colored desert bustard (*Afrotis atra*) was "disgusting." Even though the final proof (avoidance by predators) is still wanting, it is highly probable that even here the taxonomic character (a color pattern) has adaptive value.

It is hardly necessary to mention at this point such conspicuously adaptive characters as the pointed wings of migratory and high altitude birds or the bill characters of the various races of the crossbill (*Loxia curvirostra*). The bill, in particular, is one of the most plastic characters in birds and a most sensitive indicator of ecological conditions. The amount of variation within a single polytypic species or superspecies is often striking enough to simulate generic difference (e.g. *Acrocephalus luscinia-syrinx* group, *Myiagra ferrocyanea, Dicrurus hottentottus*). The extreme plasticity of the bill in the Galapagos Finches was recently discussed by Lack (1947a), in the Drepaniidae by Amadon (1947), and in the flowerpeckers (Dicaeidae) by Mayr and Amadon (1947). The bill often responds with great sensitivity to temperature differences, as demanded by Allen's rule. To include a group of mountain species of *Coracina* into a single species group on account of their short bill, as was done by Voous and Van Marle (1949) (against Ripley, 1941), seems therefore ill advised. Mountain birds often have shorter bills than their lowland relatives.

Character Geography and Faunal History

The recent comprehension of the adaptive importance of taxonomic characters warns to caution when using them for the reconstruction of faunal histories. Unfortunately, they cannot be dispensed with entirely, since we ornithologists lack a fossil record almost completely. Fortunately, they can be used if interpreted cautiously. This has not always been the case in recent studies (Chapman, 1940; Johansen, 1946; Mayr and Moynihan, 1946; Voous, 1947, 1949; Voous and Van Marle, 1949). Voous (1949), for instance, criticizes Görnitz quite rightly for considering *Pyrrhula cineracea* and *murina* as "relict" relatives, merely because they are superficially similar. But then he proceeds to reconstruct the history of the *Coracina striata* group on the basis of presence or abscence of barring. There is much in the distribution pattern of this species to suggest that this is an adaptive character and that isolated barred subspecies are not necessarily more closely related to each other than to unbarred forms.

It is possible, and indeed well documented, that a character may be gained and lost repeatedly in a chain of forms. Terminal forms, for instance, may acquire secondarily an aspect of primitiveness. The most isolated forms in some polytypic species or superspecies of Polynesian birds show this very well indeed, for instance, *Lalage sharpei* (Samoa) in the *Lalage maculosa* complex and *Todirhamphus venerata* in the *Halcyon chloris* complex. In *Pachycephala pectoralis* sexual dimorphism has been lost twice, on Rennell Island (*feminina*) and on Norfolk Island (*xanthoprocta*), giving these two subspecies a pseudo-primitive aspect. This potentiality of secondary primitiveness must be taken into consideration when reconstructing faunal histories. The genus *Junco*, for instance, has all the earmarks of an immigrant from the Old World (of the originally American Emberizinae!) (Mayr, 1942b). The Chiriqui Junco (*J. vulcani*) is therefore presumably not a primarily primitive form, but owes its primitive character to a secondary loss of characters in its specialized environment.

Students who devote their attention to the reconstruction of faunal histories with the help of a taxonomic character analysis have committed some obvious blunders by not clearly realizing the dangers of this technique. To prevent other investigators from falling into the same traps as their less fortunate predecessors, it may be useful to point out some of the most frequent fallacies.

In regard to characters

(a) that characters have no selective significance and that the occurrence of similar characters in widely separated areas must indicate former continuities.

(b) that all evolution is unidirectional, starting from a primitive stage and proceeding from there irresistibly to a specialized condition.

(c) that characters are inexorably correlated with each other so that if one character in a form is primitive, its other characters must also be primitive.

In regard to distribution

(a) that birds cannot cross barriers and that discontinuous distributions indicate a former continuity of habitat.

(b) that the habitat preference is unalterably constant through space or geological time.

In regard to evolutionary change

that there is a standard rate of evolutionary change in various species and subspecies and various characters.

It so happens that all of these statements are partly true, in fact, they are often true. But, they are not invariably true, indeed, the number of exceptions for many of these statements is very large. It is actually so large as to invalidate many of the conclusions reached in some recent zoogeographical-evolutionary essays. A little more caution in such studies would seem advisable, otherwise there is danger that this entire working method will fall in disrepute.

There is one group of characters that seems to vary independently of the environment, namely, the epigamic characters. The various plume and color characters, by which the allopatric species of *Astrapia* and *Parotia* on the mountains of New Guinea differ from each other, evolved to enhance the conspicuousness of the males during their display, rather than as a response to definite requirements of the local climate. This is, no doubt, true for most of the sexually dimorphic characters displayed in courtship.

But, even these characters have been selected (for maximum stimulatory value!).

It is very doubtful, then, whether there is any "random" variation of taxonomic characters. Rensch (1947), for instance, quotes the geographical variation of the colorful tropical lory *Trichoglossus haematodus* as an instance of random variation. However, Cain (MS) has shown that there are clines and other regularities in this variation and has established correlations between prevailing colors and the environment of the various subspecies. No definite evidence for randomness could be found.

The Role of Ecological Factors

The realization that conspecific populations replace each other in space became the most productive concept in bird taxonomy. Every isolated bird species was suspected of being a disguised member of an allopatric species. The tremendous heuristic value of this method is well known. It permitted the reduction of the 19 000 species of Sharpe's *Handlist* to the 8 600 species of birds recognized at present (Mayr, 1946). However, beyond these purely practical results, it led to a new and deeper understanding of the nature of species and of the process of species formation. This period of taxonomic research was clearly dominated by the geographical approach; species were defined as complexes of geographical races and the new concept of speciation was called the theory of geographic speciation. Although this terminology is quite correct, it failed to bring out sufficiently the equally or even more important ecological aspects of species and speciation.

This neglect of the ecological factors is historically understandable since the role of the environment was almost consistently misinterpreted in the decades before 1940. Again and again attempts were made to explain the ecological differences between species as evidence for sympatric speciation, that is, for the origin of species through ecological specialization without geographical isolation. To counteract such views, it was necessary to stress the geographical aspects of speciation. When, by 1942, geographical speciation was universally accepted as the only mode of speciation in birds (Mayr, 1942) the time had come for a re-evaluation of the ecological factors.

This was done independently and simultaneously in Germany by Stresemann and in England by Lack. While Lack (1944 ff.) studied the effects of competition, Stresemann (1943 a) was primarily concerned with the evolutionary role of minor or major ecological differences among local populations. This may then be the starting point of our considerations.

Ecological characteristics of populations

Although all the members of a species or at least of a subspecies agree in their essential ecological requirements, it must never be forgotten that there are differences not only among individuals but also among local populations. Every field ornithologist is familiar with such cases. The Savanna Sparrow (*Passerculus sandwichensis*) of the eastern United States is found in coastal salt marshes and also on dry uplands in the interior. The Redwing (*Agelaius phoeniceus*) normally lives in cattail (*Typha*) swamps, but breeding colonies have been found in alfalfa fields and even in young pine plantations. The Swainson's Warbler (*Limnothlypis swainsoni*) lives in the cane brakes of the coastal marshes in the southern United States but also in the southern Appalachian highlands above 3000 feet in thickets of rhododendron, mountain laurel, hemlock, and American holly. There are no connecting populations between the two "ecotypes" (Brooks, 1942). Similar cases have been described by Peitzmeier (1943) and others and partial reviews of the literature have been given by Stresemann (1943a) and by Mayr (1949a). There is no doubt that there is a great deal of ecological variation not only on the level of individual differences but also on the population level. What interests us primarily in this connection is the role which this ecological variability plays in the process of speciation.

There has been much discussion in the literature (Stresemann, 1943a; Timoféeff-Ressovsky, 1943; Cushing, 1941, 1944; Mayr, 1947) about whether or not these ecological differences have a genetic basis. Actually this is a rather sterile discussion, since it is evident that non-genetic conditioning as well as genetic differences are involved. But, the relation between ecological variability and speciation is not yet clearly understood. A species, which is so narrow in its requirements that it can survive anywhere only in the same narrow niche, will have a very limited range and little chance to break up into several species. The same is true, at the other extreme, for a species with high dispersal faculties that is ecologically so tolerant that it will prosper in every conceivable niche, as for instance, man. A species that has rather specific requirements at any one given locality but which is sufficiently plastic to be able to shift into a new niche when colonizing new geographical regions, has the best chance for speciation.

One of the important subjects of speciation research, then, is the mode by which species colonize new geographical areas. Though this is, strictly speaking, a subject of zoogeographical or ecological research, it has significant bearing on speciation. The most important results can be summarized in these statements: (1) new areas are normally settled by young birds that are either forced out of their birthplace by population pressure or whose homing faculty (or *Ortstreue*) is not perfectly developed, (2) the optimum biotopes (habitats) in the new district are occupied first and the inferior habitats are filled only gradually and with much delay, and (3) for every successful colonization there are many that are only temporarily successful or completely unsuccessful.

Peitzmeier (1942) is particularly impressed by the ecological conservativism of most species. When the mistle thrush (*Turdus viscivorus*) spread in recent years from northern France and Belgium into Holland and the lowlands of northwestern Germany, it occupied open deciduous woods and groves of trees near human habitations, the very habitats occupied in the area of origin of these populations of mistle thrushes. On the other hand, the mistle thrush populations of the coniferous woods of the hill country of northwestern Germany were apparently unable to change their habitat preference and invade the otherwise so favorable adjacent lowlands.

There has been one great clarification in our thoughts on this subject. Preadaptation, in the sense of Cuénot, does not exist. Cuénot, who was dominated by a typological species and subspecies concept and by a De Vriesian concept of macromutations, thought that the colonization of new territories was possible only if through mutation new "types" originated that were inferior in their home range, but preadapted for the new area. These preadapted individuals would then become in this area the progenitors of a new type.

Stresemann (1943a), Cushing (1944), Thorpe (1945), Gause (1947), and other recent authors emphasize a different process. They believe that most populations are started by individuals that are able to adjust themselves (non-genetically) to new conditions when colonizing new regions. Once such a population has become conditioned to a new niche, a new set of selective factors will operate to reconstruct this population genetically. This hypothesis is sound as long as it does not include the concept that two such ecologically slightly differing populations could co-exist sympatrically without reproductive isolating mechanisms. However, in all cases of explosive range expansion such as have occurred in the Serin Finch (*Serinus serinus*) and in the Ring Dove (*Streptopelia decaocto*), there is reason to believe that this expansion was initiated by a genetic alteration of the peripheral populations.

Some recent authors have advocated a dualistic speciation process, either via geographical races or via ecological races. It cannot be stressed too emphatically that there is no such dualism. There is no ecological race that is not spatially isolated and no geographical race that does not show certain ecological differences. The apparent discrepancy is most easily resolved if we think of these races as populations. Each of these populations has its own area of distribution—no matter how small—and is adapted to a specific niche within this area. It is the only population in this area and is therefore both geographical and ecological. The difficulties usually arise from the more or less arbitrary combining of such populations into the subspecies of the taxonomist.

An excellent study by Marshall (1948) of the ecological races of the Song Sparrow (*Melospiza melodia*) in the San Francisco Bay region of California illustrates the interdependence of geographical and ecological factors in an almost diagrammatic manner. Three subspecies occur here in salt water or brackish marshes and a fourth one near fresh water in the uplands. The upland subspecies is separated in most areas from the salt marsh subspecies by an ecologically unsuitable dry plain. Wherever this isolation exists there is a sharply defined taxonomic difference between upland and marsh birds while the marsh populations intergrade with each clinally. Where isolation is absent, the morphological difference between populations of different habitats is slight or absent: "Without this isolation the Tomales Bay salt marsh birds are identical in morphological attributes with the birds in fresh-water habitats surrounding them. Wherever it is lacking between bay marsh and upland populations as at Richardson Bay, Corte Madera and San Pablo, the respective populations become practically indistinguishable" (Marshall, l.c.). In a few places the marsh and upland birds have come into secondary contact and the area and amount of intergradation is governed entirely by the configuration of the habitat. At San Francisquito Creek, where two contrasting habitat types meet abruptly, there is little mixing of the neighboring subspecies, apparently due to habitat selection rather than to non-random mating. These ecological races of the Song Sparrow owe their origin and maintenance to a large degree to geographical isolation. They are not only ecological races, they are also geographical races.

Ecological characteristics of subspecies and species

The differences in ecological requirements among individuals and local populations are usually believed to be of minor degree within the "typical" ecology of the species. This is indeed normally the case. However, in an unexpectedly great number of species there is striking geographical variability in ecological characters. Such instances have been listed by Brooks (1942), Stresemann (1943a), and Mayr (1942, 1949a). The Island Thrush (*Turdus poliocephalus*), for instance, feeds on Rennell Island near the seashore among coral boulders, while nearby on Kulambangra and Bougainville in the Solomon Islands it lives in the upper mountain forest at 2 000–3 000 meters. The Red Crossbill (*Loxia curvirostra*) feeds on larch in the Himalayas, on spruce in the Alps, and on pine in other restricted areas. The European cormorant (*Phalacrocorax carbo*) nests in inland areas of Europe and Asia (subspecies *sinensis*) on trees, while the Atlantic race (*carbo*) nests on the coast on rocks. The nominate subspecies of the Wilson's Warbler (*Wilsonia pusilla*) is found in Canadian bogs, while another race (*chryseola*) is found in warmer and drier habitats in the chaparral country of the west. These examples of striking changes in ecological characteristics could be matched by hundreds of additional ones, affecting food, habitat, nest site, altitudinal range, and others. Each of these changes involves populations within a single species.

The geographical variability of ecological preferences has been pointed out by many authors. In the Canada Jay (*Perisoreus canadensis*) two subspecies (*griseus* and *bicolor*) come into very abrupt contact apparently on account of ecological differences (Aldrich, 1943). Similar cases were described by Dementieff (1938) for Palearctic birds, although he apparently was in part deceived by allopatry of good species (see below). *Zonotrichia capensis* with its twenty-two subspecies lives under a remarkable range of environmental conditions, which varies from the hot and arid islands Curaçao and Aruba, the steamy lowlands of French Guiana and lower Amazonia to the mountains of Haiti, Costa Rica, and the Andes south to Tierra del Fuego. Many striking instances of geographical variation of ecological characters in the species *Dicrurus adsimilis, caerulescens, paradiseus, leucophaeus*, and *hottentottus* have been described by Vaurie (1949a). Many species of Palearctic birds that occur both in Britain and in the Canary Islands (Teneriffe) differ ecologically. This difference often consists in an apparently broader tolerance of the Teneriffe birds, which are able to occupy niches that are otherwise unoccupied in these islands with their much impoverished avifauna (Lack and Southern, 1949). The geographical variability of ecological requirements in the *Rhipidura rufifrons* group is remarkable (Mayr and Moynihan, 1946).

It is easy to see how such ecological diversification could permit populations of a species to re-enter the geographical range of other populations of the same species or species group without coming into actual contact. Johansen (1947) shows this for the *Phylloscopus trochilus* group which twice invaded the Caucasus. The western *abietinus*, which invaded the area from the north, is restricted to the forest of the lower slopes, the eastern *lorenzii* of the *neglectus* group, which invaded the Caucasus from the southeast is restricted to the subalpine forest zone. In the *Halcyon chloris* group there is an overlap on Palau between *chloris teraokai* which lives in the open country (near human habitations, etc.) and *cinnamomina pelewensis*, which is restricted to the true forest. In all these cases the original ecological differences had been acquired in geographical isolation, although no doubt they were further perfected during the period of contact.

These examples demonstrate the danger of reconstructing with too much confidence the past history of species on the assumption that a species is irrevocably tied to a single habitat. Animal species are not necessarily fixed components of biotic associations to the extent that the adherents of the biome concept want us to believe. This is abundantly proven by the striking variability of habitat preferences in many species of birds. And, if different subspecies in space can differ in their ecology, why not also subspecies in time?

In fact, there is good evidence that subspecies may change their ecology in time. There are numerous instances in birds where the range of a species was split by extrinsic factors into a western and eastern portion or into a northern and southern one. If there is a climatic difference in two sections, a gradual transformation of the isolated populations will occur with respect to their ecological tolerance and requirements. For instance, the species, which was the common ancestor of Nightingale (*Luscinia megarhynchos*) and Sprosser (*L. luscinia*) became separated into a southwestern (Mediterranean) and a northeastern (Russian–west Siberian) population. The southwestern birds became adapted to a dry, open, and hot kind of habitat, the northeastern birds to a more moist, more heavily forested and cooler habitat (Stresemann, 1948). The alteration occurred as a gradual change of the entire population. When subsequently the two populations expanded their ranges and came again into physical contact, they had not only acquired reproductive isolation but also ecological compatibility. In the zone of overlap *L. megarhynchos* occurs in open, dry locations, like gardens, parks, and cemeteries, while *L. luscinia* is restricted to lake shores, river bottoms, and swampy woods (Stresemann, l.c.).

Numerous similar examples have been recorded in the literature. The *Geokichla* of the Lesser Sunda Islands became separated into a western and an eastern group. The western group on the humid islands of Lombok, Sumbawa, and Flores became the humidity-loving forest species *G. dohertyi*, while the eastern group became the aridity-tolerant species *peronii* with headquarters on Timor. When rather recently *dohertyi* spread to Timor, it settled in the humid mountain forests where it is out of competition with *peronii*, inhabitant of the open monsoon forests of the lowlands.

Finally, one more case. During a late Pleistocene or post-Pleistocene arid period most of the bird life of Australia (except some desert species) was confined to a number of drought refuges along the coast (Gentilli, 1949; Mayr, 1950). In the tree runner *Neositta chrysoptera*, the two populations that were isolated in the west (southwestern Australia and Port Darwin region) were exposed to more severe aridity than the populations along the east coast of Australia. As a result, these two populations became much more drought tolerant than the eastern populations, and when conditions became again more humid the populations from these two western refuges were able to spread over most of Australia, while the eastern populations expanded their ranges only slightly. Here again, some of the populations changed materially during their isolation from others.

Competition and Speciation

While Stresemann's studies had as their point of departure the minor or major ecological differences that are found among local populations, Lack in his studies started out from the ecological differences between species. He showed that these are not merely a negligible byproduct of speciation but rather that they are an indispensable prerequisite. Two species normally cannot co-exist in the same area unless they have sufficiently diverged in their ecological requirements so as not to be severe competitors (Lack, 1944, 1947a, 1949, etc.).

A consequence of this consideration is that ecological competition may prevent two perfectly good species from invading each other's ranges. The resulting geographical relationship of strict allopatry has often been interpreted as proving conspecificity. Stresemann (1939) cites many instances where complete or virtual allopatry does not prove conspecificity. In certain cases the presence of one species actually seems to keep out (or even drive out) another species from the area. For instance, *Ducula rosacea* seems to prevent *D. bicolor* from colonizing the small islets in the Lesser Sunda Islands, and *Tanygnathus megalorhynchus* seems to have crowded *T. sumatranus* off the small islands between Celebes and Sangir. In the African parrot genus *Agapornis* all the species are completely or almost completely allopatric (Moreau, 1948b). The realization that allopatry does not necessarily prove conspecificity has forced the taxonomist to re-examine all the more heterogeneous-appearing polytypic species. The result is that it was found, indeed, that some of them were composed of pseudo-conspecific allopatric species. Of the many cases of this sort I shall list here only a few that have recently come to my attention (Appendix). These cases are characterized by contiguity of populations without signs of intergradation, and often even a slight geographical overlap in the zone of contact.

Taxonomic consequences

From the point of view of evolution the most important aspect of the ecological differences between local populations, subspecies, and species is that they set up selection pressures that lead to evolutionary divergence. This, in turn, results in an evolutionary divergence of taxonomic characters as discussed in the previous section.

Here we shall discuss the taxonomic effects of only one factor, namely competition. Lack (1947a) showed how competition determined which ecological niche the species *Geospiza magnirostris, difficilis, fuliginosa*, and *conirostris* occupy on a given island and how this, in turn, affects the size of the bill. The same phenomenon was shown by Amadon (1947) for two species pairs on Kauai Island. Of two species of the thrush *Phaeornis* that were thrown into competition one became a berry eater, the other an insect eater, and corresponding changes occurred in bill shape. Likewise, in the genus *Loxops* where the species *virens* comes into competition with the small-billed *parva* it has developed a subspecies with a noticeably heavier bill and different feeding habits. The two rock nuthatches (*Sitta tephronota* and *neumayer*) are very much alike in size outside their area of overlap, but one is much larger and the other much smaller in Iran where the two species are sympatric (Vaurie, 1950b).

In this manner ecological competition is one of the most potent factors in the modification of old and in the evolution of new taxonomic characters.

It is now evident that the study of the role of ecological factors has become one of the most important objects of taxonomy. Bird systematics is no longer merely the study of specimens in the museum, but rather, as the new systematics, it has shifted its main attention to the biological aspects of the taxonomic categories and it has expanded to include the study of animal populations in nature, of their evolution and relationships.

The Significance of Ecological Factors for the Taxonomist

It is now realized that ecological characters are among the most important systematic characters of animals, in fact, that even morphological characters are usually merely indicators of genetic differences correlated with specific ecological conditions. The findings on ecological factors can then be summarized as follows:

(1) Every local population is slightly different from every other one in its ecological tolerance and requirements, just as every locality is slightly different from every other one.

(2) Within each population there is individual variability in many characters including ecological tolerance.

(3) Ecological characters appear to have largely a genetic basis, but there is also a considerable amount of non-genetic variability which is subject to conditioning.

(4) Species are rarely uniform ecologically. Subspecies often differ strikingly in ecological characteristics. Sometimes even morphologically monotypic species reveal a certain amount of geographical variablilty of ecological characters.

(5) Species must differ in their ecological requirements in order to be able to co-exist. The amount of difference that existed at the time of first overlap is augmented through natural selection.

(6) Competition affects the expression of morphological characters causing divergence in the zone of overlap of closely related species.

Conclusion

Work on the geographical races of birds has led to the establishment of polytypic species and has helped to solve the riddle of the process of speciation. Emphasis is now shifting to the ecological factors that are active during this process of speciation and to the effect they have on taxonomic characters. These seem to be the most promising fields of future research in bird taxonomy, together with a study of the historical factors that have produced present distribution.

Discussion:

J. S. Huxley: I would like first of all to express my admiration for Dr Mayr's important and interesting paper, and my regret that he was unable to attend this Congress in person.

I would like to make a few comments.

(1) He stated that polymorphism always had a genetic basis, in the greater viability of a heterozygote, as suggested by Mr Fischer. This, however, is not the case. In the guillemot (*Uria aalge* var. *ringvia*) Prof. Fischer himself in a recent discussion expressed the opinion that the two forms

concerned were kept in balance by climatic factors, *ringvia* having an advantage in colder and moister climates. In other cases there may be an ecological balance of a different sort as with polymorphism in mimetic butterflies or in the balance between the greater hardiness of dominant melanics in moths, and the greater cryptic advantage of the normal form in non-industrial surroundings. Lloyd Morgan christened it "organic selection". This is not a very abbreviate phrase, but some brief term is required for the phenomenon, and I hope that evolutionists will find one.

(2) Dr Mayr referred to the importance of modifications of habit as often paving the way for genetic evolutionary change. I would hope that full credit will be given to Mr Baldwin and, C. Lloyd Morgan, who first drew attention to this phenomenon.[1]

(3) Dr Mayr stressed the adaptive nature of almost all characters, and the ever-present moulding force of natural selection. I would like to remind ornithologists of a beautiful set of instances, best illustrated in birds, where we can make an estimate of the selective forces at work, and then compare the results.

In truly polygamous birds, characters promoting successful mating by males ensure a multiple reproductive advantage: the sucessful male may have offspring by several females, while less successful males mate with only one or two or even none. Characters promoting success in holding territory in monogamous territorial species confer a unitary reproductive advantage, all successful males securing one mate, all unsuccessful no mate. Finally, purely epigamic (display) characters in such species confer a fractional reproductive advantage, the better equipped males having on the average a larger progeny. These differences in the strength of selection are clearly reflected in the degree of development of the characters concerned, exaggerated or so-called dystelic display characters only being seen with multiple advantage, while with fractional advantage they are never more than slightly developed. The distinctive characters advertising territory and conferring unitary advantage, whether they be visual or auditory, are always conspicuous but never exaggerated.

[1] The literature cited in this report covers the period since the 1938 Congress. — E. M.

Appendix

Pseudo-conspecific pairs of allopatric species

1. *Psittacula cyanocephala*
   "   *rosea*
   Biswas, 1951, Amer. Mus. Novitates, no. 1500: 1–12.

2. *Anthracoceros malabaricus*
   "   *coronatus*
   Delacour, 1946, Zoologica, 31: 2.

3. *Merops superciliosus*
   "   *philippinus*
   Marien, 1950, Jour. Bombay Nat. Hist. Soc., 49 (2): 151–164.

4. *Dendrocopos hardwickii*
   "   *canicapillus*
   Greenway, 1943, Auk, 60: 564.

5. *Ammomanes cinctura*
   "   *phoenicura*
   Vaurie, 1951, Bull. Amer. Mus. Nat. Hist., 97: 431–526.

6. *Melanocorypha calandra*
   "   *bimaculata*
   Vaurie, 1951, Bull. Amer. Mus. Nat. Hist., 97: 431–526.

7. *Pycnonotus aurigaster*
   "   *cafer*
   Deignan, 1949, Jour. Washington Acad. Sci., 39: 273.

8. *Coracina melaschista*
   "   *polioptera*
   Deignan, MS.

9. *Pericrocotus cantonensis*
   "   *roseus*
   Deignan, MS.

10. *Oenanthe hispanica*
    "   *pleschanka*
    Vaurie, 1949, Amer. Mus. Novitates, no. 1425: 14–24.

11. *Acrocephalus arundinaceus*
    "   *stentoreus*
    Stresemann, E., and J. Arnold, 1949, Jour. Bombay Nat. Hist. Soc., 48: 428–443.

12. *Turdoides caudatus*
    "   *altirostris*
    Vaurie, MS.

13. *Heterophasia capistrata*
    "   *melanoleuca*
    Smythies, 1949, Ibis, 91: 637.

14. *Lanius schach*
    "   *tephronotus*
    Biswas, 1950, Jour. Bombay Nat. Hist. Soc., 49: 444–455.

15. *Rhodopechys mongolica*
    "   *githaginea*
    Vaurie, 1949, Amer. Mus. Novitates, no. 1424: 30–35.

16. *Lonchura leucogaster*
    "   *leucogastroides*
    Delacour, 1943, Zoologica, 28: 82.

17. *Acridotheres cristatellus*
    "   *fuscus*
    Marien, 1951, Jour. Bombay Nat. Hist. Soc., 49: 482–485.

18. *Oriolus chinensis*
    "   *tenuirostris*
    Delacour, 1951, L'OOiseau (in press).

References

ALDRICH, J. W., 1943. Relationships of the Canada Jays in the northwest. Wilson Bull., 55: 217–222.
— 1946 a. Significance of racial variation in birds to wildlife management. Jour. Wildlife Management, 10: 86–93.
— 1946 b. Speciation in the white-cheeked geese. Wilson Bull., 58: 94–103.
AMADON, D., 1942. Notes on non-passerine genera. 2. Amer. Mus. Novitates, no. 1176: 1–12.
— 1943. Notes on non-passerine genera. 3. Amer. Mus. Novitates, no. 1237: 1–22.
— 1947. Ecology and the evolution of some Hawaiian birds. Evolution, 1: 63–68.
— 1949. The seventy-five per cent rule for subspecies. Condor, 51: 250–258.
— 1950. The Hawaiian honeycreepers (Aves, Drepaniidæ). Bull. Amer. Mus. Nat. Hist. 95: 157–262.
BAILEY, A. M., 1948. Birds of Arctic Alaska. Colorado Mus. Popular Ser., no. 8: 151–155.
BAUER, H. and N. W. TIMOFÉEFF-RESSOVSKY, 1943. Genetik und Evolutionsforschung bei Tieren. In: Heberer, G., Die Evolution der Organismen, 335–429.
BEHLE, W. H., 1942. Distribution and variation of the Horned Larks (*Otocoris alpestris*) of western North America. Univ. California Publ. Zool., 46: 201–316.
BENSON, C. W., 1946. On the change of coloration in *Lybius zombae* (Shelley). Bull. Brit. Ornith. Club, 67: 33–35.
— 1948. Geographical voice-variation in African birds. Ibis, 90: 48–71.
BLANCHARD, B. D., 1941. The white-crowned sparrows (*Zonotrichia leucophrys*) of the Pacific seaboard: environment and annual cycle. Univ. California Publ. Zool., 46: 1–178.
BOND, R. M., 1943. Variation in western sparrow hawks. Condor, 45: 168–185.
BROOKS, M., 1942. Birds at the extremities of their ranges. Wilson Bull., 54: 12–16.
BULLOUGH, W. S., 1942. On the external morphology of the British and continental races of starling (*Sturnus vulgaris* Linnaeus). Ibis, 84: 225–239.
CAIN, A. J., 1952. Variation, selection, and speciation in the genus *Trichoglossus* (Aves, Psittaciformes).
CHAPIN, J. P., 1947. Color variation in shrikes of the genus *Chlorophoneus*. Auk, 64: 53–64.
— 1948. Variation and hybridization among the Paradise Flycatchers of Africa. Evolution, 2: 111–126.
— 1949. Relationship and voice in the genus *Calamocichla*. Ornith. Biol. Wiss., 7–16.
CHAPMAN, F. M., 1940. The post-glacial history of *Zonotrichia capensis*. Bull. American Mus. Nat. Hist., 77: 381–438.
COTT, H. B., 1947. The edibility of birds. Proc. Zool. Soc. London, 116: 371–524.
CUSHING, J. E., Jr., 1941. Non-genetic mating preference as a factor in evolution. Condor, 43: 233–236.
— 1944. The relation of non-heritable food habits to evolution. Condor, 46: 265–271.
DELACOUR, J. and E. MAYR, 1945. The family Anatidae. Wilson Bull., 57: 1–55.
DOBZHANSKY, TH., 1941. Genetics and the origin of species. Columbia Univ. Press, New York, 446 pp.
DUNAJEWSKI, A., 1939. Gliederung und Verbreitung des Formenkreises *Lanius schach* L. Jour. Ornith., 87: 28–53.
FISHER, H. I., 1947. The skeleton of recent and fossil Gymnogyps. Pacific Science, 1: 227–236.
GAUSE, G. F., 1947. Problems of evolution. Trans. Conn. Acad. Arts, Sci., 37: 17–68.
GENTILLI, J., 1949. Foundations of Australian bird geography. Emu, 49: 85–129.

HAARTMAN, L. v., 1949. Der Trauerfliegenschnäpper I. Ortstreue und Rassenbildung. Act. Zool. Fenn., 56: 1–104.
HAWBECKER, A. C., 1948. Analysis of variation in western races of the white-breasted nuthatch. Condor, 50: 26–39.
HINDWOOD, K. A. and E. MAYR, 1946. A revision of the stripe-crowned pardalotes. Emu, 46: 49–67.
HOESCH, W. and G. NIETHAMMER, 1940. Die Vogelwelt Deutsch-Südwestafrikas. Jour. Ornith., Sonderh. 75–84.
HOWARD, H., 1947 a. An ancestral Golden Eagle raises a question in taxonomy. Auk, 64: 287–291.
—— 1947 b. A preliminary survey of trends in avian evolution from Pleistocene to recent time. Condor, 49: 10–13.
HUXLEY, J., 1942. Evolution. The modern synthesis. George Allen and Unwin Ltd., London, 645 pp.
IRWIN, M. R., 1949. On immunogenetic relationships between the antigenic characters specific to two species of Columbidae. Genetics, 34: 586–606.
JANY, E., 1948. L'Influence de l'humidité du climat sur la coloration du plumage chez les pies-grieches grises de l'Afrique du Nord (*Lanius excubitor* L.). L'Oiseau, 18: 117–132.
JEPSEN, G. L., G. G. SIMPSON, and E. MAYR, 1949. Genetics, paleontology, and evolution. Princeton University Press, 474 pp.
JOHANSEN, H., 1945. Races of bean-geese. Dansk Ornith. Foren. Tidsskr., 39: 106–127.
—— 1946. Notes on systematics and distribution of the Yellow Wagtails (*Motacilla flava* L.). Dansk Ornith. Foren. Tidsskr., 40: 121–142.
—— 1947. Notes on the geographical variation of the chiffchaff (*Phylloscopus collybita* [Vieill.]). Dansk Ornith. Foren. Tidsskr., 41: 212–215.
KULLENBERG, B., 1947. Über Verbreitung und Wanderungen von 4 *Sterna* Arten. Ark. Zool., 38 A, 17: 1–80.
LACK, D., 1944 a. Ecological aspects of species formation. Ibis, 86: 260–286.
—— 1944 b. Correlation between beak and food in the crossbill. Ibis, 86: 552–553.
—— 1945. The Galapagos Finches (Geospizinae). A study in variation. Occ. Pap. California Acad. Sci., 21: 1–158.
—— 1946. The taxonomy of the robin *Erithacus rubecula* (Linn.). Bull. Brit. Ornith. Club, 66: 55–65.
—— 1947 a. Darwin's finches. Cambridge Univ. Press, 208 pp.
—— 1947 b. A further note on the taxonomy of the robin *Erithacus rubecula* (Linnaeus). Bull. Brit. Orn. Club, 67: 51–54.
—— 1947 c. The significance of clutch size. Ibis, 89: 302–352.
—— 1949. The significance of ecological isolation. In: Genetics, paleontology, and evolution by G. L. Jepsen, G. G. Simpson, and E. Mayr, 1949, Princeton University Press, 299–308.
LACK, D. and H. N. SOUTHERN, 1949. Birds on Tenerife. Ibis, 91: 607–626.
LEOPOLD, A. S., 1944. The nature of heritable wildness in turkeys. Condor, 46: 133–197.
LØPPENTHIN, B., 1943. Systematic and biologic notes on the Long-tailed Skua, *Stercorarius longicaudus* Vieill. Medd. Grønland, 131 (12): 1–26.
MANNING, T. H., 1942. Blue and Lesser Snow Geese on Southampton and Baffin Islands. Auk, 59: 158–175.
MARSHALL, J. T., Jr., 1948. Ecological races of song sparrows in the San Francisco Region. Condor, 50: 193–215, 233–256.
MAYAUD, N., 1940. Considérations sur les affinités et la systématique de *Larus fuscus* et *Larus argentatus*. Alauda, 12: 89–98.

MAYR, E., 1941. Die geographische Variation der Färbungstypen von *Microscelis leucocephalus*. Jour. f. Ornith., 89: 377–392.
—— 1942 a. Systematics and the origin of species. Columbia University Press, New York, 334 pp.
—— 1942 b. Speciation in the junco. Review of speciation in the avian genus Junco by Alden H. Miller. Ecology, 23 (3): 378–379.
—— 1943. Criteria of subspecies, species, and genera in ornithology. Ann. New York Acad. Sci., 44: 133–140.
—— 1944. The birds of Timor and Sumba. Bull. Amer. Mus. Nat. Hist., 83: 123–194.
—— 1946. The number of species of birds. Auk, 63: 64–69.
—— 1947. Ecological factors in speciation. Evolution, 1 (4): 263–288.
—— 1948 a. The bearing of the new systematics on genetical problems. The nature of species. Advances in Genetics, 2: 205–237.
—— 1948 b. Geographic variation in the Reed-Warbler. Emu, 47: 205–210.
—— 1948 c. Climatic races in plants and animals. Evolution, 2 (4): 375–376.
—— 1949 a. Speciation and systematics. Genetics, Paleontology, and Evolution, Princeton University Press, 281–298.
—— 1949 b. Geographical variation in *Accipiter trivirgatus*. Amer. Mus. Novitates, no. 1415: 1–12.
—— 1949 c. Artbildung und Variation in der *Halcyon-chloris*-Gruppe. Ornith. Biol. Wiss., 54–60.
—— 1949 d. Speciation and selection. Proc. Amer. Phil. Soc., 93 (6): 514–519.
—— 1950. Notes on the genus *Neositta*. Emu, 49 (4): 282–291.
—— 1951. Report of the chairman of the Standing Committee on Terrestrial Faunas of the Inner Pacific. Proc. Pacific Sci. Cong. (in press).
MAYR, E., and D. AMADON, 1947. A review of the Dicaeidae. Amer. Mus. Novitates, no. 1360: 1–32.
MAYR, E., and M. MOYNIHAN, 1946. Evolution in the *Rhipidura rufifrons* group. Amer. Mus. Novitates, no. 1321: 1–21.
MAYR, E., and C. VAURIE, 1948. Evolution in the family Dicruridae. Evolution, 2: 238–265.
MAYR, E., and E. STRESEMANN, 1950. Polymorphism in the chat genus *Oenanthe* (Aves). Evolution 4 (4): 291–300.
MEINERTZHAGEN, R., 1950. Some problems connected with Arabian birds. Ibis, 92: 336–340.
MILLER, A. H., 1940. Climatic conditions of the Pleistocene reflected by the ecologic requirements of fossil birds. Proc. Sixth Pac. Sci. Congress Berkeley, 2: 807–810.
—— 1941. Speciation in the avian genus Junco. Univ. California Publ. Zool., 44: 173–434.
—— 1947. Panmixia and population size with reference to birds. Evolution, 1: 186–190.
—— 1949. Some ecologic and morphologic considerations in the evolution of higher taxonomic categories. Ornith. Biol. Wiss., 84–88.
MOREAU, R. E., 1948 a. Some recent terms and tendencies in bird taxonomy. Ibis, 90: 102–111.
—— 1948 b. Aspects of evolution in the parrot genus *Agapornis*. Ibis, 90: 206–239.
PALMGREN, P., 1942. Die Populationsgrösse der Vögel als Evolutionsfaktor. Naturwissenschaften, 30: 217–220.
PALUDAN, K., 1940. Contributions to the ornithology of Iran. Danish Sci. Investig. Iran, pt. 2: 11–54.
PATEFF, P., 1947. On the systematic position of the starlings inhabiting Bulgaria and the neighbouring countries. Ibis, 89: 494–507.

PEITZMEIER, J., 1942. Die Bedeutung der oekologischen Beharrungstendenz für faunistische Untersuchungen. Jour. f. Ornith., 90: 311–322.

PETERSEN, N., and K. WILLIAMSON, 1949. Polymorphism and breeding of Rock Dove in the Faeroe Islands. Ibis, 91: 17–23.

PITELKA, F. A., 1945 a. Differentiation of the Scrub Jay, *Aphelocoma coerulescens*, in the Great Basin and Arizona. Condor, 47: 23–26.

—— 1945 b. Pterylography, molt, and age determination of American jays of the genus *Aphelocoma*. Condor, 47: 229–260.

RAND, A. L., 1943. *Larus kumlieni* and its allies. Canadian Field Nat., 56: 123–126.

—— 1948. Glaciation, an isolating factor in speciation. Evolution, 2: 314–321.

RAND, A. L., and M. A. TRAYLOR, 1950. The amount of overlap allowable for subspecies. Auk, 67: 169–183.

RENSCH, B., 1943. Die biologischen Beweismittel der Abstammungslehre. In: Heberer, G., Die Evolution der Organismen, 57–85.

—— 1947. Neuere Probleme der Abstammungslehre. Die Transspezifische Evolution, Stuttgart, 407 pp.

RIPLEY, S. D., 1941. Notes on the genus *Coracina*. Auk, 58: 381–395.

—— 1944. The bird fauna of West Sumatra islands. Bull. Mus. Comp. Zoöl., 94: 305–430.

—— 1945. Suggested terms for the interpretation of speciation phenomena. Jour. Washington Acad. Sci., 35: 337–341.

—— 1949. Avian relicts and double invasions in peninsular India and Ceylon. Evolution, 3: 150–159.

RIPLEY, S. D., and H. BIRCKHEAD, 1942. Birds collected during the Whitney South Sea Expedition 51. On the fruit pigeons of the *Ptilinopus purpuratus* group. Amer. Mus. Novitates, no. 1192: 1–14.

SALOMONSEN, F., 1944. The Atlantic Alcidae. Göteborgs Vetensk. Samh. Handl. (6) 3 B 5: 1–138.

SERVENTY, D. L., 1943. The white phase of the Giant Petrel in Australia. Emu, 42: 167–169.

SIMPSON, G. G., 1944. Tempo and mode in evolution. Columbia Univ. Press, New York, 237 pp.

SNYDER, L. L., and T. M. SHORTT, 1946. Variation in *Bonasa umbellus*, with particular reference to the species in Canada east of the Rockies. Canadian Jour. Res., D 24: 118–133.

SOUTHERN, H. N., 1943. The two phases of *Stercorarius parasiticus* (Linnaeus). Ibis, 85: 443–485.

—— 1944. Dimorphism in *Stercorarius pomarinus* (Temminck). Ibis, 86: 1–16.

—— 1946. The inheritance of head colour in the Gouldian Finch (*Poephila gouldiae* Gould). Avicult. Mag. London, 52: 126–131.

SOUTHERN, H. N., and E. C. R. REEVE, 1942. Quantitative studies on the geographical variation of birds — The Common Guillemot (*Uria aalge* Pont.). Proc. Zool. Soc. London 111 A: 255–276.

SOUTHERN, H. N., and D. L. SERVENTY, 1947. The two phases of *Astur novae-hollandiae* (Gm.) in Australia. Emu, 46: 331–347.

STEINBACHER, G., 1948. Der Einfluss der Eiszeit auf die europäische Vogelwelt. Biologisches Zentralblatt, 67: 444–456.

STEINBACHER, G., and J. STEINBACHER, 1943. Über die Entstehung und das Alter von Vogelrassen. Zool. Anz., 141: 141.

STRESEMANN, E., 1939. Die Vögel von Celebes. Teil I. Jour. f. Ornith., 88: 1–135; 389–487.

—— 1940. Zur Kenntnis der Wespenbussarde. Arch. Naturg., NF. 9: 137–193.

—— 1943 a. Oekologische Sippen-, Rassen- und Artunterschiede bei Vögeln. Jour. f. Ornith., 91: 305–324.

—— 1943 b. *Cracticus rufescens* De Vis. Ornith. Monatsb., 51: 68–72.

STRESEMANN, E., 1948. Nachtigall und Sprosser: ihre Verbreitung und Ökologie. Ornith. Berichte, 1: 193–222.

STRESEMANN, E., and N. W. TIMOFÉEFF-RESSOVSKY, 1947. Artenstehung in geographischen Formenkreisen I. Der Formenkreis *Larus argentatus-cachinnans-fuscus*. Biol. Zentralbl., 66: 57–76.

STRESEMANN, E., and J. ARNOLD, 1949. Speciation in the group of Great Reed-Warblers. Jour. Bombay Nat. Hist. Soc., 48 (3): 428–443.

SUTTER, E., 1946. Die Flügellänge junger und mehrjähriger Grünfinken und Gartenrötel. Ornith. Beob., 43: 72–85.

SUTTON, G. M., 1938. Oddly plumaged orioles from western Oklahoma. Auk, 55: 1–6.

THOMSON, A. L. et al., 1943. Physiological races in birds. Bull. Brit. Ornith. Club, 63: 73–80.

THORPE, W. H., 1945. The evolutionary significance of habitat selection. Jour. Animal Ecol., 14: 67–70.

TIMOFÉEFF-RESSOVSKY, N. W., 1943. [Discussion of Stresemann 1943 a.] Jour. f. Ornith., 91: 326–327.

TODD, W. E. D., and H. FRIEDMANN, 1947. A study of the gyrfalcons with particular reference to North America. Wilson Bull., 59: 139–150.

VAURIE, C., 1949 a. A revision of the bird family Dicruridae. Bull. Amer. Mus. Nat. Hist., 92: 205–342.

—— 1949 b. Notes on the bird genus *Oenanthe* in Persia, Afghanistan, and India. Amer. Mus. Novitates, no. 1425: 1–47.

—— 1950 a. Notes on some Asiatic titmice. Amer. Mus. Novitates, no. 1459: 1–66.

—— 1950 b. Notes on Asiatic nuthatches and creepers. Amer. Mus. Novitates, no. 1472: 1–39.

VOOUS, K. H., 1947. The history of the distribution of the genus *Dendrocopos*. Amsterdam, 142 pp.

—— 1949. Distributional history of Eurasian bullfinches, genus *Pyrrhula*. Condor, 51: 52–81.

VOOUS, K. H., and J. G. VAN MARLE, 1949. The distributional history of *Coracina* in the Indo-Australian Archipelago. Bijdragen tot de Dierk., 28: 513–529.

WHITE, C. M. N., 1948. Size as a racial character. Bull. Brit. Ornith. Club, 69: 11–13.

ZEUNER, F. E., 1943. Studies in the systematics of Troides Hübner (Lepidoptera Papilionidae) and its allies, distribution, and phylogeny in relation to the geological history of the Australian Archipelago. Trans. Zool. Soc. London, 25: 107–184.

SOME EFFECTS OF EXTERNAL CONDITIONS UPON THE WHITE MOUSE

BY

FRANCIS B. SUMNER

WITH FOURTEEN FIGURES

The production of definite modifications in the structure, color or size of animals or plants by artificial changes in the conditions of life has been successfully accomplished over and over again by a large number of investigators. I need only allude to such classical instances as the experiments of Dorfmeister, Weismann and others on butterflies, Schmankewitsch on Artemia, Cunningham on flounders, Naegeli and Bonnier on Alpine plants, or to the more recent work of Tower[1] upon beetles and Beebe[2] upon birds. The fact that such considerable modifications may be produced within the lifetime of an individual by physical or chemical means is in itself interesting if regarded simply as an illustration of the plasticity displayed by many organisms. But when we push our inquiry beyond the merely descriptive plane, we are brought face to face with some of the most fundamental problems of biology. Are these modifications adaptive in their character? Do they in any instance correspond to the features which distinguish one natural species or geographical variety from one another? And finally, do such artificially produced modifications reappear in offspring which have not themselves been subjected to the conditions of the experiment? These are some of the questions which demand an answer from the investigator. They are not so simple and easy of solution as may appear on the surface: the first

---

[1] An Investigation of Evolution in Chrysomelid Beetles of the Genus Leptinotarsa. Carnegie Institution, 1906, pp. 320, pl. 30.

[2] Geographic Variations in Birds, with Special Reference to the Effects of Humidity. Zoölogica! N. Y. Zoöl. Soc., Sept. 25, 1907, pp. 41, pl. 6.

because we cannot always declare with confidence whether or not a given character is adaptive; the second, because the average experimentalist is as little conversant with the literature and methods of the systematist as the systematist is with those of the experimentalist; and the last, for reasons so numerous that they cannot even be outlined within the limits of the present paper.

Herein are presented some of the results of an inquiry into the effects of differences of temperature and humidity upon the postnatal development of the white mouse.[3] In all, upwards of 400 individuals have been subjected to the experimental conditions during the past three winters,[4] though it must be confessed that the number of mice in any one series has been relatively small. Due allowance has been made for this fact, however, in considering the probability of the various conclusions which are offered below. As will later be pointed out, differences which are great enough and constant enough are to be regarded as significant, even though the number of individuals is small. And it is scarcely necessary to remind my readers that the statistical treatment of even such small numbers as are here under discussion requires a great amount of extremely tedious work.

It must be acknowledged at once that I have been actuated by ulterior motives in pursuing these experiments. My primary object has been to test the question of the transmission of certain characters, or, more correctly, of their reappearance in the offspring. Thus far, it is true, no satisfactory or at all convincing test of this point has been made by the writer, though the experiments are still far from being ended. Large, obvious, and readily measurable changes have, however, been produced in the generations immediately subjected to the conditions employed. Thus one of the primary requirements for the fulfilment of such a test has been realized. It is not the purpose of this paper, accordingly, to present any evidence in favor of *transmission*. I shall content

---

[3] Some of the earlier results of this work were presented briefly before the American Society of Zoölogists, New Haven, December, 1907 (abstracted in Science, March 20, 1908).

[4] To this number must be added nearly 300 others, the data for which were not available when the present paper was being prepared. A brief mention of this most recent series has, however, been inserted on a later page.

Effects of External Conditions

myself with recounting some of the modifications produced during the lifetime of the individual.

ENVIRONMENTAL CONDITIONS

Early in life, in each experiment, the mice were divided into two lots, one of which was transferred to a room artificially heated,[5] the other to a room readily accessible to the winter atmosphere. The source of heat employed was a large steam coil which was ordinarily very effective. The room used throughout most of the work was, however, extremely pervious to draughts of air, and a strong wind from the proper quarter would sometimes bring down the temperature as much as 5 or 10° C. in the course of an hour or two. In the fall and early part of the winter, the steam was available during the daytime only; commencing about January 1, it was turned on night and day. The cold room used in the two earlier series of experiments was situated in another part of the same building as the warm one. During much of the time a window was kept open in the former, though this was not necessary when the wind blew from the right direction, at which times the temperature would fall nearly to that of the atmosphere outside. An unfavorable wind, on the contrary, frequently forced in warm air from other parts of the building. Thus in the case of neither of these rooms did conditions favor the maintenance of a very uniform temperature, and at times the fluctuations were rather violent.[6] Nevertheless the mean temperature throughout the experiments was very much higher in one room than in the other (Fig. 1), and the fluctuations resulted in little if any harm to the animals. Necessarily the humidity likewise differed to an enormous degree. In the cold room, the air was frequently near the saturation point. In the warm room, on the contrary, this

---

[5] Certain rooms in the U. S. Fisheries Laboratory at Woods Hole, Mass., not otherwise in demand during the winter months, were employed for this purpose. The material equipment has been provided almost wholly by the author himself. I must except a considerable number of cages kindly lent by Professor Morgan during the third winter of the work.

[6] During the present winter my facilities have been much better. Through the kindness of the director, Dr. F. R. Lillie, I have had the privilege of locating my cold room in a small unheated building belonging to the Marine Biological Laboratory. Here, therefore, the only fluctuations are those due to changes of weather.

same air, being heated to a temperature of 20° or 30° C., its humidity (degree of saturation) fell to 50 per cent or even considerably less. There is a popular fallacy to the effect that moist air may be *dried* by heating, after the manner of one's damp clothes. The notion contains this element of truth, of course—namely that heating, except when the same degree of saturation is maintained by evaporation, does increase the capacity of the air for taking up moisture from other objects. Accordingly, in the present experiments, the sawdust upon the floor of the cages and the cotton waste used for bedding were much damper in the cold room than in the warm one. Indeed there was commonly no perceptible dampness at all in the latter, while in the former it became distinctly moist if the changing were long neglected. The relative humidity was determined at rather irregular intervals by the use of an ordinary psychrometer or wet-and-dry-bulb thermometer. The percentage values were obtained from the scale attached to the "hygrophant," manufactured by J. S. F. Huddleston of Boston, though the instrument actually used was a similar one of another make. These percentages are probably to be regarded as rather rough approximations. They are, however, believed to suffice for present purposes. The temperature, in each room, was recorded continually by means of a thermograph, and thus the daily and hourly fluctuations of temperature could be followed.[7]

Since the conditions and methods of treatment differed considerably in the different series of experiments, a further account of these is deferred to the separate discussions of the latter. No account of the feeding nor of the general care of the mice is regarded as necessary here.[8] It is sufficient to state that, except for the differences in temperature and humidity above mentioned, it was my endeavor to maintain all conditions as similar as possible for the two contrasted lots of animals.

---

[7] The laborious task of computing the mean temperatures, etc., from the tracings upon the thermograph sheets has been mainly performed by my wife and my mother.

[8] I must acknowledge my indebtedness to Prof. W. E. Castle and Prof. T. H. Morgan for valuable suggestions relative to the care of the mice. The stock was all obtained from Miss Abbie E. C. Lathrop, the well-known animal breeder, of Granby, Mass.

CHARACTERS CHOSEN FOR MEASUREMENT

In the selection of these characters it was of course desirable to choose one or more which might be supposed to be influenced adaptively under the conditions of the experiment. *Quantity of hair* was therefore chosen as one of the values to be determined. Theoretically the hair might be subjected to quantitative treatment in three different ways: (1) by determining the average *length* of the hairs on each pelt, based upon a considerable number of individual hairs, so taken as to be representative of the entire lot; (2) by finding the *number* of hairs on a given unit of skin area; (3) by ascertaining the *weight* of the entire pelage or hair-coat of each mouse. The first method was very soon found to be impracticable, owing to the fact that the hairs are seldom straight enough for the purpose, being commonly curved or zigzag. The second and later the third methods were, however, employed more or less advantageously. For the determination of the number of hairs in a given unit of skin, the latter, immediately after removal from the freshly killed animal, was subjected to a stretching process which was as far as possible identical for each pelt. Pinch clamps were fastened at six points on the margin of the skin, and to these were attached fine cords which passed outward over pulleys arranged around the skin in a circle. Each cord bore a 100-gram weight. The pelt was thus gently stretched for five minutes, after which a large cork was pressed lightly against its central region, and the skin pinned to it by a circle of 16 pins. The outlying portions of the pelt were now cut away, and the circular area which was left pinned to the cork was allowed to dry. After a variable interval, depending on convenience, a small disk of the skin was cut out by means of a tool devised for the purpose. This consisted of a steel tube, $1\frac{1}{2}$ mm. in diameter, sharpened at one end into an edge. In use, it was pressed rather lightly against the inner surface of the skin and rotated until the latter was completely cut through. The resulting disk of skin was fastened to a black tile —hair uppermost, of course—by means of glue. The hairs were now shaved off with a sharp knife and counted under a low power lens. Two disks from symmetrically placed points on the pelt of each mouse were used, and it was sought to take them from corresponding regions of the skin in the case of every animal. The entire process required much time, and the counting of the hairs proved to be extremely fatiguing to the eyes. Moreover the great individual differences in the density of the hair and even the differences in density on the two sides of the same individual showed that satisfactory figures could not be obtained without making a large number of determinations. For these and for other reasons the counting of hairs was abandoned after the first year's series of experiments. The results, so far as obtained, will be presented later in the paper.

The method of *weighing* the total pelage of each mouse has proved to be by far the most satisfactory one. Aside from the greater ease and accuracy of technique, this method has the advantage of showing whether or not the total *quantity* of hair has varied under the different conditions. It has the disadvantage, however, of not showing how the quantity has been affected, i. e., whether by the augmentation of the number of developed hairs, or of their length or their diameter. The process employed was as follows:[9] The pelt after drying was placed in a saturated solution of calcium hydrate, and left for three days (occasionally longer).[10] After rinsing, the hair was scraped off with a moderately sharp scalpel. Most of the hair, in the majority of pelts, could be removed very readily; the remainder sometimes required considerable scraping. The epidermal cells were unavoidably scraped off at the same time, but these furnished very little material, and that was for the most part removed by washing. In any case their presence constituted a source of error which was probably practically constant throughout the series. The hair, after a thorough washing in water, was treated for ten to twenty minutes with dilute HCl (5 per cent) in order to remove any calcium which might remain in association with it. After further repeated washings, it was dehydrated in 95 per cent alcohol, subjected to ether for one hour,

---

[9] Acknowledgment is due to Dr. C. L. Alsberg, of the Bureau of Plant Industry, for valuable suggestions relating to certain steps in this process

[10] An equal number of skins of the two lots were always treated at the same time, so that constant differences in treatment were avoided.

Effects of External Conditions 103

to insure the removal of fats, and then dried in the air. Each lot (the total pelage derived from a single pelt) was now put into a weighing bottle and transferred to a vacuum desiccator, until the weight was found to have become constant (usually for about four days). The final weight was recorded to the ten-thousandth of a gram, though only thousandths have been regarded in the present paper. Since a number of specimens were dealt with at once, the process did not require as much time as might be inferred from the number of steps involved. The results obtained are presented below (pp. 129-131).

Other characters chosen for measurement were: *weight, length of body* (from snout to anus), *length of tail* (from anus to tip), *length of left ear* (from the lower extremity of the incisura intertragica[11] to the tip of the pinna), and *length of left hind foot* (from heel to tip of longest toe). Certain additional determinations, which were made in special cases, will be referred to at the proper time. A uniform method of procedure was adopted in making each of these measurements. In order that my figures may be of service for comparison with those of other workers in this field, it seems desirable to detail these methods a little more fully. The weight was taken by means of a torsion balance sensitive to a few hundredths of a gram. The tail length was obtained in two ways, according to whether the animal was living or dead when measured. When measured living, the mouse was suspended by the tip of the tail, the forefeet being allowed to rest upon the table. A certain degree of stretching generally occurred, the amount of which was found to average somewhat over 2 mm. in the case of full-grown mice. Weight and tail length were the only characters which it was possible to determine accurately with living animals.[12] In measuring the tail of the dead mouse, and likewise the body length, the freshly killed animal was laid with the ventral surface uppermost upon a wooden board. A pin was passed through the roof of the mouth, thus securing the snout to the board. The latter was then held upright and the body allowed to dangle

---

[11] Or at least a point which would seem to be homologous with the part so named in the human ear.

[12] I have since found it practicable to measure the ear and foot, in the living mouse, by the use of ether (April, 1909.)

for a few seconds, after which a second pin was passed through the basal part of the tail, which was thus likewise fastened securely in place. The distance from the tip of the snout to the anus and that from the latter to the tip of the straightened tail (exclusive of terminal hairs) were found by means of a graduated sliding caliper provided with a vernier. When the ear length was taken the mouse was laid freely upon the right side. In measuring the length of the right hind foot, the animal's body was held in the left hand, the sole of its foot pressed lightly against the board, and a pin passed through the instep into the wood. The tarsal joint was bent at about right angles, and the heel allowed to rise above the surface of support as in life. The caliper spanned the distance from the tip of the heel, which was touched rather lightly, to the tip of the nail of the middle (third) toe. It was found important to make all of these measurements as soon after death as possible. In the later series, two determinations were made of each part measured, and the reading was taken in tenths of a millimeter.

Throughout the whole of this work, the practice was followed, so far as possible, of measuring warm-room and cold-room individuals alternately, or at least of alternating small groups of individuals. Only thus would such gradual changes in one's standards of judgment as would inevitably result from growing experience, affect equally the two groups to be compared.

The characters selected are few in number, we must allow, and are, for the most part, such as would not be expected to respond *adaptively* to the treatment accorded to the animals. One must, however, accept the limits imposed by brevity of time and the nature of the material at hand. When subjected to statistical treatment a very few characters are found to involve a very great amount of labor. Moreover, aside from thickness of fur, such adaptations as might be conceived of as resulting from temperature conditions would probably be histological or chemical in their nature and therefore not accessible to ready methods of quantitative treatment.

The measurements which I have chosen as being applicable to these mice are, with the exception of those relating to the hair, ones which are employed by mammalogists who concern themselves with rodents. Coues and Allen,[13] for example, give the following external measurements for Muridæ, Leporidæ, Sciuridæ, etc.: From tip of nose to (1) eye, (2) ear, (3) occiput, (4) tail; length of tail to end of (1) vertebræ, (2) hairs; length of fore-foot, and hind foot "from the tuberosity of the heel to the end of the longest claw"; height of ear. Merriam,[14] likewise, presents figures for "total length" (i. e., body plus tail); "tail vertebræ" (i. e., length of the vertebral portion of the tail); "pencil" (tuft of hair at tip of tail); "hind foot;" "ear from crown" (sometimes from "anterior root" or from "notch"); and occasionally some others, including weight. Various dental and skeletal features are of course included in the complete diagnosis of a species, as well as differences in the color, texture and length of the pelage. But many of these characters are such as do not lend themselves to measurement, while others require the preparation of cleaned skeletons.

STATISTICAL METHODS

Since it is to be presumed that many biologists are still unfamiliar with the methods of biometry, the following statement as to those employed in the present paper may not be out of place.[15] For a really earnest and, on the whole, successful endeavor to render this difficult subject intelligible to the non-mathematical mind, the reader is referred to Thorndike's "Introduction to the Theory of Mental and Social Measurements" (Science Press, 1904). Davenport's "Statistical Methods" is of course indispensable to those who are already sufficiently familiar with the use of these methods.

The *mean* or *average* of a series of values (in the present case,

---

[13] Monographs of North American Rodentia. Report of the U. S. Geological Survey, vol. xi, 1877, pp. 1–1091.

[14] "North American Fauna" series, published by the Biological Survey of the Department of Agriculture.

[15] My thanks are due to Messrs. E. L. Thorndike and R. P. Bigelow for criticising certain portions of my manuscript and to Prof. C. B. Davenport for important information relative to biometrical methods.

measurements) is obtained by dividing the sum of all the terms of the series by the number of these terms.

The *standard deviation*, which is, at present, most frequently employed as the measure of the variability of a series, is obtained by squaring all the individual "deviations" or departures from the mean, finding the sum, and then the average, of these squared deviations, and extracting the square root of the resulting average, i. e.,

$$\sqrt{\frac{\text{sum of squared deviations}}{\text{number}}}$$

The *reliability* of the average or mean has, in each case, been indicated by the *probable error*, which is the number preceded by the ± sign and annexed to the mean in the following tables. The value of this number is such that there is an even chance that the true mean (i. e., such as would be obtained from averaging an infinite number of terms) lies within the limits indicated by: *given mean ± probable error*. The probable error of an average or mean is obtained by the formula

$$\frac{.6745 \times \text{standard deviation}}{\sqrt{\text{number of terms in the series}}}$$

The reliability of an average thus varies inversely as the variability of the series and directly as the square root of the number of terms. The *probable error of the mean*, here employed as an index of reliability, is not to be confused with the *probable error of a series*, sometimes employed as an index of its variability. This latter is a number of such magnitude that it is exceeded by exactly one-half of the deviations. It has the value: *.6745 × standard deviation*.

The reliability of the standard deviation, or figure denoting the variability of each series, is indicated by the *probable error of the standard deviation*. This is obtained by the formula:

$$\frac{.6745 \times \text{standard deviation}}{\sqrt{2 \times \text{number of terms}}}$$

Effects of External Conditions

Since one of the primary objects of such investigations as the present is an inquiry into the effects of differences in the conditions of life upon supposedly homogeneous material, one of the principal points to be determined is the significance of any differences which may be discovered between the average values of a given character in two groups of individuals whose history has differed. For this purpose it is necessary to compare the *probable error of the difference* with the actually obtained difference between the two averages in question. The probable error of the difference is expressed by the formula:

$$\sqrt{(\text{probable error of the first average})^2 + (\text{probable error of the second average})^2}$$

i. e, the square root of the sum of the squares of the probable errors of the two respective averages. Now the actually obtained difference is the most probably true difference and it is as likely to be too small as too large. Nevertheless the true difference may possibly equal 0, i. e., be non-existent, in which case the obtained difference would be regarded as wholly "accidental." From the table of the values of the "probability integral" it may be calculated that the chances that a difference between two averages is due to mere accident are:

250 out of 1000 when difference between averages = 1 × probable error of the difference.
156 out of 1000 when difference between averages = 1.5 × probable error of the difference.
89 out of 1000 when difference between averages = 2 × probable error of the difference.
46 out of 1000 when difference between averages = 2.5 × probable error of the difference.
21 out of 1000 when difference between averages = 3 × probable error of the difference.
9 out of 1000 when difference between averages = 3.5 × probable error of the difference.
3 out of 1000 when difference between averages = 4 × probable error of the difference.
less than 1 out of 1000 when difference between averages = 4.5 × probable error of the difference,

In proportion as the probability *decreases* that such a difference has been due to mere chance or accident (i. e., that it is the result of a multitude of independent causes having no relation to the conditions of the experiment), it is obvious that the probability *increases* that some constant modifying influence has been operative in differentiating the two groups. It must be admitted however, that the probabilities here stated apply in full strictness only to cases where we are dealing with large numbers of individuals. Davenport suggests 200 as the minimum number of individuals to be gathered for statistical treatment when the material is available; though he grants that much smaller numbers may be employed to advantage, where we are restricted by circumstances. But it must be borne in mind, he says[16] "that the rules for determination of averages, probable errors, standard deviations and all the rest become less and less significant as the number of variants becomes smaller. Finally, in the region of twenty or so, the results can no longer be treated by mass statistics; twenty hardly makes a mass." To the experimentalist it must often happen, as in the present work, that the use of a large number of individuals, in any single series, is excluded by reason of the laboriousness of the methods employed. In such cases, we are told, no exact mathematical equivalent can be offered for the probability of a given result, even though the frequency distributions afford strong evidence on the subject. Of course the cumulative testimony of several independent series of experiments is of high value. In general, it would seem that the experimentalist demands a somewhat different statistical technique from the student of variation *per se*, and it is to be regarded as unfortunate that the methods at our disposal, have, thus far been developed mainly by the latter type of investigator, and with very little reference to the special needs of the former. To the experimentalist, who is studying the effects of artificial conditions, it is the *significance of differences*, and scarcely anything-else in the whole field of statistical theory, that is likely to be of interest.

Regarding the accuracy of my computations, I can only say that every step has been gone over at least twice, and that, wherever possible, a different method of calculation has been employed in the repetition.

RESULTS IN DETAIL
Series of 1906–1907

Owing to the small number of individuals used and the tentative character of my methods at the outset, this first series will not be

[16] In a letter to the author

discussed at all fully. The mean temperatures in the two rooms, during the period of the experiment, were 24.9° C. and 9.1° C. (76.8° F. and 48.4° F.) respectively. No further analysis of the temperature conditions seems worth while in the present experiment. The humidity was not at any time determined. Twenty-one mice (13 males and 8 females) were reared in the cold room; 20 mice (12 males and 8 females) in the warm room. The animals were not subjected to the differing temperatures until they were about three weeks old (21 ± days). Previous to that time, the undivided lots had been reared under similar conditions. Each lot comprised individuals from 7 different broods, each of the latter having been divided into two portions destined for the warm and cold rooms respectively. About half the stock, consisting of the broods earliest obtained, were subjected to the experimental conditions for a period of 106 days, the remainder for a period of 83 days. At the expiration of these terms the mice were paired for breeding purposes, and the two contrasted lots were transferred to a single room having a temperature somewhat intermediate between those previously employed. None of the animals were killed immediately after this transfer, while the females were kept until they had reared their broods. The interval between the discontinuance of the temperature differences and the killing of the animals varied from 15 to 55 days. Thus the material was far from homogeneous. Nevertheless the figures obtained seem worth recording. (Table 1.)[17]

The difference in tail length, in the males, at least, was often obvious without measurement, and it must be regarded as a significant one, even without such overwhelming corroborative evidence as is offered later. In the case of the females, the difference is much smaller, though it is greatly in excess of the probable errors. No further analysis of this table seems called for.

In the present series, the number of hairs per unit of skin area was computed for each mouse, according to the method described

---

[7] Here and elsewhere the number of individuals measured for any given character is indicated by the figure in parentheses at the head of each column. In some cases individual tail measurements have been thrown out, where the tip of the organ had obviously been lost through accident. I trust, however, that it is not necessary to urge that no merely "exceptional" figures have been rejected

TABLE 1
Series of 1906–1907

	WEIGHT IN GRAMS				BODY LENGTH				TAIL LENTGH			
	Males		Females		Males		Females		Males		Females	
	Warm (12)	Cold (13)	Warm (8)	Cold (8)	Warm (12)	Cold (12)	Warm (8)	Cold (8)	Warm (10)	Cold (12)	Warm (6)	Cold (8)
Mean..........	24.742	24.792	23.437	24.437	94.08	93.67	92.25	93.25	88.10	77.67	84.50	80.20
Standard deviation........	1.170	1.456	2.232	2.531	1.75	1.70	2.28	2.86	3.21	3.04	2.29	4.02

TABLE 2

Series of 1906–1907: Number of hairs per unit of skin surface

	MALES		FEMALES	
	Warm (12)	Cold (13)	Warm (8)	Cold (8)
	165 132 } 297	164 146 } 310	190 204 } 394	242 220 } 462
	164 137 } 301	144 174 } 318	140 174 } 314	174 182 } 356
	138 135 } 273	154 178 } 332	189 149 } 338	165 150 } 315
	113 137 } 250	159 191 } 350	164 162 } 326	136 159 } 295
	95 110 } 205	150 160 } 310	144 148 } 292	127 136 } 263
	157 123 } 280	143 95 } 238	145 161 } 306	158 147 } 305
	128 97 } 225	161 196 } 357	160 217 } 377	152 139 } 291
	109 123 } 232	148 187 } 335	111 109 } 220	151 159 } 310
	159 163 } 322	192 118 } 310		
	137 130 } 267	161 170 } 331		
	135 152 } 287	140 113 } 253		
	122 144 } 266	159 135 } 294		
		114 116 } 230		
Mean..................	267.08 ± 6.12	305.23 ± 7.37	320.87 ± 11.96	324.62 ± 13.68
Standard deviation........	31.37 ± 4.32	39.43 ± 5.22	50.12 ± 8.45	57.40 ± 9.67

above (p. 101). It seems worth while to present the individual figures obtained for these animals, since this is the only series with which the counting method was employed. In Table 2 each figure in the left-hand columns represents the number of hairs on a circular disk of skin 1.5 mm. in diameter. Two figures are given for each individual, based upon two disks of skin taken at a distance of 1 cm. to the right and the left respectively of the mid-dorsal line. The sum of these two is stated in heavy type.

It will be noted that the mean number of hairs for the cold room males is 305.23 upon the two disks, that for the warm-room males being 267.08.[18] Here, then, is a difference of 14.3 per cent. The number of individuals is small, it is true, and the probable errors are large. Even granting the constancy, however, of such a difference, between two lots of mice thus treated, it is not necessary to conclude that there has been an actual increase in the number of (developed) hairs per unit of skin surface. If the warm-room individuals be supposed to have slightly *thinner* skins than those of the cold-room lot, the greater degree of *stretching* in the former (see p. 101) would result in a less dense distribution of hair upon its surface. But whether a difference so produced would be as high as 14 per cent may well be questioned. Again, it must not be supposed that I am urging this difference in the density of the coat of hair as an instance of permanent morphological change. It may be due merely to a difference in the rate at which the hair is shed. This point will be discussed later.

The averages for the hairs of the female mice are not far from equal in the two contrasted lots. It has already been noted that the females exhibited a much smaller difference in tail length than did the males.

Series of 1907–1908

During the winter of 1907–1908 the experiments were conducted on a much larger scale than previously, the conditions employed were such as were calculated to result in the production of greater

---

[18] The number of hairs per square millimeter of skin may be readily computed, since the area of each disk was (approximately) 1.767 sq. mm. Thus the mean number for the cold room males is 86.4, that for all the mice comprised in the table is 85.0, etc.

Effects of External Conditions

modifications, and the number of measurements applied to the animals was considerably increased.

Temperature. The mean temperature[19] of the warm room for the entire season was 26.30° C. (79.34° F.), that for the cold room 6.16° C. (43.08° F.). These figures correspond roughly to those for the mean annual temperatures of Key West or Porto Rico, on the one hand, and those for Eastport, Maine, or Minneapolis, Minn., on the other.[20] But it would not, of course, be at all fair to compare the temperature conditions of the experimental rooms with those of the points named, still less to compare the climatic conditions as a whole.

The mean daily range of temperature (i. e., mean difference between the maximum and minimum for each day) was 11.9° C. (21.4° F.) in the warm room, 6.7° C. (12° F.) in the cold room. The maximum temperature reached at any time in the warm room was about 40° C. (for very brief periods), the minimum in the cold room being $-14.4°$. But these figures represent exceptional occurrences and have little significance. Curves have been constructed (Fig. 1) showing the mean daily temperature in each room during the entire period of the experiment.

It is plain, of course, that none of these figures can represent the actual temperatures to which the mice themselves were most of the time exposed. When huddled together in large numbers in a nest of cotton-waste, the temperature of the air immediately in contact with them, at least in the case of the cold-room animals, was doubtless very much higher than that in the room outside, i. e., that recorded by the thermograph. Nevertheless, we all know by experience the difference between sleeping in a cold room and sleeping in a warm one, even when the amount of bedding is varied to suit the circumstances. And it must be remembered that during part of the time the mice were feeding, exploring the cages, etc., and were then wholly exposed to the air.

Humidity. The relative humidity in the warm room (see p. 100 above) ranged from about 22 per cent to about 66 per cent, the

---

[19] The mean temperatures here given are based upon four figures daily, these being taken from the thermograph sheets.

[20] Report U. S. Weather Bureau for 1906–1907.

mean for 12 determinations made during a period of four months being 39 per cent. This is somewhat less than the mean relative humidity at Phoenix, Arizona, for the year 1906, as stated by the U. S. Weather Bureau, that point having the dryest atmosphere, with a single exception, of any place in the United States for which records are given. In general, the humidity of the warm room varied inversely as the temperature, since no compensation was made by the evaporation of water; but the degree of saturation of the outside air must have been a factor. The humidity of the cold room varied from about 49 per cent to about 90 per cent, the mean of nine determinations being 67 per cent. This figure is a trifle less than that given for the mean humidity of Philadelphia during the year 1906.

The physiological results of such differences in humidity must be far reaching. The quality of the air inspired must affect the processes of respiration, and the rate of evaporation from the body surface must differ widely in the two cases. This last was shown by the eagerness for water displayed by the warm-room individuals. To what degree the results which I offer have been brought about by differences in temperature and to what degree by differences in humidity it is impossible to state. In these experiments it has not been practicable to control the humidity, independently of the temperature, and thus it has been impossible to decide this question definitely. The subject will be referred to later.

Disposition of Stock. Twenty broods of young mice, aggregating 135 individuals, were employed for this experiment. In order to insure, so far as possible, a division into two lots of a similar hereditary endowment, one-half the individuals of each litter were subjected to the high temperature, the other half to the low. For this purpose, exchanges were made between the offspring of different mothers according to the following plan: One-half of brood A, plus one-half of brood B, were consigned to the care of the A mother in the cold room, the other half of each brood being consigned to the care of the B mother in the warm room, and so on through the series. This disposition of the young was made while the latter were 2 to 4 days old (in a few exceptional cases as much as 11 days). The members of each litter were marked at the

Effects of External Conditions 115

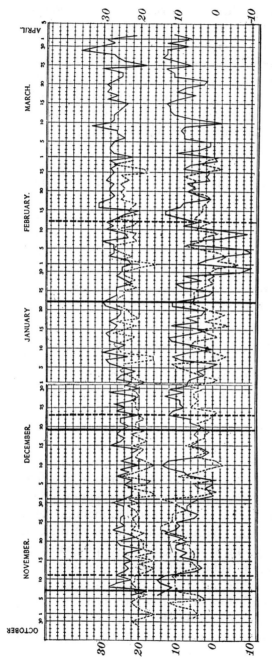

Fig. 1 Chart showing temperature conditions in warm and cold rooms during experiments of 1907–1908 and 1908–1909. These curves are based upon the mean temperatures for each day, as computed from four daily readings at equal intervals. The continuous lines represent the temperatures for the earlier winter's experiments, the broken lines those for the later year. The heavy vertical lines (continuous and broken) denote (1) mean date of birth of each year's mice; (2) mean date at which the 42-day measurements were made in each year; (3) mean date at which the 2½-month and 3-month measurements were made.

outset by a system of clipping the right ear and various fingers and toes of the right foot. Rather contrary to my expectations, the alien offspring were accepted by the mothers as readily as their own, so that little if any loss of life resulted from this procedure.

A certain proportion of deaths occurred here as always during the rearing of these mice. The number of deaths in the warm room during the first six weeks was 6, giving a mortality of about 9 per cent. The number dying in the cold room was considerably greater, being 13, or about 20 per cent of the lot. During the next month of life no deaths occurred among the warm room lot, while 6 were recorded in the cold room; but scarcely any further deaths from natural causes occurred during the next four months, i. e., until the end of the experiment. The cold room individuals, throughout the earlier part of their life, at least, were much less active than those in the warm room. During the first few weeks they kept to their nests almost constantly. Nevertheless, when mature, they were of decidedly better appearance than the warm room lot, and, when paired, they reared a much higher percentage of offspring. It must be added, however, that the reproductive capacity of both lots was found to be distressingly slight—so slight, in fact, as to render futile any attempt to make a satisfactory test of the transmissibility of the modifications which had been produced. Among the 21 females in the cold room lot, 31 pregnancies are recorded for the 15 weeks during which they were kept with the males, while in the aggregate only 48 young were reared to the age of six weeks. Indeed, the majority of the litters either consisted entirely of stillborn young, or of ones which died during the first few days after birth. In other cases, the young were apparently born healthy, but the mothers seemed unable to suckle them or perhaps lacked the instinct to do so. With the warm room lot the case was even worse. Of the young resulting from 50 pregnancies (doubtless over 200) only 35 individuals, or about 15 per cent, survived to the age of six weeks, while in the great majority of litters all the individuals died either before or shortly after birth. I am still almost wholly at a loss to account for this failure of the powers of reproduction. The mice were paired rather too young, it is true, being

2½ months old at the time. Many of them did not become pregnant till they were much older than this, however, while it is well known that female mice may bear healthy young at an even earlier age. Again, when judged by most other standards, the animals appeared to be in perfect health. They were active enough, and the fur was commonly in good condition, though the size of the females, at least, even when fully grown, was probably somewhat below normal. Moreover, after the earlier weeks of life, their mortality had been slight. That this damage to the generative powers must be set down as one of the results of the experimental conditions seems, nevertheless, probable.

Measurements at 42 Days. The weight and tail length of these mice was taken at the age of six weeks.[21] No other measurements were at that time believed to be practicable with the living animals. The following table (no. 3) presents the mean and the index of variability for each of these measurements, the two sexes being treated separately.

TABLE 3

Series of 1907-1908: Measurements at six weeks of age

	WEIGHT				TAIL LENGTH			
	Males		Females		Males		Females	
	Warm (29)	Cold (31)	Warm (33)	Cold (23)	Warm (29)	Cold (31)	Warm (33)	Cold (23)
Mean	12.997 ±0.270	13.123 ±0.374	12.282 ±0.179	11.400 ±0.301	68.83 ±0.35	54.16 ±0.68	69.06 ±0.43	51.91 ±0.83
Standard deviation	2.156 ±0.191	3.088 ±0.265	1.517 ±0.126	2.138 ±0.212	2.80 ±0.25	5.63 ±0.48	3.69 ±0.31	5.89 ±0.59

Fig. 2 shows the distribution of weights for the cold and warm room groups (sexes combined); Fig. 3 shows the distribution of tail lengths for the two groups. From the table and polygons collectively the following facts may be gathered:

[21] Owing to difference in the date of birth, this age was not attained simultaneously. The great majority, however, were born within the space of a week.

1   The tails of the warm room individuals are much longer than those of the cold room individuals. This difference amounts to 27.1 per cent for the males, 33.0 per cent for the females, and 29.7 per cent for the sexes combined. Indeed, the two types are so distinct that, but for a single individual, there would be no overlapping of the polygons; i. e. (barring this single exception), the longest tail in the cold room lot was shorter than the shortest in the warm room lot. These differences were so patent to the eye that, had the two lots of mice been mixed together accidentally, I am

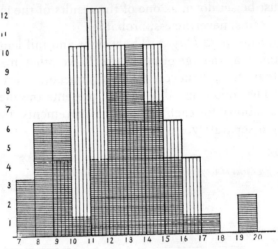

Fig. 2   Series of 1907–1908: weight at six weeks of age (sexes combined). Abscissas denote weight in grams; ordinates denote number of individuals. Vertically shaded areas represent warm-room animals; horizontally shaded areas represent cold-room animals.

sure that I should have been able to separate them again with comparatively few mistakes. Contrary to the condition in the first year's series, a greater difference is here shown by the females.

2   The warm room males were on the average 1 per cent lighter than the cold room males; the warm room females, on the contrary, were 7.7 per cent heavier. Quite similar relations in respect to weight will be found in the series of the following year. The frequency polygons for weight show that two groups of individuals, having two different "modes," are comprised in the cold room lot—a lighter and a heavier group. An analysis of the individual

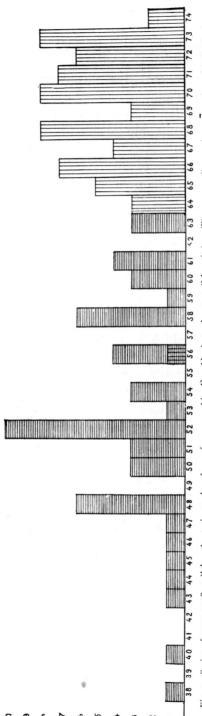

Fig. 3 Series of 1907–1908: tail length at six weeks of age (sexes combined). Abscissas denote tail length in millimeters; ordinates denote number of individuals. Shading as before.

figures shows that both the males and the females are divided quite sharply into these two groups. The number of animals is so small, however, that this phenomenon may be accidental. It will be referred to later.

3 In the case of both sexes, the variability both in weight and in tail length is considerably greater for the cold room lot than for the warm room lot.

4 The mean weight of the males (irrespective of history) is considerably greater (9.6 per cent) than that of the females; the tail length seems to bear no constant relation to sex, nor does the variability of either character.

Measurements at $2\frac{1}{2}$ Months. At the age of $2\frac{1}{2}$ months the mice were mated. At about the same time (when 75 to 78 days old) the measurements previously made (weight and tail length) were repeated. The number of females surviving to this age was 54 (33 warm + 21 cold). The number of males reserved for breeding was 13 (7 warm + 6 cold). There remained 43 males (22 warm + 21 cold) which were not needed for breeding purposes. They were therefore killed at this time, and subjected to a considerably greater number of measurements than had hitherto been employed. At this period, therefore, there are to be considered three groups of individuals (each group consisting, of course, of a warm room and a cold room half) : (1) the females; (2) the mated males; (3) the unmated males.

Referring to Table 4, it will be seen that among the female mice the warm room lot are 4.2 per cent heavier than the cold room lot, while the tail length of the former is 26.6 per cent greater. Comparing the mean figures here given with those of the table for 42 days, it will be found that in the cold room lot the mean weight has undergone an increase of 33.1 per cent during the interval between the two measurements, while the mean tail length has undergone an increase of 16 per cent.[22] In the warm room lot,

---

[22] Of course this *increase in the mean* has not exactly the same value as the *average individual increase*. The latter figure cannot be given for the present series, since the mice had not been marked so as to be individually distinguishable, although the members of each litter were identified by a brood number. Inasmuch as there had been but two deaths in the interval between the measurements, we are dealing with practically the same group in each case.

on the other hand, the increase in weight and in tail length have been 28.8 per cent and 10.3 per cent respectively. There has thus been manifested a tendency toward equalization in respect to both of these characters, but especially in respect to tail length. This appendage has undergone a percentage increase which has been more than half again as rapid in the case of the cold room (i. e., the shorter tailed) as in the warm room (i. e., longer tailed) animals. Further evidence for such a general tendency toward the neutralization of early differences will be offered later. As regards variability, the standard deviation for tail length, both in the warm and cold room lots, has undergone a slight absolute decrease,

TABLE 4

Series of 1907–1908: Females 2½ months old

	WEIGHT		TAIL LENGTH	
	Warm (33)	Cold (21)	Warm (33)	Cold (21)
Mean...............	15.821 ± 0.261	15.176 ± 0.330	76.18 ± 0.38	60.19 ± 0.76
Standard deviation....	2.221 ± 0.184	2.236 ± 0.233	3.20 ± 0.27	5.10 ± 0.53

notwithstanding a considerable *increase* in the average for each lot. The standard deviation for weight, on the contrary, has undergone an increase in each lot. The relative variability (i. e., ratio of standard deviation to mean) has increased in one case (warm room), decreased in the other (cold room).

TABLE 5

Series of 1907–1908: Mated males 2½ months old

	WEIGHT		TAIL LENGTH	
	Warm (7)	Cold (6)	Warm (7)	Cold (6)
Mean	17.843 ± 0.568	20.117 ± 0.824	78.00 ± 1.17	66.00 ± 0.62

No discussion of the "mated males" (Table 5)[23] is worth while, owing to the small number of individuals comprised. These

---

[23] A certain degree of selection was exercised in picking out these males, the larger individuals of a brood being chosen for breeding purposes. This is shown by the difference in mean weight between the "mated males" and the "unmated males."

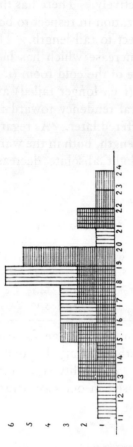

Fig. 4 Series of 1907–1908: weight of "unmated males" at 2½-months of age.

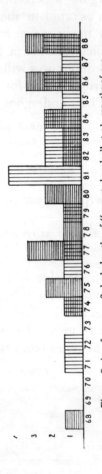

Fig. 5 Series of 1907–1908: body lengths of "unmated males" at 2½ months of age.

Fig. 6 Series of 1907–1908: tail length of "unmated males" at 2½ months of age.

TABLE 6 A

Series of 1907–1908: Males (unmated) killed at 2½ months.

WARM ROOM (22)

	Weight	Body length	Tail	Caudal vertebrae	Ear	Foot	Hair in mg.
	15.5	82	74	32	13.2	16.9	328
	14.3	76	72	33	12.6	17.1	146
	15.1	79	82	32	13.6	17.3	350
	13.8	72	72	33	12.7	18.2	198
	18.8	83	74	32	13.0	17.0	343
	17.7	81	78	33	12.4	17.2	341
	13.0	77	73	34	12.5	16.4	228
	18.3	86	77	32	13.0	17.0	271
	12.3	71	74	31	13.0	17.3	198
	16.6	83	77	32	12.3	17.5	289
	21.6	88	79	32	13.2	17.5	371
	21.7	84	79	31	13.5	16.9	318
	17.3	81	74	31	13.0	17.9	233
	16.8	81	72	33	12.3	17.2	258
	18.3	87	77	33	13.6	17.6	279
	18.7	85	72	31	12.9	17.2	252
	16.1	81	76	32	13.2	17.6	229
	11.8	74	71	32	13.3	17.5	130
	18.8	84	79	32	13.5	18.1	261
	17.2	82	78	32	13.2	17.4	234
	18.7	86	81	32	13.1	17.5	258
	20.8	88	75	31	13.0	17.6	306
Mean.....	16.964 ±0.346	81.41 ±0.69	75.73 ±0.45	32.09 ±0.11	13.005 ±0.056	17.359 ±0.055	264.59 ±8.98
Standard deviation..	2.724 ±0.277	4.78 ±0.49	3.11 ±0.32	0.74 ±0.08	0.389 ±0.040	0.382 ±0.039	62.48 ±6.35

TABLE 6 B

Series of 1907–1908: (*Males unmated*) *killed in* 2½ *months.*

COLD ROOM (21)

Weight	Body length	Tail	Caudal vertebræ	Ear	Foot	Hair in mg.
21.4	83	59	31	12.6	16.4	462
19.8	80	60	31	12.8	16.8	349
14.9	74	59	31	12.3	16.5	237
18.5	77	59	31	12.5	16.4	328
17.6	79	60	32	12.6	17.1	330
15.7	75	61	31	12.9	16.6	257
15.2	80	57	[28]*	12.5	16.2	228
13.2	75	62	32	12.2	17.4	233
16.0	78	58	[29]*	12.0	16.0	205
18.3	84	64	33	12.6	16.9	293
15.5	77	60	32	12.3	16.7	239
13.2	68	62	32	12.4	16.8	227
23.1	88	64	33	13.5	17.5	364
22.7	88	65	32	13.3	17.5	502
14.7	77	51	30	12.2	16.3	211
17.2	82	53	32	12.2	16.3	225
21.8	86	60	31	13.0	16.9	284
19.5	86	70	33	14.0	17.2	255
19.3	88	67	30	13.4	17.3	284
19.0	84	65	32	13.5	16.7	383
19.4	86	67	31	13.7	17.3	294
Mean.... 17.905 ±0.425	80.71 ±0.78	61.10 ±0.65	31.58 ±0.14	12.786 ±0.082	16.800 ±0.065	294.76 ±11.66
Standard deviation... 2.888 ±0.301	5.33 ±0.55	4.41 ±0.46	0.88 ±0.09	0.558 ±0.058	0.441 ±0.046	79.17 ±8.24

TABLE 6 C

Series of 1907–1908: Males (unmated) killed at 2½ months. Figures arranged for comparison

	WEIGHT		BODY LENGTH		TAIL LENGTH		NUMBER CAUDAL VERTEBRAE		EAR		FOOT		WEIGHT OF HAIR IN MILLIGRAMS	
	Warm	Cold	Warm	Cold	Warm	Cold	Warm	Cold	Warm	Cold	Warm	Cold	Warm	Cold
Mean.....	16.964 ±0.346	17.905 ±0.425	81.41 ±0.69	80.71 ±0.78	75.73 ±0.45	61.10 ±0.65	32.09 ±0.11	31.58 ±0.14	13.005 ±0.056	12.786 ±0.082	17.359 ±0.055	16.800 ±0.065	264.59 ±8.98	294.76 ±11.66
Standard deviation	2.724 ±0.277	2.888 ±0.301	4.78 ±0.49	5.33 ±0.55	3.11 ±0.32	4.41 ±0.46	.74 ±0.08	0.88 ±0.09	0.389 ±0.040	0.558 ±0.058	0.382 ±0.039	0.441 ±0.046	62.48 ±6.35	79.17 ±8.24

figures will be later combined, however, with certain of those given for the next group.

The *unmated males*, measured at the age of $2\frac{1}{2}$ months, constitute the most important group in the second year's series. Table 6 (A, B, and C) presents the measurements for these 43 mice. These measurements were all made after killing.

Comparison of these figures with those given for the males at 42 days is of course only possible in respect to two characters—weight and tail length. In order to determine accurately what changes have occurred in these, however, we must first combine the figures for the present group with the preceding group of "mated males," since the two together comprise the entire collection of males which had been measured earlier in life.[24] Table 7 accordingly represents the mean weight and tail length for all males at the age of $2\frac{1}{2}$ months.

TABLE 7

Series of 1907–1908: All males at $2\frac{1}{2}$ months of age

	WEIGHT		TAIL LENGTH	
	Warm (29)	Cold (27)	Warm (29)	Cold (27)
Mean............	17.176	18.396	77.41	63.33
	±0.331	±0.396	±0.44	±0.56
Standard deviation....	2.646	3.053	3.54	4.28
	±0.234	±0.281	±0.31	±0.39

Comparing these figures with those of the males at six weeks of age, we note that whereas the tails of the cold room lot have increased 16.9 per cent in the interval between the measurements, those of the warm room lot have increased only 12.5 per cent. There is thus seen to be a tendency to "catch up" on the part of the less developed organs, which has already been pointed out for

---

[24] In combining these figures an allowance is first necessary. The tail length of the dead mice was, as stated above (p. 103), obtained by a different method from that practised upon the living ones. I have found that in living mice of this size the tail is stretched on the average about 1.5 mm. during the suspension. This amount has accordingly been added to the mean tail length of the unmated males before combining with that of the mated males. The resulting figure represents approximately the tail length which would have been obtained had all been measured alive.

the females and will be discussed later. It must be added, however, that in the present case the difference in weight between the two groups has increased rather than diminished. In respect to variability, two of the four standard deviations comprised in the table show an absolute decrease, one of the others indicates a slightly lessened variability, while in the fourth case there is an increase, both absolute and relative. Not much importance is to be attached to these latter comparisons, however.

Returning to a consideration of the figures presented in Table 6 it is seen that the weight is 5.3 per cent less in the warm room lot than in the cold room lot; while the body is seven-tenths of one per cent longer and the tail 23.9 per cent longer. Passing to the new measurements (not applicable to living animals), the average length of the (left) ear is 1.7 per cent greater in the warm room lot; that of the foot 3.3 per cent greater. The average weight of hair (see p. 102) for the cold room mice is 11.4 per cent greater than that of the warm room individuals. It has seemed worth while to represent graphically the frequency distributions of these characters (Figs. 4 to 9). The difference in weight between the two sets of mice (Fig. 4) cannot with any certainty be regarded as a significant one. The difference in *body length* (Fig. 5) is too trivial to be taken into consideration.

In striking contrast, however, is the case for *tail length* (Fig. 6). Here there is no overlapping whatever of the polygons, while the modes are removed by a distance of 13 mm. Two questions present themselves respecting this difference of tail length: (1) Does it involve an actual difference in the volume of the organ? And (2) does it involve a difference in the number of vertebræ? In order to test the first question the diameter of the tail at its widest point was obtained by means of calipers for all the mice of this group. While this is a difficult measurement to make with any great accuracy, the figures are probably exact enough for present purposes. The averages for warm room and cold room animals are 2.94 mm. and 3.02 mm., respectively. There is thus the possibility of a slightly greater attenuation of the tail, accompanying its more rapid increase in length, but the former is certainly nothing like proportionate to the latter. There is

therefore an actual difference in the volume of the tail. The
number of caudal vertebræ was likewise counted for all of this

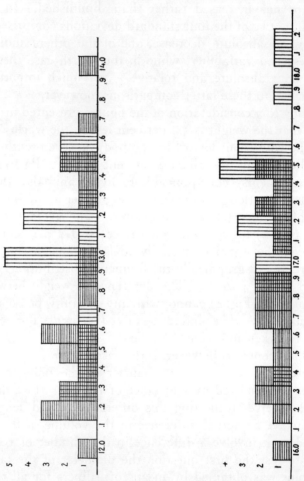

Fig. 7 Series of 1907–1908: ear length of "unmated males" at 2½ months of age.

Fig. 8 Series of 1907–1908: foot length of "unmated males" at 2½ months of age.

group. The mean figures are given in the table[25] (6), where it
is seen that the difference between the two groups (if significant

[25] The number of vertebræ was counted, from the sacrum to the last nodule of bone distinguishable under a low power objective. The task was comparatively simple and admitted of tolerable accuracy. The average error of observation was probably less than one vertebra per mouse. Since the skele-

Effects of External Conditions 129

at all) amounts to only about half a vertebra, i. e., it is not sufficient to make any appreciable difference in the tail length. In fact there is very little deviation from the mean throughout the entire series. It would be surprising indeed if the number of vertebræ had been found to be altered by temperature conditions, first because this number does not in all probability vary very widely in most species of mammals, and, secondly, because the definitive division of the embryonic tissues into vertebræ is probably complete long before birth.

The present difference in *ear length* is not convincing, but *foot length*, has, with practical certainty, been affected by the tempeature conditions.

It must be remembered that we are here dealing with mice which differ among themselves widely in size, and that neither the mean weight nor the mean body length is quite the same for the two groups. In Table 8 are presented the *relative* magnitudes of certain characters. Here we have the mean ratios between the length of tail, ear and foot in each individual and the length of the body; likewise the ratios between the weight of the hair and the square of the body length. In the case of the ear and the foot, the variability of the ratios is found to be much greater than that of the absolute measurements. This is due to the fact that these parts vary but slightly as compared with the size of the animals. Indeed their length is remarkably constant throughout each series, irrespective of the body length of the individual.

The case of the *hair* deserves a rather full discussion, since this is the character, in particular, whose modification may be supposed to be of an adaptive nature. It is seen from Table 6 that the mean weight of the pelage for the cold room individuals is 11.4 per cent greater than that of the warm room individuals. The variability is very high, to be sure, partly because the animals vary much in size, partly because they actually vary in the density of their hair

---

ton was not thoroughly denuded of muscles, etc., (alcoholic specimens were used) it was not always easy to distinguish the termination of the sacrum and the commencement of the caudal series, and an error of one or two vertebræ perhaps resulted occasionally from this cause. In a few instances, some of the minute terminal vertebræ were lacking, owing to obvious injury to the tail. In such cases, the figures have been enclosed in brackets and have not been included in making up the averages.

TABLE 8

Series of 1907–1908: Unmated males—ratios, expressed in percentages of body length

	TAIL LENGTH: BODY LENGTH		EAR LENGTH: BODY LENGTH		FOOT LENGTH: BODY LENGTH		WEIGHT OF HAIR IN GRAMS: (BODY LENGTH)^2	
	Warm (22)	Cold (21)	Warm (22)	Cold (21)	Warm (22)	Cold (21)	Warm (22)	Cold (21)
Mean............	93.23 ±0.71	75.86 ±0.83	16.023 ±0.141	15.871 ±0.115	21.391 ±0.206	20.895 ±0.191	0.03959 ±0.00111	0.04501 ±0.00145
Standard deviation.....	4.95 ±0.50	5.67 ±0.59	0.978 ±0.099	0.778 ±0.081	1.435 ±0.146	1.297 ±0.135	0.00771 ±0.0078	0.00975 ±0.00101

coat. I have therefore made an endeavor to compute the *relative* amount of hair, making allowance for the area of the skin—that is, I have obtained the ratio between the hair weight of each mouse and the square of its body length. The mean of these figures is 0.04501 mg. per sq. mm. for the cold room lot, 0.03959 for the warm room lot. According to this computation, the cold room mice have a relative amount of hair 13.6 per cent greater (heavier) than the warm room ones. It must be admitted, however, that such a method of computation is open to criticism. To estimate the relative skin areas of these mice by comparing the squares of their body lengths presupposes that they are, in the language of geom-

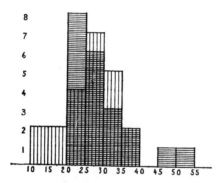

Fig. 9 Series of 1907–1908: weight of hair (absolute) of "unmated males" at 2½ months (expressed in hundredths of a gram).

etry, "similar solids," which they are not. As a matter of fact, while the warm room mice have a slightly greater body-length than the contrasted group, they are lighter, on the average, by nearly one gram, i. e., a difference of over 5 per cent. They are probably somewhat less plump, therefore. I must add, however, that I do not believe any such slight difference of shape to be accountable for the difference in the amount of hair which is shown by the two groups of mice. The most serious criticism of these figures relates to the number of individuals, which is confessedly too small to permit of our drawing any final conclusions in the presence of such high variability.

Until further data are available, however, it seems worth while to offer them for what they are worth.

Fig. 9 shows the distribution of hair weight for the entire lot, both the hot room and the cold room individuals. No fair comparison can be drawn from these polygons, as has already been stated, since mice of very different sizes are represented. In Fig. 10 I have substituted in each case for the absolute weight the ratio of the hair weight to the square of the body length. While the modes lie at the same point, the centers of gravity are considerably separated.

In view of statements cited by Lydekker[26] regarding certain modifications which are alleged to have been produced in cats by

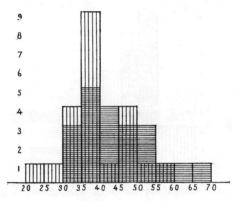

Fig. 10 Series of 1907–1908: weight of hair (relative) of "unmated males" at 2½ months (= ratio of hair-weight to square of body length).

life in a cold-storage warehouse, I endeavored to determine for the present group of mice whether there was any appreciable difference in the length of the vibrissæ between cold room and warm room individuals. The measurements have necessarily been rough in the extreme. The longest hair among the "whiskers" of one side was measured by calipers, without straightening or removing it. The mean figures obtained were 26.8 mm. for the warm room lot; 25.4 for the cold room lot. Considering the crudity of the method, the individual measurements vary compara-

[26] A Handbook to the Carnivora. Part I: Cats, etc.; pp. 158–163.

tively little, though enough, probably, to deprive this difference of any significance. No such obvious modifications as has been alleged for cats [27] is here evident.

Measurements at 7 Months. The subsequent history of the female mice varied considerably with different individuals, according to the exigencies of the breeding experiments. Each female from either lot, as she became pregnant, was transferred to a room kept at a temperature somewhat intermediate between the hot and cold rooms. If, as was commonly the case (see p. 116), her brood did not survive, she was taken back to the room from which she came. Thus, in the interval between February 6 and April 1, a considerable proportion of the females were transferred back and forth between the warm or cold room and the "intermediate" room, in some cases more than once. I have not thought it worth while to compute the average duration of each set of conditions for the lot. On April 1, all the mice were moved to the "intermediate" room and the sexes separated. On May 1, they were paired again, but with the same unfortunate results. The entire lot of females was killed and measured between June 4 and July 6. All were about 7 months of age at the time of killing, save for a few mothers of broods, which were allowed to remain with the latter till they were old enough to take care of themselves. These somewhat older individuals ($7\frac{1}{2}$ to 8 months) have, however, been included in the table herewith given. As they were, with little doubt, all fully grown, this proceedure seems fair.

The mean figures obtained for each lot of females is given in Table 9. It herewith appears that the weight in the warm room lot is 2.4 per cent greater than in the cold room lot, the body two-tenths per cent longer, the tail 14.9 per cent longer,[28] the foot 4.1 per cent longer, while the average ear length is practically equal in the

---

[27] It is true that in the case cited by Lydekker the elongation of the vibrissæ was attributed to the *darkness*, rather than to the cold.

[28] It seems probable that, so far as the tail at least is concerned, the cold-room mice have departed from the more usual or normal condition, while the warm-room individuals have been little if any modified. Fifty-nine adult female mice, of unknown history, which were received by me during the present winter, had a mean tail length of 92.8 mm.; i. e., their tails were considerably longer than even the warm-room females of Table 7. It must be noted, however, that they were larger mice, having a mean weight of 26.6 gms.

TABLE 9

Series of 1907–1908: Females about 7 months old

	WEIGHT		BODY LENGTH		TAIL LENGTH (LIVING)		EAR		FOOT	
	Warm (27)	Cold (18)	Warm (27)	Cold (18)	Warm (27)	Cold (18)	Warm (27)	Cold (18)	Warm (27)	Cold (18)
Mean	23.059 ±0.449	22.522 ±0.514	94.03 ±0.56	93.84 ±0.58	86.96 ±0.38	75.67 ±0.78	13.743 ±0.048	13.703 ±0.074	17.322 ±0.050	16.651 ±0.091
Standard deviation	3.457 ±0.317	3.233 ±0.363	4.35 ±0.40	3.62 ±0.41	2.93 ±0.27	4.90 ±0.55	0.368 ±0.034	0.464 ±0.052	0.382 ±0.035	0.573 ±0.064

TABLE 10

Series of 1907–1908: Males "mated," about 7 months old

	WEIGHT		BODY LENGTH		TAIL LENGTH (LIVING)		EAR		FOOT	
	Warm (7)	Cold (6)	Warm (7)	Cold (6)	Warm (7)	Cold (6)	Warm (7)	Cold (6)	Warm (7)	Cold (6)
Mean	23.657 ±0.375	25.167 ±0.440	95.11 ±0.86	95.69 ±0.58	87.00 ±0.88	77.83 ±0.56	14.043 ±0.800	14.067 ±0.141	17.771 ±0.053	17.283 ±0.162

Effects of External Conditions

two lots. None of these differences except those in the length of the tail and foot are to be regarded as having any significance.

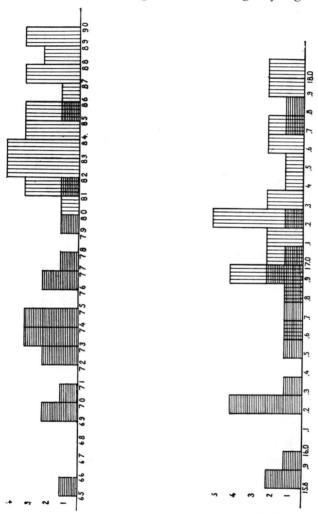

Fig. 11. Series of 1907–1908: tail length of females at 7 months of age (based upon measurements *after* death—see text.)

Fig. 12 Series of 1907–1908: foot length of emales at 7 months of age.

Comparing the measurements for this age with those (weight and tail) made upon these same mice[29] when $2\frac{1}{2}$ months old, we find

[29] Six of the warm-room lot and three of the cold-room lot had died in the meantime.

that the weight has increased during this interval 48.4 per cent in the cold room (i. e., lighter) lot; 45.8 per cent in the warm room (heavier) lot. The gain in tail length for the cold room lot has been 25.7 per cent, as compared with an increase of only 14.2 per cent on the part of the warm room individuals. In respect to each character, therefore, but more especially in respect to the tail, there has been a very obvious tendency toward a diminution of the differences between the two contrasted sets.

Referring to variability, a comparison of the standard deviations for tail length shows that there has been a slight reduction in the absolute and a considerable reduction in the relative variability in both the warm room and cold room lots. The standard deviations for weight have undergone an increase in each lot, though the relative variability has remained practically unchanged.

Frequency polygons (Figs. 11 and 12) have been plotted for tail length and foot length in this group of females. The polygons for the former overlap to a very slight extent; those for the latter are sufficiently distinct to admit of no doubt as to their significance.

The males used for breeding were kept under the extreme temperature conditions until April 1, at which time they were 4 to 5 months old. They were then transferred to the "intermediate room" along with the females, and were killed upon reaching the age of 7 months.

The measurements for this group are given in Table 10.

Comparing the present figures with those for the same mice at the age of $2\frac{1}{2}$ months, we find that the lighter warm room lot has gained to the extent of 32.7 per cent of its former weight, while the heavier cold room lot has gained only 25.1 per cent, again showing a tendency toward equalization. The same fact is evident in the growth of the tail. This has amounted to 17.9 per cent in the case of the cold room lot; 11.5 per cent in the case of the warm room lot.

Levelling Down of Early Differences.

Reference has more than once been made, in discussing particular sets of measurements, to a tendency for these experimentally produced differences to diminish with growth. A table has been prepared (Table 11) which includes all the cases in which this point can be tested.

TABLE 11

Series of 1907–1908: percentages of increase* in the intervals between successive measurements

		6 WEEKS ABSOLUTE MEASUREMENTS		2½ MONTHS PERCENTAGES OF INCREASE		7 MONTHS † PERCENTAGES OF INCREASE	
		Weight	Tail length	Weight	Tail length	Weight	Tail length
Males	Warm	12.997	68.83	**32.1%**	12.5%	**32.7%**	11.5%
	Cold	13.123	54.16	40.2%	**16.9%**	25.1%	**17.9%**
Females	Warm	12.282	69.06	28.8%	10.3%	45.8%	14.2%
	Cold	11.400	51.91	**33.1%**	**16.0%**	**48.4%**	**25.7%**

*See foot note on p. 120 above.

†In considering the figures for 7 months, it must be recalled that only 13 males (7 warm + 6 cold) have been kept till this time. In figuring percentages of increase for these males, therefore, the later figures have been compared with those for this same group of "mated males," and not with those for the males in general. In the column for 2½ months, on the contrary, the figures for "all males" are given.

Under the "6 weeks" column, we have the absolute measurements of weight and tail length for males and females belonging to the warm room and cold room lots. In the column for 2½ months is given for each group the percentage of increase of each of these characters, during the interval between the two measurements. In the "7 months" column are given the percentages of increase over the measurements for 2½ months. For each sex are presented two horizontal rows of figures: those for the "warm" and the "cold" lots respectively. It is thus easy to compare the contrasted figures for any one character. In each pair of these contrasted figures, that one has been printed in heavy type which represents the rate of increase for the group which had previously shown a *lower* mean value for the character in question. This group should, according to the hypothesis, be expected to have undergone a more rapid rate of increase. As a matter of fact, it will be noted that in 7 out of the 8 pairs of contrasted figures, the

TABLE 12
Series of 1908–1909: Weight and tail length at six weeks

	WEIGHT				TAIL LENGTH			
	Males		Females		Males		Females	
	Warm (55)	Cold (50)	Warm (43)	Cold (38)	Warm (53)	Cold (47)	Warm (42)	Cold (35)
Mean..........	12.604 ±0.283	13.180 ±0.241	12.663 ±0.217	11.889 ±0.195	67.19 ±0.55	60.11 ±0.51	68.95 ±0.48	59.49 ±0.53
Standard deviation....	3.119 ±0.201	2.522 ±0.170	2.107 ±0.153	1.783 ±0.138	5.98 ±0.39	5.20 ±0.36	4.61 ±0.34	4.63 ±0.37

number in heavy type is the larger number.[30] This decrease in bodily differences originally brought about by differences of temperature has not been due to a withdrawal of the latter. Indeed, during the interval between the "6 weeks" measurements and the "$2\frac{1}{2}$ months" measurements the temperature differences in the two rooms have increased rather than diminished. (See Fig. 1.) Later, it is true, the temperature differences gradually diminished, and commencing with April 2 they were abolished altogether. At the latter date, however, the mice averaged nearly five months old, and their growth was probably not far from complete.

While the tendency toward a reduction of the original differences between the warm and the cold room groups is thus pretty clear, the evidence for a reduction of variability within each group is not so certain. An inspection of the tables shows us twelve cases in which we can compare a later standard deviation with an earlier one for the same character. In six of these cases the later standard deviation is smaller, i. e., the decrease in variability has been absolute as well as relative. In two cases there has been a relative decrease, though not an absolute one; while in two others, the relative variability has remained practically unchanged. In only two of the twelve cases has there been any appreciable increase in the relative variability. In view of the lack of uniformity in these results, however, and the commonly high probable errors, too much significance must not be attached to them. It is worth pointing out, however, that the variability for *tail length* has decreased absolutely as well as relatively in five out of six cases.

Series of 1908–1909

At the date of writing, the experiments of the third winter have not been carried very far. A first and second series of measurements upon the living mice have, however, been made, and the results seem well worth comparing with those of the preceding

---

[30] It is likewise true that in all of these seven cases there has been a greater *absolute* gain as well as a greater percentage increase. Rate of growth seems more fairly expressed, however, in terms of *proportionate* increase.

year.[31] The temperature conditions during the first four months are indicated in Fig. 1, in which a certain degree of comparison with those of the previous winter is made possible. The mean temperature of the warm room during the first three months of

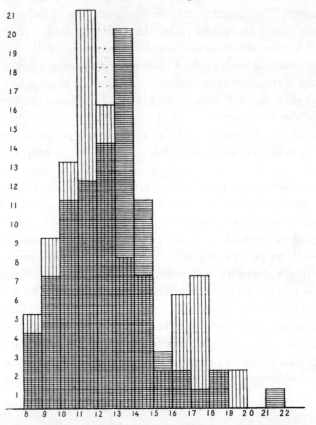

Fig. 13 Series of 1908–1909: weight at six weeks (sexes combined).

the animals' lives was 21.3° C., with a mean daily range of 12° C. In the cold room the mean temperature during this same period was 3.6° C., the range of daily fluctuation being 6°. The temper-

---

[31] The mice here used are from an entirely new stock, none of them being descendants of those raised during the preceding year. Animals from several different localities, having independent pedigree, are comprised in the lot. In the present series individual litters were not divided into a "warm room" and "cold room" half, as had been done with those of the preceding year.

ature conditions will be further discussed after the first table of measurements has been presented. The mean relative humidity in the warm room (12 determinations) was 38.5 per cent, a figure very close to that of the preceding year. The humidity of the cold room (14 determinations) was 76.5 per cent, being thus considerably greater than that of the preceding year. This is accounted for by the fact that the cold room used during the present year has been situated in a building which is entirely unheated.

With respect to weight there is a remarkably close agreement between each of the four averages here presented and the corresponding one of the preceding year. In each year's series, the males are larger in the cold room lot, while the females are larger in the warm room lot. From the magnitude of the probable errors, however, we cannot feel sure of the validity of these differences.[32]

Regarding tail length, it will be noted that the differences, while considerable, are very much smaller than those to be found in Table 3. The warm room males have tails which are 11.8 per cent longer than those of the cold room lot. This difference, in the case of the 1907–1908 lot, was 27.1. The warm room females show a mean tail length which is 15.9 per cent greater than that of the cold room individuals, as compared with a difference of 33 per cent in the earlier lot. Thus, as in the preceding year's measurements, we find that it is the females which have been modified most in this respect. But the amount of modification for each sex has been less than half that which was earlier observed. Regarding variability no deductions of interest are to be drawn. No such evidence of a higher variability in the cold room lot as was previously noted is here to be found. In fact, the reverse is possibly true.

The salient fact brought out by this comparison is the relatively small degree of modification in the tail length of the present lot of animals. This fact is not satisfactorily explained by a comparison of the temperature conditions for the two years. We find, it is true, in the later series a somewhat smaller difference in temperature between the two rooms during the first six weeks of the

---

[32] See supplementary note below.

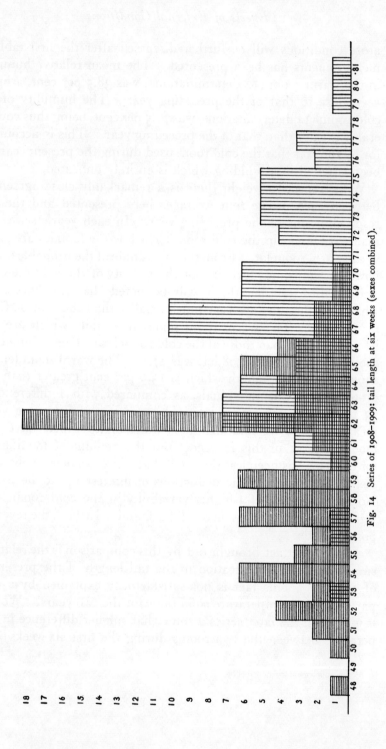

Fig. 14 Series of 1908–1909: tail length at six weeks (sexes combined).

animals' lives. This difference here amounts to 14.5° C., as compared with 16.6° during the winter of 1907–1908; and it might be at once inferred that in this fact we had a key to the difference of results. It must be noted, however, that, while the warm room temperature has been considerably lower for this period, during the later year (19.8° as compared with 23.7°), the cold room temperature has likewise been somewhat lower (5.3°[33] as compared with 7.2°). Now a comparison of Tables 3 and 12 shows us that while the *warm room* tails agree pretty closely in length in the two series, the *cold room* tails are much shorter in the earlier one

TABLE 13

Series of 1908–1909: percentages of increase in the interval between successive measurements

		6 WEEKS* ABSOLUTE MEASUREMENTS		3 MONTHS PERCENTAGES OF INCREASE	
		Weight	Tail	Weight	Tail
Males	Warm	12.604	67.19	65.58 ± ?	20.31 ± 0.61
	Cold	13.180	60.11	58.86 ± ?	22.79 ± 0.46
Females	Warm	12.663	68.95	40.09 ± ?	15.81 ± 0.42
	Cold	11.889	59.49	40.75 ± ?	18.53 ± 0.66

* See Table 12.

Were these modifications simple functions of the temperature differences, we should not have expected such a state of things. Indeed the author has no explanation to offer for the striking difference between the two years' results.

A second set of measurements was made upon the same mice at the age of three months. The statistical treatment of these later figures is not yet quite complete. I have determined, however, the mean percentage increase in weight and tail length for each sex, both in the warm room and the cold room lots. Since every

[33] This is the figure for the room in which the thermograph was kept, and in which most of the mice lived during the greater part of this period. For certain reasons, however, all of the animals were kept for a period of varying length (8 to 42 days—the last in the case of only one brood) in another room, having a mean temperature about 4° higher than that of the room first referred to. This fact of course complicates the situation somewhat.

mouse in the present year's experiments is identified by a mark of its own, it has been possible to compute the rate of increase for each animal individually. In Table 13 are presented the mean percentages of increase for each character during the period in question.

It will be seen from this table that in all four cases the figure expressing the increase in a character is *larger* in that group in which the previous absolute measurement had been *smaller*. It must be added, however, that in neither case is the figure for *weight* of any significance in this connection, despite the differences between the averages. The variability in the weight-increase has been enormous (ranging from 9 per cent to 143 per cent), so that the probable errors (not yet computed) are undoubtedly very large. For the growth of the tail, however, the case seems fairly certain. It will be recalled that in the preceding series it was the figures for tail length which bore the strongest testimony to the principle of the leveling down of original differences. Reference to the temperature curves shows that here, as previously, the differences in the conditions have increased rather than diminished during this period. As regards the increase or decrease in variability within each lot, nothing can be said here, since the standard deviations have not been computed for the 3-months measurements.

Supplementary. A yet later series of animals (born March 1909), consisting of a larger number of individuals than any of the preceding lots, has yielded, after similiar teatment, the following results: (1) a difference in tail length somewhat greater than that shown in the preceding series (16.7 per cent in the present case); (2) an indubitable difference (both absolute and relative) in foot length, which applies equally to both sexes and fully confirms the earlier conclusions on this point; (3) a similar difference in ear length, which, however, is of far less certain significance; (4) a difference in weight, both sexes being heavier in the warm room lots (cf. pp. 118, 138 above), although this difference was greater for the females (11.3 per cent) than for the males (9.1 per cent). The temperature differences to which the animals were subjected had already begun to diminish at the time of the first measurement (42 days), and both lots were transferred to the same room when at the mean age of 11 weeks. When measured later, at the mean age of 3 months, the difference in tail length had diminished to 6.7 per cent. The data for this series have not yet been fully compiled.

SUMMARY OF STATISTICAL DATA

The more significant facts which may be disentangled from this mass of data may be briefly summarized as follows:

1   The *tail length* (whether absolute or relative) was found to be very much less in the case of those mice which were reared at the lower temperature.

2   This difference was not accompanied by any appreciable difference in the number of *vertebræ*.

3   The *foot length* was likewise considerably less in all of the cold room lots, though the differences were not so large as in the case of the tail.

4   In two series, at least, the mean *ear length* was found to be smaller at the lower temperature, but these differences are perhaps to be regarded as "accidental."

5   *Body length* was not affected with any degree of certainty; while the influence of temperature upon *weight*, although evident in certain cases, was inconstant, and seemed to depend upon sex.

6   Differences in the average quantity (weight) of *hair* have been demonstrated for the only series in which the point has been tested. The cold room lot were found to have an average amount of hair which was 11.4 per cent greater absolutely and 13.6 per cent greater relatively (i. e., allowing for the dimensions of the animal) than that of the warm room lot. The number of individuals was, however, small (43) and the variability was high.

7   In another experiment a considerable difference in the average *number* of hairs per unit of skin area was found among the male mice, the number being greater in the cold room individuals. Here, likewise, the high variability and small number of individuals render any conclusions doubtful.

8   No constant differences in *variability* were observable between the warm room and the cold room mice.

9   To what extent the modifications cited above have resulted

from differences in *temperature* and to what extent from differences in *humidity* is not certain under the conditions of the experiments.

10 Comparisons of earlier and later measurements upon the same animals show that there is a distinct tendency toward the *reduction of these experimentally produced differences* during subsequent growth, even when the conditions remain unchanged.

11 A *diminution in variability* within each of the contrasted groups, during the course of growth, was shown to be probable for tail length, at least, and possible in the case of weight.

In order to complete this summary, I will add by way of anticipation:

12 The modifications thus artificially produced *are such as have long been known to distinguish northern from southern races of mammals.*

GENERAL DISCUSSION OF RESULTS

In the following comment upon the results of these studies, I shall commence with the last statement in the summary. It is a most significant fact that the experimentally produced differences which have been discussed in the present paper are found to be of just such a nature as are recognized by mammalogists and ornithologists as distinguishing the northern from the southern representatives (individuals or geographical races) of some species. J. A. Allen has repeatedly called attention to the "marked tendency to enlargement of peripheral parts under high temperature or toward the Tropics."[34] Baird and other writers had previously made incidental mention "of the larger size of the bills of southern representatives of northward ranging species" of birds, but Allen offers some detailed examples of this.[35] He concludes that "an increase in the length of the bill is most frequent in long-billed species, while in short-billed ones the increase is in general size, without material change in its proportions" (p. 231). A greater curvature sometimes accompanies the increase in length. Like-

---

[34] The influence of physical conditions in the genesis of species. Radical Review, I, 1877, pp. 108–140. (Reprinted with annotations in Annual Report of the Smithsonian Institution for 1905.)

[35] On the Mammals and Winter Birds of East Florida, etc. Bulletin of the Museum of Comparative Zoölogy, vol. 2, 1871, pp. 161–450, pl. iv–viii.

wise, "there are well-known instances of an increase in the length of the tail" (meaning the *tail feathers*). Much more explicit statements are offered regarding mammals, both by Allen and by Coues, though it is stated by the former that the responses to climatic conditions are less evident in this class than among the birds. "In mammals which have the external ear largely developed, as the wolves, foxes, some of the deer, and especially the hares, the larger size of this organ in southern as compared with northern individuals of the same species is often strikingly apparent. . . . . In *Lepus callotus* ['*Lepus texianus* and its subspecies' —later note], for example, which ranges from Wyoming southward far into Mexico, the ear is about one-fourth to one-third larger in the southern examples than in the northern. . . . . Among the domestic races of cattle those of the warm temperate and intertropical regions have much larger and longer horns than those of northern countries. . . . . Naturalists have also recorded the existence of larger feet in many of the smaller North American mammalia at the southward than at the northward among individuals of the same species. . . . ."[36] In his monograph on the Muridæ[37] Coues repeatedly makes similar statements. Referring to a mouse, "Hesperomys leucopus" (now Peromyscus leucopus) (p. 66), he says: "The arctic series averages larger than the United States specimens, and has shorter feet and ears, as well as shorter tail," and he alludes later to "the well-known law of smallness of peripheral parts in Arctic mammals" (p. 83). Comparing the red-backed vole, "Evotomys rutilus gapperi," a more southern "variety," with the species "E. rutilus,"[38] he finds that the vertebral part of the tail is, on the average, about a third of an inch longer in the former, while the foot is 72 hundredths of an inch in length, as compared with 70 hundredths of an inch in the northern form. Relatively, the differences are even

---

[36] Op. cit., 1905, pp. 382–384.

[37] Monographs of North American Rodentia.—Report of the U. S. Geological Survey, vol. xi, 1877, pp. 1–1091.

[38] It is quite possible that two or more distinct species are here referred to. I am not sufficiently familiar with the classsification of the Muridæ to know the present status of the various species and varieties referred to by Coues. "E. rutilus gapperi" is now regarded as a true species, Evotomys gapperi.

greater, since the northern animals are of larger size. Indeed, the authors cited dwell with equal emphasis upon the *larger size* of the northern representatives (individuals or varieties) of species, both of birds and mammals, as compared with the southern. Previously, Allen tells us, Baird had "explicitly announced a general law of geographical variation in size; namely a gradual decrease in size in individuals of the same species with the decrease in the latitude and altitude of their birthplaces.[39] And Allen further affirms that "this is true not only of the individual representatives of each species, but generally the largest species of each genus and family are northern.[40]

The foregoing statements were made before the days of exact biometry, and an examination of the tables of measurements offered us shows that in most cases they comprise relatively few individuals and that the material used was not homogeneous, i. e., it includes alcoholic and fresh specimens, as well as dried skins. For this reason most of these tables are not likely to be of very great use to the modern student of variation. In more recent years, a very extensive mass of similar measurements has been gathered by a considerable number of naturalists, but, so far as the writer has been able to discover few if any of these have been subjected to statistical treatment with reference to testing the generalizations of Baird, Allen and Coues.[41] It would seem overskeptical, however, to reject the emphatic opinions of a number of able naturalists upon these matters, particularly as we have no more satisfactory data at our disposal.

Regarding the pelage, there can be little doubt that this likewise responds directly or indirectly to climatic conditions. "At the northward, in individuals of the same species, the hairs are longer and softer, the under fur more abundant, and the ears and the soles of the feet better clothed. This is not only true of individuals of the same species, but of northern species collectively, as compared with their nearest southern allies."[42] Both Coues

---

[39] Op. cit., 1871, p. 230.
[40] 1905, p. 378.
[41] I state this on the authority of several of our leading students of mammalian distribution to whom I have appealed for information.
[42] Allen, 1905, p. 382.

and Allen cited many specific instances of this fact for mice, hares, squirrels and other rodents. Moreover, obvious seasonal changes are to be observed in some species. Speaking of the squirrel, Sciurus hudsonius, var. hudsonius, Allen says: "In summer the soles of the feet are naked, often wholly so to the heel; in winter they are wholly thickly furred, only the tubercles at the base of the toes being naked. The general pelage is also much fuller, longer and softer in winter than in summer."[43]

It may well be that the change in the quantity of hair which appears to have been produced in the white mice during the experiments above described was comparable to these seasonal changes, i. e., that the results were purely temporary, and would have disappeared with a cessation of the conditions employed.[44] Indeed, since the life of an individual hair is comparatively brief, it would be necessary to effect some permanent change in the physiological activity of the hair follicles, in order that differences such as these should endure. Whether or not these effects are permanent, it has been believed by many that various changes in the character of the hair coat occur in domestic animals as the result of transferrence to an unaccustomed habitat. Darwin, indeed, tells us[45] that "Great heat, however, seems to act directly on the fleece; several accounts have been published of the changes which sheep imported from Europe undergo in the West Indies. Dr. Nicholson of Antigua informs me that, after the third generation, the wool disappears from the whole body, except over the loins; and the animal then appears like a goat with a dirty door-mat on its back." And again:[46] "It has been asserted on good authority [Isidore Geoffroy St. Hilaire] that horses kept during several years in the deep coal-mines of Belgium become covered with velvety hair, almost like that on the mole." The "classical" Porto Santo rabbit may be cited as another and perhaps more authentic instance of the modification of mammals through changed cli-

---

[43] Monographs of N. A. Rodentia, p. 675.

[44] It is uncertain, to be sure, in how far the season changes of the hair coat of mammals are *direct* responses to climatic conditions.

[45] Variation of Animals and Plants under Domestication, vol. i, p. 124.

[46] Op. cit., vol. ii, p. 336.

matic conditions. Here, not only the hair, but other features, were affected.

So far as the present writer is aware, however, no such differences as have formed the principal theme of this paper have been previously brought about by direct experiment or even produced under such circumstances as would warrant one in stating positively that they were the immediate results of external conditions. Lydekker, in the work already referred to, cites a case on the authority of "an American newspaper" (so notoriously infallible in matters scientific!) which would certainly be important if true. It is worthy of mention only because the modifications alleged accord so well, in some respects, with those which have been demonstrated for mice. In order to combat the rats in a cold-storage warehouse at Pittsburgh—so the story runs—cats were introduced. The first of these died. "One cat was finally introduced . . . . which was able to withstand the low temperature. She was a cat of unusually thick fur, and she thrived and grew fat in quarters where the temperature was below 30°. By careful nursing, a brood of seven kittens was developed in the warehouse into sturdy thick-furred cats that loved an Icelandic climate. They have been distributed among the other cold-storage warehouses of Pittsburgh, and have created a peculiar breed of cats, adapted to the conditions under which they must exist to find their prey. These cats are *short-tailed* [italics mine], chubby pussies, with hair as thick and full of under-fur as the wild cats of the Canadian woods. One of the remarkable things about them is the development of their 'feelers.' . . . . In the cold warehouses the feelers grow to a length of five and six inches. This is probably because the light is dim in these places, and all movements must be the result of the feeling sense."

I am informed by Dr. A. E. Ortmann, who has kindly taken the trouble to make some inquiries regarding this story, that he can find no foundation for it whatever. Those who had heard of it at all did not take it seriously. Moreover, as Dr. Ortmann points out, it seems quite unlikely that cats could be forced to live in a cold-storage warehouse unless caged. It has, nevertheless, seemed worth while to cite this account, owing to the prominence given to it by Lydekker.

Passing to the question of the *adaptiveness* of these experimentally produced modifications, that of the hair would surely seem to fall within this category. A complementary physiological explanation for the change would doubtless be likewise possible, had we a sufficient knowledge of the various processes concerned. The shrinkage or "drawing in" of the peripheral parts under the influence of a cold climate might also be regarded as adaptive, for the reduction of these thinly clothed surfaces would diminish, at least theoretically, the radiation of heat from the body. Here again a simple physiological explanation is likewise possible. We might either appeal to the effect upon the peripheral circulation (as does Allen) or to the direct influence of temperature upon the protoplasm of the growing parts. In the case of the feet of the mice in the above experiments, the greater activity of the young animals in the warm room, and the greater consequent exercise of the limbs, may possibly have played some part in bringing about the difference.

To what degree the modifications which I have described have been due to temperature and to what degree they have been due to humidity is not clear under the conditions of the experiments. As has been stated, the two have varied inversely. Allen and Coues seem to regard such differences, when presented by mammals in nature, as due chiefly to the temperature factor. Nevertheless, the former writer tells us, speaking of hares, that "there is also a marked tendency to an enlargement of the ears in proportion to the aridity of the habitat. . . . . In this connection, also, attention may be called to the fact that all of the long-eared species of American hares are found exclusively over the most arid portions of the continent. . . . ."[47] And it may be added that the color of the pelage of mammals and that of the plumage of birds is well known to vary with the hygrometric conditions. In many species of birds the degree of pigmentation is said to be a function of the mean humidity of the habitat. Tower, indeed, regards the humidity as being much more important than the temperature in the production of color changes in beetles. Until,

---

[47] Monographs of N. A. Rodentia, p. 272.

therefore, it is possible to separate these two factors in our experiments, we cannot state with any certainty to what degree each has been operative. *A priori*, it would seem, perhaps, that the changes in the mice have been such as could more reasonably be attributed to temperature.

The fact that the same sort of differences as those which sometimes obtain in nature between northern and southern species or varieties of animals have been produced by artificial conditions acting within the individual lifetime will be taken by some as evidence that these differences in nature are likewise entirely "ontogenetic" or acquired independently by each individual. Conversely, the neo-Lamarckian will perhaps argue—and with equal right—that here we have evidence that natural varieties and species have resulted from the accumulated effects of external conditions since the reality of such effects has been palpably demonstrated by the present experiments. Neither conclusion is justified by the facts before us. It remains to be settled experimentally (and thus only!) whether or not such modifications are transmissible.

It has already been stated that no constant difference in size between the warm room and cold room individuals has been found to obtain throughout my series. Here, then, the reputed effects of natural climatic conditions have not been paralleled. It is quite possible that the cold was so severe during the early growth stages that some individuals were stunted. Indeed it has been pointed out for the 1907–1908 series that there was considerable mortality amongt the cold room lot in early life. Reference to the frequency polygons in Fig. 2 shows that there are two distinct modes among the cold room individuals; and I have determined that this is equally true of each of the sexes taken separately. The impression conveyed is that there are two pretty distinct groups, one of which was stunted by exposure to the cold, the other being favorably affected, so as even to surpass the warm room lot in size. It must be added, however, that no such effect is manifest in the 1908–1909 series.

One of the most important general conclusions which seem warranted from an analysis of the foregoing results is the principle of the levelling down of experimentally produced inequalities, even while the conditions which gave rise to them remain in full force. A diminution in initial differences of size has been demonstrated by Minot[48] in the case of growing guinea-pigs. His findings upon this point are thus summed up: "The study of the individual variations yields two important conclusions: *First*, that any irregularity in the growth of an individual tends to be followed by an opposite compensating irregularity. *Second*, the variability diminishes with the age." Thus, "if an individual grows for a period excessively fast, there immediately follows a period of slower growth, and *vice versa*, those that remain behind for a time, if they remain in good health, make up the loss (at least in great part if not always completely) soon after. . . . . It is probable that the same is true for man and that therefore the usual and even the severer illnesses of childhood and youth do not greatly affect the ultimate size of the adult." Pearson,[49] likewise, has shown that the variability both of weight and of stature in man diminishes from infancy to adult life. And indeed it is a matter of common experience that an early handicap in the size or strength of a child is frequently "outgrown," wholly or in part.

The variability which the above-named writers have considered is doubtless in part due to blastogenic differences, in part to somatogenic ones, resulting from fetal or post-natal conditions of nutrition, etc. In my own results, however, the most noteworthy fact is not a reduction in the general variability of my stock, but the diminution of differences whose cause is known to be external—*and this while the effective conditions remain unchanged*. The foregoing statement applies to the growth of the tail, both of the male and the female mice, between the age of six weeks and the age of $2\frac{1}{2}$ (or 3) months. It likewise holds, with some qualification, for the growth of the tail during the next interval between the measurements, i. e., between $2\frac{1}{2}$ and 7 months. In the latter case, however, the data are fewer, and the allowance is necessary that about midway during this third period the temperature conditions were equalized

---

[48] Senescence and Rejuvenation. Journal of Physiology, vol. xii, no, 2. 1891, pp. 97–153, pl. ii–iv.
[49] Proceedings of the Royal Society, lxvi, 1900, p. 23 (cited by Vernon, in "Variation in Animals and Plants").

for the entire lot of animals. The diminution of differences in weight between the contrasted groups of animals is less certain, though it appears tolerably clear in the case of the 1907–1908 females. It will be recalled, however, that the differences in weight were only very doubtfully regarded as results of the temperature conditions.

This tendency toward a reduction of experimentally produced differences in the relative size of parts, should it prove to be general, is of considerable theoretic interest. It adds another to the many well-known examples of a "regulative" tendency in living things.[50] After the initial shock of change, with its resulting effect in deflecting the organism away from its individual norm, there would seem to be a continuous effort to regain the latter. Here we have a principle which might be said to bear the same relation to individual growth as Galton's "law of filial regression" bears to stem-history, though the analogy may be merely superficial. In either case, however, we have to do with a "reversion to mediocrity." The process in question is directly opposed to that conceived of by Weismann, in his theory of "germinal selection," as occurring among the determinants of the germ-plasm. According to this hypothesis, a given determinant, if once handicapped by unfavorable nutrition, is more and more pushed to the wall by its more fortunate competitors until it may be totally annihilated. The disappearance of useless structures in phylogeny and the frequent orthogenetic trend of evolution is thus explained. If it be objected that this analogy of mine is out of place I can only reply that Weismann's whole conception of a struggle among the determinants of the germ-plasm was derived from what was assumed to occur among the parts of the organism as a whole. Some evidence has been offered above for the existence of a tendency in the growing body quite at variance with the demands of that theory. To many readers, on the other hand, it will doubt-

---

[50] Vernon's principle that "the permanent effect of environment on the growth of a developing organism diminishes rapidly and regularly from the time of impregnation onwards" (op. cit., p. 199) would account for the failure of these differences to augment with the growth of the organism. But it certainly would not in itself account for the *absolutely* greater increase shown by the more retarded organs (or organisms) mentioned on p. 137 above.

less seem quite frivolous to attempt any serious refutation of the "germinal selection" hypothesis. In justification, I will but call attention to the fact that this theory is not only treated respectfully but is ably defended in the most recent general treatise on heredity.[51]

---

[51] I refer to Thompson's admirable work, Heredity (G. P. Putnam's Sons, 1908), which should be in the hands of every student of this province.

THE RÔLE OF ISOLATION IN THE FORMATION OF A NARROWLY LOCALIZED RACE OF DEER–MICE (PEROMYSCUS).*

DR. F. B. SUMNER

SCRIPPS INSTITUTION, LA JOLLA, CALIF.

No one who has critically examined large numbers of specimens, belonging to such a widely distributed and diversified genus as *Peromyscus,* can fail to be impressed with two facts. First, the differences upon which the so-called "subspecies" are based are real and obvious ones. But, secondly, the actual subspecies which are recognized and named are necessarily highly artificial groups. On the one hand, each subspecies intergrades with others to such an extent that the assignment of a given specimen to one or the other group is often quite arbitrary. And on the other hand, even these "subspecies" themselves are far from being elementary. They are composite groups, comprising, in many cases, a number—perhaps a great number—of distinguishable local types. The word *distinguishable* is here used in a qualified sense. It is likely that the distinctions would commonly be obvious just in proportion as the collections were made at points which were remote from one another.

Indeed, it has been said by one who has monographed this genus of mice[1] that "classification becomes . . . like dividing the spectrum and depends largely upon the standards set, for, theoretically at least, the possibilities of subdivision are unlimited (p. 17)."

None the less, it is generally believed that where well-marked physical or other barriers are interposed between two groups of individuals, this continuous intergradation

* Read before Ecological Society of America, San Diego meeting, August, 1916.

[1] Osgood, "Revision of the Mice of the American Genus, *Peromyscus,*" North American Fauna, No. 28. Washington, 1909.

of racial characters may be largely interrupted. It is the object of this paper to discuss a case of this sort which I have had the opportunity of studying during the past year.

The subspecies *Peromyscus maniculatus rubidus*, according to Osgood,[2] who first described it, occupies a strip of varying width on the "coast of California and Oregon from San Francisco Bay to the mouth of the Columbia River." In discussing certain local variations shown by this subspecies throughout its range, the same writer states that "six specimens from the Outer Peninsula, near Samoa, Humboldt Bay, are decidedly paler than others from the neighboring redwoods. They evidently represent an incipient and very local subspecies, and well illustrate the plasticity of the group to which they belong." Osgood further remarks that "a careful study of this variation and the local conditions doubtless would prove instructive" (p. 66).

During the latter part of May, 1916, I trapped on two consecutive nights in the neighborhood from which Osgood obtained his six "aberrant" specimens of *rudibus*.[3] About one hundred live-traps were set on each occasion. Twenty-eight specimens were taken, of which twenty-one were later available for skinning and for careful measurement. These last were all in either mature or adolescent pelage, and were about evenly divided in respect to sex.

The distinctness of this race from the *rubidus* of the redwood forests on the mainland was evident from a casual inspection of the living mice. A more careful comparison of freshly killed specimens from the two localities, and later of their prepared skins, justifies the following generalizations. These impressions were formed independently by several other persons to whom I showed the specimens, and were confirmed by more careful ex-

[2] *Loc. cit.*, p. 65.

[3] The trapping was done between one and two miles northwest of the village of Samoa. Besides these *Peromyscus*, the only other animal caught was a single specimen of *Microtus*.

amination and measurement. (1) The Samoa lot, as a whole, were paler than the redwood lot; (2) the tails of the former were shorter, and (3) the ears were longer.

To consider first the coat color, the mean difference between the two series of skins is evident at a glance. Likewise, it is plain that the palest Samoa specimen is paler than the palest Eureka (redwood) specimen, and that the darkest among the former is paler than the darkest among the latter. It must be admitted, however, that the two series overlap rather broadly,[4] the darker skins of the Samoa stock being as dark as or darker than the paler ones of the Eureka stock.

An attempt to express the color of a mammal's pelage in terms of any set of "standard" colors is beset with great difficulties. Instead of a uniformly tinted, plane surface, we have to do with a mixture of variously colored hairs, further diversified by minute shadows and reflections. I have, nevertheless, endeavored, in a rough way, to "match" the colors of these two races with those of Ridgeway's "Color Standards and Color Nomenclature."[5] In the Samoa race, the general tone of the lateral regions of the body lies between the "tawny olive" and "Saccardo's umber," that of the dorsal darker stripe being not far from "sepia." In the Eureka mice, the lateral regions range from "Saccardo's umber" to "sepia," the dorsal stripe being of a depth somewhere between "sepia" and black. These comparisons will at least enable the reader to judge of the degree of difference between the two racs.[6]

As regards the tail, it was plain without measurement that the average length of this member was greater in the

[4] I have at present for reference twenty-one skins of the Samoa lot and thirty skins of wild adults from the redwoods. Ten of the latter individuals were trapped and skinned at about the same time as the former, so that the factor of season may be disregarded.

[5] Washington, 1912. Published by the author.

[6] In my further studies of *Peromyscus* I plan to employ two revolving color-wheels, on one of which the skin itself will be rotated, on the other sectors of black, white and various primary colors. This apparatus is now being tested by Mr. H. H. Collins and myself.

Eureka than in the Samoa race, though here again the difference related to averages and did not hold for all individual cases.

A comparison of the mean figures for *absolute* tail length in two series of mice is not entirely justifiable, particularly if the two lots of individuals differ somewhat in mean body size. But the *relative* tail lengths (expressed as percentages of body-length) may be fairly compared, since there is good evidence that these ratios remain nearly constant after the first few months of life. The following table allows of a comparison between the two races, in respect to this character:

	Number of Cases	Mean (Percentage)	Standard Deviation
Eureka (males)	83	104.39 ± 0.37	4.95
Eureka (females)	53	103.60 ± 0.54	5.85
Samoa (sexes combined)	21	97.48 ± 0.94	6.38

The differences between the Samoa lot (sexes combined) and the Eureka males and females are 6.91 per cent. and 6.12 per cent., respectively. These differences are about seven and six times their probable errors, respectively. Their significance may therefore be regarded as fairly certain, despite the small numbers comprised in the Samoa series.

As regards foot-length, the two races do not differ significantly. But the ear, as already stated, is appreciably longer in the Samoa mice, this difference being perceptible, even without measurement. Here, as in the case of tail-length, a simple comparison of gross averages for the two groups would be unjustifiable. But in the present instance, the conversion of the absolute values into percentages of body-length would be equally unjustifiable, since the growth of the ear is not at all proportionate to that of the body as a whole. We must therefore resort to the method of "size groups," *i. e.*, we must divide each of our two lots of animals into small groups comprising individuals of nearly equal size.

In the case at hand, we have fifteen groups, or rather

pairs of groups, within which a comparison of average ear-length is possible. In twelve cases the mean figure is greater for the Samoa mice, in two cases it is greater for the Eureka mice, while in one case the two figures do not differ appreciably. The probabilities against such a preponderance being due to chance are of course high. The mean difference in ear-length between the two lots, computed according to a method described by me in an earlier paper,[7] is 0.87 mm. Those who have made careful measurements of mice will regard such a difference in the length of this appendage as far from trivial.

Let me now say something as to the environmental conditions under which these two races of *rubidus* live. Those which I have designated as the "Eureka" or "redwood" race were trapped by me during two different years, within a distance of two miles from the southern limits of the city of Eureka, California. The region is one covered in large part by redwood forest, most of which is of second growth, although there are some small areas that have never been logged. The predominant tree is the redwood (*Sequoia sempervirens*), but several other conifers are common, the most abundant of these being the Sitka spruce (*Picea sitchensis*), Douglas fir (*Pseudotsuga taxifolia*), and lowland fir (*Abies grandis*). The red alder (*Alnus rubra*), cascara (*Rhamnus purshiana*), waxberry (*Myrica californica*), red elderberry (*Sambucus racemosa*), and a willow (*Salix hookeriana*) appear to be the chief non-coniferous trees of this district.[8] The "wild lilac" (*Ceanothus thyrsiflorus*) is likewise common in some of the more open areas, often reaching the proportions of a small tree.

Except in recently cleared tracts, the region is one of dense underbrush, the shrubbery and vines forming, in fact, a veritable jungle which is frequently hard to pene-

[7] *Journal of Experimental Zoology,* Vol. 18, April, 1915, particularly, pp. 341 et seq.

[8] For the determination of many of the plants referred to in this paper I am indebted to Professor H. M. Hall, of the University of California, and to Mr. J. P. Tracy, of Eureka.

trate. Here we meet with the thimble-berry (*Rubus parviflorus* var. *velutinus*), the salmon-berry (*Rubus spectabilis* var. *menziesii*), huckleberry (*Vaccinium ovatum*), red bilberry (*V. parvifolium*), salal (*Gaultheria shallon*), and in the more open areas the blackberry (*Rubus vitifolia*). Two ferns (*Aspidium munitum* and *Pteris aquilina*) are extremely abundant, the latter in particular forming dense growths higher than a man's head. In the

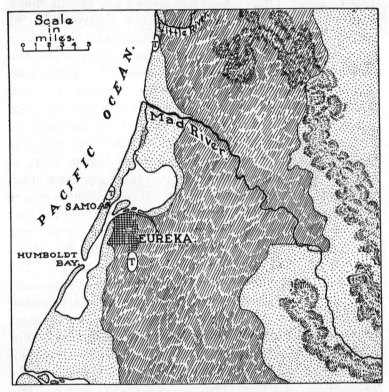

FIG. 1. Map of the vicinity of Humboldt Bay, California, based upon J. N. Lentell's map of Humboldt County. The three principal trapping stations are designated by the letter T. Area occupied by redwood forests is indicated by oblique shading.

more open areas a tall annual of the evening primrose family (*Epilobium angustifolium*) constitutes an important element in the vegetation.

One coming from the more arid parts of California can

not fail to be impressed by the prevailing humidity of both soil and atmosphere in this region. In the dense shade of the great redwoods the ground is damp, even during the summer months, and the fallen logs are covered with mosses and fungi.

When we cross Humboldt Bay to the narrow peninsulas separating this body of water from the ocean (Fig. 1), we enter a quite different environment. No redwoods are found, the woods, where present, are open, and the ground is prevailingly dry and sandy. In the wooded area, extending down the axis of the northern peninsula, the predominant tree is a small pine (*Pinus contorta*), though the waxberry and willow (*Salix hookeriana*) are likewise abundant, and small specimens of the Sitka spruce are fairly common. Among the more frequent shrubs are the huckleberry (*V. ovatum*), the twinberry (*Lonicera involucrata*) and silk tassel bush (*Garrya elliptica*). The ground is largely covered by two plants of trailing habit, the bearberry (*Arctostaphylos uva-ursi*) and the beach strawberry (*Fragaria chilensis*).

On its ocean side, the peninsula is bordered by a wide strip of shifting sand. Here the process of dune formation may be witnessed to perfection, the dunes often reaching a height of forty or fifty feet. In places the encroachments of the sand upon the hard-pressed vegetation are evidently rapid, solid ramparts of willows and spruces being steadily engulfed by an advancing wall, frequently as high as the trees themselves. Nevertheless, even on the open sands of the dunes, certain trailing plants maintain a precarious foothold. Among the commonest of these are to be mentioned the yellow sand verbena (*Abronia latifolia*), the beach strawberry (*F. chilensis*), beach pea (*Lathyrus littoralis*), and two species of Franseria (*F. chamissonis* and *F. bipinnatifida*), while the succulent *Mesembryanthemum aequilaterale* is occasionally met with.

Despite the nearness to the ocean and the high atmospheric humidity, the peninsula region seems dry in comparison with the redwood forests. This is due in part to the loose, sandy character of the soil—where, indeed, any real soil exists—and to the comparative lack of shelter from the prevailing westerly winds. Evaporation here is doubtless more rapid than in the comparatively stagnant air of the forests.

To my surprise, the footprints of mice and other small mammals were abundant, even on the shifting sands, in the areas of sparsest vegetation. Since these tracks, for the most part, were effaced every day by the wind, the animals must have been present in large numbers. Indeed, it was in or close to the dune region that I trapped most of the twenty-eight *Peromyscus*. It seems more than possible, therefore, that the predominantly paler shade of the mice dwelling here may be due to the same causes which are operative in producing the yet paler hues of many of the desert rodents.

What the effective factors are can not yet be stated with certainty in either case. Protective coloration is of course an obvious explanation, but it is one of doubtful applicability in the case of animals which are almost wholly nocturnal in their habits. For this and other reasons it seems more likely that the pale coloration of these mice stands in some more direct relation to the humidity of their immediate surroundings. That it is not, however, a strictly ''somatic'' phenomenon, called forth anew in each generation, I have already shown for the desert race, *P. m. sonoriensis*.[9]

Whether or not the peculiar color of the pelage in the Samoa race is likewise hereditary I have endeavored to test experimentally. Seven living females and a number of males were brought to La Jolla in June, 1916. Unfortunately, it was not possible to obtain more than two broods of young, comprising three individuals, one male and two females. These animals were carefully examined at the age of five months, in comparison with over forty individuals, derived from the redwood stock, which were

[9] AMERICAN NATURALIST, Nov., 1915. I have since reared this race in Berkeley as far as the third (in one instance the fourth) cage-born generation, without any certain modification in color.

mainly of the same age or older, and likewise reared from birth at La Jolla. Not a single individual of the latter stock was as pale as either of the two females of Samoa parentage. The male of the Samoa race was, however, of about the average shade of the redwood descendants. As stated above, some of the wild parents, trapped on the peninsula, were likewise as dark as many of the redwood series.

No certain conclusions can, of course, be based upon these three individuals. But the condition of the two females certainly lends support to the belief that the peculiar coat color of the Samoa race, however it was acquired, has become fixed germinally.

Reference to the map shows that the northern peninsula of Humboldt Bay is largely isolated, so far as land-living rodents are concerned. In addition to the ocean and the bay, a marshy tract extends from the latter to the Mad River, which, in turn, interposes a further barrier on the north, and nearly converts the peninsula into an island. Beyond the mouth of Mad River, this same type of sand-dune formation extends uninterruptedly to the mouth of Little River, about six miles to the north, where it ends abruptly and the shore line becomes precipitous.

Now this northward extension of the sand-dune region is not isolated by any physical barrier from the redwood forest, which here comes near to the coast. It occurred to me, therefore, to attempt the collection of *Peromyscus* from a point somewhere within this region. The locality chosen was close to the ocean, about two miles south of Little River and four to five miles north of Mad River. Here the conditions were found to be closely similar to those on the exposed side of the northern peninsula of Humboldt Bay. The dunes were on the whole lower, however, and some minor differences were noted in the flora. The belt of shifting sand here ranges from five or six hundred feet to perhaps a fourth of a mile, giving place on the landward side to a narrow meadow or marshy area, succeeded by a high, steep, wooded ridge.

About ninety traps were set on two consecutive nights,

yielding in all forty-eight *Peromyscus,* all belonging to the subspecies *rubidus*. Many of these were still in juvenile pelage and such individuals were kept and allowed to mature in captivity.

A hasty comparison of the living Little River animals (as I shall call them) with those from the Samoa and Eureka trapping grounds made it plain that, in respect to color, they belonged with the latter group rather than the former. Careful comparisons of series of dead mice and of skins were made later and the bodies were subjected to the customary measurements. Owing to numerous deaths, however, only twenty-eight individuals were available for these purposes.

This more critical examination confirmed my earlier belief that the Little River mice agreed pretty closely, in average color, with the redwood stock, but that they differed widely from those taken on the peninsula. It seemed probable, however, that the mean shade was slightly lighter than that of the former animals, making them, to this extent, intermediate.

One conclusion then seemed plain. The peninsula race, exposed to certain modifying conditions, was enabled to differentiate from the mainland stock, owing to the almost insuperable barriers to migration. The Little River stock, exposed to practically the same conditions, have not formed a distinguishable race, because the rate of differentiation has been far exceeded by the rate of diffusion, or intermingling with the great body of more typical "*rubidus*," dwelling in the redwood forests which extend back from the coast. We might seem to have, therefore, a particularly clear cut example of the effectiveness of isolation in the formation of a local race.

Now, I am not yet prepared to admit that such conclusions would be groundless. But here, as so often happens, a further study of the data has shown that the problem is more complex than was at first suspected. It is true that the mice of the more northern sand dunes have not formed a distinct race as regards *color*. But it is none the less certain that they differ from those of the

Eureka region in regard to both the length of the tail and that of the ear. In respect to the former character, they agree pretty closely with the Samoa race, the difference from the redwood stock being statistically even more certain in this case. To still further complicate the situation, we find that the ear, instead of being longer, is shorter than that of the redwood mice by about half a millimeter, and thus averages about one and one half millimeters shorter than in the peninsula race. Here, too, the differences are even more certain statistically than those which distinguish the Eureka and Samoa series.

The numbers are small, of course, only twenty-eight of the Little River mice having been available for measurement. But as regards tail length, the difference between the averages is seven to nine[10] times its probable error, so that the likelihood of its being due to random sampling is very small.

Have we, then, here merely another example of inconclusive data, which might best have been left unpublished? I do not think so. The mere existence of these local differences in color and in the size of parts deserves careful description, whatever interpretation we may place upon them.

Moreover, I am disposed to believe that the case of coat color is not entirely comparable with that of the length of the appendages. In another article[11] I have given reasons for thinking that some of the differences in the former may have arisen in nature as more or less direct effects of environmental conditions. On the other hand, I have shown that such an explanation would be of very difficult application as regards some of the measurable differences in the parts of the body, even though the latter are known to be readily influenced by various experimental agencies.

Now the evidence at hand is sufficient to show that any environmentally produced modifications of coat color are

---

[10] Depending on whether the comparison is made with the Eureka males or females, the sexes being combined in the case of the Little River group.
[11] AMERICAN NATURALIST, Nov., 1915.

at best rather gradual. *Rubidus* remains *rubidus* and *sonoriensis* remains *sonoriensis*, after several generations of captivity in changed climates. But even the first cageborn generation of each of my subspecies is found to be highly modified by confinement, in respect to the mean length of certain of the appendages. That this somatic plasticity would be accompanied by a high degree of germinal instability, as regards these parts, could not, of course, be predicted in advance. But the frequent appearance of local differences of type renders it probable that this is true. Whether or not these local peculiarities are due in some indirect way to environmental factors, or whether they are due to "spontaneous" mutation, need not concern us here. The main point to bear in mind is the probability that the pelage color is somewhat more stable in these mice than are the bodily proportions, despite the fact that it is the former, rather than the latter, which gives the clearest evidence of a definite correlation with known factors of the environment.

For the reason just stated, it is possible that the differentiation of a new color race might require fairly rigid isolation; whereas local differences in some of the measurable parts might arise in the presence of no other barrier than the naturally slow rate of diffusion of a nonmigratory animal. As was remarked earlier, we have reason to suppose that representative collections from an indefinite number of localities would reveal the existence of statistically certain differences between the mice of many of these localities. In most cases, it would probably be unjustifiable to assign these series to distinct *races*, or other definite taxonomic groups, since it is likely that perfect intergradation would be found between most of them, and that the degree of difference would be largely a function of the distance apart of their respective habitats.

These last remarks are, of course, largely conjectural. Part of the author's present program consists in a careful study of local differences of the sort here discussed.

It is hoped that this will render possible more definite answers to some of these difficult questions.

It seems to be held by certain zoologists that any discernible difference between two local types, if at all constant, ought to be in some way recognized in the nomenclature. Indeed, I have been advised to name this modified race of *rubidus* from the northern peninsula of Humboldt Bay. Such a practise, if carried out consistently, would lead either to an endless multiplication of subspecies, or else to the introduction of quadrinomial names. Either procedure would, I think, be deplorable. The actual needs of the situation can commonly be met, I believe, by stating the locality from which a given specimen or collection was taken. The bestowing of formal names creates the false impression of a multitude of well-defined entities which do not, in reality, exist. Moreover, it is my firm conviction that nomenclature should have for its object the recognition of resemblances as well as the recognition of differences. The first of these functions is all too frequently overlooked.

GENETIC, DISTRIBUTIONAL, AND EVOLUTIONARY STUDIES OF THE SUBSPECIES OF DEER MICE (PEROMYSCUS)

by

F. B. SUMNER

GENETIC, DISTRIBUTIONAL, AND EVOLUTIONARY STUDIES OF THE SUBSPECIES OF DEER MICE (PEROMYSCUS)

by

F. B. SUMNER

(The Scripps Institution of Oceanography of the University of California)

TABLE OF CONTENTS

	Page
INTRODUCTION	2
GENERAL CHARACTERISTICS, LIFE HISTORY, HABITS, ETC.	6
METHOD OF COLLECTING	12
CARE OF STOCK	14
LINEAR AND COLORIMETRIC MEASUREMENTS	20
NATURE OF RACIAL DIFFERENCES	25
THE HERITABILITY OF RACIAL DIFFERENCES	27
THE HERITABILITY OF INDIVIDUAL DIFFERENCES	28
"MUTATIONS"	30
Albinism 31; *Pallid* 31; *Hairless* 33; *"Yellow"* 34; *Grizzled* 35; *Miscellaneous heritable peculiarities* 37.	
RESULTS OF EXPERIMENTS IN HYBRIDIZATION	39
CORRELATION	51
RACIAL DIFFERENCES IN RELATION TO ENVIRONMENT	56
OTHER LINES OF INVESTIGATION UPON PEROMYSCUS	86

	Page
BIBLIOGRAPHY	98
REGISTER	105

INTRODUCTION

In the following pages are summarized the results of an inquiry devoted primarily to the genetic and evolutionary status of geographic races, as exemplified in certain species of mice, belonging to the genus *Peromyscus*. Despite the high scientific interest which attaches to the geographic races or subspecies of animals, and the preoccupation of taxonomists, for many years past, with the task of naming and describing these, it is a curious fact that until rather recently there was little or no critical evidence that the differences involved were hereditary at all. By some zoologists (J. A. ALLEN, for example) it was believed that such was the case, but the evidence offered was mainly circumstantial. On the other hand, it was queried by JORDAN whether we might not be dealing here merely with "ontogenetic species," whose characters were direct responses to local conditions, and were not fixed germinally. Even as late as 1921, two British ornithologists [1], writing in a standard journal, did not hesitate to assert their belief that by far the majority of present-day subspecific forms among birds were unstable environmental modifications which would quickly disappear in a new environment.

That in the mammalian genus *Peromyscus*, at least, and presumably in a wide range of other forms, the differences between subspecies are chiefly or wholly hereditary is now shown by abundant experimental evidence. Subspecies from widely different climatic regions maintain their characteristic differences of form and color, after being reared for a number of generations in a common environment.

Of course, such evidence as the last throws no light upon the phylogenetic origin of these differences. It is not inconsistent with the belief that climatic influences may have a cumulative effect in the course of sufficiently great periods of time. The present writer made it plain that he commenced this program of investigation with a

---

[1] LOWE and MACKWORTH-PRAED (1921).

distinct bias in favor of such an interpretation of the phenomena of geographic variation. This was due in part to previous experiments of his own, upon white mice (1909, 1915a), in the course of which certain modifying effects of temperature were revealed [1]), of such a nature as to conform with ALLEN's generalization respecting the "marked tendency to enlargement of peripheral parts under high temperature or toward the Tropics" (1877).

However, the problem of the underlying causes of geographic variation has proved to be increasingly complex, as the present work has progressed. It is at once the most interesting and the most difficult of the problems which here offer themselves. And it is only fair to admit that this is probably the one upon which the least light has been thrown by the present investigations.

Another matter of high theoretical importance, in relation to which little previous experimental work had been done, was the question whether the characters by which subspecies differ from one another belong to the same category as the ones which distinguish the various color varieties of domestic animals, or the various "mutant" stocks which have been bred so extensively for many years by students of Mendelism. Are the differences between these two classes of characters merely differences of complexity, or are there fundamental differences in their mode of origin, method of hereditary transmission, and significance for the theory of evolution?

At first glance, it would seem fairly obvious that the latter alternative is the correct one. Variations of the "mutant" sort have long been known to systematic zoologists and botanists, under such names as "sports", "freaks", "aberrations", or sometimes simply as "varieties." Some of these variations — we may take albinism as an extreme example — were known to ignore taxonomic lines entirely, and to crop out sporadically in species belonging to widely different groups. They were inherited according to the all-or-nothing principle: there were no intermediates. Their incidence commonly bore no definite relation to geographic locality. And for the most part they

---

[1]) Reference is here made only to the effects upon the generation immediately treated, not to the more than doubtful inheritance of such effects. The former result was obtained in four different series of experiments and has been obtained independently by PRZIBRAM (1909 and later).

represented no new acquisition, but merely the loss of color or of some other characteristic.

On the other hand, the characters distinguishing species and other natural groups did not make their appearance in any such erratic manner. So far as known, they were not inherited in alternative fashion, when species or subspecies were crossed, but presented all possible intermediate conditions. They frequently bore very definite relations to habitat or geographic range. And in some cases, at least, they appeared to be characters of a positive sort, having probable adaptive value.

Without going into the history of this controversy, it is sufficient to note that systematic zoologists and botanists, and field naturalists in general, have stressed the differences between these two classes of characters, while students of heredity, since the advent of Mendelism, have tended to minimize the differences and to interpret them as due to differing degrees of complexity in the genetic basis of the characters concerned.

The preceding paragraphs will serve to indicate some of the problems which seemed to call for investigation at the time when these studies of *Peromyscus* were undertaken. They may also make clear why I was, at the outset, sympathetic with a Lamarckian explanation of the phenomena of geographic variation, and why I was skeptical of the endeavor to interpret all hereditary characters in terms of Mendelian genes. To what extent the results of these investigations have forced me to change my viewpoint will appear later. I may say here, however, that I have come to regard the "multiple factor" extension of the Mendelian principle as providing a fairly adequate interpretation of the inheritance of subspecific characters, while I am less disposed than formerly to admit the efficacy of environmental stimuli to bring about hereditary changes in the course of a few generations.

Owing to these considerable changes in my own viewpoint, and to the further fact that the results of the *Peromyscus* investigations have been published in about a dozen different journals, throughout a period of fifteen years, it now seems worth while to present a rather full recapitulation of these results and of the conclusions which at present appear to be warranted. In a few cases, hitherto unpublished data have been made use of. In preparing this summary, it has seemed desirable to include a brief account of the methods which have been employed by me in dealing with these animals. It may be expected that the genus *Peromyscus* will figure prominently in genetic and ecological studies in the future [1]).

The investigations herewith detailed were commenced late in the year 1913, under the auspices of the Scripps Institution for Biological Research of the University of California. A preliminary year was spent in Berkeley, during which considerable time was devoted to an examination of the invaluable series of skins in the Museum of Vertebrate Zoology, which had been accumulated through the labors of Professor JOSEPH GRINNELL and his staff. I here received my first real introduction to the richly diversified phenomena of geographic variation among mammals and birds, which are so strikingly shown in the Pacific Coast region of North America.

After about a year at Berkeley, the *Peromyscus* project was transferred to the laboratories of the Scripps Institution at La Jolla, where it was conducted continuously until the fall of 1930. The ultimate discontinuance of this project was due to a radical change in administrative policy, in accordance with which the Scripps Institution for Biological Research was transformed into the Scripps Institution of Oceanography.

In concluding this brief historical statement, acknowledgment should be made to Dr. WILLIAM E. RITTER, first director of the Scripps Institution, under whom these studies were commenced and for years liberally supported; and to Dr. T. WAYLAND VAUGHAN, his successor in the directorship, under whom this support was continued for some further years, even after the change which was made in the scope of this institution's activities. To the friendly interest of Dr. GRINNELL was due the cooperation and hospitality of the Museum of Vertebrate Zoology, during the first as well as many subsequent years of this enterprize. Provision for much of the all-important field work, and some other phases of the program, was due to the generosity of the late E. W. SCRIPPS. Finally, my profound thanks are due to the Carnegie Institution of Washington, and to its president, Dr. J. C. MERRIAM, for a liberal grant which made possible the continuance of this work through the years 1927—1930.

---

[1]) MORGAN (1911) commenced the rearing of these mice with a view to their possible use in genetic studies, but he soon abandoned this attempt.

GENERAL CHARACTERISTICS, LIFE HISTORY, HABITS, ETC.

The genus *Peromyscus* comprizes small, mouse-like rodents, which range throughout the North American continent, from Labrador and arctic Alaska to Florida and Central America. OSGOOD, in his "Revision" (1909) divides the genus into 6 subgenera and 43 species, while he recognizes 143 subspecies or geographic races [1]. The taxonomic characters of this genus and of its major subdivisions are irrelevant to the present discussion. They may be found by reference to OSGOOD. Only subspecific differences concern us here, and among these only such as are adapted to the purposes of the present studies. At this point, it is sufficient to state that these differences relate to many of the measurable parts of the body, as well as to the depth and the extent of the pigmentation, both of the hair and the skin.

The rodents belonging to this genus range in size from forms smaller than the common house-mouse to ones which might more appropriately by termed rats than mice. Unfortunately, they are not known by any generally accepted common name, being variously called "field mice", "wood mice", "deer mice" and "white-footed mice."

The last of these names is descriptive, since not only the feet, but the entire ventral half of the body, up to a rather sharply defined level, are white, or whitish in appearance, the remainder of the body being brown, buff or gray. The hairs on the feet, and in a few cases on the belly as well, are completely white, i.e., they are unpigmented from base to tip. More commonly, however, the hairs of the latter region are darkly pigmented throughout their basal half or more, the remainder of their length being white. On the colored portions of the pelage, most of the hairs are of the banded type ("agouti" of the breeders). Throughout the greater part of their length, they contain the same black or chocolate colored ("sepia") pigment, as the basal region of the ventral hairs. But this dark pigmentation is interrupt-

---

[1] Including with these such species as are represented, so far as known, by only a single local form.

I learn from Dr. OSGOOD that since the publication of his revision, six new species have been described, and six new subspecies added to species already known. These are all from Mexico and Central America.

ed, not far from the distal end of the hair, by a band of varying width, containing a yellowish or orange pigment. The marked differences in the shade and color of the various subspecies, as well as variations of the "mutant" type, are due in part to the relative lengths of these differently colored zones in the single hairs, in part to the proportional numbers of different types of hair, in part to the state of concentration of the pigments themselves.

The tail of *Peromyscus* is clad with short, stiff hairs, and here, as on the body, we commonly find a dorsal region, of varying width, covered with dark hairs, and a ventral, or ventro-lateral, region covered with white ones (commonly entirely white). In many species, the division between these regions is a sharp one, the dorsal hairs forming a definite longitudinal dark stripe, the "tail stripe" of this and previous papers.

Some of the species of white-footed mice are extremely abundant. My own record catch is 165 "mice", which were taken in 100 live-traps, in the course of a single night. Of these, all but 13 were *Peromyscus*, of three species, while 134 belonged to the single subspecies, *P. maniculatus rufinus*. Likewise, some of the species have a very wide geographic range. The most ubiquitous of these, *Peromyscus maniculatus*, to which I have devoted more time than to any other, is known to range from the Atlantic to the Pacific Coast of North America, and from the Arctic Circle nearly to Yucatan. In this territory, it is represented by about 35 subspecies (figure 1). Of this species Osgood remarks that "it is probable that a line, or several lines, could be drawn from Labrador to Alaska and thence to southern Mexico throughout which not a single square mile is not inhabited by some form of this species."

Peromyscus, in nature, is largely, if not wholly, nocturnal in its major activities. The following test which I made illustrates this characteristic, for one subspecies at least. One hundred live-traps were set between 10 : 30 a.m. and 12 : 30 p.m. They were visited again between 4 : 45 and 5 : 55 p.m. (the sun set about 5 : 30). Not a single trap contained a mouse. On the following morning, these 100 traps contained exactly 100 mice, 95 of which were *P. maniculatus rufinus*. (SUMNER and SWARTH, 1924). In captivity, likewise, these mice are predominantly nocturnal in their habits, though by no means exclusively so. They are frequently seen to leave the nest compartment of the cage for food, during the daylight hours, and the activity records likewise show that some individuals are far

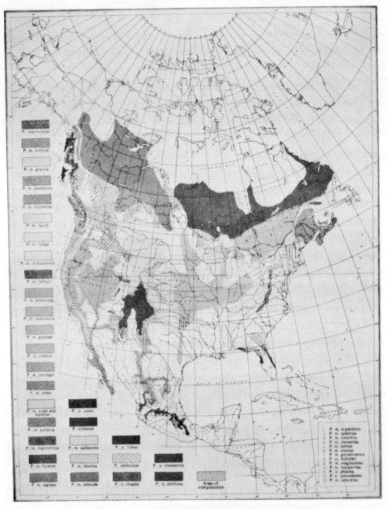

FIGURE 1. Distribution of subspecies of *Peromyscus maniculatus* and a few closely related species. (From OSGOOD, 1909. In the original map, these areas are indicated in color.)

from quiescent during this period (p. 95)[1]). Such behavior may,

---

[1]) See also JOHNSON (1926).

however, result from the highly unnatural conditions of captivity.

It is likely that many, perhaps most, of the white-footed mice inhabit holes in the ground. One species (*P. polionotus*) at least, is known to dig readily recognizable burrows, having interesting peculiarities (HOWELL, 1921; SUMNER and KAROL, 1929). In the case of other species, it is not easy to discover to what extent they dig their own burrows [1]. Several subspecies of *P. maniculatus* were found to burrow in the ground, in the large inclosed yards which were used in the long-time transplantation experiment (SUMNER, 1924a). This was particularly true of *P. m. dubius*, which tunneled everywhere through the ground.

Much of value could undoubtedly be learned regarding the behavior, particularly the "social" behavior, of these mice in captivity, if systematic studies were made by competent students [2]. In the absence of these, a few random observations may be worth recording.

Probably the most striking feature of the behavior of *Peromyscus* in captivity is the back handspring movement, which many individuals execute with extraordinary persistence. This may be watched in the open compartment of the cage, particularly at night, though it has frequently been observed in the daytime. In its most perfect form, the movement consists of a complete back somersault through the air, the mouse returning to its original position. Sometimes the animal appears to scramble up one wall of the cage before hurling itself backward. On the other hand, they have been frequently seen to stand in the middle of a large container and execute the handspring without touching its walls. It is this type of movement which is chiefly responsible for the larger strokes of the pen in the kymograph experiments (p. 96). This curious habit is not confined to any single race or species, but has been displayed by a considerable proportion of our stock. Teleologically speaking, it is perhaps fair to call it "exercise." To what extent, if at all, it is practiced in nature cannot be stated.

---

[1] The burrowing and nesting habits of some other members of the genus have been discussed by SETON (1909, 1920), and JOHNSON (1926).

[2] Casual observations have been recorded by a number of writers, particularly SETON (1909, 1920).

Another characteristic performance of these animals consists in "drumming" upon a hard surface, by means of a rapid vibratory movement of one of the front feet. It has been observed by us particularly among recently caught mice, while still in the tin-bottomed traps (p. 13). Whether the sound thus produced serves as a means of communication with other mice, as has been claimed, certainly demands proof.

Problems of "social" behavior related to the reactions of individual to one another. The following statements are based upon generalized impressions, rather than carefully controlled observations, but I believe that they are substantially accurate. They may at least stimulate others to undertake more careful investigations.

Mice of the same or opposite sex, shortly after capture, may, in many cases, be placed together in the same cage, with little danger of fatal quarrels, and they are likely, in such cases to live together amicably thereafter. Animals which have been kept in captivity for a considerable period are much more likely to show hostile reactions if brought together [1]).

Adult males which are "strangers" frequently attack one another with fatal consequences. On the other hand, the killing or serious injury of a female by a male is a very rare event.

Females frequently attack males, with which they have been brought together for breeding purposes, and not rarely kill them. The loss of males, in this way, is often a serious handicap to breeding experiments. In at least one case, a female is known to have killed four intended mates in succession.

Females much less commonly kill one another.

When the "getting acquainted" process is ended, fatalities are rare, though they may occur even after the lapse of months. Sick or injured animals are frequently killed by their cage-mates, and cannibalism in such cases is not rare.

---

[1]) It has always been our practice, in bringing together mice (regardless of sex) which are strangers to one another, to start with a freshly cleaned cage, or preferably one which is new to all of the animals concerned. The introduction of one mouse directly into the cage of another has been avoided. This procedure has probably reduced the number of fatalities, but it has by no means eliminated them.

The age attained by *Peromyscus* in nature is, of course, impossible to state. In captivity, one is known to have reached an age of nearly six years (5 yrs., 8 mo.), though some individuals considerably younger than this have shown obvious symptoms of senility.

Males have been known to be capable of effective copulation at the age of six weeks, while several females have given birth to broods at the age of ten weeks or less, one having become pregnant when 44 days old. Gravid females in juvenile pelage (thus probably less than two months old) are occasionally trapped. Such individuals, however, are probably exceptionally precocious. Fertility is known to have been retained, in several cases, nearly to the age of three years, though there is no good reason to suppose that it may not persist considerably longer than this.

Despite the great numbers of effective matings during the course of the *Peromyscus* experiments, a successful copulation was not once observed, though unwelcome advances on the part of the male were not rarely seen. This is in strong contrast to the frequency with which the process may be observed among white mice. It is probable that copulation in *Peromyscus* occurs at night, or perhaps entirely in the nest compartment of the cage.

While it probably varies somewhat, the period of gestation may be regarded as approximately 22 days. This is the s h o r t e s t interval which has been known to elapse between the placing together of a male and female and the birth of a brood [1]). It is likewise the l o n g e s t known interval between the removal of a male and the birth of a brood.

Pregnant females have been trapped throughout the year, and in captivity, broods may be obtained at any season [2]). Birth records (SUMNER, McDANIEL and HUESTIS, 1922) seemed to indicate the existence of a period of slack breeding during November, December and January. Since, however, the number of broods born in any month has depended largely upon the requirements of the breeding experiments, it cannot be used as a reliable index of seasonal variation in reproductive activity.

---

[1]) SUMNER (1916). In examining a large number of subsequent records, I find 11 instances in which broods were born 23 days after the first opportunity for copulation, but no case in which the interval was less.

[2]) Cf. SCHEFFER (1924).

The number of young in a litter has ranged from one to nine, the latter number having been recorded but once. The average, based upon 2,321 broods, believed to be complete, is 3.25 [1]).

The young are born blind, naked, and devoid of pigment in the skin. Dark pigment appears on the dorsal surface, on the second day after birth, and the emergence of the hair may be detected at about the same time. Within four or five weeks, the first coat of hair is well established, and the skin pigment had disappeared from view. The first coat of hair, the so-called juvenile (or "juvenal") pelage is commonly darker than the succeeding coats, and approaches a neutral gray hue. Subsequent pelages are much richer in color, showing a considerable measure of brown or yellow. Replacement of the first pelage usually commences at four weeks after birth, and is completed by the age of about three months. The first post-juvenile coat, while more richly colored than the juvenile one, is less so than the succeeding pelages. The details of the process of molting, under normal, as well as experimental conditions, have been observed and recorded by COLLINS (1918, 1923).

METHOD OF COLLECTING

Nearly all of the mice used in these studies were caught in the "Delusion" mouse-trap [2]) (figure 2). This is a live-trap, and inflicts no injury in the process of capture. It is likewise so constructed that the trap is not closed and thrown out of further use by the first arrival. This is a great advantage. One frequently catches several animals (my record is seven) without resetting. The bait ordinarily used has been uncooked rolled oats, such as is so generaly employed in trapping small rodents. This has sometimes been mixed with ground, uncooked peanuts.

Commonly, a considerable number of these traps (50 to 150 or more) have been set on each day of a collecting trip. One learns to re-

---

[1]) KAROL (1928). The earlier record (1922) for 1, 567 broods was 3.22.

[2]) The company which formerly produced this trap has discontinued its manufacture. A trap of identical pattern, made by another company, was found to be of inferior workmanship, and practically useless. It might be possible, however, to reconstruct this somewhat. This type of trap has the additional advantage of packing fairly compactly, ten in a bag.

cognize favorable territory only by experience. Much of the time, there is very little "sign" to guide one in trapping *Peromyscus*. Only in the case of the *polionotus* group were recognizable burrows met with.

In hot weather, traps must be gathered up in the morning before the sun is high, particularly if they are unshaded; while in cold weather it is desirable to provide a warm bed of grass or other material in the inner compartment. One may readily lose the greater part of

FIGURE 2. Delusion mouse-trap. The front door of the ante-chamber closes when the mouse enters, but opens again when the animal passes through a second gateway into the inner compartment.

his catch from cold, even on nights when the temperature remains considerably above the freezing-point.

After much experience, I came to regard it as unprofitable to leave the traps set in the same locality for two nights in succession. The extent of the catch falls off rapidly after the first night. Commonly, the entire line should be moved a considerable distance.

After visiting the trap line, those which contained mice were brought back to field headquarters. Here the animals could be killed, measured and skinned, if desired, but more commonly they were transferred to the compartments of a special chest to be shipped to the home station.

There are numerous advantages in using live-traps for collecting small mammals, even when the latter are to be killed immediately. (1) More accurate measurements can be made, and better skins prepared, from specimens which are intact until killed by chloroform or ether, instead of being mangled in a spring-trap and measured while in rigor mortis. (2) Dead specimens frequently undergo putrefaction, or are damaged by carnivorous animals, before the traps are reached. (3) It is frequently far more convenient to defer skinning the animals for some hours or even days, after capture, or if immature, it may be desirable to keep them for months. (4) As already stated, it is frequently possible, with a live-trap, to catch several specimens at one setting.

In a large proportion of cases, my own field procedure has included the listing, or even collection, of the prevailing plants of a region, and sometimes the photographing of habitats.

CARE OF STOCK

The following procedure has been gradually developed in the course of our years of experience in rearing these animals. It is possible that it could be considerably improved [1]). A standard type of cage (figure 3) was early adopted. This was built of wood of medium hardness [2]), and was painted inside and out with asphalt varnish. The total inside dimensions of this cage were: length 15 inches, width $8\frac{1}{2}$, height $8\frac{1}{4}$. It was divided into two equal compartments by a transverse vertical partition, having a small passage-way at the bottom. The right-hand compartment was covered in front by galvanized-iron screen ($\frac{1}{8}$ to $\frac{1}{4}$-inch mesh). Here the food was placed. The left-hand compartment was the nest chamber. It was closed, except for the small doorway, just referred to, and a $\frac{3}{4}$-inch auger-hole, guarded by screen, near the top of the opposite wall. Each compart-

---

[1]) DICE (1929) has adopted a very different type of cage from that employed by us, and a somewhat richer, though much more artificial, ration.

[2]) "Oregon pine" (= Douglas fir) has been generally used. Redwood is too soft.

ment was provided with an independent cover. The covers were not hinged, or otherwise fastened to the body of the cage. Each was constructed of two layers of wood, the lower being of a size to fit readily into the top of the cage, the upper one resting on the rim. The floor of both compartments was covered to a depth of about an inch with sawdust, while the left-hand compartment, in addition, was half filled with straw, and a small bunch of cotton batting was added when broods of young were present or expected.

FIGURE 3. Cage used for keeping and rearing *Peromyscus*.

These materials were changed, and the cage cleaned, once a week, except in cases where only one animal was present. During this process, the mice were removed and placed in a large glass jar. Care was necessary to avoid being bitten during this process. Larger species were commonly lifted by the tail. For this purpose, a pair of rubber-tipped forceps was used. Smaller mice, particularly those of the *polionotus* group, were commonly herded into a narrow tin can, held against the passage-way between the two chambers.

When mating was desired, one male was usually kept with two to five females. The latter were commonly isolated when pregnant.

Great care has been necessary in mating an individual known to be vicious. The killing of the male was not a rare occurrence, as already noted. Any animals of particular value should be closely watched after being mated, and separated if the female is seriously pugnacious.

A cardboard shipping tag was loosely tacked to the front of each cage, on the left-hand side. On this were entered, in soft pencil, all relevant date, including particularly the registry number of each mouse present, the date of admitting or removing the male — when matings were being made — and the date of the last cleaning of the cage. Various memoranda were also frequently written conspicuously with chalk upon the front of the cage.

Each mouse carried an identification mark or combination of such marks, corresponding to a registration number in a "stock register" book. It is hardly worth while to describe in detail the rather complicated system of marking which we have employed. It consisted of amputating various toes and fingers, and perforating one ear by means of a leather-punch. The generation ("C_1", "C_2", [1] or F_1, F_2, etc.) could be thus indicated, as well as the individual number. Sometimes it was necessary to give distinctive "racial" marks to the members of certain local collections, where confusion would otherwise have been possible. In all cases, it was necessary that the left ear and left hind foot should remain intact, since these were among the important parts to be measured. During the marking process, the mice were etherized in a special jar.

Separate books have been kept for s t o c k r e g i s t r y (giving parentage, date of birth, and some other entries), b i r t h s o f b r o o d s (with subsequent history of each up to time of marking), m a t i n g s, g e n e r a l d e s c r i p t i v e n o t e s, f i e l d r e c o r d s, and m e a s u r e m e n t s. For all but the last two, loose-leaf note-books were found to be preferable.

The dietary has varied considerably in the course of these studies, though few changes were made during the last six years. It was perhaps far from perfect, even at the end. As will appear shortly, abnormalities of growth, of a sort which are frequently attributable to deficiencies of diet, were by no means completely eliminated. However,

---

[1] „C_1", „C_2", etc. are the designations which were given to the first, second (etc.) cage-born generations.

persons with considerable knowledge of animal nutrition were consulted at various stages of the work, and the diet, as finally adopted, would seem to furnish all of the necessary food principles, including vitamines. It is possible that the protein content was too low.

The rations, during the later years [1]), consisted of a mixture of oats and sunflower-seed (6 : 1); a mush, consisting of alfalfa-leaf meal and wheat-germ meal (about 4 : 1), well moistened with milk into which cod-liver oil (one to four tablespoons per quart) had been thoroughly beaten; and a raw succulent vegetable. To meet the last requirement, spineless cactus [2]), cut into cubes, and lettuce were given on alternate days. To meet the desire of the mice to gnaw hard food, a small oval dog-biscuit was put into each cage every week. No free water has been given throughout the course of these experiments, aside from that contained in the milk and the succulent vegetables. The practice of providing water in special containers was abandoned almost at the outset, owing partly to its inconvenience, partly to a realization that *Peromyscus*, throughout much of its range, does not commonly have access to drinking water.

It was early noted (SUMNER, 1915) that cage-bred *Peromyscus* differed, on the average, from wild ones in total size and in bodily proportions. The cage-bred animals were found to average smaller than the wild ones, and their tails, feet and certain bones were shorter, both absolutely and relatively. Thus, in one of the earlier series (SUMNER, 1918) the relative tail length (ratio to body length), in the "C_2" generation of *P. m. rubidus* from Eureka, was 92.49, as compared with 104.01, in the "wild" material; that of *gambelii* from La Jolla being 79.39, as compared with 83.95; and that of *sonoriensis* from Victorville being 75.18, as compared with 81.30. It must here be noted, however, that the three races maintain the same relative positions as in nature, even when subjected to conditions which result in such abnormal growth. This is a circumstance of importance in its bearing upon the genetic character of these subspecific differences.

I was at first disposed to attribute these abnormal characteristics of cage-bred animals to the restriction of activity in confinement.

---

[1]) Only one change (the addition of lettuce, in 1928) was made since the fall of 1924.

[2]) This was adopted by reason of its inexpensiveness, high water content, and acceptability to the stock. Raw carrots serve the purpose very well.

But they were soon found to involve, in some cases, changes of an obviously pathological nature. Deformation of the bones, apparently rachitic in character, was not infrequent. This included shortening of the femurs, warping of the skull or pelvis, and other abnormalities. Many variations in diet were tried, with the endeavor to correct any vitamin deficiencies, should these be responsible. In 1924, cod-liver oil was regularly added to the dietary.

With a single exception, no controlled experiments were conducted to determine the effects of any of these ingredients of the dietary [1]). Whenever improvements (real or supposed) were made in the latter, the entire stock was served with the new rations. Comparisons were possible only with previous results, although other conditions besides diet may have changed in the meantime. Thus, the first use of cod-liver oil nearly coincided with the substitution of *P. polionotus* for *P. maniculatus* as the chief object of experimentation. It is none the less worth recording that the average differences in tail and foot length between wild and cage-bred stocks of the three subspecies of *polionotus* were slight compared with those recorded for three subspecies of *maniculatus* in 1918. For the *polionotus* series alone it was likewise pointed out (1930) that "scarcely any of the more extreme cases of stunting and deformation, such as appeared in each of the series of animals before administration of cod-liver oil to the mothers, are to be found among those reared after the treatment was commenced." Nevertheless, neither stunting nor deformation was ever entirely eliminated.

Captive animals, in general, have tended to be fatter than those freshly caught. In some cases, this fattening has been excessive. Cage-bred individuals have been encountered, in which the weight has been more than twice that of normal animals of the same length. In occasional specimens, the entire shape of the body has been transformed, great pads of subcutaneous fat appearing in the shoulder re-

---

[1]) In this case, chopped boiled liver (administered daily) was added to the rations of one half of certain series, the other half serving as controls. The treatment was commenced with the parents some months before birth of the experimental animals, and it was continued for more than a year. While some benefit seems to have resulted, both as regards size and proportions, this benefit was slight and inconstant, and the treatment was not adopted for all the stock.

gion and elsewhere. Fortunately such cases have been relatively infrequent.

Another serious handicap to the breeding experiments which has not been overcome is the prevalence of sterility. While some strains presented a high proportion of fertility, others have been encountered in which a large majority of the mice were completely infertile. Any exact estimate of the relative numbers of such sterile individuals is out of the question. Probably at least half of the total stock was thus afflicted, even at the close of the experiments [1]. The practical result of this situation has been the necessity, in many cases, of rearing far greater numbers of animals than would otherwise be required, in order to obtain the desired number of breeders. In a few instances, indeed, a line of experimentation has been brought to a close through the total sterility of essential individuals.

The incidence of sterility has been highly erratic. Mice of reduced size or abnormal proportions, or excessively fat individuals, have doubtless been more likely to be sterile than those of normal appearance, but the correlation has been far from absolute. Nor has the trouble been confined to inbred, or even to cage-reared, animals. Many "wild" specimens of both sexes have persistently refused to breed, even in the prime of life.

Wheat-germ meal and lettuce were added to the dietary as a result of EVANS's findings with "vitamin E", but these have had little or no effect in reducing sterility in our stock.

Thus, despite all of our efforts, we never entirely solved the problem of rearing fully normal *Peromyscus* in captivity. It is likely that sustained efforts, directed exclusively to this end, would overcome these difficulties [2]. It is evident that the morphological changes here referred to must be constantly reckoned with in dealing with meas-

---

[1] Not including those which had become sterile owing to age.

[2] It was found that mice reared in large open yards, with extensive areas of soil to dig in, were of distinctly larger size and better proportions than those reared in cages (1924a). It is obvious, however, that studies involving pedigree breeding could not be conducted with mass cultures such as these.

The fact that throughout our experiments, occasional descent lines were met with which retained a high degree of fertility, for a number of generations, suggests the possibility that genetic factors might be partly responsible for the differences. If so, selection might afford a solution of the difficulty.

urements based upon cage-bred stock, particularly since these effects of captivity involve some of the very parts which are of interest in the comparison of subspecies. It is hardly necessary to say that full allowance has been made for this disturbing element, in dealing with all results which depend upon these measurements. The matter has received attention in many of the published reports upon these studies, commencing with the first. It is obvious, too, that no results based upon "wild" material are in any way affected by the factors here discussed.

Fortunately, pigmental characters, at least those which concern coat color, are little if at all influenced by captivity, even in otherwise abnormal individuals. Cage-bred mice differ little, if at all, in color characters, from those of the wild generation (SUMNER, 1930). The mean width of the tail stripe has been found to diminish somewhat, however, in some captive series.

LINEAR AND COLORIMETRIC MEASUREMENTS

Since I have published rather recently (1927), a detailed account of the procedure which has been employed by me, both in the preparation of skins and in the quantitative determination of pigmental characters and the measurement of body parts, these methods will be dealt with here, only in so far as is necessary to an understanding of the "characters" to be considered below.

The linear measurements are, in part, ones which have been previously employed by taxonomic mammalogists. They comprize determinations of total length, length of tail, foot, ear, femur, and pelvis, and the length and breadth of the skull. To these may be added weight (a not very useful "character"), and the number of caudal vertebrae. My measurements have been made to the nearest tenth of a millimeter, and a standard procedure has been developed with a view to greater accuracy and objectivity.

Skins have, from the first, been prepared in quite a different manner from that employed in making the conventional "study skins," to be found in museums. My own skins are exclusively flat ones, which have been stretched to a uniform degree of tension, while still fresh, and cleaned with a fat solvent when dry. The advantages of flat skins for colorimetric studies are obvious. Indeed photometric determinations would hardly be possible with any others. Nor would the accurate measurement of the different areas of the pelage. Add to this advantage the possibility of controlling the degree of stretching (an important factor as affecting the apparent shade of the pelage), the possibility of extracting the fat (another cause of apparent color differences), and the great economy of space in storing, and it is difficult to see why the old-time method of preparing "study skins" should be retained at all.

These prepared pelts have been subjected to two types of measurement, *areal* and *colorimetric*.

(1) Areal determinations, by the use of a planimeter, have been made of the entire skin, as thus prepared, as well as of the more limited area which is clad with colored hairs. The value employed in the statistical treatment has been the ratio of the latter to the former. But it must be pointed out that the "colored area", as dealt with in these studies, does not in all cases refer to precisely the same region of the skin. In the great majority of species

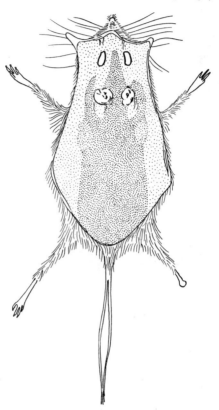

FIGURE 4. Skin of *P. p. albifrons*, seen from inner side, showing region of hair pigmented at base ("a" + "b"), and region of unpigmented hair ("c"). (From SUMNER, 1926).

and subspecies of *Peromyscus* the only "colored area" to be reckoned with is the visibly colored dorso-lateral half of the pelage, the remaining ventro-lateral half being constituted by white-tipped hairs. In a few subspecies, on the other hand, as well as in certain mutant types, three areas of the pelage may be distinguished: (a) a dorsal

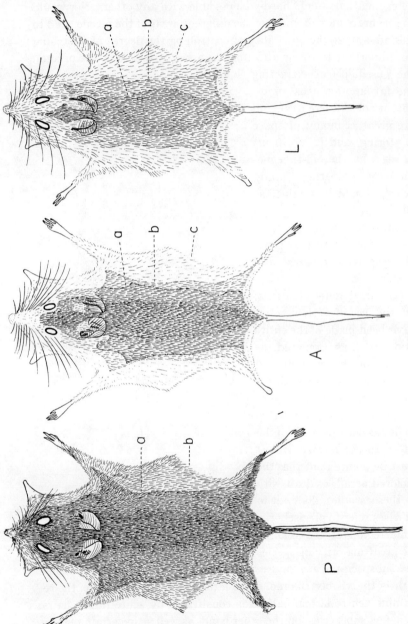

FIGURE 5. Pelage areas of three subspecies of *Peromyscus polionotus*. P = *polionotus*; A = *albifrons*; L = *leucocephalus*. a, b, and c are the areas thus designated in the text. The reader should not be confused by the dark mid-dorsal stripe shown in P and A. This forms part of area a. (From SUMNER, 1926).

area constituted by hairs which are pigmented from base to tips; (b) a narrow intermediate zone, constituted by hairs which are pigmented in the basal portion only (such as cover the entire ventral surface of the more typical *Peromyscus*); and (c) a ventral area of entirely unpigmented hairs. By transmitted light, regions "a" and "b" together constitute a more or less opaque area, which stands in sharp contrast with the semi-transparent "c", and may generally be measured with fair accuracy by the planimeter.

(2) Colorimetric determinations have been made by means of the Ives Tint-Photometer. Since commencing the use of this instrument, in 1922, various changes have been made in the procedure, such that the figures presented in my earlier and later papers are not at all comparable with one another. The last published account of my technique (1927) details the procedure finally adopted and used during the later years of the work.

The readings with this instrument indicate the amount of light, of certain wave-lengths, which is reflected from a chosen area of the pelage, expressed as percentages of the amount reflected from an equal area of pure white magnesium carbonate. Three color filters in succession are commonly interposed between the objects (both the skin and standard white) and the eye, these being red, green, and blue-violet. When using each filter, the light from the standard white surface is cut down by a diaphragm until it balances that from the object, and the reading is then taken.

In my later discussions, I have confined myself, for the most part, to the values for "red" (R), and for the "index of saturation" $\left(\dfrac{R-V}{R}\right)$ The first of these represents merely the reading obtained with the red filter. Any one of these three readings, taken by itself, might serve very well as an index of the paleness or darkness of the pelage, regardless of its color values. The reading for red has been so used, merely because it is the largest of the three. The "index of saturation" $\left(\dfrac{R-V}{R}\right)$ is a measure of the richness of coloration, without reference to whether the skin is pale or dark [1]). High values indicate

---

[1]) This fraction consists of the reading for red, minus the reading for blue-violet (V), divided by the reading for red. There is, of course, no blue or vio-

richly colored (brownish or yellowish) skins, low values an approach to neutral gray.

Another derived value has been employed, in order to indicate, in a roughly approximate way, the spectral position of the dominant color. This is the fraction $\dfrac{R-V}{G-V}$, and has been referred to in a number of my papers as the "red-green ratio." The mean value of this fraction may differ significantly in two races, one of which, for example, inclines toward a yellower, the other toward a redder hue. On the whole, however, the ratio agrees fairly closely in the various species and subspecies which have been dealt with. The large, obvious differences in the appearance of their pelages depend primarily upon differences in shade and in saturation, rather than in wave-length.

From the nature of the case, absolute values, or even very precise ones, are impossible to obtain with material of this sort, whether we are dealing with measurements of area or color. But when taken by the same observer, after sufficient practice, the values recorded are sufficiently comparable to express fairly well the differences either between individuals or races. In any case, no other satisfactory quantitative method of evaluating the pigmentation of mammalian pelages is known to me [1]).

Certain other measurements of pigmental characters must be added to the foregoing. Reference has already been made to the longitudinal stripe of black hairs which runs along the dorsal surface of the tail in most members of this genus. In species in which this stripe is sharply outlined, its width may be determined fairly accurately,

---

let in a brownish or yellowish skin, except in so far as the latter likewise reflects considerable white light. The fraction here considered represents the proportion of "free" red to the total red rays reflected from the object.

[1]) COLLINS (1923) employed two revolving color-wheels, on one of which were placed the customary sectors of colored paper, on the other the skin to be tested. With this apparatus, COLLINS obtained some valuable results, and the method used by him was probably more accurate than any which had been previously employed. But the method is extremely time-consuming, and its accuracy is limited both by the purity of the colors in the paper disks, and by the accuracy of the color perceptions of the observer. The customary method of comparing hair or feathers with the colored papers of RIDGWAY's "Standards", is extremely crude, even if frequently useful. (This subject is fully discussed in my 1927 paper).

and this width, taken as a percentage of the total circumference of the tail, is a character of considerable importance and interest. In a few geographic races, this stripe is rudimentary, extending only part of the way from the base of the tail to its tip. In such cases, the proportionate length of this stripe is determined.

The pigmentation of the sole of the hind foot likewise presents wide racial and individual differences, and the degree of this pigmentation may be assigned an approximate numerical value by reference to an arbitrary set of standards [1]).

We must here likewise include the measurement of certain microscopic characters for single hairs (length of subterminal band, etc.) as determined by HUESTIS (1925).

NATURE OF RACIAL DIFFERENCES

In using the term "race" I here commonly refer to a geographic race, and this, in turn, I employ as synonymous with subspecies. It is, however, desirable to use the term race in a somewhat more elastic sense, so as to include any local population which differs significantly from another, even though it is not expedient to assign the two to distinct subspecies.

Two subspecies may differ so widely that no individual of one group would be assigned to the other, even by an untrained person. In such cases, the differences may relate to a considerable number of characters, and some of these may display no overlapping of values when the measurements are plotted. *Peromyscus maniculatus rubidus* and *P. m. sonoriensis* are two such forms, as are also *P. polionotus polionotus* and *P. p. leucocephalus*.

In other cases, two subspecies may differ by very few independent characters, and in respect to these the differences may not be absolute, but may reveal themselves only when series of specimens of the two forms are compared. While the means and the total ranges of the measurements for the two races may differ considerably, there may be a considerable overlap when the values for any single charac-

---

[1]) Part of my procedure, in the preparation of a specimen, consisted in the preservation of the left hind foot. This was first placed in alcohol, later transferred to glycerin.

ter are plotted. *P. maniculatus sonoriensis* and *P. m. gambelii* are related in this way.

It will appear below that a subspecies is not a homogeneous collection of animals, throughout its entire geographic range. The uniformity is merely relative. I have never compared two local collections from points at all remote from one another without finding significant differences between them. And sometimes the differences between two such sub-races are evident to the eye. To create new subspecies whenever such undoubted local differences are observed, would, however, plunge us into endless confusion.

Finally, we have wide and very obvious individual differences within the population of any single locality.

The differences just considered, whether those between subspecies or minor local races, or between individuals of the same race, relate to the same structures or "characters." They include quantitative or qualitative differences of every organ or part which has been measured or carefully observed. The total size of the body, length of tail, feet and ears, skull, pelvis and femur, number of caudal vertebrae, width (or length) of the dorsal tail stripe, area of the colored portion of the pelage, shade and color of the latter, proportionate numbers of the different types of hairs, total length of hair, length of subterminal band and of the dark tip in the colored hairs, character of the pigment in these hairs, depth of the melanic pigment in the skin of the feet, ears, snout, scrotum, and even in the internal tissues of the tail and the histological structure of the thyroid gland. To these structural and chemical differences must be added marked individual and racial differences in temperament, propensity to fattening, and doubtless other characters of a physiological nature.

Neither observation nor measurement reveals any distinction between racial differences and individual ones. I may anticipate by saying that, so far as these differences are heritable, they probably depend upon the same genetic factors in one case as the other. It will be found, however, that racial differences are to a far greater extent genetic than are individual ones. Indeed, the differences between an average *rubidus* and an average *sonoriensis*, for example, are probably wholly genetic, while the differences between a dark and a pale individual, within either of these races, are demonstrably non-genetic, in part.

THE HERITABILITY OF RACIAL DIFFERENCES

From the beginning, our procedure has involved the collection of mice from various climatically different regions, and the rearing of them together, under identical conditions, at La Jolla [1]). As soon as offspring were obtained from these imported strains, it became evident (SUMNER, 1915) that the subspecific differences were hereditary, for the most part, at least. This experience has been confirmed throughout the entire course of these experiments, during which nine subspecies, belonging to three different species, have been reared in large numbers through two or more cage-bred generations. There has never been any evidence of a convergence of characters, under the influence of a common environment and identical food.

In order to subject this question to a more searching test, and reveal any slight changes, if such occurred, a long time experiment was conducted. I will quote from my summarized statement at the close of this experiment (1924a):

"(I). Mice belonging to the subspecies *Peromyscus maniculatus sonoriensis*, from the Mojave Desert, were reared for more than eight years at La Jolla, the resulting stock representing a minimum of seven and a maximum of twelve or more generations. During this period, they did not, in respect to any single measured character, undergo a modification in the direction of the La Jolla subspecies *gambeli*. On the contrary, the mice of the later generations were in some regards less like *gambeli* than were their ancestors trapped in the desert.

"(II). Mice belonging to the subspecies *P. m. rubidus*, from the northwest coast of California, were reared at La Jolla for six years, the resulting stock (with a very few exceptions) representing a minimum of four and a maximum of six or seven generations. Here again, the slight differences between the ancestral stock and its descendants were not such as to indicate a modification in the direction of the local race. As in the case of *sonoriensis*, the mice of the later generations of *rubidus* were, on the whole, less like *gambeli* than were their wild ancestors from Humboldt County.

"(III). Comparing the two introduced strains, *sonoriensis* and *rubidus*, there was no tendency towards convergence, under the influence of a common environment. To judge from the samples at our disposal, there was actually a slight divergence in respect to all but one of the characters which were measured.

"(IV). The nature of these slight differences between the

---

[1]) Preliminary experiments were conducted for one year at Berkeley.

transplanted and ancestral series of a given race renders it highly improbable that they have been due to changed climatic conditions. To some extent, they are known to be the results of captivity, irrespective of climate."

In regard to the foregoing experiment, it must be said that the numbers of individuals employed, although not great, were adequate for the comparisons, and that both linear and colorimetric measurements were made.

Less extensive, though fairly strong evidence is also available respecting the hereditary character of the differences between certain highly localized sub-races. A few specimens which were reared at La Jolla from a pale sub-race of *rubidus*, trapped upon an isolated sandy peninsula on the coast of Humboldt County, California, proved to display the paler tones characteristic of their parents (SUMNER, 1917, 1918). Likewise, the rather pronounced differences between certain local populations of *P. polionotus albifrons* were evident to the eye in the limited number of offspring reared from these (1929).

The mode of inheritance of these various racial differences, as revealed by a study of subspecific hybrids, will be considered after we have dealt with the inheritance of certain individual differences.

THE HERITABILITY OF INDIVIDUAL DIFFERENCES

That individual variations in the value of any character which can be determined quantitatively are in part genetic could have been predicted with high confidence. Early in these studies (1918), I reported parent-offspring correlation coefficients which had been determined for two characters, relative tail length (expressed as a percentage of body length) and width of the tail stripe. The figures for these two characters, in a number of series, averaged about 0.3. Later, more extensive, determinations (SUMNER, 1923b, 1930; HUESTIS, 1925; SUMNER and HUESTIS, 1925) have revealed the heritability, in varying degrees, of individual differences in nearly every measurable character which has thus far been tested for the purpose. In addition to the characters just named, parent-offspring, and in some cases fraternal, correlations have been determined for foot length, the indices of asymmetry of certain paired bones, foot pigmentation, extent of the colored area of the pelage, various values which were employed to express the shade (relative paleness or darkness), the dominant color, and the degree of saturation of this color, in the pelage, the proportions of different types of hair, total length of the hair, length of the subterminal ("agouti") band and of the dark tip.

All of the foregoing have given mean positive coefficients of undoubted significance, except the indices of asymmetry (sinistro-dextral ratios) and two indices ("red: green ratio" and $\frac{\text{"R} - \text{V"}}{\text{R}}$), which represent rather crudely the spectral position of the dominant color, and the degree of saturation of the latter, respectively. The first named coefficients are probably based upon differences of a non-genetic sort (SUMNER and HUESTIS, 1921). The latter relate to "characters," the individual values of which are only very rough approximations, owing to the method of determination (SUMNER, 1927). Hence the lack of correlation here may be only apparent [1]).

The following table gives the mean values of the coefficients of parent-offspring and fraternal correlation, for the various classes of characters, in all of the series which have been studied [2]). The considerable differences in the correlations indicated for these various characters depend in part upon the precision with which the characters may be measured, in part upon the proportion of genetic to non-genetic variability, in part to chance. The numbers (particularly of parents) upon which some of the correlations are based are small.

---

[1]) The coefficient for $\frac{R - V}{R}$ between the 22 selected F_2 parents and their 65 offspring, derived from the *leucocephalus-albifrons* cross, is + 0.381. In general, the F_2—F_3 coefficients, being based to a larger degree upon genetic variations, exceed any of the others, for all of the characters here considered.

[2]) For the sake of simplification, "characters" of a similar nature have been thrown together. As a rule, the means are not weighted, the means of different series being given equal value in the averages, regardless of the number of individuals in a series. The parent-offspring and fraternal correlations are not always comparable, since the latter were not computed for all series represented in the former. The figures for the correlations between selected F_2 parents and their F_3 offspring have not been included in the computing the present averages, since they are not comparable with the others. Their inclusion would considerably increase some of the figures in the table.

	Parent-offspring	Fraternal
Foot length and tail length, figures combined.	+ 0.243	+ 0.514
Foot pigmentation.	+ 0.245	+ 0.372
Colored area of pelage and breadth (or length) of tail stripe	+ 0.299	+ 0.351
Shade of pelage, based on averaged values for "black," "white," and "color" (SUMNER, 1923b, SUMNER and HUESTIS, 1925), combined with those for "red" (SUMNER, 1930).	+ 0.236	+ 0.360
Index of saturation $\left(\dfrac{R-V}{R}\right)$	+ 0.078	
Red: green ratio.	+ 0.069	− 0.011
A, B, C and D hairs, figures combined (HUESTIS)	+ 0.212	+ 0.279
Total length of hair, length of "agouti" band, and length of dark tip, figures combined (HUESTIS).	+ 0.317	+ 0.349
Indices of asymmetry.	+ 0.047	

COLLINS (1923) reared offspring and later descendants from two contrasted series of *P. maniculatus gambelii*, which had been selected with reference to coat color as "buff" and "dark." COLLINS's table and histograms show that the offspring of these series differed quite significantly in the same direction as the selected parents. Similar differences were observed, though not measured, in later generations.

Unpublished experiments of my own show that exceptionally pale strains, both of *P. polionotus* and *P. p. albifrons*, may be perpetuated and that, in the latter race at least, the condition may be intensified by selection.

"MUTATIONS"

Individual differences of the sort thus far discussed differ merely in degree from the differences which distinguish geographic races. Likewise, in their inheritance, they do not conform to any obvious Mendelian scheme. We have, however, encountered another type of intra-racial variability, which is plainly inherited in alternative fashion, and which manifestly belongs in the same class as the familiar "mutations" of our breeders (SUMNER, 1917*a*, 1918, 1922*a*, 1924*b*, 1928; SUMNER and COLLINS, 1922).

Albinism. — A case of albinism, in *Peromyscus leucopus noveboracensis*, was reported by CASTLE (1912). Some albino descendants were obtained by Professor CASTLE from this original wild specimen, which was caught in Michigan. The strain was, however, brought to an end by sterility.

In view of the extent of our breeding operations at La Jolla, throughout so long a period, it is perhaps surprising that albinism did not appear but once "spontaneously." Nor have I or my collaborators ever trapped a wild specimen. Two albinos appeared in 1919, in a brood belonging to the subspecies, *P. maniculatus gambelii*. The parents were sibs, belonging to the first cage-born generation. They likewise had six normally colored offspring, at least one of which proved to be heterozygous for albinism. From this original stock an indefinite number of albino descendants were reared [1]. The albinism of this strain was typical and complete. Genetically, it was a simple recessive. Owing to the circumstances of its appearance, I cannot, of course, state whether the actual genetic change occurred subsequent to the capture of the ancestral pair, or whether one of these chanced to be a heterozygote for this factor. The latter appears to be the more probable alternative.

Pallid. — Another striking color mutation likewise appeared "spontaneously" only among the descendants of a single pair. Like albinism, the "pallid" mutation [2] consisted in a loss of pigment, though the loss was far less complete.

---

[1] This, and the other chief mutant strains here discussed (with the exception of "grizzled") were sent, in 1925, to Dr. LEE R. DICE, of the University of Michigan. They were later transferred by him to Dr. W. H. FELDMAN. The "grizzled" stock was sent to Michigan in 1927.

[2] In my original account of these mice (1917*a*) they were inappropriately termed "partial albinos."

"It is characterized primarily by the lack of most of the black (or sepia) pigment found in normal mice.... This lack appears in the absence of all-black (i.e., non-banded) hairs from the pelage, and the extreme reduction of pigment in the basal zone of the others. The latter is of a pale ashy hue instead of slate-colored. Furthermore, the eyes are dark red instead of black, the ears are not appreciably pigmented, and the dorsal tail stripe (normally due to dark hairs) is scarcely perceptible. A further peculiarity of this strain is the fact that the eyes are smaller, or at least less protruding, than in the wild type. The pallid mice are pale gray when young, developing a considerable admixture of yellow or orange when adult."

Unlike the albinos, the pallids presented a considerable range of variation in color. They differed considerably in shade, as well as in the richness (yellowness) of the coat color. Likewise, the eyes ranged in shade from a condition nearly as pale as the pink eyes of the albinos to one nearly as dark as those of the normal animal. However, there were no true intergrades between the pallid and normal condition, and there was never the least doubt respecting the nature of one of these individuals. The variations in shade were doubtless due to other color factors which were introduced by the various stocks with which the pallid strain was crossed.

The pallid mutation appeared in 1915, in the F_2 generation of a cross between *P. maniculatus rubidus* and *P. m. sonoriensis*. Four "pallids" and seven of the wild type resulted from the mating of an F_1 male and his two sisters, each of these last giving birth to two pallids. It is certain that this was no simple segregation phenomenon, due to the recombination of factors regularly present in the two subspecies which were crossed. Several hundred F_2 and F_3 hybrids between these two subspecies have been reared in the course of these experiments with no other independent appearance of this aberration As in the case of albinism, there is no means of determining whether or not the gene mutation took place subsequent to captivity.

The pallid strain proved to be highly fertile, and was reared for a number of years. It has behaved consistently as a simple one-factor recessive.

The pallid and albino mutations are interesting in showing a high degree of linkage between the genes primarily concerned. While the material was not extensive enough to make possible any exact estimate of cross-over values, the existence of such a linkage is indubitable.

"Matings have been made (1) between 'extracted' albinos and 'pure' pallids (i.e., those known to be free from the factor for albinism), (2) between extracted pallids and pure albinos, and (3) between extracted pallids and extracted albinos. There were likewise a number of matings, in which the pedigrees were less simple. On the assumption of a wholly independent segregation of these factors, our F_2 pallids (of simple pedigree) should have a $\frac{2}{3}$ chance of being heterozygous for albinism, while our F_2 albinos should have a $\frac{3}{4}$ chance of being either homozygous or heterozygous for pallid [1])."

"Eighteen F_2 mice were involved in these tests.... The cumulative testimony of all these matings is overwhelming. Not a single pallid mouse and only two albinos have appeared among the 135 young which have thus far been born. Had there been a normal proportion of 'carriers' among the parents, these matings should have yielded 37 albinos and 18 pallids, as the most probably 'expected' numbers. That all of the offspring with two exceptions (these being sibs) were of the wild type is evidence of a high degree of linkage (in this case 'repulsion') between the albino and the pallid factors."

In view of the seemingly close similarity between my "pallids" and the "red-eyed yellow" rats described by CASTLE and others [2]), and likewise the existence, in each case, of a high degree of linkage with albinism, I have suggested the possibility of an actual homology between the genes here concerned.

Hairless. — This bizarre mutation appeared twice in the history of our stock. Individuals destined to be hairless acquired the first coat of hair, but this began to thin out at the age of two to three weeks, and thereafter it was gradually lost, though the vibrissae persisted throughout life, and scattered hairs remained for many months on other parts of the body. As a rule, the claws of the hairless mice ultimately became much elongated, in extreme cases assuming a spiral twist. Some individuals, when fully grown, displayed an extreme corrugation of the skin, similar to that which has been described for hairless *Mus musculus*, under the name of "rhinoceros mice."

Both albinic and colored hairless *Peromyscus* were reared, but no linkage between the hairless and color factors was detected. "Hairless" behaved as a simple recessive, though it is not improbable that some variations of this character (time of complete loss of hair, de-

---
[1]) It is a safe assumption that the double recessive form would be albino.
[2]) CASTLE (1914, 1919), CASTLE and WRIGHT (1915), DUNN (1920).

gree of skin corrugation, etc.) may have been dependent on modifying factors.

This mutation, like most of those described in this section, appeared in *P. maniculatus gambelii*. It appeared in two apparently independent descent lines. In one of these cases, the first hairless individual was distant only one generation from one of its wild ancestors; in the other case, there was a lapse of four generations. As in the mutations previously discussed, the actual genetic change may have occurred many generations earlier.

Hairless mammals, belonging to various groups, have, of course, long been known. In my paper dealing with the hairless *Peromyscus* (1924b), I cited a considerable number of previous descriptions of hairless rats and mice. Since then, many additional cases have been recorded by others, and some genetic studies have been conducted with these. Apparently, there is more than one hairless mutation now known in mice. Similar conditions seem to result from changes in a number of distinct genetic factors.

"*Yellow*". — A reddish or yellowish condition of the pelage was reported and discussed under the not very appropriate term "yellow." This appeared on six distinct occasions, in strains which were, so far as known, unrelated. The condition referred to resulted from a considerable decrease in the amount of black pigment in the pelage, and a corresponding increase in the amount of yellow pigment. The proportion of all-black (unbanded) hairs was diminished, and the breadth of the subterminal colored band, in the other hairs, was increased. The ventro-lateral white area was more intensely white, owing to an increase in the length of the unpigmented tips [1]). The eyes, ears and feet displayed, however, a normal amount of dark pigment.

All of the six independent appearances of "yellow" were among the stock of *gambelii* trapped at La Jolla.

This character commonly behaved as a simple recessive, in crosses with the normal condition, and it was at first believed to be such. But there appeared a considerable proportion of darker "atypical" yellows, whose genetic status was not evident. This last expression was used quite arbitrarily, however, since it was possible to arrange,

---

[1]) The basal pigmented zone of the hair was commonly entirely lacking on the mid-ventral line of the body.

among the offspring of "yellow" parents, a graded series between the most "typical" yellows and specimens closely resembling the paler and more buff-tinted individuals of the wild type.

But the most confusing situation arose, following a considerable series of crosses between "yellow" and "pallid." The F_1 generation, as expected, comprized animals which closely resembled normal *gambelii* of a medium shade. In the F_2 generation, along with normals and pallids, which appeared in proportions very close to the expected ones for a di-hybrid cross, there occurred a motley group of real and doubtful "yellows," some of which showed by the breeding test, that they were not yellows at all.

Likewise, at least two of the six independent "yellow" strains in our stock [1] differed from one another quite appreciably in color, and this difference was shown to be genetic [2]. Each bred fairly true to its own type, and crosses between the "a" and "b" strains were of intermediate hue.

The situation is one which could readily be explained by the existence of one major factor difference, variously complicated by the presence of modifying factors. It was not practicable, in the course of our work, to isolate these factors.

Grizzled. — This is another, often striking, color variation, having a somewhat complex genetic basis. It was found in at least three independent lines of *P. maniculatus gambelii*, while two specimens apparently showing the same condition were observed in *P. m. sonoriensis*.

The grizzled condition displayed an indefinite series of gradations, ranging from the presence of a small proportion of white hairs on the dorsal surface of the snout to a condition in which the entire body was nearly white (Figure 6).

In even the most extreme cases, however, there were numerous black hairs in the dorsal region of the pelage, while the eyes, in every case, were fully black, and there was abundant black pigment in the skin of the ears, soles of the feet and some other parts.

The tail stripe was commonly affected, even in low-grade cases,

---

[1] Only two were investigated at all carefully.

[2] The colors roughly approximated RIDGWAY's "clay color" and "ochraceous tawny," respectively.

while in the more pronounced ones the black hairs were almost entirely replaced by white. These two manifestations were not, how-

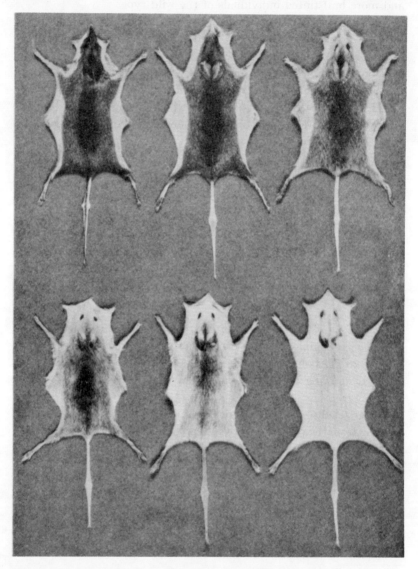

FIGURE 6. Normal *P. maniculatus gambelii* (upper left) and five grades of the multifactor color variation described as "grizzled." (From SUMNER, 1928).

ever, inseparable. The tail stripe was occasionally normal, even in cases where the grizzling of the body was well marked, while it was largely lacking in some specimens which were otherwise normal.

One interesting peculiarity of the grizzled condition was that it did not commonly manifest itself early in life, its appearance being sometimes deferred for as much as a year after birth. In the more extreme grades, on the other hand, the condition might be obvious at the age of a few months, though the degree of the aberration continued to increase for many months thereafter.

The results of my rather extended, though by no means exhaustive breeding tests with this "sport" may be summarized as follows: (1) every grizzled parent here considered bore (or sired) some grizzled offspring, even when mated with normal, unrelated individuals; (2) every grizzled parent likewise bore (or sired) some non-grizzled offspring, even in matings of grizzled × grizzled; (3) in no case, did any of the offspring of wild stock × grizzled display a higher grade of grizzling than the grizzled parent; (4) the mean grade of the progeny was correlated with the mean grade of the parents; (5) eight offspring of extracted normals were all themselves normal.

From these circumstances, it was concluded. first, that we had to do with a character dependent upon more than one genetic factor, and second, that these factors (or some of them) were dominant, though perhaps incompletely so.

Thus, we have encountered in *Peromyscus* a number of marked variations in color or other characters which behaved as simple Mendelian recessives, along with some others which appeared to depend upon a number of independently segregating factor differences. These last conformed less obviously to Mendelian requirements. Such cases may be regarded as bridging, to some extent, the apparent gap between "mutant" characters and those which distinguish subspecies.

Miscellaneous heritable peculiarities. — A number of interesting individual variations were encountered, concerning which little was learned regarding the mode of transmission. Either the characters were such as did not seem to warrant extensive breeding tests, or the tests failed to give decisive results. Some of these peculiarities are listed herewith:

(1) Incomplete tail stripe (lack at posterior end). This was noted

in *P. m. rubidus*, belonging to several independent descent lines, coming both from California and Oregon. In one of these, the incidence of the aberration was such that it seemed to be unmistakably hereditary, though the manner of its transmission was not determined. The same peculiarity was noted in three *sonoriensis*, all sibs of the fourth cage-born generation, and in two *gambelii* from near Calistoga, California [1]).

(2) White spots ("star" or "blaze", or others) on head. These were met with in various races. Sometimes they occurred in related individuals, but no serious effort was made to trace their heredity.

(3) A pigmentless condition of the tip of the snout in *P. m. rubidus* due both to white hairs and to lack of skin pigment. Inspection of the pedigrees of the affected individuals made it likely that this character was a simple recessive.

(4) Entirely white ventral hair (unpigmented from base to tip). This condition has been observed once, in a male specimen of *rubidus* from Duncan Mills, Sonoma County. Unfortunately, the specimen was killed before the peculiarity was noted. A condition resembling this in appearance, though probably genetically different, is normal in certain other subspecies.

(5) HUESTIS (1925) has discussed several discontinuous variations in minor pelage characters, which seemed to be hereditary [2]).

(6) Well developed claws on the rudimentary first digits (thumbs) of the forefoot. This condition, noted first in *gambelii*, was repeated in some of the offspring of affected individuals, though not in all. Breeding tests were not carried far.

(7) A very pronounced case of epilepsy was observed in a female *gambelii*, of the second cage-bred generation. This mouse, throughout a period of more than a year and a half, almost invariably had a typical epileptic seizure when removed from her cage to a glass battery-jar. This abnormality reappeared, in a less pronounced form, in several of the offspring of this female and of her normal sister by a

---

[1]) LLOYD (1912) records the occurrence of a similar condition in the house rat, in India, and gives evidence for the existence of much restricted local strains, in which the character seems to have arisen independently.

[2]) CASTLE (1916) records that "the production of white-spotted races from small beginnings observed in wild stocks has been accomplished in the laboratory by CASTLE and PHILLIPS in the case of *Peromyscus*."

normal brother. In at least one instance it persisted to a third generation. In the case of the offspring, however, considerable auditory stimulation was usually necessary to induce the fits. Further breeding was discontinued, owing to the high sterility of the strain. The original female seems to have outgrown the malady. No seizures were noted after the age of about 29 months.

(8) A "waltzing" habit, similar to that of the Japanese waltzing mouse, was observed in one male belonging to the F_2 generation of a cross between the albino and pallid strains. This male was mated to two of his sisters, and later to his daughters. Of the rather numerous offspring resulting from this continued inbreeding, only two showed the abnormality unmistakably, so that the evidence for the heredity of this characters is not very strong. It is of interest that in all three cases the rotation was clockwise.

RESULTS OF EXPERIMENTS IN HYBRIDIZATION

In this and the next section will be discussed the genetic status of subspecies, so far as this is indicated by experimental hybridization and by studies of inter-racial and intra-racial correlations. By genetic status is here intended the nature of the hereditary differences and their manner of transmission.

It was early found (SUMNER, 1917b, 1918) that certain, at least, among the subspecies of *Peromyscus* could be crossed very readily and would give rise to fertile hybrids. Preliminary experiments made it evident that no clear segregation of subspecific types, or even of any single character difference distinguishing such types was to be observed. Furthermore, these earlier experiments failed to reveal any general increase of variability in the second hybrid generation, even as regards characters which manifested marked differences in the races crossed. Such results tended to confirm an interpretation to which I already inclined, namely, that the characters which distinguish subspecies and other "natural" groups are in some essential way different from the "mutant" characters which are dealt with by most geneticists, and that they are not inherited in Mendelian fashion.

As more adequate material (both quantitatively and qualitatively) became available, it was evident that the familiar increase in variability occurred preponderantly, though by no means uniformly, following a subspecific cross. But one circumstance rendered difficult any satisfactory interpretation of this fact. Analysis seemed to show that the degree of this increase bore little relation to the degree of difference between the two parent races in respect to a given character. Characters which differed widely in the parent subspecies might yield standard deviations which were no greater in the F_2 generation than in the F_1, while some characters in respect to which there was little or no racial difference showed a marked increase of variability. Indeed, the greatest increase of all was shown in the variability of a series of characters (indices of asymmetry of certain paired bones) which were shown rather conclusively not to be hereditary at all (SUMNER and HUESTIS, 1921) [1]. Various reasons were given (1923b, 284—287) for rejecting an interpretation earlier suggested, namely, that the higher variability of the F_2 generation might have been due to a progressive increase in the degree of abnormality resulting from the conditions of captivity. It was therefore concluded that the F_2 hybrids, in these experiments, were more variable than the F_1 for the reason that they were F_2 hybrids and not because of any extraneous circumstances. But the facts also seemed to indicate that increases of variability, following a cross, might be due to some other cause than the generally accepted one of the segregation of multiple genetic factors.

Later studies have led to further changes in the viewpoint adop-

---

[1] Determinations (not hitherto published) have recently been made of the indices of asymmetry of the *Peromyscus polionotus* series of hybrids. The number of individuals here comprized is much smaller than the aggregate number of series on which the previous report was based, and only two of the four indices previously considered are here included. Although the results, in themselves, would certainly not be regarded as conclusive, they have a certain confirmatory value. Standard deviations have been computed for the sinistro-dextral ratios of the femurs and the innominate bones, in the F_1 and F_2 generations of the *leucocephalus-albifrons* and *leucocephalus-polionotus* crosses. Since the sexes, in each case, were dealt with separately, there are available for comparison between the two hybrid generations eight pairs of standard deviations. In four of these cases the F_2 figure is higher (differences 1.6, 2.3, 2.5, and 3.8 times their probable errors, respectively); in three cases the difference between the two was virtually 0; while, in the remaining case, the F_2 deviation was smaller, the difference, however, being only 1.3 times the probable error.

ted. While the difficulties raised in the preceding paragraph have not been satisfactorily disposed of, the evidence for genetic segregation, as regards some of the subspecific characters, at present seems overwhelming.

Additional data, together with a more searching examination of the earlier evidence, made it plain that there was a closer relation than had hitherto been recognized between the increased variability of the F_2 generation and the degree of difference between the parent races, in respect to a given character (SUMNER and HUESTIS, 1925). But the most convincing evidence for a factorial interpretation of such differences has been derived from crosses between subspecies belonging to the *Peromyscus polionotus* group. Analyses of the racial differences among these mice, together with an account of their distributional relations, have been presented in a number of former papers (1926, 1928a, 1929, 1929a) while results from inter-racial crosses in this group have recently been treated rather fully (1930).

Before proceeding, I shall summarize the character differences which distinguish the races here considered. Their distributional relations will be dealt with below. For the former, I cannot do better than quote directly from a statement already published (SUMNER, 1926, pp. 182—183):

> "These races comprise a very pale, largely white form, *leucocephalus*, which is restricted to an island of intensely white sand; an intermediate form, *albifrons*, which is found, under closely similar environmental conditions, upon the adjacent mainland [1]), and which probably represents the parent type from which *leucocephalus* sprang; and a dark form, *polionotus*, found in Alabama, Georgia and northern Florida. The last race is nearer, in several respects, to the more typical conditions in the genus *Peromyscus*, and, if not actually ancestral to the various subspecies of *P. polionotus*, it has probably departed less from the primitive condition than have any of these others.
>
> "The three races here considered agree closely in weight and in body length. They present considerable sexual differences in size, however, the females averaging larger than the males.
>
> "The races present characteristic differences in the length of the tail, foot and ear [2]); and the foot, as in other species of *Peromyscus*, averages larger in males than in females.

[1]) Also for considerable distances inland. (See below.)

[2]) The tail and foot of *leucocephalus* are distinctly longer than those of *polionotus*. The coast dwelling populations of *albifrons* are intermediate

"The dorsal tail-stripe is fully developed only in *polionotus*. It is rudimentary or lacking (rarely of full length) in *albifrons*, and wanting altogether in *leucocephalus*.

"The soles of the hind feet are frequently pigmented in *polionotus*, rarely and feebly in *albifrons*, quite unpigmented in *leucocephalus*.

"In the dark race, *polionotus*, the hairs throughout the ventral whitish region are pigmented except near the distal ends. In *albifrons* and *leucocephalus*, much of the ventral and lateral hair is white from base to tips. Between this region of pure white hair and that of the dorsal region, which is pigmented from base to tips, is an intermediate zone of varying width, in which the hair is pigmented at the base only.

"The extent of the pigmented region of the hair (as seen from the surface) is greatest in *polionotus* and least in *leucocephalus*, the condition in *albifrons* being intermediate. These areas have been measured with approximate accuracy by means of the planimeter.

"Analysis of pelages with the tint-photometer shows that these races differ in shade (whether light or dark), and in purity of color (whether highly colored or tending toward neutral gray). They do not differ appreciably in the quality (spectral position) of the color itself."

It may be added that a few of these subspecific differences are absolute, in the sense that one race may possess, in full measure, a characteristic which another entirely lacks. For the most part, however, the differences are ones of degree. In such cases, the values for a given character may differ so widely in two races that there is no overlapping of the frequency "polygons"; or they may differ merely in respect to averages, there being a more or less broad overlap of the individual values.

The chief results of the hybridization experiments with the *polionotus* group I shall present in a series of excerpts from the "Summary" of my recent extended report upon these studies (1930).

"In respect to all of these racial differences, both linear and colorimetric, the first as well as later hybrid generations show an intermediate condition. For most characters, the mean value in the F_1 and F_2 generations is approximately midway between the parental means, *i.e.*, there is no appreciable dominance.

"Dominance is strikingly shown, however, in the case of one

---

in these respects, although inland collections of this subspecies do not differ materially from *polionotus*.

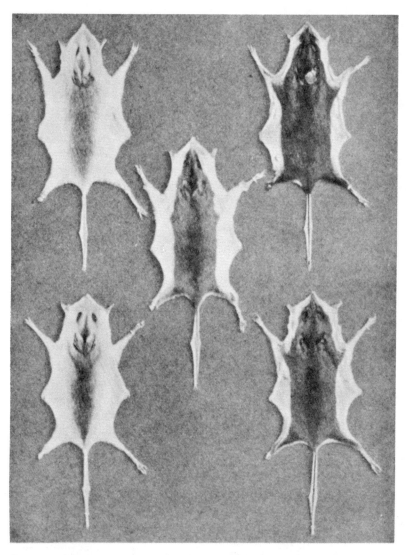

FIGURE 7. Upper left: skin of *P. polionotus leucocephalus*, close to average in respect to all pelage characters. Upper right: *P. p. polionotus* (average). Center: *P. p. albifrons* (average). Lower left: extreme pale segregant in F_2 generation of *leucocephalus* × *polionotus* cross. Lower right: extreme dark segregant in same cross. (Same specimens as shown in color in plate VIII, SUMNER, 1930).

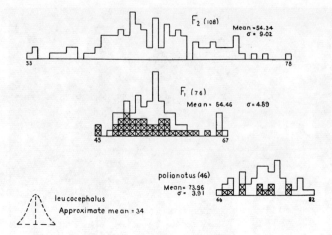
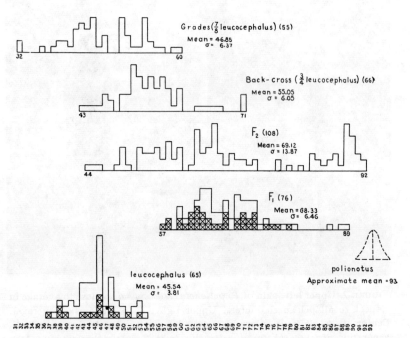

FIGURE 8. Genetic behavior of the colored area of the pelage (both determinations) in the *leucocephalus* × *polionotus* cross. Cross-hatched squares indicate parents of the next generation. (From SUMNER, 1930).

character, tail stripe. Lack of tail stripe is dominant, though incompletely so, over its presence, this phenomenon being clearly

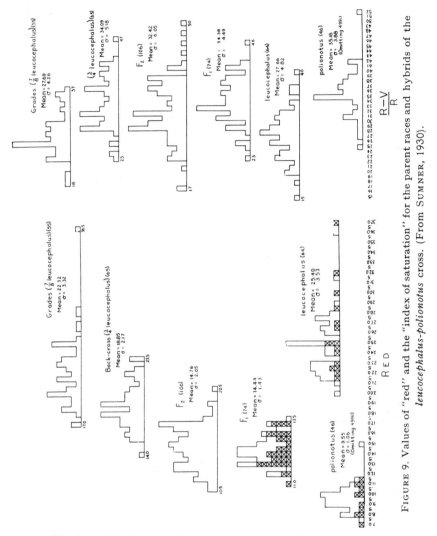

FIGURE 9. Values of "red" and the "index of saturation" for the parent races and hybrids of the *leucocephalus-polionotus* cross. (From SUMNER, 1930).

illustrated in the two interracial crosses, and in one interspecific cross [1]. This dominance is "of the 'fluctuating' type, there

---

[1] A single hybrid resulted from a mating of *P. maniculatus sonoriensis* ♀ × *P. polionotus leucocephalus* ♂. Although this was the only hybrid obtained from numerous attempted matings between these species, it proved to be fertile with both parent forms.

being an enormous range of variability in the first hybrid generation....

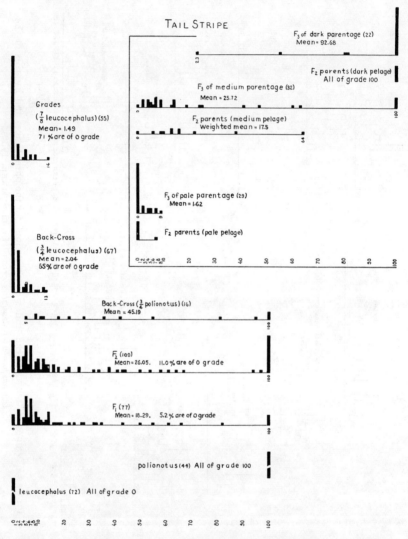

FIGURE 10. Values of "tail stripe" (here proportionate length — see text) for *leucocephalus* (in all cases 0), *polionotus* (in all cases 100), and various generations of hybrids. Insert (upper right) shows values for selected F_2 parents and their F_3 offspring. The parents had been selected according to shade of pelage, as "pale", "medium" and "dark". (From SUMNER, 1930).

"Dominance is less strikingly shown in the case of another pigmental character, relative pallor or darkness of the pelage, represented by the value for 'red' (R) [1].... A dark pelage (indicated by a low value for R) is incompletely dominant over a paler one....

"The facts cited in the two preceding paragraphs are rather unexpected, inasmuch as the degree of development of the tail stripe is negatively correlated with 'red' (*i.e.*, positively corre-

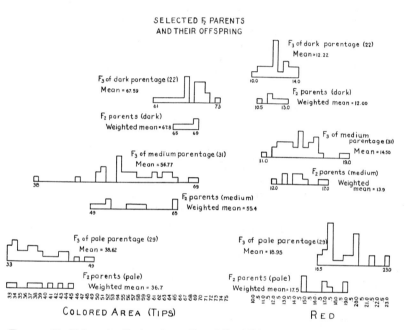

FIGURE 11. Values for "colored area" and "red," in selected F_2 parents and their F_3 offspring. (From SUMNER, 1930).

lated with depth of pigmentation). Yet presence of tail stripe is recessive, while depth of pigmentation tends to be dominant.

"In the case of.... these characters which display a partial dominance, there is a shifting of the mean in the recessive direction in the F_2 generation, as compared with the F_1.

"In no case does our evidence indicate that a racial difference in respect to any distinguishable character is dependent upon a single pair of Mendelian allelomorphs....

---

[1] See p. 23.

"That genetic segregation occurs, none the less, in respect to one important class of characters at least, is conspicuously shown by the graphs for the various measurements of the intensity and extensity of pigmentation, as well as by the relative magnitudes of the standard deviations for the F_1 and F_2 hybrid generations, as shown in the tables. It is most conspicuous in the widest of these crosses (*leucocephalus-polionotus*), though quite pronounced in the *leucocephalus-albifrons* cross. The two backcross generations show rather erratic relations in this respect.

"As regards the linear measurements of body parts, there is, on the contrary, little or no evidence of segregation. The standard deviations of the F_1 and F_2 generations have been compared for seven sets of linear measurements (tail, foot, ear, certain bones), considered separately for the two sexes. In the *leucocephalus-albifrons* cross the F_2 figure is actually more often smaller than larger.... In the *leucocephalus-polionotus* cross, on the other hand, in which the differences between the parent races are much more pronounced, the F_2 figure is larger than the F_1 in eleven cases out of fourteen.... In both of these crosses the standard deviations for number of caudal vertebrae are greater in the F_2 than in the F_1 generation. .... The differences, taken singly, are in most of these cases trivial.

"The number of genetic factors commonly concerned in any single character difference is probably considerable. Various estimates of these numbers are obtained by various methods of calculation. It is likely that the lowest of these are erroneous.

"In the narrower *leucocephalus-albifrons* cross several individuals in an F_2 generation of seventy-four reach or surpass the mean of one or the other parent race in respect to the value of colored area or of red. Even in the wider *leucocephalus-polionotus* cross two or three individuals out of 106 reach the mean of each parent race in respect to the value of colored area, though not of red. One individual, however, reaches the value of red of its own *polionotus* grandparent.

"If we provisionally consider those individuals which reach or surpass the mean of one parent race, with respect to a given character, as 'pure' segregants for that character [1]), and base our computations upon the F_2, and the first and second backcross generations of the *leucocephalus-albifrons* and *leucocephalus-polionotus* crosses, respectively, we reach the following estimates. The difference between *leucocephalus* and *albifrons*, in respect to the magnitude of the colored area, is determined by about four factors (two to six), the difference in their values for

---

[1]) This is on the assumption that the chance for an incomplete segregant to exceed the mean value of a given parent race would be balanced by the chance that a pure segregant would fall below this level. No such exact correspondence could, of course, be expected.

red being determined by two to three factors. On the other hand, the relative magnitudes of colored area in *leucocephalus* and *polionotus* would seem to depend, according to this method of computation, upon between three and four factor differences, while the relative magnitudes of red would depend upon three to five factor diferences. Similar estimates are obtained if we rank as a 'pure' segregant any individual which reaches or surpasses its own particular ancestor (or ancestors) of one or another race, in respect to a given character.

"That most of the estimates above given, aside from their manifest inconsistencies, are too small would seem to be indicated by a comparison of the histograms for the 'grades' ($\frac{7}{8}$ *leucocephalus*) with the theoretical distribution of values in a generation of this composition [1]....

"If we seek for individuals which measure up to the 'average' condition of one or another of our pure races, in respect to the ensemble of pigment characters (*i.e.*, a standard which would include half of the population of a given pure race) [2]), we do not find a single such case in the F_2 generation of either of our two principal crosses.... Five out of fifty-eight back-cross individuals, and nine out of forty-one, among the grades, in the *leucocephalus-albifrons* cross, conform, however, to the standards set for an 'average' *leucocephalus*. In the *leucocephalus-polionotus* cross we do not meet with any cases of this sort until we reach the grades, among which we find eleven out of fifty- five which may be rated as 'average' *leucocephalus*.'"

A consideration of all the evidence led to the conclusion that

"if we are to retain the multiple factor interpretation at all,.... we must attribute any one of the subspecific differences here considered to a much greater number of factors than three or four. This means that there is something wrong with the argument based upon the number of apparently 'pure' segregants. Our argument contained the implicit assumption that all of the

---

[1]) The theoretical distributions, in a second back-cross with *leucocephalus* (here called "grades") were plotted for each possible number of factor differences from two to ten. The asymmetry of distribution is so marked in all of the polygons up to those for five or six factors that it should manifest itself even in rather small populations. As a matter of fact, no undoubted asymmetry is displayed in the values plotted for this generation, in either of the crosses represented, or for any of the characters so treated.

[2]) The procedure followed, in setting this standard, was detailed in my 1930 paper (pages 306, 307). Unfortunately, two misprints on the latter page, tend to render my entire argument unintelligible. For "$(4/5)^1$" on the 10th line, should be substituted $(4/5)^3$; while for "$(5/6)^2$", on line 19, should be substituted $(5/6)^4$.

allelomorphs tending to increase the degree of pigmentation....
were contained in one of the two subspecies entering a cross,
while all of the allelomorphs tending to decrease the degree of
pigmentation were contained in the other [1]). But this, of course,
is not necessarily true. *Leucocephalus*, for example, might have
the constitution aabbccddeeFF, in respect to a given character
difference, while *albifrons* had the constitution AABBCCDD
EEff. In such an event, it is plain that the chance that an F_2
individual would equal or exceed one or the other of the parent
races would be very much greater than if all of the 'lower case'
genes were on one side, and all the 'capitals" on the other."

The early failure to detect any subspecific differences in *Peromyscus* which behaved as if dependent upon single Mendelian factors led to the suggestion that the "characters" considered were not well chosen for the purpose. Possibly some of the more elementary components which contributed to coat color, for example, might be found to segregate in relatively simple fashion.

Partly with this in view, HUESTIS undertook an investigation of the inheritance of some of the more "elementary" characters contributing to the color and texture of the pelage. Counts were made of four different types of hairs which could be distinguished in the pelage, and measurements were made of the total length of the hair, the length of the sub-terminal colored ("agouti") band, and of the dark terminal region. All of these "characters" were found to differ more or less in the various geographic races which were studied, as well as in certain marked color varieties ("mutants"), and these differences were largely responsible for the racial differences in the pelage (HUESTIS, 1923, 1925; SUMNER and HUESTIS, 1925).

HUESTIS found abundant evidence of segregation for all of these more "elementary" characters contributing to coat color, as judged by the relative magnitudes of the standard deviations in the successive hybrid generations, and by certain other critera. But the increase of variability, thus revealed, was no greater, with a single exception, than that shown by values representing the aggregate shade of these hairs, when combined in the pelage. "While the data at hand do not warrant any such simple inference, we may safely affirm that

---

[1]) This same objection, among others, applies to the formula of CASTLE, and WRIGHT (CASTLE, 1921). The latter leads to utterly impossible conclusions when applied to my own material (1930, p. 357).

microscopic analysis has thus far helped us little in the search for unit factors underlying the coat color of *Peromyscus*."

Thus, if coat color is capable of resolution into genetically more elementary characters, it is evident that the analysis must proceed along different lines from those chosen. In the next section it will be shown with a high probability that various pigmental characters, relating to different regions of the body, and affecting both the extensity and intensity of pigmentation, depend, to a large degree, upon common genetic factors. But it has already been shown that these factors are all of the cumulative or polymeric type. Thus we have a situation in which every character difference between two races is determined by a polymeric group of factors, but also one in which each such group manifests itself in manifold ways in the body of the organism. In order to isolate the effects of single factors, therefore, it would be necessary to split up one of these factor groups, rather than to analyze a complex of bodily effects into its components. But the outlook for success does not seem to be very promising.

CORRELATION

Some of the most important light upon the genetic nature of subspecific differences has been derived from a study of correlations among the various measured characters. Correlations have been determined for as many characters as possible. Two types of correlation have been considered, inter-racial and intra-racial, respectively. The first of these indicates the degree to which two given characters are found to vary together, when we study a series of closely related races. The second reveals the tendency for these characters to vary together within the same population (here including populations of hybrids, as well as of pure-race individuals). The computation of coefficients has been restricted almost wholly to the second type of correlation (intra-racial). Inter-racial correlations are frequently of high interest, but the number of races in any comparative series is commonly too small for the computation of coefficients, while the presence of some degree of correlation is often sufficiently indicated by comparisons of mean values in the races forming the series under consideration.

To discuss intra-racial correlation first, it has been found necessary to deal separately with two classes of characters: (a), body parts subjected to linear measurement (tail, foot, ear and various bones), and (b) pigmental characters (shade of pelage, photometrically determined, index of saturation, extent of colored area of pelage, foot-pigmentation, tail stripe [1]).

In general, it may be said: (1) that many of the body parts tend to vary together in their relative magnitudes [2]; (2) nearly all of the pigmental characters (all of those involving quantity of pigment) vary together in a positive sense; (3) there is no clear evidence of a biologically significant correlation between a character of the first class and one of the second.

Inter-racial correlations, so far as these concern the length of body parts, are altogether erratic. While, within single populations, certain parts (e.g., tail and foot) tend to vary together in their relative size, such concomitant variation may or may not be encountered when we examine a series of geographic races. Throughout considerable tracts, a positive correlation may hold; in other territories the correlation may be entirely dissolved.

Inter-racial correlations in pigmental characters, on the other hand, are even more pronounced than are intra-racial ones. Darker races, like darker individuals, tend to have more extended colored areas in their pelages, deeper pigmentation in the skin of their feet, broader (and longer) tail stripes, etc. At an earlier stage of these studies, indeed, I noted the obvious correlation between certain pigmental characters, when several geographic races were compared, although I failed to find adequate evidence of a correlation between some of these same characters, within each race, taken singly [3]. This supposed fact led me to certain theoretical conclusions which I now believe to be untenable. The fact that inter-racial correlations among these characters are commonly much higher than intra-racial ones now appears to me intelligible on the ground that the differences

---

[1]) Actually, the measurement of tail stripe is linear.

[2]) That is, when the influence of total body size has been eliminated. That there is, in a population of mixed size, a large degree of correlation among all the measurable parts, goes without saying.

[3]) SUMNER (1918, p. 195, 1923a, 1925). COLLINS (1923) also records that "pigmentation of the feet was found to vary independently of the color of the pelage."

between races are almost wholly genetic in their nature, while the differences among the individuals of a race are largely of the non-genetic sort. These non-hereditary modifications of the pigmental characters (however they may be produced) appear to be largely independent of one another and thus tend to obscure the correlations among them [1].

In view of these relations among our geographic races, in respect to characters based on pigmentation, it is highly interesting to consider the relative magnitudes of the correlation coefficients, in the various generations of hybrids derived from the crossing of the subspecies of *P. polionotus*. These

> "are commonly lowest for the pure races and F_1 hybrids (lowest of all, frequently, in the latter); higher for the F_2 generation, and highest of all in the F_3 generation, derived from selected F_2 parents. The coefficients for the back-crosses are variable, though much more often higher than lower, in comparison with those for the F_1. These relations are, for the most part, such as might be expected, on the supposition that the characters concerned are in some way genetically connected, and that they therefore tend to segregate together.... The higher degree of correlation among these characters which is found in the F_2 generation, as compared with the F_1, is due to an increase in the proportion of the total variability which is genetic, combined with the fact that the characters are in some way bound together genetically. The still further increase in these correlations which is manifested in the F_3 generation is due to the procedure adopted in the selection and mating of the parents." (SUMNER, 1930).

> "Regarding the nature of this genetic connection among our correlated characters, two hypotheses are possible: (1) The characters in question may be merely different manifestations of the same genetic factors,"

e.g., both the extent and the depth of the pigmentation may be determined by factors which control the total amount of pigment produced,

> "(2) The factors underlying these characters may be more or less closely linked, due to their presence in the same chromosome...."

Another alternative, that of parallel modification, through envi-

---

[1] SUMNER (1929a, 1930). In this recognition of the difference between intra-racial and inter-racial (intra-specific and inter-specific) variation, and likewise in the essential features of the interpretation here adopted, I was anticipated by BUBNOFF (1919) in a discussion of fossil *Ammonites*. I am not, however, in agreement with that writer on all points.

ronmental influences acting during ontogeny, could hardly account for correlations which increase in the "segregating" generations of hybrids [1]).

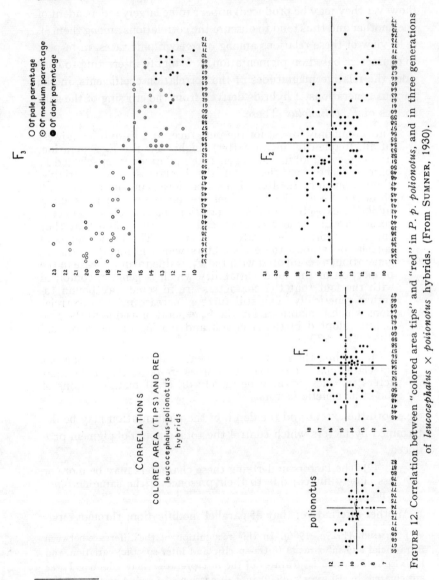

FIGURE 12. Correlation between "colored area tips" and "red" in *P. p. polionotus*, and in three generations of *leucocephalus* × *polionotus* hybrids. (From SUMNER, 1930).

---

[1]) It has been shown (SUMNER, 1915a, 1918) that parallel modifications

Reasons are given (SUMNER, 1930) for believing that these correlations among the various pigmental characters result from their being determined in part by a genetic basis common to all of them. Important among these reasons is

> "the frequent association between intensity and extensity of pigmentation, upon various parts of the body, and in animals belonging to widely different groups.... In connection with this class of facts, the hypothesis of linkage can hardly be invoked as an explanation. For studies of linkage thus far made have surely revealed no general tendency toward the close propinquity, within the same chromosome, of genes which bring about similar physiological or morphological conditions." (Pp. 360—361).

On the linkage hypothesis, we should have to suppose that the genes belonging to a considerable number of different polymeric series were linked together more or less closely, while those within the same polymeric series were genetically quite independent of one another [1]). Such a situation might be intelligible on the basis of polyploidy, but it is, at best, entirely hypothetical, and is not in harmony with the fact mentioned at the close of the preceding paragraph.

> "This common genetic basis for all the various pigmental characters is not, however, absolute. There is a considerable degree of independent variability among these characters, and it may be shown [2]) that a large fraction of this independent variability is genetic. Thus, two pigmental characters may be supposed to have certain factors in common and certain ones peculiar to themselves." (Ibid., p. 355).

Finally,

> "there is some evidence for the existence of correlations between the bodily appendages (tail and foot) and the pigmental characters, in certain of the segregating generations of hybrids, in all three crosses. The coefficients are preponderantly of the 'expecpected' sign, on the assumption that the character differences of a subspecies should segregate together. While the considerable

may be produced in the length of the tail, foot, and some other parts, by certain physical agencies (temperature, dietary (?) deficiencies, leading to a rachitic condition). On the other hand, pigmental characters, with few exceptions, have shown themselves to be uninfluenced by environmental factors which call forth marked changes in bodily proportions.

[1]) See TAMMES (1912).

[2]) By the method of partial correlation and in other ways.

series of coincidences here displayed can hardly be credited to random sampling, there are circumstances which render the foregoing interpretation somewhat questionable." (Ibid. p. 355).

Several years ago I devoted two papers (1923c, 1924) to a matter which has obvious relations to some of the questions just considered, as well as to the genetic status of subspecific characters in general. Arguments were offered for "the partial genetic independence in size of the various parts of the body," and exception was taken to CASTLE'S contention (1922) "that the genetic agencies affecting size in rabbits are general in their action, influencing in the same direction all parts of the body," and furthermore, that these factors probably are not only "*general* in their action," but "*exclusively* so." In support of my viewpoint, there were offered evidences of the independent variability of various measurable parts of the body in *Peromyscus*, when the influence of total body size was eliminated. The existence was likewise shown, both in mice and rabbits, of a considerable degree of "net" correlation between various body parts, *i.e.*, such correlation as would (theoretically) be obtained in series of animals having constant body length. It was made clear, moreover, that part, at least, of the independent variability here indicated was of the genetic sort. Further evidence for the same general conclusion was based upon the known occurrence within groups of closely related animals, of striking differences in the proportional size of almost every bodily part. In other words, the genetic factors underlying these last must have undergone independent modification in the course of phylogeny, a thing which would have been quite impossible had not some of these size factors been specific, rather than general, in their action.

RACIAL DIFFERENCES IN RELATION TO ENVIRONMENT

The term subspecies is commonly employed as synonymous with geographic race. By definition, therefore, such groups occupy different territories, and it would accordingly seem inexpedient to try to distinguish representatives of two or more subspecies (belonging to the same species) within the population of any single locality and habitat [1]). The extreme variants, in a single population, may show as wide differences as those between the means of two subspecies, and these differences, as already pointed out, may be largely genetic ones. But it would seem more reasonable to credit such differences to *intra-racial* variation, rather than to the coexistence of more than one geographic race in the same locality. To be consistent, the boundaries between subspecies should be regarded as geographic, rather than morphological ones, although morphological considerations must, of course, determine whether or not two geographically separated populations shall be recognized as distinct subspecies.

To the foregoing statements, one possible qualification should be admitted. Cases have been reported (OSGOOD, 1909) [2]) in which the more highly divergent races of a "Rassenkreis" appear to have come together at some time subsequent to their differentiation, and have failed to blend in this common territory, thus behaving toward one another like distinct species. In such cases, one would wish to know how recent the meeting of these two unlike races has been; likewise whether, within the territory in question, the two occupy the same habitat. So far as I have found in *Peromyscus*, the most unlike subspecies appear to be perfectly fertile *inter* se [3]).

In viewing a map which portrays the geographic races of a widely ranging and "plastic" species (OSGOOD's plate 1 (1909) is an excellent example), one cannot fail to be struck by the largely arbitrary manner in which the territory seems to be divided. To quote from one of my earlier papers (1918, pp. 180, 181):

> "Looking at the distribution maps in such publications as those mentioned, one is impressed by a seeming analogy between the boundaries of these various subspecific ranges and those of the political subdivisions of the earth's surface. In considerable degree these last are bounded by geographic features, but to a large extent, also, the lines of demarcation seem to be drawn quite arbitrarily — the territories merely bound one another.... And, in any case, it is doubtful whether any geographic barrier, save a continuous body of water or a lofty and unbroken range of mountains could prevent the free diffusion of such rodents."

---

[1]) This statement does not, of course, apply to birds during migration.

[2]) See also KLAUBER (1930, pp. 132, 137) for a similar case among rattlesnakes.

[3]) I. e., as fertile as members of the same race. Infertility, in captivity, is of frequent occurrence. (See page 19).

On the other hand, the ranges of single subspecies

"frequently comprise territory having a wide diversity of physical conditions. For example, *Peromyscus maniculatus gambeli* is represented as ranging from the foggy coastal area of central and southern California across the hot, semi-arid San Joaquin Valley to the snowy heights of the Sierra Nevada. And in latitude, its range is said to extend roughly from the 31st to the 48th parallel.... We might well be puzzled to discover any common elements of the physical environment which were responsible for the presence of the same subspecies under such widely divergent conditions of life. Particularly is this true when the environmental differences, as in the present case, far exceed those between the habitats of certain quite distinct subspecies."

One not familiar with the facts of the case might be disposed to doubt the validity of such subdivisions of a species, and to attribute the anomalous situations here described to the overwrought imaginations of our taxonomists. But any such general interpretation of the facts may certainly be rejected. I will illustrate by a single case. Mice from Calistoga, in northern California, within the stated range of "*gambelii*", were found to much more closely resemble "*gambelii*" from La Jolla, 500 miles to the southward, than they did mice from Duncan Mills, in "*rubidus*" territory, only 27 miles to the westward (SUMNER, 1920). These facts are the more striking, since no real geographic barrier intervenes between Calistoga and Duncan Mills, while the entire section of California in which Calistoga lies is isolated from the central and southern parts of the state by San Francisco Bay and the Sacramento River (Figure 13).

Whether two forms shall be regarded as distinct species, or as subspecies within the same species, is commonly determined by reference to the "criterion of intergradation." If two populations merge and intergrade with one another at contiguous points in their ranges, or if they are connected by one or more intermediate forms, each of which intergrades with its neighbors, they are regarded as subspecies. The whole aggregation is commonly designated as a species, though the terms "Formenkreis" (KLEINSCHMIDT), "Rassenkreis" (RENSCH) and some other have been employed. The principle of "intergradation" has been extended so as to include insular forms, which are now geographically isolated, but whose measurable characters overlap, to some extent, those of races on the nearby mainland.

In general, it must be insisted that the distinction between species and subspecies is not absolute, but is more or less conventional. A satisfactory explanation of the origin of subspecies would likewise account for the origin of species.

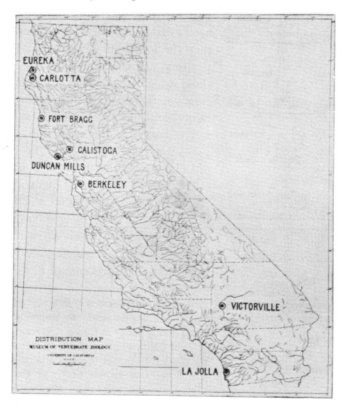

FIGURE 13. Chief stations in California, at which collections of the subspecies of *Peromyscus maniculatus* were obtained. (From SUMNER, 1918).

As already stated, these subspecies are far from being homogeneous throughout their respective ranges. The *"gambelii"* of Berkeley is not identical with the *"gambelii"* of La Jolla, and the *"rubidus"* of Duncan Mills differs rather widely from that of Eureka or Carlotta (figure 14) [1]).

---

[1]) The Duncan Mills population may perhaps be regarded as showing in-

Nor is either *P. polionotus polionotus* or *P. p. albifrons* uniform within its range, as will be shown later.

The recognition of a definite number of subspecies, and their assignment to definite territories implies, however, that each is *relatively* homogeneous within its range; or in other words that the local differences within its own territory are small in comparison with those which distinguish one subspecies from another. This requirement is certainly fulfilled in very many cases.

The nature of intergradation between two subspecies I have made the subject of an intensive study in one particular case (1929, 1929a). In this case (figures 15, 16, 17), one of the subspecies (*P. p. albifrons*) proved to undergo a progressive increase in both the intensity and extensity of its pigmental characters, as we passed northward from the Gulf Coast of Florida into

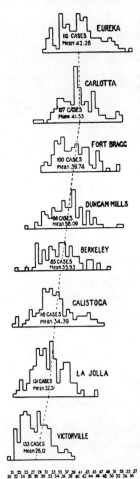

FIGURE 14. Distribution of values of the width of the tail stripe (ratio to circumference of tail) for subspecies of *Peromyscus maniculatus*, at eight California stations (see figure 13). The broken lines connect the means of the series. The order corresponds nearly or quite with that for mean annual rainfall. Specimens from Victorville are assigned to subspecies *sonoriensis*; those from La Jolla, Calistoga and Berkely to *gambelii*; those from Duncan Mills, Fort Bragg, Carlotta and Eureka to *rubidus*. (From SUMNER, 1918).

the interior [1]). At a point about forty miles inland, a fairly abrupt

---

tergradation with *gambelii*. But this could hardly be claimed for the Fort Bragg specimens, which were quite distinct from those taken at the more northern points.

[1]) This is the order in which our investigation proceeded, and it seems best to adhere to it here, although it is doubtless the reverse of the evolutionary order.

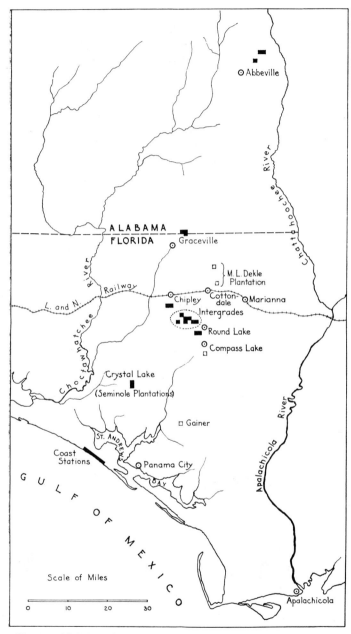

FIGURE 15. Map of portions of Florida and Alabama, showing stations at which *P. p. albifrons* and *P. p. polionotus* were trapped in 1927. (From SUMNER, 1929a).

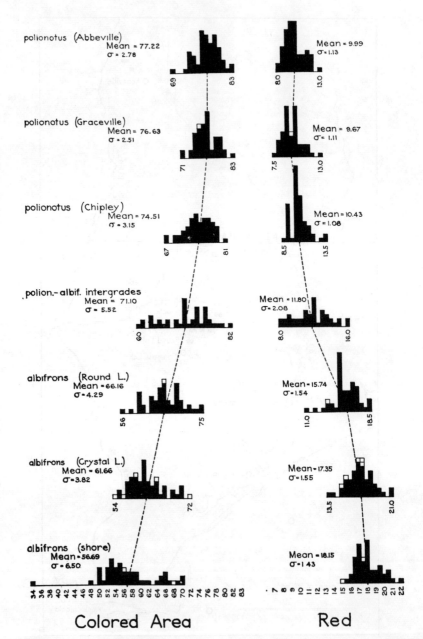

FIGURE 16. Values of "colored area" and "red" at each of seven chief collecting stations shown in map (fig. 15). The broken lines connect the means for the local collections. The unshaded squares represent a few individuals, which, for certain reasons, are not included in the regular series upon which the averages are based. (From SUMNER, 1929).

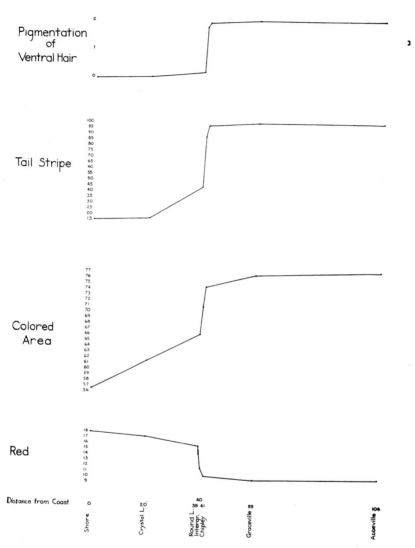

FIGURE 17. Graphs showing mean values for four pigmental characters in each of the seven collections (figure 16). In the present figure, unlike the preceding, the stations are plotted with reference to their actual distances from the coast. The condition as regards "red" is by no means contradictory to that shown by the other characters, as might be inferred from this figure, since it is *low* values which indicate *heavy* pigmentation and *high* values which represent *light* pigmentation. (From SUMNER, 1929).

change took place within a "zone of intergradation" only a few miles in width. To the northward of this, we encountered populations which could fairly be assigned to *P. p. polionotus*, though evidence of the same gradient, as regards some of the pigmental characters, could be detected for a further distance of perhaps sixty miles [1]). In the zone of intergradation, the population consisted of some extreme types which reached or exceeded the average of one or the other "pure" race, along with a much larger proportion of intermediate forms, the mean of which lay closer to *polionotus* than to *albifrons*.

Thus, in the present case, there is manifested a pronounced gradient, in respect to both the intensity and extensity of pigmentation, extending throughout a distance of perhaps a hundred miles. In the "zone of intergradation," this gradient merely becomes steeper, but the trend appears to be the same, with minor exceptions, throughout the entire line [2]). Whether the foregoing situation is a typical one, in all respects, is questionable. However that may be, gradients of this general nature, in the values of various measurable "characters," both linear and colorimetric, are not infrequent. Furthermore, there are certain undoubted correlations between these gradients and known factors of the environment. The most striking and indubitable of these correlations is the long recognized association between dark pigmentation and high atmospheric humidity, or conversely between pale coloration and an arid climate. This association has been chiefly discussed in relation to mammals and birds, but abundant examples may be found among reptiles, insects and

---

[1]) Whether this gradient actually held as far as Abbeville, 104 miles from the coast, cannot be stated. No trapping was done between Graceville (59 miles) and Abbeville, owing to the slight differences between the collections obtained at these two stations. That the gradient extended to some point beyond Graceville is, however, evident from the charts.

[2]) This picture, and the explanation which follows, have been oversimplified by the omission of one complicating circumstance. This is the fact that no such gradient is manifested in the linear measurements of tail, foot, etc. While these are distinctly longer in the coast collections than in those taken at any of the inland stations, the latter do not appear to show any consistent difference among themselves. However, it has been pointed out elsewhere that these members are subject to highly erratic local variations, a fact which may be related to the ease with which they may be modified artificially through physical agencies.

perhaps some other groups of animals. Its existence was first brought into general notice by ALLEN (1871).

The reality of this correlation is strikingly illustrated by local variations in coat color and numerous other pigmental characters in *Peromyscus*. In general, throughout the Pacific Coast area, we meet with an increase of dark pigment as we pass from south to north, and from the interior toward the coast. This corresponds, in each case, with an increase in both rainfall and atmospheric humidity.

The nature of the causal relation between these two correlated variables remains to be discovered. Several hypotheses have been offered: (1) Atmospheric humidity has, in some more or less direct way, influenced the metabolism of the body, and thus the formation of pigment, without regard to the utility of the result. In this case, we should have to suppose that the germ cells have undergone parallel (or subsequent) changes. (2) The degrees of pigmentation may be merely visible indices of physiological differences, in such a manner that the type of organism best adapted to desert life has a low capacity for pigment formation, while that best adapted to coastal fogs has a high capacity. (3) The optical conditions of the environment, acting through the eyes or skin, have in some direct way influenced pigment formation. (4) We have here a simple case of "concealing coloration," brought about through the selective elimination of individuals which chanced to be least in harmony with the color tone of their surroundings. In this connection, it is hardly necessary to state that both soil and vegetation, in arid regions, are prevailingly pale, in comparison with those of more humid regions.

For the first of these alternatives we have no direct evidence. Attempts to test the question experimentally have thus far failed to give convincing results. The chief arguments which have been offered in support of this hypothesis are based upon the inadequacy of other explanations. (SUMNER, 1921, 1925; SUMNER and SWARTH, 1924).

The second alternative is thus far purely hypothetical. It is supported, so far as I know, by no evidence, either direct or circumstantial. On the contrary, the presence of extremely depigmented beach-dwelling races, in regions of high humidity, would render such a supposition very doubtful.

The third alternative would probably be dismissed by most biologists without a hearing. However, we have fairly conclusive evidence that the pelage color of mammals may be influenced in some way, direct or indirect, by the shade or color tone of the background, inde-

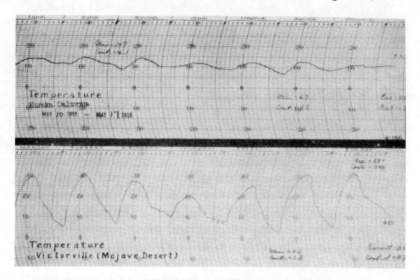

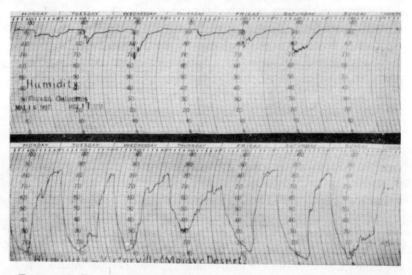

FIGURE 18. Thermograph and hygrograph records for a week in May, at Eureka, California, in a humid region of the northwest coast, and at Victorville, in the Mojave Desert.

pendently of any atmospheric factors. It has long been known that among fishes and animals of several other groups the entire colora-

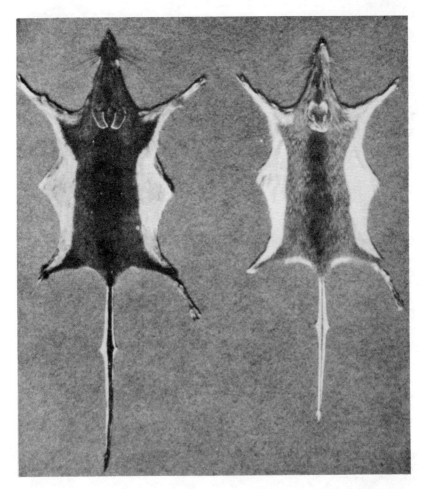

FIGURE 19. Skin of *P. maniculatus rubidus* (left) from vicinity of Eureka, California (humid), and one of *P. m. sonoriensis* (right), from Mojave Desert, California. (In these and the skins next figured, the ventral cut was not carried forward through the head, according to the procedure later adopted).

tion of the body may undergo rapid changes in response to optic stimuli received from underlying or surrounding surfaces. Furthermore, it is now certain that the actual production and destruction

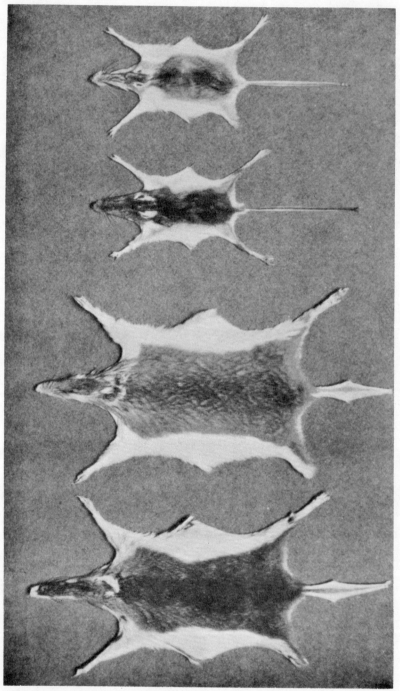

FIGURE 20. Larger skins: one of *Onychomys leucogaster fuliginosus* (left) from San Francisco Mountain region, Arizona, and one of *O. l. melanophrys* (right), from the nearby Painted Desert. Smaller skins: *Perognathus flavus fuliginosus* (left) and *P. f. bimaculatus* (right), from same two general localities. (See SUMNER and SWARTH, 1924).

of pigment may be influenced greatly, though less rapidly, by stimuli received through the eyes [1]). Such phenomena are, I believe, utterly unknown among mammals, but the possibility should be borne in mind. Even highly improbable things occasionally prove to be true!

Thus, in certain cases at least, we seem almost driven to the acceptance of the fourth of these alternative hypotheses. Mice inhabiting sandy beaches, especially when these are isolated from the mainland, are in some instances conspicuously pale (JAMESON, 1898; BANGS, 1905; OSGOOD, 1909; G. M. ALLEN, 1914, 1920; HOWELL, 1920; SUMNER, 1917, 1926). The extraordinarily depigmented form, *Peromyscus polionotus leucocephalus*, already discussed, occurs upon an island reef of pure white quartz sand, situated close to the Florida coast. In this race, most of the hair is white from base to tip, while the pigment of the skin is likewise greatly reduced.

On the adjacent mainland, among beaches and dunes of equally white sand, is found another very pale race, *P. p. albifrons*, which has not, however, been modified to such an extreme degree. This last circumstance is doubtless due to the continuity of this coastal, sandy zone with very different territory, a short distance inland. The graded differences in the character of the *albifrons* populations, as we pass from the coast to the interior, have already been discussed.

It should be stated that both of the last mentioned races dwell in a region of high rainfall and high atmospheric humidity. We must accordingly look to optical, rather than atmospheric factors for an explanation of these peculiarities of coat color. And since the direct effect of visual stimuli upon pigment formation among mammals can hardly as yet be considered seriously, there seems to remain in this case only the rather threadbare hypothesis of "protective coloration", brought about through the differential survival of those individuals which chanced to be least conspicuous against their pale background. This hypothesis has considerable inherent probability, though there is little direct evidence in its favor, for any group of animals, and none, so far as I know, for mammals.

That selective elimination of this sort commonly requires long pe-

---

[1]) BABÀK (1913), KUNTZ (1917), MURISIER (1920—21), SUMNER and WELLS (unpublished notes).

riods of time to work in is certain. I have pointed out one case (1926, p. 158), in which a thriving colony of *albifrons*, inhabiting an almost isolated reef of white sand, and apparently having very little oppor-

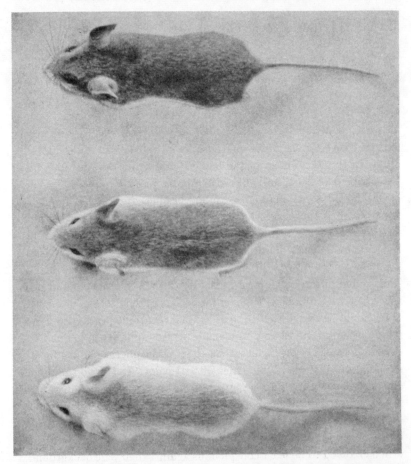

FIGURE 21. Specimens of *leucocephalus, albifrons* and *polionotus*, photographed against background of white sand, taken from region inhabited by *leucocephalus*. (From SUMNER, 1926).

tunity to diffuse with the mainland population, has none the less undergone no perceptible change in the direction of *leucocephalus*. This at least shows the improbability that competition, on the basis of color adaptation, is at all keen among these animals.

I had earlier described a case (1921), in which a typically pale desert subspecies, *Peromyscus crinitus stephensi*, was found to occur in great

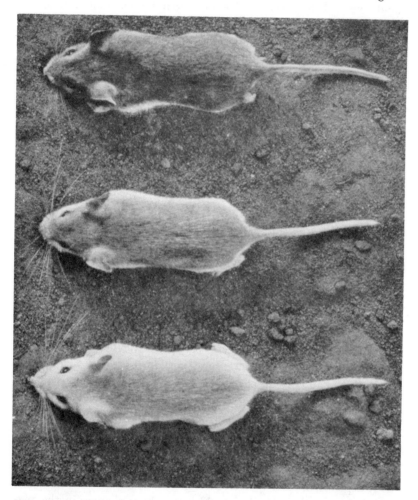

FIGURE 22. Same specimens as in figure 21, on background of dark soil, from region inhabited by *P. p. polionotus*. (From SUMNER, 1926).

numbers [1]) upon an area of very dark lava in the Mojave Desert. Since this lava flow is surrounded on all sides by pale sand, and

---

[1]) It was by far the most abundant mouse trapped here.

since the mouse in question is restricted very largely to rocky regions, this colony appears to be isolated, much as if it dwelt upon an island. Nevertheless, these mice were found to average no darker than those trapped upon an area of pale, non-volcanic rock, situated in another part of the same desert [1]). This case was cited by me as one of the reasons for attributing the peculiar hues of desert animals to atmospheric rather than optical causes. It must be admitted, however, that if this instance is to be used as an argument against the necessity of concealing coloration among these animals, the case cited in the preceding paragraph tends toward the same conclusion. Yet in relation to these pale, beach-dwelling types, the atmospheric factor can not be invoked as an alternative explanation. It would, perhaps, be fairer to cite both of these cases as evidence that selective elimination, on the basis of coat color, is very far from rigid in *Peromyscus*, and that thousands of generations may be necessary before its effects are manisfested. However that may be, such cases are sufficient to cast serious doubts upon various claims which have been made respecting the pronounced effects of highly local conditions upon the pelage color of rodents [2]).

I am now prepared to admit that certain of the arguments which I offered in earlier papers (1921, 1923a, 1925) for the direct responsibility of atmospheric factors in bringing about the pallid hues of desert animals have lost some of their cogency. A further illustration of this relates to the depigmentation of certain parts (e.g., soles of the feet) which can play no part in concealment. I had earlier believed that the pigmentation of these and some other parts had no genetic connection with that of the hair. The parallel modification of these various pigmental regions seemed therefore to imply the simultaneous effects of some environmental agent. More recent evidence (see pages 53—55) has shown that pigmentation throughout the entire body depends, in part, upon a common genetic basis. Thus, selection, with reference to coat color, could bring about changes in the pigmentation of invisible parts.

There remains, however, the circumstance that desert races of

---

[1]) The alleged effect of lava flows, in darkening the color of desert rodents, was also studied carefully in another locality, likewise with negative results (SUMNER and SWARTH, 1924).

[2]) For example, OSGOOD (1909), p. 16.

certain animals which would seem to have no need for concealing coloration (skunks and bats) nevertheless give pronounced evidence of depigmentation.

It is not improbable, therefore, that both of these explanations must be reckoned with. Pronounced differences of atmospheric humidity may be capable of bringing about genetic changes in pigment formation, independently of selection, while selection, on the basis of concealing coloration, may enhance these effects in the case of many desert animals. This is a "compromise verdict" of a decidedly banal character, but it none the less seems to accord best with our present knowledge.

Let us return to a consideration of the relation between character gradients and environmental gradients. In the case which I have most carefully investigated, the pigmental characters were found to undergo parallel changes, in the direction of increasing pigmentation, from the sea-coast to a distance of perhaps a hundred miles inland. The rate of change was not uniform throughout this distance, however, the gradient becoming extremely steep for the space of a few miles. A really adequate interpretation of such phenomena is greatly to be desired. Unfortunately, this case, and perhaps all similar cases, are far from simple (SUMNER, 1929, 1929a). Had we any reasons, in the present instance, for attributing these changes to climatic factors, the pigmental gradients could perhaps be plausibly explained, since we are passing from south to north, and likewise from, the coast to the interior. Also, the abrupt change encountered in the "area of intergradation" might be ascribed to the existence of a "critical point" in some environmental factor, such as those which are believed to be concerned in the delimitation of the "life zones", "faunas", "floras", "associations", etc. of the students of distribution.

Unfortunately, the only physical agency which we can say with any confidence has been concerned in the present case presents no gradient parallel with these biological changes. Unless all appearances are illusory, the depigmentation of these coast-dwelling races stands in some direct causal relation to the bare white sand on which they dwell. But this white sand is limited to a narrow zone close to the sea, and there is no pronounced gradient in soil color as we pass inland from this littoral fringe. Throughout the range of *albifrons*, it is true, the sandy soil is paler, in comparison with that encountered farther north, this difference depending upon a difference in the respective geological formations. Furthermore, these soil types succeed one another rather abruptly, though it is important to add that this occurs at a level which does not correspond with the boundary between the territories of *albifrons* and *polionotus*. But the ground is largely covered with vegetation, and the differences in soil color are not very evident except in cultivated areas and along roadways. In any case, it would be difficult to find here an explanation for the extreme depigmentation of the inland populations of *albifrons*, a process which has been carried farther here than in any of our desert types, except a few of the most highly modified ones.

It may be added that there are no geological data which warrant the conjecture that the present characteristics of these local populations have been due to conditions which have long since vanished (1929a).

The two most perplexing questions here raised are: (1) Why have the bleaching effects of the coastal sands spread so far beyond the confines of the latter? and (2) Why do we encounter such an abrupt change in our mouse population in a region where no adequate environmental cause is discernible, and in spite of the complete fertility of these races *inter se*, and the apparently complete fertility of the hybrids?

One hypothesis, if tenable, would answer both of these questions fairly satisfactorily. We may assume with some confidence that the more typical and widespread of these subspecies, *P. p. polionotus* (or a form closely similar to the latter) was ancestral to the other races of this group, including *P. p. albifrons*. We may next suppose that *polionotus* extended its range from the interior to the white sand-dunes and beaches of the Florida coast, if indeed it was not already at hand when the latter first emerged from the sea; that it there underwent an extensive process of depigmentation, in response to the color tone of its background; that owing to particularly favorable conditions in this littoral region, the new race (*albifrons*) multiplied rapidly; and finally, that the population pressure thus engendered led to a return wave of migration back toward the interior. The present "zone of intergradation, "forty miles from the coast, may be supposed to represent the line along which the centrifugal movement of the *albifrons* population meets and is arrested by an oppositely directed movement on the part of the *polionotus* population.

The existence of pigmental gradients within the *albifrons* territory may be attributed either to the continuous absorption of the previous inhabitants of this territory, in the course of its occupation by *albifrons,* or to a reversed selection, resulting from the return of the latter to the much darker soil of the interior, or perhaps to both of these causes combined. Furthermore, the assumed biological barrier must be far from absolute, so that a certain amount of invasion and mutual contamination would affect the composition of both populations. Reference to the histograms (figure 16) shows that such contamination is not unlikely. And the failure of the beach-dwelling populations of *albifrons* to reach the extreme condition of pallor shown by the island race *leucocephalus* is further evidence of a certain amount of reversed migration from the interior.

It is evident that a number of unproved assumptions underly such an explanation as the foregoing. On the other hand, none of these assumptions seem to be inherently improbable. It can hardly be doubted that there are considerable local differences in population pressure, due to differential net fecundity, with a consequent expansion of certain populations at the expense of others. It seems equally probable that, periodically at least, the competition for food and other conditions of life among these animals is very keen. Under such conditions, the wanderings of each individual would tend to be in the direction of a falling gradient of population pressure. Thus, the territory of one race would be unlikely to be invaded by animals of a neighboring race having the same food habits. This pressure gradient would not, however, have any relation to differences of race, as such, for the animals of each race would be equally unlikely to turn back toward the interior of their own territory. I shall return to this subject later.

Such an explanation would account for the maintenance of the rather sharp line of division between these mutually fertile populations, in the absence of any probable ecological barrier, and it would not involve us in the impossible assumption that each of these races is adapted so completely to the peculiarities of its own territory that it could not survive in that of its neighbor. For even if the division between the ranges of *polionotus* and *albifrons* corresponded to pronounced differences in the color of the soil [1]), this would surely not suffice to prevent the extensive migration of individuals of one race into the territory of the other. It has been shown above that selective elimination, on the basis of coat-color, is far from keen.

Would not this same type of explanation account for some of the anomalies earlier mentioned in the ranges of subspecies in general? It would relieve us from the hopeless task of searching for common elements in the enormously diversified environment of *P. maniculatus gambelii*, for example, to which that race is peculiarly adapted; while it would likewise relieve us of the necessity of finding some potent, if intangible, barrier dividing a seemingly uniform and continuous area which chanced to be tenanted by two subspecies [2]).

Despite obvious flaws in the analogy, we seem justified, within certain limits, in comparing the relations shown by our distribution maps with ones which would result if a collection of spherical rubber bags were placed in a rigid container, and then strongly but unequally inflated. The bags would come to bound one another, without the need of any other agent to mould their outlines. Fluctuations in relative pressure would lead to the continual shifting of boundaries, whether in our physical model or in an assemblage of subspecies. It is quite unlikely that the distribution patterns which we now see are definitive ones.

This last picture is quite a different one from that offered us by WILLIS (1922), who finds the expanding ranges of species to overlap one another broadly, or even to be included one within the other ("wheels within wheels"). Such an arrangement as the last would obviously be impossible to maintain in the case of subspecies of animals, which are, so far as known, perfectly fertile *inter se*.

WILLIS has given convincing reasons for attributing the distribution of plant and animal species largely to their inherent tendency to expand, with little regard to their special "fitness" for the particu-

---

[1]) See p. 74.

[2]) It is possible that light might occasionally be thrown upon these anomalies by a knowledge of recent geological history. But I should not expect much help from this source, even if our knowledge were vastly more complete. One need only ask the question: Why should two mutually fertile races maintain their distinctness after the withdrawal of an earlier barrier between them?

lar geographic areas which they come to occupy. I am, however, disposed to attach more importance than does WILLIS to adaptation as a factor both in the origin and perpetuation of local types. While this is impossible to prove in the vast majority of cases, the demonstrable occurrence of adaptive modifications of color suggests that other adaptations of a less obvious sort may be not infrequent.

In the example which I have most fully discussed, it has been assumed that the pigmental gradient in question has been primarily due to a modifying influence of an extreme sort, operating on one margin of the range of a subspecies. The trend which is manifest as we progress inland is attributed to the diminishing effects of this unilateral influence. Cases of this sort are perhaps exceptional. Such an interpretation is hardly applicable where we have a close correspondence between environmental and character gradients throughout an extensive geographic area. Thus, along the Pacific Coast of North America, we find the species *Peromyscus maniculatus* represented by a series of forms, having a steadily increasing pigmentation, from the peninsula of Lower California to Puget Sound and perhaps much farther north. Here we likewise have a definite climatic gradient, in respect to rainfall, cloudiness and atmospheric humidity [1]). That these climatic factors may all operate through their effect upon the color tone of the soil, and other elements of the background, is a possibility which has already been fully admitted.

Whatever the causal relations here involved, however, we can hardly doubt that they have been operative throughout a large part of the territory in question. Nevertheless, we have here the same picture as before, of a series of relatively discontinuous types, succeeding one another along this climatic gradient, instead of a uniformly graded trend of variation throughout its entire extent. It is here that the hypothesis of discontinuous centers of dispersal, based upon local differences of population pressure, would render the situation more intelligible.

But the existence of geographic trends of variation is not limited to characters concerned with color. Abundant instances could be cited, both among birds and mammals, in which the values of various bodily measurements are arranged for considerable distances in

---

[1]) Also temperature, of course, but this factor may probably be omitted from the present discussion.

geographic sequence, such that the mean difference between two local populations is, roughly speaking, a function of their separation in space. RENSCH, for example (1929, p. 119), presents wing measurements for five subspecies of a bird, *Parus atricapillus*, showing that these structures display a steady increase in length, from the Rhine to northern Siberia. *Peromyscus* furnishes examples of the same sort. Along the Pacific Coast of North America, from San Francisco Bay northward, we have a graded series, in respect to the length of the tail and foot. Thus at Berkeley the subspecies *gambelii* shows a tail length about 82 per cent of that of the head and trunk combined, while in Alaska, in the subspecies *hylaeus*, the tail is 10 per cent longer than the remainder of the body.

As regards these bodily measurements, however, it would be hard to discover any general correlation between such trends of variation and known environmental gradients. One might at first be disposed to look upon the two cases just cited as affording an example of such a correlation. There is doubtless a fairly steady decline in mean temperature as we pass from south to north. However, the facts in the present instance are exactly the reverse of those found in certain other parts of the continent, upon which J. A. ALLEN (1877) based his generalization of the "marked tendency to enlargement of peripheral parts under high temperature or toward the Tropics." They are likewise quite out of harmony with my own experimental data (1909, 1915*a*), respecting the effects of temperature upon the growth of mice and those of PRZIBRAM (1909 and later) upon rats.

It is of interest, furthermore, to point out that both the tail and foot of *Peromyscus* become longer, as we pass from the interior to the coast, whether in northern California, Oregon, or Alaska [1]). Since, in the Pacific Coast area, we likewise have a progressive increase in atmospheric humidity and rainfall, whether we pass from the interior to the seashore, or from south to north, it might be suggested that humidity, rather than temperature, has been the effective factor here. However, the fact that no such trend holds south of San Francisco Bay, and the existence of various other anomalies in the geo-

---

[1]) Along one Alaskan river, SWARTH (1922) found the length of both tail and foot, in *Peromyscus*, to increase from the interior to the coast, and a similar relation was observed in respect to tail length in one species of *Microtus*.

graphic distribution of tail and foot lengths in California (SUMNER, 1923a), render this extremely doubtful.

Despite the absence of any general law governing the direction of such trends, and our resulting inability to correlate these last with known factors of the physical environment, we can hardly doubt the existence of some sort of a causal relation between the two, at least in certain cases. At present, we cannot even conjecture whether most of these bodily differences are adaptive, or whether they are merely the non-adaptive effects of environmental impacts. So far as tested, they have proved to be hereditary, irrespective of immediate peculiarities of the environment. In any case, one thing seems certain. The mere process of random mutation is utterly inadequate to account for such an accumulation of genetic differences, without the aid of some directive agent.

The geographic extent of some of these gradients, both of structure and color, rules out of consideration one rather obvious explanation. This is the supposition that, in general, we have to do in such cases with only two original divergent populations, which have given rise, through hybridization, to all possible intermediate combinations. The only advantage of such a supposition is the escape which it would afford from the necessity of admitting some unknown relation between organic and environmental gradients.

It is obvious that the validity of the hypothesis here adopted, relative to the mutual limitation of subspecific ranges, depends entirely upon the correctness of one assumption, namely, that there are radially directed gradients of population pressure, within the territory of each subspecies, such that competition for the necessities of life would be found to be increasingly keen by an animal which endeavored to pass inward from the periphery. For the existence of such gradients within the range of a subspecies we admittedly have scarcely a shred of direct evidence. The occasional mass migrations of lemmings and other rodents, and perhaps those of certain insects, seem to result from population pressures which have reached the explosion point [1]. And it does not seem unlikely that the same principle is in constant operation, on a less spectacular scale, among great numbers of animal species. But the correspondence of subspe-

---

[1] ELTON (1927), chap. IX.

cific distribution areas with such areas of centrifugally directed migration would be difficult to prove. The hypothesis derives its chief strength from the fact that it accounts for a situation for which no other solution now seems available.

The causes which lead to an increased population pressure in any locality are doubtless very obscure. The fact that increases of this sort are sometimes periodic in their occurrence, and that they may occur simultaneously at rather widely separated points, suggests that they may be influenced by climatic cycles (ELTON, l.c.). But the latter would be operative largely through their effect upon the food supply. Increased food supply, *if permanent*, would not, however, necessarily result in increased population pressure. The ratio of population to the means of subsistence might remain unaltered. However that may be, it is likely that the ultimate effect of more favorable conditions of life [1]) would frequently be an increase of population pressure. Rapid multiplication would entail a diminution in food supply and the optimum density of population would soon be exceeded. Centrifugal migration might well be one of the consequences of such a condition [2]). Theoretically, therefore, the most productive "center of distribution" would seem to be one in which periods of optimum life conditions alternated with periods of stringency. Since the latter would, in large degree, result automatically from the former, it may be surmised that any region of particularly favorable life conditions would tend to become, periodically, at least, a center of dispersal for the surrounding territory.

Variations in physiological fertility may also contribute to differences in population pressure, but in respect to these we are on even more uncertain ground. The term "reproductive selection" was proposed by PEARSON, years ago (1896), to designate the results of the supposed differential fertility of certain types, and it would seem *a priori* highly probable that some races might be able to outbreed others without having any specific advantage as regards adaptation to their particular environments.

The fact that in many cases two or more *species* occupy territories that are not mutually exclusive does not seem inconsistent with the

---

[1]) Including here the scarcity of predatory or parasitic enemies.

[2]) Our mouse "plagues" seem to depend upon an exaggeration of such conditions.

view here upheld. For (1) different species may be presumed not to compete so closely, as a rule, as do related subspecies. Thus they frequently occupy different "ecological niches" within the same geographic range. (2) Among animals, species are commonly completely infertile *inter se*. Hence they can interpenetrate one another's ranges without blending. Representatives of two subspecies, on the contrary, would freely hybridize, if they made their way into one another's territory, and if this process continued unchecked, the two races might be replaced by one. The reported occurrence of immensely wide "areas of intergradation," between two subspecific ranges, may indicate that such complete fusion is actually in progress there. Finally, cases are not infrequent in which species actually do occupy mutually exclusive territories, and may (save for the absence of intergrading) present much the same picture as do two contiguous subspecies.

The line of reasoning here followed, in relation to the differentiation and distribution of subspecies, is not, of course, entirely new. Some of its essential features were presented years ago by GRINNELL (1904) [1]), in a very suggestive paper on "The origin and distribution of the chestnut-backed chickadee." GRINNELL there makes considerable use of hypothetical "centers of distribution," in accounting for the differentiation of subspecies, following the invasion of new territory. Such centers, according to him, represent areas of most rapid multiplication, due to the presence of exceptionally favorable life conditions. Gradients of population pressure are likewise recognized implicitly, though without being so designated. The virtual isolation afforded the populations at the centers of these areas of centrifugal migration, he believes, makes possible the differentiation of local races. I find no indication, however, that GRINNELL is prepared to attribute the boundaries of subspecific territories, in any degree, to the mere meeting of these oppositely directed currents of population movement, in the absence of any form of physical barrier or of any abrupt change in environmental conditions. This is an essential feature of the hypothesis which I have here advocated. The only alternative would seem to call for an impossible degree of specfic adjustment, on the part of each race, to its immediate local conditions.

---

[1]) See also, GRINNELL, DIXON and LINSDALE (1930), pp. 108—109.

A more recent paper, dealing with this same question of the boundaries of subspecific ranges, is the interesting one of MEISE (1928). This writer argues that subspecies cannot be regarded as adapted, each to the special conditions of its environment, since each may live under such utterly diverse conditions. Present distributions are unintelligible, he believes, without reference to the past history of the races in question. MEISE recognizes the possibility that two oppositely directed "waves of differentiation" may meet and block one another. In cases where the zone of blending ("Mischgebiet") between such races is narrow, he believes that it must be accounted for by a partial infertility of the hybrids.

If, by "historical factors" MEISE has reference to the immediate past, his contention that a knowledge of this would be necessary to a full understanding of existing distribution limits is undeniable. These limits are doubtless shifting continually, and there is no reason for regarding the present situation as definitive. It may be doubted, however, whether much light would be thrown on present subspecific boundaries by a knowledge of former geological or geographic conditions. These boundaries could not be expected long to outlive the conditions which brought them into being. Nor do I believe that we are justified in assuming any degree of infertility, on the part of subspecific hybrids, to account for a narrow zone of intergradation. In addition to the fact that such an assumption is opposed by a certain amount of direct experimental evidence, it is plain that neither hybrid sterility nor sexual antipathy between the races would prevent the free invasion of one another's territory, on the part of contiguous races.

In the foregoing attempt to account for certain spatial relationships encountered in the study of geographic variation, very little consideration has been given to the more fundamental problems of the origin of the initial differences from which such variation proceeds, and the manner of their accumulation into definite racial "characters." This is inevitable, in view of the slight progress which has thus far been made in the solution of these problems. The extraordinarily persistent analysis of mutations and their mode of inheritance, during the past three decades, still leaves us almost wholly in the dark regarding the agencies which call forth these mutations in nature, while the now generally accepted viewpoint regarding the magnitude and incidence of these mutations has brought us back to a conception of natural selection not very different from that held by Darwin. These "mutations" are, for the most part, changes of slight magnitude; they are still "random," so far as any adaptive significance is concerned, and they show not the least tendency to maintain variational trends, of an "orthogenetic" sort, independently of selection. They would thus seem to leave the natural selection theory open to the same chief objections which confronted it in Darwin's day, with certain added difficulties due to a curtailment in the amount of variation which is available for selection. However, this last source of difficulty may be more apparent than real. There seems to be a widespread tendency to look to such phenomena as those revealed by *Drosophila* for a key to the entire evolutionary process. Thus, it is natural to believe that gene mutations, similar to those which have been studied so successfully in the fruit-fly, are occurring in nature, in every group of organisms. It likewise seems reasonable to assume (despite the lack of any direct evidence) that an occasional mutation of this sort occurs which is calculated to give the species an advantage, instead of a handicap. The next assumption to be made is that these advantageous gene mutations are gradually accumulated through natural selection, until a perceptible divergence occurs. Such a conception is obviously open to many of the earlier objections to the natural selection hypothesis — ones which were supposed to be avoided by recourse to Lamarckism, orthogenesis, or "mutation" (in DE VRIES's sense). Moreover, such an assumed accumulation of single gene mutations, one at a time, fails to square with the fact that not only the least distinguishable local races, but even individuals within a race, display numerous character differences, each of which commonly depends upon a considerable series of genetic factors, cumulative in their action.

It is hardly necessary to point out that selection is not concerned, directly, with single genes, but with the aggregate fitness of entire organisms, and that this fitness commonly depends upon a large number of varying genetic elements. Unlike the more or less pure cultures which are frequently reared in the laboratory, natural populations are genetically highly mixed. This condition is maintained through the free interbreeding of diverse strains, probably, in extreme cases, through the crossing of widely different races or even species. Thus, while every one of the underlying unitary differences may be supposed to have originated, at some time, as a single mutation, the material upon which selection acts has arisen through the endless shuffling of these elements as a result of sexual reproduction. The processes of mutation and selection thus need not go on *pari passu*. Much genetic change might come about, and be widely disseminated throughout a population, before it became subject to selection at all. Large genetic differences might thus be available at the very commencement of a given selective process, and the gap between the "pre-useful" and the "useful" stages of a structure might be bridged by purely fortuitous gametic combinations.

This viewpoint is not, of course, offered as new in any single detail. But it stresses a class of facts which are sometimes overlooked. It is the service of the exponents of the theory of evolution through hybridization [1]) to have emphasized the important part played by crossing (in the broad sense) in the production of genetic "variation" in nature, and to have thrown the burden of proof upon those who may be disposed to attribute every novelty to "mutation."

Finally, it is only fair to state, that I cannot myself regard the natural selection of purely random mutations, even with the qualifications set forth in the last few paragraphs, as furnishing an adequate or satisfying account of the mechanism of evolution. While admitting the paucity, if not the total lack of direct evidence in this field, I still lean strongly toward the view that the process of natural selection must be supplemented by adaptive responses of a more direct nature.

To recapitulate briefly the views set forth above respecting the origin of geographic races, I believe it likely:

(1) That a new geographic race, if not actually isolated, commonly has its origin at some point more or less remote from the main mass of the parent stock;

(2) That, in order to insure its perpetuation, a rather rapid expansion must ensue, owing either to particularly favorable environmental conditions, or to a high inherent fertility of the new race or to both;

(3) That the expansion resulting from this high population pres-

---

[1]) LOTSY (1916), HAGEDOORN (1921).

sure continues, in the absence of a physical barrier, until an equally high population pressure from another direction is encountered arising either from the parent race or another closely related race;

(4) That the races blend through hybridization in a "zone of intergradation," the width of which varies inversely with the strength of the population pressures from the two sides;

(5) That, in the absence of some physical barrier, two contiguous races would remain distinct only so long as there was a preponderance of centrifugal movement from both of the "centers of dispersal;"

(6) That physical barriers, or abrupt changes of environmental conditions, oftentimes account in part for the boundaries of the ranges of subspecies;

(7) That in no cases do subspecies (at least among mammals) appear to have arisen through a single act of mutation, since, so far as known, the differences between even the most closely related geographic races depend upon considerable numbers of genetic factors;

(8) That the frequent occurrence of serial gradations in the magnitude of one or more characters, sometimes correlated, throughout extensive regions, with some obvious gradient in the physical environment, shows that the latter may exert a directive influence either upon the origin or the survival of the elementary differences which underly these characters;

(9) That such an accumulation of factorial differences must have resulted either from selection on the basis of special adaptedness to local conditions, or from the direct effect of environmental conditions, or perhaps from both of these combined;

(10) That, however much adaptation to local conditions may have figured in the differentiation of a given race, the extent of the territory ultimately occupied by it is by no means an index of the special biological needs of the race in question, but that the range is determined largely by extrinsic causes, i.e., physical barriers and opposed population pressures from other races [1]);

(11) That, in the case of a subspecies occupying a highly diversified range, the condition ultimately assumed by it may represent the mean condition of races which have arisen in a number of places,

---

[1]) Cf. GRINNELL (1914), pp. 105—106.

and have subsequently blended; or it seems even possible that some of those subspecies which occupy immense geographic ranges (e.g., *P. m. gambelii*) may represent a polyphyletic assemblage of types, indistinguishable morphologically, though genetically distinct in origin.

OTHER LINES OF INVESTIGATION UPON PEROMYSCUS

I shall here discuss briefly several lines of investigation which have little immediate relation to genetics, along with certain others which were not carried far enough to yield decisive results. It is to be hoped that some of these lines will be pursued much farther by investigators having adequate time and equipment.

(1) Studies of the process of molting were conducted for a number of years by H. H. COLLINS (1918, 1923). These consisted of detailed observations of the successive pelage phases of the growing mouse and of the annual molt of the adult, in several species of *Peromyscus*, supplemented by experimental studies of the replacement of the pelage after its removal from various areas. COLLINS's investigations likewise comprized a description and classification of different types of hair, based upon the microscopic examination of these. The latter studies have been supplemented by those of HUESTIS (1925) and APGAR (1930).

(2) Secondary sexual differences were discussed more or less incidentally in a number of papers (1918, 1920, 1923b, 1923c, 1930). In nearly all collections of any size, the mean body length (head plus trunk) was distinctly greater in females than in males, while the mean relative (and likewise commonly the absolute) foot length was consistently greater in the males. For four local collections of *P., maniculatus*, belonging to three subspecies (1920), it was found that the length of the pelvic bones (relative as well as absolute) was greater in the females, but this relation does not seem to hold for the *P. polionotus* series (1930). Other possible sexual differences in the mean values of various characters may be noted in limited collections, but they do not appear to have any general validity.

(3) The number of each sex in the broods of young has been recorded, as a general rule, throughout the entire course of these ex-

periments. The results of these enumerations have been tabulated and published in two papers (SUMNER, McDANIEL, and HUESTIS, 1922; KAROL, 1928). As a whole, the results from these time-consuming studies illustrate the familiar contradictoriness of data concerning the sex ratio, and the difficulty of discovering significant correlations between this and any external or internal factors of possible influence. For example, the aggregate ratio reported in the first paper, based upon more than 4,600 animals, was 97.37; while that found by KAROL, for more than 2400 young, born subsequently to the earlier report was 114.93. This difference may be due, in part, to a change in the species chiefly used during the later years, but it must be noted that the ratios obtained during these earlier years varied enormously from year to year. The extreme figures are those for 1916 and 1917, the ratios for which were 125.36 ± 7.82 and 70.56 ± 4.70. These were based upon 471 and 423 individuals, respectively. The difference between the ratios is six times its probable error. An analysis of these two years' populations, with reference to a number of the factors suspected of influencing the sex ratio, failed to reveal any clue. All that can be asserted with confidence is that these surprising relations are not due to errors either in the records or the computations.

A seasonal rhythm was displayed by the material first analyzed, and is retained with some modification, when the two groups are thrown together. Two annual maxima are shown, in the spring and fall (or late summer) respectively, while minima occur in winter and in midsummer [1]). However, an analysis of the behavior, in this respect, of various component groups within these populations, makes it doubtful whether this apparent seasonal rhythm has any biological significance [2]).

One wholly consistent result was, however, obtained from both the earlier and the later studies. When the number of each possible combination of males and females, in broods of each size, was compared with the number expected according to chance, the conformity was

---

[1]) By an extraordinary error in diction, this rhythm was referred to in both reports as "biennial."

[2]) The most nearly consistent feature of these seasonal relations is the presence of a fall maximum, both in the earlier and later material. There is no indication of a spring maximum in the latter.

found to be, on the whole, very close. "There is thus no preponderant tendency toward the production of homosexual litters, and thus no likelihood that polyembryony or true twinning is at all common in these animals."

(4) Under the heading "superfetation and deferred fertilization," I described years ago (1916), some cases in which the birth of broods was delayed considerably beyond the normal period of gestation, along with two, in which the records seemed to show that a later brood had resulted from the insemination of the mother during or before an earlier pregnancy.

In one of these latter cases, a female *gambelii* gave birth, on August 12, to a single male offspring. This was followed, on or about August 25, by the birth of three more young [1]). The father was removed on or before August 12, but this fact is immaterial, in the present case. In this instance, if we leave aside the possibility of error, there are several possible explanations. Either (a) a second effective copulation, accompanied by ovulation, occurred during the first pregnancy; or (b) ova liberated during this period were fertilized by sperm held in the fallopian tubes from a previous copulation; or (c) the ova for both broods were all fertilized at the same time, but for some reason the development of the second lot was prolonged.

In the second of the instances cited above, a female *rubidus* is recorded as giving birth to a brood of two on June 5, and a second brood of three on or about July 3, 28 days later. The records indicate that the father was removed on May 17. Here, likewise, the second brood developed from ova which could only have been fertilized by sperm received during or prior to the first pregnancy. We must also suppose either that the actual fertilization of the second batch of eggs was deferred until nearly a week after the birth of the first brood, or, if it occurred earlier, that the period of development was prolonged. It may be repeated that the normal period of gestation in *Peromyscus* is about 22 days.

It should be stated, in connection with the foregoing two cases, that only two of a similar nature have occurred since the publication of the 1916 report, although they have been rather carefully

---

[1]) Miss KING (1913) had earlier recorded two cases in which a second brood of white rats had followed an earlier one after an interval of 12 to 14 days.

watched for. In one of these later cases, the records indicate that the birth of one brood (1 ♂, 1 ♀) was followed, after an interval of 43 days, by the birth of a second brood (2 ♀ + 1 ?), although it is recorded that the father of the first brood was removed on the day before their birth. In the second case, the mother gave birth to a brood of five, nine days (or less) after the removal of her mate. These died shortly after birth. She was remated in three days to the same male, and 17 days thereafter (29 days after last previous opportunity for copulation and 20 days (+ ?) after birth of earlier brood) delivered nine young.

Granting the accuracy of the foregoing records [1]), we have in the first of these cases two alternatives. Either the second brood of young were due to the insemination of the mother by the same mate, at least 44 days prior to their birth; or the father of the second brood was the male belonging to the first brood, which, at the time of copulation, would have been 21 days (43 minus 22) old. In the second of the foregoing cases, we must conclude, either that the insemination which resulted in the later brood took place during or before the earlier pregnancy, or that the period of gestation, for the later brood, was 17 days or less.

It must be admitted that the paucity of the evidence here offered, covering a period of nearly 17 years, considerably weakens its force, despite the precautions taken to guard against error [2]).

One unquestionable fact of interest was, however, revealed by the data reported at that time, as well as by later records. In many cases in which the male parent was removed from the cage on the date of birth (or discovery) of the earlier brood, a second brood has appeared after a considerably longer interval than the normal period of gestation. In the cases previously recorded (1916, p. 275), the interval has ranged from 26 to 39 (±) days, and I have apparently reliable later records of 23 to 33 days, along with somewhat questionable ones, as high as 34 to 47 days. In each of these more extreme cases, no male was present in the first brood.

---

[1]) In the first case, the record of the removal of the father was probably made by an assistant. In the second case, all the essential records were entered by myself, and later examination of these reveals no likelihood of error.

[2]) In the great majority of cases, the date of removal of the male was recorded upon the cage label.

Similar cases had earlier been reported by DANIEL (1910) and others, and had been attributed to a prolongation of embryonic development, due to the fact that the mother was already suckling another brood. However, this retardation amounted to 15 and 17 days, in two of my cases, and indeed it ran as high as 25 days (47 days after possible copulation), if we accept my "case 5" as authentic in this respect. According to DANIEL, the retardation of development amounts to about one day for each young mouse which is suckled at the time by the mother. In the three cases cited above where the retardation has been greatest (15, 17, 25? days), the numbers of young comprized in the first broods were 3, 3 and 2, respectively. For these reasons, it seemed likely that postponed fertilization, effected by spermatozoa which had been received some time previously, was a more probably explanation than prolongation of embryonic development [1]).

(5) What may be termed an "autotomy" of the tail was described for several rodents (SUMNER and COLLINS, 1918). This was chiefly encountered in members of the genus *Perognathus* (pocket mice), belonging to an entirely distinct family from *Peromyscus*, but somewhat similar cases were observed in one species of the latter genus. Many specimens of *Perognathus*, when held by the tail detach a part of the latter member by a rapid gyratory movement of the body. The tail may be broken at any point in its length, and the break occurs through a vertebra rather than between two of these. These is no true regeneration, though a dense terminal brush of long hairs, more pronounced than the normal one, is usually formed upon the stump. In the only species of *Peromyscus* which was found to display a habit at all similar to this, the skin of a large part of the tail slipped off when this member was held by the hand, and the mouse thus readily escaped. The exposed axial part of the tail was doubtless lost later. Whether either of these methods of escape is of any considerable service in nature may well be doubted, although mice with tails reduced to stumps are not rarely met with.

(6) Despite the absolutely negative character of the results, it seems worth while to mention two series of experiments, to which

---

[1]) The interval between two successive broods was not, however, prolonged in all cases. I now have records for a probable interval as low as 22 days, with more certain ones for 23 days.

considerable time was devoted in the earlier days of the *Peromyscus* investigations. Both were attempts to influence the germ-cells in specific ways. One of these experiments was suggested by those of GUYER and SMITH, which had seemed to demonstrate the production of hereditary eye defects in rabbits by injecting blood serum which was supposed to contain an anti-lens substance. The attempt was made by us to produce an anti-hair substance in the blood of fowls, by injecting a suspension of finely ground albino hair from mice. The serum of these fowls was injected into pregnant mice, following GUYER's procedure. No hairless mice resulted!

The other experiment of this sort was a rather naive endeavor to test the inheritance of functional modifications. A considerable series of mice were subjected to treatment which crippled one hind leg (either left or right). In some cases part of the leg was amputated; in others the sciatic nerve was cut. Numerous young were reared from both right and left-legged parents (SUMNER, 1917b). Of these, 87 individuals were tested. In some cases, only the parents had been subjected to the operation, in others, the grandparents also, and in a few cases the three preceding generations. The young were tested by a measurement of the relative strength of pull of the two hind legs, as recorded by dynamometers. No correlation was evident between the side giving the stronger pull and the treatment to which the parents had been subjected.

(7) An important phase of the *Peromyscus* investigations was only fairly commenced, at the time when these investigations were discontinued. I refer to studies of physiological and behavioristic phenomena among these animals, undertaken particularly with a view to revealing specific and subspecific differences of a physiological nature. Since much time, in the aggregate, was devoted to the perfection of technique and to preliminary experiments along these lines, it seems worth while to describe these briefly. The investigations actually commenced comprized (a) a study of total bodily activity, as revealed by recording wheel-cages; (b) the distribution of activity (diurnal rhythm, etc.), as recorded on slowly moving kymograph sheets; (c) comparative studies of water consumption.

Ten wheel-cages were constructed, following the general pattern of those used by RICHTER and WANG (1926) for rats, but adapted to the special requirements of *Peromyscus* (Figure 23). As was to be expected, the greater part of the activity thus recorded was found to occur at night. It was soon evident too that the amount of daily

FIGURE 23. Recording wheel-cage, used for measuring general bodily activity of *Peromyscus*.

exercise was in many cases very great, and in fact, altogether excessive. In the first series of experiments, a number of the animals actually ran themselves to death. Thereafter, each individual was allowed to run only on alternate days, the intervening time being spent in a nest-box, detached from the wheel-cage [1]). Even so, there occurred a number of deaths which can only be attributed to over activity.

The most obvious feature in comparing these records was the enormous individual difference in the energy output of these animals. While occasional apparently healthy individuals remained continuously in the nest boxes, and seldom or never revolved the wheels, others appear to have kept the latter in almost continual revolution throughout the entire night, and even to a considerable extent during the daytime. The highest record was 90,049 revolutions for the 24-hour period, of which nearly 65,000 were scored for the period between 4 : 20 p. m. and 8 : 15 a.m. The next greatest number was 84,701, while records of 20,000 to 40,000 were not uncommon [2]).

In view of these great individual variations in running capacity, it is evident that racial or specific differences, if such exist, would be impossible to demonstrate without the use of large numbers. However, I shall present a few figures for what they are worth. Ten adult male *P. m. sonoriensis*, for a period of ten days, gave a mean daily record of 10,747 revolutions, while ten adult male *rubidus*, during the same period, gave a mean reading of 7,196. The four highest records, but also the two lowest, were among the *sonoriensis*. Nine adult female *sonoriensis* and ten *rubidus*, for a similar period, gave averages of 9,363 and 3,493, revolutions, respectively. Here, the three largest records, as well as the single lowest, were scored by *sonoriensis*. Comparing the records, day by day, it is found that for each sex, the *sonoriensis* daily run was, after the first day, consistently higher than the *rubidus* run [3]).

[1]) These small boxes were so made that they could be bolted to the wheel-cages on the active day, and removed (another box, containing a fresh mouse being substituted) on the day of rest. Thus, the mice could be readily transferred without being disturbed. In a twenty-day experiment, each mouse made a record with each of the ten recording wheels, thus eliminating the effect of differences in the friction of the mechanism.

[2]) This does not, however, imply that the animals ran throughout the entire period while the wheel was in rotation. After starting the latter in active motion, a mouse would frequently cling to the periphery of the wheel and be carried around by it. These figures, likewise, are subject to a considerable margin of possible error, due to peculiarities of the recording mechanism, which need not be discussed here.

[3]) There was a fairly steady increase in the daily run throughout the ten days, i. e., as long as the experiment was continued.

Whether these figures actually represent a racial difference in running capacity is, of course, questionable. Aside from other elements of uncertainty, the mice here used were of the "wild" generation. and therefore of unknown age. The *sonoriensis* specimens were, on the average, about one gram heavier than the *rubidus*. On the other hand, this apparently superior·activity of *sonoriensis*, in comparison with *rubidus*, is confirmed by the limited results of the kymograph experiments, to be discussed shortly, and is doubly suggestive, in view of the histological differences in the thyroid glands of the two races (YOCOM and HUESTIS, 1928).

In the next series of experiments [1]), only cage-bred individuals, of approximately the same age, were employed. The mean daily runs of four [2]) male *P. maniculatus sonoriensis* and five male *P. eremicus eremicus*, for a period of five days, were 33,305 and 8,993 respectively Those of five females of each race, for the same period, were 14,178 and 11,038, respectively. Four male *P. maniculatus gambelii* and five male *P. eremicus fraterculus* gave mean daily records of 17,396 and 14, 417 revolutions, respectively, while five females of each of these subspecies gave records of 22,085 and 11,762, respectively. The fact that in each of the four comparisons between *Peromyscus maniculatus* and *P. eremicus* (i.e., *P. m. sonoriensis* with *P. e. eremicus*, and *P. m. gambelii* with *P. e. fraterculus*, the sexes being dealt with separately), the first named species gave a higher record than the second is a circumstance of possible significance. It is of particular interest, in view of the seeming specific differences in water consumption to be discussed presently.

For recording the periodicity of activity a quite different type of apparatus was employed [3]). Each mouse under observation was placed in a triangular cage, of galvanized iron, mounted upon three legs, terminating below in flat disks or feet. Each cage was stood upon three tambour cups, the feet resting upon the rubber diaphragms of the latter. The three tambour cups under each cage were connected

---

[1]) These were conducted with the collaboration of Mr. L. G. Ross.

[2]) The number of males in this lot, and that of *gambelii*, was reduced by death, resulting doubtless from over-activity. A second one of these male *sonoriensis* died the day after having made a score of 90,049.

[3]) This was likewise adapted from an apparatus employed by RICHTER (1927) for white rats.

with a single rubber tube, which passed to a recording tambour, fitted with a pen filled with the usual recording ink. In each experiment, four cages and four pens were employed, the latter being arranged in vertical series, and recording simultaneously. The kymograph used was of the SHERRINGTON-STARLING type, and carried a sheet 2.45 meters in length. It was operated by an electric motor, with speed-reducing gears, commonly so adjusted that a complete revolution required between three and four days.

The pens responded appreciably, even to slight movements of the animals, while the familiar vigorous handsprings (page 9) were frequently registered in pen tracings several centimeters in amplitude.

This apparatus was in operation almost continuously for several months, and more than thirty records (in each case, with four mice at a time) are available. As in the case of the wheel-cages, results were obtained which would be of high interest had they been based upon a sufficiently great number of animals. Here, as before, individual differences were very great. Frequently each of the four mice which were run simultaneously recorded a different activity pattern, to which it adhered fairly consistently. In the case of some very vigorous animals the pen-strokes, during the night, were repeated in such rapid succession as to produce an almost continuous band of ink upon the paper. The daytime performance was commonly restricted to infrequent and less active movements, though this was not true in all cases. Occasional specimens were observed in which a low degree of activity was distributed fairly uniformly throughout the entire 24 hours [1]).

The following figures are based upon twelve series of four animals each. Each of these quartets consisted of two *sonoriensis* and two *rubidus*, each race being represented by a male and female specimen. The mice used were all young cage-bred animals and, so far as possible, the members of a quartet were approximately of the same age. The members of each quartet have been graded, by simple inspection of the tracings, according to the relative degree of activity there in-

---

[1]) JOHNSON, at Illinois (1926), found by the use of a simple recording device that the regular diurnal rhythm of activity in *P. leucopus noveboracensis* persisted for a month in total darkness; and that a daily periodicity (though no longer coinciding with night and day) was still evident after more than seven months.

dicated. Thus a *sonoriensis* held the first place in ten cases, a *rubidus* in only two cases, etc.

Grade	1	2	3	4
Sonoriensis	10	9	2	3
Rubidus	2	3	10	9

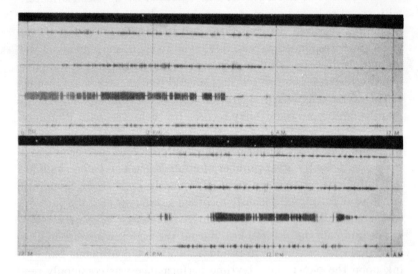

FIGURE 24. Kymograph record showing diurnal rhythm of activity of four *Peromyscus* for two days (Reading from top to bottom, are *sonoriensis*, ♂ and ♀, and *rubidus*, ♂ and ♀). More often, *sonoriensis* showed a higher activity than *rubidus*.

These results, meager as they are, are in obvious accord with the records of these two subspecies in the wheel-cages. No differences in the activity patterns of the sexes were noted, and no periodic fluctuations in the activity of females, due to oestrus [1]).

The comparative study of water consumption was untertaken by L. G. Ross (1930). Mr. Ross employed a graduated drinking fountain, with which it was possible to obtain daily readings of the water withdrawn. Only dry food was given during the period of the experiment. Three subspecies of *P. maniculatus* (*sonoriensis*, *gambelii* and *rubidus*) were tested; likewise two subspecies of *P. eremicus* (*P. eremicus eremicus* and *P. eremicus fraterculus*).

---

[1]) SLONAKER (1924), RICHTER and WANG (1926).

No undoubted subspecific differences in water consumption were encountered within either species. This was somewhat unexpected, in view of the considerable differences in the life conditions to which these races are accustomed in nature. *P. m. sonoriensis* and *P. e. eremicus*, on the one hand, are typically desert dwelling races, while *P. m. gambelii* and *rubidus*, and *P. e. fraterculus* are denizens of the far more humid coastal plain.

But one highly interesting difference was found. Comparing the *P. maniculatus* representative with the *P. eremicus* one, for each locality (i.e., *sonoriensis* with *eremicus*, and *gambelii* with *fraterculus*), it was found that the former species consumed appreciably more water, in proportion to body weight, than the latter. This fact is perhaps significantly related to two facts in the distribution of these species. (1) *P. maniculatus* is a widely distributed form, its geographic races ranging over almost the whole continent, including chiefly non-arid regions, while *P. eremicus* is typically a denizen of northern Mexico and our arid southwest. (2) In regions where both occur together, within the same geographic range, *P. maniculatus* tends to be distinctly less xerophilous than *P. eremicus*. Thus, the physiological differences apparently here found may have their explanation in racial history.

Any such experiment as the foregoing should, of course, be repeated with material from different sources. While the results cited were statistically of fairly high "significance," this merely means that they are probably not due to random sampling. It is not impossible that independent stocks from other localities would fail to confirm these findings [1]).

As bearing upon possible subspecific differences in activity, reference may here be made to the comparative studies of the thyroid gland by YOCOM and HUESTIS at Oregon (1928), in which certain characteristic differences of histological structure were found between the thyroids of *P. m. rubidus* and *P. m. sonoriensis*. In the

---

[1]) Previous studies by DICE (1922), with the "prairie deermouse" (*P. maniculatus bairdii*) and the "forest deer-mouse" (*P. leucopus noveboracensis*), indicated that the former consumed more water, in proportion to body weight, than the latter, although the absolute water consumption did not differ significantly in the two cases. Here, the physiological difference, if real, bears no obvious relation to habitat preferences.

former, the follicles were found to be thin-walled and distended with secretion. The latter, on the other hand, "show relatively little storage of secretion and the epithelial cells give the appearance of an actively secreting gland." I am informed by Dr. HUESTIS that the reality of these histological differences has recently been confirmed by Dr. YOCOM and himself (using, in this case, *rubidus* and *gambelii*), and that it is just as easy to distinguish these two subspecies by the thyroids as by the pelages.

BIBLIOGRAPHY

This has been divided into two sections, the first of which is limited to papers based upon the program of research with *Peromyscus* which was conducted under the direction of the present author during the years 1913 to 1930. The second section comprizes all other papers referred to in the foregoing pages, along with a few which are not specifically mentioned.

In connection with the papers of the first group, it seems desirable to list the more serious errata. These are, for the most part 'printers' errors, for which the authors are not responsible. In some cases, they result in misstatement of fact, or render an important discussion unintelligible.

Errata

SUMNER (1917), p. 178, fourth line from bottom: "annual" should read "perennial."

SUMNER (1920), p. 385, fifth line from bottom; "σ and σ_y" should read "σ_x and σ_y."

SUMNER (1923b), p. 262, foot-note 20: "p. 253" should read "p. 279"; p. 270, sixth line from bottom: "F_1" should read "F_2."

SUMNER (1923c), p. 394, in table: "Gross σ" and "Net σ" should read "Gross r" and "Net r."

SUMNER (1924a), p. 483, foot-note 4: "Ibid" should read "Ibis."

SUMNER (1929), p. 114, fifth paragraph: statements regarding figures 1 and 2 do not correspond to present numbering of figures.

SUMNER (1930), p. 307, line 10: "$(4/5)^1$" should read "$(4/5)^3$"; line 19: "$(5/6)^2$" should read "$(5/6)^4$."

SUMNER and HUESTIS (1921), p. 448, second foot-note: "amblyomyopic" should read "amblyopic," "fraction" should read "refraction;" p. 453, fifth line from bottom: delete matter in parenthesis.

SUMNER, McDANIEL and HUESTIS (1922), pp. 128 and 160: "biennial" should read "semiannual."

SUMNER and HUESTIS (1925), p. 43, fourth line below table: "0.28" should read "0.238."

COLLINS, H. H. (1918). "Studies of normal moult and of artificially induced regeneration of pelage in Peromyscus." Journal of Experimental Zoology, vol. 27, pp. 73—99.

COLLINS, H. H. (1923). "Studies of the pelage phases and of the nature of color variations in mice of the genus Peromyscus." Journal of Experimental Zoology, vol. 38, pp. 45—107.

HUESTIS, R. R. (1923). "The heredity of microscopic hair characters in Peromyscus." Proceedings National Academy of Sciences, vol. 9, pp. 352—355.

HUESTIS, R. R. (1925). "A description of microscopic hair characters and of their inheritance in Peromyscus." Journal of Experimental Zoology, vol. 41, pp. 429—470.

KAROL, J. J. (1928). "The sex ratio in Peromyscus." Biological Bulletin, vol. 55, pp. 151—162.

ROSS, L. G. (1930). "A comparative study of daily water-intake among certain taxonomic and geographic groups within the genus Peromyscus." Biological Bulletin, vol. 59, pp. 326—338.

SUMNER, F. B. (1915). "Genetic studies of several geographic races of California deer-mice." American Naturalist, vol. 49, pp. 688—701.

SUMNER, F. B. (1916). "Notes on superfetation and deferred fertilization among mice." Biological Bulletin, vol. 30, pp. 271—285.

SUMNER, F. B. (1917). "The rôle of isolation in the formation of a narrowly localized race of deer-mice (Peromyscus)." American Naturalist, vol. 51, pp. 173—185.

SUMNER, F. B. (1917a). "Several color ,mutations' in mice of the genus Peromyscus. "Genetics, vol. 2, pp. 291—300.

SUMNER, F. B. (1917b). "Modern conceptions of heredity and genetic studies at the Scripps Institution." Bulletin Scripps Institution for Biological Research, no. 3, pp. 1—24.

SUMNER, F. B. (1918). "Continuous and discontinuous variations and their inheritance in Peromyscus." American Naturalist, vol. 52, pp. 177—208, 290—301, 439—454.

SUMNER, F. B. (1920). "Geographic variation and Mendelian inheritance." Journal of Experimental Zoology, vol. 30, pp. 369—402.

SUMNER, F. B. (1921). "Desert and lava-dwelling mice, and the problem of protective coloration in mammals." Journal of Mammalogy, vol. 2, pp. 75—86.

SUMNER, F. B. (1922). "Longevity in Peromyscus." Journal of Mammalogy, vol. 3, pp. 79—81.

SUMNER, F. B. (1922a). "Linkage in Peromyscus." American Naturalist, vol. 56, pp. 412—417.

SUMNER, F. B. (1923). "Studies of subspecific hybrids in Peromyscus." Proceedings National Academy of Sciences, vol. 9, pp. 47—52.

SUMNER, F. B. (1923a). "Some facts relevant to a discussion of the origin and inheritance of specific characters." American Naturalist, vol. 57, pp. 238—254.

Sumner, F. B. (1923b). "Results of experiments in hybridizing subspecies of Peromyscus." Journal of Experimental Zoology, vol. 38, pp. 245—292.

Sumner, F. B. (1923c). "Size-factors and size-inheritance." Proceedings National Academy of Sciences, vol. 9, pp. 391—397.

Sumner, F. B. (1924). "The partical genetic independence in size of the various parts of the body." Proceedings National Academy of Sciences, vol. 10, pp. 178—180.

Sumner, F. B. (1924a). "The stability of subspecific characters under changed conditions of environment." American Naturalist, vol. 58, pp. 481—505.

Sumner, F. B. (1924b). "Hairless mice." Journal of Heredity, vol. 15, pp. 475—481.

Sumner, F. B. (1925). "Some biological problems of our southwestern deserts." Ecology, vol. 6, pp. 352—371.

Sumner, F. B. (1926). "An analysis of geographic variation in mice of the Peromyscus polionotus group from Florida and Alabama." Journal of Mammalogy, vol. 7, pp. 149—184.

Sumner, F. B. (1927). "Linear and colorimetric measurements of small mammals." Journal of Mammalogy, vol. 8, pp. 177—206.

Sumner, F. B. (1928). "Observations on the inheritance of a multifactor color variation in white-footed mice (Peromyscus)." American Naturalist, vol. 62, pp. 193—106.

Sumner, F. B. (1928a). "Continuation of ecological and genetical studies with Peromyscus." Carnegie Institution Year Book no. 27 (1927—28), pp. 335—339.

Sumner, F. B. (1929). "The analysis of a concrete case of intergradation between two subspecies." Proceedings National Academy of Sciences, vol. 15, pp. 110—120.

Sumner, F. B. (1929a). "The analysis of a concrete case of intergradation between two subspecies. II. Additional data and interpretations." Proceedings National Academy of Sciences, vol. 15, pp. 481—493.

Sumner, F. B. (1929b). "Continuation of ecological and genetical studies with Peromyscus." Carnegie Institution Year Book no. 28 (1928—29), pp. 346—347.

Sumner, F. B. (1930). "Genetic and distributional studies of three subspecies of Peromyscus." Journal of Genetics, vol. 23, pp. 275—376.

Sumner, F. B. (1930a). "Continuation of ecological and genetical studies with Peromyscus." Carnegie Institution Year Book no. 29 (1929—30), pp. 360—365.

Sumner, F. B., and H. H. Collins, (1918). "Autotomy of the tail in rodents.' Biological Bulletin, vol. 34, pp. 1—6.

Sumner, F. B., and H. H. Collins, (1922). "Further studies of color mutations in mice of the genus Peromyscus." Journal of Experimental Zoology, vol. 36, pp. 289—321.

Sumner, F. B., and R. R. Huestis, (1921). "Bilateral asymmetry and its relation to certain problems of genetics." Genetics, vol. 6, pp. 445—485.

SUMNER, F. B., and R. R. HUESTIS, (1925). "Studies of coat-color and foot pigmentation in subspecific hybrids of Peromyscus eremicus." Biological Bulletin, vol. 48, pp. 37—55.

SUMNER, F. B. and J. J. KAROL, (1929). "Notes on the burrowing habits of Peromyscus polionotus." Journal of Mammalogy, vol. 10, pp. 213—215.

SUMNER, F. B., M. E. MCDANIEL, and HUESTIS, R. R. (1922). "A study of influences which may affect the sex-ratio of the deer-mouse (Peromyscus)." Biological Bulletin, vol. 43, pp. 123—165.

SUMNER, F. B. and H. S. SWARTH, (1924). "The supposed effects of the color tone of the background upon the coat color of mammals." Journal of Mammalogy, vol. 5, pp. 81—113.

ALLEN, G. M. (1914). "Pattern development in mammals and birds." American Naturalist, vol. 48, pp. 467—484.

ALLEN, G. M. (1920). "An insular race of cotton rat from the Florida Keys." Journal of Mammalogy, vol. 1, pp. 235—236.

ALLEN, J. A. (1871). "On the mammals and winter birds of East Florida." Bulletin Museum Comparative Zoology, vol. 2, pp. 161—450.

ALLEN, J. A. (1877). "The influence of physical conditions in the genesis of species." Radical Review, vol. 1 (reprinted, with additions, in Annual Report of Smithsonian Institution for 1905 (1906)), pp. 375—402).

APGAR, C. S. (1930). "A comparative study of the pelage of three forms of Peromyscus." Journal of Mammalogy, vol. 11, pp. 485—493.

BATESON, W. (1913). "Problems of genetics." New Haven, 258 pp.

BABÁK, E. (1913). "Ueber den Einfluss des Lichtes auf die Vermehrung der Hautchromatophoren." Pflüger's Archiv für die gesammte Physiologie, Bd. 149, pp. 462—470.

BANGS, O. (1905). (Title not known). Proceedings of the New England Zoological Club, pp. 14—15.

BONHOTE. (1915). "Vigour and heredity". London, 263 pp.

BUBNOFF, S. VON (1919). "Ueber einige grundlegende Prinzipien der paläontologischen Systematik." Zeitschrift für induktive Abstammungs- und Vererbungslehre, Bd. 21, pp. 158—168.

BUXTON, P. A. (1923). "Animal life in deserts; a study of the fauna in relation to the environment." London, xv + 176 pp.

CASTLE, W. E. (1912). "On the origin of an albino race of deer mouse." Science, n. s., vol. 35, pp. 346—348.

CASTLE, W. E. (1914). "Some new varieties of rats and guinea-pigs and their relation to problems of color inheritance." American Naturalist, vol. 48, pp. 65—73.

CASTLE, W. E. (1916). "Genetics and eugenics." Cambridge, 353 pp. (Revised editions in 1920, 1930).

CASTLE, W. E. (1919). "Studies of heredity in rabbits, rats, and mice. Part III. Observations on the occurrence of linkage in rats and mice." Publication no. 288, Carnegie Institution of Washington, pp. 29—36.

CASTLE, W. E. (1921). "An improved method of estimating the number of genetic factors concerned in cases of blending inheritance." Science, n. s., vol. 54, p. 223.

CASTLE, W. E. (1922). "Genetic studies of rabbits and rats." Publication no. 320, Carnegie Institution of Washington, pp. 1—55.

CASTLE, W. E. (1924). "Does the inheritance of differences in general size depend upon general or special size factors?" Proceedings National Academy of Sciences, vol. 10, pp. 19—22.

CASTLE, W. E. (1924). "Are the various parts of the body genetically independent in size?" Proceedings National Academy of Sciences, vol. 10, pp. 181—182.

CASTLE, W. E., and S. WRIGHT, (1915). "Two color mutations in rats which show partial coupling." Science, n.s., vol. 42, pp. 193—195.

DANIEL, J. F. (1910). "Observations on the period of gestation in white mice." Journal of Experimental Zoology, vol. 9, pp. 865—870.

DICE, L. R. (1922). "Some factors affecting the distribution of the prairie vole, forest deer mouse, and prairie deer mouse." Ecology, vol. 3, pp. 29—47.

DICE, L. R. (1929). "A new laboratory cage for small mammals, with notes on methods of rearing Peromyscus." Journal of Mammalogy, vol. 10, pp. 116—124.

DUNN, L. C. (1920). "Linkage in mice and rats." Genetics, vol. 5, pp. 325—343.

DUNN, L. C. (1921). "Unit character variation in rodents." Journal of Mammalogy, vol. 2, pp. 121—140.

ELTON, C. A. (1927). "Animal ecology". London, XVII + 207 pp.

GRINNELL, J. (1904). "The origin and distribution of the chestnut-backed chickadee." Auk, vol. 21, pp. 364—382.

GRINNELL, J. (1914). "An account of the mammals and birds of the lower Colorado valley." University of California Publications in Zoology, vol. 12, pp. 51—294.

GRINNELL, J. (1922). "A geographical study of the kangaroo rats of California." University of California Publications in Zoology, vol. 24, pp. 1—124.

GRINNELL, J., DIXON, J., and J. M. LINSDALE, (1930). "Vertebrate natural history of a section of Northern California through the Lassen Peak region". University of California Publications in Zoology, vol. 35, pp. 1—594.

HAGEDOORN, A. L., and A. C. (1921). "The relative value of the processes causing evolution." The Hague, 294 pp.

HOWELL, A. H. (1920). "Description of a new species of beach mouse from Florida." Journal of Mammalogy, vol. 1, pp. 237—240.

HOWELL, A. H. (1921). "A biological survey of Alabama." U. S. Department of Agriculture. North American Fauna, no. 45.

HUESTIS, R. R., and H. B. YOCOM, (1930). "Effects of thyroxin upon the thyroid gland and the regeneration and pigmentation of hair in Peromyscus." Archiv für Entwicklungsmechanik der Organismen, 121 Bd., pp. 128—134.

JAMESON, H. L. (1898). "On a probable case of protective coloration in the house-mouse (Mus musculus Linn.)." Journal of the Linnean Society (Zoology), vol. 26, pp. 465—473.

JOHNSON, M. S. (1926). "Activity and distribution of certain wild mice in relation to biotic communities." Journal of Mammalogy, vol. 7, pp. 245—277.

KING, H. D. (1913). "Some anomalies in the gestation of the albino rat (Mus norvegicus albinus)." Biological Bulletin, vol. 24, pp. 377—391.

KLAUBER, L. M. (1930). "New and renamed subspecies of Crotalus confluentus Say, with remarks on related species." Transactions of the San Diego Society of Natural History, vol. 6, pp. 95—104.

KLEINSCHMIDT, O. (1926). "Die Formenkreislehre und das Weltwerden des Lebens." Halle, x + 188 pp.

KUNTZ, A. (1917). "The histological basis of adaptive shades and colors in the flounder Paralichthys albiguttus." Bulletin U.S. Bureau of Fisheries, vol. 35, pp. 1—29.

LLOYD, R. E. (1912). "The growth of groups in the animal kingdom." London, 185 pp.

LOTSY, J. P. (1916). "Evolution by means of hybridization." The Hague, 166 pp.

LOWE, P. R., and C. MACKWORTH-PRAED. (1921). "The last phase of the subspecies." Ibis, 11th ser., vol. 3, pp. 344—347.

MORGAN, T. H. (1911). "The influence of heredity and of environment in determining the coat colors in mice." Annals of the New York Academy of Sciences, vol. 21, pp. 87—117.

MEISE, W. (1928). "Rassenkreuzungen an den Arealgrenzen." Verhandlungen der Deutschen Zoologischen Gesellschaft, pp. 96—104.

MURISIER, P. (1920—21). "Le pigment mélanique de la truite (Salmo lacustris L.)." Revue Suisse de Zoologie, vol. 28, pp. 45—97, 149—195, 243—299.

OSGOOD, W. H. (1909). "Revision of the mice of the American genus Peromyscus." U. S. Department of Agriculture, North American Fauna, no. 28, 285 pp.

OSBORN, H. F. (1915). "Origin of single characters as observed in fossil and living animals and plants." American Naturalist, vol. 49, pp. 193—239.

OSBORN, H. F. (1927). "The origin of species, V: speciation and mutation." American Naturalist, vol. 61, pp. 5—42.

PEARSON, KARL. (1896). "Reproductive selection." Natural Science, vol. 8, pp. 321—325.

PRZIBRAM, H. (1909). "Uebertragungen erworbener Eigenschaften bei Säugetieren: Versuche mit Hitze-Ratten." Verhandlungen der Gesellschaft deutscher Naturforscher und Aertzte, 81 Versammlung, 2 Teil, 1 Hälfte, pp. 179—180.

RENSCH, B. (1929). "Das Prinzip geographischer Rassenkreise und das Problem der Artbildung." Berlin, 206 pp.

Richter, C. P. (1927). "Animal behavior and internal drives." Quarterly Review of Biology, vol. 2, pp. 307—343.

Richter, C. P. and G. H. Wang, (1926). "New apparatus for measuring the spontaneous motility of animals." Journal of Laboratory and Clinical Medicine, vol. 12, p. 289.

Ridgway, R. (1912). "Color standards and color nomenclature." Washington, 43 pp., 53 colored plates.

Robson, G. C. (1928). "The species problem, an introduction to the study of evolutionary divergence in natural populations." Edinburgh and London, 283 pp.

Scheffer, T. H. (1924). "Notes on the breeding of Peromyscus." Journal of Mammalogy, vol. 5, pp. 258—260.

Seton, E. T. (1909). "Life-histories of northern animals." New York, 2 vols. (particularly, vol. 1, chapt. XIX, XIXa).

Seton, E. T. (1920). "Notes on the breeding habits of captive deermice." Journal of Mammalogy, vol. 1, pp. 134—138.

Slonaker, J. R. (1924). "The effect of pubescence, oestruation and menopause on the voluntary activity in the albino rat." American Journal of Physiology, vol. 68, pp. 284—315.

Sumner, F. B. (1909). "Some effects of external conditions upon the white mouse." Journal of Experimental Zoology, vol. 7, pp. 97—155.

Sumner, F. B. (1915a). "Some studies of environmental influence, heredity, correlation and growth, in the white mouse." Journal of Experimental Zoology, vol. 18, pp. 325—432.

Swarth, H. S. (1922). "Birds and mammals of the Stikine River region of northern British Columbia and southeastern Alaska." University of California Publications in Zoology, vol. 24, pp. 125—314.

Tammes, Tine. (1912). "Einige Korrelationserscheinungen bei Bastarden." Recueil des Travaux Botaniques Néerlandais, vol. 9, pp. 69—84.

Willis, J. C. (1922). "Age and area." Cambridge (England), X + 259 pp.

Yocom, H. B., and R. R. Huestis, (1928). "Histological differences in the thyroid glands from two subspecies of Peromyscus maniculatus." Anatomical Record, vol. 39, pp. 57—62.

REGISTER

Abnormalities of cage-breed mice 17, 18.
Abundance of deer mice 7.
Activity, nocturnal and diurnal 7, 95, 96.
Albinos 31.
ALLEN, G. M. 69, 101.
ALLEN, J. A. 2, 3, 65, 78, 101.
APGAR, C. S. 86, 101.
Areal measurements of pelages 21.
Asymmetry, indices of 40.
Autotomy of tail 90.

BABÁK, E. 69, 101.
BANGS, O. 69, 101.
Beach-dwelling mice, pale color of 69.
Behavior 9, 10.
Breeding 11, 12.
BUBNOFF, S. 53, 101.

Cages 14—16.
Carnegie Institution of Washington 5.
CASTLE, W. E. 31, 33, 38, 50, 56, 101, 102.
Chickadee, chestnut-backed 81.
COLLINS, H. H. 12, 24, 30, 31, 52, 86, 90, 99.
Color determinations 23, 24.
Concealing coloration, in relation to *Peromyscus* 65, 69—74.
Correlation between racial characters and environment 64—79.
Correlation, inter-racial 51.
Correlation, intra-racial 51—56.
Correlation, parent-offspring 30.

DANIEL, J. F. 90, 102.
Desert animals, pale color of 71—73.
DICE, L. R. 14, 31, 97, 102.
Differences, racial 25, 26.
Differences, racial, heritability of 27, 28.
Differences, individual 26.
Differences, individual, heritability of 28—30.
Distribution, geographic 7, 8, 57—64.
DIXON, J. 81.
Dominance 42, 45, 47.
DUNN, L. C. 33, 102.

ELTON, C. A. 79, 80, 102.
Epilepsy 38, 39.
EVANS, H. M. 19.

Feeding 17, 18.
FELDMAN, W. H. 31.
Fertilization, deferred 88—90.

Gradients, parallel, of environmental factors and characters of animals 64, 65, 73—75, 77—79.
GRINNELL, J. 5, 81, 85, 102.
Grizzled mutant 35—37.
GUYER, M. F. 91.

HAGEDOORN, A. L. and A. C. 84, 102.
Hairless mutant 33, 34.
Heritable peculiarities, miscellaneous 37—39.
HOWELL, A. H. 9, 69, 102.
HUESTIS, R. R. 11, 25, 28, 29, 30, 38, 40, 41, 50, 86, 87, 94, 97, 98, 99.
Humidity, supposed effect on color of mammals 65—67, 77.
Hybridization of species 45.
Hybridization of subspecies 39—51.

Inheritance of functional modifications, tests of 90, 91.
Intergradation between subspecies 58, 60—64.

JAMESON, H. L. 69, 103.
JOHNSON, M. S. 8, 9, 95, 103.
JORDAN, D. S. 2.

KAROL, J. J. 9, 12, 87, 99.
KING, H. D. 88, 103.
KLAUBER, L. M. 57, 103.
KLEINSCHMIDT, O. 58, 103.
KUNTZ, A. 69, 103.
Kymograph 95, 96.

Lava, alleged effects upon color of rodents 71, 72.

Linkage between albino and pallid genes 32, 33.
LINSDALE, J. M. 81.
Live-traps, advantages of 14.
LLOYD, R. E. 38, **103**.
LOTSY, J. P. 84, **103**.
LOWE and MACWORTH-PRAED 2, **103**.

MCDANIEL, M. E. 11, 87.
Measurements, colorimetric 23, 24.
Measurements, linear 20.
MEISE 82, **103**.
MERRIAM, J. C. 5.
Molting 12, 86.
MORGAN, T. H. 5, **103**.
MURISIER, P. 69, **103**.
Museum of Vertebrate Zoology 5.

Number of factors, attempts to estimate 48—50.

Onychomys leucogaster 68.
OSGOOD, W. H. 6, 7, 57, 69, 72, **103**.

Pallid mutant 31—33.
Parus atricapillus 78.
PEARSON, K. 80, **103**.
Pelages 12.
Perognathus 68, 90.
Peromyscus crinitus 71.
Peromyscus eremicus 94, 96, 97.
Peromyscus leucopus 31, 97.
Peromyscus maniculatus (subspecies of) 7—9, 17, 18, 25—27, 30—32, 34, 35, 38, 45—49, 59, 60, 67, 76—78, 86, 88, 93, 94, 96, 97.
Peromyscus polionotus (subspecies of) 13, 15, 18, 21, 25, 28—30, 40—50, 53, 54, 60, 61, 69—71, 73—75, 86.
PHILLIPS, J. C. 38.
Physiological differences between subspecies 91—98.
Pigment of hair 6.
Population pressure 74—76, 79—81.
PRZIBRAM, H. 3, 78, **103**.

Red-eyed yellow 33.
RENSCH, B. 58, 78, **103**.

Rhinoceros mice 33.
RICHTER, C. P. 91, 94, 96, **103**.
RIDGWAY, R. 24, 35, **104**.
RITTER, W. E. 5.
Ross, L. G. 94, 96, 99.

SCHEFFER, T. H. 11, **104**.
SCRIPPS, E. W. 5.
Secondary sexual differences 86.
Segregation, genetic 40, 41, 48, 53.
Selection theory, in relation to subspecies 83, 84.
SETON, E. T. 9, **104**.
Sex ratios 86—88.
Size 6.
Skins, method of preparing 20.
SLONAKER, J. R. 96, **104**.
Sterility in captivity 19.
SUMNER, F. B. 3, 9, 11, 17, 20—22, 27—31, 34, 36, 39, 41—47, 52—56, 58—65, 68—70, 73, 79, 86—88, 91, 98, 99, **100**, **104**.
Superfetation, possible cases of 88—90.
SWARTH, H. S. 7, 65, 68, 72, 78.

TAMMES, T. 55, **104**.
Thyroid, apparent racial differences in 97, 98.
Tint-photometer, Ives 23.
Transplantation of races 27, 28.
Trapping 12—14.
Trends, geographic (see gradients).

VAUGHAN, T. W. 5.

Waltzing habit 39.
WANG, G. H. 91, 96, **103**.
Water consumption, apparent specific differences in 96, 97.
WELLS, N. A. 69.
Wheel-cages 91—94.
WILLIS, J. C. 76, 77, **104**.
WRIGHT, S. 33, 50, **102**.

Yellow mutant 34, 35.
YOCOM, H. B. 94, 97, 98, **102**.

CRITERIA FOR VERTEBRATE SUBSPECIES, SPECIES AND GENERA: THE MAMMALS

By

E. RAYMOND HALL

University of California, Berkeley, California

Mr. Chairman, members of the American Society of Ichthyologists and Herpetologists, members of the American Society of Mammalogists, and guests: We had expected as a speaker at this time one of the senior mammalogists who now is unable to attend. I am glad to appear as a substitute because the subject under discussion is one in which I am especially interested. In these extemporaneous remarks I propose: (1) to indicate some steps which I think useful to take in classifying mammalian specimens as to subspecies; (2) to express my personal views as to criteria for subspecies, species, and genera of mammals; (3) to illustrate how some of these criteria for subspecies and species may be applied to closely related insular kinds of mammals; and (4) to suggest a way in which subspecies may disappear without becoming extinct.

When I undertake to classify mammalian specimens as to subspecies or species, or when I present a series to a beginning student for classification, I like to observe the following steps: (a) select for initial, intensive study a large series, 30 or more individuals, from one restricted locality; (b) segregate these by sex; (c) arrange specimens of each sex from oldest to youngest; (d) divide these into age-groups and within a given group, of one sex, from one locality, of what is judged to be one species, measure the amount of so-called individual variation; (e) with this measurement as a "yardstick," compare individuals, and if possible series, comparable as to sex and age (and seasons where characteristics of the pelage are involved) from this and other localities. The differences found are usually properly designated as geographic variations and form the basis for recognition of subspecies, which in turn comprise one of the tools used by some students of geographic variation.

As to criteria for the recognition of genera, species and subspecies of mammals, it seems to me that if crossbreeding occurs freely in nature where the geographic ranges of two kinds of mammals meet, the two kinds should be treated as subspecies of one species. If at this and all other places where the ranges of the two kinds meet or overlap, no crossbreeding occurs, then the two kinds are to be regarded as two distinct, full species. The concept of a species, therefore, is relatively clear-cut and precise; the species is a definite entity. Furthermore, if a zoologist knows the morphological characteristics diagnostic of the species, he has no difficulty in identifying a particular individual as of one species or another. In identification of subspecies, difficulty is frequently encountered, especially with individuals which originate in an area of intergradation.

The category next higher than the species, namely, the genus, is less definite and more subjective as regards its limits than is the species. As the species is the definite, clear-cut starting point for defining subspecies, the species is likewise the starting point for consideration of genera. Degree of difference is the criterion for a genus. The genus lies about midway between the species and the family. Because the limits of the family, like those of the genus, are subjective, it follows that the criterion for recognition of genera, although precise enough at the lower point of beginning, the species, is elastic at the upper end—namely, at the level of the family.

In summary, the criterion for subspecies is intergradation, that for species is lack of intergradation, and that for genera is degree of difference. These ideas agree in general with the ideas expressed by the previous speakers.

One of the situations in which it is difficult, or impossible, to apply these criteria to conditions actually existing in nature is comprised in some insular populations. Frequently the populations on two islands near each other differ enough to warrant subspecific or possibly specific distinction. A means of deciding on specific versus subspecific status for these populations is to find on the adjacent mainland a continuously distributed, related kind of mammal which there breaks up into subspecies. Ascertain the degree of difference between each pair of mainland subspecies which intergrade directly. If the maximum degree of difference between the insular kinds is greater than the difference between the two subspecies on the mainland, which intergrade directly, and greater than that between either insular kind and the related population on the nearby mainland, the two insular kinds may properly be treated as full species. If the maximum degree of difference between the insular kinds is no greater than, or less than, the difference found on the mainland between pairs of subspecies which intergrade directly, the insular kinds may properly be treated as subspecies of one species. In fine, the criterion is degree of difference with the limitation of geographic adjacency, rather than intergradation or lack of it.

Now to my fourth point, namely the suggestion that many subspecies disappear without becoming extinct. Permit me first to observe that although species and subspecies seem to have the same kinds of distinguishing characters, which appear to be inherited by means of essentially the same kinds of mechanisms in the germ plasm, there are two noteworthy differences between species and subspecies. One already implied is that, in a species which is continually distributed over a given area, its characters at the boundaries of its range are sharp, definite, and precise. Some of its characters comprised in size, shape and color, at any one place are either those of one species or instead unequivocally those of some other, whereas the characters of a subspecies, particularly at or near the place where two subspecies meet, more often than not are various combinations of those of the two subspecies and in many individual characters there is blending.

Second, through a given epoch of geological time while a species is in existence, one or more of its subspecies may disappear and one or several new subspecies may be formed. Subspecies, therefore, on the average are shorter-lived than species.

Now the disappearance of subspecies is to be expected on *a priori* grounds if we suppose that new subspecies are formed in every geological epoch. There is reason to believe that in the Pleistocene, the epoch of time immediately preceding the Recent, there were even more species of mammals than there are now. In each of several successively corresponding periods of Tertiary time before the Pleistocene, probably there were as many species as now. Probably too, these species then were about as productive of subspecies as species are now. Had even half of these subspecies persisted, either as subspecies unchanged or in considerable part by becoming full species, there would now be an array of species and subspecies many times as numerous as actually does exist. It is obvious therefore that many disappeared.

In accounting for this adjustment of numbers of kinds of mammals, I have spoken of the disappearance of subspecies rather than of their extinction because I can imagine how a species, say, the pocket gopher *Thomomys townsendii*, in the middle Pleistocene with three subspecies (geographic races) could have come down to the present by means of each of the three subspecies having gradually changed its characters into those of one of the three subspecies existing today in the area of northern Nevada that I have in mind. In this way, disappearance of subspecies living in the Pleistocene has been accomplished, without their having become extinct in the sense that the subspecies left no living descendants. Of course this has to be true for some of the subspecies of each successively preceding epoch if any animals at all persist, but what I wish to emphasize is the strong probability that many, perhaps more than 50 per cent, disappeared thus without actually becoming extinct, when, for example, two successive stages of the Pleistocene, south of the ice sheet, are considered. In this regard it is pertinent to recall that each of three Pleistocene kinds of pocket gophers, *Thomomys* (probably species *talpoides*) *gidleyi*, *Thomomys* (probably species *townsendii*) *vetus*, and *Thomomys* (probably species *bottae*) *scudderi*, from a short distance over the northern boundary of Nevada, differs from living representatives corresponding to it (several subspecies of one species) in greater width labially of the individual cheek teeth of the lower jaw. Significant for the thesis being defended is the point that each and all of these *Thomomys* in the Pleistocene differed, at least as regards the shape of the teeth, in the same way from the three living species which I feel confident are their descendants.

Let us suppose that three hypothetical subspecies of *Thomomys townsendii* in middle Pleistocene time each gradually changed into three different subspecies inhabiting about the same areas in upper Pleistocene time, and that these in turn were the ancestors of the three subspecies living in those same general areas today. A total of nine kinds is thus accounted for. At any one time there was geographical intergradation, which has reference to horizontal direction. Also there was intergradation up through time, which has reference for present purposes to a vertical direction. If I had before me all the material necessary to substantiate this or a similar case, I would be inclined to recognize nine subspecies of one species. This hypothetical case emphasizes the importance of intergradation, the criterion for subspecies.

In review: I have mentioned some preliminary steps useful for a person to take when he aims to analyze variation in mammals and to establish species and subspecies thereon; intergradation is the criterion for subspecies and degree of difference is the criterion for genera; degree of difference with the limitation of geographic adjacency may be used as the criterion for insular populations (the classification of which is doubtful as between subspecies and species); and, finally, I have sought to stress the importance of intergradation as a criterion for subspecies by showing how subspecies may disappear without becoming extinct.

NATURAL SCIENCES IN AMERICA

An Arno Press Collection

Allen, J[oel] A[saph]. **The American Bisons,** Living and Extinct. 1876

Allen, Joel Asaph. **History of the North American Pinnipeds:** A Monograph of the Walruses, Sea-Lions, Sea-Bears and Seals of North America. 1880

American Natural History Studies: The Bairdian Period. 1974

American Ornithological Bibliography. 1974

Anker, Jean. **Bird Books and Bird Art.** 1938

Audubon, John James and John Bachman. **The Quadrupeds of North America.** Three vols. 1854

Baird, Spencer F[ullerton]. **Mammals of North America.** 1859

Baird, S[pencer] F[ullerton], T[homas] M. Brewer and R[obert] Ridgway. **A History of North American Birds:** Land Birds. Three vols., 1874

Baird, Spencer F[ullerton], John Cassin and George N. Lawrence. **The Birds of North America.** 1860. Two vols. in one.

Baird, S[pencer] F[ullerton], T[homas] M. Brewer, and R[obert] Ridgway. **The Water Birds of North America.** 1884. Two vols. in one.

Barton, Benjamin Smith. **Notes on the Animals of North America.** Edited, with an Introduction by Keir B. Sterling. 1792

Bendire, Charles [Emil]. **Life Histories of North American Birds** With Special Reference to Their Breeding Habits and Eggs. 1892/1895. Two vols. in one.

Bonaparte, Charles Lucian [Jules Laurent]. **American Ornithology:** Or The Natural History of Birds Inhabiting the United States, Not Given by Wilson. 1825/1828/1833. Four vols. in one.

Cameron, Jenks. **The Bureau of Biological Survey:** Its History, Activities, and Organization. 1929

Caton, John Dean. **The Antelope and Deer of America:** A Comprehensive Scientific Treatise Upon the Natural History, Including the Characteristics, Habits, Affinities, and Capacity for Domestication of the Antilocapra and Cervidae of North America. 1877

Contributions to American Systematics. 1974

Contributions to the Bibliographical Literature of American Mammals. 1974

Contributions to the History of American Natural History. 1974

Contributions to the History of American Ornithology. 1974

Cooper, J[ames] G[raham]. **Ornithology.** Volume I, Land Birds. 1870

Cope, E[dward] D[rinker]. **The Origin of the Fittest:** Essays on Evolution and **The Primary Factors of Organic Evolution.** 1887/1896. Two vols. in one.

Coues, Elliott. **Birds of the Colorado Valley.** 1878

Coues, Elliott. **Birds of the Northwest.** 1874

Coues, Elliott. **Key To North American Birds.** Two vols. 1903

Early Nineteenth-Century Studies and Surveys. 1974

Emmons, Ebenezer. **American Geology:** Containing a Statement of the Principles of the Science. 1855. Two vols. in one.

Fauna Americana. 1825-1826

Fisher, A[lbert] K[enrick]. **The Hawks and Owls of the United States in Their Relation to Agriculture.** 1893

Godman, John D. **American Natural History:** Part I — Mastology and **Rambles of a Naturalist.** 1826-28/1833. Three vols. in one.

Gregory, William King. **Evolution Emerging:** A Survey of Changing Patterns from Primeval Life to Man. Two vols. 1951

Hay, Oliver Perry. **Bibliography and Catalogue of the Fossil Vertebrata of North America.** 1902

Heilprin, Angelo. **The Geographical and Geological Distribution of Animals.** 1887

Hitchcock, Edward. **A Report on the Sandstone of the Connecticut Valley,** Especially Its Fossil Footmarks. 1858

Hubbs, Carl L., editor. **Zoogeography.** 1958

[Kessel, Edward L., editor]. **A Century of Progress in the Natural Sciences: 1853-1953.** 1955

Leidy, Joseph. **The Extinct Mammalian Fauna of Dakota and Nebraska,** Including an Account of Some Allied Forms from Other Localities, Together with a Synopsis of the Mammalian Remains of North America. 1869

Lyon, Marcus Ward, Jr. **Mammals of Indiana.** 1936

Matthew, W[illiam] D[iller]. **Climate and Evolution.** 1915

Mayr, Ernst, editor. **The Species Problem.** 1957

Mearns, Edgar Alexander. **Mammals of the Mexican Boundary of the United States.** Part I: Families Didelphiidae to Muridae. 1907

Merriam, Clinton Hart. **The Mammals of the Adirondack Region,** Northeastern New York. 1884

Nuttall, Thomas. **A Manual of the Ornithology of the United States and of Canada.** Two vols. 1832-1834

Nuttall Ornithological Club. **Bulletin of the Nuttall Ornithological Club:** A Quarterly Journal of Ornithology. 1876-1883. Eight vols. in three.

[Pennant, Thomas]. **Arctic Zoology. 1784-1787.** Two vols. in one.

Richardson, John. **Fauna Boreali-Americana;** Or the Zoology of the Northern Parts of British America, Containing Descriptions of the Objects of Natural History Collected on the Late Northern Land Expeditions Under Command of Captain Sir John Franklin, R. N. Part I: Quadrupeds. 1829

Richardson, John and William Swainson. **Fauna Boreali-Americana:** Or the Zoology of the Northern Parts of British America, Containing Descriptions of the Objects of Natural History Collected by the Late Northern Land Expeditions Under Command of Captain Sir John Franklin, R. N. Part II: The Birds. 1831

Ridgway, Robert. **Ornithology.** 1877

Selected Works By Eighteenth-Century Naturalists and Travellers. 1974

Selected Works in Nineteenth-Century North American Paleontology. 1974

Selected Works of Clinton Hart Merriam. 1974

Selected Works of Joel Asaph Allen. 1974

Selections From the Literature of American Biogeography. 1974

Seton, Ernest Thompson. **Life-Histories of Northern Animals: An Account of the Mammals of Manitoba.** Two vols. 1909

Sterling, Keir Brooks. **Last of the Naturalists:** The Career of C. Hart Merriam. 1974

Vieillot, L. P. **Histoire Naturelle Des Oiseaux de L'Amerique Septentrionale,** Contenant Un Grand Nombre D'Especes Decrites ou Figurees Pour La Premiere Fois. 1807. Two vols. in one.

Wilson, Scott B., assisted by A. H. Evans. **Aves Hawaiienses:** The Birds of the Sandwich Islands. 1890-99

Wood. Casey A., editor. **An Introduction to the Literature of Vertebrate Zoology.** 1931

Zimmer, John Todd. **Catalogue of the Edward E. Ayer Ornithological Library.** 1926